ALGEBRA

an intermediate course

SECOND EDITION

ALGEBRA

an intermediate course

SECOND EDITION

Raymond A. Barnett

Merritt College

Thomas J. Kearns

Northern Kentucky University

McGraw-Hill, Inc.

New York St. Louis San Francisco Auckland Bogotá
Caracas Lisbon London Madrid Mexico Milan
Montreal New Delhi Paris San Juan Singapore
Sydney Tokyo Toronto

ALGEBRA: An Intermediate Course

5 6 7 8 9 0   WEBWEB   9 2

ISBN 0-07-00374 9-3

The editors were Robert A. Weinstein and Phillip A. Butcher; the production supervisor was Leroy A. Young; the designer was Janet Bollow. Project supervision was done by Phyllis Niklas. Webcrafters, Inc., was printer and binder. New drawings were done by Carl Brown.
Cover photo was provided by Melvin L. Prueitt, Motion Picture/Video Production, Los Alamos National Laboratory.

Library of Congress Cataloging-in-Publication Data

Barnett, Raymond A.
  Algebra: an intermediate course.

  Includes index.
  1. Algebra.   I. Kearns, Thomas J.   II. Title.
QA154.2.B347   1987   512.9   86-21452
ISBN 0-07-00374 9-3

Contents

Tables   757

Preface

This is a second course in algebra for those who have had basic algebra or for those who need a review before proceeding further. The book is designed for use in self-paced classes, math laboratories, and lecture–discussion classes.

Improvements contained in this second edition of *Algebra: An Intermediate Course* evolved out of the generous response from the many users of the first edition. How do you improve a book that is already working well in the classroom? You seek substantial feedback from those who have actually used the text in the classroom.

■ Principal Changes from the First Edition

1. **Topic organization** has been improved. This was achieved by compressing **elementary algebra review topics**, absorbing and compressing the **applications** chapter from the first edition into the new Chapter 4 (applications are still liberally distributed throughout the book), and moving some topics to the **appendix**. **Natural number exponents** and properties have been placed at the beginning of Chapter 2 for use in basic polynomial operations. **Polynomial division** has been moved to Chapter 3 ("Algebraic Fractions"). **Functions** are introduced earlier (Chapter 8 instead of 10), and **exponential** and **logarithmic functions** immediately follow (Chapter 9 instead of 11).
2. **Substantially rewritten chapters** include: Chapter 1 ("Preliminaries"), Chapter 4 ("First-Degree Equations and Inequalities in One Variable"), Chapter 8 ("Relations and Functions"), and Chapter 9 ("Exponential and Logarithmic Functions").
3. **New material** includes an optional Section 3-3 on **synthetic division**, Section 10-3 on **systems and augmented matrices**, Section 10-4 on **systems of linear inequalities**, Section 8-3 on **graphing polynomial functions**, Chapter 11 on **sequences and series**, Appendix A on **significant digits**, and Appendixes F and G on **matrices and matrix operations**.
4. **Exercise sets** have been expanded, and more difficult problems have been added to many B-level and C-level exercises.
5. The **Appendix** has been expanded. All **table evaluation** of logarithmic functions is included there. (The text emphasizes calculator evaluation of these functions.) Also, the Appendix now contains two new sections on **matrices** for those who want to cover these topics in this course.
6. The use of **hand calculators** has been increased throughout the book. All use of logarithmic tables has been moved to the Appendix.
7. **Interval notation** has been introduced relative to the study of inequalities (Sections 4-6 and 4-7). For example,

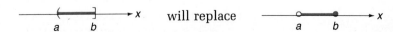

   and $(a, b]$ will be used along with $a < x \le b$.
8. **Restrictions on variables** have been reworded for clarity and consistency.

9. The use of **diagrams** and **figures** has been expanded.
10. There are more worked-out **examples** with matched problems.
11. There is more **boxed material** for emphasis.
12. **Applications** have been made current.
13. More **common errors** are highlighted.
14. **Photographs** associated with applications have been added to increase motivation.
15. Short **historical comments** have been added where appropriate.
16. **Chapter review** sections have been improved.
17. **Check Exercises** have been converted from multiple-choice to short answer problems.

▪ Important Features Retained from the First Edition

1. **Exposition:** Considerable effort has been directed toward making the exposition clear, informal, accurate, and nonthreatening. Vocabulary is controlled. Technical terms and statements are presented only with substantial motivation and clear, concrete illustrations.
2. **Word problems and applications:** The translation of verbal statements into symbolic forms receives much attention, and the process is gradually expanded to significant applications. By the end of Chapter 4 students will have had much experience in both areas. Important applications are liberally distributed throughout the text.
3. **Provisions for substantial student involvement and feedback:** Students are substantially involved in the instructional process throughout.
   (A) Following each **example** is a **matched problem** with room for its solution right in the text (see page 30). Complete solutions to the matched problems are found near the end of each section in a specially screened box (see page 32).
   (B) At the end of each section is a comprehensive, carefully graded, **Practice Exercise** set (see page 32). The problems are presented in matched pairs and the student is directed to work through the odd-numbered problems first, check answers, and then to work through even-numbered problems in areas of weakness. Answers to all Practice Exercises are given in the back of the book.
   (C) At the end of each chapter is grouped a series of five- or ten-problem **Check Exercise** sets (see page 53). Each individual Check Exercise corresponds to one section in the chapter. Space is provided for students to show their work. These check sets can be turned in (pages are perforated), easily corrected, and returned to the student. Answers are in an answer key for instructors. A color stripe that runs down the right-hand edge of each Check Exercise page is designed to allow students or instructors to turn quickly to a Check Exercise set for a given section.
   (D) Near the end of each chapter is a **Diagnostic (Review) Exercise** (see page 47). A student should work all the problems, check answers in the back of the book, and then review text sections corresponding to problems missed (corresponding section numbers are given in italics following each answer).
   (E) Following each Diagnostic (Review) Exercise is a **Practice Test** for the chapter (see page 51). Students should take this Practice Test as if it were

an in-class test, allowing 50 minutes or less, check answers in the back of the book, and then review text sections where weaknesses still prevail (corresponding section numbers are given in italics following each answer).

■ Student Aids

1. **Common student errors** are clearly identified at places where they naturally occur (see Sections 3-1, 4-1, 4-5, 5-2).
2. **Think boxes** (dashed boxes) are used to enclose steps that are usually performed mentally (see Sections 1-4, 2-2, 3-6, 4-1).
3. **Annotation** of examples and developments is found throughout the text to help students through critical stages (see Sections 1-4, 2-4, 3-2, 3-6, 4-1).
4. **Functional use of a second color** guides students through critical steps (see Sections 2-1, 2-4, 3-3, 3-6, 4-1).
5. **Chapter review sections** include a review of all important terms and symbols and a comprehensive Diagnostic (Review) Exercise. Answers to all Diagnostic (Review) Exercises are included in the back of the book and are keyed (with numbers in italics) to the corresponding text sections.
6. **Summaries** of formulas and symbols (keyed to the sections in which they are introduced) and the metric system are inside the front and back covers of the text for convenient reference.
7. A **solutions manual** is available at a nominal cost through a bookstore. The manual includes detailed solutions to all odd-numbered problems and all chapter Diagnostic (Review) Exercises and Practice Tests.
8. **Calculator steps** are often included (see Sections 1-4, 8-3, 9-1, 9-4).

■ Instructor Aids

1. A unique **computer-generated random test system** is available to instructors without cost. The system, using Apple II or IBM-PC computers and any one of more than 30 dot matrix printers, will generate an almost unlimited number of chapter tests and final examinations, each different from the other, quickly and easily. At the same time, the system produces an answer key and a student worksheet with an answer column that exactly matches the answer column on the answer key. Graphing grids are included on the answer key and on the student worksheet for problems requiring graphs.
2. **Section checks** can be handled in a simple way by making use of the Check Exercises grouped at the end of each chapter. These Check Exercises contain either five or ten problems with ample space to show work. The whole Check Exercise is arranged on the front and back of a single page for easy handling. The perforated page can be torn out, turned in, easily graded, and returned to the student in a short time. (These Check Exercises also provide an effective and easy-to-use **homework** control in lecture-discussion classes.)
3. A **printed and bound test battery** is also available to instructors without cost. The battery contains several tests for each chapter and several final examinations. Included are easy-to-grade answer keys along with student worksheets with answer columns that exactly match the answer columns on the answer keys. Graphing grids are included on the answer keys and on the student worksheets for problems requiring graphs.

4. A **student solutions manual,** containing worked out solutions to all chapter Diagnostic (Review) Exercises and Practice Tests and all odd-numbered problems in the text, is available to instructors without charge.

5. An **instructor's Check Exercise Answer Key** is available to instructors without cost. The key is easy to use and will slip easily into the back of the book for ready availability.

■ Error Check

Because of the careful checking and proofing by a number of very competent people (acting independently), the authors and publisher believe this book to be substantially error-free. If any errors remain, the authors would be grateful if corrections were sent to: Mathematics Editor, College Division, 27th floor, McGraw-Hill Book Company, 1221 Avenue of the Americas, New York, New York 10020.

■ Acknowledgments

The authors wish to thank Peter A. Lindstrom, North Lake College, for his helpful suggestions for this edition. In addition, we wish to thank the following for their detailed reviews: David W. Bange, University of Wisconsin—La Crosse; Carol A. Edwards, St. Louis Community College; Robert Elmore, Santa Barbara City College; Art Molner, Meramec Community College; Richard Reese, Pensacola Junior College; Beverly S. Rich, Illinois State University; Kenneth J. Schabell, Riverside City College; Lynn Tooley, Bellevue Community College; and Paige Yuhn, Santa Barbara City College.

Special thanks are due to Debra C. Calloway, Thomas More College, and Daniel J. Curtin, Northern Kentucky University, for their careful checking of the whole manuscript, including answers to all exercises, examples, and matched problems.

Raymond A. Barnett
Thomas J. Kearns

To the Student

The following suggestions will help you get the most out of this book and your efforts.

As you study the text we suggest a five-step process. For each section:

1. Read the mathematical development.
2. Work through the illustrative example.   } Repeat the 1-2-3 cycle until the section is finished.
3. Work the matched problem.
4. Review the main ideas in the section.
5. Work the assigned exercise at the end of the section.

Even though this book is designed for you to work in the text as you read, you should have extra paper, pencils, and a wastebasket at hand. In fact, no mathematics text should be read without pencil and paper in hand; mathematics is not a spectator sport. Just as you cannot learn to swim by watching someone else, you cannot learn mathematics simply by reading worked examples—you must work problems, lots of them.

If you have difficulty with the course, then, in addition to doing the regular assignments, spend more time on the examples and matched problems and work more A exercises, even if they are not assigned. If the A exercises continue to be difficult for you, you probably should take elementary algebra before attempting this course. If you find the course too easy, then work more C exercises, even if they are not assigned. If the C exercises are consistently easy for you, you are probably ready to start college algebra. See your instructor.

Raymond A. Barnett
Thomas J. Kearns

Preliminaries ■ 1

INSTRUCTIONS FOR STUDENTS IN A SELF-PACED CLASS OR LAB

YES — **HAVE YOU HAD INTERMEDIATE ALGEBRA BEFORE THIS COURSE?** — NO

1. Work Diagnostic (Review) Exercise 1-6 on page 47. Check answers in back of book; then work through text sections corresponding to problems missed. (Section numbers are in italics following each answer.)
2. When finished with step 1, take Practice Test Chapter 1 on page 51 as a final check of your understanding of the chapter. Check answers in the back of the book; then review sections where weakness still prevails. (Corresponding section numbers are in italics following each answer.)
3. When you think you are ready, ask your instructor for a graded test for Chapter 1.
4. If your instructor approves, after the test is corrected, go to the next chapter.

1. Work through each section in the chapter as follows:
   (A) Read discussion.
   (B) Read each example and work the corresponding matched problem. Check your solutions to the matched problem in Solutions to Matched Problems on the indicated page.
   (C) At the end of a section work the odd-numbered problems in the Practice Exercise and check answers; then work even-numbered problems in areas of weakness. (Answers to *all* Practice Exercise sets are in the back of the book.)
   (D) Work Check Exercise as instructed. Tear out and turn in as directed by your instructor. (Answers are not in the text.)
2. Repeat each step in item 1 for each section in the chapter.
3. After the instructional part of the chapter is completed, proceed with steps 1 to 4 in the box above this one.

Chapter 1 ■ Preliminaries

Section 1-1   Basic Concepts

- ■ Sets
- ■ The Set of Real Numbers
- ■ The Real Number Line

Algebra is often referred to as "generalized arithmetic." In arithmetic we deal with basic arithmetic operations (addition, subtraction, multiplication, and division) on specific numbers. In algebra we continue to use all that we know in arithmetic, but, in addition, we reason and work with symbols that represent or are placeholders for one or more numbers. The rules for manipulating and reasoning with these symbols depend, in large part, on certain properties of numbers (since the symbols represent numbers). In this chapter we will review important number systems and some of their basic properties. To make our discussions here and elsewhere in the text precise, we first introduce a few useful notions about sets.

■ Sets

Our use of the word "set" will not differ appreciably from the way it is used in everyday language. Words such as "set," "collection," "bunch," and "flock" all convey the same idea. Thus, we think of a **set** as a collection of objects with the important property that given any object we can tell whether it is or is not in the set. Capital letters, such as A, B, and C, are often used to designate particular sets. For example,

$$A = \{3, 5, 7\} \qquad B = \{4, 5, 6\}$$

specify sets A and B.

Each object in a set is called a **member** or **element** of the set. Symbolically:

$a \in A$	means	"a is an element of set A"
$a \notin A$	means	"a is not an element of set A"

Referring to sets A and B above, we see that

$5 \in A$ 5 is an element of set A.

$3 \notin B$ 3 is not an element of set B.

A set is usually described in one of two ways:

1. By **listing** the elements between braces $\{\ \ \}$:

$$\{3, 5, 7\}$$

Note: The order in which elements are listed is irrelevant. Also, a given element is not listed more than once.

2. By enclosing a **rule** within braces $\{\ \ \}$ that determines the elements in the set:

$$\{x \mid x^2 = 81\}$$ Read: "The set of all x such that $x^2 = 81$."

- rule
- such that
- all x
- The set of

EXAMPLE 1   Let A be the set of all numbers x such that $x^2 = 25$. The set A may be specified as follows:

Listing method: $A = \{-5, 5\}$
Rule method: $A = \{x \mid x^2 = 25\}$ Read: "The set of all x such that $x^2 = 25$."

PROBLEM 1[†]   Let B be the set of all x such that $x^2 = 49$.

(A) Specify B by the listing method.
(B) Specify B by the rule method.
(C) Indicate true (T) or false (F): $7 \in B$, $7 \notin B$, $49 \notin B$.

Solution   (A)
(B)
(C) $7 \in B$ _____ $7 \notin B$ _____ $49 \notin B$ _____

The letter x introduced above is a variable. In general, a **variable** is a symbol used as a placeholder for elements out of a set with two or more elements. (This set is called the **replacement set** for the variable.) A **constant**, on the other hand, is a symbol that names exactly one object. The symbol "8" is a constant, since it always names the number eight.

The introduction of variables into mathematics occurred about A.D. 1600. A French mathematician, François Vieta (1540–1603), is singled out as the one mainly responsible for this new idea. Many mark this point as the beginning of modern mathematics.

Some sets are **finite** (there is one counting number that indicates the total number of elements), some sets are **infinite** (in counting the elements, we never come to an end), and some sets are **empty** (the set contains no elements). Empty sets are also called **null** sets. Symbolically:

$\varnothing$ represents "the empty set."

[†] Answers to matched problems following examples are located at the end of each section before the exercise set.

EXAMPLE 2 $A = \{2, 4, 6\}$ A finite set

$N = \{1, 2, 3, 4, 5, \ldots\}$ An infinite set

This is the set of **natural**,
or **counting**, **numbers**.
(The three dots indicate the
pattern continues without
end.)

$B = \{x \mid x \text{ is a counting number between 2 and 3}\} = \varnothing$ An empty or null set

PROBLEM 2   Indicate whether the set is finite, infinite, or empty:

(A) $U = \{x \mid x \text{ is a counting number less than } 0\}$
(B) $E = \{2, 4, 6, \ldots\}$
(C) $G = \{x \mid x^2 = 25\}$

Solution   **(A)**   **(B)**   **(C)**

If each element of set A is also an element of set B, we say that A is a **subset** of set B. The set of all women in a class is a subset of the whole class. (Note that the definition of a subset allows a set to be a subset of itself.) If two sets have exactly the same elements (the order of listing does not matter), the sets are said to be **equal**. Set A is equal to set B if and only if A is a subset of B and B is a subset of A.

Symbolically:

Subsets

$A \subset B$ means "A is a subset of B." $\{3, 5\} \subset \{3, 5, 7\}$
$A = B$ means "A is equal to B." $\{4, 6\} = \{6, 4\}$

It is useful and interesting to note that:

$\varnothing$ is a subset of every set.

It is certainly true that every element of $\varnothing$ is an element of any given set, since $\varnothing$ has no elements. (Note that the symbol $\varnothing$ is not enclosed by braces.)

Do not confuse the two symbols $\subset$ and $\in$. The former is used only between two sets and the latter is used only between an element of a set and a set. Note the following:

$4 \in \{2, 4, 6\}$ Correct usage of $\in$

~~$4 \subset \{2, 4, 6\}$~~   Incorrect usage of $\subset$

~~$\{4\} \in \{2, 4, 6\}$~~   Incorrect usage of $\in$

$\{4\} \subset \{2, 4, 6\}$ Correct usage of $\subset$

EXAMPLE 3   Let $A = \{-3, 0, 5\}$, $B = \{0, 5, -3\}$, and $C = \{0, 5\}$. Then each of the following statements is true:

$$C \subset A \qquad C \not\subset B \qquad A = B \qquad A \subset B \qquad \emptyset \subset A \qquad A \neq C$$

PROBLEM 3   Let $M = \{-4, 6\}$, $N = \{6, -4\}$, and $P = \{-4\}$. Indicate true (T) or false (F):

(A) $M \neq N$ **(B)** $P \subset N$ **(C)** $P \in N$
(D) $N \subset M$ **(E)** $\emptyset \subset P$ **(F)** $M \subset P$

Solution   **(A)**   **(B)**   **(C)**   **(D)**   **(E)**   **(F)**

■ The Set of Real Numbers

The real number system is the number system you have used most of your life. In algebra we are interested in manipulating symbols in order to change or simplify algebraic expressions and to solve algebraic equations. Because many of these symbols represent real numbers, it is important to briefly review the set of real numbers and some of its important subsets (see Table 1). Figure 1 illustrates how these sets of numbers are related to one another.

TABLE 1
THE SET OF REAL NUMBERS

SYMBOL	NUMBER SYSTEM	DESCRIPTION	EXAMPLES
N	Natural numbers	Counting numbers (also called positive integers)	$1, 2, 3, \ldots$
J	Integers	Set of natural numbers, their negatives, and 0	$\ldots, -2, -1, 0, 1, 2, \ldots$
Q	Rational numbers	Any number that can be represented as a/b, where a and b are integers and $b \neq 0$	$-4; \frac{-3}{5}; 0; 1; \frac{2}{3}; 3.67$
R	Real numbers	Set of all rational and irrational numbers (the irrational numbers are all the real numbers that are not rational)	$-4; \frac{-3}{5}; 0; 1; \frac{2}{3}; 3.67; \sqrt{2}; \pi; \sqrt[3]{5}$

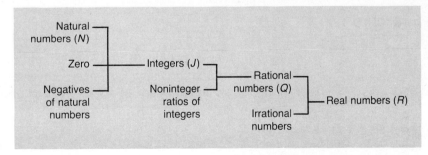

FIGURE 1 The real number system

The set of integers contains all the natural numbers and something else (their negatives and 0). The set of rational numbers contains all the integers (for example, 5 can be expressed as the ratio of two integers, $\frac{5}{1}$) and something else (noninteger ratios of integers). And the set of real numbers contains all the rational numbers and something else (irrational numbers). In short, N is a subset of J, J is a subset of Q, and Q is a subset of R. Symbolically,

$$N \subset J \subset Q \subset R$$

Rational numbers have repeating decimal representations, whereas irrational numbers have infinite nonrepeating decimal representations.[†] For example, the decimal representations of the rational numbers 2, $\frac{4}{3}$, and $\frac{5}{11}$ are, respectively,

$$2 = 2.0000\ldots \qquad \frac{4}{3} = 1.333\ldots \qquad \frac{5}{11} = 0.454545\ldots$$

whereas those of the irrational numbers $\sqrt{2}$ and π are, respectively,

$$\sqrt{2} = 1.41421356\ldots \qquad \pi = 3.14159265\ldots$$

■ The Real Number Line

A one-to-one correspondence exists between the set of real numbers and the set of points on a line; that is, each real number corresponds to exactly one point, and each point to exactly one real number. A line with a real number associated with each point, and vice versa, as in Figure 2, is called a **real number line** or simply a **real line**. Each number associated with a point is called the **coordinate** of the point. The point with coordinate 0 is called the **origin**. The arrow indicates a positive direction; the coordinates of all points to the right of the origin are called **positive real numbers** and those to the left of the origin are called **negative real numbers**.

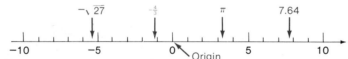

FIGURE 2 A real number line

Solutions to Matched Problems

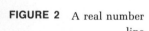

1. **(A)** $\{-7, 7\}$ **(B)** $\{x \mid x^2 = 49\}$ **(C)** T, F, T
2. **(A)** Empty   **(B)** Infinite   **(C)** Finite
3. **(A)** F   **(B)** T   **(C)** F   **(D)** T   **(E)** T   **(F)** F

Practice Exercise 1-1 ■

Work odd-numbered problems first, check answers, and then work even-numbered problems in areas of weakness. Answers to all problems are in the back of the book. Make every effort to work a problem yourself before you look at an answer.

A   *In Problems 1–12, indicate true (T) or false (F).*

1. $3 \in \{2, 3, 4\}$ _____   2. $4 \in \{3, 5, 7\}$ _____

[†] It is instructive to experiment with your hand calculator by dividing any integer by another nonzero integer. Also, see what happens when you take the square root of any number that is not a perfect square. More will be said about irrational numbers in Chapter 5.

3. $5 \notin \{2, 3, 4\}$ _____

4. $6 \notin \{2, 4, 6, 8\}$ _____

5. $\{2, 3\} \subset \{1, 2, 3\}$ _____

6. $\{2, 4\} \subset \{1, 2, 3\}$ _____

7. $\{7, 3, 5\} = \{3, 5, 7\}$ _____

8. $\{3, 1, 2\} \subset \{1, 2, 3\}$ _____

9. $\varnothing \subset \{2, 5\}$ _____

10. $\varnothing \subset \{1, 3\}$ _____

11. $\{7\} \in \{3, 5, 7\}$ _____

12. $\{7\} \subset \{3, 5, 7\}$ _____

B **13.** Give an example of a negative integer, an integer that is neither positive nor negative, and a positive integer. _____

14. Give an example of a negative rational number, a rational number that is neither positive nor negative, and a positive rational number. _____

15. Give an example of a rational number that is not an integer. _____

16. Give an example of an integer that is not a natural number. _____

Write each set in Problems 17–28 using the listing method; that is, list the elements between braces.

17. $\{x \mid x \text{ is a whole number between 5 and 10}\}$ _____

18. $\{x \mid x \text{ is a whole number between 4 and 8}\}$ _____

19. $\{x \mid x \text{ is a letter in "}status\text{"}\}$ _____

20. $\{x \mid x \text{ is a letter in "}Illinois\text{"}\}$ _____

21. $\{x \mid x \text{ was a woman president of the United States}\}$ _____

22. $\{x \mid x \text{ is a month starting with the letter } B\}$ _____

23. $\{x \mid x - 5 = 0\}$ _____

24. $\{x \mid x + 3 = 0\}$ _____

25. $\{x \mid x + 9 = x + 1\}$ _____

26. $\{x \mid x - 3 = x + 2\}$ _____

27. $\{x \mid x^2 = 4\}$ _____

28. $\{x \mid x^2 = 9\}$ _____

29. Indicate which of the following are true:

 (A) All natural numbers are integers. _____

 (B) All real numbers are irrational. _____

 (C) All rational numbers are real numbers. _____

30. Indicate which of the following are true:

 (A) All integers are natural numbers. _____

(B)   All rational numbers are real numbers. _____

(C)   All natural numbers are rational numbers. _____

31.   Each of the following real numbers lies between two successive integers on a real number line. Indicate which two:

(A) $\frac{15}{4}$ _____   **(B)** $-\frac{4}{3}$ _____   **(C)** $-\sqrt{7}$ _____

32.   Each of the following real numbers lies between two successive integers on a real number line. Indicate which two:

(A) $\frac{25}{7}$ _____   **(B)** $-\frac{11}{4}$ _____   **(C)** $-\sqrt{19}$ _____

C **33.**   If $A = \{1, 2, 3, 4\}$ and $B = \{2, 4, 6\}$, find:

(A) $\{x \mid x \in A \text{ or } x \in B\}^{\dagger}$ _____   **(B)** $\{x \mid x \in A \text{ and } x \in B\}$ _____

34.   If $M = \{-2, 0, 2\}$ and $N = \{-1, 0, 1, 2\}$, find:

(A) $\{x \mid x \in M \text{ or } x \in N\}$ _____   **(B)** $\{x \mid x \in M \text{ and } x \in N\}$ _____

35.   Given the sets of numbers N (natural numbers), J (integers), Q (rational numbers), and R (real numbers), indicate to which sets each of the following numbers belongs:

(A) -3 _____   **(B)**   3.14 _____

(C) π _____   **(D)** $\frac{2}{3}$ _____

36.   Given the sets of numbers N, J, Q, and R (see Problem 35), indicate to which sets each of the following numbers belongs:

(A)   8 _____   **(B)** $\sqrt{2}$ _____

(C) $-1.414\cdot$ _____   **(D)** $\frac{-5}{2}$ _____

37.   If $c = 0.151515\ldots$, then $100c = 15.1515\ldots$ and

$$100c - c = (15.1515\ldots) - (0.151515\ldots)$$
$$99c = 15$$
$$c = \frac{15}{99} = \frac{5}{33}$$

Proceeding similarly, convert the repeating decimal $0.090909\ldots$ into a fraction. (All repeating decimals are rational numbers, and all rational numbers have repeating decimal representations.) _____

38.   Repeat Problem 37 for $0.181818\ldots$. _____

---

† Unless otherwise stated, we will use "or" as it is usually used in mathematics—as an inclusive or. Thus, in this case, x is an element of the set if $x \in A$ or $x \in B$, or both.

CALCULATOR PROBLEMS

Express each number as a decimal fraction to the capacity of your calculator. Observe the repeating decimal representation of the rational numbers and the apparent nonrepeating decimal representation of the irrational numbers.

39. **(A)** $\frac{8}{9}$ _____ **(B)** $\frac{3}{11}$ _____

(C) $\sqrt{5}$ _____ **(D)** $\frac{11}{8}$ _____

40. **(A)** $\frac{13}{6}$ _____ **(B)** $\sqrt{21}$ _____

(C) $\frac{7}{16}$ _____ **(D)** $\frac{29}{111}$ _____

APPLICATIONS[†]

41. The executive committee of a student council consists of a president, vice president, secretary, and treasurer and is denoted by the set $\{P, V, S, T\}$. How many two-person subcommittees are possible; that is, how many two-element subsets can be formed? _____

The Check Exercise for this section is on page 53.

42. How many three-person subcommittees are possible in Problem 41?

_____

Section 1-2 Equality and Inequality

■ Algebraic Expressions
■ Equality Relation
■ Inequality Relation

■ Algebraic Expressions

An **algebraic expression** is a meaningful symbolic form involving constants, variables, mathematical operations, and grouping symbols. For example,

$$2 + 8 \qquad 4 \cdot 3 - 7 \qquad 16 - 3(7 - 4)$$

$$5x - 3y \qquad 7(x + 2y) \qquad 4\{u - 3[u - 2(u + 1)]\}$$

are all algebraic expressions.

Two or more algebraic expressions (each taken as a single entity) joined by plus or minus signs are called **terms**. (For reasons that will become clear in Chapter 2, a term includes the sign that precedes it.) Two or more algebraic expressions joined by multiplication are called **factors**. For example,

$$\underbrace{3(x - y)}_{\text{Term}} + \underbrace{(x + y)(x - y)}_{\text{Term}}$$

has two terms, $3(x - y)$ and $(x + y)(x - y)$, and each term has two factors. The first term has factors 3 and $(x - y)$, and the second term has factors $(x + y)$

[†] Applications will often be placed after the C-level exercises or calculator problems and may vary in level of difficulty from easy (no asterisks) and moderately difficult (marked *) to difficult (marked **). In short, they are not necessarily C-level problems in difficulty. Calculator problems will also vary in level of difficulty.

and $(x - y)$. A term may contain several factors, and a factor may contain several terms.

■ Equality Relation

The use of an **equality sign** (=) between two expressions asserts that the two expressions are names or descriptions of exactly the same object. The symbol $\neq$ means **is not equal to**. Statements involving the use of an equality or an inequality sign may be true or they may be false:

$$15 - 3 = 4 \cdot 3 \quad \text{True statement}$$

$$7 - 2 = \tfrac{8}{2} \quad \text{False statement}$$

$$7 - 2 \neq \tfrac{8}{2} \quad \text{True statement}$$

It is interesting to note that the equality sign did not appear until rather late in history—the sixteenth century. It was introduced by the English mathematician Robert Recorde (1510–1558).

If two algebraic expressions involving at least one variable are joined with an equal sign, the resulting form is called an **algebraic equation**. The following are algebraic equations in one or more variables:

$$2x - 3 = 3(x - 5) \qquad a + b = b + a \qquad 3x + 5y = 7$$

Since a variable is a placeholder for constants from a given replacement set, an equation is neither true nor false as it stands; it does not become so until the variables have been replaced by constants. Formulating algebraic equations is an important first step in solving many practical problems using algebraic methods.

EXAMPLE 4   Translate each statement into an algebraic equation using x as the only variable:

(A) 5 times a number is 3 more than twice the number.
(B) 4 times a number is 5 less than twice the number.

Solution   **(A)** Let x = The unknown number; then the statement translates as follows:

5 (times) (a number) (is) 3 (more than) (twice) (the number)

$$5 \quad \cdot \quad x \quad = \quad 3 \quad + \quad 2 \quad x$$

Thus,

$$5x = 3 + 2x$$

(B) Let x = The unknown number; then the statement translates as follows:

4 (times) (a number) (is) (5) (less than) (twice the number)

$$4 \quad \cdot \quad x \quad = \quad 2x \quad - \quad 5$$

[*Note:* "Less than" in this context means "subtracted from."] Thus,

$$4x = 2x - 5 \quad \text{Not } 4x = 5 - 2x$$

PROBLEM 4   Translate each statement into an algebraic equation using x as the only variable:

(A) 7 is 3 more than a certain number.
(B) 12 is 9 less than a certain number.
(C) 3 times a certain number is 6 less than twice that number.
(D) If 6 is subtracted from a certain number, the difference is twice a number that is 4 less than the original number.

Solution   **(A)**

     (B)

     (C)

     (D)

Several important properties of the equality symbol ($=$) follow directly from its logical meaning. These properties must hold any time the symbol is used.

Basic Properties of Equality
If a, b, and c are names of objects, then:
1. $a = a$.   REFLEXIVE PROPERTY
2. If $a = b$, then $b = a$.   SYMMETRIC PROPERTY
3. If $a = b$ and $b = c$, then $a = c$.   TRANSITIVE PROPERTY
4. If $a = b$, then either may replace the other in   SUBSTITUTION PRINCIPLE any statement without changing the truth or falsity of the statement.

The properties of equality are used extensively throughout mathematics. For example, using the symmetric property, we may reverse the left and right sides of an equation any time we wish. That is:

$$\text{If} \quad A = P + Prt, \quad \text{then} \quad P + Prt = A.$$

Using the transitive property, we find that if

$$2x + 3x = (2 + 3)x \quad \text{and} \quad (2 + 3)x = 5x$$

then

$$2x + 3x = 5x$$

And, finally, if we know that

$$C = \pi D \qquad \text{and} \qquad D = 2R$$

then, using the substitution principle, D in the first formula may be replaced by $2R$ from the second formula to obtain

$$C = \pi(2R) = 2\pi R$$

■ Inequality Relation

We now turn to the **inequality** or **order relation**. This relation has to do with "less than" and "greater than." If a and b are any real numbers and $a \neq b$, then a must be "less than" b or a must be "greater than" b.

Just as we use $=$ to replace the words "is equal to," we will use the **inequality symbols** $<$ and $>$ to represent "is less than" and "is greater than," respectively. Thus, we can write the following symbolic forms and their corresponding equivalent verbal forms:

Inequality Symbols

$a < b$	a is less than b
$a > b$	a is greater than b
$a \leq b$	a is less than or equal to b
$a \geq b$	a is greater than or equal to b

It no doubt seems obvious to you that

$$5 < 8$$

is true, but it may not seem equally obvious that

$$-8 < -5 \qquad 0 > -10 \qquad -30{,}000 < -1$$

To make the inequality relation precise so that we can interpret it relative to *all* real numbers, we need a precise definition of the concept.

Definition of $a < b$ and $b > a$

For a and b real numbers, we say that **a is less than b** or **b is greater than a** and write

$$a < b \qquad \text{or} \qquad b > a$$

if there exists a positive real number p such that $a + p = b$ (or equivalently, $b - a = p$.)

We would certainly expect that if a positive number is added to *any* real number, the sum will be larger than the original. That is essentially what the definition states. When we write

$$a \leq b$$

we mean **a is less than or equal to b**, and when we write

$$a \geq b$$

we mean that **a is greater than or equal to b**.

The inequality symbols $<$ and $>$ have a very clear geometric interpretation on the real number line. If $a < b$, then a is to the left of b; if $c > d$, then c is to the right of d (Figure 3).

FIGURE 3 $a < b, c > d$

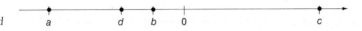

EXAMPLE 5   Indicate true (T) or false (F). Letters a, b, c, and d refer to Figure 3.

(A) $-3{,}000 > 0$ **(B)** $-10 \leq 2$ **(C)** $-5 \geq -5$
(D) $0 < -25$ **(E)** $a < c$ **(F)** $d \geq b$

Solution   **(A)** F   **(B)** T   **(C)** T   **(D)** F   **(E)** T   **(F)** F

PROBLEM 5   Indicate true (T) or false (F). Letters a, b, c, and d refer to Figure 3.

(A) $0 < 25$ **(B)** $-35 \geq 3$ **(C)** $-5 \leq -5$
(D) $-25 > 0$ **(E)** $b \leq c$ **(F)** $d < b$

Solution   **(A)**   **(B)**   **(C)**   **(D)**   **(E)**   **(F)**

We assume that the following two important inequality properties hold for all real numbers:

Basic Inequality Properties
For any real numbers a, b, and c: **1.** Either $a < b$, $a = b$, or $a > b$.   TRICHOTOMY PROPERTY **2.** If $a < b$ and $b < c$, then $a < c$.   TRANSITIVE PROPERTY

The **double inequality** $a < x \leq b$ means that $a < x$ and $x \leq b$; that is, x is between a and b, including b but not including a. Similar interpretations are given to forms such as $a \leq x < b$, $a \leq x \leq b$, and $a < x < b$.

Let us now turn to simple **inequality statements** (inequality forms involving at least one variable) of the form

$$x > 2 \qquad\qquad -2 < x \leq 3$$
$$x \leq -3 \qquad\qquad 0 \leq x \leq 5$$

We are interested in graphing such statements on a real number line. In general, to **graph an inequality statement** in one variable on a real number line is to graph the set of all real number replacements of the variable that make the statement true. This set is called the **solution set** of the inequality statement.

EXAMPLE 6   Graph on a real number line: **(A)** $x > 2$ **(B)** $-2 < x \le 3$

Solution   **(A)** The solution set for $x > 2$ is the set of *all* real numbers greater than 2. Graphically, this set includes *all* the points to the right of 2:

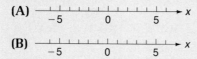

The parenthesis through 2 indicates that 2 is not included.

(B) The solution set for $-2 < x \le 3$ is the set of *all* real numbers between -2 and 3, including 3 but not -2. Graphically:

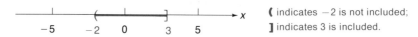

(indicates -2 is not included;
] indicates 3 is included.

PROBLEM 6   Graph on a real number line: **(A)** $x \le -3$ **(B)** $0 < x < 5$

Solution   **(A)**

(B)

More will be said about solving equations and inequalities of a more complicated nature in Chapter 4. Just as formulating algebraic equations is an important step in solving many practical problems, formulating appropriate algebraic inequality statements is an important step in solving other types of practical problems.

EXAMPLE 7   Translate each statement into an algebraic inequality statement using x as the only variable:

(A) 8 times a number is greater than or equal to 10 more than the number.
(B) 4 less than twice a number is less than 6 times the number.

Solution   **(A)** Let x = The unknown number(s); then the statement translates as follows:

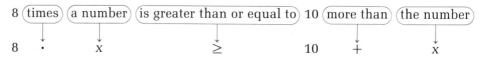

Or, more compactly,

$$8x \ge 10 + x$$

(B) Let x = The unknown number(s); then the statement translates as follows:

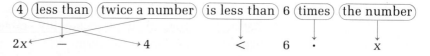

Or, more compactly,

$$2x - 4 < 6x$$

Note: "Less than" in the first part of the statement means "subtract from"; "is less than" in the middle of the statement means $<$.

PROBLEM 7 | Translate each statement into an algebraic inequality statement using x as the only variable:

(A) 7 less than twice a number is greater than 5 times the number.
(B) 3 more than a number is less than or equal to 5 less than twice the number.

Solution | (A)

(B)

Solutions to Matched Problems

4. (A) $7 = 3 + x$ (B) $12 = x - 9$ (C) $3x = 2x - 6$
   (D) $x - 6 = 2(x - 4)$
5. (A) T   (B) F   (C) T   (D) F   (E) T   (F) T
6. (A)
   (B)
7. (A) $2x - 7 > 5x$ (B) $3 + x \leq 2x - 5$

Practice Exercise 1-2 ■

Work odd-numbered problems first, check answers, and then work even-numbered problems in areas of weakness. Answers to all problems are in the back of the book. Make every effort to work a problem yourself before you look at an answer.

A *Write in symbolic form.*

1. $11 - 5$ is equal to $\frac{12}{2}$. _____
2. $\frac{36}{2}$ is equal to $3 \cdot 6$ _____

3. 4 is greater than -18. _____
4. 3 is greater than -7. _____

5. -12 is less than -3. _____
6. -20 is less than 1. _____

7. x is greater than or equal to -8. _____

8. x is less than or equal to 12. _____

9. x is between -2 and 2. _____

10. x is between 0 and 10. _____

Replace equality or inequality symbols with appropriate verbal form.

11. $\frac{48}{4} = 15 - 3$ _____   **12.** $24 + 3 = 9 \cdot 3$ _____

13. $12 > 11$ _____   **14.** $5 > -2$ _____

15. $11 < 12$ _____   **16.** $-2 < 5$ _____

17. $x \leq 4$ _____   **18.** $x \leq -3$ _____

19. $6 - 4 \neq \frac{9}{3}$ _____   **20.** $\frac{15}{3} \neq \frac{10}{5}$ _____

21. $2x - 1 \geq 3$ _____   **22.** $3(x + 4) \geq 25$ _____

Replace each question mark with $=$, $<$, or $>$ to form a true statement.

23. $-3 \; ? \; 7$ _____   **24.** $1 \; ? \; -6$ _____

25. $-2 \; ? \; -5$ _____   **26.** $0 \; ? \; -100$ _____

27. $-1.35 \; ? \; -1$ _____   **28.** $-6.33 \; ? \; -6$ _____

29. $\frac{55}{5} \; ? \; \frac{22}{2}$ _____   **30.** $10 + 3 \; ? \; \frac{39}{3}$ _____

31. $6 - 4 \; ? \; \frac{24}{6}$ _____   **32.** $6 \cdot 2 \; ? \; 4 + 6$ _____

33. $\frac{3}{2} \; ? \; \frac{3}{4}$ _____   **34.** $-\frac{3}{2} \; ? \; -\frac{3}{4}$ _____

35. $-10 \; ? \; -2 \; ? \; -1$ _____   **36.** $-2 \; ? \; 0 \; ? \; 2$ _____

Referring to the number line below, replace each question mark in Problems 37–42 with either $<$ or $>$ to form a true statement.

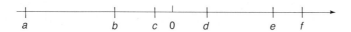

37. $e \; ? \; a$ _____   **38.** $a \; ? \; d$ _____   **39.** $c \; ? \; b$ _____

40. $e \; ? \; f$ _____   **41.** $0 \; ? \; d$ _____   **42.** $0 \; ? \; a$ _____

B **43.** If we add a positive real number to any real number, will the sum be greater than or less than the original number? _____

44. If we add a positive real number to any real number, will the sum be to the right or left of the original number on a number line? _____

Graph on a real number line for x a real number.

45. $x \leq 3$

46. $x \geq -2$

47. $-5 < x \leq -1$

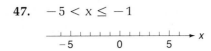

48. $-4 \leq x < 1$

49. $x > -4$

50. $x > 1$

51. $-1 < x < 3$

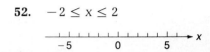

52. $-2 \leq x \leq 2$

Translate each statement into an algebraic equation or inequality statement using x as the only variable.

53. $x - 8$ is positive. _____

54. $2x - 8$ is negative. _____

55. $x + 4$ is not negative. _____

56. $4x + 1$ is not positive. _____

57. 80 is 3 more than twice a certain number. _____

58. 18 is 3 times a certain number. _____

59. x is greater than or equal to -3 and less than 4. _____

60. x is greater than 0 and less than 10. _____

61. 26 is 12 less than a certain number. _____

62. 32 is 5 less than a certain number. _____

63. x is less than 6 less than twice x. _____

64. x is greater than 9 more than 4 times x. _____

65. 6 times a number is 4 more than 3 times the number. _____

66. 7 times a number is 12 less than 4 times the number. _____

67. 6 less than a certain number is 5 times the number that is 7 more than the certain number. _____

68. 5 more than a certain number is 3 times the number that is 4 less than the certain number. _____

69. $\frac{9}{5}C + 32$ is between 63 and 72, inclusive. _____

70. $\frac{5}{9}(F - 32)$ is between 20 and 25, inclusive. _____

71. The sum of three consecutive natural numbers is 186. _____

72. The sum of three consecutive natural numbers is 372. _____

C *Replace each question mark with an appropriate symbol to make the statement an illustration of the given property of equality or inequality.*

73. Symmetry property: If $-5 = t$, then ?. _____

74. Transitive property: If $x < 3$ and $3 < y$, then ?. _____

75. Transitive property: If $5x + 7x = (5 + 7)x$ and $(5 + 7)x = 12x$, then ?.

_____

76. Trichotomy property: If $x \neq 3$, then either $x < 3$ or ?. _____

77. Reflexive property: $3 - x = $? _____

78. Substitution principle: If $2x + 3y = 5$ and $y = x - 3$, then $2x + 3(?) = 5$.

_____

79. What is wrong with the following argument? Four is an even number and 8 is an even number; hence we can write "4 = Even number" and "8 = Even number." By the symmetric law for equality, we can write "Even number = 8," and we can conclude, using the transitive law for equality (since 4 = Even number and Even number = 8), that 4 = 8. _____

80. What is wrong with the following argument? Rod is a human and Jan is a human; hence we can write "Rod = Human" and "Jan = Human." By the symmetric law for equality, we can write "Human = Jan," and we can conclude, using the transitive law for equality (since Rod = Human and Human = Jan), that Rod = Jan. _____

APPLICATIONS **81.** In a rectangle with area 90 square meters, the length is 3 meters less than twice the width. Write an equation relating the area with the length and width, using x as the only variable. [*Note:*   Area = (Length)(Width).]

_____

82. In a rectangle with area 500 square inches, the length is 5 inches longer than the width. Write an equation relating the area with the length and the width, using x as the only variable. _____

*83. In a rectangle with perimeter 210 centimeters the length is 10 centimeters less than 3 times the width. Write an equation relating the perimeter with the length and width, using x as the only variable. _____

*84. In a rectangle with perimeter 186 feet, the width is 7 feet less than the length. Write an equation relating the perimeter with the length and the width, using x as the only variable. _____

**85. A rectangular lot is to be fenced with 400 feet of wire fencing.

   (A) If the lot is x feet wide, write a formula for the area of the lot in terms of x. _____

(B) What real number values can x assume? Write the answer in terms of a double inequality statement that also allows the area to be 0.

The Check Exercise for this section is on page 55.

**86. A flat sheet of cardboard in the shape of an 18 by 12 inch rectangle is to be used to make an open-topped box by cutting an x by x inch square out of each corner and folding the remaining part appropriately.

(A) Write an equation for the volume of the box in terms of x. _____

(B) What real number values can x assume? Formulate the answer in terms of a double inequality statement that also allows the volume to be 0. _____

[*Note:* Volume = (Length)(Width)(Height).]

Section 1-3 Real Number Properties

- Overview
- Commutative Properties
- Associative Properties
- Simplifying Algebraic Expressions
- Identity Properties

■ Overview

In the last section we discussed algebraic expressions, equations, and inequality statements. We now take a closer look at some of the basic properties of the set of real numbers that enable us to convert algebraic expressions into "equivalent forms." These assumed properties, called **axioms**, become operational rules in the algebra of real numbers.

AXIOMS

Basic Properties of the Set of Real Numbers
Let R be the set of real numbers and x, y, and z arbitrary elements of R.
ADDITION PROPERTIES
CLOSURE:　　　　$x + y$ is a unique element in R.
ASSOCIATIVE:　　$(x + y) + z = x + (y + z)$
COMMUTATIVE:　$x + y = y + x$
IDENTITY:　　　　0 is the additive identity; that is, $0 + x = x + 0 = x$ for all $x \in R$, and 0 is the only element in R with this property.
INVERSE:　　　　For each $x \in R$, $-x$ is its unique additive inverse; that is, $x + (-x) = (-x) + x = 0$, and $-x$ is the only element in R relative to x with this property.

Continued

MULTIPLICATION PROPERTIES

CLOSURE: xy is a unique element in R.

ASSOCIATIVE: $(xy)z = x(yz)$

COMMUTATIVE: $xy = yx$

IDENTITY: 1 is the multiplicative identity; that is, for $x \in R$, $1x = x1 = x$, and 1 is the only element in R with this property.

INVERSE: For each $x \in R$, $x \neq 0$, $1/x$ is its unique multiplicative inverse; that is, $x(1/x) = (1/x)x = 1$, and $1/x$ is the only element in R relative to x with this property.

COMBINED PROPERTY

DISTRIBUTIVE: $x(y + z) = xy + xz$

Don't let the names of these properties frighten you. Most of the ideas presented here are quite simple. In fact, you have been using many of these properties in arithmetic for a long time. In this section we will informally consider several of the properties and will discuss others as needed in other parts of the book.

■ Commutative Properties

You are already familiar with the commutative properties for addition and multiplication. They simply indicate that the order in which addition or multiplication is performed doesn't matter: $2 + 3 = 3 + 2$ and $2 \cdot 3 = 3 \cdot 2$.

EXAMPLE 8   If x and y are real numbers, then, using the commutative properties for addition or multiplication, we know that

$$x + 9 = 9 + x \qquad\qquad 7y = y7$$
$$2x + 3y = 3y + 2x \qquad 3 + yx = 3 + xy$$

PROBLEM 8   Use an appropriate commutative property to replace each question mark with an appropriate symbol:

(A) $x + 3 = 3 + ?$ (B) $3x + 5y = 5y + ?$
(C) $5(y + x) = 5(? + y)$ (D) $y + x4 = y + 4?$

Solution   (A)   (B)

(C)   (D)

Does the commutative property hold relative to subtraction and division? That is, do $x - y = y - x$ and $x \div y = y \div x$ for all real numbers x and y, division by 0 excluded? The answer is no, since, for example, $5 - 3 \neq 3 - 5$ and $8 \div 4 \neq 4 \div 8$.

■ Associative Properties

When computing

$$4 + 3 + 5 \qquad \text{or} \qquad 4 \cdot 3 \cdot 5$$

why don't we need parentheses to show us which two numbers are to be added or multiplied first? The answer is to be found in the associative axioms. These axioms allow us to write

$$(4 + 3) + 5 = 4 + (3 + 5) \qquad \text{and} \qquad (4 \cdot 3) \cdot 5 = 4 \cdot (3 \cdot 5)$$

so it doesn't matter how we group relative to either operation.

EXAMPLE 9   If x, y, and z represent real numbers, then, using the associative axiom, we know that

$$(x + 7) + 2 = x + (7 + 2) \qquad 3(5y) = (3 \cdot 5)y$$
$$x + (x + 3) = (x + x) + 3 \qquad (2y)y = 2(yy)$$

Thus, the associative axiom tells us how parentheses can be moved relative to addition and multiplication.

PROBLEM 9   Using an appropriate associative axiom, replace each question mark with an appropriate symbol:

(A) $(x + 3) + 9 = x + (?)$ **(B)** $6(3y) = (?)y$
(C) $(5 + z) + z = 5 + (?)$ **(D)** $(3z)z = 3(?)$

Solution   **(A)**   **(B)**

(C)   **(D)**

Does the associative axiom hold for subtraction and division? The answer is no, since, for example, $(8 - 4) - 2 \neq 8 - (4 - 2)$ and $(8 \div 4) \div 2 \neq 8 \div (4 \div 2)$. Evaluate each side of each equation to see why.

Conclusion
Relative to addition, commutativity and associativity permit us to change the order of addition at will and insert or remove parentheses as we please. The same is true for multiplication, but not for subtraction and division.

■ Simplifying Algebraic Expressions

The commutative and associative axioms provide us with our first tools for transforming algebraic expressions into equivalent forms. (Two algebraic expressions are **equivalent** over given replacement sets for the variables if both yield equal numbers for each replacement of the variables by numbers from their respective replacement sets.) Let us use the associative and commutative properties to formally transform

$$(x + 8) + (y + 2)$$

into a simpler equivalent form. By formally transforming, we mean that each step will be justified by a basic axiom or an earlier stated property. We will then indicate how most of these steps are performed mentally.

$$
\begin{aligned}
(x + 8) + (y + 2) &= [(x + 8) + y] + 2 && \text{Associative property for } + \\
&= [x + (8 + y)] + 2 && \text{Associative property for } + \\
&= [x + (y + 8)] + 2 && \text{Commutative property for } + \\
&= x + [(y + 8) + 2] && \text{Associative property for } + \\
&= x + [y + (8 + 2)] && \text{Associative property for } + \\
&= x + (y + 10) && \text{Substitution property for } = \\
&= (x + y) + 10 && \text{Associative property for } +
\end{aligned}
$$

Normally, we do most of these steps mentally and simply write

$$
\begin{aligned}
(x + 8) + (y + 2) &= x + 8 + y + 2 \\
&= x + y + 10
\end{aligned}
$$

Even though we did not write each step as in the formal treatment, you should not lose sight of the fact that the associative and commutative axioms are behind the mental steps taken in the simpler version.

EXAMPLE 10   Simplify mentally, using commutative and associative axioms:

(A) $(3x + 5) + (2y + 7) = 3x + 5 + 2y + 7$
$$
= 3x + 2y + 12
$$

(B) $(6x)(3y) = 6x3y$
$$
\begin{aligned}
&= 6 \cdot 3xy \\
&= 18xy
\end{aligned}
$$

PROBLEM 10   Repeat Example 10 for:

(A) $(2a + 3) + (3b + 11)$ **(B)** $(4m)(6n)$

Solution   **(A)**

(B)

■ Identity Properties

The identity axiom for addition states that 0 is the unique real number having the property that when it is added to any real number we get that number back again. Thus,

$$
5 + 0 = 5 \qquad 0 + 3x = 3x \qquad (x + y) + 0 = x + y
$$

Similarly, the number 1 plays the same role relative to multiplication. That is, when any number is multiplied by 1, we get that number back again. Thus, we can write

$$
1 \cdot 8 = 8 \qquad 1x = x \qquad 1xy = xy
$$

The other axioms will be discussed as needed.

Solutions to Matched
Problems

> **8. (A)** $x + 3 = 3 + [x]$ **(B)** $3x + 5y = 5y + [3x]$
>   **(C)** $5(y + x) = 5([x] + y)$ **(D)** $y + x4 = y + 4[x]$
> **9. (A)** $(x + 3) + 9 = x + (3 + 9)$ **(B)** $6(3y) = (6 \cdot 3)y$
>   **(C)** $(5 + z) + z = 5 + (z + z)$ **(D)** $(3z)z = 3(zz)$
> **10. (A)** $(2a + 3) + (3b + 11) = 2a + 3 + 3b + 11 = 2a + 3b + 14$
>   **(B)** $(4m)(6n) = 4m6n = 24mn$

Practice Exercise 1-3 ■

Work odd-numbered problems first, check answers, and then work even-numbered problems in areas of weakness. Answers to all problems are in the back of the book. Make every effort to work a problem yourself before you look at an answer.

All variables represent real numbers.

A   Replace each question mark with an appropriate expression that will illustrate the use of the indicated real number property.

1.   Commutative property $(+)$: $x + 3 = ?$ _____

2.   Commutative property $(+)$: $m + n = ?$ _____

3.   Associative property $(\cdot)$: $5(7z) = ?$ _____

4.   Associative property $(\cdot)$: $(uv)w = ?$ _____

5.   Commutative property $(\cdot)$: $nm = ?$ _____

6.   Commutative property $(\cdot)$: $dc = ?$ _____

7.   Associative property $(+)$: $9 + (11 + M) = ?$ _____

8.   Associative property $(+)$: $(x + 7) + 5 = ?$ _____

9.   Identity property $(+)$: $7x + 0 = ?$ _____

10.   Identity property $(+)$: $0 + (x + z) = ?$ _____

11.   Identity property $(\cdot)$: $1(x + y) = ?$ _____

12.   Identity property $(\cdot)$: $1(uv) = ?$ _____

State the justifying axiom for each statement.

13. $12 + w = w + 12$ _____   14. $2x + 3 = 3 + 2x$

15. $m + (n + 3) = (m + n) + 3$ _____

16. $(3x + y) + 5 = 3x + (y + 5)$ _____

17. $20x = x20$ _____   **18.** $MN = NM$ _____

19. $4(8y) = (4 \cdot 8)y$ _____   **20.** $(12u)v = 12(uv)$ _____

21. $3x + 0 = 3x$ _____   **22.** $0 + (2x + 3) = 2x + 3$ _____

23. $1m = m$ _____   **24.** $uv = 1uv$ _____

Remove parentheses and simplify:

25. $(x + 7) + 2$ _____   **26.** $3 + (5 + m)$ _____

27. $4(5y)$ _____   **28.** $6(8n)$ _____

29. $12 + (u + 3)$ _____   **30.** $(4 + x) + 13$ _____

31. $(3x)7$ _____   **32.** $4(y3)$ _____

33. $0 + 1x$ _____   **34.** $(1y + 3) + 0$ _____

B   *State the justifying axiom for each statement.*

35. $2 + (y + 3) = 2 + (3 + y)$ _____

36. $(3m)n = n(3m)$ _____

37. $7(y4) = 7(4y)$ _____

38. $5 + (y + 2) = (y + 2) + 5$ _____

39. $3x + 2y = 2y + 3x$ _____

40. $3x + (2x + 5y) = (3x + 2x) + 5y$ _____

41. $(2x)(x + 3) = 2[x(x + 3)]$ _____

42. $(x + 3) + (2 + y) = x + [3 + (2 + y)]$ _____

Remove parentheses and simplify:

43. $(x + 7) + (y + 4) + (z + 1)$ _____

44. $(7 + m) + (8 + n) + (3 + p)$ _____

45. $(3x + 5) + (4y + 6)$ _____   **46.** $(3a + 7) + (5b + 2)$ _____

47. $0 + (1x + 3) + (y + 2)$ _____   **48.** $1(x + 3) + 0 + (y + 2)$ _____

49. $(12m)(3n)(1p)$ _____   **50.** $(8x)(4y)(2z)$ _____

C **51.** Indicate whether true (T) or false (F), and for each false statement find real number replacements for a and b that will illustrate its falseness. For all real numbers a and b:

(A) $a + b = b + a$ _____ (B) $a - b = b - a$ _____

(C) $ab = ba$ _____ (D) $a \div b = b \div a$ _____

52. Indicate whether true (T) or false (F), and for each false statement find real number replacements for a, b, and c that will illustrate its falseness. For all real numbers a, b, and c:

(A) $(a + b) + c = a + (b + c)$ _____

(B) $(a - b) - c = a - (b - c)$ _____

(C) $a(bc) = (ab)c$ _____

(D) $(a \div b) \div c = a \div (b \div c)$ _____

53. Supply a reason for each step.

STATEMENT		REASON
1. $(x + 3) + (y + 4) = (x + 3) + (4 + y)$	**1.** _____
2. $\qquad = x + [3 + (4 + y)]$	**2.** _____
3. $\qquad = x + [(3 + 4) + y]$	**3.** _____
4. $\qquad = x + (7 + y)$	**4.** _____
5. $\qquad = x + (y + 7)$	**5.** _____
6. $\qquad = (x + y) + 7$	**6.** _____

54. Supply a reason for each step.

STATEMENT	REASON
1. $(5x)(2y) = (x5)(2y)$	**1.** _____
2. $\qquad = x[5(2y)]$	**2.** _____
3. $\qquad = x[(5 \cdot 2)y]$	**3.** _____
4. $\qquad = x(10y)$	**4.** _____
5. $\qquad = (x10)y$	**5.** _____
6. $\qquad = (10x)y$	**6.** _____
7. $\qquad = 10(xy)$	**7.** _____

The Check Exercise for this section is on page 57.

Section 1-4   Addition and Subtraction

- The Negative of a Number
- The Absolute Value of a Number
- Addition of Real Numbers
- Subtraction of Real Numbers
- Combined Operations

Before we review addition and subtraction of real numbers we will say a few words about the operations "the negative of" and "the absolute value of." These operations are useful in describing operations related to addition, subtraction, multiplication, and division of real numbers.

■ The Negative of a Number

For each real number x, we denoted its additive inverse by

$$-x \begin{cases} \text{Additive inverse of } x \\ \text{Opposite of } x \\ \text{Negative of } x \end{cases}$$

All the names on the right describe the same thing and are used interchangeably. You will recall that the **opposite of or negative of a number x** is obtained from x by changing its sign. The opposite of or negative of 0 is 0.

EXAMPLE 11   **(A)** $-(+5) = -5$ **(B)** $-(-8) = +8$ or 8
(C) $-(0) = 0$ **(D)** $-[-(-4)] = -(+4) = -4$

Examples 11(A) and (B) illustrate the fact that:

$-x$ is not necessarily a negative number.

That is, $-x$ represents a negative number if x is positive and a positive number if x is negative.
   It is not difficult to show that

THEOREM 1 | **Double Negative Property** |
|---|
| For a any real number: $$-(-a) = a$$ |

PROBLEM 11   Find:

(A) $-(+11)$ **(B)** $-(-12)$ **(C)** $-(0)$ **(D)** $-[-(+6)]$

Solution   **(A)**   **(B)**

(C)   **(D)**

It is now important to note the three distinct uses of the minus sign.

Multiple Uses of the Minus Sign

1. As the operation "subtract": $9 \overset{\downarrow}{-} 3 = 6$
2. As the operation "the negative or opposite of": $\overset{\downarrow}{-}(-8) = 8$
3. As part of a number symbol: $\overset{\downarrow}{-}4$

■ **The Absolute Value of a Number**

The **absolute value** of a number x is an operation on x, denoted by the symbol

$$|x|$$

(not square brackets). The absolute value of a number can be thought of geometrically as the distance of the number from 0 on the real number line expressed as a positive number or 0. For example, both 5 and -5 are five units from 0 (see Figure 4). Thus, we can write $|5| = 5$ and $|-5| = 5$. Figure 4 also illustrates the fact that $|-8| = 8$ and $|7| = 7$.

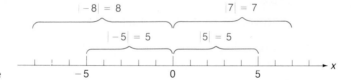

FIGURE 4 Absolute value

Symbolically, and more formally, we define absolute value as follows:

Absolute Value

$$|x| = \begin{cases} x & \text{if x is positive} \\ 0 & \text{if x is 0} \\ -x & \text{if x is negative} \end{cases}$$

Note: $-x$ is positive if x is negative.

It is important to remember that:

The absolute value of a number is never negative.

EXAMPLE 12   **(A)** $|24| = 24$ **(B)** $|-7| = 7$ **(C)** $|0| = 0$
(D) $-(|-8| + |-3|) = -(8 + 3) = -(11) = -11$

Note: $-(11)$ represents the opposite or negative of the positive number 11, while -11 represents a negative number.

PROBLEM 12   Evaluate:

(A) $|-13|$ (B) $|43|$ (C) $-|-4|$ (D) $-(|-6|-|+2|)$

Solution   (A)   (B)

(C)   (D)

■ Addition of Real Numbers

We are now ready to review addition of real numbers. We will consider three cases: addition of positive numbers, addition of negative numbers, and addition of numbers with unlike signs. From the axioms in Section 1-3, we know that $0 + x = x + 0 = x$ for all real numbers x, so 0 involved in addition is taken care of.

Addition of Positive Numbers
Add positive numbers as in arithmetic. (We assume you can do this.)

For example:

$$18 + 12 = 30 \qquad 6.32 + 1.04 = 7.36$$

We now consider the other two cases. The axioms for the real numbers (Section 1-3) and the operations "the negative of" and "absolute value of" can be used to produce operational rules for adding negative numbers and adding numbers with unlike signs. We state these rules without proof.

THEOREM 2

Addition of Negative Numbers
If a and b are both negative, their sum is the negative of the sum of their absolute values.

Mentally block out the signs of the two numbers (take absolute values), add as in arithmetic, and then attach a minus sign to the result. For example:

$$(-5) + (-3) \;\overline{\left| = -(5 + 3) \right|}^{\dagger} = -8$$

THEOREM 3

Addition of Numbers with Unlike Signs
To add two numbers with unlike signs, subtract the smaller absolute value from the larger absolute value. Then attach the sign of the number with the largest absolute value to the result.

† Throughout the book dashed boxes are used to represent steps that are usually done mentally.

Mentally block out the signs of the two numbers (take their absolute values), subtract the smaller from the larger, and then attach the sign of the number with the largest absolute value. For example:

$$(-3) + (+9) \; \boxed{= +(9 - 3)} = 6$$

$$(+3) + (-9) \; \boxed{= -(9 - 3)} = -6$$

To add three or more numbers, add all of the positive numbers together, add all of the negative numbers together (the commutative and associative axioms justify this procedure), and then add the two resulting sums as above.

EXAMPLE 13 $3 + (-6) + 8 + (-4) + (-5) = (3 + 8) + [(-6) + (-4) + (-5)]$

$$= 11 + (-15) \; \boxed{= -(15 - 11)} = -4$$

PROBLEM 13   Add: $6 + (-8) + (-4) + 10 + (-3) + 1$

Solution

■ **Subtraction of Real Numbers**

We define subtraction as follows:

Subtraction

For a and b any real numbers:

$$a - b = a + (-b)$$

To subtract b from a, add the opposite of b to a.

Recall that the opposite of b is the same as the negative of b and the additive inverse of b:

Opposite of -9

$$(-3) - (-9) = (-3) + 9 = 6$$

Change to addition

You should get to the point where you can perform this type of subtraction mentally, and simply write down the answer.

EXAMPLE 14   **(A)** $8 - (-5) \; \boxed{= 8 + 5} = 13$

(B) $(-8) - 5 \; \boxed{= (-8) + (-5)} = -13$

(C) $(-8) - (-5) \boxed{= (-8) + 5} = -3$

(D) $(0 - 5) \boxed{= 0 + (-5)} = -5$

PROBLEM 14   Subtract:

(A) $4 - 7$ **(B)** $7 - (-4)$ **(C)** $(-7) - 4$
(D) $(-7) - (-4)$ **(E)** $(-4) - (-7)$ **(F)** $0 - (-4)$

Solution   **(A)**   **(B)**

             (C)   **(D)**

             (E)   **(F)**

■ Combined Operations

When three or more terms are combined by addition and subtraction and symbols of grouping are omitted, we convert (mentally) any subtraction to addition and add. Thus,

$$8 - 5 + 3 \boxed{= 8 + (-5) + 3} = 6$$
$$\text{\emph{Think}}$$

EXAMPLE 15   **(A)** $2 - 3 - 7 + 4 \boxed{= 2 + (-3) + (-7) + 4} = -4$

$$\text{\emph{Think}}$$

             (B) $-4 - 8 + 2 + 9 \boxed{= (-4) + (-8) + 2 + 9} = -1$

$$\text{\emph{Think}}$$

PROBLEM 15   Evaluate: **(A)** $5 - 8 + 2 - 6$ **(B)** $-6 + 12 - 2 - 1$

Solution   **(A)**

       (B)

EXAMPLE 16   Evaluate each for $x = 2$, $y = -3$, and $z = -9$:

(A) $x + y$ **(B)** $y - z$ **(C)** $y - (z - x)$ **(D)** $\left| (-y) - |z| \right|$

Solution   **(A)** $x + y$

        $\boxed{(\ \ ) + (\ \ )}$ Use of parentheses as indicated prevents many sign errors.

        $(2) + (-3) = -1$

(B) $y - z$

$$(\quad) - (\quad)$$

$$(-3) - (-9)$$

$$= (-3) + (9) = 6$$

(C) $y - (z - x)$

$(-3) - [(-9) - 2]$ — Replace parentheses with brackets (another form of parentheses), substitute values, and then evaluate starting inside the square brackets.

$$= (-3) - (-11)$$

$$= (-3) + (11) = 8$$

(D) $\left| (-y) - |z| \right|$

$\left| [-(-3)] - |-9| \right|$ — Evaluate $-(-3)$ and $|-9|$ first.

$= |(3) - (9)|$ — Subtract inside absolute value signs.

$= |-6| = 6$ — Take the absolute value.

PROBLEM 16   Evaluate for $x = -4$, $y = 5$, and $z = -11$:

(A) $y + z$ **(B)** $x - y$ **(C)** $(z - x) + y$ **(D)** $\left| (x) - |z| \right|$

Solution   **(A)**   **(B)**

(C)   **(D)**

EXAMPLE 17   Using a calculator, evaluate each for $x = -504.394$, $y = 829.077$, and $z = -1,023.998$:

(A) $y - x$ **(B)** $x - (y - z)$

Solution   First store x, y, and z in memories 1, 2, and 3:

$\boxed{504.394}\,\boxed{+/-}\,\boxed{STO}\,\boxed{1}$

$\boxed{829.077}\,\boxed{STO}\,\boxed{2}$

$\boxed{1,023.998}\,\boxed{+/-}\,\boxed{STO}\,\boxed{3}$

(A) $y - x = 1,333.471$

$\boxed{RCL}\,\boxed{2}\,\boxed{-}\,\boxed{RCL}\,\boxed{1}\,\boxed{=}$

(B) $x - (y - z) = -2,357.469$

$\boxed{RCL}\,\boxed{1}\,\boxed{-}\,\boxed{(}\,\boxed{RCL}\,\boxed{2}\,\boxed{-}\,\boxed{RCL}\,\boxed{3}\,\boxed{)}\,\boxed{=}$

PROBLEM 17   Use the values of x, y, and z in Example 17 to evaluate:

(A) $x - y$ **(B)** $x - (z - y)$

Solution   **(A)**   **(B)**

11. **(A)** $-(+11) = -11$ **(B)** $-(-12) = 12$ **(C)** $-(0) = 0$
    (D) $-[-(+6)] = -(-6) = 6$
12. **(A)** $|-13| = 13$ **(B)** $|43| = 43$ **(C)** $-|-4| = -(+4) = -4$
    (D) $-(|-6| - |+2|) = -(6 - 2) = -(+4) = -4$
13. $6 + (-8) + (-4) + 10 + (-3) + 1 = 17 + (-15) = 2$
14. **(A)** $4 - 7 = -3$ **(B)** $7 - (-4) = 11$
    (C) $(-7) - 4 = -11$ **(D)** $(-7) - (-4) = -3$
    (E) $(-4) - (-7) = 3$ **(F)** $0 - (-4) = 4$
15. **(A)** $5 - 8 + 2 - 6 = 7 - 14 = -7$
    (B) $-6 + 12 - 2 - 1 = 12 - 9 = 3$
16. **(A)** $5 + (-11) = -6$ **(B)** $-4 - 5 = -9$
    (C) $[-11 - (-4)] + 5 = (-7) + 5 = -2$
    (D) $\big||(-4) - |-11|\big| = |(-4) - 11| = |-15| = 15$
17. **(A)** $-1{,}333.471$ **(B)** $1{,}348.681$

Practice Exercise 1-4 ■

Work odd-numbered problems first, check answers, and then work even-numbered problems in areas of weakness. Answers to all problems are in the back of the book. Make every effort to work a problem yourself before you look at an answer.

A   *Evaluate.*

1. $-(+7)$ _____

2. $-(+12)$ _____

3. $-(-6)$ _____

4. $-(-8)$ _____

5. $|+2|$ _____

6. $|+9|$ _____

7. $|-27|$ _____

8. $|-32|$ _____

9. $|0|$ _____

10. $-(0)$ _____

11. $(-7) + (-3)$ _____

12. $(-7) + (+3)$ _____

13. $(+7) + (-3)$ _____

14. $(-12) + (+8)$ _____

15. $(+3) - (+9)$ _____

16. $(+3) - (-9)$ _____

17. $(+9) - (-3)$ _____

18. $(-9) - (-3)$ _____

19. The negative of a number is (*always, sometimes, never*) a negative number.

_____

20. The absolute value of a number is (*always, sometimes, never*) a positive number. _____

B   *Evaluate.*

21. $-[-(-3)]$ _____   22. $-[-(+6)]$ _____

23. $-|-(+2)|$ _____   24. $-|-(-3)|$ _____

25. $-(|-9|-|-3|)$ _____   26. $-(|-14|-|-8|)$ _____

27. $(-2)+(-6)+3$ _____   28. $(-2)+(-8)+5$ _____

29. $5-7-3$ _____   30. $3-2+4$ _____

31. $-7+6-4$ _____   32. $-4+7-6$ _____

33. $-2-3+6-2$ _____   34. $-4+7-3-2$ _____

35. $6-[3-(-9)]$ _____   36. $(-10)-[(-6)+3]$ _____

37. $[6-(-8)]-[(-8)-6]$ _____   38. $[3-5]+[(-5)-(-2)]$ _____

Replace each question mark with an appropriate real number.

39. $-(?)=5$ _____   40. $-(?)=-8$ _____

41. $|?|=7$ _____   42. $|?|=-4$ _____

43. $(-3)+?=-8$ _____   44. $?+5=-6$ _____

45. $(-3)-?=-8$ _____   46. $?-(-2)=-4$ _____

Evaluate for $x=3$, $y=-8$, *and* $z=-2$.

47. $x+y$ _____   48. $y+z$ _____

49. $y-x$ _____   50. $y-z$ _____

51. $(x-z)+y$ _____   52. $y-(z-x)$ _____

53. $|(-z)-|y||$ _____   54. $||-y|-|12||$ _____

55. $-||y|-|x||$ _____   56. $-||-10|-|x||$ _____

C   *Which of the following hold for all integers a, b, and c? Illustrate each false statement with an example showing that it is false.*

57. $a+b=b+a$ _____   58. $a+(-a)=0$ _____

59. $a-b=b-a$ _____   60. $a-b=a+(-b)$ _____

61. $(a+b)+c=a+(b+c)$ _____

62. $(a-b)-c=a-(b-c)$ _____

63. $|a + b| = |a| + |b|$ _____

64. $|a - b| = |a| - |b|$ _____

65. Supply the reasons for each of the following steps:

STATEMENT		REASON
1. $b + [a + (-b)] = b + [(-b) + a]$	**1.**	_____
2. $\qquad = [b + (-b)] + a$	**2.**	_____
3. $\qquad = 0 + a$	**3.**	_____
4. $\qquad = a$	**4.**	_____

66. Supply the reasons for each step:

STATEMENT		REASON
1. $(a + b) + [(-a) + (-b)] = (b + a) + [(-a) + (-b)]$	**1.**	_____
2. $\qquad = [(b + a) + (-a)] + (-b)$	**2.**	_____
3. $\qquad = \{b + [a + (-a)]\} + (-b)$	**3.**	_____
4. $\qquad = (b + 0) + (-b)$	**4.**	_____
5. $\qquad = b + (-b)$	**5.**	_____
6. $\qquad = 0$	**6.**	_____
7. Therefore, $-(a + b) = (-a) + (-b)$.	**7.**	_____

CALCULATOR PROBLEMS

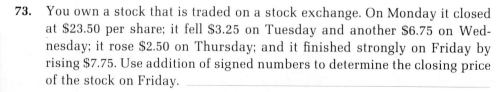

Evaluate each for $x = 23.417$, $y = -52.608$, _and_ $z = -13.012$.

67. $y + x$ _____   **68.** $x + z$ _____

69. $x - y$ _____   **70.** $z - y$ _____

71. $(x - z) + y$ _____   **72.** $(y + z) - x$ _____

APPLICATIONS

73. You own a stock that is traded on a stock exchange. On Monday it closed at $23.50 per share; it fell $3.25 on Tuesday and another $6.75 on Wednesday; it rose $2.50 on Thursday; and it finished strongly on Friday by rising $7.75. Use addition of signed numbers to determine the closing price of the stock on Friday. _____

The Check Exercise for
this section is on
page 59.

74. Find, using subtraction of signed numbers, the difference in the height between the highest point in the United States, Mount McKinley (20,270 feet) and the lowest point in the United States, Death Valley (-280 feet).

Section 1-5 Multiplication, Division, and Combined Operations

- Multiplication of Real Numbers
- Division of Real Numbers
- Combined Operations

Having discussed addition and subtraction, we now turn to multiplication and division. We will then be in a position to consider problems involving all four arithmetic operations $(+, -, \cdot, \div)$.

■ Multiplication of Real Numbers

We start with the product of two positive numbers.

Multiplication of Positive Numbers
Multiply positive numbers as in arithmetic. (We assume you can do this.) The product is a positive number.

For example:

$5 \cdot 7 = 35$ $(3.24)(6.13) = 19.8612$

THEOREM 4

Multiplication Involving Zero
For any real number a: $a \cdot 0 = 0$

This is a theorem that can be proved on the basis of the material in the preceding sections (see Problem 75 in Practice Exercise 1-5).

To get an idea of how numbers with unlike signs must be multiplied, consider the product

$(+2)(-7)$

We start with something we know is true and then proceed through a sequence of logical steps to a conclusion that must also be true:

$(+7) + (-7) = 0$	Inverse property for addition
$(+2)[(+7) + (-7)] = (+20)0$	Property of equality (if $a = b$, then $ca = cb$) that follows from properties of equality in Section 1-2
$(+2)0 = 0$	Theorem 4
$(+2)(+7) + (+2)(-7) = 0$	Distributive axiom and transitive property for equality
$(+14) + (+2)(-7) = 0$	Substitution principle for equality
$(+2)(-7) = -14$	Additive inverse axiom $[(+2)(-7)$ must be the additive inverse of $(+14)]$

Thus, we see that if the axioms of the real numbers stated in Section 1-3 hold, then the product of $(+2)$ and (-7) must be -14. There is no other choice! Similar arguments are used in the general proof of the following theorem.

THEOREM 5

Multiplication Involving a Negative Number
Numbers with unlike signs: The product of two numbers with unlike signs is a negative number and is found by taking the negative of the product of the absolute values of the two numbers. *Two negative numbers:*   The product of two negative numbers is positive and is found by multiplying the absolute values of the two numbers.

Thus, we see that:

The product of two numbers with unlike signs is negative.
The product of two numbers with like signs is positive.

EXAMPLE 18   **(A)** $2(-7) \boxed{= -(2 \cdot 7)} = -14$

(B) $(-2)(-7) \boxed{= 2 \cdot 7} = 14$

(C) $0(-7) = 0$

PROBLEM 18   Evaluate: **(A)** $4(-3)$ **(B)** $(-4)3$ **(C)** $(-4)(-3)$ **(D)** $0(-3)$

Solution   **(A)**   **(B)**

(C)   **(D)**

Several important sign properties for multiplication are summarized in the following theorem.

THEOREM 6

Sign Properties for Multiplication
For a and b any real numbers: **(A)** $(-1)a = -a$ **(B)** $(-a)b = -(ab)$ **(C)** $(-a)(-b) = ab$

EXAMPLE 19   Evaluate $(-a)b$ and $-(ab)$ for $a = -5$ and $b = 4$.

Solution $(-a)b = [-(-5)]4 = 5 \cdot 4 = 20$

$-(ab) = -[(-5)4] = -(-20) = 20$

PROBLEM 19

Evaluate $(-a)(-b)$ and ab for $a = -5$ and $b = 4$.

Solution

$(-a)(-b) =$

$ab =$

Expressions of the form

$-ab$

occur frequently and at first glance are confusing to students. If you were asked to evaluate $-ab$ for $a = -3$ and $b = +2$, how would you proceed? Would you take the negative of a and then multiply it by b, or multiply a and b first and then take the negative of the product? Actually it does not matter! Because of Theorem 6 we get the same result either way since $(-a)b = -(ab)$. If we consider other material in this section, we find that

$$-ab = \begin{cases} (-a)b \\ a(-b) \\ -(ab) \\ (-1)ab \end{cases}$$

For example,

$-3(4) = (-3)(4) = 3(-4) = -(3)(4) = (-1)(3)(4)$

and we are at liberty to replace any one of these five forms with another from the same group.

EXAMPLE 20

Evaluate $-ab$ for $a = -7$ and $b = 4$.

Solution

Most people would proceed in one of the following ways:

$$-(-7)(4) = \begin{cases} -[(-7)(4)] = -(-28) = 28 \\ [-(-7)](4) = (7)(4) = 28 \end{cases}$$

Both are correct.

PROBLEM 20

Evaluate $-ab$ for $a = 6$ and $b = -3$ two different ways.

Solution

■ Division of Real Numbers

You will recall in arithmetic that to check the division problem

$$9 \overline{\smash{)}36} \;\; \overset{4}{}$$

we multiply 9 by 4 to obtain 36. We will use this checking requirement to transform division into multiplication. Instead of asking "What is 9 divided into 36?", we ask "What number times 9 is 36?" Both questions have the same answer. The latter way of looking at division is the more useful of the two because of its generalization to other number systems.

Definition of Division

We write

$$a \div b = Q$$
$$\left.\begin{array}{c} Q \\ b \,\overline{)\, a} \end{array}\right\} \qquad \text{if and only if} \qquad Qb = a \text{ and } Q \text{ is unique}$$

The quotient Q is the number that must be multiplied times b to produce a.

Let us use the definition to find

$$12 \div (-3) = ? \qquad \text{or} \qquad -3 \,\overline{)\, 12}^{\;?}$$

We ask "What number must (-3) be multiplied by to produce 12?" From our discussion of multiplication, we know the answer to be -4. Thus, we write

$$12 \div (-3) = -4 \qquad \text{or} \qquad -3 \,\overline{)\, 12}^{\;-4}$$

since $(-4)(-3) = 12$.

What about division involving 0?

$$5 \,\overline{)\, 0}^{\;?} \qquad ? \cdot 5 = 0$$

$$0 \,\overline{)\, 5}^{\;?} \qquad ? \cdot 0 = 5$$

$$0 \,\overline{)\, 0}^{\;?} \qquad ? \cdot 0 = 0$$

In the first case the quotient is 0, since $0 \cdot 5 = 0$. In the second case we find that no real number times 0 can produce 5; hence, this quotient is not defined. In the third case we have the other extreme—any real number will produce 0 when multiplied times 0; hence, the quotient is not unique. We conclude that:

Zero cannot be used as a divisior—ever!

The two division symbols $\div$ and $\overline{)}$ from arithmetic are not used a great deal in algebra and higher mathematics. The horizontal bar (—) and slash mark (/) are the symbols most frequently used. Thus

$$a/b \qquad \frac{a}{b} \qquad a \div b \qquad \text{and} \qquad b \,\overline{)\, a} \qquad \text{In each case } b \text{ is the divisor.}$$

all name the same number (assuming the quotient is defined), and we can write

$$a/b = \frac{a}{b} = a \div b = b \,\overline{)\, a}$$

Now to the mechanics of division:

Division and Positive Numbers

Divide positive numbers as in arithmetic. (We assume you can do this.)

For example:

$$\frac{8}{4} = 2 \qquad \frac{5}{7} = 0.\overline{714285}$$

THEOREM 7

Division Involving Zero

For a any nonzero real number:

$$\frac{0}{a} = 0 \qquad \frac{a}{0} \text{ is not defined} \qquad \frac{0}{0} \text{ is not defined}$$

For example:

$$\frac{0}{-3} = 0 \qquad \frac{-3}{0} \text{ is not defined} \qquad \frac{0}{0} \text{ is not defined}$$

THEOREM 8

Division Involving a Negative Number

Numbers with unlike signs: The quotient of two numbers with unlike signs is a negative number and is found by taking the negative of the quotient of the absolute values of the two numbers.

Two negative numbers: The quotient of two negative numbers is positive and is found by dividing the absolute values of the two numbers.

We see that:

The quotient of two numbers with unlike signs is negative.
The quotient of two numbers with like signs is positive.

EXAMPLE 21 Evaluate: **(A)** $\dfrac{-22}{11} = -2$ **(B)** $\dfrac{-36}{-12} = 3$ **(C)** $\dfrac{48}{-16} = -3$

PROBLEM 21 Evaluate: **(A)** $\dfrac{-36}{-9}$ **(B)** $\dfrac{24}{-8}$ **(C)** $\dfrac{-72}{12}$

Solution **(A)** **(B)** **(C)**

Several important sign properties for division are summarized in the following theorem.

THEOREM 9

Sign Properties for Division
For all real numbers a and b, $b \neq 0$:
(A) $\dfrac{-a}{-b} = \dfrac{a}{b} = -\dfrac{-a}{b} = -\dfrac{a}{-b}$ $\dfrac{-3}{-4} = \dfrac{3}{4} = -\dfrac{-3}{4} = -\dfrac{3}{-4}$
(B) $\dfrac{-a}{b} = \dfrac{a}{-b} = -\dfrac{a}{b}$ $\dfrac{-3}{4} = \dfrac{3}{-4} = -\dfrac{3}{4}$

EXAMPLE 22   Evaluate $\dfrac{-a}{b}$, $\dfrac{a}{-b}$, and $-\dfrac{a}{b}$ for $a = -6$ and $b = 2$.

Solution

$$\frac{-a}{b} = \frac{-(-6)}{2} \qquad \frac{a}{-b} = \frac{(-6)}{-(2)} \qquad -\frac{a}{b} = -\frac{-6}{2}$$

$$= \frac{6}{2} = 3 \qquad\qquad = \frac{-6}{-2} = 3 \qquad\qquad = -(-3) = 3$$

PROBLEM 22   Evaluate $\dfrac{-a}{-b}$, $\dfrac{a}{b}$, and $-\dfrac{-a}{b}$ for $a = -6$ and $b = 2$.

Solution

$$\frac{-a}{-b} =$$

$$\frac{a}{b} =$$

$$-\frac{-a}{b} =$$

■ Combined Operations

We now consider problems involving various combinations of the **arithmetic operations** $+$, $-$, $\cdot$, and $\div$ as well as **grouping symbols** such as **parentheses** ( ), **brackets** [ ], **braces** { }, and **fraction bars** —.

To start, suppose several people were asked to evaluate

$$6 - 4(-3) + \frac{-6}{2}$$

To get the same result from all (a reasonable request), we need an agreement indicating the order in which the operations should be performed.

Order of Operations

(A) *If no grouping symbols are present:*
  1. Perform any multiplication and division, proceeding from left to right.
  2. Perform any addition and subtraction, proceeding from left to right.

(B) *If symbols of grouping are present:*
  1. Simplify above and below any fraction bars following the steps in (A).
  2. Simplify within other symbols of grouping, generally starting with the innermost and working outward, following the steps in (A).

Thus,

$$6 - 4(-3) + \frac{-6}{2} = 6 - (-12) + (-3) = 6 + 12 + (-3) = 15$$

EXAMPLE 23   Evaluate for $x = -24$, $y = 2$, and $z = -3$:

(A) $2x - 3yz + \dfrac{x}{z}$ **(B)** $\dfrac{x}{y} - \dfrac{16z + xy}{y + z}$ **(C)** $-x - y(x - 5yz)$

Solution   **(A)** $2x - 3yz \quad + \dfrac{x}{z}$

$$2(\ \ ) - 3(\ \ )(\ \ ) + \frac{(\ \ )}{(\ \ )}$$

Using parentheses as indicated will help to reduce sign errors.

$$2(-24) \quad - 3(2)(-3) + \frac{(-24)}{(-3)}$$

Multiplication and division precede addition and subtraction.

$$= (-48) - (-18) \quad + 8$$
$$= (-48) + 18 \quad\quad + 8 = -22$$

(B) $\dfrac{x}{y} - \dfrac{16z + xy}{y + z}$

$$\frac{(-24)}{2} - \frac{16(-3) + (-24)(2)}{2 + (-3)}$$

$$= (-12) - \frac{(-48) + (-48)}{-1}$$

$$= (-12) - \frac{(-96)}{(-1)} = (-12) - 96 = -108$$

(C) $-x - y(x - 5yz)$

$$-(-24) - 2[(-24) - 5(2)(-3)]$$

Notice how brackets and parentheses are used. Brackets are just another kind of parentheses; each may replace the other as desired.

$$= 24 - 2[(-24) - (-30)]$$
$$= 24 - 2(6) = 24 - 12 = 12$$

PROBLEM 23   Evaluate for $u = 36$, $v = -4$, and $w = -3$:

(A) $3vw - \dfrac{u}{3w} + 4v$ **(B)** $\dfrac{9w - 8v}{v - w} - \dfrac{u}{v}$ **(C)** $u - [7 - 2(u - 4vw)]$

Solution   **(A)**

(B)

(C)

Solutions to Matched Problems

18. **(A)** $4(-3) = -12$ **(B)** $(-4)3 = -12$ **(C)** $(-4)(-3) = 12$
(D) $0(-3) = 0$

19. $(-a)(-b) = [-(-5)][-(4)] = (5)(-4) = -20;\ ab = (-5)(4) = -20$

20. $-[(6)(-3)] = -(-18) = 18$ and $(-6)(-3) = 18$

21. **(A)** $\dfrac{-36}{-9} = 4$ **(B)** $\dfrac{24}{-8} = -3$ **(C)** $\dfrac{-72}{12} = -6$

22. $\dfrac{-a}{-b} = \dfrac{-(-6)}{-(2)} = \dfrac{6}{-2} = -3;\ \dfrac{a}{b} = \dfrac{-6}{2} = -3;$

$-\dfrac{-a}{b} = -\dfrac{-(-6)}{2} = -\dfrac{6}{2} = -3$

23. **(A)** $3(-4)(-3) - \dfrac{36}{3(-3)} + 4(-4) = 36 - (-4) - 16 = 24$

(B) $\dfrac{9(-3) - 8(-4)}{(-4) - (-3)} - \dfrac{36}{-4} = \dfrac{-27 - (-32)}{-1} - (-9)$

$= -5 - (-9) = 4$

(C) $36 - [7 - 2(36 - 4(-4)(-3))] = 36 - [7 - 2(36 - 48)]$

$= 36 - [7 - (-24)]$

$= 36 - 31 = 5$

Practice Exercise 1-5 ■

Work odd-numbered problems first, check answers, and then work even-numbered problems in areas of weakness. Answers to all problems are in the back of the book. Make every effort to work a problem yourself before you look at an answer.

A   *Evaluate, performing the indicated operations.*

1. $(-3)(-5)$ _____   **2.** $(-7)(-4)$ _____

3. $(-18) \div (-6)$ _____   **4.** $(-20) \div (-4)$ _____

5. $(-2)(+9)$ _____   **6.** $(+6)(-3)$ _____

7. $\dfrac{-9}{+3}$ _____   **8.** $\dfrac{+12}{-4}$ _____   **9.** $0(-7)$ _____

10. $(-6)0$ _____   **11.** $0/5$ _____   **12.** $0/(-2)$ _____

13. $3/0$ _____   **14.** $-2/0$ _____   **15.** $0 \div 0$ _____

16. $\dfrac{0}{0}$ _____   **17.** $\dfrac{-21}{3}$ _____   **18.** $\dfrac{-36}{-4}$ _____

19. $(-4)(-2)+(-9)$ _____   **20.** $(-7)+(-3)(+2)$ _____

21. $(+5)-(-2)(+3)$ _____   **22.** $(-7)-(-3)(-4)$ _____

23. $5-\dfrac{-8}{2}$ _____   **24.** $7-\dfrac{-16}{-2}$ _____

25. $(-1)(-8)$ and $-(-8)$ _____   **26.** $(-1)(+3)$ and $-(+3)$ _____

27. $-12+\dfrac{-14}{-7}$ _____   **28.** $\dfrac{-10}{5}+(-7)$ _____

29. $\dfrac{6(-4)}{-8}$ _____   **30.** $\dfrac{5(-3)}{3}$ _____

31. $\dfrac{22}{-11}-(-4)(-3)$ _____   **32.** $3(-2)-\dfrac{-10}{-5}$ _____

33. $\dfrac{-16}{2}-\dfrac{3}{-1}$ _____   **34.** $\dfrac{27}{-9}-\dfrac{-21}{-7}$ _____

35. $(+5)(-7)(+2)$ _____   **36.** $(-6)(-3)(+4)$ _____

37. $(-22)(+36)(0)$ _____   **38.** $(+19)(0)(-35)$ _____

B   *Evaluate, performing the indicated operations.*

39. $[(+2)+(-7)][(+8)-(+10)]$ _____

40. $[(-3)-(+8)][(+4)+(-2)]$ _____

41. $12-7[(-4)(5)-2(-8)]$ _____

42. $9-5[(-2)-3]$ _____

43. $\dfrac{9}{-3}-\dfrac{3+9(-2)}{-2-(-3)}$ _____

44. $\dfrac{4(-2) - (-5)}{(-9) - (-6)} - \dfrac{-24}{-8}$ _____

45. $\{[8/(-2)] - [21 + 5(-3)]\} - (-2)(-4)$ _____

46. $7 - \{9 - [5 - 2(-3)] - (8/2)\}$ _____

Evaluate Problems 47–68 for $w = 2$, $x = -3$, $y = 0$, *and* $z = -24$.

47. z/w _____   **48.** z/x _____   **49.** w/y _____

50. y/x _____   **51.** $\dfrac{z}{x} - wz$ _____   **52.** $wx - \dfrac{z}{w}$ _____

53. $\dfrac{xy}{w} - xyz$ _____   **54.** $wxy - \dfrac{y}{z}$ _____   **55.** $-|w||x|$ _____

56. $(|x||z|)$ _____   **57.** $\dfrac{|z|}{|x|}$ _____   **58.** $-\dfrac{|z|}{|w|}$ _____

59. $(wx - z)(z - 8x)$ _____   **60.** $(5x - z)(wx - 3w)$ _____

61. $wx + \dfrac{z}{wx} + wz$ _____   **62.** $xyz + \dfrac{y}{z} + x$ _____

63. $\dfrac{8x}{z} - \dfrac{z - 6x}{wx}$ _____   **64.** $\dfrac{w - x}{w + x} - \dfrac{z}{2x}$ _____

65. $\dfrac{24}{3w - 2x} - \dfrac{24}{3w + 2x}$ _____   **66.** $\dfrac{48}{z + 8x} - \dfrac{48}{z - 8x}$ _____

67. $\dfrac{z}{wx} - 2[z + 3(2x - w)]$ _____

68. $\dfrac{8wx}{-z} - x[5 + 2(z + 9w)]$ _____

69. Any integer divided by 0 is (*always, sometimes, never*) 0. _____

70. Zero divided by *any* integer is (*always, sometimes, never*) 0. _____

71. A product made up of an odd number of negative factors is (*sometimes, always, never*) negative. _____

72. A product made up of an even number of negative factors is (*sometimes, always, never*) negative. _____

C **73.** If the quotient $\dfrac{x}{y}$ exists, and neither x nor y is 0, when is it equal to $\dfrac{-|x|}{|y|}$? _____

74. If the quotient $\dfrac{x}{y}$ exists, and neither x nor y is 0, when is it equal to $\dfrac{|x|}{|y|}$? _____

75. Provide the reasons for each step in the proof that $a0 = 0$ for all real numbers a (Theorem 4).

STATEMENT		REASON
1.	$a0 = a(0 + 0)$	**1.** _____
2.	$a0 = a0 + a0$	**2.** _____
3.	$a0 + [-(a0)] = (a0 + a0) + [-(a0)]$	**3.** _____
4.	$0 = a0 + \{a0 + [-(a0)]\}$	**4.** _____
5.	$0 = a0 + 0$	**5.** _____
6.	$0 = a0$	**6.** _____
7.	$a0 = 0$	**7.** _____

76. Provide the reasons for each step in the proof that $(-1)a = -a$ (Theorem 6A).

STATEMENT		REASON
1.	$a + (-1)a = 1a + (-1)a$	**1.** _____
2.	$= a[1 + (-1)]$	**2.** _____
3.	$= a \cdot 0$	**3.** _____
4.	$= 0$	**4.** _____
5. Therefore, $(-1)a = -a$.		**5.** _____

The Check Exercise for this section is on page 61.

Section 1-6 Chapter Review

A **set** is a collection of objects called **members** or **elements** of the set. Sets are usually described by **listing** {list of elements} or by a **rule** {x|rule that determines that x is a member}. That an object a belongs to set A is denoted $a \in A$; that it does not is denoted by $a \notin A$. The set containing no elements is called the **empty set** or the **null set** and is denoted $\varnothing$. Sets are **finite** if the elements can be counted (and there is an end); they are **infinite** otherwise. If each element of set A is also in set B, we say A is a **subset** of B and write $A \subset B$. A **variable** is a symbol that represents unspecified elements from a **replacement set**; a **constant** is a symbol for one object in a set. *(1-1)*

The **real number** system consists of

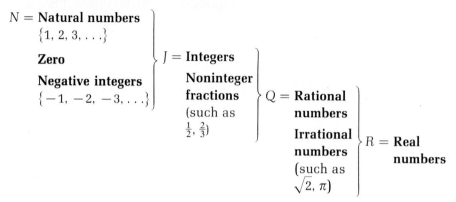

The real numbers can be represented as points on a **real number line (real line)** where each point is associated with a single real number called the **coordinate** of the point. The point with coordinate 0 is called the **origin**, with **positive real numbers** to the right and **negative real numbers** to the left. *(1-1)*

An **algebraic expression** is made up from variables, constants, mathematical operation signs, and grouping symbols. Expressions joined by plus or minus signs are called **terms**; those joined by multiplication are **factors**. Algebraic expressions joined by an equal sign are called **algebraic equations**. Equality satisfies these **equality properties**:

Reflexive property:	$a = a$
Symmetric property:	If $a = b$, then $b = a$.
Transitive property:	If $a = b$ and $b = c$, then $a = c$.
Substitution property:	If $a = b$, then either may be substituted for the other.

The **inequality symbols** $<$, $>$, $\leq$, $\geq$ denote **less than, greater than, less than or equal to**, and **greater than or equal to**, respectively; $a < b$ and $b > a$ mean $a + p = b$ for some positive number p. For an **inequality statement**—that is, an inequality involving a variable—the numbers that make it true are called the **solution set**. The solution set can be represented by a **graph of the inequality statement**. *(1-2)*

The real numbers under addition and multiplication satisfy several basic **axioms** that are listed and named in Section 1-3. Among these are the **commutative properties** $[a + b = b + a, \quad a \cdot b = b \cdot a]$, the **associative properties** $[(a + b) + c = a + (b + c)$ and $(ab)c = a(bc)]$, and the **identity properties** $[a + 0 = 0 + a = a, \ 1 \cdot a = a \cdot 1 = a]$. These properties allow us to manipulate algebraic expressions. Two algebraic expressions are **equivalent** if they yield the same number for all possible values of the variables. *(1-3)*

For any real number x, $-x$ denotes its **additive inverse**, the number we add to x to get 0; this is also called the **negative of the number**. For any real number x, its absolute value $|x|$ is the distance from x to the origin, a nonnegative quantity. Rules for addition of real numbers are given in Section 1-4. Subtraction is accomplished in terms of addition by $a - b = a + (-b)$. *(1-4)*

Rules for multiplication of real numbers are given in Section 1-5. Division is defined in terms of multiplication: $a \div b$ is equal to a **quotient** Q if $bQ = a$ and Q is unique. Division by 0 is not defined. Multiplication and division satisfy these **sign properties**:

$$(-a)b = -(ab) = a(-b) \qquad (-a)(-b) = ab$$

$$\frac{-a}{b} = \frac{a}{-b} = -\frac{a}{b} \qquad \frac{-a}{-b} = \frac{a}{b} = -\frac{-a}{b} = -\frac{a}{-b}$$

The **order of operations** for arithmetic operations, unless **grouping symbols** indicate otherwise, is first to do multiplications and divisions left to right and then to do additions and subtractions left to right. *(1-5)*

Diagnostic (Review) Exercise 1-6 ∎

Work through all the problems in this chapter review and check answers in the back of the book. (Answers to all problems are there, and following each answer is a number in italics indicating the section in which that type of problem is discussed.) Where weaknesses show up, review appropriate sections in the text. When you are satisfied that you know the material, take the practice test following this review.

A 1. For $A = \{1, 2, 3, 4, 5\}$, $B = \{1, 3, 5\}$, and $C = \{5, 1, 3\}$, indicate true (T) or false (F):

(A) $4 \in A$ _____

(B) $4 \notin C$ _____

(C) $B \in A$ _____

(D) $A \subset B$ _____

(E) $C \subset A$ _____

(F) $B \subset C$ _____

(G) $\varnothing \subset B$ _____

(H) $B \neq C$ _____

2. $3.127127\ldots$ represents a (*rational, irrational*) number. _____

Evaluate.

3. $3 \cdot 7 - 4$ _____

4. $7 + 2 \cdot 3$ _____

5. $(-8) + 3$ _____

6. $(-9) + (-4)$ _____

7. $(-3) - (-9)$ _____

8. $4 - 7$ _____

9. $0 - (-3)$ _____

10. $(-12) - 0$ _____

11. $(-7)(-4)$ _____

12. $3(-6)$ _____

13. $(-16)/4$ _____

14. $(-12)/(-2)$ _____

15. $(-6)/0$ _____

16. $0/(-3)$ _____

17. $10 - 3(6 - 4)$ _____

18. $(-8) - (-2)(-3)$ _____

19. $(-9) - [(-12)/3]$ _____

20. $(4 - 8) + 4(-2)$ _____

21. $|-8|$ _____   **22.** $-(-5)$ _____

23. $-[-(-3)]$ _____   **24.** $-|-(-2)|$ _____

25. $-(|-3| + |-2|)$ _____   **26.** $-(|-8| - |3|)$ _____

Remove parentheses and simplify using commutative and associative axioms mentally.

27. $7 + (x + 3)$ _____   **28.** $(3x)5$ _____

29. $(2x)(4y)$ _____

30. $(y + 7) + (x + 2) + (z + 3)$ _____

31. $0 + (1x + 2)$ _____   **32.** $(x + 0) + 0y$ _____

Replace each question mark with $<$ or $>$ to form a true statement.

33. $7 \; ? \; 2$ _____   **34.** $-7 \; ? \; -2$ _____

35. $-12 \; ? \; 0$ _____   **36.** $-342 \; ? \; -3$ _____

37. $0 \; ? \; -45$ _____   **38.** $-50 \; ? \; 20$ _____

B   39. For $N =$ The set of natural numbers, $J =$ The set of integers, $Q =$ The set of rational numbers, $R =$ The set of real numbers, indicate true (T) or false (F):

(A) $-5 \in N$ _____   **(B)** $-5 \in R$ _____

(C) $Q \subset R$ _____   **(D)** $Q \subset J$ _____

(E) $1.43 \in Q$ _____   **(F)** $-\frac{2}{3} \in R$ _____

(G) $\sqrt{2} \in Q$ _____   **(H)** $\pi \in R$ _____

40. Graph on a real number line.

(A) $x < -1$ **(B)** $-4 \leq x < 3$

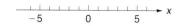   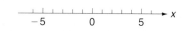

Evaluate.

41. $2[9 - 3(3 - 1)]$ _____   **42.** $6 - 2 - 3 - 4 + 5$ _____

43. $[(-3) - (-3)] - (-4)$ _____   **44.** $[-(-4)] + (-|-3|)$ _____

45. $[(-16)/2] - (-3)(4)$ _____

46. $(-2)(-4)(-3) - \dfrac{-36}{(-2)(9)}$ _____

47. $2\{9 - 2[(3 + 1) - (1 + 1)]\}$ _____

48. $(-3) - 2\{5 - 3[2 - 2(3 - 6)]\}$ _____

49. $\dfrac{12 - (-4)(-5)}{4 + (-2)} - \dfrac{-14}{7}$ _____   **50.** $\dfrac{24}{(-4) + 4} - \dfrac{24}{(-4) + 4}$ _____

51. $3[14 - x(x + 1)]$ for $x = 3$ _____

52. $-(-x)$ for $x = -2$ _____

53. $-(|x| - |w|)$ for $x = -2$ and $w = -10$ _____

54. $(x + y) - z$ for $x = 6$, $y = -8$, $z = 4$ _____

55. $\left(2x - \dfrac{z}{x}\right) - \dfrac{w}{x}$ for $w = -10$, $x = -2$, $z = 0$ _____

56. $\dfrac{(xyz + xz) - z}{z}$ for $x = -6$, $y = 0$, $z = -3$ _____

57. $\dfrac{x - 3y}{z - x} - \dfrac{z}{xy}$ for $x = -3$, $y = 2$, and $z = -12$ _____

Translate each statement into an algebraic equation or inequality statement using x as the only variable.

58. $x - 1$ is positive. _____

59. $2x + 3$ is not negative. _____

60. 50 is 10 less than twice a certain number. _____

61. x is less than 12 less than twice x. _____

62. x is greater than or equal to -5 and less than 5. _____

63. 8 more than a certain number is 5 times the number that is 6 less than the certain number. _____

64. In a rectangle with area 1,200 square centimeters, the width is 10 centimeters less than the length x. Write an equation relating the area with the length and width, using x as the only variable. _____

65. If the length of a rectangle is 5 meters longer than its width x and the perimeter is 43 meters, write an algebraic equation relating the sides and the perimeter. _____

Replace each question mark with an appropriate symbol to form a true statement.

66. $-a + ? = 0$ _____   **67.** $a + (-a) = ?$ _____

68. $a + ? = a$ _____   **69.** $0 + ? = a$ _____

70. $? \cdot a = a$ _____   **71.** $a \cdot ? = 1, a \neq 0$ _____

72. If $a + p = b$ for some positive number p, then a is (greater than, less than) b. _____

73. If $a - p = b$ for some positive number p, then a is (greater than, less than) b. _____

Replace each question mark with an appropriate symbol to make the statement an illustration of the stated property. (All variables represent real numbers.)

74. Symmetry property: If $P + Prt = A$, then ?. _____

75. Transitive property: If $x + y < 5$ and $5 < z$, then ?. _____

76. Commutative property: $P + Q = ?$ _____

77. Inverse property: $x + ? = 0$ _____

78. Substitution principle: If $3u - 2v = 5$ and $v - u + 4$, then $3u - 2(?) = 5$.

_____

79. Identity property: $(?)x = x$ _____

80. Associative property: $(x + 3) + 5 = ?$ _____

81. Trichotomy property: If $y \neq 5$, then either $y > 5$ or ?. _____

State the real number axiom that justifies each statement.

82. $5 + (x + 3) = 5 + (3 + x)$ _____

83. $5 + (3 + x) = (5 + 3) + x$ _____

84. $5(x3) = 5(3x)$ _____   **85.** $5(3x) = (5 \cdot 3)x$ _____

86. $(x + y) + 0 = x + y$ _____   **87.** $(ab) + [-(ab)] = 0$ _____

C   **88.** Evaluate $uv - 3\{x - 2[(x + y) - (x - y)] + u\}$ for $u = -2$, $v = 3$, $x = 2$, and $y = -3$. _____

89. Evaluate $\dfrac{5w}{x - 7} - \dfrac{wx - 4}{x - w}$ for $w = -4$ and $x = 2$. _____

90. Replace the question marks with appropriate symbols:

$$\{3, 4, 5, 6\} = \{x \in N \mid ? \leq x \,?\, 7\} \,\underline{\hspace{5cm}}$$

91. Write each set by the listing method:

(A) $\{x \in J \mid -2 < x \leq 2\}$ _____   **(B)** $\{x \in N \mid 5 < x < 6\}$ _____

92. If $M = \{3, 4, 5, 7\}$ and $N = \{4, 5, 6\}$, find:

(A) $\{x \mid x \in M \text{ or } x \in N\}$ _____   **(B)** $\{x \mid x \in M \text{ and } x \in N\}$ _____

93. Describe the elements in each set:

(A) $\{x \in R \mid |x| = x\}$ _____   **(B)** $\{x \in R \mid |x| = -x\}$ _____

Practice Test Chapter 1 ■

Take this as if it were a graded test by working the problems within a 50-minute time period. Do not look back in the chapter. Choose one of three levels of difficulty: least difficult, Problems 1–12; more difficult, add Problem 13; most difficult, add Problems 13 and 14. Use the answers in the back of the book to correct your work. The answers are keyed to appropriate text sections so that you can easily locate and review sections where difficulties still persist.

Evaluate Problems 1–6.

1. $(-3)\{4 - 2[3 - (5 - 8)]\}$ _____

2. $(-4)(-8) - \dfrac{-30}{5}$ _____

3. $-(|-3| - |8|)$ _____

4. $\left(3x - \dfrac{y}{z}\right) - \dfrac{z}{x}$ for $x = -1, y = 6, z = -2$ _____

5. $\dfrac{-(-x)(y + z)}{xy - yz - 1}$ for $x = 3, y = -2, z = -1$ _____

6. $\dfrac{xy + 1}{y + 2x}$ for $x = 3, y = -6$ _____

Translate Problems 7 and 8 into algebraic equations or inequality statements using x as the only variable.

7. Three times a certain number is greater than the quantity that is 6 less than 4 times the certain number. _____

8. Four more than twice a number is 5 less than 3 times the number.

    _____

9. Graph $-4 < x \leq 3$ on a real number line. _____

10. A rectangle with length 3 more than twice its width has perimeter 54 centimeters. Write an equation for the perimeter in terms of the width x.

    _____

11. For $N =$ the set of natural numbers, $J =$ the set of integers, $Q =$ the set of rational numbers, and $R =$ the set of real numbers, which of the following statements are false? _____

    (A) $1.4001 \in Q$ (B) $-5 \in R$ (C) $\sqrt{2} \in Q$
    (D) $11 \in N$ (E) $0 \in J$ (F) $-1 \in N$

12. Which of the following statements illustrate a commutative property?

    _____

    (A) $xy + z = yx + z$ (B) $xy + z = z + xy$
    (C) $x(y + z) = xy + xz$ (D) $x(y + z) = (y + z)x$

13. Which of the following statements illustrate an associative property?

    _____

    (A) $(a + b) + (c + d) = [(a + b) + c] + d$
    (B) $(a + b) + (c + d) = (c + d) + (a + b)$
    (C) $(ab)(cd) = a[b(cd)]$
    (D) $(ab)(cd) = (ba)(dc)$

14. Write the set $A = \{x \in J \mid -1 \leq x < 4\}$ by the listing method. _____

Check Exercise 1-1 ■

Work the following problems without looking at any text examples. Show your work in the space provided. Write your answer in the answer column.

1. _____

2. _____

3. _____

4. _____

5. _____

6. _____

7. _____

8. _____

9. _____

10. _____

1. Which of the statements below are false?

   (A) $\{-2, 7, 3\} = \{3, 7, -2\}$
   (B) $-2 \in \{-2, 7, 3\}$
   (C) $3 \notin \{-2, 7, 3\}$
   (D) $5 \notin \{-2, 7, 3\}$

2. Which of the statements below are false?

   (A) $\{2, 5\} \subset \{1, 2, 3, 5\}$
   (B) $\{2, 3, 4\} \subset \{1, 2, 3, 5\}$
   (C) $\varnothing \subset \{1, 2, 3, 5\}$
   (D) $\{1, 2, 3, 5\} \subset \{1, 2, 3, 5\}$

3. Let M be the set of all numbers x such that $x^2 = 1$. Write M, using the listing method.

4. Let M be the set of all numbers x such that $x^2 = 1$. Write M, using the rule method.

5. Which of the following numbers belong to the set of rational numbers?

   $$3.1415, \qquad -\sqrt{2}, \qquad \frac{5}{7}, \qquad 0, \qquad \pi$$

6. Which of the following numbers belong to the set of integers?

$$\frac{2}{3}, \quad \pi, \quad 0, \quad -7, \quad 3.14, \quad 45$$

7. Which of the following numbers belong to the set of natural numbers?

$$-2, \quad 3.05, \quad 13, \quad \frac{12}{19}, \quad 1{,}985$$

8. If $A = \{3, 5, 7\}$ and $B = \{5, 7, 9\}$, list the elements of $\{x \mid x \in A \text{ or } x \in B\}$.

9. If $A = \{3, 5, 7\}$ and $B = \{5, 7, 9\}$, list the elements of $\{x \mid x \in A \text{ and } x \in B\}$.

10. How many subsets does the set $\{a, b, c\}$ have?

Check Exercise 1-2 ■

Work the following problems without looking at any text examples. Show your work in the space provided. Write your answer or draw your final graph in the answer column.

1. _____

2. _____

3. _____

$-5 \quad 0 \quad 5 \quad x$

4. _____

$-5 \quad 0 \quad 5 \quad x$

5. _____

6. _____

7. _____

8. _____

9. _____

10. _____

1. Which of the statements below are false?

   (A) $-3 < 7$
   (B) $-6.33 > -6$
   (C) $\frac{2}{3} > \frac{3}{4}$
   (D) $-5 > -7$

2. Which of the statements below are false for this number line?

   (A) $b < d$
   (B) $a < 0$
   (C) c is a positive number.
   (D) $d < b$

3. Write "x is less than or equal to 5" in symbolic form.

4. Graph $x < 3$ on a real number line.

5. Graph $-3 \le x < 4$ on a real number line.

Translate Problems 6–9 into algebraic equations or inequality statements using x as the only variable.

6. Twice a number is greater than 9 less than 3 times the number.

7. Two more than 3 times a number is less than or equal to 5 less than 4 times the number.

8. 7 times a number is 5 less than 3 times the number.

9. 3 subtracted from a certain number is twice the quantity which is 2 less than the certain number.

10. In a rectangle with perimeter 100 centimeters the length is 5 centimeters less than twice the width. Write an equation relating the perimeter with the length and width, using x as the only variable. ($P = 2a + 2b$)

Check Exercise 1-3 ■

Work the following problems without looking at any text examples. Show your work in the space provided. Write your answer in the answer column.

1. _____

2. _____

In Problems 1–6 state the justifying real number property for each statement.

1. $5 + 7y = 7y + 5$

3. _____

4. _____

5. _____

2. $m(x + y) = mx + my$

6. _____

7. _____

8. _____

3. $5 + (A + 3) = 5 + (3 + A)$

9. _____

10. _____

4. $(x + 2)(x + 3) = (x + 3)(x + 2)$

5. $7m + m + 3m = (7 + 1 + 3)m$

6. $5x + 10y + 15z = 5(x + 2y + 3z)$

In Problems 7–9 remove parentheses and simplify.

7. $(5u + 3) + (2v + 7)$

8. $(a + 3) + (b + 2) + (c + 1)$

9. $(4u)(3v)(2w)$

10. Which of the statements below are true for all real numbers a, b, and c?

   (A) $a + b = b + a$
   (B) $(ab)c = a(bc)$
   (C) $a - b = b - a$
   (D) $a(b + c) = ab + ac$

Check Exercise 1-4 ∎

ANSWER COLUMN

Work the following problems without looking at any text examples. Show your work in the space provided. Write your answer in the answer column.

1. _____

2. _____   *Evaluate Problems 1–8.*

    1. $-[-(-4)]$

3. _____

4. _____

5. _____

    2. $8 + (-7)$

6. _____

7. _____

8. _____

    3. $-|-8|$

9. _____

10. _____

    4. $(-12) - (-9)$

    5. $6 - 10$

6. $-7 - 4 + 6 - 8 + 9$

7. $-(|-9| - |-12|)$

8. $[(-8) - 3] - [3 - (-9)]$

9. Replace the question mark with an appropriate number:

$(-7) - (?) = 5$

10. Evaluate for $x = -9$ and $y = 2$: $|x - y| - |x|$

Check Exercise 1-5 ■

ANSWER COLUMN

Work the following problems without looking at any text examples. Show your work in the space provided. Write your answer in the answer column.

1. _____

2. _____   *Evaluate Problems 1–8.*

    1. $(-3)(-2) - 8$

3. _____

4. _____

5. _____   **2.** $(-5)(0)(-3)(2)$

6. _____

7. _____

8. _____   **3.** $\dfrac{-45}{9} - (-3)(3)$

9. _____

10. _____

    4. $\dfrac{8 - 5}{0}$

    5. $\dfrac{0}{(-2)(-5)}$

6. $\dfrac{27}{-3} - \dfrac{-8}{-4}$

7. $[(-8) - (-2)][9 - (-1)]$

8. $\dfrac{(-4) - (+8)}{(-2)(3)} - \dfrac{0}{-12}$

9. Evaluate for $u = -12$, $v = 0$, and $w = 4$:

$$uvw - \dfrac{v}{u} - |u|\,|w|$$

10. Evaluate for $x = -30$, $y = 4$, and $z = -3$:

$$\dfrac{2x}{yz} - \dfrac{x - 6}{z - 3}$$

Polynomials

**INSTRUCTIONS FOR STUDENTS IN A
SELF-PACED CLASS OR LAB**

YES — **HAVE YOU HAD INTERMEDIATE ALGEBRA BEFORE THIS COURSE?** — NO

1. Work Diagnostic (Review) Exercise 2-8 on page 112. Check answers in back of book; then work through text sections corresponding to problems missed. (Section numbers are in italics following each answer.)
2. When finished with step 1, take Practice Test Chapter 2 on page 114 as a final check of your understanding of the chapter. Check answers in the back of the book; then review sections where weakness still prevails. (Corresponding section numbers are in italics following each answer.)
3. When you think you are ready, ask your instructor for a graded test for Chapter 2.
4. If your instructor approves, after the test is corrected, go to the next chapter.

1. Work through each section in the chapter as follows:
   (A) Read discussion.
   (B) Read each example and work the corresponding matched problem. Check your solutions to the matched problem in Solutions to Matched Problems on the indicated page.
   (C) At the end of a section work the odd-numbered problems in the Practice Exercise and check answers; then work even-numbered problems in areas of weakness. (Answers to *all* Practice Exercise sets are in the back of the book.)
   (D) Work Check Exercise as instructed. Tear out and turn in as directed by your instructor. (Answers are not in the text.)
2. Repeat each step in item 1 for each section in the chapter.
3. After the instructional part of the chapter is completed, proceed with steps 1 to 4 in the box above this one.

Chapter 2 ■ Polynomials

Section 2-1　Natural Number Exponents

- ■ Definition and Five Properties
- ■ Use of the Five Properties

■ Definition and Five Properties

Recall that

$$a^5 = a \cdot a \cdot a \cdot a \cdot a \qquad \text{Five factors of } a$$

and, in general:

Natural Number Exponent

For n a natural number and a a real number:

$$a^n = a \cdot a \cdot \cdots \cdot u \qquad n \text{ factors of } a$$

Exponent ↗　Base ↙

$$2^3 = 2 \cdot 2 \cdot 2 = 8$$

Exponent forms are encountered so frequently in algebra that it is essential for you to become completely familiar with the following five properties of exponents and their uses. These properties follow primarily from the definition of natural number exponent in the box and the commutative and associative properties of real numbers. We will give only informal arguments for each property. In all cases a and b represent real numbers and m and n represent natural numbers.

Consider:

$$a^3 a^4 = \overbrace{(a \cdot a \cdot a)}^{3 \text{ factors}} \overbrace{(a \cdot a \cdot a \cdot a)}^{4 \text{ factors}} = \overbrace{(a \cdot a \cdot a \cdot a \cdot a \cdot a \cdot a)}^{3 + 4 \text{ factors}} = a^{3+4} = a^7$$

which suggests:

Property 1

$$a^m a^n = a^{m+n}$$

$$a^5 a^2 \boxed{= a^{5+2}} = a^7$$

Consider:

$$4 \text{ groups of 3 factors each}$$

$$(a^3)^4 = a^3 \cdot a^3 \cdot a^3 \cdot a^3 = \overbrace{(a \cdot a \cdot a)(a \cdot a \cdot a)(a \cdot a \cdot a)(a \cdot a \cdot a)}$$

$$4 \cdot 3 \text{ factors}$$

$$= \overbrace{(a \cdot a \cdot a \cdot a \cdot a \cdot a \cdot a \cdot a \cdot a \cdot a \cdot a \cdot a)} = a^{4 \cdot 3} = a^{12}$$

which suggests:

Property 2

$$(a^n)^m = a^{mn}$$

$$(a^2)^5 \begin{array}{|c|} \hline = a^{5 \cdot 2} \\ \hline \end{array} = a^{10}$$

Consider:

$$4 \text{ factors of } (ab) \qquad 4 \text{ factors of } a \quad 4 \text{ factors of } b$$

$$(ab)^4 = \overbrace{(ab)(ab)(ab)(ab)} = \overbrace{(a \cdot a \cdot a \cdot a)}\overbrace{(b \cdot b \cdot b \cdot b)} = a^4 b^4$$

which suggests:

Property 3

$$(ab)^m = a^m b^m$$

$$(ab)^7 = a^7 b^7$$

Consider:

$$5 \text{ factors of } a/b \qquad\qquad 5 \text{ factors of } a$$

$$\left(\frac{a}{b}\right)^5 = \left(\overbrace{\frac{a}{b} \cdot \frac{a}{b} \cdot \frac{a}{b} \cdot \frac{a}{b} \cdot \frac{a}{b}}\right) = \frac{\overbrace{a \cdot a \cdot a \cdot a \cdot a}}{\underbrace{b \cdot b \cdot b \cdot b \cdot b}} = \frac{a^5}{b^5}$$

$$5 \text{ factors of } b$$

which suggests:

Property 4

$$\left(\frac{a}{b}\right)^m = \frac{a^m}{b^m}$$

$$\left(\frac{a}{b}\right)^3 = \frac{a^3}{b^3}$$

Consider:

(A) $\dfrac{a^7}{a^3} = \dfrac{a \cdot a \cdot a \cdot a \cdot a \cdot a \cdot a}{a \cdot a \cdot a}$

$$= \frac{\cancel{(a \cdot a \cdot a)}(a \cdot a \cdot a \cdot a)}{\cancel{(a \cdot a \cdot a)}} = a^{7-3} = a^4$$

(B) $\dfrac{a^3}{a^3} = \dfrac{a \cdot a \cdot a}{a \cdot a \cdot a} = 1$

(C) $\dfrac{a^4}{a^7} = \dfrac{a \cdot a \cdot a \cdot a}{a \cdot a \cdot a \cdot a \cdot a \cdot a \cdot a}$

$= \dfrac{(a \cdot a \cdot a \cdot a)}{(a \cdot a \cdot a \cdot a)(a \cdot a \cdot a)} = \dfrac{1}{a^{7-4}} = \dfrac{1}{a^3}$

which suggests:

Property 5

$$\dfrac{a^m}{a^n} = \begin{cases} a^{m-n} & \text{if } m \text{ is larger than } n \\ 1 & \text{if } m = n \\ \dfrac{1}{a^{n-m}} & \text{if } n \text{ is larger than } m \end{cases}$$

$\dfrac{a^8}{a^3} \boxed{= a^{8-3}} = a^5$ $\dfrac{a^8}{a^8} = 1$ $\dfrac{a^3}{a^8} \boxed{= \dfrac{1}{a^{8-3}}} = \dfrac{1}{a^5}$

COMMON ERROR   Remember that the properties of exponents apply to products and quotients, not to sums and differences. Many mistakes are made in algebra by people applying a property of exponents to the wrong algebraic form. For example:

$(ab)^2 = a^2 b^2$ but $(a + b)^2 \neq a^2 + b^2$

Recall that in the order of operations agreed to in Section 1-5, multiplications and divisions are to be done before additions and subtractions unless grouping symbols indicate otherwise. This scheme is extended to include the taking of powers (and later roots) by agreeing that:

Powers take precedence over multiplication and division.

Thus, for example,

$\dfrac{2^3}{5} = \dfrac{8}{5} = 1.6$

and   The exponent applies only to the 2.

$5 \cdot 2^3 = 5 \cdot 8 = 40$

but

$\left(\dfrac{2}{5}\right)^3 = (0.4)^3 = 0.064$

and   Here the parentheses indicate that the exponent applies to the expression within.

$(5 \cdot 2)^3 = 10^3 = 1{,}000$

COMMON ERROR   Particular care is required when applying exponents to expressions involving negative numbers:

$-4^2 \neq (-4)^2$ because $\begin{cases} -4^2 = -(4 \cdot 4) = -16 \\ (-4)^2 = (-4)(-4) = 16 \end{cases}$

The exponent properties are summarized here for m and n natural numbers:

Exponent Properties

For m and n positive integers:

1. $a^m a^n = a^{m+n}$

2. $(a^n)^m = a^{mn}$

3. $(ab)^m = a^m b^m$

4. $\left(\dfrac{a}{b}\right)^m = \dfrac{a^m}{b^m}$

5. $\dfrac{a^m}{a^n} = \begin{cases} a^{m-n} & \text{if } m \text{ is larger than } n \\ 1 & \text{if } m = n \\ \dfrac{1}{a^{n-m}} & \text{if } n \text{ is larger than } m \end{cases}$

■ Use of the Five Properties

As before, the dashed boxes in the following examples indicate steps that are usually carried out mentally.

EXAMPLE 1   Simplify, using natural number exponents only; that is, rewrite the expression so that each variable occurs only once and has only a single natural number exponent applied to it.

(A) $x^{12} x^{13} \boxed{= x^{12+13}} = x^{25}$ **(B)** $(t^7)^5 \boxed{= t^{5 \cdot 7}} = t^{35}$

(C) $(xy)^5 = x^5 y^5$ **(D)** $\left(\dfrac{u}{v}\right)^3 = \dfrac{u^3}{v^3}$

(E) $\dfrac{x^{12}}{x^4} \boxed{= x^{12-4}} = x^8$ **(F)** $\dfrac{t^4}{t^9} \boxed{= \dfrac{1}{t^{9-4}}} = \dfrac{1}{t^5}$

PROBLEM 1   Simplify, using natural number exponents only:

(A) $x^8 x^6$ **(B)** $(u^4)^5$ **(C)** $(xy)^9$

(D) $\left(\dfrac{x}{y}\right)^4$ **(E)** $\dfrac{x^{10}}{x^3}$ **(F)** $\dfrac{x^3}{x^{10}}$

Solution   **(A)**   **(B)**   **(C)**

(D)   **(E)**   **(F)**

EXAMPLE 2   Simplify, using natural number exponents only:

(A) $(x^2y^3)^4 \boxed{= (x^2)^4(y^3)^4} = x^8y^{12}$ **(B)** $\left(\dfrac{u^3}{v^4}\right)^3 \boxed{= \dfrac{(u^3)^3}{(v^4)^3}} = \dfrac{u^9}{v^{12}}$

(C) $\dfrac{2x^9y^{11}}{4x^{12}y^7} \boxed{= \dfrac{2}{4} \cdot \dfrac{x^9}{x^{12}} \cdot \dfrac{y^{11}}{y^7} = \dfrac{1}{2} \cdot \dfrac{1}{x^3} \cdot \dfrac{y^4}{1}} = \dfrac{y^4}{2x^3}$

PROBLEM 2   Simplify, using natural number exponents only:

(A) $(u^3v^4)^2$ **(B)** $\left(\dfrac{x^4}{y^3}\right)^2$ **(C)** $\dfrac{9x^7y^2}{3x^5y^3}$ **(D)** $\dfrac{(2x^2y)^3}{(4xy^3)^2}$

Solution   **(A)**   **(B)**

(C)   **(D)**

Knowing the rules of the game of chess doesn't make you good at playing chess; similarly, memorizing the properties of exponents doesn't necessarily make you good at using these properties. To acquire skill in their use, you must use these properties in a fairly large variety of problems. The following exercises should help you acquire this skill.

Solutions to Matched
Problems

1. (A) $x^8x^6 = x^{14}$ **(B)** $(u^4)^5 = u^{20}$ **(C)** $(xy)^9 = x^9y^9$

(D) $\left(\dfrac{x}{y}\right)^4 = \dfrac{x^4}{y^4}$ **(E)** $\dfrac{x^{10}}{x^3} = x^7$ **(F)** $\dfrac{x^3}{x^{10}} = \dfrac{1}{x^7}$

2. (A) $(u^3v^4)^2 = u^6v^8$ **(B)** $\left(\dfrac{x^4}{y^3}\right)^2 = \dfrac{x^8}{y^6}$ **(C)** $\dfrac{9x^7y^2}{3x^5y^3} = \dfrac{3x^2}{y}$

(D) $\dfrac{(2x^2y)^3}{(4xy^3)^2} = \dfrac{2^3x^6y^3}{4^2x^2y^6} = \dfrac{x^4}{2y^3}$

Practice Exercise 2-1 ■

Work odd-numbered problems first, check answers, and then work even-numbered problems in areas of weakness. Answers to all problems are in the back of the book. Make every effort to work a problem yourself before you look at an answer.

A   *Replace the question marks with appropriate symbols.*

1. $y^2y^7 = y^?$ _____   **2.** $x^7x^5 = x^?$ _____

3. $y^8 = y^3 y^?$ _____

4. $x^{10} = x^? x^6$ _____

5. $(u^4)^3 = u^?$ _____

6. $(v^2)^3 = ?$ _____

7. $x^{10} = (x^?)^5$ _____

8. $y^{12} = (y^6)^?$ _____

9. $(uv)^7 = ?$ _____

10. $(xy)^5 = x^5 y^?$ _____

11. $p^4 q^4 = (pq)^?$ _____

12. $m^3 n^3 = (mn)^?$ _____

13. $\left(\dfrac{a}{b}\right)^8 = ?$ _____

14. $\left(\dfrac{x}{y}\right)^4 = \dfrac{x^?}{y^4}$ _____

15. $\dfrac{m^3}{n^3} = \left(\dfrac{m}{n}\right)^?$ _____

16. $\dfrac{x^7}{y^7} = \left(\dfrac{x}{y}\right)^?$ _____

17. $\dfrac{n^{14}}{n^8} = n^?$ _____

18. $\dfrac{x^7}{x^3} = x^?$ _____

19. $m^6 = \dfrac{m^8}{m^?}$ _____

20. $x^3 = \dfrac{x^?}{x^4}$ _____

21. $\dfrac{x^4}{x^{11}} = \dfrac{1}{x^?}$ _____

22. $\dfrac{a^5}{a^9} = \dfrac{1}{a^?}$ _____

23. $\dfrac{1}{x^8} = \dfrac{x^4}{x^?}$ _____

24. $\dfrac{1}{u^2} = \dfrac{u^?}{u^9}$ _____

Simplify, using appropriate properties of exponents.

25. $(5x^2)(2x^9)$ _____

26. $(2x^3)(3x^7)$ _____

27. $\dfrac{9x^6}{3x^4}$ _____

28. $\dfrac{4x^8}{2x^6}$ _____

29. $\dfrac{6m^5}{8m^7}$ _____

30. $\dfrac{4u^3}{2u^7}$ _____

31. $(xy)^{10}$ _____

32. $(cd)^{12}$ _____

33. $\left(\dfrac{m}{n}\right)^5$ _____

34. $\left(\dfrac{x}{y}\right)^6$ _____

B   35. $(4y^3)(3y)(y^6)$ _____

36. $(2x^2)(3x^3)(x^4)$ _____

37. $(5 \times 10^8)(7 \times 10^9)$ _____

38. $(2 \times 10^3)(3 \times 10^{12})$ _____

39. $(10^7)^2$ _____

40. $(10^4)^5$ _____

41. $(x^3)^2$ _____

42. $(y^4)^5$ _____

43. $(m^2 n^5)^3$ _____

44. $(x^2 y^3)^4$ _____

45. $\left(\dfrac{c^2}{d^5}\right)^3$ _____

46. $\left(\dfrac{a^3}{b^2}\right)^4$ _____

47. $\dfrac{9u^8 v^6}{3u^4 v^8}$

48. $\dfrac{2x^3y^8}{6x^7y^2}$ _____

49. $(2s^2t^4)^4$ _____

50. $(3a^3b^2)^3$ _____

51. $6(xy^3)^5$ _____

52. $2(x^2y)^4$ _____

53. $\left(\dfrac{mn^3}{p^2q}\right)^4$ _____

54. $\left(\dfrac{x^2y}{2w^2}\right)^3$ _____

55. $\dfrac{(4u^3v)^3}{(2uv^2)^6}$ _____

56. $\dfrac{(2xy^3)^2}{(4x^2y)^3}$ _____

57. $\dfrac{(9x^3)^2}{(-3x)^2}$ _____

58. $\dfrac{(-2x^2)^3}{(2^2x)^4}$ _____

59. $\dfrac{-x^2}{(-x)^2}$ _____

60. $\dfrac{-2^2}{(-2)^2}$ _____

61. $\dfrac{(-x^2)^2}{(-x^3)^3}$ _____

62. $\dfrac{-2^4}{(-2a^2)^4}$ _____

63. $\left(-\dfrac{x}{y}\right)^3\left(\dfrac{y^2}{w}\right)^2\left(\dfrac{w}{x^2}\right)^3$ _____

64. $\left(-\dfrac{a^2b}{c}\right)^2\left(\dfrac{c}{b^2}\right)^3\left(\dfrac{1}{a^3}\right)^2$ _____

65. $\dfrac{3(x+y)^3(x-y)^4}{(x-y)^26(x+y)^5}$ _____

66. $\dfrac{10(u-v+w)^8}{5(u-v+w)^{11}}$ _____

C   *Simplify, assuming n is restricted so that each exponent represents a positive integer.*

67. $x^{5-n}x^{n+2}$ _____

68. $y^{2n+2}y^{n-2}$ _____

69. $\dfrac{x^{2n}}{x^n}$ _____

70. $\dfrac{x^{n+2}}{x^n}$ _____

71. $(x^{n+1})^2$ _____

72. $(x^{n+1})^n$ _____

The Check Exercise for this section is on page 117.

73. $\dfrac{u^{n+3}v^n}{u^{n+1}v^{n+4}}$ _____

74. $\dfrac{(x^ny^{n+1})^2}{x^{2n+1}y^{2n}}$ _____

Section 2-2   Addition and Subtraction of Polynomials

- Polynomials; Degree
- Distributive Axiom
- Combining Like Terms
- Removing Symbols of Grouping
- Addition and Subtraction

Polynomials; Degree

An algebraic expression involving only the operations of addition, subtraction, and multiplication on variables and constants is called a **polynomial**. Here are some examples:

Polynomials

$$3x - 1 \qquad x \qquad 2x^2 - 3x + 2 \qquad 5$$

$$x^3 - 3x^2y - 4y^2 \qquad 0 \qquad x^2 - \tfrac{2}{3}xy + 2y^2$$

In a polynomial, a variable cannot appear in a denominator, as an exponent, within a radical, or within absolute-value bars. The following expressions are, therefore, not polynomials:

Nonpolynomials

$$\frac{2x + 1}{3x^2 - 5x + 7} \qquad 3^x \qquad x^3 - 2\sqrt{x} + \frac{1}{x^3} \qquad |2x^3 - 5| \qquad \frac{1}{x} \qquad \sqrt{x}$$

We see that a polynomial in one variable x is constructed by adding or subtracting constants and terms of the form ax^n, where a is a real number and n is a natural number. A polynomial in two variables x and y is constructed by adding or subtracting constants and terms of the form $ax^m y^n$, where again a is a real number and m and n are natural numbers.

EXAMPLE 3 Which of the following expressions are polynomials?

(A) $3x^2 - 2xy + y^2$ **(B)** $\sqrt{2x} - \dfrac{3}{x} + 5$ **(C)** $\dfrac{x^2 - 3x + 2}{x - 3}$

(D) $6x^3 - \sqrt{2}x - \tfrac{1}{3}$ **(E)** $\sqrt{x^2 - 3x + 1}$ **(F)** $4x^3y^2 - \sqrt{3}xy^2z^5$

Solution Expressions A, D, and F are polynomials.

PROBLEM 3 Which of the following expressions are polynomials?

(A) $3x^2 - 2x + 1$ **(B)** $\sqrt{x - 3}$ **(C)** $x^2 - 2xy + y^2$

(D) $\dfrac{x - 1}{x^2 + 2}$ **(E)** $10^x + 5$ **(F)** $2xyz$

Solution

It is convenient to identify certain types of polynomials for more efficient study. The concept of degree is used for this purpose. The **degree of a term** in a polynomial is the sum of the powers of the variables in the term. Thus, if a term has only one variable, the degree of the term is the power of the variable. The degree 0 is assigned to a nonzero constant term—that is, a term without any variables. The **degree of a polynomial** is the degree of its term with the highest degree. The constant 0 is not assigned a degree, either as a term or as a polynomial.

EXAMPLE 4 What is the degree of each term in the following polynomials? What is the degree of the polynomial?

(A) $4x^3$ **(B)** $3x^3y^2$ **(C)** $3x^5 - 2x^4 + x^2 - 3$
(D) $x^2 - 2xy + y^2 + 2x - 3y + 2$

Solution **(A)** This polynomial and its only term are of degree 3.
(B) This polynomial and its only term are of degree 5.

(C) The degrees of the four terms, in order, are 5, 4, 2, and 0; the degree of the polynomial is 5.

(D) The degrees of the six terms, in order, are 2, 2, 2, 1, 1, and 0; the degree of the polynomial is 2.

PROBLEM 4   What is the degree of each term in the following polynomials? What is the degree of the polynomial?

(A) $7x^5$ (B) $3x^3y^4$ (C) $5x^3 - 7x^2 + x - 9$
(D) $4x^2y - xy^2 + xy + y^2 + x + 6$

Solution   (A)   (B)

(C)   (D)

We also call a one-term polynomial a **monomial**, a two-termed polynomial a **binomial**, and a three-termed polynomial a **trinomial**.

For example,

$4x^3 - 3x + 7$ $5x - 2y$ $6x^4y^3$ 7

Trinomial   Binomial   Monomial   Monomial

degree 3   degree 1   degree 7   degree 0

■ **Distributive Axiom**

Basic to adding and subtracting polynomials, as well as many other operations on algebraic expressions, is the distributive axiom

$$a \cdot (b + c) = a \cdot b + a \cdot c$$

where a, b, and c are any real numbers. (The times dot · is included for emphasis.) In words, this axiom states that multiplication distributes over addition.

EXAMPLE 5   Multiply, using the distributive axiom:

(A) $3(x + y) = 3x + 3y$ (B) $5(2x^2 + 3) = 10x^2 + 15$
(C) $2x(x + 1) = 2x^2 + 2x$ (D) $-3(5x^2 - 2) = -15x^2 + 6$

PROBLEM 5   Multiply, using the distributive axiom:

(A) $5(m + n)$ (B) $4(3y^3 + 5)$ (C) $3y(2 + y)$ (D) $-2(y^2 - y)$

Solution   (A)   (B)

(C)   (D)

It should be clear from the properties of equality in Chapter 1 that we can also write the distributive axiom in the form

$$ab + ac = a(b + c)$$

That is, a factor common to each term can be taken out. Again, recall that we multiply factors and add or subtract terms.

EXAMPLE 6   Take out factors common to both terms:

(A) $6x + 6y = \mathbf{6}x + \mathbf{6}y = \mathbf{6}(x + y)$
(B) $7x^2 + 14 = 7x^2 + \mathbf{7} \cdot 2 = \mathbf{7}(x^2 + 2)$
(C) $ax + ay = \mathbf{a}x + \mathbf{a}y = \mathbf{a}(x + y)$

PROBLEM 6

Take out factors common to both terms:

(A) $9m + 9n$ **(B)** $6y^3 + 12$ **(C)** $du + dy$

Solution   **(A)**   **(B)**   **(C)**

Several other useful distributive forms follow from the basic version given above and from other properties discussed in the first chapter:

Additional Distributive Properties

$$(ab + ac) = (b + c)a$$
$$(ab - ac) = a(b - c) = (b - c)a$$
$$a(b + c + d + \cdots + f) = ab + ac + ad + \cdots + af$$
$$ab + ac + ad + \cdots + af = a(b + c + d + \cdots + f)$$
$$= (b + c + d + \cdots + f)a$$

■ Combining Like Terms

A constant that is present as a factor in a term is called the **numerical coefficient** (or simply the **coefficient**) of the term. If no constant appears in the term, then the coefficient is understood to be 1. The coefficient of a term in a polynomial includes the sign that precedes it, so that a term like $-3x^2$ should be thought of as $(-3)x^2$.

EXAMPLE 7   What is the coefficient of each term in the polynomial $3x^4 - 2x^3 + x^2 - x + 3$?

Solution

$$3x^4 - 2x^3 + x^2 - x + 3 = 3x^4 + (-2)x^3 + 1 \cdot x^2 + (-1)x + 3$$

where: Coefficient of x^4, Coefficient of x^3, Coefficient of x^2, Coefficient of x.

The coefficient of x^4 is 3, that of x^3 is -2, that of x^2 is 1, and that of x is -1.

PROBLEM 7

What is the coefficient of each term in the polynomial
$5x^4 - x^3 - 3x^2 + x - 7$?

Solution

First term: Second term: Third term:

Fourth term: Fifth term:

Two terms are called **like terms** if they have exactly the same variable factors to the same powers. The numerical coefficients may or may not be the same. Since constant terms involve no variables, all constant terms are like terms. If an algebraic expression contains two or more like terms, these terms can be combined into a single term by making use of the distributive law.

EXAMPLE 8   Combine like terms:

(A) $3x + 7x \;\boxed{= (3 + 7)x}\; = 10x$

(B) $6m - 9m \;\boxed{= (6 - 9)m}\; = -3m$

(C) $3z + 5 - z + 2 \;\boxed{\begin{aligned}&= 3z - z + 5 + 2\\&= (3 - 1)z + (5 + 2)\end{aligned}}$
$= 2z + 7$

(D) $5x^3y - 2xy - x^3y - 2x^3y \;\boxed{\begin{aligned}&= 5x^3y - x^3y - 2x^3y - 2xy\\&= (5 - 1 - 2)x^3y - 2xy\end{aligned}}$
$= 2x^3y - 2xy$

PROBLEM 8

Combine like terms:

(A) $5y + 4y$ (B) $2u - 6u$ (C) $4x - 1 + 3x + 2$
(D) $6mn^2 - m^2n - 3mn^2 - mn^2$

Solution

(A)   (B)

(C)   (D)

It should be clear that free use has been made of the axioms discussed in the first chapter. Most of the steps illustrated in the dashed boxes are done mentally. The process is quickly mechanized as follows:

Like terms are combined by adding their numerical coefficients.

EXAMPLE 9   Combine like terms mentally:

(A) $3x - 5y + 6x + 2y \;\boxed{= 3x + 6x - 5y + 2y}\; = 9x - 3y$

(B) $x^3y^2 - 2x^2y^3 + 5x^2y^2 - 4x^2y^3 - x^3y^2 - 5x^2y^2$

$$= x^3y^2 - x^3y^2 - 2x^2y^3 - 4x^2y^3 + 5x^2y^2 - 5x^2y^2$$

$$= -6x^2y^3$$

PROBLEM 9

Combine like terms mentally:

(A) $7m + 8n - 5m - 10n$

(B) $2u^4v^2 - 3uv^3 - u^4v^2 + 6u^4v^2 + 2uv^3 - 6u^4v^2$

Solution

(A)

(B)

■ Removing Symbols of Grouping

How can we simplify expressions such as

$$2(3x - 5y) - 2(x + 3y)$$

You no doubt would guess that we could rewrite this expression, using the various forms of the distributive property, as

$$6x - 10y - 2x - 6y$$

and combine like terms to obtain

$$4x - 16y$$

and your guess would be correct.

EXAMPLE 10

Remove parentheses and simplify:

(A) $2(3x^2 - 2x + 5) + (x^2 + 3x - 7)$

$$= 2(3x^2 - 2x + 5) + 1(x^2 + 3x - 7)$$
Think

$$= 6x^2 - 4x + 10 + x^2 + 3x - 7$$

$$= 7x^2 - x + 3$$

(B) $(x^3 - 2x - 6) - (2x^3 - x^2 + 2x - 3)$

$$= 1(x^3 - 2x - 6) + (-1)(2x^3 - x^2 + 2x - 3)$$ Be careful with the sign
Think   here.

$$= x^3 - 2x - 6 - 2x^3 + x^2 - 2x + 3$$

$$= -x^3 + x^2 - 4x - 3$$

(C) $[3x^2 - (2x + 1)] - (x^2 - 1)$

$$= [3x^2 - 2x - 1] - (x^2 - 1)$$ Remove inner parentheses first.

$$= 3x^2 - 2x - 1 - x^2 + 1$$

$$= 2x^2 - 2x$$

(D) $y - \{x - [2y - (x + y)]\}$

$\qquad = y - \{x - [2y - x - y]\}$ Remove innermost parentheses first.

$\qquad = y - \{x - [y - x]\}$ Simplify within parentheses.

$\qquad = y - \{x - y + x\}$ Remove inner parentheses.

$\qquad = y - x + y - x$

$\qquad = 2y - 2x$

PROBLEM 10   Remove parentheses and simplify:

(A) $3(u^2 - 2v^2) + (u^2 + 5v^2)$
(B) $(m^3 - 3m^2 + m - 1) - (2m^3 - m + 3)$
(C) $(x^3 - 2) - [2x^3 - (3x + 4)]$
(D) $\{x - [y - (z + x)] + x\} - [y - (x + z)]$

Solution   **(A)**

(B)

(C)

(D)

■ Addition and Subtraction

Addition and subtraction of polynomials can be thought of in terms of removing parentheses and combining like terms, as illustrated in Example 10. Horizontal and vertical arrangements are illustrated in the next two examples. You should be able to work either way, letting the situation dictate the choice.

EXAMPLE 11   Add:

$\qquad x^4 - 3x^3 + x^2 \qquad -x^3 - 2x^2 + 3x \qquad \text{and} \qquad 3x^2 - 4x - 5$

Solution   Add horizontally:

$\qquad (x^4 - 3x^3 + x^2) + (-x^3 - 2x^2 + 3x) + (3x^2 - 4x - 5)$

$\qquad\qquad = x^4 - 3x^3 + x^2 - x^3 - 2x^2 + 3x + 3x^2 - 4x - 5$

$\qquad\qquad = x^4 - 4x^3 + 2x^2 - x - 5$

or vertically by lining up like terms and adding their coefficients:

$$
\begin{array}{r}
x^4 - 3x^3 + x^2 \\
- x^3 - 2x^2 + 3x \\
3x^2 - 4x - 5 \\
\hline
x^4 - 4x^3 + 2x^2 - x - 5
\end{array}
$$

PROBLEM 11   Add horizontally and vertically:

$$3x^4 - 2x^3 - 4x^2 \qquad x^3 - 2x^2 - 5x \qquad \text{and} \qquad x^2 + 7x - 2$$

Solution

EXAMPLE 12   Subtract: $4x^2 - 3x + 5$ from $x^2 - 8$

Solution
$$(x^2 - 8) - (4x^2 - 3x + 5) \qquad \text{or}$$
$$= x^2 - 8 - 4x^2 + 3x - 5$$
$$= -3x^2 + 3x - 13$$

$$\begin{array}{r} x^2 \qquad\;\; - 8 \\ 4x^2 - 3x + \; 5 \\ \hline -3x^2 + 3x - 13 \end{array}$$ $\leftarrow$ Change signs and add.

PROBLEM 12   Subtract: $2x^2 - 5x + 4$ from $5x^2 - 6$

Solution

Solutions to Matched Problems

3. (A), (C), and (F) are polynomials.
4. **(A)** 5   **(B)** 7
   (C) Degrees of terms are 3, 2, 1, and 0; degree of polynomial is 3
   (D) Degrees of terms are 3, 3, 2, 2, 1, and 0; degree of polynomial is 3
5. **(A)** $5(m + n) = 5m + 5n$ **(B)** $4(3y^3 + 5) = 12y^3 + 20$
   (C) $3y(2 + y) = 6y + 3y^2$ **(D)** $-2(y^2 - y) = -2y^2 + 2y$
6. **(A)** $9m + 9n + 9(m + n)$ **(B)** $6y^3 + 12 = 6(y^3 + 2)$
   (C) $du + dy = d(u + y)$
7. The coefficients in order are 5, -1, -3, 1, and -7.
8. **(A)** $5y + 4y \begin{array}{|c|} \hline = (5 + 4)y \\ \hline \end{array} = 9y$

   (B) $2u - 6u \begin{array}{|c|} \hline = (2 - 6)u \\ \hline \end{array} = -4u$

   (C) $4x - 1 + 3x + 2 \begin{array}{|c|} \hline = 4x + 3x - 1 + 2 \\ \hline \end{array} = 7x + 1$

8. (D) $6mn^2 - m^2n - 3mn^2 - mn^2$ $\boxed{= 6mn^2 - 3mn^2 - mn^2 - m^2n \atop = (6 - 3 - 1)mn^2 - m^2n}$

$$= 2mn^2 - m^2n$$

9. (A) $7m + 8n - 5m - 10n = 2m - 2n$
  (B) $2u^4v^2 - 3uv^3 - u^4v^2 + 6u^4v^2 + 2uv^3 - 6u^4v^2 = u^4v^2 - uv^3$

10. (A) $3(u^2 - 2v^2) + (u^2 + 5v^2) = 3u^2 - 6v^2 + u^2 + 5v^2 = 4u^2 - v^2$
  (B) $(m^3 - 3m^2 + m - 1) - (2m^3 - m + 3)$
$$= m^3 - 3m^2 + m - 1 - 2m^3 + m - 3$$
$$= -m^3 - 3m^2 + 2m - 4$$
  (C) $(x^3 - 2) - [2x^3 - (3x + 4)] = x^3 - 2 - (2x^3 - 3x - 4)$
$$= x^3 - 2 - 2x^3 + 3x + 4$$
$$= -x^3 + 3x + 2$$
  (D) $\{x - [y - (z + x)] + x\} - [y - (x + z)]$
$$= [x - (y - z - x) + x] - (y - x - z)$$
$$= x - y + z + x + x - y + x + z = 4x - 2y + 2z$$

11. Horizontal: $(3x^4 - 2x^3 - 4x^2) + (x^3 - 2x^2 - 5x) + (x^2 + 7x - 2)$
$$= 3x^4 - 2x^3 - 4x^2 + x^3 - 2x^2 - 5x + x^2 + 7x - 2$$
$$= 3x^4 - x^3 - 5x^2 + 2x - 2$$
  Vertical: $3x^4 - 2x^3 - 4x^2$
$$\begin{array}{r} x^3 - 2x^2 - 5x \\ x^2 + 7x - 2 \\ \hline 3x^4 - x^3 - 5x^2 + 2x - 2 \end{array}$$

12. $(5x^2 - 6) - (2x^2 - 5x + 4)$
$$= 5x^2 - 6 - 2x^2 + 5x - 4 = 3x^2 + 5x - 10$$

Practice Exercise 2-2 ■

Work odd-numbered problems first, check answers, and then work even-numbered problems in areas of weakness. Answers to all problems are in the back of the book. Make every effort to work a problem yourself before you look at an answer.

A Indicate the degree of each monomial.

1. $3x^5$ _____ **2.** $-7x^3$ _____

3. $-2xy^3$ _____ **4.** $5x^2y$ _____

5. $21x^2yz^3$ _____ **6.** $-13xy^3z^4$ _____

Identify each of the following as a monomial, binomial, or trinomial and give its degree.

7. $3x^2 + 7$ _____ **8.** $1 + 2y + 3y^2$ _____

9. $-u^5 + u^4v^2 - u^3v$ _____ **10.** $-7a^2 + 2b^5$ _____

11. $-t^3 + 8$ _____   **12.** $13x^3y^2z$ _____

13. $5p^2r^3st^2$ _____   **14.** $xyz + x^2z^2 + yz^2$ _____

Given the polynomial $7x^4 - 3x^3 - x^2 + x - 3$, indicate the following.

15. The coefficient of the second term _____

16. The coefficient of the third term _____

17. The exponent of the variable in the second term _____

18. The exponent of the variable in the fourth term _____

19. The coefficient of the fourth term _____

20. The coefficient of the first term _____

Simplify by removing parentheses, if any, and combining like terms.

21. $9x + 8x$ _____   **22.** $7x + 3x$ _____

23. $9x - 8x$ _____   **24.** $7x - 3x$ _____

25. $5x + x + 2x$ _____   **26.** $3x + 4x + x$ _____

27. $4t - 8t - 9t$ _____   **28.** $2x - 5x + x$ _____

29. $4y + 3x + y$ _____   **30.** $2x + 3y + 5x$ _____

31. $8 + 4x - 4$ _____   **32.** $-3 - x + 5$ _____

33. $5m + 3n - m - 9n$ _____   **34.** $2x + 8y - 7x - 5y$ _____

35. $3(u - 2v) + 2(3u + v)$ _____   **36.** $2(m + 3n) + 4(m - 2n)$ _____

37. $4(m - 3n) - 3(2m + 4n)$ _____

38. $2(x - y) - 3(3x - 2y)$ _____

39. $(2u - v) + (3u - 5v)$ _____   **40.** $(x + 3y) + (2x - 5y)$ _____

41. $(3x + 2) + (x - 5)$ _____   **42.** $(y + 7) - (4 - 2y)$ _____

43. $2(x + 5) - (x - 1)$ _____   **44.** $3(z - 2) - 2(z - 3)$ _____

Add.

45. $6x + 5$ and $3x - 8$ _____   **46.** $3x - 5$ and $2x + 3$ _____

47. $7x - 5$, $-x + 3$, and $-8x - 2$ _____

48. $2x + 3,\ -4x - 2,$ and $7x - 4$ _____

49. $5x^2 + 2x - 7,\ 2x^2 + 3,$ and $-3x - 8$ _____

50. $2x^2 - 3x + 1,\ 2x - 3$ and $4x^2 + 5$ _____

Subtract.

51. $3x - 8$ from $2x - 7$ _____   **52.** $4x - 9$ from $2x + 3$ _____

53. $2y^2 - 6y + 1$ from $y^2 - 6y - 1$ _____

54. $x^2 - 3x - 5$ from $2x^2 - 6x - 5$ _____

B *Simplify by removing symbols of grouping, if any, and combining like terms.*

55. $-x^2y + 3x^2y - 5x^2y$ _____

56. $-4r^3t^3 - 7r^3t^3 - 7r^3t^3 + 9r^3t^3$ _____

57. $y^3 + 4y^2 - 10 + 2y^3 - y + 7$ _____

58. $3x^2 - 2x + 5 - x^2 + 4x - 8$ _____

59. $a^2 - 3ab + b^2 + 2a^2 + 3ab - 2b^2$ _____

60. $2x^2y + 2xy^2 - 5xy + 2xy^2 - xy - 4x^2y$ _____

61. $x - 3y - 4(2x - 3y)$ _____   **62.** $a + b - 2(a - b)$ _____

63. $y - 2(x - y) - 3x$ _____   **64.** $x - 3(x + 2y) + 5y$ _____

65. $-2(-3x + 1) - (2x + 4)$ _____

66. $-3(-t + 7) - (t - 1)$ _____

67. $2(x - 1) - 3(2x - 3) - (4x - 5)$ _____

68. $-2(y - 7) - 3(2y + 1) - (-5y + 7)$ _____

69. $4t - 3[4 - 2(t - 1)]$ _____

70. $3x - 2[2x - (x - 7)]$ _____

71. $3[x - 2(x + 1)] - 4(2 - x)$ _____

72. $2(3x + y) - 2[y - (3x + 2)]$ _____

Replace each question mark with an appropriate algebraic expression.

73. $5 + m - 2n = 5 - (?)$ _____   **74.** $2 + 3x - y = 2 - (?)$ _____

75. $w^2 - y - z = w^2 - (?)$ _____

76. $x - y - z + 5 = x - (?)$ _____

Add.

77. $2x^4 - x^2 - 7, 3x^3 + 7x^2 + 2x,$ and $x^2 - 3x - 1$ _____

78. $3x^3 - 2x^2 + 5, 3x^2 - x - 3,$ and $2x + 4$ _____

Subtract.

79. $5x^3 - 3x + 1$ from $2x^3 + x^2 - 1$ _____

80. $3x^3 - 2x^2 - 5$ from $2x^3 - 3x + 2$ _____

81. Subtract the sum of the first two polynomials from the sum of the last two: $3m^3 - 2m + 5, 4m^2 - m, 3m^2 - 3m - 2,$ and $m^3 + m^2 + 2$ _____

82. Subtract the sum of the last two polynomials from the sum of the first two: $2x^2 - 4xy + y^2, 3xy - y^2, x^2 - 2xy - y^2,$ and $-x^2 + 3xy - 2y^2$

_____

C *Remove symbols of grouping and combine like terms.*

83. $2t - 3\{t + 2[t - (t + 5)] + 1\}$ _____

84. $x - \{x - [x - (x - 1)]\}$ _____

85. $w - \{x - [z - (w - x) - z] - (x - w)\} + x$ _____

86. $3x^2 - 2\{x - x[x + 4(x - 3)] - 5\}$ _____

87. $\{[(x - 1) - 1] - x\} - 1$ _____

88. $1 - \{1 - [1 - (1 - x)]\}$ _____

89. $x - \{1 - [x - (1 - x)]\}$ _____

90. $x - \{[1 - (x - 1)] - x\}$ _____

APPLICATIONS   **91.** The width of a rectangle is 5 meters less than its length. If x is the length of the rectangle, write an algebraic expression that represents the perimeter P of the rectangle and simplify the expression. _____

92. Repeat Problem 91 if the length of the rectangle is 3 meters more than twice its width. _____

93. A pile of coins consists of nickels, dimes, and quarters. There are 5 fewer dimes than nickels and 2 more quarters than dimes. If x equals the number of nickels, write an algebraic expression that represents the value of the pile in cents. Simplify the expression. [*Hint:* If x represents the number of nickels, then what do $x - 5$ and $(x - 5) + 2$ represent?] _____

94. A parking meter contains dimes and quarters only. There are 4 fewer quarters than dimes. If x represents the number of dimes, write an algebraic expression that represents the total value of all coins in the meter in cents. Simplify the expression. _____

*95. A board is to be cut into four pieces. The largest piece is to be 3 times as long as the smallest; the other two pieces are each to be twice as long as the smallest. If x represents the length of the smallest piece, write an algebraic expression for the total length of the board. Simplify the expression.

_____

*96. A wire is to be cut into three pieces so that the second piece is 3 feet longer than the first and the third is 8 feet longer than the sum of the other two. Let x represent the length of the smallest piece. Write an algebraic expression for the total length of the wire in feet. Simplify the expression.

_____

*97. A jogger runs for some time at a rate of 8 kilometers per hour and then runs twice as long at a rate of 12 kilometers per hour. Let t be the time, in hours, run at the slower pace. Write an algebraic expression for the total distance run. Simplify the expression. (Recall that Distance = Rate × Time.) _____

The Check Exercise for this section is on page 119.

*98. A racer drives for 2 hours at one speed and then for another hour and a half at a speed 12 miles per hour slower. Let s be her initial speed. Write an algebraic expression for the total distance driven. Simplify the expression. _____

Section 2-3   Multiplication of Polynomials

■ Multiplication of Monomials
■ Multiplication of Polynomials
■ Mental Multiplication of Binomials
■ Squaring Binomials

■ Multiplication of Monomials

You have already had experience in multiplying monomials in Section 2-1 on exponents. We will review the process in the following example.

EXAMPLE 13   Multiply:

(A) $x^3x^5 \;\boxed{= x^{3+5}}\; = x^8 \quad x^3x^5 \neq x^{3 \cdot 5}$

(B) $(3m^{12})(5m^{23}) \;\boxed{= 3 \cdot 5m^{12+23}}\; = 15m^{35}$

(C) $(-3x^3y^4)(2x^2y^3) \;\boxed{= (-3)(2)x^{3+2}y^{4+3}}\; = -6x^5y^7$

PROBLEM 13	Multiply: **(A)** $y^4 y^7$	**(B)** $(9x^4)(3x^2)$	**(C)** $(4u^3 v^2)(-3uv^3)$
Solution	**(A)**	**(B)**	**(C)**

■ Multiplication of Polynomials

How do we multiply polynomials with more than one term? The distributive property plays a central role in the process and leads directly to the following mechanical rule:

Mechanics of Multiplying Two Polynomials

To multiply two polynomials, multiply each term of one by each term of the other. Then add like terms.

EXAMPLE 14   Multiply:

(A) $3x^2(2x^2 - 3x + 4) = 6x^4 - 9x^3 + 12x^2$

(B) $(2x - 3)(3x^2 - 2x + 3)$ or

$$= 2x(3x^2 - 2x + 3) - 3(3x^2 - 2x + 3)$$

$$= 6x^3 - 4x^2 + 6x - 9x^2 + 6x - 9$$

$$= 6x^3 - 13x^2 + 12x - 9$$

$$\begin{array}{r} 3x^2 - 2x + 3 \\ 2x - 3 \\ \hline 6x^3 - 4x^2 + 6x \\ - 9x^2 + 6x - 9 \\ \hline 6x^3 - 13x^2 + 12x - 9 \end{array}$$

Note that either way, each term in $3x^2 - 2x + 3$ is multiplied by each term in $2x - 3$. In the vertical arrangement, by multiplying by $2x$ first the like terms line up more conveniently. Students usually prefer a vertical arrangement for this type of problem.

PROBLEM 14

Multiply:

(A) $2m^3(3m^2 - 4m - 3)$ **(B)** $2x^2 + 3x - 1$ **(C)** $2x^2 + 3x - 2$
                                                    $\underline{3x - 4}$ $\underline{3x^2 - 2x + 1}$

Solution   **(A)**   **(B)**   **(C)**

■ Mental Multiplication of Binomials

For reasons that will become clear shortly, it is essential that you learn to multiply first-degree polynomials of the type $(3x + 2)(2x - 1)$ and $(3x - y)(x + 2y)$ mentally. To discover relationships that will make this possible, let us first multiply $(3x + 2)$ and $(2x - 1)$ using a vertical arrangement:

$$
\begin{array}{r}
3x + 2 \\
2x - 1 \\
\hline
6x^2 + 4x \\
- 3x - 2 \\
\hline
6x^2 + x - 2
\end{array}
$$

Now let us use a horizontal arrangement and try to discover a method that will enable us to carry out the multiplication mentally. We start by multiplying each term in the first binomial times each term in the second binomial:

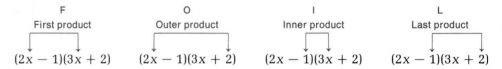

F	O	I	L
First product	Outer product	Inner product	Last product

$(2x - 1)(3x + 2)$ $(2x - 1)(3x + 2)$ $(2x - 1)(3x + 2)$ $(2x - 1)(3x + 2)$

Performing these four operations on one line, we obtain

F	O	I	L
First	Outer	Inner	Last
product	product	product	product

$$(2x - 1)(3x + 2) = 6x^2 \quad + 4x \quad - 3x \quad - 2$$

The inner and outer products are like terms and hence combine into one term. Thus,

$$(2x - 1)(3x + 2) = 6x^2 + x - 2$$

To speed up the process we combine the inner and outer products mentally. The method just described is called the **FOIL method**. We note again that the product of two first-degree polynomials is a second-degree polynomial. A simple three-step process for carrying out the FOIL method is illustrated in Example 15.

EXAMPLE 15   Mentally multiply:

(A) $(2x - 1)(3x + 2) = 6x^2 + x - 2$

The like terms are obtained in step 2 by multiplying the inner and outer products, and they are combined mentally.

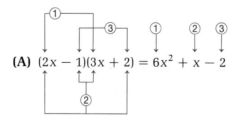

(B) $(2a - b)(a + 3b) = 2a^2 + 5ab - 3b^2$

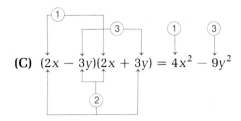

(C) $(2x - 3y)(2x + 3y) = 4x^2 - 9y^2$

Notice that a middle term does not appear since its coefficient is 0.

PROBLEM 15

Mentally multiply:

(A) $(3x - 2)(2x - 1)$ **(B)** $(a - 3b)(2a + 3b)$
(C) $(5x - y)(5x + y)$ **(D)** $(4x - 3y)(2x + 5y)$

Solution

(A)   **(B)**

(C)   **(D)**

In Section 2-5 we will consider the reverse problem: given a second-degree polynomial, such as $2x^2 - 5x - 3$, find first-degree factors with integer coefficients that will produce this second-degree polynomial as a product. To be able to factor second-degree polynomial forms with any degree of efficiency, you must know how to mentally multiply quickly and accurately first-degree factors of the types illustrated in this section.

The FOIL method can be applied to any binomials, not just those of degree 1. In such cases, however, the inner and outer products may not be like terms.

EXAMPLE 16

Mentally multiply:

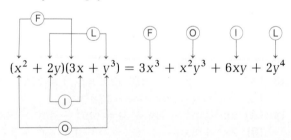

$(x^2 + 2y)(3x + y^3) = 3x^3 + x^2y^3 + 6xy + 2y^4$

PROBLEM 16

Mentally multiply: $(2a^3 + b^2)(a - 2b^2)$

Solution

■ **Squaring Binomials**

Since

$$(a + b)^2 = (a + b)(a + b) = a^2 + 2ab + b^2$$
$$(a - b)^2 = (a - b)(a - b) = a^2 - 2ab + b^2$$

we can formulate a simple mechanical rule for squaring any binomial directly.

Mechanical Rule for Squaring Binomials

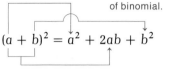

Square first term of binomial.   Square second term of binomial.

$$(a + b)^2 = a^2 + 2ab + b^2$$

Double the product of the two terms in the binomial.

We do the same thing for $(a - b)^2$, except that the sign of the middle term on the right becomes negative.

EXAMPLE 17   Square each binomial, using the mechanical rule:

(A) $(2x + 3)^2 = (2x)^2 + 2[(2x)(3)] + (3)^2 = 4x^2 + 12x + 9$

(B) $(3x - 2y)^2 = (3x)^2 - 2[(3x)(2y)] + (2y)^2 = 9x^2 - 12xy + 4y^2$

PROBLEM 17   Square each binomial, using the mechanical rule:

(A) $(3y + 4)^2$ **(B)** $(2u - 3v)^2$

Solution   **(A)**   **(B)**

Solutions to Matched Problems

13. (A) $y^4y^7 = y^{11}$ **(B)** $(9x^4)(3x^2) = 27x^6$
    (C) $(4u^3v^2)(-3uv^3) = -12u^4v^5$
14. (A) $2m^3(3m^2 - 4m - 3) = 6m^5 - 8m^4 - 6m^3$

(B)

$$
\begin{array}{r}
2x^2 + 3x - 1 \\
3x - 4 \\
\hline
6x^3 + 9x^2 - 3x \\
- 8x^2 - 12x + 4 \\
\hline
6x^3 + x^2 - 15x + 4
\end{array}
$$

(C)

$$
\begin{array}{r}
2x^2 + 3x - 2 \\
3x^2 - 2x + 1 \\
\hline
6x^4 + 9x^3 - 6x^2 \\
- 4x^3 - 6x^2 + 4x \\
2x^2 + 3x - 2 \\
\hline
6x^4 + 5x^3 - 10x^2 + 7x - 2
\end{array}
$$

15. (A) $(3x - 2)(2x - 1) = 6x^2 - 7x + 2$
    (B) $(a - 3b)(2a + 3b) = 2a^2 - 3ab - 9b^2$
    (C) $(5x - y)(5x + y) = 25x^2 - y^2$
    (D) $(4x - 3y)(2x + 5y) = 8x^2 + 14xy - 15y^2$
16. $(2a^3 + b^2)(a - 2b^2) = 2a^4 - 4a^3b^2 + ab^2 - 2b^4$
17. (A) $(3y + 4)^2 = 9y^2 + 24y + 16$
    (B) $(2u - 3v)^2 = 4u^2 - 12uv + 9v^2$

Practice Exercise 2-3 ■

Work odd-numbered problems first, check answers, and then work even-numbered problems in areas of weakness. Answers to all problems are in the back of the book. Make every effort to work a problem yourself before you look at an answer.

A *Multiply.*

1. y^2y^3 _____

2. x^3x^2 _____

3. $(5y^4)(2y)$ _____

4. $(2x)(3x^4)$ _____

5. $(8x^{11})(-3x^9)$ _____

6. $(-7u^9)(5u^7)$ _____

7. $(-3u^4)(2u^5)(-u^7)$ _____

8. $(2x^3)(-3x)(-4x^5)$ _____

9. $(cd^2)(c^2d^2)$ _____

10. $(a^2b)(ab^2)$ _____

11. $(-3xy^2z^3)(-5xyz^2)$ _____

12. $(-2xy^3z)(3x^3yz)$ _____

13. $y(y + 7)$ _____

14. $x(1 + x)$ _____

15. $5y(2y - 7)$ _____

16. $3x(2x - 5)$ _____

17. $3a^2(a^3 + 2a^2)$ _____

18. $2m^2(m^2 + 3m)$ _____

19. $2y(y^2 + 2y - 3)$ _____

20. $2x(2x^2 - 3x + 1)$ _____

21. $7m^3(m^3 - 2m^2 - m + 4)$ _____

22. $3x^2(2x^3 + 3x^2 - x - 2)$ _____

23. $5uv^2(2u^3v - 3uv^2)$ _____

24. $4m^2n^3(2m^3n - mn^2)$ _____

25. $2cd^3(c^2d - 2cd + 4c^3d^2)$ _____

26. $3x^2y(2xy^3 + 4x - y^2)$ _____

B **27.** $(3y + 2)(2y^2 + 5y - 3)$ _____

28. $(2x - 1)(x^2 - 3x + 5)$ _____

29. $(m + 2n)(m^2 - 4mn - n^2)$ _____

30. $(x - 3y)(x^2 - 3xy + y^2)$ _____

31. $(2m^2 + 2m - 1)(3m^2 - 2m + 1)$ _____

32. $(x^2 - 3x + 5)(2x^2 + x - 2)$ _____

33. $(a + b)(a^2 - ab + b^2)$ _____

34. $(a - b)(a^2 + ab + b^2)$ _____

35. $(2x^2 - 3xy + y^2)(x^2 + 2xy - y^2)$ _____

36. $(a^2 - 2ab + b^2)(a^2 + 2ab + b^2)$ _____

Multiply mentally.

37. $(x + 3)(x + 2)$ _____   **38.** $(m - 2)(m - 3)$ _____

39. $(a + 8)(a - 4)$ _____   **40.** $(m - 12)(m + 5)$ _____

41. $(t + 4)(t - 4)$ _____   **42.** $(u - 3)(u + 3)$ _____

43. $(m - n)(m + n)$ _____   **44.** $(a + b)(a - b)$ _____

45. $(4t - 3)(t - 2)$ _____   **46.** $(3x - 5)(2x + 1)$ _____

47. $(3x + 2y)(x - 3y)$ _____   **48.** $(2x - 3y)(x + 2y)$ _____

49. $(2m - 7)(2m + 7)$ _____   **50.** $(3y + 2)(3y - 2)$ _____

51. $(6x - 4y)(5x + 3y)$ _____   **52.** $(3m + 7n)(2m - 5n)$ _____

53. $(2s - 3t)(3s - t)$ _____   **54.** $(2x - 3y)(3x - 2y)$ _____

55. $(x^3 + y^3)(x + y)$ _____   **56.** $(x^2 + y^3)(x^3 + y)$ _____

57. $(2x - y^2)(x^2 + 3y)$ _____   **58.** $(x^3 - y^2)(2x + y^2)$ _____

Square each binomial, using the mechanical rule.

59. $(3x + 2)^2$ _____   **60.** $(4x + 3y)^2$ _____

61. $(2x - 5y)^2$ _____   **62.** $(2x - 7)^2$ _____

63. $(6u + 5v)^2$ _____   **64.** $(7p + 2q)^2$ _____

65. $(2m - 5n)^2$ _____   **66.** $(4x - 1)^2$ _____

C   *Simplify.*

67. $(x + 2y)^3$ _____

68. $(2m - n)^3$ _____

69. $(3x - 1)(x + 2) - (2x - 3)^2$ _____

70. $(2x + 3)(x - 5) - (3x - 1)^2$ _____

71. $2(x - 2)^3 - (x - 2)^2 - 3(x - 2) - 4$ _____

72. $(2x - 1)^3 - 2(2x - 1)^2 + 3(2x - 1) + 7$ _____

73. $-3x\{x[x - x(2 - x)] - (x + 2)(x^2 - 3)\}$ _____

74. $2\{(x - 3)(x^2 - 2x + 1) - x[3 - x(x - 2)]\}$ _____

75. $(2x - 1)(2x + 1)(3x^3 - 4x + 3)$ _____

76. $(x - 1)(x - 2)(2x^3 - 3x^2 - 2x - 1)$ _____

77.   If you are given two polynomials, one of degree m and another of degree n, and $m > n$, then what is the degree of their product? _____

78.   What is the degree of the sum of the two polynomials in Problem 77?

_____

APPLICATIONS   **79.**   The length of a rectangle is 8 meters more than its width. If y is the length of the rectangle, write an algebraic expression that represents its area. Change the expression to a form without parentheses. _____

**The Check Exercise for
this section is on
page 121.**   **80.**   Repeat Problem 79 if the length of the rectangle is 3 meters less than twice the width. _____

Section 2-4   Factoring Out Common Factors; Factoring by Grouping

■   Factoring Out Common Factors
■   Factoring by Grouping

■   Factoring Out Common Factors

You have already had experience in factoring out common factors from polynomials. The distributive property of real numbers in the form

$$ab + ac = a(b + c)$$

is the important property behind the process. Note that factoring is just the opposite of multiplication. That is, to factor a polynomial means to find factors whose product is the given polynomial.

EXAMPLE 18   Factor out factors common to all terms:

(A) $6x^2 + 15x \boxed{= 3x \cdot 2x + 3x \cdot 5} = 3x(2x + 5)$

(B) $2u^3v - 6u^2v^2 + 8uv^3 \boxed{= 2uv \cdot u^2 - 2uv \cdot 3uv + 2uv \cdot 4v^2}$

$$= 2uv(u^2 - 3uv + 4v^2)$$

PROBLEM 18   Factor out factors common to all terms:

(A) $12y^2 - 28y$ (B) $6m^4 - 15m^3n + 9m^2n^2$

Solution   (A)   (B)

Now look closely at the following four examples and try to determine what they all have in common.

$$2xy + 3y = y(2x + 3)$$
$$2xA + 3A = A(2x + 3)$$
$$2x(x - 4) + 3(x - 4) = (x - 4)(2x + 3)$$
$$2x(3x + 1) + 3(3x + 1) = (3x + 1)(2x + 3)$$

Because of the commutative property, common factors can be taken out either on the left or on the right.

The factoring involved in each example is essentially the same. The only difference is the nature of the common factors being taken out. In the last two examples think of $(x - 4)$ and $(3x + 1)$ as single numbers, just as A represents a single number in the second example.

EXAMPLE 19   Factor out factors common to all terms:

(A) $5x(x - 1) - (x - 1)\ \boxed{= 5x(x - 1) - 1(x - 1)} = (x - 1)(5x - 1)$

(B) $3x(2x - y) - 2y(2x - y) = (2x - y)(3x - 2y)$

PROBLEM 19   Factor out factors common to all terms:

(A) $3m(m + 2) - (m + 2)$ (B) $2u(u + 3v) - 3v(u + 3v)$

Solution   (A)   (B)

■ **Factoring by Grouping**

Some polynomials can be factored by grouping terms in such a way that we obtain results that look like Example 19. We can then complete the factoring following the procedures used there. This process will prove useful in Section 2-6, where an efficient method is developed for factoring a second-degree polynomial as the product of two first-degree polynomials.

EXAMPLE 20   Factor by grouping:

(A) $2x^2 - 8x + 3x - 12$ (B) $5x^2 - 5x - x + 1$
(C) $6x^2 - 3xy - 4xy + 2y^2$

Solution   (A) $2x^2 - 8x + 3x - 12$ Group the first two and last two terms.

$= (2x^2 - 8x) + (3x - 12)$ Remove common factors from each group.

$= 2x(x - 4) + 3(x - 4)$ The common factor $(x - 4)$ can be taken out.

$= (x - 4)(2x + 3)$ The factoring is complete.

(B) $5x^2 - 5x - x + 1$ Group the first two and last two terms.

$= (5x^2 - 5x) - (x - 1)$ Remove common factors from each group. Note that the signs change in the second set of parentheses.

$= 5x(x - 1) - 1(x - 1)$ The common factor $(x - 1)$ can be taken out.

$= (x - 1)(5x - 1)$ The factoring is complete.

(C) $6x^2 - 3xy - 4xy + 2y^2$

$= (6x^2 - 3xy) - (4xy - 2y^2)$ Signs change inside the second set of parentheses.

$= 3x(2x - y) - 2y(2x - y)$

$= (2x - y)(3x - 2y)$

PROBLEM 20

Factor by grouping:

(A) $6x^2 + 2x + 9x + 3$ **(B)** $3m^2 + 6m - m - 2$
(C) $2u^2 + 6uv - 3uv - 9v^2$

Solution   **(A)**

(B)

(C)

In Example 20, the polynomials were arranged in such a way that grouping the first two terms and the last two terms led to common factors. The process is not always this neat, however, and you will sometimes have to rearrange terms in order to group them profitably for factoring (if the polynomial can be factored at all).

EXAMPLE 21   Factor $y^2 + xz + xy + yz$ by grouping.

Solution $y^2 + xz + xy + yz$ If we proceed as in Example 20, no common factor can be factored out to complete the factoring.

$= (y^2 + xz) + (xy + yz)$

$= (y^2 + xz) + y(x + z)$

$y^2 + xz + xy + yz$ Rearrange the terms and proceed again as in Example 20.

$= y^2 + xy + xz + yz$

$= y(y + x) + (x + y)z$

$= y(x + y) + z(x + y)$

$= (y + z)(x + y)$

PROBLEM 21   Factor $ac + bd + bc + ad$ by grouping.

Solution

Solutions to Matched   **18. (A)** $12y^2 - 28y = 4y(3y - 7)$
Problems   **(B)** $6m^4 - 15m^3n + 9m^2n^2 = 3m^2(2m^2 - 5mn + 3n^2)$
                            19. (A) $3m(m + 2) - (m + 2) = (m + 2)(3m - 1)$
                                 (B) $2u(u + 3v) - 3v(u + 3v) = (u + 3v)(2u - 3v)$
                            20. (A) $6x^2 + 2x + 9x + 3 = (6x^2 + 2x) + (9x + 3)$
$$= 2x(3x + 1) + 3(3x + 1)$$
$$= (3x + 1)(2x + 3)$$
   (B) $3m^2 + 6m - m - 2 = (3m^2 + 6m) - (m + 2)$
$$= 3m(m + 2) - 1(m + 2)$$
$$= (m + 2)(3m - 1)$$
   (C) $2u^2 + 6uv - 3uv - 9v^2 = (2u^2 + 6uv) - (3uv + 9v^2)$
$$= 2u(u + 3v) - 3v(u + 3v)$$
$$= (u + 3v)(2u - 3v)$$
21. $ac + bd + bc + ad = ac + bc + ad + bd = c(a + b) + (a + b)d$
$$= (a + b)c + (a + b)d = (a + b)(c + d)$$

Practice Exercise 2-4 ∎

Work odd-numbered problems first, check answers, and then work even-numbered problems in areas of weakness. Answers to all problems are in the back of the book. Make every effort to work a problem yourself before you look at an answer.

Write in factored form by removing factors common to all terms.

A   **1.** $2xA + 3A$ _____   **2.** $xM - 4M$ _____

    3. $10x^2 + 15x$ _____   **4.** $9y^2 - 6y$ _____

    5. $14u^2 - 6u$ _____   **6.** $20m^2 + 12m$ _____

    7. $6u^2 - 10uv$ _____   **8.** $14x^2 - 21xy$ _____

    9. $10m^2n - 15mn^2$ _____   **10.** $9u^2v + 6uv^2$ _____

    11. $2x^3y - 6x^2y^2$ _____   **12.** $6x^2y^2 - 6xy^3$ _____

    13. $3x(x + 2) + 5(x + 2)$ _____   **14.** $4y(y + 3) + 7(y + 3)$ _____

    15. $3m(m - 4) - 2(m - 4)$ _____   **16.** $x(x - 1) - 4(x - 1)$ _____

    17. $x(x + y) - y(x + y)$ _____   **18.** $m(m - n) + n(m - n)$ _____

B   **19.** $6x^4 - 9x^3 + 3x^2$ _____   **20.** $6m^4 - 8m^3 - 2m^2$ _____

     21. $8x^3y - 6x^2y^2 + 4xy^3$ _____

     22. $10u^3v + 20u^2v^2 - 15uv^3$ _____

     23. $8x^4 - 12x^3y + 4x^2y^2$ _____   **24.** $9m^4 - 6m^3n - 6m^2n^2$ _____

     25. $3x(2x + 3) - 5(2x + 3)$ _____   **26.** $2u(3u - 8) - 3(3u - 8)$ _____

     27. $x(x + 1) - (x + 1)$ _____   **28.** $3u(u - 1) - (u - 1)$ _____

     29. $4x(2x - 3) - (2x - 3)$ _____   **30.** $3y(4y - 5) - (4y - 5)$ _____

Replace each question mark with an algebraic expression that will make both sides equal.

     31. $3x^2 - 3x + 2x - 2 = (3x^2 - 3x) + (?)$ _____

     32. $2x^2 + 4x + 3x + 6 = (2x^2 + 4x) + (?)$ _____

     33. $3x^2 - 12x - 2x + 8 = (3x^2 - 12x) - (?)$ _____

     34. $2y^2 - 10y - 3y + 15 = (2y^2 - 10y) - (?)$ _____

     35. $8u^2 + 4u - 2u - 1 = (8u^2 + 4u) - (?)$ _____

     36. $6x^2 + 10x - 3x - 5 = (6x^2 + 10x) - (?)$ _____

Factor out common factors from each group, and then complete the factoring if possible.

     37. $(3x^2 - 3x) + (2x - 2)$ _____

     38. $(2x^2 + 4x) + (3x + 6)$ _____

     39. $(3x^2 - 12x) - (2x - 8)$ _____

     40. $(2y^2 - 10y) - (3y - 15)$ _____

     41. $(8u^2 + 4u) - (2u + 1)$ _____

     42. $(6x^2 + 10x) - (3x + 5)$ _____

Factor as the product of two first-degree polynomials using grouping. (These problems are related to Problems 31–42.)

     43. $3x^2 - 3x + 2x - 2$ _____   **44.** $2x^2 + 4x + 3x + 6$ _____

     45. $3x^2 - 12x - 2x + 8$ _____   **46.** $2y^2 - 10y - 3y + 15$ _____

     47. $8u^2 + 4u - 2u - 1$ _____   **48.** $6x^2 + 10x - 3x - 5$ _____

Factor as the product of two first-degree factors using grouping.

49. $2m^2 - 8m + 5m - 20$ _____ **50.** $5x^2 - 10x + 2x - 4$ _____

51. $6x^2 - 9x - 4x + 6$ _____ **52.** $12x^2 + 8x - 9x - 6$ _____

C **53.** $3u^2 + 4 - 12u - u$ _____ **54.** $6m^2 - 2 + 4m - 3m$ _____

55. $6u^2 - 2v^2 + 3uv - 4uv$ _____

56. $2x^2 + 2y^2 - 4xy - xy$ _____

57. $6x^2 - 5y^2 + 3xy - 10xy$ _____

58. $4u^2 + 12v^2 - 3uv - 16uv$ _____

The Check Exercise for **59.** $3a^2 + 3b^2 + 9ab + ab$ _____ **60.** $a^2 + b^2 + ab + ab$ _____
this section is on
page 123. **61.** $uw + vx - vw - ux$ _____ **62.** $2ab + 12 + 6b + 4a$ _____

Section 2-5 Factoring Second-Degree Polynomials

We now turn our attention to factoring second-degree polynomials of the form

$$2x^2 - 5x - 3 \quad \text{and} \quad 2x^2 + 3xy - 2y^2$$

into the product of two first-degree polynomials with integer coefficients. By now it should be very easy for you to obtain (by mental multiplication) the products

$$(x - 3)(x + 2) = x^2 - x - 6 \quad \text{and} \quad (x - 3y)(x + 2y) = x^2 - xy - 6y^2$$

but can you reverse the process? Can you, for example, find integers a, b, c, and d so that

$$2x^2 - 5x - 3 = (ax + b)(cx + d)$$

Representing a second-degree polynomial with integers as coefficients as the product of two first-degree polynomials with integer coefficients is not as easy as multiplying first-degree polynomials. In this section we will develop a method of attack that is relatively easy to understand, but not always easy to apply. In the next section we will develop an approach to the problem that builds on factoring by grouping discussed in the last section. This approach is a little harder to understand, but once the method is understood, it is fairly easy to apply.

Let us start with a very simple polynomial whose factors you will likely guess at once:

$$x^2 + 6x + 8$$

Our problem is to find two first-degree factors with integer coefficients, if they exist. To start we write

$$x^2 + 6x + 8 = (x + \quad)(x + \quad) \quad \text{Why must both signs be positive?}$$

Now what are the constant terms? Since they are positive-integer factors of 8, we write

$$\frac{8}{\begin{array}{l} 1 \cdot 8 \\ 2 \cdot 4 \end{array}}$$

If we try the first pair (mentally), we obtain the first and last terms, but not the middle term. We next try 2 and 4, which gives us the middle term as well as the first and last terms. Thus,

$$x^2 + 6x + 8 = (x + 2)(x + 4)$$

Because of the commutative property, we could also write

$$x^2 + 6x + 8 = (x + 4)(x + 2)$$

Let us try another polynomial. Find first-degree factors with integer coefficients for

$$2x^2 - 7x + 6$$

Again, we write

$$2x^2 - 7x + 6 = (2x - \quad)(x - \quad) \qquad \text{Why must both signs be negative?}$$

The constant terms both must be factors of 6. The possibilities are

$$\frac{6}{\begin{array}{l} 1 \cdot 6 \\ 6 \cdot 1 \\ 2 \cdot 3 \\ 3 \cdot 2 \end{array}}$$

All pairs give the first and last terms in $2x^2 - 7x + 6$, but will any give the middle term, $-7x$?

Testing each pair (this is why you need to do binomial multiplication mentally), we find that the last pair gives the middle term. Thus,

$$2x^2 - 7x + 6 = (2x - 3)(x - 2)$$

Before you conclude that all second-degree polynomials with integer coefficients have first-degree factors with integer coefficients, consider the following simple polynomial:

$$x^2 + x + 2$$

Proceeding as above, we write

$$x^2 + x + 2 = (x + \quad)(x + \quad)$$

$$\frac{2}{\begin{array}{l} 1 \cdot 2 \\ 2 \cdot 1 \end{array}}$$

and find that neither pair produces the middle term, x. Hence, we conclude that

$$x^2 + x + 2$$

has no first-degree factors with integer coefficients, and we say the polynomial is not factorable using integers. (Any polynomial can be factored as itself times

1 or its negative times -1; we will say that a polynomial with integer coefficients is **not factorable using integers** if it cannot be factored into two polynomials of lower degree with integer coefficients.)

EXAMPLE 22　Factor each polynomial, if possible, using integer coefficients:

(A) $2x^2 + 3xy - 2y^2$　　**(B)** $x^2 - 3x + 4$　　**(C)** $6x^2 + 5xy - 4y^2$

Solution　**(A)** $2x^2 + 3xy - 2y^2$

$$= (2x + \quad y)(x - \quad y)$$

Put in what we know. Signs must be opposite. (We can reverse this choice if we get $-3xy$ instead of $+3xy$ for the middle term.)

Now what are the factors of 2 (the coefficient of y^2)?

$$\frac{2}{\begin{array}{l}1 \cdot 2 \\ 2 \cdot 1\end{array}}$$

The first choice gives us $-3xy$ for the middle term—close, but not there—so we reverse our choice of signs to obtain

$$2x^2 + 3xy - 2y^2 = (2x - y)(x + 2y)$$

(B) $x^2 - 3x + 4 = (x - \quad)(x - \quad)$

$$\frac{4}{\begin{array}{l}2 \cdot 2 \\ 1 \cdot 4 \\ 4 \cdot 1\end{array}}$$

No choice produces the middle term; hence,

$$x^2 - 3x + 4$$

is not factorable using integer coefficients.

(C) $6x^2 + 5xy - 4y^2 = (\quad x + \quad y)(\quad x - \quad y)$

The signs must be opposite in the factors, since the third term is negative. (We can reverse our choice of signs later if necessary.)

We now write all factors of 6 and of 4:

$$\frac{6}{\begin{array}{l}2 \cdot 3 \\ 3 \cdot 2 \\ 1 \cdot 6 \\ 6 \cdot 1\end{array}} \qquad \frac{4}{\begin{array}{l}2 \cdot 2 \\ 1 \cdot 4 \\ 4 \cdot 1\end{array}}$$

and try each choice on the left with each on the right—a total of 12 combinations that give us the first and last terms in the polynomial $6x^2 + 5xy - 4y^2$. The question is: does any combination also give us the middle term, $5xy$? After trial and error and, perhaps, some educated guessing among the choices, we find that $3 \cdot 2$ matched with $4 \cdot 1$ gives us the correct middle term. Thus,

$$6x^2 + 5xy - 4y^2 = (3x + 4y)(2x - y)$$

If none of the 24 combinations (including reversing our sign choice) had produced the middle term, then we would conclude that the polynomial is not factorable using integers.

PROBLEM 22   Factor each polynomial, if possible, using integer coefficients:

(A) $x^2 - 8x + 12$ (B) $x^2 + 2x + 5$
(C) $2x^2 + 7xy - 4y^2$ (D) $4x^2 - 15xy - 4y^2$

Solution   (A)   (B)

(C)   (D)

CONCLUDING REMARKS   In problems like those found in Example 22(C), if the coefficients of the first and last terms get larger and larger with more and more factors, the number of combinations that need to be checked increases very rapidly. And it is quite possible in most practical situations that none of the combinations will work. It is important, however, that you understand the approach presented here, since it will work for most of the simpler factoring problems you will encounter. The next (optional) section introduces a systematic approach to the problem of factoring that will reduce the amount of trial and error substantially and even tell you whether a polynomial can be factored before you proceed too far.

In conclusion, we point out that if a, b, and c are selected at random out of the integers, the probability that

$$ax^2 + bx + c$$

is not factorable in the integers is much greater than the probability that it is. But even being able to factor some second-degree polynomials leads to marked simplification of some algebraic expressions and an easy way to solve some second-degree equations, as will be seen later.

Solution to Matched   **22.** (A) $x^2 - 8x + 12 = (x - 2)(x - 6)$ (B) Not factorable
Problem   (C) $2x^2 + 7xy - 4y^2 = (2x - y)(x + 4y)$
(D) $4x^2 - 15xy - 4y^2 = (4x + y)(x - 4y)$

Practice Exercise 2-5 ∎

Work odd-numbered problems first, check answers, and then work even-numbered problems in areas of weakness. Answers to all problems are in the back of the book. Make every effort to work a problem yourself before you look at an answer.

Factor in the integers, if possible. If the expression is not factorable, say so.

A 1. $x^2 + 5x + 4$ _____ 2. $x^2 + 4x + 3$ _____

3. $x^2 + 5x + 6$ _____ 4. $x^2 + 7x + 10$ _____

5. $x^2 - 4x + 3$ _____ 6. $x^2 - 5x + 4$ _____

7. $x^2 - 7x + 10$ _____ 8. $x^2 - 5x + 6$ _____

9. $y^2 + 3y + 3$ _____ 10. $y^2 + 2y + 2$ _____

11. $y^2 - 2y + 6$ _____ 12. $x^2 - 3x + 5$ _____

13. $x^2 + 8xy + 15y^2$ _____ 14. $x^2 + 9xy + 20y^2$ _____

15. $x^2 - 10xy + 21y^2$ _____ 16. $x^2 - 10xy + 16y^2$ _____

17. $u^2 + 4uv + v^2$ _____ 18. $u^2 + 5uv + 3v^2$ _____

19. $3x^2 + 7x + 2$ _____ 20. $2x^2 + 7x + 3$ _____

21. $3x^2 - 7x + 4$ _____ 22. $2x^2 - 7x + 6$ _____

B 23. $3x^2 - 14x + 8$ _____ 24. $2y^2 - 13y + 15$ _____

25. $3x^2 - 11xy + 6y^2$ _____ 26. $2x^2 - 7xy + 6y^2$ _____

27. $n^2 - 2n - 8$ _____ 28. $n^2 + 2n - 8$ _____

29. $x^2 - 4x - 6$ _____ 30. $x^2 - 3x - 8$ _____

31. $3x^2 - x - 2$ _____ 32. $6m^2 + m - 2$ _____

33. $x^2 + 4xy - 12y^2$ _____ 34. $2x^2 - 3xy - 2y^2$ _____

35. $3u^2 - 11u - 4$ _____ 36. $8u^2 + 2u - 1$ _____

37. $6x^2 + 7x - 5$ _____ 38. $2m^2 - 3m - 20$ _____

39. $3s^2 - 5s - 2$ _____ 40. $2s^2 + 5s - 3$ _____

41. $3x^2 + 2xy - 3y^2$ _____ 42. $2x^2 - 3xy - 4y^2$ _____

43. $5x^2 - 8x - 4$ _____   **44.** $12x^2 + 16x - 3$ _____

45. $6u^2 - uv - 2v^2$ _____   **46.** $6x^2 - 7xy - 5y^2$ _____

47. $8x^2 + 6x - 9$ _____   **48.** $6x^2 - 13x + 6$ _____

49. $3u^2 + 7uv - 6v^2$ _____   **50.** $4m^2 + 10mn - 6n^2$ _____

51. $4u^2 - 19uv + 12v^2$ _____   **52.** $12x^2 - xy - 6y^2$ _____

C **53.** $12x^2 - 40xy - 7y^2$ _____   **54.** $15x^2 + 17xy - 4y^2$ _____

55. $12x^2 + 19xy - 10y^2$ _____   **56.** $24x^2 - 31xy - 15y^2$ _____

A polynomial such as $x^4 + 5x^2 + 6$ is not second degree but can still be factored using the techniques of this section. We can think of x^2 as the variable and rewrite the polynomial as $(x^2)^2 + 5(x^2) + 6$, and factor it as $(x^2 + 2)(x^2 + 3)$ as in Example 22(B). Use this method to factor the following polynomials.

57. $x^4 + 4x^2 + 3$ _____   **58.** $x^4 - x^2 - 2$ _____

The Check Exercise for
this section is on
page 125.
59. $2x^4 - x^2 - 3$ _____   **60.** $6x^4 - 7x^2 - 3$ _____

61. $x^6 + x^3 - 6$ _____   **62.** $2x^6 - 7x^3 - 4$ _____

Section 2-6   *ac* Test and Factoring (Optional)

In Example 22(C), the polynomial $6x^2 + 5xy - 4y^2$ was factored by trial and error as $6x^2 + 5xy - 4y^2 = (3x + 4y)(2x - y)$. The following process leads to the same result:

$$6x^2 + 5xy - 4y^2 \qquad \text{Rewrite } 5xy \text{ as } 8xy - 3xy.$$
$$= 6x^2 + 8xy - 3xy - 4y^2 \qquad \text{Group the first two and last two terms.}$$
$$= (6x^2 + 8xy) - (3xy + 4y^2) \qquad \text{Factor out the common factors.}$$
$$= 2x(3x + 4y) - y(3x + 4y) \qquad \text{Factor out the common factor } (3x + 4y).$$
$$= (3x + 4y)(2x - y)$$

The first step in this process—rewriting $5xy$ as $8xy - 3xy$—is what makes it work. To see why the choice of 8 and -3 works, consider the process again:

$$6x^2 + 5xy - 4y^2 \qquad \text{Rewrite } 5xy \text{ as a sum with unknown coefficients; note}$$
$$= 6x^2 + pxy + qxy - 4y^2 \qquad \text{that } p + q \text{ must be 5.}$$
$$= (6x^2 + pxy) + (qxy - 4y^2)$$
$$= x(6x + py) + y(qx - 4y)$$

For $x(6x + py)$ and $y(qx - 4y)$ to have a common factor, $6x + py$ must be a multiple of $qx - 4y$. Thus, the ratio of 6 to q must be the same as p to -4;

that is,

$$\frac{6}{q} = \frac{p}{-4}$$

so that $pq = -24$. Therefore, we need two integers p and q such that $p + q = 5$ and $pq = -24$; 8 and -3 work.

This process can be generalized and applied to second-degree polynomials of the following type:

$$ax^2 + bx + c \quad \text{and} \quad ax^2 + bxy + cy^2 \tag{1}$$

The process provides a test, called the **ac test for factorability**, that not only tells us if the polynomials of type (1) can be factored using integer coefficients, but, in addition, leads to a direct way of factoring those that are factorable.

---

ac Test for Factorability

If in polynomials of type (1) the product ac has two integer factors p and q whose sum is the coefficient of the middle term b; that is, if integers p and q exist so that

$$pq = ac \quad \text{and} \quad p + q = b \tag{2}$$

then polynomials of type (1) have first-degree factors with integer coefficients. If no integers p and q exist that satisfy Equations (2), then polynomials of type (1) will not have first-degree factors with integer coefficients.

Once we find integers p and q in the ac test, if they exist, our work is almost finished, since we can then write polynomials of type (1), splitting the middle term, in the forms

$$ax^2 + px + qx + c \quad \text{and} \quad ax^2 + pxy + qxy + cy^2 \tag{3}$$

Then the factoring can be completed in a couple of steps using factoring by grouping discussed at the end of Section 2-4.

Let us make the discussion concrete through several examples.

EXAMPLE 23   Factor, if possible, using integer coefficients:

(A) $2x^2 + 11x - 6$ **(B)** $4x^2 - 7x + 4$ **(C)** $6x^2 + 5xy - 4y^2$

Solution   **(A)** *Step 1:*   Test $2x^2 + 11x - 6$ for factorability using the ac test:

$$ac = (2)(-6) = -12$$

Try to find two integer factors of -12 whose sum is $b = 11$. We write (or think) of all two-integer factors of -12:

$$pq$$
$$(-3)(4)$$
$$(3)(-4)$$
$$(2)(-6)$$
$$(-2)(6)$$
$$(-1)(12)$$
$$(1)(-12)$$

and test to see if any of these pairs add up to 11, the coefficient of the middle term. We see that the next to the last pair works; that is,

$$\overset{p}{(-1)}\overset{q}{(12)} = \overset{ac}{-12} \quad \text{and} \quad \overset{p}{(-1)} + \overset{q}{(12)} = \overset{b}{11}$$

We can conclude, because of the *ac* test, that $2x^2 + 11x - 6$ can be factored using integer coefficients.

Step 2: Now split the middle term in the original equation using $p = -1$ and $q = 12$ as coefficients for x. This is possible, since $p + q = b$.

$$2x^2 + \overset{b}{11}x - 6 = 2x^2 - \overset{p}{1}x + \overset{q}{12}x - 6$$

Step 3: Factor the result obtained in step 2 by grouping. (This will always work if we can get to step 2, and it doesn't matter whether the values for p and q are reversed.)

$2x^2 - x + 12x - 6$	Group the first two and last two terms.
$= (2x^2 - x) + (12x - 6)$	Factor out the common factors.
$= x(2x - 1) + 6(2x - 1)$	Factor out the common factor $(2x - 1)$.
$= (2x - 1)(x + 6)$	The factoring is complete.

Thus,

$$2x^2 + 11x - 6 = (2x - 1)(x + 6)$$

This process can be reduced to a few key operational steps when all of the commentary is eliminated and some of the process is done mentally. The only trial and error occurs in step 1, and with a little practice that step will go fairly fast.

(B) Compute *ac* for $4x^2 - 7x + 4$:

$$ac = (4)(4) = 16$$

Write (or think) all two-integer factors of 16, and try to find a pair whose sum is -7, the coefficient of the middle term.

$$pq$$
$$(4)(4)$$
$$(-4)(-4)$$
$$(2)(8)$$
$$(-2)(-8)$$
$$(1)(16)$$
$$(-1)(-16)$$

None of these adds up to $-7 = b$; thus, according to the ac test,

$$4x^2 - 7x + 4$$

is not factorable using integer coefficients.

(C) Compute ac for $6x^2 + 5xy - 4y^2$:

$$ac = (6)(-4) = -24$$

Does -24 have two integer factors whose sum is $5 = b$? A little trial and error (either mentally or by listing) will turn up

$$\overset{p}{} \overset{q}{} \quad\quad \overset{ac}{} \quad\quad\quad\quad \overset{p}{} \quad \overset{q}{} \quad \overset{b}{}$$
$$(8)(-3) = -24 \quad\quad \text{and} \quad\quad 8 + (-3) = 5$$

We now split the middle term, $5xy$, into two terms and write

$$6x^2 + \overset{b}{5xy} - 4y^2 = 6x^2 + \overset{p}{8xy} - \overset{q}{3xy} - 4y^2$$

Then complete the factoring by grouping:

$$\begin{aligned}
6x^2 + 8xy - 3xy - 4y^2 &= (6x^2 + 8xy) - (3xy + 4y^2) \\
&= 2x(3x + 4y) - y(3x + 4y) \\
&= (3x + 4y)(2x - y)
\end{aligned}$$

Thus,

$$6x^2 + 5xy - 4y^2 = (3x + 4y)(2x - y)$$

PROBLEM 23

Factor, if possible, using integer coefficients. Proceed as above.

(A) $4x^2 + 4x - 3$ **(B)** $6x^2 - 3x - 4$ **(C)** $6x^2 - 25xy + 4y^2$

Solution   **(A)**   **(B)**

    (C)

Solution to Matched
Problem

23. (A) $ax^2 + bx + c = 4x^2 + 4x - 3$; thus, $a = 4$, $b = 4$, and $c = -3$.

$ac = (4)(-3) = -12$

$\dfrac{pq}{}$

$(-2)(6)$ ⟵— This pair works; that is, $(-2)(6) = -12$ and $(-2) + 6 = 4 = b$.
$(2)(-6)$
$(3)(-4)$
$(-3)(4)$
$(1)(-12)$
$(-1)(12)$

Thus, we write

$$4x^2 + 4x - 3 = 4x^2 - 2x + 6x - 3$$
$$= (4x^2 - 2x) + (6x - 3)$$
$$= 2x(2x - 1) + 3(2x - 1)$$
$$= (2x - 1)(2x + 3)$$

(B) $ax^2 + bx + c = 6x^2 - 3x - 4$; thus, $a = 6$, $b = -3$, and $c = -4$.

$ac = (6)(-4) = -24$

No two-integer factors of -24 add up to $-3 = b$. We conclude that the polynomial cannot be factored using integer coefficients.

(C) $ax^2 + bx + c = 6x^2 - 25xy + 4y^2$; thus $a = 6$, $b = -25$, and $c = 4$.

$ac = (6)(4) = 24$

Testing two-integer factors of 24, we find that $(-1)(-24) = 24$ and $(-1) + (-24) = -25 = b$; thus, the polynomial is factorable.

$$6x^2 - 25xy + 4y^2 = 6x^2 - xy - 24xy + 4y^2$$
$$= (6x^2 - xy) - (24xy - 4y^2)$$
$$= x(6x - y) - 4y(6x - y) = (6x - y)(x - 4y)$$

Practice Exercise 2-6 ■

Work odd-numbered problems first, check answers, and then work even-numbered problems in areas of weakness. Answers to all problems are in the back of the book. Make every effort to work a problem yourself before you look at an answer.

Factor, if possible, using integer coefficients. Use the ac test and proceed as in Example 23.

A   **1.** $3x^2 - 7x + 4$ _____   **2.** $2x^2 - 7x + 6$ _____

3. $x^2 + 4x - 6$ _____

4. $x^2 - 3x - 8$ _____

5. $2x^2 + 5x - 3$ _____

6. $3x^2 - 5x - 2$ _____

7. $3x^2 - 5x + 4$ _____

8. $2x^2 - 11x + 6$ _____

B 9. $3x^2 - 14x + 8$ _____

10. $2y^2 - 13y + 15$ _____

11. $6x^2 + 7x - 5$ _____

12. $5x^2 - 8x - 4$ _____

13. $6x^2 - 4x - 5$ _____

14. $5x^2 - 7x - 4$ _____

15. $2m^2 - 3m - 20$ _____

16. $12x^2 + 16x - 3$ _____

17. $3u^2 - 11u - 4$ _____

18. $8u^2 + 2u - 1$ _____

19. $6u^2 - uv - 2v^2$ _____

20. $6x^2 - 7xy - 5y^2$ _____

21. $3x^2 + 2xy - 3y^2$ _____

22. $2x^2 - 3xy - 4y^2$ _____

23. $8x^2 + 6x - 9$ _____

24. $6x^2 - 5x - 6$ _____

25. $4m^2 + 10mn - 6n^2$ _____

26. $3u^2 + 7uv - 6v^2$ _____

27. $3u^2 - 8uv - 6v^2$ _____

28. $4m^2 - 9mn - 6n^2$ _____

29. $4u^2 - 19uv + 12v^2$ _____

30. $12x^2 - xy - 6y^2$ _____

C 31. $12x^2 - 40xy - 7y^2$ _____

32. $15x^2 + 17xy - 4y^2$ _____

33. $18x^2 - 9xy - 20y^2$ _____

34. $15m^2 + 2mn - 24n^2$ _____

35. Find all integers b such that $x^2 + bx + 12$ can be factored. _____

36. Find all positive integers c less than 15 so that $x^2 - 7x + c$ can be factored.

Refer to the approach introduced in Exercise 2-5 (Problems 57–62) for factoring certain polynomials of degree higher than 2. Apply the ac test and factor, if possible, using integer coefficients.

37. $6x^4 + 7x^2 + 2$ _____

38. $5x^4 - 11x^2 + 2$ _____

The Check Exercise for
this section is on
page 127.

39. $2x^4 + 5x^2y^2 + 3y^4$ _____

40. $3x^4 + 5x^2y^2 - 2y^4$ _____

41. $2x^6 - x^3 - 3$ _____

42. $3x^6 - 10x^3 - 8$ _____

Section 2-7   More Factoring

- Sum and Difference of Two Squares
- Sum and Difference of Two Cubes
- Combined Forms
- More Factoring by Grouping
- Higher-Degree Polynomials
- General Strategy

■ Sum and Difference of Two Squares

If we multiply $(A - B)$ and $(A + B)$, we obtain

$$(A - B)(A + B) = A^2 - B^2$$

which is a difference of two squares. Writing this result in reverse order, we obtain a very useful factoring formula. If we try to factor the sum of two squares, $A^2 + B^2$, we find that it cannot be factored using integer coefficients unless A and B have common factors. (Try it to see why.)

Sum and Difference of Two Squares	
$A^2 + B^2$ $\quad$ cannot be factored using integer coefficients unless A and B have common factors.	(1)
$A^2 - B^2 = (A - B)(A + B)$	(2)

EXAMPLE 24   **(A)** $x^2 - y^2 = (x - y)(x + y)$ $\quad$ **(B)** $9x^2 - 4 = (3x - 2)(3x + 2)$
(C) $m^2 + n^2$ is not factorable using integer coefficients
(D) $9u^2 - 25v^2 = (3u - 5v)(3u + 5v)$

PROBLEM 24   Factor, if possible, using integer coefficients:

(A) $x^2 - 25$ $\quad$ **(B)** $16u^2 - v^2$ $\quad$ **(C)** $9x^2 + y^2$ $\quad$ **(D)** $7x^2 - 3$

Solution   **(A)**   **(B)**

(C)   **(D)**

■ Sum and Difference of Two Cubes

It is easy to verify, by direct multiplication of the right sides, the following factoring formulas for the sum and difference of two cubes:

Sum and Difference of Two Cubes
$A^3 + B^3 = (A + B)(A^2 - AB + B^2)$ (3)
$A^3 - B^3 = (A - B)(A^2 + AB + B^2)$ (4)

These formulas are used in the same way as the factoring formula for the difference of two squares. (Notice that neither $A^2 - AB + B^2$ nor $A^2 + AB + B^2$ factor further using integer coefficients.) Both formulas should be memorized.

To factor

$$y^3 - 27$$

we first note that it can be written in the form

$$y^3 - 3^3$$

Thus, we are dealing with the difference of two cubes. If in the factoring formula (4) we let $A = y$ and $B = 3$, we obtain

$$y^3 - 27 = y^3 - 3^3 = (y - 3)(y^2 + 3y + 9)$$

EXAMPLE 25   Factor as far as possible using integer coefficients:

(A) $8x^3 + 27 = (2x)^3 + 3^3 = (2x + 3)(4x^2 - 6x + 9)$

(B) $t^3 - 1 = (t - 1)(t^2 + t + 1)$

PROBLEM 25   Factor as far as possible using integer coefficients:

(A) $u^3 + 8$ **(B)** $x^3 - 8y^3$

Solution   **(A)**   **(B)**

■ Combined Forms

We now consider some examples that involve removing common factors as well as factoring second-degree and third-degree forms.

General Factoring Principle
Remove common factors first before proceeding further.

EXAMPLE 26   Factor as far as possible using integer coefficients:

(A) $18x^3 - 8x$ **(B)** $6x^3 + 12y^2 + 12y$
(C) $3x^4 - 24xy^3$ **(D)** $8x^3y + 20x^2y^2 - 12xy^3$

Solution   **(A)** $18x^3 - 8x$ Remove common factors first.

$\qquad = 2x(9x^2 - 4)$ Factor the difference of two squares.

$\qquad = 2x(3x - 2)(3x + 2)$ Factoring is complete.

(B) $6y^3 + 12y^2 + 12y$ Remove common factors.

$\qquad = 6y(y^2 + 2y + 2)$ Cannot be factored further using integer coefficients. Note that the constant factor 6 need not be factored into primes.

(C) $3x^4 - 24xy^3$ Remove common factors.

$\qquad = 3x(x^3 - 8y^3)$ Factor the difference of two cubes.

$\qquad = 3x(x - 2y)(x^2 + 2xy + 4y^2)$

(D) $8x^3y + 20x^2y^2 - 12xy^3$

$\qquad = 4xy(2x^2 + 5xy - 3y^2)$

$\qquad = 4xy(2x - y)(x + 3y)$

PROBLEM 26   Factor as far as possible using integer coefficients:

(A) $3x^3 - 48x$ **(B)** $3x^3 - 15x^2y + 18xy^2$
(C) $4x^3 + 12x^2 + 12x$ **(D)** $2u^4 - 16u$

Solution   **(A)**

(B)

(C)

(D)

■ More Factoring by Grouping

Occasionally, polynomial forms of a more general nature than we considered in Section 2-4 can be factored by appropriate grouping of terms. The following example illustrates the process.

EXAMPLE 27   **(A)** $x^2 + xy + 2x + 2y$ Group the first two and last two terms.

$\qquad = (x^2 + xy) + (2x + 2y)$ Remove common factors.

$\qquad = x(x + y) + 2(x + y)$ Remove the common factor $(x + y)$.

$\qquad = (x + y)(x + 2)$ Notice that the factors are not first-degree factors of the same type.

(B) $x^3 + 3x^2 + 4x + 12$ Group the first two and last two terms.

$\qquad = (x^3 + 3x^2) + (4x + 12)$ Remove the common factor.

$\qquad = x^2(x + 3) + 4(x + 3)$ Remove the common factor $x + 3$.

$\qquad = (x^2 + 4)(x + 3)$

PROBLEM 27

Factor by grouping terms:

(A) $x^2 - xy + 5x - 5y$ **(B)** $2x^3 - x^2 + 4x - 2$

Solution

(A)

(B)

■ Higher-Degree Polynomials

The techniques and examples of factoring introduced so far have dealt mainly with polynomials of degree 2 or 3. Some higher-degree polynomials can be factored in the same ways.

EXAMPLE 28

Factor: **(A)** $x^4 - 4$ **(B)** $x^4 + 6x^2 + 8$ **(C)** $x^6 + y^3$

Solution

(A) $x^4 - 4$ Think of x^4 as $(x^2)^2$.

       $= (x^2)^2 - 2^2$ Factor as the difference of two squares.

       $= (x^2 - 2)(x^2 + 2)$

(B) $x^4 + 6x^2 + 8$ Think of x^4 as $(x^2)^2$.

       $= (x^2)^2 + 6(x^2) + 8$ Factor as a second-degree polynomial in x^2.

       $= (x^2 + 4)(x^2 + 2)$

(C) $x^6 + y^3$ Think of x^6 as $(x^2)^3$.

       $= (x^2)^3 + y^3$ Factor as the sum of two cubes.

       $= (x^2 + y)[(x^2)^2 - (x^2)y + y^2)]$ Simplify the second factor.

       $= (x^2 + y)(x^4 - x^2y + y^2)$

PROBLEM 28

Factor: **(A)** $x^4 - y^4$ **(B)** $x^4 - 3x^2 + 2$ **(C)** $8x^6 - 27$

Solution

(A)

(B)

(C)

■ General Strategy

There is no standard procedure (**algorithm**) for factoring a polynomial. However, the following strategy may be helpful.

General Strategy for Factoring

---

1. Remove common factors (Section 2-4).
2. If the polynomial has two terms, look for a difference of two squares or a sum or difference of two cubes (Section 2-7).
3. If the polynomial has three terms:
   (A) Try trial and error (Section 2-5).
   (B) Or use the ac test (Section 2-6).
4. If the polynomial has more than three terms, try grouping (Sections 2-4 and 2-7).

Factoring requires skill, creativity, and perseverance. Often the appropriate technique is not immediately apparent and practice is necessary to develop your recognition of what might work on a given problem. Exercises 2-7 and 2-8 contain numerous problems for this purpose.

Solutions to Matched Problems

24. (A) $x^2 - 25 = (x - 5)(x + 5)$ **(B)** $16u^2 - v^2 = (4u - v)(4u + v)$
(C) $9x^2 + y^2$ not factorable **(D)** $7x^2 - 3$ not factorable

25. (A) $u^3 + 8 = (u + 2)(u^2 - 2u + 4)$
(B) $x^3 - 8y^3 = (x - 2y)(x^2 + 2xy + 4y^2)$

26. (A) $3x^3 - 48x = 3x(x^2 - 16) = 3x(x - 4)(x + 4)$
(B) $3x^3 - 15x^2y + 18xy^2 = 3x(x^2 - 5xy + 6y^2) = 3x(x - 2y)(x - 3y)$
(C) $4x^3 + 12x^2 + 12x = 4x(x^2 + 3x + 3)$
(D) $2u^4 - 16u = 2u(u^3 - 8) = 2u(u - 2)(u^2 + 2u + 4)$

27. (A) $x^2 - xy + 5x - 5y = (x^2 - xy) + (5x - 5y)$
$$= x(x - y) + 5(x - y) = (x - y)(x + 5)$$
(B) $2x^3 - x^2 + 4x - 2 = (2x^3 - x^2) + (4x - 2)$
$$= x^2(2x - 1) + 2(2x - 1) = (x^2 + 2)(2x - 1)$$

28. (A) $x^4 - y^4 = (x^2)^2 - (y^2)^2 = (x^2 + y^2)(x^2 - y^2)$
$$= (x^2 + y^2)(x + 1)(x - 1)$$
(B) $x^4 - 3x^2 + 2 = (x^2)^2 - 3(x^2) + 2 = (x^2 - 1)(x^2 - 2)$
$$= (x - 1)(x + 1)(x^2 - 2)$$
(C) $8x^6 - 27 = (2x^2)^3 - 3^3 = (2x^2 - 3)[(2x^2)^2 + (2x^2) \cdot 3 + 3^2]$
$$= (2x^2 - 3)(4x^4 + 6x^2 + 9)$$

Practice Exercise 2-7 ■

Work odd-numbered problems first, check answers, and then work even-numbered problems in areas of weakness. Answers to all problems are in the back of the book. Make every effort to work a problem yourself before you look at an answer.

Factor as far as possible using integer coefficients.

A **1.** $v^2 - 25$ _____ **2.** $x^2 - 81$ _____ **3.** $9x^2 - 4$ _____

4. $4m^2 - 1$ _____ **5.** $x^2 + 49$ _____ **6.** $y^2 + 64$ _____

7. $9x^2 - 16y^2$ _____ **8.** $25u^2 - 4v^2$ _____

9. $x^3 + 1$ _____

10. $y^3 - 1$ _____

11. $m^3 - n^3$ _____

12. $p^3 + q^3$ _____

13. $8x^3 + 27$ _____

14. $u^3 - 8v^3$ _____

15. $6u^2v^2 - 3uv^3$ _____

16. $2x^3y - 6x^2y^3$ _____

17. $2x^2 - 8$ _____

18. $3y^2 - 27$ _____

19. $2x^3 + 8x$ _____

20. $3x^4 + 27x^2$ _____

21. $12x^3 - 3xy^2$ _____

22. $2u^3v - 2uv^3$ _____

23. $2x^4 + 2x$ _____

24. $xy^3 + x^4$ _____

25. $6x^2 + 36x + 48$ _____

26. $4x^2 - 4x - 24$ _____

27. $3x^3 - 6x^2 + 15x$ _____

28. $2x^3 - 2x^2 + 8x$ _____

B **29.** $x^2y^2 - 16$ _____

30. $m^2n^2 - 36$ _____

31. $a^3b^3 + 8$ _____

32. $27 - x^3y^3$ _____

33. $4x^3y + 14x^2y^2 + 6xy^3$ _____

34. $3x^3y - 15x^2y^2 + 18xy^3$ _____

35. $4u^3 + 32v^3$ _____

36. $54x^3 - 2y^3$ _____

37. $60x^2y^2 - 200xy^3 - 35y^4$ _____

38. $60x^4 + 68x^3y - 16x^2y^2$ _____

39. $xy + 2x + y^2 + 2y$ _____

40. $x^2 + 3x + xy + 3y$ _____

41. $x^2 - 5x + xy - 5y$ _____

42. $x^2 - 3x - xy + 3y$ _____

43. $ax - 2bx - ay + 2by$ _____

44. $mx + my - 2nx - 2ny$ _____

45. $15ac - 20ad + 3bc - 4bd$ _____

46. $2am - 3an + 2bm - 3bn$ _____

47. $x^3 - 2x^2 - x + 2$ _____

48. $x^3 - 2x^2 + x - 2$ _____

49. $(y - x)^2 - y + x$ _____

50. $x^2(x - 1) - x + 1$ _____

51. $x^2y^2 - xy - 6$ _____

52. $a^2b^2 - 7ab + 12$ _____

53. $z^4 - z^2 - 6$ _____

54. $x^4 + 4x^2 + 4$ _____

C   **55.** $x^8 - 4$ _____   **56.** $a^6 + 8a^3 - 20$ _____

57. $r^4 - s^4$ _____   **58.** $16a^4 - b^4$ _____

59. $x^4 - 3x^2 - 4$ _____   **60.** $x^4 - 7x^2 - 18$ _____

61. $(x - 3)^2 - 16y^2$ _____   **62.** $(x + 2)^2 - 9y^2$ _____

63. $(a - b)^2 - 4(c - d)^2$ _____   **64.** $(x^2 - x)^2 - 9(y^2 - y)^2$ _____

65. $25(4x^2 - 12xy + 9y^2) - 9a^2b^2$ _____

66. $18a^3 - 8a(x^2 + 8x + 16)$ _____

67. $x^6 - 1$ _____   **68.** $a^6 - 64b^6$ _____

69. $2x^3 - x^2 - 8x + 4$ _____   **70.** $4y^3 - 12y^2 - 9y + 27$ _____

71. $25 - a^2 - 2ab - b^2$ _____   **72.** $x^2 - 2xy + y^2 - 9$ _____

73. $16x^4 - x^2 + 6xy - 9y^2$ _____   **74.** $x^4 - x^2 + 4x - 4$ _____

75. $x^3 - 2x^2 + 3x - 6$ _____   **76.** $x^3 + 2x^2 - 5x - 10$ _____

77. $x^5 - x^4 + x - 1$ _____   **78.** $x^5 + x^3 + 2x^2 + 2$ _____

The Check Exercise for   **79.** $3x^3 - x^2 + 12x - 4$ _____   **80.** $2x^3 + x^2 + 4x + 2$ _____
this section is on
page 129.   **81.** $x^2 + 4x - y^2 + 4$ _____   **82.** $x^2 + y^2 - z^2 + 2xy$ _____

Section 2-8   Chapter Review

For a natural number n, the **base** a raised to the **exponent** n is $a^n = a \cdot a \cdot \cdots \cdot a$ (n factors). Natural number exponents satisfy these **five properties of exponents**:

1. $a^m a^n = a^{m+n}$
2. $(a^n)^m = a^{mn}$
3. $(ab)^m = a^m b^m$
4. $\left(\dfrac{a}{b}\right)^m = \dfrac{a^m}{b^m}$

5. $\dfrac{a^m}{a^n} = \begin{cases} a^{m-n} & \text{if } m > n \\ 1 & \text{if } m = n \\ \dfrac{1}{a^{n-m}} & \text{if } n > m \end{cases}$

In combined operations, powers take precedence over multiplication and division.   *(2-1)*

A **polynomial** is an algebraic expression involving only the operations of addition, subtraction, and multiplication on variables and constants. The constant factor in a term is called the **coefficient**. The **degree of a term** in a polynomial is the sum of the powers of the variables; constants are assigned degree

0. The **degree of a polynomial** is the highest degree of its terms. Polynomials with one, two, and three terms are called **monomials**, **binomials**, and **trinomials**, respectively. The **distributive axiom** $a \cdot (b + c) = ab + ac$ is used to combine **like terms**—that is, terms with identical variables to the same power. Addition and subtraction of polynomials are accomplished by **grouping** and combining like terms.   *(2-2)*

Monomials are multiplied by using the first property of exponents. Polynomials are multiplied by multiplying each term of one by each term of the other and adding. Binomials are mentally multiplied by the FOIL method

$$(A + B)(C + D) = AC + AD + BC + BD$$

with First, Last, Inner, Outer and F O I L labels,

and squared by

$$(A + B)^2 = A^2 + 2AB + B^2 \qquad (A - B)^2 = A^2 - 2AB + B^2 \quad (2\text{-}3)$$

To **factor** a polynomial means to find factors whose product is the given polynomial. Factors common to all terms can be factored out by the distributive law. **Grouping** may lead to factoring out common factors.   *(2-4)*

Second-degree polynomials can be factored by trial and error or the **ac test**.   *(2-5)*

The *ac* test to factor $ax^2 + bx + c$ or $ax^2 + bxy + cy^2$ involves finding two integers p and q such that $pq = ac$, $p + q = b$, rewriting the middle term as $px + qx$ or $pxy + qxy$, and grouping.   *(2-6)*

The sum $A^2 + B^2$ of two squares cannot be factored unless there are common factors. The difference of two squares is factored

$$A^2 - B^2 = (A - B)(A + B)$$

The sum and difference of two cubes are factored

$$A^3 + B^3 = (A + B)(A^2 - AB + B^2)$$
$$A^3 - B^3 = (A - B)(A^2 + AB + B^2) \quad (2\text{-}7)$$

A general strategy for factoring is given in Section 2-7.

Diagnostic (Review) Exercise 2-8 ■

Work through all the problems in this chapter review and check answers in the back of the book. (Answers to all problems are there, and following each answer is a number in italics indicating the section in which that type of problem is discussed.) Where weaknesses show up, review appropriate sections in the text. When you are satisfied that you know the material, take the practice test following this review.

A   *Simplify, using natural number exponents only.*

1. $\dfrac{x^8}{x^3}$ _____   **2.** $(xy)^3$ _____   **3.** $\left(\dfrac{x}{y}\right)^3$ _____

4. $\dfrac{x^3}{x^8}$ _____   **5.** $(x^3)^8$ _____   **6.** $\dfrac{x^3}{x^3}$ _____

7. $x^3 x^8$ _____   **8.** $(-2x)^3$ _____

9. $(-2x^3)(3x^8)$ _____   **10.** $\dfrac{6x^4 y}{8x^2 y^3}$ _____

11. Given the polynomial $3x^5 - 2x^3 + 7x^2 - x + 2$:

    (A) What is its degree? _____

    (B) What is the degree of the second term? _____

12. Consider the product $(3x^3 y^2)(-2xy^2 z^3)$ of two monomials.

    (A) What is the degree of the first factor? _____

    (B) What is the degree of the product? _____

Add, subtract, or multiply as indicated.

13. $(2x + 5) + (x^2 - 4)$ _____

14. $(2x + 5) - (x^2 - 4)$ _____

15. $(2x + 5)(x^2 - 4)$ _____

16. $(3x^2 + 2x + 1) - (2x^2 - 3x + 4)$ _____

17. $(3x^2 + 2x + 1) + (2x^2 - 3x + 4)$ _____

18. $(3x^2 + 2x + 1)(2x^2 - 3x + 4)$ _____

19. $[x^2 - (1 - x - x^2)] - (x + 1)$ _____

20. $-2x\{(x^2 + 2)(x - 3) - x[x - x(3 - x)]\}$ _____

Factor as far as possible in the integers.

21. $3x^4 + 9x^3$ _____   **22.** $2x^2 y^3 + x^3 y$ _____

B   **23.** $2x^2 - 10x + x - 5$ _____   **24.** $4x^2 + 11x - 3$ _____

25. $4x^2 - 25$ _____   **26.** $x^2 + 3xy + 2y^2$ _____

27. $a^2 + 9$ _____   **28.** $x^2 + 2x + 4 + 2x$ _____

29. $2x^2 - 5x - 3$ _____   **30.** $x^2 - 3x + 9 - 3x$ _____

31. $3x^2 + 2x - 3x - 2$ _____   **32.** $a^3 - 8$ _____

33. $x^2 + 6x + 8$ _____   **34.** $2x^2 - 5xy + 2y^2$ _____

35. $x^3 + 64$ _____   **36.** $a^2 - 9b^2$ _____

Simplify, using natural number exponents only.

37. $\left(\dfrac{2x^3}{y^8}\right)^2$ _____   **38.** $(-x^2y)^2(-xy^2)^3$ _____

39. $\dfrac{-4(x^2y)^3}{(-2x)^2}$ _____   **40.** $(3xy^3)^2(x^2y)^3$ _____

41. $\left(\dfrac{-2x}{y^2}\right)^3$ _____   **42.** $\left(\dfrac{3x^3y^2}{2x^2y^3}\right)^2$ _____

C   *Factor as far as possible in the integers.*

43. $x^6 + 1$ _____   **44.** $2x^4 + 6x^3 + 3x^2 + x^3$ _____

45. $x^2a^3 - 4xa^3 + 4a^3$ _____   **46.** $3x^3y - 3xy^3$ _____

47. $x^3 - 3x^2 + x - 3$ _____   **48.** $x^6 + x^5 + x^2 + x$ _____

49. $2x^3 - 2x^2 - 4x$ _____   **50.** $x^2 + 2xy + y^2 - 1$ _____

51. $x^4 - 1$ _____   **52.** $8x^3 - 125$ _____

53. $3x^3 + 2x^2 - 15x - 10$ _____   **54.** $x^4 + 2x^3 - 3x^2$ _____

55. $-x^5y^3 - 2x^4y^2 - x^3y$ _____   **56.** $2a^4 + 2a$ _____

Practice Test Chapter 2 ■

Take this as if it were a graded test by working the problems within a 50-minute time period. Do not look back in the chapter. Choose one of three levels of difficulty: least difficult, Problems 1–12; more difficult, add Problem 13; most difficult, add Problems 13 and 14. Use the answers in the back of the book to correct your work. The answers are keyed to appropriate text sections so that you can easily locate and review sections where difficulties still persist.

1.   Which of the following statements are false? _____

   (A)   The degree of $7x^5y^2z$ is 7.
   (B)   The coefficient of the last term in $x^2 + 2xy + 3y^2$ is 3.
   (C) $7x^3 + y^2$ is a binomial.
   (D)   The degree of $4x^3 + 3x^2 + 2x + 1$ is 4.

Simplify Problems 2 and 3 using natural number exponents only.

2. $\dfrac{(2x^2)^2(3y)}{x(2y)^3}$ _____   **3.** $(-xy)^3(-2x^2y^3)^2$ _____

4. Add $2x + 3$, $4x^2 - x + 3$, and $2x^2 - 5$. _____

5. Subtract $x^2 - 2x + 1$ from $2x^2 + 3$. _____

6. Multiply $3x^2 + 2x - 1$ and $4x - 5$. _____

7. Add the product of $(2x + 3)$ and $(4x - 1)$ to $-3x^2 + 5x - 7$.

_____

Factor Problems 8–14 as far as possible in the integers.

8. $3x^2 + 18x + 24$ _____

9. $12x^3y^2 - 3x^2y^3$ _____

10. $3x^2 - y^2 + 3xy - xy$ _____

11. $2x^3y^2 - 4x^2y^2 - 6xy^2$ _____

12. $27x^3 - 8$ _____

13. $6x^2 + x - 15$ _____

14. $x^4 + x^2 - x^2y^2 - y^2$ _____

Check Exercise 2-1 ∎

ANSWER COLUMN

1. _____

2. _____

3. _____

4. _____

5. _____

6. _____

7. _____

8. _____

9. _____

10. _____

Work the following problems without looking at any text examples. Show your work in the space provided. Write your answer in the answer column.

Simplify, using appropriate properties of exponents.

1. $2^3 \cdot 2^2$

2. $(3x^4)(2x^5)$

3. $(2u^3)(4u)(u^5)$

4. $\dfrac{12x^6}{8x^4}$

5. $\left(\dfrac{m^4}{n^3}\right)^3$

6. $(2 \times 10^6)(4 \times 10^7)$

7. $\left(\dfrac{x^2 y}{u^3 v^2}\right)^3$

8. $\dfrac{(2x^3 y)^2}{(4xy^2)^2}$

9. $\dfrac{-3^2}{(-3)^2}$

10. $\dfrac{(a+b)^7 (a-b)^4}{(a+b)^9 (a-b)}$

Check Exercise 2-2 ■

ANSWER COLUMN

Work the following problems without looking at any text examples. Show your work in the space provided. Write your answer in the answer column.

1. _____

2. _____

Simplify Problems 1–5 by removing symbols of grouping, if any, and combining like terms.

1. $(x + 3y) - (2x - y)$

3. _____

4. _____

5. _____

2. $2m^3n - 4mn^3 + 3mn^3 - m^3n$

6. _____

7. _____

8. _____

3. $4(y + 3) - (y - 2) + 3(2y - 1)$

9. _____

10. _____

4. $4[3 - 2(x - 3) - 3x]$

5. $2\{3x - [x - (2x - 1)] - 5\}$

6. Which of the following statements are false?
   (A) $3x + 1$ is a binomial.
   (B) The degree of $3x^4y^5$ is 9.
   (C) The degree of $3x^4 + 2x^2 + 1$ is 3.
   (D) The coefficient of the first term in $3x^4 + 2x^2 + 1$ is 3.

7. Replace the question mark with an appropriate algebraic expression.

   $$3(x + y) - x - y = 3(x + y) - (?)$$

Problems 8–10 refer to the following three polynomials:

$$2x^2 - 3x + 1 \qquad 3x - 5 \qquad x^2 + 4x - 2$$

8. Add all three polynomials.

9. Subtract the sum of the last two polynomials from the first.

10. Subtract the first polynomial from the third.

Check Exercise 2-3 ∎

Work the following problems without looking at any text examples. Show your work in the space provided. Write your answer in the answer column.

1. _____

2. _____

3. _____

4. _____

5. _____

6. _____

7. _____

8. _____

9. _____

10. _____

Multiply in Problems 1–9.

1. $(-2x^2y)(3xy^2 - 2x^2y)$

2. $(3x - 2)(x + 4)$

3. $(5x - y)(2x - 3y)$

4. $(2m - 5n)(3m + 2n)$

5. $2x^2 - 3x + 2$
  $\underline{3x - 2}$

6.
$$2x^2 - xy + 3y^2$$
$$\underline{x^2 + 2xy - y^2}$$

7. $(5a - 2)^2$

8. $(6m - n)(6m + n)$

9. $(4x + 5y)^2$

10. Simplify:

$(4x - 1)(2x + 3) - (2x - 3)^2$

Check Exercise 2-4 ▪

ANSWER COLUMN

Work the following problems without looking at any text examples. Show your work in the space provided. Write your answer in the answer column.

1. _____

Write Problems 1–5 in factored form by removing all factors common to all terms.

2. _____

1. $8u^3v - 12uv^2$

3. _____

4. _____

5. _____

2. $6m^2n - 6mn - 8mn^2$

6. _____

7. _____

8. _____

3. $7x(x - 3) + 2(x - 3)$

9. _____

10. _____

4. $5y(2y + 1) - (2y + 1)$

5. $3m(2n - 1) - 2(2n + 1)$

6. Fill in the open parentheses with an appropriate algebraic expression:

$$6x^2 + 9x - 2x - 3 = (6x^2 + 9x) - (\qquad)$$

7. Remove all common factors from each group; then complete the factoring, if possible.

$$(3x^2 - 3x) - (2x - 2)$$

Factor Problems 8–10 as the product of two first-degree factors, using grouping.

8. $6x^2 - 9x + 4x - 6$

9. $6x^2 + 15x - 2x - 5$

10. $10m^2 + 15mn - 4mn - 6n^2$

Check Exercise 2-5 ■

Work the following problems without looking at any text examples. Show your work in the space provided. Write your answer in the answer column.

1. _____

Factor, using integer coefficients. If not factorable, say so.

2. _____

1. $2x^2 - 7x + 3$

3. _____

4. _____

5. _____

2. $x^2 - 4x + 6$

6. _____

7. _____

8. _____

3. $x^2 + x - 6$

9. _____

10. _____

4. $x^2 - 5xy + 6y^2$

5. $2u^2 + uv + 3v^2$

6. $8x^2 + 27x + 12$

7. $4x^2 + 4x - 3$

8. $4m^2 + 15mn - 4n^2$

9. $12x^2 + xy - 6y^2$

10. $16w^2 + 14w - 15$

Check Exercise 2-6 ■

ANSWER COLUMN

Work the following problems without looking at any text examples. Show your work in the space provided. Write your answer in the answer column.

1. _____

2. _____

Factor, if possible, using integer coefficients. Use the ac test.

1. $2x^2 + 7x - 4$

3. _____

4. _____

5. _____

2. $4x^2 + 5x - 6$

3. $6u^2 - 3u - 4$

4. $3x^2 - 11xy + 6y^2$

5. $4u^2 + 9uv - 4v^2$

Check Exercise 2-7 ■

ANSWER COLUMN

ANSWER COLUMN

1. _____

2. _____

3. _____

4. _____

5. _____

6. _____

7. _____

8. _____

9. _____

10. _____

Work the following problems without looking at any text examples. Show your work in the space provided. Write your answer in the answer column.

Factor as far as possible, using integer coefficients.

1. $4u^2 - 1$

2. $m^3 + n^3$

3. $2u^2 + 8v^2$

4. $8y^3 - 28y^2 - 16y$

5. $2x^3y - 6x^2y + 12xy$

6. $2m^4 - 16m$

7. $ac + ad - bc - bd$

8. $x^4 - y^4$

9. $(u - 2)^2 - 4v^2$

10. $x^3 - 3x^2 - 4x + 12$

Algebraic Fractions ■ 3

**INSTRUCTIONS FOR STUDENTS IN A
SELF-PACED CLASS OR LAB**

(YES) ── **HAVE YOU HAD INTERMEDIATE ALGEBRA BEFORE THIS COURSE?** ── (NO)

1. Work Diagnostic (Review) Exercise 3-7 on page 162. Check answers in back of book; then work through text sections corresponding to problems missed. (Section numbers are in italics following each answer.)
2. When finished with step 1, take Practice Test Chapter 3 on page 164 as a final check of your understanding of the chapter. Check answers in the back of the book; then review sections where weakness still prevails. (Corresponding section numbers are in italics following each answer.)
3. When you think you are ready, ask your instructor for a graded test for Chapter 3.
4. If your instructor approves, after the test is corrected, go to the next chapter.

1. Work through each section in the chapter as follows:
   (A) Read discussion.
   (B) Read each example and work the corresponding matched problem. Check your solutions to the matched problem in Solutions to Matched Problems on the indicated page.
   (C) At the end of a section work the odd-numbered problems in the Practice Exercise and check answers; then work even-numbered problems in areas of weakness. (Answers to *all* Practice Exercise sets are in the back of the book.)
   (D) Work Check Exercise as instructed. Tear out and turn in as directed by your instructor. (Answers are not in the text.)
2. Repeat each step in item 1 for each section in the chapter.
3. After the instructional part of the chapter is completed, proceed with steps 1 to 4 in the box above this one.

Chapter 3 ■ Algebraic Fractions

Section 3-1   Rational Expressions

- ■ Rational Expressions
- ■ Fundamental Principle of Fractions
- ■ Reducing to Lowest Terms
- ■ Raising to Higher Terms

■ Rational Expressions

Fractional forms in which the numerator and denominator are polynomials are called **rational expressions**. For example,

$$\frac{1}{x} \qquad \frac{-6}{y-3} \qquad \frac{x-2}{x^2-2x+5} \qquad \frac{x^2-3xy+y^2}{3x^3y^4} \qquad \frac{3x^2-8x-4}{5}$$

are all rational expressions. (Recall that a nonzero constant is a polynomial of degree 0.) More generally, a rational expression is an algebraic expression involving only the operations of addition, subtraction, multiplication, and division on variables and constants.

Each rational expression names a real number for real number replacements of the variables, division by 0 excluded. Hence, all properties of the real numbers apply to those expressions.

■ Fundamental Principle of Fractions

You will recall from arithmetic that

$$\frac{8}{12} \left[= \frac{8 \div 4}{12 \div 4} \right] = \frac{2}{3} \qquad \text{and} \qquad \frac{3}{5} \left[= \frac{2 \cdot 3}{2 \cdot 5} \right] = \frac{6}{10}$$

The first example illustrates reducing a fraction to lowest terms, while the second example illustrates raising a fraction to higher terms. Both illustrate the use of the **fundamental principle of fractions**.

The Fundamental Principle of Fractions

For all polynomials P, Q, and K with Q, $K \neq 0$:

$\dfrac{PK}{QK} = \dfrac{P}{Q}$ We may divide out a common factor K from both the numerator and denominator. This is called **reducing to lower terms** or **canceling** the factor K.

$\dfrac{P}{Q} = \dfrac{PK}{QK}$ We may multiply the numerator and denominator by the same nonzero factor K. This is called **raising to higher terms**.

■ Reducing to Lowest Terms

To reduce a fraction to **lowest terms** is to divide out or cancel *all* common factors from the numerator and denominator.

COMMON ERROR   It is important to keep in mind when reducing fractions to lowest terms that it is only common *factors* in products that can be divided out or canceled; common *terms* in sums or differences cannot be canceled. For example:

Common factors may be canceled: $\dfrac{2y}{3y} = \dfrac{2\overset{1}{\cancel{y}}}{3\underset{1}{\cancel{y}}} = \dfrac{2}{3}$

Common terms may not be canceled: $\dfrac{2+y}{3+y} \neq \dfrac{2+\cancel{y}}{3+\cancel{y}} \neq \dfrac{2}{3}$

EXAMPLE 1   Reduce to lowest terms:

(A) $\dfrac{8x^2y}{12xy^2} = \dfrac{(4xy)(2x)}{(4xy)(3y)}$ Cancel common factors.

$= \dfrac{2x}{3y}$

(B) $\dfrac{6x^2 - 3x}{3x} = \dfrac{\overset{1}{\cancel{3x}}(2x - 1)}{\underset{1}{\cancel{3x}}}$ Factor the top, then cancel common factors.

$= 2x - 1$ *Note:* $\dfrac{6x^2 - \overset{1}{\cancel{3x}}}{\underset{1}{\cancel{3x}}}$ is wrong. (Why?)

(C) $\dfrac{x^2y - xy^2}{x^2 - xy} = \dfrac{\overset{1}{\cancel{xy}}(\overset{1}{\cancel{x - y}})}{\underset{1}{\cancel{x}}(\underset{1}{\cancel{x - y}})}$ Factor the top and bottom, then cancel common factors.

$= y$

(D) $\dfrac{3x^3 - 2x^2 - 8x}{x^4 - 8x} = \dfrac{\overset{1}{\cancel{x}}(3x^2 - 2x - 8)}{\underset{1}{\cancel{x}}(x^3 - 8)}$

$= \dfrac{(3x + 4)(\overset{1}{\cancel{x - 2}})}{(\underset{1}{\cancel{x - 2}})(x^2 + 2x + 4)}$

$= \dfrac{3x + 4}{x^2 + 2x + 4}$

PROBLEM 1   Reduce to lowest terms:

(A) $\dfrac{16u^5v^3}{24u^3v^6}$ (B) $\dfrac{4x}{8x^2 - 4x}$ (C) $\dfrac{x^2 - 3x}{x^2y - 3xy}$ (D) $\dfrac{2x^3 - 8x}{2x^4 + 16x}$

Solution   **(A)**   **(B)**

            (C)   **(D)**

■ Raising to Higher Terms

The following example shows how to use the fundamental principle of fractions to raise fractions to higher terms.

EXAMPLE 2   **(A)** $\dfrac{3}{2x} = \dfrac{(4xy)(3)}{(4xy)(2x)} = \dfrac{12xy}{8x^2y}$ **(B)** $\dfrac{2x}{3y} = \dfrac{(x-y)2x}{(x-y)3y} = \dfrac{2x^2 - 2xy}{3xy - 3y^2}$

(C) $\dfrac{x - 2y}{2x + y} = \dfrac{(2x-y)(x-2y)}{(2x-y)(2x+y)} = \dfrac{2x^2 - 5xy + 2y^2}{4x^2 - y^2}$

PROBLEM 2   Complete the raising-to-higher-terms process by replacing the question marks with appropriate expressions:

(A) $\dfrac{3u^2}{4v} = \dfrac{?}{20u^2v^2}$ **(B)** $\dfrac{3m}{2n} = \dfrac{3m^3 - 3m}{?}$

(C) $\dfrac{2x - 3}{x - 4} = \dfrac{?}{3x^2 - 14x + 8}$

Solution   **(A)**   **(B)**

            (C)

Solutions to Matched Problems

1. (A) $\dfrac{16u^5v^3}{24u^3v^6} = \dfrac{2u^2}{3v^3}$ **(B)** $\dfrac{4x}{8x^2 - 4x} = \dfrac{\overset{1}{\cancel{4x}}}{\underset{1}{\cancel{4x}}(2x - 1)} = \dfrac{1}{2x - 1}$

(C) $\dfrac{x^2 - 3x}{x^2y - 3xy} = \dfrac{\overset{1}{\cancel{x}}\overset{1}{\cancel{(x - 3)}}}{\underset{1}{\cancel{xy}}\underset{1}{\cancel{(x - 3)}}} = \dfrac{1}{y}$

1. (D) $\dfrac{2x^3 - 8x}{2x^4 + 16x} = \dfrac{2x(x^2 - 4)}{2x(x^3 + 8)} = \dfrac{\overset{1}{2x}(x - 2)\overset{1}{(x + 2)}}{\underset{1}{2x}(x + 2)(x^2 - 2x + 4)} = \dfrac{x - 2}{x^2 - 2x + 4}$

2. (A) $\dfrac{3u^2}{4v} = \dfrac{(5u^2v)(3u^2)}{(5u^2v)(4v)} = \dfrac{15u^4v}{20u^2v^2}$ **(B)** $\dfrac{3m}{2n} = \dfrac{3m(m - 1)}{2n(m - 1)} = \dfrac{3m^2 - 3m}{2mn - 2n}$

(C) $\dfrac{2x - 3}{x - 4} = \dfrac{(3x - 2)(2x - 3)}{(3x - 2)(x - 4)} = \dfrac{6x^2 - 13x + 6}{3x^2 - 14x + 8}$

Practice Exercise 3-1 ■

Work odd-numbered problems first, check answers, and then work even-numbered problems in areas of weakness. Answers to all problems are in the back of the book. Make every effort to work a problem yourself before you look at an answer.

A   *Reduce to lowest terms.*

1. $\dfrac{3x^3}{6x^5}$ _____

2. $\dfrac{9u^5}{3u^6}$ _____

3. $\dfrac{14x^3y}{21xy^2}$ _____

4. $\dfrac{20m^4n^6}{15m^5n^2}$ _____

5. $\dfrac{15y^3(x - 9)^3}{5y^4(x - 9)^2}$ _____

6. $\dfrac{2x^2(x + 7)}{6x(x + 7)}$ _____

7. $\dfrac{(2x - 1)(2x + 1)}{3x(2x + 1)}$ _____

8. $\dfrac{(x + 3)(2x + 5)}{2x^2(2x + 5)}$ _____

9. $\dfrac{x^2 - 2x}{2x - 4}$ _____

10. $\dfrac{2x^2 - 10x}{4x - 20}$ _____

11. $\dfrac{m^2 - mn}{m^2n - mn^2}$ _____

12. $\dfrac{a^2b + ab^2}{ab + b^2}$ _____

Complete the raising-to-higher-terms process by replacing the question marks with appropriate expressions.

13. $\dfrac{3}{2x} = \dfrac{?}{8x^2y}$ _____

14. $\dfrac{5x}{3} = \dfrac{10x^3y^2}{?}$ _____

15. $\dfrac{7}{3y} = \dfrac{?}{6x^3y^2}$ _____

16. $\dfrac{5u}{4v^2} = \dfrac{20u^3v}{?}$ _____

B   *Reduce to lowest terms.*

17. $\dfrac{x^2 + 6x + 8}{3x^2 + 12x}$ _____

18. $\dfrac{x^2 + 5x + 6}{2x^2 + 6x}$ _____

19. $\dfrac{x^2 - 9}{x^2 + 6x + 9}$ _____

20. $\dfrac{x^2 - 4}{x^2 + 4x + 4}$ _____

21. $\dfrac{4x^2 - 9y^2}{4x^2y + 6xy^2}$ _____

22. $\dfrac{a^2 - 16b^2}{4ab - 16b^2}$ _____

23. $\dfrac{x^2 - xy + 2x - 2y}{x^2 - y^2}$ _____

24. $\dfrac{u^2 + uv - 2u - 2v}{u^2 + 2uv + v^2}$ _____

25. $\dfrac{6x^3 + 28x^2 - 10x}{12x^3 - 4x^2}$ _____

26. $\dfrac{12x^3 - 78x^2 - 42x}{16x^4 + 8x^3}$ _____

27. $\dfrac{x^3 - 8}{x^2 - 4}$ _____

28. $\dfrac{y^3 + 27}{2y^3 - 6y^2 + 18y}$ _____

29. $\dfrac{x^2 + 10x + 21}{x^2 + 5x - 14}$ _____

30. $\dfrac{x^2 - 4x + 3}{x^2 + 2x - 15}$ _____

31. $\dfrac{6x^2z + 4xyz}{3xyz + 2y^2z}$ _____

32. $\dfrac{2x^2 + x - 3}{4x^2 + x - 5}$ _____

33. $\dfrac{x^3 - 3x^2 + 2x - 6}{x^2 - 5x + 6}$ _____

34. $\dfrac{x^3 - x^2 + x - 1}{x^2 - 1}$ _____

Complete the raising-to-higher-terms process by replacing the question marks with appropriate expressions.

35. $\dfrac{3x}{4y} = \dfrac{?}{4xy + 4y^2}$ _____

36. $\dfrac{4m}{5n} = \dfrac{4m^2 - 4mn}{?}$ _____

37. $\dfrac{x - 2y}{x + y} = \dfrac{x^2 - 3xy + 2y^2}{?}$ _____

38. $\dfrac{2x + 5}{x - 3} = \dfrac{?}{3x^2 - 8x - 3}$ _____

C   Reduce to lowest terms.

39. $\dfrac{x^3 - y^3}{3x^3 + 3x^2y + 3xy^2}$ _____

40. $\dfrac{2u^3v - 2u^2v^2 + 2uv^3}{u^3 + v^3}$ _____

41. $\dfrac{ux + vx - uy - vy}{2ux + 2vx + uy + vy}$ _____

42. $\dfrac{mx - 2my + nx - 2ny}{mx - 2my - nx + 2ny}$ _____

43. $\dfrac{x^4 - y^4}{(x^2 - y^2)(x + y)^2}$ _____

44. $\dfrac{x^4 - 2x^2y^2 + y^4}{x^4 - y^4}$ _____

45. $\dfrac{x^4 + x}{x^3 - x^2 + x}$ _____

46. $\dfrac{x^5 + x^3 + x}{x^4 + x^2 + 1}$ _____

47. $\dfrac{x^2 + y^2}{x^3 + y^3}$ _____   **48.** $\dfrac{z^3 - z}{z^3 - z^2 - z + 1}$ _____

49. $\dfrac{x^2u + uxy + vxy + vy^2}{xu^2 + uvy + uvx + yv^2}$ _____

The Check Exercise for this section is on page 165.

50. $\dfrac{x^2y^2z - xyz^2 - xy + z}{x^2yz - x - xyz^2 + yz}$ _____

Section 3-2   Quotients of Polynomials

There are times when it is useful to find quotients of polynomials by a long-division process similar to that used in arithmetic. Several examples will illustrate the process.

EXAMPLE 3   **(A)** Divide: $2x^2 + 5x - 12$ by $x + 4$
(B) Divide: $x^3 + 8$ by $x + 2$
(C) Divide: $3 - 7x + 6x^2$ by $3x + 1$

Solution   **(A)** $x + 4 \, \overline{\smash{)}\, 2x^2 + 5x - 12}$ Arrange both polynomials in descending powers of the variable if this is not already done.

$$\begin{array}{r} 2x \phantom{{}+{}} \\ x + 4 \, \overline{\smash{)}\, 2x^2 + 5x - 12} \end{array}$$ Divide the first term of the divisor into the first term of the dividend. That is, what must x be multiplied by so that the product is exactly $2x^2$? Answer: $2x$.

$$\begin{array}{r} 2x \phantom{{}+{}} \\ x + 4 \, \overline{\smash{)}\, 2x^2 + 5x - 12} \\ \underline{2x^2 + 8x} \phantom{{}-12} \\ -3x - 12 \end{array}$$ Multiply the divisor by $2x$, line up like terms as indicated, subtract, and bring down -12 from above.

$$\begin{array}{r} 2x - 3 \\ x + 4 \, \overline{\smash{)}\, 2x^2 + 5x - 12} \\ \underline{2x^2 + 8x} \phantom{{}-12} \\ -3x - 12 \\ \underline{-3x - 12} \\ 0 \end{array}$$ Repeat the process until the degree of the remainder is less than that of the divisor, or the remainder is 0.

Check $(x + 4)(2x - 3) = 2x^2 + 5x - 12$

(B) $$\begin{array}{r} x^2 - 2x + 4 \\ x + 2 \, \overline{\smash{)}\, x^3 + 0x^2 + 0x + 8} \\ \underline{x^3 + 2x^2} \phantom{{}+0x+8} \\ -2x^2 + 0x \\ \underline{-2x^2 - 4x} \\ 4x + 8 \\ \underline{4x + 8} \\ 0 \end{array}$$ Note that the terms $0x^2$ and $0x$ need to be included. Now proceed as in part (A).

Can you check this problem?

$$
\begin{array}{r}
2x - 3 \\
3x + 1 \overline{) 6x^2 - 7x + 3} \\
\underline{6x^2 + 2x} \\
-9x + 3 \\
\underline{-9x - 3} \\
6 = R
\end{array}
$$

(C) at left, with the division shown above.

Arrange $3 - 7x + 6x^2$ in descending powers of x, then proceed as above until the degree of the remainder is less than the degree of the divisor.

$6 = R$ Remainder

Check

Just as in arithmetic, when there is a remainder we check by adding the remainder to the product of the divisor and quotient. Thus,

$$(3x + 1)(2x - 3) + 6 \overset{?}{=} 6x^2 - 7x + 3$$
$$6x^2 - 7x - 3 + 6 \overset{?}{=} 6x^2 - 7x + 3$$
$$6x^2 - 7x + 3 \overset{\checkmark}{=} 6x^2 - 7x + 3$$

PROBLEM 3

Divide, using the long-division process, and check:

(A) $(2x^2 + 7x + 3)/(x + 3)$ **(B)** $(x^3 - 8)/(x - 2)$
(C) $(2 - x + 6x^2)/(3x - 2)$

Solution   **(A)**   Check

(B)   Check

(C)   Check

Solution to Matched Problem

3. (A)

$$2x + 3 \overline{\smash{\big)}\ 2x^2 + 7x + 3} \quad \frac{2x + 1}{}$$

$$\underline{2x^2 + 6x}$$

$$x + 3$$

$$\underline{x + 3}$$

$$0 = R$$

Check $(x + 3)(2x + 1)$
$= 2x^2 + 7x + 3$

(B)

$$x - 2 \overline{\smash{\big)}\ x^3 + 0x^2 + 0x - 8} \quad \frac{x^2 + 2x + 4}{}$$

$$\underline{x^3 - 2x^2}$$

$$2x^2 + 0x$$

$$\underline{2x^2 - 4x}$$

$$4x - 8$$

$$\underline{4x - 8}$$

$$0 = R$$

Check $x^2 + 2x + 4$
$\underline{x\ -\ 2}$
$x^3 + 2x^2 + 4x$
$\underline{\ -\ 2x^2 - 4x - 8}$
$x^3 \qquad\qquad\ -\ 8$

(C)

$$3x - 2 \overline{\smash{\big)}\ 6x^2 -\ \ x + 2} \quad \frac{2x + 1}{}$$

$$\underline{6x^2 - 4x}$$

$$3x + 2$$

$$\underline{3x - 2}$$

$$4 = R$$

Check $(3x - 2)(2x + 1) + 4$
$= 6x^2 - x - 2 + 4$
$= 6x^2 - x + 2$

Practice Exercise 3-2 ■

Work odd-numbered problems first, check answers, and then work even-numbered problems in areas of weakness. Answers to all problems are in the back of the book. Make every effort to work a problem yourself before you look at an answer.

Divide, using the long-division process. Check the answers.

A **1.** $(3x^2 - 5x - 2)/(x - 2)$ _____

2. $(2x^2 + x - 6)/(x + 2)$ _____

3. $(2y^3 + 5y^2 - y - 6)/(y + 2)$ _____

4. $(x^3 - 5x^2 + x + 10)/(x - 2)$ _____

5. $(3x^2 - 11x - 1)/(x - 4)$ _____

6. $(2x^2 - 3x - 4)/(x - 3)$ _____

7. $(8x^2 - 14x + 3)/(2x - 3)$ _____

8. $(6x^2 + 5x - 6)/(3x - 2)$ _____

9. $(6x^2 + x - 13)/(2x + 3)$ _____

10. $(6x^2 + 11x - 12)/(3x - 2)$ _____

11. $(x^2 - 4)/(x - 2)$ _____

12. $(y^2 - 9)/(y + 3)$ _____

B **13.** $(12x^2 + 11x - 2)/(3x + 2)$ _____

14. $(8x^2 - 6x + 6)/(2x - 1)$ _____

15. $(8x^2 + 7)/(2x - 3)$ _____

16. $(9x^2 - 8)/(3x - 2)$ _____

17. $(-7x + 2x^2 - 1)/(2x + 1)$ _____

18. $(13x - 12 + 3x^2)/(3x - 2)$ _____

19. $(x^3 - 1)/(x - 1)$ _____

20. $(a^3 + 27)/(a + 3)$ _____

21. $(x^4 - 81)/(x - 3)$ _____

22. $(x^4 - 16)/(x + 2)$ _____

23. $(4a^2 - 22 - 7a)/(a - 3)$ _____

24. $(8c + 4 + 5c^2)/(c + 2)$ _____

25. $(x + 5x^2 - 10 + x^3)/(x + 2)$ _____

26. $(5y^2 - y + 2y^3 - 6)/(y + 2)$ _____

27. $(3 + x^3 - x)/(x - 3)$ _____

28. $(3y - y^2 + 2y^3 - 1)/(y + 2)$ _____

C **29.** $(9x^4 - 2 - 6x - x^2)/(3x - 1)$ _____

30. $(4x^4 - 10x - 9x^2 - 10)/(2x + 3)$ _____

31. $(8x^2 - 7 - 13x + 24x^4)/(3x + 5 + 6x^2)$ _____

32. $(16x - 5x^3 - 8 + 6x^4 - 8x^2)/(2x - 4 + 3x^2)$ _____

33. $(9x^3 - x + 2x^5 + 9x^3 - 2 - x)/(2 + x^2 - 3x)$ _____

34. $(12x^2 - 19x^3 - 4x - 3 + 12x^5)/(4x^2 - 1)$ _____

35. Given polynomials $P = x^3 - 6x^2 + 12x - 4$ and $D = x^2 - 3x + 2$, find polynomials Q and R such that $P = DQ + R$ and the degree of R is less than the degree of D, or $R = 0$. _____

**The Check Exercise for
this section is on
page 167.**

36. Repeat the preceding problem for $P = x^4 + x^3 - 4x^2 + 7x + 2$ and $D = x^2 - x + 1$. _____

Section 3-3 Synthetic Division (Optional)

Any polynomial can be divided by a first-degree polynomial of the form $x - r$ using the algebraic long-division process described in Section 3-2. In some circumstances such divisions have to be done repeatedly and a quicker, more concise method, called **synthetic division**, is useful. The method is most easily understood through an example. Let us start by dividing $P = 2x^4 + 3x^3 - x - 5$ by $x + 2$, using ordinary long division. The critical parts of the process are indicated in color.

$$
\begin{array}{r}
2x^3 - 1x^2 + 2x - 5 \qquad \text{Quotient} \\
x + 2 \,\overline{)\, 2x^4 + 3x^3 + 0x^2 - 1x - 5} \qquad \text{Dividend} \\
\underline{2x^4 + 4x^3} \\
-1x^3 + 0x^2 \\
\underline{-1x^3 - 2x^2} \\
2x^2 - 1x \\
\underline{2x^2 + 4x} \\
-5x - 5 \\
\underline{-5x - 10} \\
5 \qquad \text{Remainder}
\end{array}
$$

The numerals printed in color, which represent the essential part of the division process, are arranged more conveniently as

$$
\begin{array}{r|rrrrr}
 & \multicolumn{5}{c}{\text{Dividend coefficients}} \\
 & 2 & 3 & 0 & -1 & -5 \\
 & & 4 & -2 & 4 & -10 \\
\hline
2 & 2 & -1 & 2 & -5 & 5 \\
 & \multicolumn{4}{c}{\text{Quotient}} & \text{Remainder} \\
 & \multicolumn{4}{c}{\text{coefficients}} &
\end{array}
$$

We see that the second and third rows of numerals are generated as follows. The first coefficient 2 of the dividend is brought down and multiplied by 2 from the divisor, and the product 4 is placed under the second dividend coefficient 3 and subtracted. The difference -1 is again multiplied by the 2 from the divisor, and the product is placed under the third coefficient from the dividend and subtracted. This process is repeated until the remainder is reached. The process can be made a little faster, and less prone to sign errors, by changing $+2$ from the divisor to -2 and adding instead of subtracting. Thus,

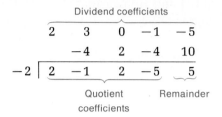

		Dividend coefficients		
2	3	0	-1	-5
	-4	2	-4	10

-2 | 2 | -1 | 2 | -5 | 5

Quotient   Remainder
coefficients

Key Steps in the Synthetic Division Process: $P \div (x - r)$

1. Arrange the coefficients of P in order of descending powers of x. (Write 0 as the coefficient for each missing power.)
2. After writing the divisor in the form $x - r$, use r to generate the second and third rows of numbers as follows. Bring down the first coefficient of the dividend and multiply it by r; then add the product to the second coefficient of the dividend. Multiply this sum by r, and add the product to the third coefficient of the dividend. Repeat the process until a product is added to the constant term of P. [*Note:*   This process is well suited to hand calculator use. Store r; then proceed from left to right recalling r and using it as indicated.]
3. The last number in the third row of numbers is the remainder; the other numbers in the third row are the coefficients of the quotient, which is of degree 1 less than P.

EXAMPLE 4   Use synthetic division to find the quotient and remainder resulting from dividing $P = 4x^5 - 30x^3 - 50x - 2$ by $x + 3$.

Solution $x + 3 = x - (-3)$; therefore, $r = -3$.

	4	0	-30	0	-50	-2
		-12	36	-18	54	-12
-3	4	-12	6	-18	4	-14

The quotient is $4x^4 - 12x^3 + 6x^2 - 18x + 4$ with a remainder of -14.

PROBLEM 4   Repeat Example 4 with $P = 3x^4 - 11x^3 - 18x + 8$ and divisor $x - 4$.

Solution

Solution to Matched Problem	**4.**	3	−11	0	−18	8	Quotient: $3x^3 + x^2 + 4x - 2$
			12	4	16	8	Remainder: 0
	4	3	1	4	−2	0	

Practice Exercise 3-3 ∎

Work odd-numbered problems first, check answers, and then work even-numbered problems in areas of weakness. Answers to all problems are in the back of the book. Make every effort to work a problem yourself before you look at an answer.

A *Divide using synthetic division. Write the quotient and indicate the remainder.*

1. $(x^3 + 2x^2 + 3x + 4) \div (x - 2)$ _____

2. $(x^3 + 2x^2 + 3x + 4) \div (x - 1)$ _____

3. $(x^3 + 2x^2 + 3x + 4) \div (x + 1)$ _____

4. $(x^3 + 2x^2 + 3x + 4) \div (x + 2)$ _____

5. $(2x^3 - x^2 + x - 2) \div (x - 3)$ _____

6. $(2x^3 - x^2 + x - 2) \div (x - 1)$ _____

7. $(2x^3 - x^2 + x - 2) \div (x + 1)$ _____

8. $(2x^3 - x^2 + x - 2) \div (x + 3)$ _____

9. $(x^4 + x^3 + 3x^2 + 3x + 5) \div (x - 4)$ _____

10. $(x^4 + x^3 + 3x^2 + 3x + 5) \div (x - 2)$ _____

11. $(x^4 + x^3 + 3x^2 + 3x + 5) \div (x + 2)$ _____

12. $(x^4 + x^3 + 3x^2 + 3x + 5) \div (x + 4)$ _____

B **13.** $(x^3 - 2x + 4) \div (x - 2)$ _____

14. $(x^3 - 2x + 4) \div (x - 1)$ _____

15. $(x^3 - 2x + 4) \div (x + 1)$ _____

16. $(x^3 - 2x + 4) \div (x + 2)$ _____

17. $(x^4 - 3x + 5) \div (x - 4)$ _____

18. $(x^4 - 3x + 5) \div (x - 2)$ _____

19. $(x^4 - 3x + 5) \div (x + 2)$ _____

20. $(x^4 - 3x + 5) \div (x + 4)$ _____

C   *If the polynomial P is written in the form $P = Q \cdot (x - r) + R$, it is easy to evaluate P for the value $x = r$: since for this value of x, $Q \cdot (x - r) = 0$, P must equal the remainder R. Use this fact and synthetic division to evaluate the polynomial $P = x^5 - 3x^3 + x^2 - 1$ for the following values of x.*

21. 3 _____   **22.** 2 _____   **23.** 1 _____

24. −1 _____   **25.** −2 _____   **26.** −3 _____

 Synthetic division can be used with any values for r, not just integer values. Use synthetic division with a hand calculator to evaluate to six decimal places the polynomial $P = x^3 - 2x^2 - 3x + 4$ for the following values of x.

27. 1.1 _____   **28.** 1.01 _____   **29.** 1.35 _____

The Check Exercise for this section is on page 169.   **30.** 2.11 _____   **31.** −3.3 _____   **32.** 1.001 _____

33. 3.102 _____   **34.** −3.141 _____

Section 3-4   Multiplication and Division

- Multiplication of Rational Expressions
- Division of Rational Expressions

Multiplication of Rational Expressions

We start with the process of multiplying rational forms, which is a direct consequence of real number properties.

Multiplication

If P, Q, R, and S are polynomials (Q, $S \neq 0$), then

$$\frac{P}{Q} \cdot \frac{R}{S} = \frac{P \cdot R}{Q \cdot S}$$

EXAMPLE 5   **(A)** $\dfrac{3a^2b}{4c^2d} \cdot \dfrac{8c^2d^3}{9ab^2} = \dfrac{(3a^2b) \cdot (8c^2d^3)}{(4c^2d) \cdot (9ab^2)} = \dfrac{24a^2bc^2d^3}{36ab^2c^2d}$

$$= \frac{(2ad^2)\overset{1}{(12abc^2d)}}{(3b)\underset{1}{(12abc^2d)}} = \frac{2ad^2}{3b}$$

This process is easily shortened to the following when it is realized that, in effect, any factor in either numerator may cancel any like factor in either denominator. Thus,

$$\frac{\overset{1\,\cdot\,a\,\cdot\,1}{\cancel{3a^2b}}}{\underset{1\,\cdot\,1\,\cdot\,1}{\cancel{4c^2d}}}\cdot\frac{\overset{2\,\cdot\,1\,\cdot\,d^2}{\cancel{8c^2d^3}}}{\underset{3\,\cdot\,1\,\cdot\,b}{\cancel{9ab^2}}}=\frac{2ad^2}{3b}$$

(B) $(x^2-4)\cdot\dfrac{2x-3}{x+2}=\dfrac{\overset{1}{\cancel{(x+2)}}(x-2)}{1}\cdot\dfrac{(2x-3)}{\underset{1}{\cancel{(x+2)}}}$ Factor where possible, cancel common factors, then multiply and write the answer.

$$=(x-2)(2x-3)$$

(C) $\dfrac{4a^2-9b^2}{4a^2+12ab+9b^2}\cdot\dfrac{6a^2b}{8a^2b^2-12ab^3}=\dfrac{\overset{1}{\cancel{(2a-3b)}}\overset{1}{\cancel{(2a+3b)}}}{\underset{(2a+3b)}{(2a+3b)^2}}\cdot\dfrac{\overset{3a}{\cancel{6a^2b}}}{\underset{2b\quad\ 1}{4ab^2\cancel{(2a-3b)}}}$

$$=\frac{3a}{2b(2a+3b)}$$

PROBLEM 5   Multiply and reduce to lowest terms:

(A) $\dfrac{4x^2y^3}{9w^2z}\cdot\dfrac{3wz^2}{2xy^4}$ 　　　　　　 **(B)** $\dfrac{x+5}{x^2-9}\cdot(x+3)$

(C) $\dfrac{x^2-9y^2}{x^2-6xy+9y^2}\cdot\dfrac{6x^2y}{2x^2+6xy}$

Solution   **(A)**

(B)

(C)

■ Division of Rational Expressions

Theorem 1 follows from the definition of division. [*Recall:* $A\div B=Q$ if and only if $QB=A$ and Q is unique.]

THEOREM 1

Division
If P, Q, R, and S are polynomials (Q, R, $S \neq 0$), then Divisor is R/S ⟶   ⟵ Reciprocal of divisor is S/R $$\frac{P}{Q} \div \frac{R}{S} = \frac{P}{Q} \cdot \frac{S}{R}$$ That is, to divide one fraction by another, multiply by the reciprocal of the divisor.

To prove that the indicated process is a valid procedure, one has only to show that the product of R/S and $(P/Q) \cdot (S/R)$ is P/Q, a problem left to the exercises.

EXAMPLE 6   **(A)** $\dfrac{6a^2b^3}{5cd} \div \dfrac{3a^2c}{10bd} = \dfrac{\overset{2 \cdot 1}{6a^2b^3}}{\underset{1 \cdot 1}{5cd}} \cdot \dfrac{\overset{2 \cdot 1}{10bd}}{\underset{1 \cdot 1}{3a^2c}} = \dfrac{4b^4}{c^2}$

(B) $(x + 4) \div \dfrac{2x^2 - 32}{6xy} = \dfrac{\overset{1}{x + 4}}{1} \cdot \dfrac{\overset{3}{6xy}}{\underset{1}{2(x - 4)(x + 4)}} = \dfrac{3xy}{x - 4}$

(C) $\dfrac{10x^3y}{3xy + 9y} \div \dfrac{4x^2 - 12x}{x^2 - 9} = \dfrac{\overset{5 \cdot x^2 \cdot 1}{10x^3y}}{\underset{1 \cdot 1}{3y(x + 3)}} \cdot \dfrac{\overset{1}{(x + 3)}\overset{1}{(x - 3)}}{\underset{2 \cdot 1 \cdot 1}{4x(x - 3)}} = \dfrac{5x^2}{6}$

PROBLEM 6

Divide and reduce to lowest terms:

(A) $\dfrac{8w^2z^2}{9x^2y} \div \dfrac{4wz}{6xy^2}$ **(B)** $\dfrac{2x^2 - 8}{4x} \div (x + 2)$

(C) $\dfrac{x^2 - 4x + 4}{4x^2y - 8xy} \div \dfrac{x^2 + x - 6}{6x^2 + 18x}$

Solution   **(A)**

(B)

(C)

Solutions to Matched
Problems

5. (A) $\dfrac{4x^2y^3}{9w^2z} \cdot \dfrac{3wz^2}{2xy^4} = \dfrac{2xz}{3wy}$

(B) $\dfrac{x+5}{x^2-9} \cdot \dfrac{x+3}{1} = \dfrac{x+5}{(x-3)(x+3)} \cdot \dfrac{x+3}{1} = \dfrac{x+5}{x-3}$

(C) $\dfrac{x^2-9y^2}{x^2-6xy+9y^2} \cdot \dfrac{6x^2y}{2x^2+6xy} = \dfrac{(x-3y)(x+3y)}{(x-3y)^2} \cdot \dfrac{6x^2y}{2x(x+3y)}$

$$= \dfrac{3xy}{x-3y}$$

6. (A) $\dfrac{8w^2z^2}{9x^2y} \div \dfrac{4wz}{6xy^2} = \dfrac{8w^2z^2}{9x^2y} \cdot \dfrac{6xy^2}{4wz} = \dfrac{4wzy}{3x}$

(B) $\dfrac{2x^2-8}{4x} \div \dfrac{x+2}{1} = \dfrac{2(x-2)(x+2)}{4x} \cdot \dfrac{1}{x+2} = \dfrac{x-2}{2x}$

(C) $\dfrac{x^2-4x+4}{4x^2y-8xy} \div \dfrac{x^2+x-6}{6x^2+18x} = \dfrac{(x-2)^2}{4xy(x-2)} \cdot \dfrac{6x(x+3)}{(x+3)(x-2)} = \dfrac{3}{2y}$

Practice Exercise 3-4 ▪

Work odd-numbered problems first, check answers, and then work even-numbered problems in areas of weakness. Answers to all problems are in the back of the book. Make every effort to work a problem yourself before you look at an answer.

Do not change improper fractions in your answers to mixed fractions; that is, write $\frac{7}{2}$, not $3\frac{1}{2}$.

A   Multiply and reduce to lowest terms.

1. $\dfrac{10}{9} \cdot \dfrac{12}{15}$ _____

2. $\dfrac{3}{7} \cdot \dfrac{14}{9}$ _____

3. $\dfrac{2a}{3bc} \cdot \dfrac{9c}{a}$ _____

4. $\dfrac{2x}{3yz} \cdot \dfrac{6y}{4x}$ _____

5. $\dfrac{3x^2}{4} \cdot \dfrac{16y}{12x^3}$ _____

6. $\dfrac{2x^2}{3y^2} \cdot \dfrac{9y}{4x}$ _____

Divide and reduce to lowest terms.

7. $\dfrac{9m}{8n} \div \dfrac{3m}{4n}$ _____

8. $\dfrac{6x}{5y} \div \dfrac{3x}{10y}$ _____

9. $\dfrac{a}{4c} \div \dfrac{a^2}{12c^2}$ _____

10. $\dfrac{2x}{3y} \div \dfrac{4x}{6y^2}$ _____

11. $\dfrac{x}{3y} \div 3y$ _____   12. $2xy \div \dfrac{x}{y}$ _____

Perform the indicated operations and reduce to lowest terms.

13. $\dfrac{8x^2}{3xy} \cdot \dfrac{12y^3}{6y}$ _____   14. $\dfrac{6a^2}{7c} \cdot \dfrac{21cd}{12ac}$ _____

15. $\dfrac{21x^2y^2}{12cd} \div \dfrac{14xy}{9d}$ _____   16. $\dfrac{3uv^2}{5w} \div \dfrac{6u^2v}{15w}$ _____

17. $\dfrac{9u^4}{4v^3} \div \dfrac{-12u^2}{15v}$ _____   18. $\dfrac{-6x^3}{5y^2} \div \dfrac{18x}{10y}$ _____

19. $\dfrac{3c^2d}{a^3b^3} \div \dfrac{3a^3b^3}{cd}$ _____   20. $\dfrac{uvw}{5xyz} \div \dfrac{5vy}{uwxz}$ _____

B   21. $\dfrac{3x^2y}{x-y} \cdot \dfrac{x-y}{6xy}$ _____   22. $\dfrac{x+3}{2x^2} \cdot \dfrac{4x}{x+3}$ _____

23. $\dfrac{x+3}{x^3+3x^2} \cdot \dfrac{x^3}{x-3}$ _____   24. $\dfrac{a^2-a}{a-1} \cdot \dfrac{a+1}{a}$ _____

25. $\dfrac{x-2}{4y} \div \dfrac{x^2+x-6}{12y^2}$ _____   26. $\dfrac{4x}{x-4} \div \dfrac{8x^2}{x^2-6x+8}$ _____

27. $\dfrac{6x^2}{4x^2y-12xy} \cdot \dfrac{x^2+x-12}{3x^2+12x}$ _____

28. $\dfrac{2x^2+4x}{12x^2y} \cdot \dfrac{6x}{x^2+6x+8}$ _____

29. $(t^2-t-12) \div \dfrac{t^2-9}{t^2-3t}$ _____   30. $\dfrac{2y^2+7y+3}{4y^2-1} \div (y+3)$ _____

31. $\dfrac{m+n}{m^2-n^2} \div \dfrac{m^2-mn}{m^2-2mn+n^2}$ _____

32. $\dfrac{x^2-6x+9}{x^2-x-6} \div \dfrac{x^2+2x-15}{x^2+2x}$ _____

33. $-(x^2-3x) \cdot \dfrac{x-2}{x-3}$ _____   34. $-(x^2-4) \cdot \dfrac{3}{x+2}$ _____

35. $\left(\dfrac{d^5}{3a} \div \dfrac{d^2}{6a^2}\right) \cdot \dfrac{a}{4d^3}$ _____   36. $\dfrac{d^5}{3a} \div \left(\dfrac{d^2}{6a^2} \cdot \dfrac{a}{4d^3}\right)$ _____

37. $\dfrac{2x^2}{3y^2} \cdot \dfrac{-6yz}{2x} \cdot \dfrac{y}{-xz}$ _____   38. $\dfrac{-a}{-b} \cdot \dfrac{12b^2c}{15ac} \cdot \dfrac{-10}{4b}$ _____

C   **39.** $\dfrac{9 - x^2}{x^2 + 5x + 6} \cdot \dfrac{x + 2}{x - 3}^{\dagger}$

40. $\dfrac{2 - m}{2m + m^2} \cdot \dfrac{m^2 + 4m + 4}{m^2 - 4}$

41. $\dfrac{x^2 - xy}{xy + y^2} \div \left(\dfrac{x^2 - y^2}{x^2 + 2xy + y^2} \div \dfrac{x^2 - 2xy + y^2}{x^2 y + xy^2} \right)$

42. $\left(\dfrac{x^2 - xy}{xy + y^2} \div \dfrac{x^2 - y^2}{x^2 + 2xy + y^2} \right) \div \dfrac{x^2 - 2xy + y^2}{x^2 y + xy^2}$

43. $(x^2 - x - 6)/(x - 3) = x + 2$, except for what values of x?

The Check Exercise for this section is on page 171.

44. $(x^2 - 1)/(x - 1)$ and $x + 1$ name the same real number for (*all*, *all but one*, *no*) replacements of x by real numbers.

Section 3-5   Addition and Subtraction

Addition and subtraction of rational expressions are based on the corresponding properties of real fractions. Thus,

---

Addition and Subtraction

If P, D, and Q are polynomials ($D \neq 0$), then

$$\frac{P}{D} + \frac{Q}{D} = \frac{P + Q}{D} \tag{1}$$

$$\frac{P}{D} - \frac{Q}{D} = \frac{P - Q}{D} \tag{2}$$

In words: if the denominators of two rational expressions are the same, we may either add or subtract the expressions by adding or subtracting the numerator and placing the result over the common denominator. If the denominators are not the same, we use the fundamental principle of fractions to change the form of each fraction so they have a common denominator, then use either Property 1 or 2.

    Even though any common denominator will do, the problem will generally become less involved if the least common denominator (LCD) is used. Recall that the least common denominator is the least common multiple (LCM) of all the denominators; that is, it is the "smallest" quantity exactly divisible by each denominator.

---

† Note that $9 - x^2 = -(x^2 - 9)$, or, in general, $b - a = -(a - b)$.

If the LCD is not obvious (often it is), then it is found as follows:

Finding the Least Common Denominator (LCD)

Step 1: Factor each denominator completely using integer coefficients.

Step 2: The LCD must contain each *different* factor that occurs in all of the denominators to the highest power it occurs in any one denominator.

EXAMPLE 7   Find the least common multiple for:

(A) $18x^3$, $15x$, $10x^2$ **(B)** $6(x-3)$, x^2-9, $4x^2+24x+36$

Solution   **(A)** Write each expression in completely factored form, including coefficients:

$$18x^3 = 2 \cdot 3^2 x^3 \qquad 15x = 3 \cdot 5x \qquad 10x^2 = 2 \cdot 5x^2$$

The LCM must contain each different factor (2, 3, 5, and x) to the highest power it occurs in any one denominator. Thus,

$$LCM = 2 \cdot 3^2 \cdot 5x^3 = 90x^3$$

(B) Factor each expression completely:

$$6(x-3) = 2 \cdot 3(x-3) \qquad x^2 - 9 = (x-3)(x+3)$$
$$4x^2 + 24x + 36 = 4(x^2 + 6x + 9) = 2^2(x+3)^2$$

Thus,

$$LCM = 2^2 \cdot 3(x-3)(x+3)^2 = 12(x-3)(x+3)^2$$

PROBLEM 7   Find the LCM for:

(A) $15y^2$, $12y$, $9y^4$ **(B)** $3x^2 - 12$, $x^2 - 4x + 4$, $12(x+2)$

Solution   **(A)**   **(B)**

EXAMPLE 8   Combine into a single fraction and simplify:

(A) $\dfrac{x+1}{x-2} + \dfrac{3x-2}{x-2}$ **(B)** $\dfrac{1}{x-3} - \dfrac{x-2}{x-3}$

Solution **(A)** $\dfrac{(x+1)}{x-2} + \dfrac{(3x-2)}{x-2}$

When a numerator has more than one term, place terms in parentheses before proceeding. Since denominators are the same, use Property 1 to add.

$= \dfrac{(x+1)+(3x-2)}{x-2}$

Simplify the numerator.

$= \dfrac{x+1+3x-2}{x-2}$

$= \dfrac{4x-1}{x-2}$

(B) $\dfrac{1}{x-3} - \dfrac{(x-2)}{x-3}$

Use Property 2 to subtract.

$= \dfrac{1-(x-2)}{x-3}$

Simplify the numerator. Watch signs.

$= \dfrac{1-x+2}{x-3}$

Sign errors are frequently made where the arrow points.

$= \dfrac{3-x}{x-3}$

$3-x \neq x-3;\; 3-x = -(x-3)$

$= \dfrac{-(x-3)}{(x-3)}$

Reduce to lowest terms.

$= -1$

PROBLEM 8

Combine into a single fraction and simplify:

$$\dfrac{x+3}{2x-5} - \dfrac{3x-2}{2x-5}$$

Solution

EXAMPLE 9 Combine into a single fraction and simplify:

$$\dfrac{3}{2y} - \dfrac{1}{3y^2} + 1$$

$\text{LCD} = 6y^2$

$= \dfrac{3y \cdot 3}{3y \cdot 2y} - \dfrac{2 \cdot 1}{2 \cdot 3y^2} + \dfrac{6y^2}{6y^2}$

Use the fundamental principle of fractions to make each denominator $6y^2$.

$= \dfrac{9y}{6y^2} - \dfrac{2}{6y^2} + \dfrac{6y^2}{6y^2}$

$= \dfrac{9y-2+6y^2}{6y^2}$

Arrange the numerator in descending powers of y.

$= \dfrac{6y^2+9y-2}{6y^2}$

PROBLEM 9   Combine into a single fraction and simplify:

$$\frac{5}{4x^3} - \frac{1}{3x} + 2$$

Solution

EXAMPLE 10 $\dfrac{4}{3x^2 - 27} - \dfrac{x-1}{4x^2 + 24x + 36}$ Factor each denominator completely.

$$= \frac{4}{3(x-3)(x+3)} - \frac{(x-1)}{2^2(x+3)^2} \qquad \text{LCD} = 12(x-3)(x+3)^2$$

$$= \frac{4(x+3)\cdot 4}{4(x+3)\cdot 3(x-3)(x+3)} - \frac{3(x-3)\cdot(x-1)}{3(x-3)\cdot 2^2(x+3)^2}$$

$$= \frac{16(x+3)}{12(x-3)(x+3)^2} - \frac{3(x-3)(x-1)}{12(x-3)(x+3)^2}$$

$$= \frac{16(x+3) - 3(x-3)(x-1)}{12(x-3)(x+3)^2}$$

$$= \frac{16x + 48 - 3(x^2 - 4x + 3)}{12(x-3)(x+3)^2}$$

$$= \frac{16x + 48 - 3x^2 + 12x - 9}{12(x-3)(x+3)^2}$$

$$= \frac{-3x^2 + 28x + 39}{12(x-3)(x+3)^2}$$

Use the fundamental
principle of fractions
to make each
denominator
$12(x-3)(x+3)^2$.

PROBLEM 10   Combine into a single fraction and simplify:

$$\frac{3}{2x^2 - 8x + 8} - \frac{x+1}{3x^2 - 12}$$

Solution

Solutions to Matched
Problems

7. (A) $15y^2 = 3 \cdot 5y^2$

$\qquad 12y = 2^2 \cdot 3y$

$\qquad 9y^4 = 3^2 y^4$

$\qquad \text{LCM} = 2^2 \cdot 3^2 \cdot 5y^4 = 180y^4$

(B) $\quad 3x^2 - 12 = 3(x - 2)(x + 2)$

$\qquad x^2 - 4x + 4 = (x - 2)^2$

$\qquad 12(x + 2) = 2^2 \cdot 3(x + 2)$

$\qquad\qquad \text{LCM} = 2^2 \cdot 3(x - 2)^2(x + 2) = 12(x - 2)^2(x + 2)$

8. $\dfrac{x + 3}{2x - 5} - \dfrac{3x - 2}{2x - 5} = \dfrac{(x + 3) - (3x - 2)}{2x - 5} = \dfrac{x + 3 - 3x + 2}{2x - 5} = \dfrac{5 - 2x}{2x - 5}$

$\qquad\qquad\qquad\qquad\qquad\qquad\qquad\qquad = \dfrac{-(2x - 5)}{2x - 5} = -1$

9. $\dfrac{5}{4x^3} - \dfrac{1}{3x} + \dfrac{2}{1} = \dfrac{3(5)}{3(4x^3)} - \dfrac{4x^2(1)}{4x^2(3x)} + \dfrac{12x^3(2)}{12x^3(1)} \quad \text{LCD} = 12x^3$

$\qquad\qquad = \dfrac{15}{12x^3} - \dfrac{4x^2}{12x^3} + \dfrac{24x^3}{12x^3} = \dfrac{24x^3 - 4x^2 + 15}{12x^3}$

10. $\dfrac{3}{2x^2 - 8x + 8} - \dfrac{x + 1}{3x^2 - 12} = \dfrac{3}{2(x - 2)^2} - \dfrac{(x + 1)}{3(x - 2)(x + 2)}$

$\qquad = \dfrac{3(x + 2)(3)}{3(x + 2)[2(x - 2)^2]} - \dfrac{2(x - 2)(x + 1)}{2(x - 2)[3(x - 2)(x + 2)]} \quad \text{LCD} = 6(x - 2)^2(x + 2)$

$\qquad = \dfrac{9(x + 2)}{6(x - 2)^2(x + 2)} - \dfrac{2(x - 2)(x + 1)}{6(x - 2)^2(x + 2)} = \dfrac{9x + 18 - (2x^2 - 2x - 4)}{6(x - 2)^2(x + 2)}$

$\qquad = \dfrac{9x + 18 - 2x^2 + 2x + 4}{6(x - 2)^2(x + 2)} = \dfrac{-2x^2 + 11x + 22}{6(x - 2)^2(x + 2)}$

Practice Exercise 3-5 ■

Work odd-numbered problems first, check answers, and then work even-numbered problems in areas of weakness. Answers to all problems are in the back of the book. Make every effort to work a problem yourself before you look at an answer.

A   *Find the least common multiple (LCM) for each group of expressions.*

1. $3, x$ _____

2. $4, y$ _____

3. $x, 1$ _____

4. $y, 1$ _____

5. v^2, v, v^3 _____

6. x, x, x^2 _____

7. $3x, 6x^2, 4$ _____

8. $8u^3, 6u, 4u^2$ _____

9. $x + 1, x - 2$ _____

10. $x - 2, x + 3$ _____

11. $y + 3, 3y$ _____

12. $x - 2, 2x$ _____

Combine into single fractions and simplify.

13. $\dfrac{7x}{5x^2} + \dfrac{2}{5x^2}$ _____

14. $\dfrac{3m}{2m^2} + \dfrac{1}{2m^2}$ _____

15. $\dfrac{4x}{2x-1} - \dfrac{2}{2x-1}$ _____

16. $\dfrac{5a}{a-1} - \dfrac{5}{a-1}$ _____

17. $\dfrac{y}{y^2-9} - \dfrac{3}{y^2-9}$ _____

18. $\dfrac{2x}{4x^2-9} + \dfrac{3}{4x^2-9}$ _____

19. $\dfrac{5}{3k} - \dfrac{6x-4}{3k}$ _____

20. $\dfrac{1}{2a^2} - \dfrac{2b-1}{2a^2}$ _____

21. $\dfrac{3x}{y} + \dfrac{1}{4}$ _____

22. $\dfrac{2}{x} - \dfrac{1}{3}$ _____

23. $\dfrac{2}{y} + 1$ _____

24. $x + \dfrac{1}{x}$ _____

25. $\dfrac{u}{v^2} - \dfrac{1}{v} + \dfrac{u^3}{v^3}$ _____

26. $\dfrac{1}{x} - \dfrac{y}{x^2} + \dfrac{y^2}{x^3}$ _____

27. $\dfrac{2}{3x} - \dfrac{1}{6x^2} + \dfrac{3}{4}$ _____

28. $\dfrac{1}{8u^3} + \dfrac{5}{6u} - \dfrac{3}{4u^2}$ ___ ___

29. $\dfrac{2}{x+1} + \dfrac{3}{x-2}$ _____

30. $\dfrac{1}{x-2} + \dfrac{1}{x+3}$ _____

31. $\dfrac{3}{y+3} - \dfrac{2}{3y}$ _____

32. $\dfrac{2}{x-2} - \dfrac{3}{2x}$ _____

B *Find the LCM for each group of expressions.*

33. $12x^3, 8x^2y^2, 3xy^2$ _____

34. $9u^3v^2, 6uv, 12v^3$ _____

35. $15x^2y, 25xy, 5y^2$ _____

36. $18m^4n^2, 12m^2n^4, 9mn$ _____

37. $6(x-1), 9(x-1)^2$ _____

38. $8(y-3)^2, 6(y-3)$ _____

39. $6(x-7)(x+7), 8(x+7)^2$ _____

40. $3(x-5)^2, 4(x+5)(x-5)$ _____

41. x^2-4, x^2+4x+4 _____

42. x^2-6x+9, x^2-9 _____

43. $3x^2 + 3x, 4x^2, 3x^2 + 6x + 3$ _____

44. $3m^2 - 3m, m^2 - 2m + 1, 5m^2$ _____

Combine into a single fraction and simplify.

45. $\dfrac{2}{9u^3v^2} - \dfrac{1}{6uv} + \dfrac{1}{12v^3}$ _____   **46.** $\dfrac{1}{12x^3} + \dfrac{3}{8x^2y^2} - \dfrac{2}{3xy^2}$ _____

47. $\dfrac{4t - 3}{18t^3} + \dfrac{3}{4t} - \dfrac{2t - 1}{6t^2}$ _____   **48.** $\dfrac{3y + 8}{4y^2} - \dfrac{2y - 1}{y^3} - \dfrac{5}{8y}$ _____

49. $\dfrac{t + 1}{t - 1} - 1$ _____   **50.** $2 + \dfrac{x + 1}{x - 3}$ _____

51. $5 + \dfrac{a}{a + 1} - \dfrac{a}{a - 1}$ _____   **52.** $\dfrac{1}{y + 2} + 3 - \dfrac{2}{y - 2}$ _____

53. $\dfrac{2}{3(x - 5)^2} - \dfrac{1}{4(x + 5)(x - 5)}$ _____

54. $\dfrac{1}{6(x - 7)(x + 7)} + \dfrac{3}{8(x + 7)^2}$ _____

55. $\dfrac{5}{6(x - 1)} + \dfrac{2}{9(x - 1)^2}$ _____

56. $\dfrac{3}{8(y - 3)^2} - \dfrac{1}{6(y - 3)}$ _____

57. $\dfrac{3}{x + 3} - \dfrac{3x + 1}{(x - 1)(x + 3)}$ _____

58. $\dfrac{4}{2x - 3} - \dfrac{2x + 1}{(2x - 3)(x + 2)}$ _____

59. $\dfrac{3s}{3s^2 - 12} + \dfrac{1}{2s^2 + 4s}$ _____

60. $\dfrac{2t}{3t^2 - 48} + \dfrac{t}{4t + t^2}$ _____

61. $\dfrac{3}{x^2 - 4} - \dfrac{1}{x^2 + 4x + 4}$ _____

62. $\dfrac{2}{x^2 - 6x + 9} - \dfrac{1}{x^2 - 9}$ _____

63. $\dfrac{2}{x+3} - \dfrac{1}{x-3} + \dfrac{2x}{x^2-9}$ _____

64. $\dfrac{2x}{x^2-y^2} + \dfrac{1}{x+y} - \dfrac{1}{x-y}$ _____

C **65.** $\dfrac{x}{x^2-x-2} - \dfrac{1}{x^2+5x-14} - \dfrac{2}{x^2+8x+7}$ _____

66. $\dfrac{m^2}{m^2+2m+1} + \dfrac{1}{3m+3} - \dfrac{1}{6}$ _____

67. $\dfrac{1}{3x^2+3x} + \dfrac{1}{4x^2} - \dfrac{1}{3x^2+6x+3}$ _____

68. $\dfrac{1}{3m(m-1)} + \dfrac{1}{m^2-2m+1} - \dfrac{1}{5m^2}$ _____

69. $\dfrac{xy^2}{x^3-y^3} - \dfrac{y}{x^2+xy+y^2}$ _____

70. $\dfrac{x}{x^2-xy+y^2} - \dfrac{xy}{x^3+y^3}$ _____

For the next six problems note that $b - a = -(a - b)$; thus, $3 - y = -(y - 3)$, $1 - x = -(x - 1)$, and so on.

71. $\dfrac{5}{y-3} - \dfrac{2}{3-y}$ _____ **72.** $\dfrac{3}{x-1} + \dfrac{2}{1-x}$ _____

73. $\dfrac{3}{x-3} + \dfrac{x}{3-x}$ _____ **74.** $\dfrac{-2}{2-y} - \dfrac{y}{y-2}$ _____

75. $\dfrac{1}{5x-5} - \dfrac{1}{3x-3} + \dfrac{1}{1-x}$ _____

The Check Exercise for this section is on page 173.

76. $\dfrac{x+7}{ax-bx} + \dfrac{y+9}{by-ay}$ _____

Section 3-6　Complex Fractions

A fractional form with fractions in its numerator or denominator is called a **complex fraction**. It is often necessary to represent a complex fraction as a **simple fraction**—that is (in all cases we will consider), as the quotient of two polynomials. The process does not involve any new concepts. It is a matter of applying old concepts in the right way. In particular, we will find the fundamental principle of fractions,

$$\frac{PK}{QK} = \frac{P}{Q} \qquad Q, K \neq 0 \tag{1}$$

of considerable use. Several examples should clarify the process.

EXAMPLE 11   Express as simple fractions: **(A)** $\dfrac{\frac{2}{3}}{\frac{3}{4}}$ **(B)** $\dfrac{1\frac{1}{2}}{3\frac{2}{3}}$

Solution   **(A)** Use the fundamental principle of fractions (1) and multiply numerator and denominator by a number divisible by both 3 and 4—that is, 12, the LCD of the two internal fractions.

$$\frac{\frac{2}{3}}{\frac{3}{4}} = \frac{12 \cdot \frac{2}{3}}{12 \cdot \frac{3}{4}}$$ Multiply top and bottom by 12, then cancel internal denominators.

$$= \frac{4 \cdot 2}{3 \cdot 3}$$

$$= \frac{8}{9}$$

Note: We can also work this problem by inverting the denominator and multiplying. Thus,

$$\tfrac{2}{3} \cdot \tfrac{4}{3} = \tfrac{8}{9}$$

(B) Recall that $1\frac{1}{2}$ and $3\frac{2}{3}$ represent sums and not products; that is, $1\frac{1}{2} = 1 + \frac{1}{2}$ and $3\frac{2}{3} = 3 + \frac{2}{3}$. Thus,

$$\frac{1\frac{1}{2}}{3\frac{2}{3}} = \frac{1 + \frac{1}{2}}{3 + \frac{2}{3}}$$ Write mixed fractions as sums.

$$= \frac{6\left(1 + \frac{1}{2}\right)}{6\left(3 + \frac{2}{3}\right)}$$ Multiply top and bottom by 6, the LCD of all fractions within the main fraction.

$$= \frac{6 \cdot 1 + 6 \cdot \frac{1}{2}}{6 \cdot 3 + 6 \cdot \frac{2}{3}}$$ The denominators 2 and 3 cancel.

$$= \frac{6 + 3}{18 + 4} = \frac{9}{22}$$ A simple fraction.

PROBLEM 11   Express as simple fractions: **(A)** $\dfrac{\frac{3}{5}}{\frac{1}{4}}$ **(B)** $\dfrac{2\frac{3}{4}}{4\frac{1}{3}}$

Solution   **(A)**   **(B)**

EXAMPLE 12   Express as simple fractions:

(A) $\dfrac{1 - \dfrac{1}{x^2}}{1 + \dfrac{1}{x}}$

Multiply top and bottom by x^2, the LCD of all internal fractions.

$$= \frac{x^2\left(1 - \dfrac{1}{x^2}\right)}{x^2\left(1 + \dfrac{1}{x}\right)}$$

$$= \frac{x^2 \cdot 1 - x^2 \cdot \dfrac{1}{x^2}}{x^2 \cdot 1 + x^2 \cdot \dfrac{1}{x}}$$

$$= \frac{x^2 - 1}{x^2 + x}$$

Factor top and bottom to reduce to lowest terms.

$$= \frac{(x - 1)(\overset{1}{\cancel{x + 1}})}{x(\underset{1}{\cancel{x + 1}})}$$

$$= \frac{x - 1}{x}$$

(B) $\dfrac{\dfrac{a}{b} - \dfrac{b}{a}}{\dfrac{a}{b} + 2 + \dfrac{b}{a}}$

LCD $= ab$

$$= \frac{ab\left(\dfrac{a}{b} - \dfrac{b}{a}\right)}{ab\left(\dfrac{a}{b} + 2 + \dfrac{b}{a}\right)}$$

Use the fundamental principle of fractions; that is, multiply top and bottom by ab to clear internal fractions.

$$= \frac{ab \cdot \dfrac{a}{b} - ab \cdot \dfrac{b}{a}}{ab \cdot \dfrac{a}{b} + ab \cdot 2 + ab \cdot \dfrac{b}{a}}$$

$$= \frac{a^2 - b^2}{a^2 + 2ab + b^2}$$

Reduce to lowest terms.

$$= \frac{(a - b)(a + b)}{(a + b)^2}$$

$$= \frac{a - b}{a + b}$$

A simple fraction.

PROBLEM 12   Express as simple fractions:

(A) $\dfrac{1 - \dfrac{1}{3x}}{1 - \dfrac{1}{9x^2}}$ (B) $\dfrac{\dfrac{x}{y} + 1 - \dfrac{2y}{x}}{\dfrac{x}{y} - \dfrac{y}{x}}$

Solution   **(A)**   **(B)**

Solutions to Matched
Problems

11. **(A)** $\dfrac{\frac{3}{5}}{\frac{1}{4}}$ $\boxed{= \dfrac{20\left(\frac{3}{5}\right)}{20\left(\frac{1}{4}\right)} = \dfrac{4 \cdot 3}{5 \cdot 1}}$ $= \dfrac{12}{5}$ **(B)** $\dfrac{2 + \frac{3}{4}}{4 + \frac{1}{3}} = \dfrac{\mathbf{12}\left(2 + \frac{3}{4}\right)}{\mathbf{12}\left(4 + \frac{1}{3}\right)} = \dfrac{24 + 9}{48 + 4} = \dfrac{33}{52}$

12. **(A)** $\dfrac{1 - \dfrac{1}{3x}}{1 - \dfrac{1}{9x^2}} = \dfrac{\mathbf{9x^2}\left(1 - \dfrac{1}{3x}\right)}{\mathbf{9x^2}\left(1 - \dfrac{1}{9x^2}\right)} = \dfrac{9x^2 - 3x}{9x^2 - 1}$

$= \dfrac{3x(\cancel{3x - 1})}{(\cancel{3x - 1})(3x + 1)} = \dfrac{3x}{3x + 1}$

(B) $\dfrac{\dfrac{x}{y} + 1 - \dfrac{2y}{x}}{\dfrac{x}{y} - \dfrac{y}{x}} = \dfrac{\mathbf{xy}\left(\dfrac{x}{y} + 1 - \dfrac{2y}{x}\right)}{\mathbf{xy}\left(\dfrac{x}{y} - \dfrac{y}{x}\right)} = \dfrac{x^2 + xy - 2y^2}{x^2 - y^2}$

$= \dfrac{(x + 2y)(\cancel{x - y})}{(\cancel{x - y})(x + y)} = \dfrac{x + 2y}{x + y}$

Practice Exercise 3-6 ■

Work odd-numbered problems first, check answers, and then work even-numbered problems in areas of weakness. Answers to all problems are in the back of the book. Make every effort to work a problem yourself before you look at an answer.

Express as simple fractions reduced to lowest terms.

A   **1.** $\dfrac{\frac{1}{2}}{\frac{2}{3}}$ _____   **2.** $\dfrac{\frac{1}{4}}{\frac{2}{3}}$ _____   **3.** $\dfrac{\frac{3}{8}}{\frac{5}{12}}$ _____

4. $\dfrac{\frac{4}{15}}{\frac{5}{6}}$ _____

5. $\dfrac{1\frac{1}{3}}{2\frac{1}{6}}$ _____

6. $\dfrac{3\frac{1}{10}}{2\frac{1}{5}}$ _____

7. $\dfrac{1\frac{2}{9}}{2\frac{5}{6}}$ _____

8. $\dfrac{2\frac{4}{15}}{1\frac{7}{10}}$ _____

9. $\dfrac{\frac{x}{y}}{\frac{1}{y^2}}$ _____

10. $\dfrac{\frac{1}{b^2}}{\frac{a}{b}}$ _____

11. $\dfrac{\frac{y}{2x}}{\frac{1}{3x^2}}$ _____

12. $\dfrac{\frac{2x}{5y}}{\frac{1}{3x}}$ _____

B **13.** $\dfrac{1+\frac{3}{x}}{x-\frac{9}{x}}$ _____

14. $\dfrac{1-\frac{2}{x}}{x-\frac{4}{x}}$ _____

15. $\dfrac{1-\frac{y^2}{x^2}}{1-\frac{y}{x}}$ _____

16. $\dfrac{\frac{a^2}{b^2}-1}{\frac{a}{b}-1}$ _____

17. $\dfrac{\frac{1}{x}+\frac{1}{y}}{\frac{y}{x}-\frac{x}{y}}$ _____

18. $\dfrac{b-\frac{a^2}{b}}{\frac{1}{a}-\frac{1}{b}}$ _____

19. $\dfrac{\frac{x}{y}-2+\frac{y}{x}}{\frac{x}{y}-\frac{y}{x}}$ _____

20. $\dfrac{1+\frac{2}{x}-\frac{15}{x^2}}{1+\frac{4}{x}-\frac{5}{x^2}}$ _____

21. $\dfrac{\frac{a^2}{a-b}-a}{\frac{b^2}{a-b}+b}$ _____

22. $\dfrac{n-\frac{n^2}{n-m}}{1+\frac{m^2}{n^2-m^2}}$ _____

23. $\dfrac{\frac{m}{m+2}-\frac{m}{m-2}}{\frac{m+2}{m-2}-\frac{m-2}{m+2}}$ _____

24. $\dfrac{\frac{y}{x+y}-\frac{x}{x-y}}{\frac{x}{x+y}+\frac{y}{x-y}}$ _____

C **25.** $1-\dfrac{1}{1-\frac{1}{x}}$ _____

26. $2-\dfrac{1}{1-\frac{2}{x+2}}$ _____

27. $1-\dfrac{x-\frac{1}{x}}{1-\frac{1}{x}}$ _____

28. $\dfrac{t-\dfrac{1}{1+\frac{1}{t}}}{t+\dfrac{1}{t-\frac{1}{t}}}$ _____

29. $1 + \cfrac{1}{1 + \cfrac{1}{1 + \cfrac{1}{1 + x}}}$ _____

30. $1 - \cfrac{1}{1 - \cfrac{1}{1 - \cfrac{1}{1 - x}}}$ _____

APPLICATIONS

31. A formula for the average rate r for a round trip between two points, where the rate going is r_G and the rate returning is r_R, is given by the complex fraction

$$r = \cfrac{2}{\cfrac{1}{r_G} + \cfrac{1}{r_R}}$$

Express r as a simple fraction. _____

32. The airspeed indicator on a jet aircraft registers 500 miles per hour. If the plane is traveling with an airstream moving at 100 miles per hour, then the plane's ground speed would be 600 miles per hour—or would it? According to Einstein, velocities must be added according to the following formula:

$$v = \cfrac{v_1 + v_2}{1 + \cfrac{v_1 v_2}{c^2}}$$

The Check Exercise for this section is on page 175.

where v is the resultant velocity, c is the speed of light, and v_1 and v_2 are the two velocities to be added. Convert the right side of the equation into a simple fraction. _____

Section 3-7   Chapter Review

A **rational expression** is an algebraic expression involving only the operations of addition, subtraction, multiplication, and division on variables and constants. A rational expression A/B can be **reduced to lowest terms** or **raised to higher terms** by using the **fundamental principle of fractions**: $A/B = AK/BK$. *(3-1)*

A polynomial P can be divided by a polynomial **divisor** D by using **algebraic long division** to yield a **quotient** Q and **remainder** R so that $P = D \cdot Q + R$ and either R is 0 or the degree of R is smaller than that of D.   *(3-2)*

The **synthetic division process** is an abbreviated long-division method used when the divisor is of the form $x - r$.   *(3-3)*

Rational expressions are multiplied and divided as follows:

$$\frac{A}{B} \cdot \frac{C}{D} = \frac{A \cdot C}{B \cdot D} \qquad \frac{A}{B} \div \frac{C}{D} = \frac{A}{B} \cdot \frac{D}{C} = \frac{A \cdot D}{B \cdot C} \quad \text{(3-4)}$$

Rational expressions with a common denominator are added and subtracted as follows:

$$\frac{A}{D} + \frac{B}{D} = \frac{A + B}{D} \qquad \frac{A}{D} - \frac{B}{D} = \frac{A - B}{D}$$

Rational expressions with different denominators are converted to ones with a common denominator and then added or subtracted. *(3-5)*

A **complex fraction** is a fraction in which the numerator or denominator is a fraction. The fundamental principle of fractions and the operations of addition, subtraction, multiplication, and division are used to convert complex fractions to **simple fractions**—that is, to quotients of two polynomials. *(3-6)*

Diagnostic (Review) Exercise 3-7 ■

Work through all the problems in this chapter review and check answers in the back of the book. (Answers to all problems are there, and following each answer is a number in italics indicating the section in which that type of problem is discussed.) Where weaknesses show up, review appropriate sections in the text. When you are satisfied that you know the material, take the practice test following this review.

A   *Perform the indicated operation and simplify. Express each answer as a simple fraction in lowest terms.*

1. $\dfrac{18x^3y^2(z+3)}{12xy^2(z+3)^3}$ _____

2. $\dfrac{x^2+2x+1}{x^2-1}$ _____

3. $1+\dfrac{2}{3x}$ _____

4. $\dfrac{2}{x}-\dfrac{1}{6x}+\dfrac{1}{3}$ _____

5. $\dfrac{1}{6x^3}-\dfrac{3}{4x}-\dfrac{2}{3}$ _____

6. $\dfrac{4x^2y^3}{3a^2b^2}\div\dfrac{2xy^2}{3ab}$ _____

7. $\dfrac{6x^2}{3(x-1)}-\dfrac{6}{3(x-1)}$ _____

8. $1-\dfrac{m-1}{m+1}$ _____

9. $\dfrac{3}{x-2}-\dfrac{2}{x+1}$ _____

10. $(d-2)^2\div\dfrac{d^2-4}{d-2}$ _____

11. $\dfrac{x+1}{x+2}-\dfrac{x+2}{x+3}$ _____

12. $\dfrac{\frac{1}{4}}{\frac{2}{3}}$ _____

13. $\dfrac{2\frac{3}{4}}{1\frac{1}{2}}$ _____

14. $\dfrac{1-\dfrac{2}{y}}{1+\dfrac{1}{y}}$ _____

B   *Divide to find the quotient and remainder.*

15. $(x^3-3x^2+x-3)\div(x-1)$ _____

16. $(x^3+x)\div(x^2+1)$ _____

17. $(x^4+2x^3+3x^2+4x+5)\div(x^2+2)$ _____

18. $(x^4+x^2-1)\div(x+2)$ _____

19. $(x^4 - 1) \div (x - 1)$ _____

20. $(x^4 + x + x^3) \div (1 + x + x^2)$ _____

Perform the indicated operation and simplify. Express each answer as a simple fraction in lowest terms.

21. $\dfrac{2}{5b} - \dfrac{4}{3b^3} - \dfrac{1}{6a^2b^2}$ _____

22. $\dfrac{2}{2x - 3} - 1$ _____

23. $\dfrac{4x^2y}{3ab^2} \div \left(\dfrac{2a^2x^2}{b^2y} \cdot \dfrac{6a}{2y^2} \right)$ _____

24. $\dfrac{x}{x^2 + 4x} + \dfrac{2x}{3x^2 - 48}$ _____

25. $\dfrac{x^3 - x}{x^2 - x} \div \dfrac{x^2 + 2x + 1}{x}$ _____

26. $\dfrac{\dfrac{x}{y} - \dfrac{y}{x}}{\dfrac{x}{y} + 1}$ _____

27. $\dfrac{x}{x^3 - y^3} - \dfrac{1}{x^2 + xy + y^2}$ _____

28. $\dfrac{\dfrac{y^2}{x^2 - y^2} + 1}{\dfrac{x^2}{x - y} - x}$ _____

29. $\dfrac{x^3 - 1}{x^2 + x + 1} \div \dfrac{x^2 - 1}{x^2 + 2x + 1}$ _____

30. $\dfrac{1}{3x^2 - 27} - \dfrac{x - 1}{4x^3 + 24x^2 + 36x}$ _____

C **31.** $\dfrac{4}{s^2 - 4} + \dfrac{1}{2 - s}$ _____

32. $\dfrac{y^2 - y - 6}{(y + 2)^2} \cdot \dfrac{2 + y}{3 - y}$ _____

33. $\dfrac{y}{x^2} \div \left(\dfrac{x^2 + 3x}{2x^2 + 5x - 3} \div \dfrac{x^3y - x^2y}{2x^2 - 3x + 1} \right)$ _____

34. $\dfrac{1 - \dfrac{1}{1 + \dfrac{x}{y}}}{1 - \dfrac{1}{1 - \dfrac{x}{y}}}$ _____

35. $\left(x - \dfrac{1}{1 - \dfrac{1}{x}} \right) \div \left(\dfrac{x}{x + 1} - \dfrac{x}{1 - x} \right)$ _____

Practice Test Chapter 3 ∎

Take this as if it were a graded test by working the problems within a 50-minute time period. Do not look back in the chapter. Choose one of three levels of difficulty: least difficult, Problems 1–12; more difficult, add Problem 13; most difficult, add Problems 13 and 14. Use the answers in the back of the book to correct your work. The answers are keyed to appropriate text sections so that you can easily locate and review sections where difficulties still persist.

Divide. Write the quotient and indicate the remainder.

1. $(x^4 - 3x^3 + x^2 - 2x + 1) \div (x^2 + 1)$ _____

2. $(2x^3 - 3x + 5) \div (x - 4)$ _____

Perform the indicated operations. Reduce to lowest terms and express any complex fractions as simple fractions in lowest terms.

3. $\dfrac{2x}{x^2 + 1} + \dfrac{3x - 5}{x^2 + 1}$ _____

4. $3 - \dfrac{x + 2}{x - 2}$ _____

5. $\dfrac{x - 2}{x + 2} - \dfrac{x - 3}{x + 3}$ _____

6. $\dfrac{4x - 12}{2x^2 - 10x + 12} \cdot \dfrac{x^2 - 3x + 2}{x - 4}$ _____

7. $\dfrac{a^4}{3b^3} \cdot \left(\dfrac{b^4}{2a^3} \div \dfrac{-3b}{4a} \right)$ _____

8. $\dfrac{1}{xy} + \dfrac{2}{yz} + \dfrac{3}{xz}$ _____

9. $\dfrac{1}{x + 1} - \dfrac{2}{x - 1} + \dfrac{4x}{x^2 - 1}$ _____

10. $\dfrac{a^3 + b^3}{a^2 + 2ab + b^2} \div \dfrac{a^3 - a^2 b + ab^2}{a^2 - ab - 2b^2}$ _____

11. $\dfrac{1 + \dfrac{2a}{b} + \dfrac{a^2}{b^2}}{\dfrac{b^2}{a^2} + \dfrac{2b}{a} + 1}$ _____

12. $\dfrac{2x - 1}{x} \cdot \left(x - \dfrac{x}{1 + \dfrac{1}{1 - \dfrac{1}{x}}} \right)$ _____

13. $\left(\dfrac{1}{x - 1} - \dfrac{1}{1 - x} \right) \div \left(\dfrac{1}{x - 1} + \dfrac{1}{x + 1} \right)$ _____

14. $\dfrac{a^2 + 2ab}{ab - b^2} \div \left(\dfrac{ab + 2b^2}{a^2 - ab} \div \dfrac{a^2 b + ab^2 - 2b^3}{a^3 - 2a^2 b + ab^2} \right)$ _____

Check Exercise 3-1 ■

Work the following problems without looking at any text examples. Show your work in the space provided. Write your answer in the answer column.

1. _____

In Problems 1–5 reduce to lowest terms.

2. _____

1. $\dfrac{6x^5y^2}{8xy^4}$

3. _____

4. _____

5. _____

2. $\dfrac{(x-y)^2}{(x+y)(x-y)}$

6. _____

7. _____

8. _____

3. $\dfrac{3x^2(x-1)}{6x(x-1)^2}$

9. _____

10. _____

4. $\dfrac{u^2v-uv}{u^2-uv}$

5. $\dfrac{a^2-4b^2}{a^2-4ab+4b^2}$

Complete the raising-to-higher-terms process in Problems 6–8 by replacing the question marks with appropriate expressions.

6. $\dfrac{3x}{5y} = \dfrac{?}{15xy^2}$

7. $\dfrac{2}{x} = \dfrac{2x - 2}{?}$

8. $\dfrac{x + y}{x - 2y} = \dfrac{?}{x^2 - 4y^2}$

In Problems 9 and 10 reduce to lowest terms.

9. $\dfrac{ac + bc - ad - bd}{c^2 + dc - 2d^2}$

10. $\dfrac{x^3 - y^3}{3x^3y + 3x^2y^2 + 3xy^3}$

Check Exercise 3-2 ∎

ANSWER COLUMN

Work the following problems without looking at any text examples. Show your work in the space provided. Write your answer in the answer column.

1. _____

2. _____   *Divide, using the long-division process, and check answers.*

1. $(6x^2 - 5x - 2)/(2x - 3)$

3. _____

4. _____

5. _____

2. $(x^2 + 4)/(x - 2)$

3. $(8x^3 - 8x + 3)/(2x - 1)$

4. $(1 - x^2 - 7x + x^3)/(x - 3)$

5. $(2x^4 - 3 + 5x + 3x^2 - 3x^3)/(3 - 2x + x^2)$

Check Exercise 3-3 ■

ANSWER COLUMN

Work the following problems without looking at any text examples. Show your work in the space provided. Write your answer in the answer column.

1. _____

2. _____

In Problems 1–4 divide using synthetic division. Write the quotient and indicate the remainder.

3. _____

1. $(x^3 - x^2 - x - 1) \div (x - 3)$

4. _____

5. _____

2. $(x^3 - 2x^2 - 22x - 21) \div (x + 3)$

3. $(x^4 - 4x^2 - 5x + 6) \div (x - 2)$

4. $(2x^3 + 5x^2 + 7x - 1) \div (x + 1)$

5. Evaluate $x^5 - 2x^4 + x^3 - 3x^2 + x - 4$ at $x = 2$ using synthetic division.

Check Exercise 3-4 ∎

ANSWER COLUMN

1. _____

2. _____

3. _____

4. _____

5. _____

Work the following problems without looking at any text examples. Show your work in the space provided. Write your answer in the answer column.

Perform the indicated operations and reduce to lowest terms.

1. $\dfrac{8x^2y}{6z^3} \div \dfrac{4y^3}{3x^2z^2}$

2. $\dfrac{3x^2 + 9x}{x^2 - 2x - 15} \cdot (x - 5)$

3. $\dfrac{2x^2 + 5x - 3}{16x^3y - 8x^2y} \div \dfrac{x^2 + 6x + 9}{6x^2y^2 + 18xy^2}$

4. $\dfrac{m^3}{ab} \div \left(\dfrac{4m^2}{b^2} \div \dfrac{a^2}{m} \right)$

5. $\dfrac{3-x}{3x+x^2} \cdot \dfrac{x^2+6x+9}{x^2-9}$

Check Exercise 3-5 ■

ANSWER COLUMN

1. _____

2. _____

3. _____

4. _____

5. _____

6. _____

7. _____

8. _____

9. _____

10. _____

Work the following problems without looking at any text examples. Show your work in the space provided. Write your answer in the answer column.

In Problems 1–4 find the least common multiple (LCM) for each set of algebraic expressions.

1. $8, \quad 3x, \quad 2y$

2. $9xy^2, \quad 4x^2y, \quad 6xy$

3. $4(x - 1), \quad 12(x - 1)^2, \quad 6$

4. $6x^2 - 6, \quad 4x^2 - 8x + 4$

In Problems 5–10 combine into a single fraction and simplify.

5. $\dfrac{1}{8} + \dfrac{1}{3x} - \dfrac{1}{2y}$

6. $\dfrac{2}{9xy^2} - \dfrac{3}{4x^2y} + \dfrac{1}{6xy}$

7. $\dfrac{x+2}{x-3} - 1$

8. $\dfrac{1}{x^2 - 10x + 25} - \dfrac{1}{x^2 - 25}$

9. $\dfrac{1}{2u^2 - 4u} - \dfrac{1}{6u^2} - \dfrac{2u - 4}{12u^3 - 48u^2 + 48u}$

10. $\dfrac{x}{x-4} + \dfrac{x-1}{4-x}$

Check Exercise 3-6 ■

Work the following problems without looking at any text examples. Show your work in the space provided. Write your answer in the answer column.

1. _____

Express as simple fractions reduced to lowest terms.

2. _____

1. $\dfrac{1\frac{2}{3}}{3\frac{1}{2}}$

3. _____

4. _____

5. _____

2. $\dfrac{\dfrac{m}{3n}}{\dfrac{1}{2n^2}}$

3. $\dfrac{\dfrac{b}{a} - \dfrac{a}{b}}{\dfrac{1}{a} - \dfrac{1}{b}}$

4. $\dfrac{1 - \dfrac{b}{a}}{1 - \dfrac{b^2}{a^2}}$

5. $1 - \dfrac{1}{1 + \dfrac{1}{x}}$

First-Degree Equations and Inequalities in One Variable

■ **4**

INSTRUCTIONS FOR STUDENTS IN A SELF-PACED CLASS OR LAB

YES — **HAVE YOU HAD INTERMEDIATE ALGEBRA BEFORE THIS COURSE?** — NO

1. Work Diagnostic (Review) Exercise 4-8 on page 248. Check answers in back of book; then work through text sections corresponding to problems missed. (Section numbers are in italics following each answer.)
2. When finished with step 1, take Practice Test Chapter 4 on page 251 as a final check of your understanding of the chapter. Check answers in the back of the book; then review sections where weakness still prevails. (Corresponding section numbers are in italics following each answer.)
3. When you think you are ready, ask your instructor for a graded test for Chapter 4.
4. If your instructor approves, after the test is corrected, go to the next chapter.

1. Work through each section in the chapter as follows:
   (A) Read discussion.
   (B) Read each example and work the corresponding matched problem. Check your solutions to the matched problem in Solutions to Matched Problems on the indicated page.
   (C) At the end of a section work the odd-numbered problems in the Practice Exercise and check answers; then work even-numbered problems in areas of weakness. (Answers to *all* Practice Exercise sets are in the back of the book.)
   (D) Work Check Exercise as instructed. Tear out and turn in as directed by your instructor. (Answers are not in the text.)
2. Repeat each step in item 1 for each section in the chapter.
3. After the instructional part of the chapter is completed, proceed with steps 1 to 4 in the box above this one.

Chapter 4 ▪ First-Degree Equations and Inequalities in One Variable

▪ Introduction

In this section we will review methods of solving equations such as

$$2(3x - 5) - 2 = 5 - (3x + 2) \qquad \frac{x}{3} - \frac{x-1}{2} = \frac{1}{6} \qquad \frac{5}{2x-1} - \frac{1}{2} = \frac{3x}{4x-2}$$

A **solution** or **root** of an equation involving a single variable is a replacement of the variable by a constant that makes the left side of the equation equal to the right side. For example, $x = 4$ is a solution of

$$2x - 1 = x + 3$$

since

$$2(4) - 1 = 4 + 3 \quad \text{That is, } 7 = 7.$$

The set of all solutions is called the **solution set**. To **solve an equation** is to find its solution set.

Knowing what we mean by a solution set for an equation is one thing, but finding it is another. Our objective is to develop a systematic method of solving equations that is free of guesswork. We start by introducing the idea of equivalent equations. We say that two equations are **equivalent** if they both have the same solution set—any solution of one is a solution of the other.

The basic idea in solving equations is to perform operations on equations that produce *simpler* equivalent equations and to continue the process until we reach an equation whose solution is obvious—generally an equation such as

$$x = -5$$

The following properties of equality produce equivalent equations when applied.

THEOREM 1

> **Equality Properties**
>
> For a, b, and c any real numbers:
>
> **1.** If $a = b$, then $a + c = b + c$.   ADDITION PROPERTY
>
> If $x - 2 = 3$, then $(x - 2) + 2 = 3 + 2$.
>
> **2.** If $a = b$, then $a - c = b - c$.   SUBTRACTION PROPERTY
>
> If $x + 4 = 5$, then $(x + 4) - 4 = 5 - 4$.
>
> **3.** If $a = b$, then $ca = cb$, $c \neq 0$.   MULTIPLICATION PROPERTY
>
> If $\dfrac{x}{2} = 3$, then $2 \cdot \dfrac{x}{2} = 2 \cdot 3$.
>
> **4.** If $a = b$, then $\dfrac{a}{c} = \dfrac{b}{c}$, $c \neq 0$.   DIVISION PROPERTY
>
> If $5x = 10$, then $\dfrac{5x}{5} = \dfrac{10}{5}$.

These properties of equality follow directly from the basic equality properties discussed in Section 1-2.

We can think of the process of solving an equation as a game. The objective of the game is to isolate the variable (with a coefficient of 1) on one side of the equation (usually the left), leaving a constant on the other side. We are now ready to solve equations. The following strategy might prove helpful.

> **Equation-Solving Strategy**
>
> **1.** Use the multiplication property in Theorem 1 to remove fractions if present.
> **2.** Simplify the left and right sides of the equation by removing grouping symbols and combining like terms.
> **3.** Use the equality properties in Theorem 1 to get all variable terms on one side (usually the left) and all constant terms on the other side (usually the right). Combine like terms in the process.
> **4.** Isolate the variable (with a coefficient of 1), using the division or multiplication property of equality.

■ **Equations with Integer Coefficients**

EXAMPLE 1   Solve $3x - 2(2x - 5) = 2(x + 3) - 8$ and check.

Solution

$$3x - 2(2x - 5) = 2(x + 3) - 8 \qquad \text{Clear parentheses.}$$

$$3x - 4x + 10 = 2x + 6 - 8 \qquad \text{Combine like terms.}$$

$$-x + 10 = 2x - 2 \qquad \text{Isolate } x \text{ on the left side.}$$

$$-x + 10 - 10 = 2x - 2 - 10 \qquad \text{Subtraction property}$$

$$-x = 2x - 12$$

$$-x - 2x = 2x - 12 - 2x \qquad \text{Subtraction property}$$

$$-3x = -12$$

$$\frac{-3x}{-3} = \frac{-12}{-3} \qquad \text{Division property}$$

$$x = 4$$

You should soon be performing these three steps mentally.

We have produced a string of simpler equivalent equations using Theorem 1. Since 4 is a solution to the last equation it must be a solution to the original equation.

Check $$3x - 2(2x - 5) = 2(x + 3) - 8$$
$$3 \cdot 4 - 2(2 \cdot 4 - 5) \overset{?}{=} 2(4 + 3) - 8$$
$$12 - 2 \cdot 3 \overset{?}{=} 2 \cdot 7 - 8$$
$$6 \overset{\checkmark}{=} 6$$

PROBLEM 1

Solution

Solve $8x - 3(x - 4) = 3(x - 4) + 6$ and check.

Check:

If all terms in Example 1 had been transferred to the left side (leaving 0 on the right) and like terms combined we would have obtained

$$-3x + 12 = 0$$

which is a special case of

$$ax + b = 0 \tag{1}$$

Any equation in which both sides are first-degree polynomials in x is equivalent to Equation (1). Such an equation is called a **first-degree equation in one variable**, or a **linear equation**. Such an equation equivalent to (1) with $a \neq 0$ has a unique solution, as can be seen here:

$$ax + b = 0 \qquad a \neq 0$$
$$ax = -b$$
$$x = -\frac{b}{a}$$

If $a = 0$ in Equation (1), then the original equation equivalent to (1) may have no solution at all or an infinite number of solutions. Consider the following equation:

$$2(x - 2) = 4x - (2x + 3)$$
$$2x - 4 = 4x - 2x - 3$$
$$2x - 4 = 2x - 3$$
$$0x = 1 \quad \text{$0x - 1 = 0$ is form $ax + b = 0$.}$$
$$0 = 1 \quad \text{Not possible!}$$

Since $0 = 1$ can never be true, the original equivalent equation has no solution. Now consider the following equation:

$$4x + 2 = 2(2x + 1)$$
$$4x + 2 = 4x + 2$$
$$0x = 0 \quad \text{$0x + 0 = 0$ is form $ax + b$.}$$
$$0 = 0$$

Since $0 = 0$ is always true, the same is true of the original equivalent equation; that is, every real number is a solution and the equation has infinitely many solutions.

In general, it is important to know under what conditions a particular type of equation has a solution and how many solutions are possible. We have now answered both questions for equations equivalent to

$$ax + b = 0 \qquad a \neq 0$$

Each has exactly one solution. Other types of equations will be studied later that have more than one solution. For example,

$$x^2 - 4 = 0 \quad \text{Second-degree equation in one variable}$$

has two solutions: -2 and 2.

■ Equations Involving Fractions: No Variables in Denominators

To solve equations involving fractions with no variables in denominators, we can start by using the multiplication property of equality to clear the fractions and then proceed as above. What do we multiply both sides by to clear the fractions? We use any common multiple, preferably the LCM, of all denominators present in the equation.

EXAMPLE 2   Solve: $\dfrac{x}{3} - \dfrac{1}{2} = \dfrac{5}{6}$

Solution

$$\frac{x}{3} - \frac{1}{2} = \frac{5}{6} \qquad \text{Clear fractions by multiplying both sides by 6,}$$
$$\text{the LCM of all the denominators.}$$

$$6 \cdot \left(\frac{x}{3} - \frac{1}{2} \right) = 6 \cdot \frac{5}{6} \qquad \text{Clear parentheses. Wrong: } \overset{2}{6} \cdot \left(\frac{x}{3} - \frac{1}{2} \right)$$
$$\overset{1}{}$$

$$6 \cdot \frac{x}{3} - 6 \cdot \frac{1}{2} = 6 \cdot \frac{5}{6}$$

$$2x - 3 = 5 \qquad \text{The equation is now free of fractions.}$$
$$2x = 8$$
$$x = 4$$

PROBLEM 2   Solve: $\frac{1}{4}x - \frac{2}{3} = \frac{5}{12}x$

Note: $\frac{1}{4}x = \frac{x}{4}$ and $\frac{5}{12}x = \frac{5x}{12}$

Solution

EXAMPLE 3   Solve: $0.2x + 0.3(x - 5) = 13$

Solution   Some equations involving decimal-fraction coefficients are readily solved by first clearing decimals:

$$0.2x + 0.3(x - 5) = 13 \quad \text{Multiply by 10 to clear decimals.}$$

$$10(0.2x) + 10[0.3(x - 5)] = 10 \cdot 13$$

$$2x + 3(x - 5) = 130$$

$$2x + 3x - 15 = 130$$

$$5x = 145$$

$$x = 29$$

PROBLEM 3   Solve: $0.3(x + 2) + 0.5x = 3$

Solution

EXAMPLE 4   Solve: $5 - \dfrac{2x - 1}{4} = \dfrac{x + 2}{3}$

Solution   Before multiplying both sides by 12, the LCM of the denominators, enclose any numerator with more than one term in parentheses:

$$5 - \frac{(2x - 1)}{4} = \frac{(x + 2)}{3} \qquad \text{Multiply both sides by 12.}$$

$$12 \cdot 5 - \overset{3}{12} \cdot \frac{(2x - 1)}{\underset{1}{4}} = \overset{4}{12} \cdot \frac{(x + 2)}{\underset{1}{3}} \qquad \text{12 is exactly divisible by each denominator.}$$

$$60 - 3(2x - 1) = 4(x + 2)$$

$$60 - 6x + 3 = 4x + 8$$

$$-6x + 63 = 4x + 8$$

$$-10x = -55$$

$$x = \frac{11}{2} \quad \text{or} \quad 5.5$$

PROBLEM 4   Solve: $\dfrac{x + 3}{4} - \dfrac{x - 4}{2} = \dfrac{3}{8}$

Solution

■ A Common Error

A very common error occurs about now—students tend to confuse *algebraic expressions* involving fractions with *algebraic equations* involving fractions. Consider the two problems:

(A) Solve: $\dfrac{x}{2} + \dfrac{x}{3} = 10$ **(B)** Add: $\dfrac{x}{2} + \dfrac{x}{3} + 10$

The problems look very much alike but are actually very different. To solve the equation in (A) we multiply both sides by 6 (the LCM of 2 and 3) to clear the fractions. This works so well for equations, students want to do the same thing for problems like (B). The only catch is that (B) is not an equation and the multiplication property of equality does not apply. If we multiply (B) by 6, we obtain an expression 6 times as large as the original. To add in (B) we find the LCD and proceed as in Section 3-5.

Compare the following:

(A) $\dfrac{x}{2} + \dfrac{x}{3} = 10$ **(B)** $\dfrac{x}{2} + \dfrac{x}{3} + 10$

$$6 \cdot \frac{x}{2} + 6 \cdot \frac{x}{3} = 6 \cdot 10$$ $$= \frac{3 \cdot x}{3 \cdot 2} + \frac{2 \cdot x}{2 \cdot 3} + \frac{6 \cdot 10}{6 \cdot 1}$$

$$3x + 2x = 60$$ $$= \frac{3x}{6} + \frac{2x}{6} + \frac{60}{6}$$

$$5x = 60$$

$$x = 12$$ $$= \frac{5x + 60}{6}$$

■ **Equations Involving Fractions: Variables in Some Denominators**

If an equation involves a variable in one or more denominators, such as

$$\frac{2}{3} - \frac{2}{x} = \frac{4}{x}$$

we may proceed in essentially the same way as above.

But we must avoid any value of x that makes a denominator 0.

EXAMPLE 5   Solve: $\dfrac{2}{3} - \dfrac{2}{x} = \dfrac{4}{x}$

Solution

$$\frac{2}{3} - \frac{2}{x} = \frac{4}{x} \qquad x \neq 0$$

We note that $x \neq 0$, then multiply both sides by $3x$, the LCM of the denominators. If 0 turns up later as a "solution," it must be discarded.

$$3x \cdot \frac{2}{3} - 3x \cdot \frac{2}{x} = 3x \cdot \frac{4}{x}$$

$3x$ is exactly divisible by each denominator.

$$2x - 6 = 12$$
$$2x = 18$$
$$x = 9$$

PROBLEM 5   Solve: $\dfrac{3}{x} - \dfrac{1}{2} = \dfrac{4}{x}$

Solution

EXAMPLE 6   Solve: $\dfrac{3x}{x-2} - 4 = \dfrac{14 - 4x}{x-2}$

Solution

$$\frac{3x}{x-2} - 4 = \frac{14 - 4x}{x-2} \qquad x \neq 2$$

If 2 turns up as a "solution," it must be discarded.

$$(x-2)\frac{3x}{(x-2)} - 4(x-2) = (x-2)\frac{(14-4x)}{(x-2)}$$

Multiply by $(x-2)$, the LCM of the denominators. Also place all binomial numerators and denominators in parentheses.

$$3x - 4(x-2) = 14 - 4x$$
$$3x - 4x + 8 = 14 - 4x$$
$$-x + 8 = 14 - 4x$$
$$3x = 6$$
$$x = 2$$

$x = 2$ cannot be a solution to the original equation (see comments above).

The equation has no solution. (Hence, the solution set is empty.)

PROBLEM 6

Solve: $\dfrac{2x}{x-1} - 3 = \dfrac{7-3x}{x-1}$

Solution

EXAMPLE 7

Solve: $2 - \dfrac{3x}{1-x} = \dfrac{8}{x-1}$

Solution

$$2 - \frac{3x}{1-x} = \frac{8}{x-1} \qquad \text{Recall:} \quad 1-x = -(x-1)$$

$$2 - \frac{3x}{-(x-1)} = \frac{8}{x-1} \qquad \text{Recall:} \quad -\frac{a}{-b} = \frac{a}{b}$$

$$2 + \frac{3x}{x-1} = \frac{8}{x-1} \qquad \text{Multiply both sides by } (x-1),$$
$$\text{keeping in mind that } x \neq 1.$$

$$\boxed{(x-1)(2) + (x-1)\left(\frac{3x}{x-1}\right) = (x-1)\left(\frac{8}{x-1}\right)}$$

$$2x - 2 + 3x = 8$$
$$5x = 10$$
$$x = 2$$

PROBLEM 7

Solve: $\dfrac{5x}{x-2} + \dfrac{10}{2-x} = 3$

Solution

Solutions to Matched
Problems

1. $8x - 3(x - 4) = 3(x - 4) + 6$

$8x - 3x + 12 = 3x - 12 + 6$

$5x + 12 = 3x - 6$

$2x = -18$

$x = -9$

Check: $8(-9) - 3[(-9) - 4] \stackrel{?}{=} 3[(-9) - 4] + 6$

$-72 - 3(-13) \stackrel{?}{=} 3(-13) + 6$

$-72 + 39 \stackrel{?}{=} -39 + 6$

$-33 \stackrel{\vee}{=} -33$

2. $\dfrac{x}{4} - \dfrac{2}{3} = \dfrac{5x}{12}$

$\boxed{12\left(\dfrac{x}{4} - \dfrac{2}{3}\right) = 12\left(\dfrac{5x}{12}\right)}$

$3x - 8 = 5x$

$-2x = 8$

$x = -4$

3. $0.3(x + 2) + 0.5x = 3$

$\boxed{10\,[0.3(x + 2)] + 10(0.5x) = 10(3)}$

$3(x + 2) + 5x = 30$

$3x + 6 + 5x = 30$

$8x = 24$

$x = 3$

4. $\dfrac{x + 3}{4} - \dfrac{x - 4}{2} = \dfrac{3}{8}$

$\boxed{8 \cdot \dfrac{(x + 3)}{4} - 8 \cdot \dfrac{(x - 4)}{2} = 8 \cdot \dfrac{3}{8}}$

$2(x + 3) - 4(x - 4) = 3$

$2x + 6 - 4x + 16 = 3$

$-2x + 22 = 3$

$-2x = -19$

$x = \dfrac{-19}{-2} = \dfrac{19}{2} \quad \text{or} \quad 9.5$

5. $\dfrac{3}{x} - \dfrac{1}{2} = \dfrac{4}{x} \qquad x \neq 0$

$\boxed{2x\left(\dfrac{3}{x} - \dfrac{1}{2}\right) = 2x\left(\dfrac{4}{x}\right)}$

$6 - x = 8$

$-x = 2$

$x = -2$

6. $\dfrac{2x}{x - 1} - 3 = \dfrac{7 - 3x}{x - 1} \qquad x \neq 1$

$\boxed{(x - 1)\left(\dfrac{2x}{x - 1}\right) - 3(x - 1) = (x - 1)\left(\dfrac{7 - 3x}{x - 1}\right)}$

$2x - 3x + 3 = 7 - 3x$

$-x + 3 = 7 - 3x$

$2x = 4$

$x = 2$

7.

$$\frac{5x}{x-2} + \frac{10}{2-x} = 3 \qquad x \neq 2$$

$$\frac{5x}{x-2} + \frac{10}{-(x-2)} = 3$$

$$\frac{5x}{x-2} - \frac{10}{x-2} = 3$$

$$(x-2)\left(\frac{5x}{x-2}\right) - (x-2)\left(\frac{10}{x-2}\right) = (x-2)(3)$$

$$5x - 10 = 3(x-2)$$
$$5x - 10 = 3x - 6$$
$$2x = 4$$
$$x = 2 \qquad \text{This cannot be a solution. The equation has no solution.}$$

Practice Exercise 4-1 ■

Work odd-numbered problems first, check answers, and then work even-numbered problems in areas of weakness. Answers to all problems are in the back of the book. Make every effort to work a problem before you look at an answer.

A Solve and check.

1. $3(x + 2) = 5(x - 6)$ _____   **2.** $5x + 10(x - 2) = 40$ _____

3. $4(x - 2) = 4x - 8$ _____   **4.** $3y + 6 = 3(y + 2)$ _____

5. $5 + 4(t - 2) = 2(t + 7) + 1$ _____

6. $7x - (8x - 4) - 2 = 5 - (4x + 2)$ _____

7. $3x - (x + 2) = 5x - 3(x - 1)$ _____

8. $x - 2(x - 4) = 3x - 2(2x + 1)$ _____

9. $10x + 25(x - 3) = 275$ _____

10. $x + (x + 2) + (x + 4) = 54$ _____

11. $5x - (7x - 4) - 2 = 5 - (3x + 2)$ _____

12. $-3(4 - t) = 5 - (t + 1)$ _____

13. $2(3x + 1) - 8x = 2(1 - x)$ _____

14. $3(4x - 2) - 8x = 6x - 2(x + 3)$ _____

15. $x(x - 1) + 5 = x^2 + x - 3$ _____

16. $x(x + 2) = x(x + 4) - 12$ _____

17. $x(x - 4) - 2 = x^2 - 4(x + 3)$ _____

18. $t(t - 6) + 8 = t^2 - 6t - 3$ _____

B *Solve.*

19. $\dfrac{x}{5} - 2 = \dfrac{3}{5}$ _____

20. $\dfrac{x}{7} - 1 = \dfrac{1}{7}$ _____

21. $\dfrac{x}{3} + \dfrac{x}{6} = 4$ _____

22. $\dfrac{y}{4} + \dfrac{y}{2} = 9$ _____

23. $\dfrac{m}{4} - \dfrac{m}{3} = \dfrac{1}{2}$ _____

24. $\dfrac{n}{5} - \dfrac{n}{6} = \dfrac{6}{5}$ _____

25. $\dfrac{5}{12} - \dfrac{m}{3} = \dfrac{4}{9}$ _____

26. $\dfrac{2}{3} - \dfrac{x}{8} = \dfrac{5}{6}$ _____

27. $0.7x = 21$ _____

28. $0.9x = 540$ _____

29. $0.7x + 0.9x = 32$ _____

30. $0.3x + 0.5x = 24$ _____

31. $\dfrac{1}{2} - \dfrac{2}{x} = \dfrac{3}{x}$ _____

32. $\dfrac{2}{x} - \dfrac{1}{3} = \dfrac{5}{x}$ _____

33. $\dfrac{1}{m} - \dfrac{1}{9} = \dfrac{4}{9} - \dfrac{2}{3m}$ _____

34. $\dfrac{1}{2t} + \dfrac{1}{8} = \dfrac{2}{t} - \dfrac{1}{4}$ _____

35. $\dfrac{x - 2}{3} + 1 = \dfrac{x}{7}$ _____

36. $\dfrac{x + 3}{2} - \dfrac{x}{3} = 4$ _____

37. $\dfrac{2x - 3}{9} - \dfrac{x + 5}{6} = \dfrac{3 - x}{2} - 1$ _____

38. $\dfrac{3x + 4}{3} - \dfrac{x - 2}{5} = \dfrac{2 - x}{15} - 1$ _____

39. $0.1(x - 7) + 0.05x = 0.8$ _____

40. $0.4(x + 5) - 0.3x = 17$ _____

41. $0.02x - 0.5(x - 2) = 5.32$ _____

42. $0.3x - 0.04(x + 1) = 2.04$ _____

43. $\dfrac{7}{y - 2} - \dfrac{1}{2} = 3$ _____

44. $\dfrac{9}{A + 1} - 1 = \dfrac{12}{A + 1}$ _____

45. $\dfrac{3}{2x - 1} + 4 = \dfrac{6x}{2x - 1}$ _____

46. $\dfrac{5x}{x + 5} = 2 - \dfrac{25}{x + 5}$ _____

47. $\dfrac{2E}{E - 1} = 2 + \dfrac{5}{2E}$ _____

48. $\dfrac{3N}{N - 2} - \dfrac{9}{4N} = 3$ _____

49. $\dfrac{n - 5}{6n - 6} = \dfrac{1}{9} - \dfrac{n - 3}{4n - 4}$ _____

50. $\dfrac{1}{3} - \dfrac{s - 2}{2s + 4} = \dfrac{s + 2}{3s + 6}$ _____

51. $5 + \dfrac{2x}{x - 3} = \dfrac{6}{x - 3}$ _____

52. $\dfrac{6}{x - 2} = 3 + \dfrac{3x}{x - 2}$ _____

53. $\dfrac{x^2 + 2}{x^2 - 4} = \dfrac{x}{x - 2}$ _____

54. $\dfrac{5}{x - 3} = \dfrac{33 - x}{x^2 - 6x + 9}$ _____

C *Solve.*

55. $\dfrac{3x}{24} - \dfrac{2 - x}{10} = \dfrac{5 + x}{40} - \dfrac{1}{15}$ _____

56. $\dfrac{2x}{10} - \dfrac{3 - x}{14} = \dfrac{2 + x}{5} - \dfrac{1}{2}$ _____

57. $\dfrac{5t - 22}{t^2 - 6t + 9} - \dfrac{11}{t^2 - 3t} - \dfrac{5}{t} = 0$ _____

58. $\dfrac{x - 33}{x^2 - 6x + 9} + \dfrac{5}{x - 3} = 0$ _____

59. $5 - \dfrac{2x}{3 - x} = \dfrac{6}{x - 3}$ _____

60. $\dfrac{3x}{2 - x} + \dfrac{6}{x - 2} = 3$ _____

61. $\dfrac{1}{c^2 - c - 2} - \dfrac{3}{c^2 - 2c - 3} = \dfrac{1}{c^2 - 5c + 6}$ _____

The Check Exercise for
this section is on
page 253.

62. $\dfrac{5t - 22}{t^2 - 6t + 9} - \dfrac{11}{t^2 - 3t} = \dfrac{5}{t}$ _____

Section 4-2 Applications: Number, Geometric, and Proportion Problems

■ A Strategy for Solving Word Problems
■ Number Problems
■ Geometric Problems
■ Ratio and Proportion Problems

We are now ready to solve a wide variety of interesting and useful word problems and applications. In this section we introduce a strategy for solving word problems and use it to solve relatively easy number, geometric, and ratio and proportion problems. After gaining some experience in setting up and solving these easier problems, the next two sections will make use of the strategy in solving problems of a more challenging nature.

■ A Strategy for Solving Word Problems

A great many practical problems can be solved using algebraic techniques—so many, in fact, there is no one method of attack that will work for all. However, we can formulate a strategy that may help you organize your approach.

A Strategy for Solving Word Problems

1. Read the problem carefully—several times if necessary—until you understand the problem, know what is to be found, and know what is given.
2. If appropriate, draw figures or diagrams and label known and unknown parts.
3. Look for formulas connecting the known quantities with the unknown quantities.
4. Let one of the unknown quantities be represented by a variable, say x, and try to represent all other unknown quantities in terms of x. This is an important step and must be done carefully. Be sure you clearly understand what you are letting x represent.
5. Form an equation relating the unknown quantities with the known quantities. This step may involve the translation of an English sentence into an algebraic sentence, the use of relationships in a geometric figure, the use of certain formulas, and so on.
6. Solve the equation and write answers to *all* parts of the problem requested.
7. Check all solutions in the original problem.

■ Number Problems

In earlier sections you had experience in translating verbal forms into symbolic forms. We now take advantage of that experience to solve a variety of number problems.

EXAMPLE 8   Find a number such that 6 more than one-half the number is two-thirds the number.

Solution   Let

   x = The number

We symbolize each part of the problem as follows:

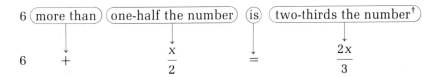

$$6 \quad + \quad \frac{x}{2} \quad = \quad \frac{2x}{3}$$

We now solve the equation as in the last section:

$$6 + \frac{x}{2} = \frac{2x}{3}$$

$$\boxed{\; 6 \cdot 6 + 6 \cdot \frac{x}{2} = 6 \cdot \frac{2x}{3} \;} \quad \text{Clear fractions.}$$

$$36 + 3x = 4x$$

$$-x = -36$$

$$x = 36$$

Check   6 more than one-half the number: $6 + \frac{36}{2} = 6 + 18 = 24$

Two-thirds the number: $\frac{2}{3}(36) = 2 \cdot 12 = 24$

PROBLEM 8   Find a number such that 10 less than two-thirds the number is one-fourth the number. Write an equation and solve.

Solution

---

† If x is a number, then two-thirds x can be written $\frac{2}{3}x$ or $\frac{2x}{3}$, since $\frac{2}{3}x = \frac{2}{3} \cdot \frac{x}{1} = \frac{2x}{3}$. The latter form will be more convenient for our purposes.

EXAMPLE 9   Find a number such that 2 less than twice the number is 5 times the quantity that is 2 more than the number.

Solution   Let

$$x = \text{The number}$$

Symbolize each part:

2 less than twice the number: $2x - 2$ Not $2 - 2x$

is: $=$

5 times the quantity
that is 2 more than the number: $5(x + 2)$ Why would $5x + 2$ be incorrect?

Write an equation and solve:

$$2x - 2 = 5(x + 2)$$
$$2x - 2 = 5x + 10$$
$$-3x = 12$$
$$x = -4$$

Checking is left to you.

PROBLEM 9   Find a number such that 4 times the quantity that is 2 less than the number is 1 more than 3 times the number. Write an equation and solve.

Solution

EXAMPLE 10   Find three consecutive even numbers such that twice the first plus the third is 10 more than the second.

Solution   Let

$$x = \text{First of three consecutive even numbers}$$
$$x + 2 = \text{Second consecutive even number}$$
$$x + 4 = \text{Third consecutive even number}$$

The difference between any two consecutive even numbers is 2.

Form an equation and solve:

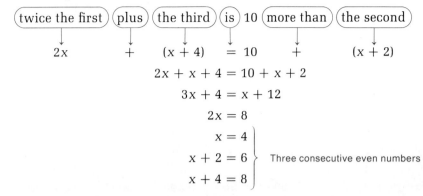

$$2x + x + 4 = 10 + x + 2$$
$$3x + 4 = x + 12$$
$$2x = 8$$
$$\left. \begin{array}{l} x = 4 \\ x + 2 = 6 \\ x + 4 = 8 \end{array} \right\} \text{ Three consecutive even numbers}$$

Thus, the three consecutive even numbers are 4, 6, and 8.

Check   Twice the first plus the third: $2 \cdot 4 + 8 = 8 + 8 = 16$

  10 more than the second: $10 + 6 = 16$

PROBLEM 10   Find three consecutive odd numbers such that 4 times the first minus the third is the same as the second. Write an equation and solve.

Solution

■ Geometric Problems

Recall that the **perimeter of a triangle or rectangle** is the distance around the figure. Symbolically:

TRIANGLE

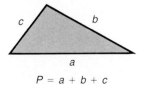

$P = a + b + c$

RECTANGLE

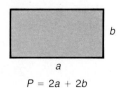

$P = 2a + 2b$

EXAMPLE 11   If one side of a triangle is one-fourth the perimeter, the second side is 7 meters, and the third side is two-fifths the perimeter, what is the perimeter?

Solution   Let

$$P = \text{Perimeter}$$

Draw a triangle and label sides, as shown. Thus,

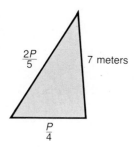

$$P = a + b + c$$

$$P = \frac{P}{4} + 7 + \frac{2P}{5}$$

$$20 \cdot P = 20 \cdot \frac{P}{4} + 20 \cdot 7 + 20 \cdot \frac{2P}{5} \quad \text{Clear fractions.}$$

$$20P = 5P + 140 + 8P$$

$$7P = 140$$

$$P = 20 \text{ meters}$$

Check

$$\text{Side 1} = \frac{P}{4} = \frac{20}{4} = \underline{} 5 \text{ meters}$$

$$\text{Side 2} = \underline{} 7 \text{ meters}$$

$$\text{Side 3} = \frac{2P}{5} = \frac{2 \cdot 20}{5} = \underline{} 8 \text{ meters}$$

$$\overline{20 \text{ meters}} \quad \text{Perimeter}$$

PROBLEM 11   If one side of a triangle is one-third the perimeter, the second side is 7 centimeters, and the third side is one-fifth the perimeter, what is the perimeter of the triangle? Set up an equation and solve.

Solution

EXAMPLE 12   Find the dimensions of a rectangle with perimeter 84 centimeters if its width is two-fifths the length.

Solution   Draw a rectangle and label sides. Let

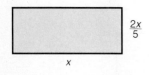

$$x = \text{Length}$$

Begin with the formula for the perimeter of a rectangle:

$$2a + 2b = P$$

$$2x + 2 \cdot \frac{2x}{5} = 84$$

$$2x + \frac{4x}{5} = 84$$

$$\boxed{5 \cdot 2x + 5 \cdot \frac{4x}{5} = 5 \cdot 84} \qquad \text{Clear fractions.}$$

$$10x + 4x = 420$$

$$14x = 420$$

$$x = 30 \text{ centimeters} \quad \text{Length}$$

$$\frac{2x}{5} = 12 \text{ centimeters} \quad \text{Width}$$

Checking is left to you.

PROBLEM 12   Find the dimensions of a rectangle with perimeter 176 centimeters if its width is three-eighths its length. Write an equation and solve.

Solution

■ **Ratio and Proportion Problems**

The ratio of two quantities is the first quantity divided by the second quantity. Symbolically:

The Ratio of a to b
The ratio of a to b, $b \neq 0$, is $\dfrac{a}{b}$.

EXAMPLE 13   If a parking meter has 45 nickels, 30 dimes, and 15 quarters, then the ratio of nickels to quarters is

$$\frac{45}{15} = \frac{3}{1}$$

(which is also written 3:1 or 3/1 and is read "3 to 1").

PROBLEM 13   In Example 13 what is the ratio of quarters to nickels? Of dimes to nickels?

Solution   Quarters to nickels: Dimes to nickels:

In addition to providing a way of comparing known quantities, ratios also provide a way of finding unknown quantities.

EXAMPLE 14   Suppose you are told that the ratio of quarters to dimes in a parking meter is 3/5 and that there are 40 dimes in the meter. How many quarters are in the meter?

Solution   Let

$$q = \text{Number of quarters}$$

Then the ratio of quarters to dimes is $q/40$. Thus,

$$\frac{q}{40} = \frac{3}{5} \qquad \text{To isolate } q, \text{ multiply both sides by 40.}$$

$$q = 40 \cdot \frac{3}{5}$$

$$q = 24 \text{ quarters}$$

PROBLEM 14   If the ratio of dimes to quarters in a meter is 3/2 and there are 24 quarters in the meter, how many dimes are there?

Solution

A statement of equality between two ratios, as in Example 14, is called a **proportion**; that is,

A Proportion
$$\frac{a}{b} = \frac{c}{d} \qquad b, d \neq 0$$

EXAMPLE 15 | If a car can travel 192 kilometers on 32 liters of gas, how far will it go on 60 liters?

Solution | Let

$$x = \text{Distance traveled on 60 liters}$$

Then

$$\frac{x}{60} = \frac{192}{32} \qquad \frac{\text{km}}{\text{L}} = \frac{\text{km}}{\text{L}} \quad \text{(kilometers per liter)}$$

$$x = 60 \cdot \frac{192}{32} \qquad \text{We isolate } x \text{ by multiplying both sides by 60—we do not need to use the LCM of 60 and 32.}$$

$$= 360 \text{ kilometers}$$

PROBLEM 15 | If there are 24 milliliters of sulfuric acid in 64 milliliters of solution, how many milliliters are in 48 milliliters of the same solution? Set up a proportion and solve.

Solution |

We will show how the concept of proportion can be used to convert metric to English and vice versa. (A summary of metric units is located inside the back cover of the text.)

EXAMPLE 16 | If there is 0.45 kilogram in 1 pound, how many pounds are in 90 kilograms?

Solution | Let x = Number of pounds in 90 kilograms. Set up a proportion (preferably with x in the numerator on the left side, since you have a choice). That is, set up a proportion of the form

$$\frac{\text{Pounds}}{\text{Kilograms}} = \frac{\text{Pounds}}{\text{Kilograms}} \qquad \text{Each ratio represents pounds per kilogram.}$$

Thus

$$\frac{x}{90} = \frac{1}{0.45}$$

$$x = 90 \cdot \frac{1}{0.45}$$

$$x = 200 \text{ pounds}$$

PROBLEM 16   If there are 2.2 pounds in 1 kilogram, how many kilograms are in 100 pounds? Set up a proportion and solve to two decimal places.

Solution

EXAMPLE 17   If there are 3.76 liters in 1 gallon, how many gallons are in 50 liters? Set up a proportion (with the variable in the numerator on the left side) and solve to two decimal places.

Solution   Let x = Number of gallons in 50 liters. We set up a proportion of the form

$$\frac{\text{Gallons}}{\text{Liters}} = \frac{\text{Gallons}}{\text{Liters}}$$ Each side gives gallons per liter.

Thus,

$$\frac{x}{50} = \frac{1}{3.76}$$

$$x = 50 \cdot \frac{1}{3.76}$$

$$x = 13.30 \text{ gallons}$$

PROBLEM 17   If there are 1.09 yards in 1 meter, how many meters are in 80 yards? Set up a proportion and solve to two decimal places.

Solution

SUGGESTION   If you are having trouble with word problems (many people do), return to the worked-out examples. Cover the solutions, proceed with your own solution until you get stuck, and then uncover only enough of the solution to get you started again. After completing an example in this way, immediately work the matched problem following the example and then work similar problems in the exercise set.

Solutions to Matched Problems

Checks are left to the reader.

8. Let x = The number

$$\frac{2x}{3} - 10 = \frac{x}{4}$$

$$\boxed{12 \cdot \frac{2x}{3} - 12 \cdot 10 = 12 \cdot \frac{x}{4}}$$

$$8x - 120 = 3x$$
$$5x = 120$$
$$x = 24$$

9. Let x = The number

$$4(x - 2) = 3x + 1$$
$$4x - 8 = 3x + 1$$
$$x = 9$$

10. Let x, x + 2, and x + 4 represent three consecutive odd numbers.

$$4x - (x + 4) = x + 2$$
$$4x - x - 4 = x + 2$$
$$3x - 4 = x + 2$$
$$2x = 6$$

$$\left.\begin{array}{l} x = 3 \\ x + 2 = 5 \\ x + 4 = 7 \end{array}\right\} \begin{array}{l}\text{Three consecutive} \\ \text{odd numbers}\end{array}$$

11.

$$\frac{P}{3} + 7 + \frac{P}{5} = P$$

$$\boxed{15 \cdot \frac{P}{3} + 15 \cdot 7 + 15 \cdot \frac{P}{5} = 15P}$$

$$5P + 105 + 3P = 15P$$
$$8P + 105 = 15P$$
$$-7P = -105$$
$$P = 15 \text{ centimeters}$$

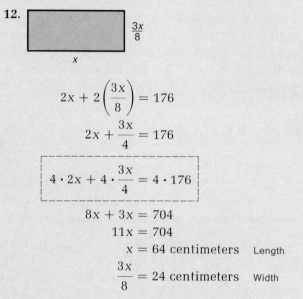

12.

$$2x + 2\left(\frac{3x}{8}\right) = 176$$

$$2x + \frac{3x}{4} = 176$$

$$4 \cdot 2x + 4 \cdot \frac{3x}{4} = 4 \cdot 176$$

$$8x + 3x = 704$$

$$11x = 704$$

$$x = 64 \text{ centimeters} \quad \text{Length}$$

$$\frac{3x}{8} = 24 \text{ centimeters} \quad \text{Width}$$

13. Quarters to nickels: $\frac{15}{45} = \frac{1}{3}$;   Dimes to nickels: $\frac{30}{45} = \frac{2}{3}$

14. Let x = Number of dimes

$$\frac{x}{24} = \frac{3}{2}$$

$$x = 24 \cdot \frac{3}{2} = 36 \text{ dimes}$$

15. Let x = Amount of acid

$$\frac{x}{48} = \frac{24}{64}$$

$$x = 48 \cdot \frac{24}{64} = 18 \text{ milliliters}$$

16. Let x = Number of kilograms

$$\frac{x}{100} = \frac{1}{2.2}$$

$$x = 100 \cdot \frac{1}{2.2}$$

$$\approx 45.45 \text{ kilograms}$$

17. Let x = Number of meters

$$\frac{x}{80} = \frac{1}{1.09}$$

$$x = 80 \cdot \frac{1}{1.09}$$

$$\approx 73.39 \text{ meters}$$

Practice Exercise 4-2 ■

Work odd-numbered problems first, check answers, and then work even-numbered problems in areas of weakness. Answers to all problems are in the back of the book. Make every effort to work a problem yourself before you look at an answer.

A   *If x represents a number, write an algebraic expression for each of the following numbers.*

1. Twice x _____

2. 3 times x _____

3. 3 less than x _____

4. 2 more than x _____

5. One-third x _____

6. One-fifth x _____

7. Three-fourths x _____

8. Two-fifths x _____

9. 5 less than two-thirds x _____

10. 11 less than three-fourths x _____

Write as a ratio.

11. 33 dimes to 22 nickels _____

12. 17 quarters to 51 dimes _____

13. 25 centimeters to 10 centimeters _____

14. 30 meters to 18 meters _____

15. 300 kilometers to 24 liters _____

16. 320 miles to 12 gallons _____

Solve each proportion.

17. $\dfrac{m}{16} = \dfrac{5}{4}$ _____   **18.** $\dfrac{n}{12} = \dfrac{2}{3}$ _____

19. $\dfrac{x}{13} = \dfrac{21}{39}$ _____   **20.** $\dfrac{x}{12} = \dfrac{27}{18}$ _____

Number Problems *Find numbers meeting each of the indicated conditions. Write an equation using x and solve.*

21. 7 times a number is 12 less than 4 times the number. _____

22. 6 times a number is 24 more than 3 times the number. _____

23. 3 more than one-sixth the number is $\frac{2}{3}$. _____

24. 2 more than one-fourth the number is $\frac{1}{2}$. _____

25. Three consecutive integers whose sum is 96 _____

26. Three consecutive integers whose sum is 78 _____

27. Three consecutive even numbers whose sum is 42 _____

28. Three consecutive even numbers whose sum is 54 _____

Geometric Problems *Set up appropriate equations and solve.*

29. A 12-foot steel rod is cut into two pieces so that one piece is 3 feet less than twice the length of the other piece. How long is each piece? _____

30. A 32-centimeter string is cut into two pieces so that one piece is 4 centimeters more than 3 times the length of the other piece. How long is each piece? _____

31. Find the dimensions of a rectangle with perimeter 36 feet if the width is 6 feet shorter than the length. _____

32. Find the dimensions of a rectangle with perimeter 54 meters if the length is 7 meters longer than the width. _____

Ratio–Proportion Problems

Set up appropriate proportions and solve. Compute decimal answers to two decimal places.

33. If in a pay telephone the ratio of quarters to dimes is 5/8 and there are 96 dimes, how many quarters are there? _____

34. If in a parking meter the ratio of pennies to nickels is 13/6 and there are 78 nickels, how many pennies are there? _____

35. If the ratio of the length of a rectangle to its width is 5/3 and its width is 24 meters, how long is it? _____

36. If the ratio of the width of a rectangle to its length is 4/7 and its length is 56 centimeters, how wide is it? _____

37. If a car can travel 108 kilometers on 12 liters of gas, how far will it go on 18 liters? _____

38. If a boat can travel 72 miles on 18 gallons of diesel fuel, how far will it travel on 15 gallons? _____

B   *The following number, geometric, and ratio and proportion problems are mixed together. Set up appropriate equations and solve. Compute decimal answers to two decimal places.*

39. Find the dimensions of a rectangle with perimeter 66 centimeters if its length is 3 centimeters more than twice its width. _____

40. Find the dimensions of a rectangle with perimeter 128 meters if its length is 6 meters less than 4 times the width. _____

41. Find a number such that 2 less than one-sixth the number is 1 more than one-fourth the number. _____

42. Find a number such that 5 less than half the number is 3 more than one-third the number. _____

43. If there are 9 milliliters of hydrochloric acid in 46 milliliters of solution, how many milliliters will be in 52 milliliters of solution? _____

44. If 0.75 cup of flour is needed in a recipe that will feed 6 people, how much flour will be needed in the recipe that will feed 9 people? _____

45. Find three consecutive odd numbers such that the sum of the first and second is 5 more than the third. _____

46. Find three consecutive odd numbers such that the sum of the second and third is 1 more than 3 times the first. _____

47. If there are 1.06 quarts in 1 liter, how many liters are in 1 gallon (4 quarts)?

_____

48. If there are 2.2 pounds in 1 kilogram, how many kilograms are in 10 pounds? _____

49. If there is 0.62 mile in 1 kilometer, how many kilometers are in 1 mile?

_____

50. If there is 0.91 meter in 1 yard, how many yards are in 1 meter?

_____

51. Find the dimensions of a rectangle with perimeter 84 meters if its width is one-sixth its length. _____

52. Find the dimensions of a rectangle with perimeter 72 centimeters if its width is one-third its length. _____

53. If a commission of $240 is charged on the purchase of 200 shares of a stock, how much commission would be charged for 500 shares of the same stock? _____

54. If the price/earnings ratio of a common stock is 8.4 and the stock earns $23.50 per share, what is the price of the stock per share? [*Note:* Ratios are often written as decimal fractions. In this case 8.4 (that is, 8.4/1) is the ratio of price per share to earnings per share. Thus, if x is the price per share, we obtain the proportion x/23.5 = 8.4.] _____

55. A 35 by 23-millimeter colored slide is used to make an enlargement whose longer side is 10 inches. How wide will the enlargement be if all of the slide is used? _____

56. A 3.25 by 4.25-inch negative is used to produce an enlargement whose shortest side is 12 inches. How long will the enlargement be if all of the negative is used? _____

57. Find a number such that 4 less than three-fifths the number is 8 more than one-third the number. _____

58. Find a number such that 5 more than two-thirds the number is 10 less than one-fourth the number. _____

59. If there is 0.26 gallon in 1 liter, how many liters are in 5 gallons?

60. If there is 0.94 liter in 1 quart, how many quarts are in 10 liters?

C   **61.** If one side of a triangle is two-fifths the perimeter P, the second side is 70 centimeters, and the third side is one-fourth the perimeter, what is the perimeter? _____

62. If one side of a triangle is one-fourth the perimeter P, the second side is 3 meters, and the third side is one-third the perimeter, what is the perimeter? _____

63. Estimate the total number of trout in a lake if a sample of 300 is netted, marked, and released, and after a suitable period for mixing, a second sample of 250 produces 25 marked trout. (Assume that the ratio of the marked trout in the second sample to the total number of the sample is the same as the ratio of those marked in the first sample to the total lake population.) _____

64. Repeat the last problem with a first (marked) sample of 400 and a second sample of 264 with only 24 marked trout. _____

65. On a trip across the Grand Canyon in Arizona, a group traveled one-third the distance by mule, 6 kilometers by boat, and one-half the distance by foot. How long was the trip? _____

66. A high diving tower is located in a lake. If one-fifth the height of the tower is in sand, 6 meters in water, and one-half the total height in air, what is the total height of the tower? _____

67. If in the figure on page 205 the diameter of the smaller pipe is 12 millimeters and the diameter of the larger pipe is 24 centimeters, how much force would be required to lift a 1,200-kilogram car? (Neglect the weight of the hydraulic lift equipment and use the proportion shown in the figure.)

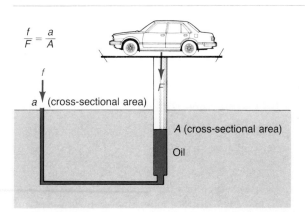

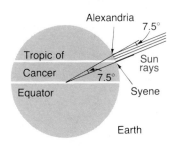

The Check Exercise for this section is on page 255.

68. Do you have any idea how one might measure the circumference of the earth? In 240 B.C. Eratosthenes measured the size of the earth from its curvature. At Syene, Egypt (lying on the Tropic of Cancer), the sun was directly overhead at noon on June 21. At the same time in Alexandria, a town 500 miles directly north, the sun's rays fell at an angle of 7.5° to the vertical. Using this information and a little knowledge of geometry (see the figure), Eratosthenes was able to approximate the circumference of the earth using the following proportion: the circumference of the earth C is to 500 as 360 is to 7.5. Compute Eratosthenes' estimate.

Section 4-3   Applications: Rate–Time and Mixture Problems

■ Rate–Time Problems
■ Mixture Problems

■ Rate–Time Problems

There are many types of rate–time problems in addition to the distance–rate–time problems with which you are probably familiar. In this section we will look at rate–time problems as a general class of problems that includes distance–rate–time problems as one of many special cases.

If a runner travels 20 kilometers in 2 hours, then the ratio

$$\frac{20 \text{ kilometers}}{2 \text{ hours}} \qquad \text{or} \qquad 10 \text{ kilometers per hour}$$

is called a **rate**. It is the number of kilometers produced (covered) in each unit of time (each hour). Similarly, if an automatic bottling machine bottles 3,200 bottles of soft drinks in 20 minutes, then the ratio

$$\frac{3,200 \text{ bottles}}{20 \text{ minutes}} = 160 \text{ bottles per minute}$$

is also a rate. It is the number of bottles produced in each unit of time (each minute).

In general, if Q is the quantity of something produced (kilometers, words, parts, and so on) in T units of time (hours, years, minutes, seconds, and so on), and R is the rate (quantity produced in one unit of time), then Q, T, and R are related by the following formula:

Quantity–Rate[†]–Time Formula

$Q = RT$ Quantity = (Rate)(Time)   (1)

If Q is distance D, then

$D = RT$ Distance = (Rate)(Time)   (2)

Formulas (1) and (2) should be memorized. [Formula (2) is simply a special case of (1).] Two other forms of Formulas (1) and (2) (one for rate R and the other for time T) are easily derived from (1) and (2) by dividing both sides of each by either T or R:

$$Q = RT \qquad\qquad D = RT$$
$$RT = Q \qquad\qquad RT = D$$
$$T = \frac{Q}{R} \qquad R = \frac{Q}{T} \qquad T = \frac{D}{R} \qquad R = \frac{D}{T}$$

EXAMPLE 18   A gas pump in a service station can deliver 36 liters in 4 minutes.

(A) What is its rate (liters per minute)?
(B) How much gas can be delivered in 5 minutes?
(C) How long will it take to deliver 54 liters?

Solution   **(A)** $R = \dfrac{Q}{T} = \dfrac{36 \text{ liters}}{4 \text{ minutes}} = 9$ liters per minute

(B) $Q = RT = (9 \text{ liters per minute})(5 \text{ minutes}) = 45$ liters

(C) $T = \dfrac{Q}{R} = \dfrac{54 \text{ liters}}{9 \text{ liters per minute}} = 6$ minutes

PROBLEM 18   A woman jogs 18 kilometers in 2 hours.

(A) What is her rate (kilometers per hour)?
(B) How far will she jog in 1.5 hours?
(C) How long will it take her to jog 12 kilometers?

Solution   **(A)**   **(B)**   **(C)**

[†] These are also referred to as **average rates**.

EXAMPLE 19  A jet plane leaves San Francisco and travels at 650 kilometers per hour toward Los Angeles. At the same time another plane leaves Los Angeles and travels at 800 kilometers per hour toward San Francisco. If the cities are 570 kilometers apart, how long will it take the jets to meet, and how far from San Francisco will they be at that time?

Solution Let T = Number of hours until both planes meet. Draw a diagram and label known and unknown parts. Both planes will have traveled the same amount of time when they meet.

$$650 \text{ km/hr} \longrightarrow \quad \overset{\text{Meeting}}{\underset{\times}{\text{point}}} \quad \longleftarrow 800 \text{ km/hr}$$

SF $\quad D_1 = 650T \quad\quad\quad\quad D_2 = 800T \quad$ LA

$$\begin{pmatrix} \text{Distance plane} \\ \text{from SF travels} \\ \text{to meeting point} \end{pmatrix} + \begin{pmatrix} \text{Distance plane} \\ \text{from LA travels} \\ \text{to meeting point} \end{pmatrix} = \begin{pmatrix} \text{Total distance} \\ \text{from} \\ \text{SF to LA} \end{pmatrix}$$

$$D_1 \quad\quad + \quad\quad D_2 \quad\quad = \quad\quad 570$$
$$650T \quad\quad + \quad\quad 800T \quad\quad = \quad\quad 570$$
$$1{,}450T = 570$$
$$T = \frac{570}{1{,}450} \approx 0.39 \text{ hour}$$

Distance from SF = (Rate from SF)(Time from SF)

$$\approx (650)(0.39) = 253.5 \text{ kilometers}$$

PROBLEM 19 If an older printing press can print 45 fliers per minute and a newer press can print 80, how long will it take both presses together to print 4,500 fliers? How many will the older press have printed by then? (Note that mathematically Problem 19 and Example 19 are the same.)

Solution

EXAMPLE 20 Find the total amount of time to print the fliers in Problem 19 if the newer press is brought on the job 10 minutes later than the older press and both continue until the job is completed.

Solution   Let

$$x = \text{Time to complete whole job}$$

Then

$$x = \text{Time old press is on the job}$$

$$x - 10 = \text{Time new press is on the job}$$

$$\begin{pmatrix} \text{Quantity} \\ \text{printed} \\ \text{by old press} \end{pmatrix} + \begin{pmatrix} \text{Quantity} \\ \text{printed} \\ \text{by new press} \end{pmatrix} = \text{Total needed}$$ *Remember:*
Quantity = (Rate)(Time)

$$45x \quad + \quad 80(x - 10) \quad = 4{,}500$$

$$45x + 80x - 800 = 4{,}500$$

$$125x = 5{,}300$$

$$x = 42.4 \text{ minutes}$$

PROBLEM 20   A car leaves a city traveling at 60 kilometers per hour. How long will it take a second car traveling at 80 kilometers per hour to catch up to the first car if it leaves 2 hours later? Set up an equation and solve. [*Hint:* Both cars will have traveled the same distance when the second car catches up to the first.]

Solution

EXAMPLE 21   A speedboat takes 1.5 times longer to go 120 miles up a river than to return. If the boat cruises at 25 miles per hour in still water, what is the rate of the current?

Solution   Let

$$x = \text{Rate of current in miles per hour}$$

$$25 - x = \text{Rate of boat upstream}$$

$$25 + x = \text{Rate of boat downstream}$$

$$\text{Time upstream} = (1.5)(\text{Time downstream})$$

$$\frac{\text{Distance upstream}}{\text{Rate upstream}} = (1.5)\frac{\text{Distance downstream}}{\text{Rate downstream}}$$ *Recall that T = D/R from D = RT.*

$$\frac{120}{25 - x} = (1.5)\frac{120}{25 + x}$$

$$\frac{120}{25 - x} = \frac{180}{25 + x}$$

$$(25 + x)120 = (25 - x)180$$

$$3{,}000 + 120x = 4{,}500 - 180x$$

$$300x = 1{,}500$$

$$x = 5 \text{ miles per hour} \quad \text{Rate of current}$$

Check

$$\text{Time upstream} = \frac{\text{Distance upstream}}{\text{Rate upstream}} = \frac{120}{20} = 6 \text{ hours}$$

$$\text{Time downstream} = \frac{\text{Distance downstream}}{\text{Rate downstream}} = \frac{120}{30} = 4 \text{ hours}$$

Thus, time upstream is 1.5 times longer than time downstream.

PROBLEM 21

A fishing boat takes twice as long to go 24 miles up a river than to return. If the boat cruises at 9 miles per hour in still water, what is the rate of the current?

Solution

EXAMPLE 22

An advertising company has an automated mailing machine that can fold, stuff, and address a particular mailing in 6 hours. With the help of a newer machine the job can be completed in 2 hours. How long would it take the new machine to do the job alone?

Solution

Let

$$x = \text{Time for new machine to do the job alone}$$

If a job can be completed in 6 hours (old machine), then the rate of completion is 1/6 job per hour. If a job can be completed in 2 hours (both machines together), then the rate of completion is 1/2 job per hour. If a job can be completed in x hours (new machine alone), then the rate of completion is 1/x job per hour. Thus,

$$\begin{pmatrix} \text{Rate of old} \\ \text{machine} \end{pmatrix} + \begin{pmatrix} \text{Rate of new} \\ \text{machine} \end{pmatrix} = \begin{pmatrix} \text{Rate} \\ \text{together} \end{pmatrix}$$

$$\frac{1}{6} \qquad + \qquad \frac{1}{x} \qquad = \qquad \frac{1}{2} \qquad \text{Multiply by 6x.}$$

$$x + 6 = 3x$$

$$2x = 6$$

$$x = 3 \text{ hours} \quad \text{New machine alone}$$

Check

$$\frac{1}{6} + \frac{1}{3} = \frac{1}{6} + \frac{2}{6} = \frac{3}{6} = \frac{1}{2}$$

PROBLEM 22   At a family cabin, water is pumped and stored in a large water tank. Two pumps are used for this purpose. One can fill the tank by itself in 6 hours, and the other can do the job in 9 hours. How long will it take both pumps operating together to fill the tank?

Solution

■ Mixture Problems

A variety of applications can be classified as mixture problems. And even though the problems come from different areas, their mathematical treatment is essentially the same.

EXAMPLE 23   A concert brought in $27,200 on the sale of 4,000 tickets. If tickets sold for $5 and $8, how many of each were sold?

Solution   Let

$$x = \text{The } number \text{ of \$5 tickets sold}$$

Then

$$4,000 - x = \text{The } number \text{ of \$8 tickets sold}$$

We now form an equation using the value of the tickets before and after mixing:

$$\text{Value before mixing} = \text{Value after mixing}$$

$$\begin{pmatrix} \text{Value of} \\ \text{\$5 tickets} \\ \text{sold} \end{pmatrix} + \begin{pmatrix} \text{Value of} \\ \text{\$8 tickets} \\ \text{sold} \end{pmatrix} = \begin{pmatrix} \text{Total value} \\ \text{of all} \\ \text{tickets sold} \end{pmatrix}$$

$$5x \qquad + 8(4,000 - x) = \qquad 27,200$$

$$5x + 32,000 - 8x = 27,200$$

$$-3x = -4,800$$

$$x = 1,600 \quad \text{\$5 tickets}$$

$$4,000 - x = 2,400 \quad \text{\$8 tickets}$$

Check $(\$5)(1,600) + (\$8)(2,400) = \$27,200$

PROBLEM 23   Suppose you receive 40 nickels and quarters in change worth $4. How many of each type of coin do you have?

Solution

Let us now consider some mixture problems involving percent. Recall that 23% in decimal form is 0.23, 6.5% is 0.065, and so on.

EXAMPLE 24   How many milliliters of distilled water must be added to 60 milliliters of 70% acid solution to obtain a 60% solution? [*Note:*   A 70% acid solution is 70% pure acid and 30% distilled water.]

Solution   Let x = Number of milliliters of distilled water added. We illustrate the situation before and after mixing, keeping in mind that the amount of acid present before mixing must equal the amount of acid present after mixing.

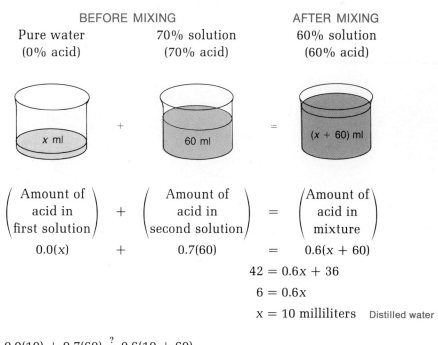

BEFORE MIXING

AFTER MIXING

Pure water (0% acid)   70% solution (70% acid)   60% solution (60% acid)

$$\begin{pmatrix} \text{Amount of} \\ \text{acid in} \\ \text{first solution} \end{pmatrix} + \begin{pmatrix} \text{Amount of} \\ \text{acid in} \\ \text{second solution} \end{pmatrix} = \begin{pmatrix} \text{Amount of} \\ \text{acid in} \\ \text{mixture} \end{pmatrix}$$

$$0.0(x) + 0.7(60) = 0.6(x + 60)$$

$$42 = 0.6x + 36$$

$$6 = 0.6x$$

$$x = 10 \text{ milliliters} \quad \text{Distilled water}$$

Check $0.0(10) + 0.7(60) \overset{?}{=} 0.6(10 + 60)$

$42 \overset{\checkmark}{=} 42$

PROBLEM 24   How many centiliters of pure alcohol must be added to 35 centiliters of a 20% solution to obtain a 30% solution?

Solution

EXAMPLE 25   A chemical storeroom has a 20% alcohol solution and a 50% alcohol solution. How many centiliters must be taken from each to obtain 24 centiliters of a 30% solution?

Solution   Let

$$x = \text{Amount of 20\% solution used}$$

Then

$$24 - x = \text{Amount of 50\% solution used}$$

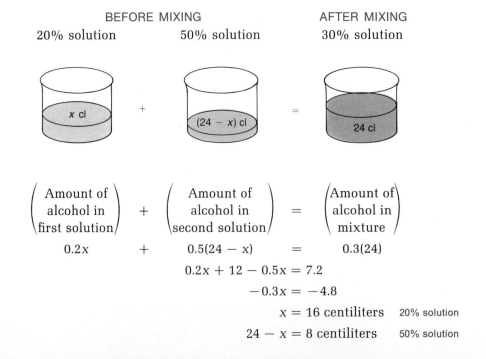

BEFORE MIXING

20% solution   50% solution

AFTER MIXING

30% solution

x cl   + $(24 - x)$ cl   =   24 cl

$$\begin{pmatrix} \text{Amount of} \\ \text{alcohol in} \\ \text{first solution} \end{pmatrix} + \begin{pmatrix} \text{Amount of} \\ \text{alcohol in} \\ \text{second solution} \end{pmatrix} = \begin{pmatrix} \text{Amount of} \\ \text{alcohol in} \\ \text{mixture} \end{pmatrix}$$

$$0.2x \quad + \quad 0.5(24 - x) \quad = \quad 0.3(24)$$

$$0.2x + 12 - 0.5x = 7.2$$

$$-0.3x = -4.8$$

$$x = 16 \text{ centiliters} \quad \text{20\% solution}$$

$$24 - x = 8 \text{ centiliters} \quad \text{50\% solution}$$

Check $0.2(16) + 0.5(8) \overset{?}{=} 0.3(24)$

$7.2 \overset{\vee}{=} 7.2$

PROBLEM 25   Repeat Example 25 using 10% and 40% stockroom solutions.

Solution

EXAMPLE 26   A coffee shop wishes to blend \$3-per-pound coffee with \$4.25-per-pound coffee to produce a blend selling for \$3.50 per pound. How much of each should be used to produce 50 pounds of the new blend?

Solution   This problem is mathematically very close to Example 25. Let

$x = $ Amount of \$3-per-pound coffee used

Then

$50 - x = $ Amount of \$4.25-per-pound coffee used

Value before blending = Value after blending

$$\begin{pmatrix} \text{Value of} \\ \text{\$3-per-pound} \\ \text{coffee used} \end{pmatrix} + \begin{pmatrix} \text{Value of} \\ \text{\$4.25-per-pound} \\ \text{coffee used} \end{pmatrix} = \begin{pmatrix} \text{Total value} \\ \text{of 50 pounds} \\ \text{of the blend} \end{pmatrix}$$

$3x \quad + \quad 4.25(50 - x) \quad = \quad 3.50(50)$

$3x + 212.5 - 4.25x = 175$

$-1.25x = -37.5$

$x = 30$ pounds $\quad$ \$3 coffee

$50 - x = 20$ pounds $\quad$ \$4.25 coffee

Check $(\$3)(30) + (\$4.25)(20) \overset{?}{=} (\$3.50)(50)$

$\$175 \overset{\vee}{=} \$175$

PROBLEM 26   Repeat Example 26 using $2.75-per-pound coffee and $4-per-pound coffee.

Solution

Solutions to Matched Problems

18. (A) $R = \dfrac{D}{T} = \dfrac{18 \text{ kilometers}}{2 \text{ hours}} = 9$ kilometers per hour

(B) $D = RT = (9 \text{ kilometers per hour})(1.5 \text{ hours}) = 13.5$ kilometers

(C) $T = \dfrac{D}{R} = \dfrac{12 \text{ kilometers}}{9 \text{ kilometers per hour}} = 1\frac{1}{3}$ hours

19. Let T = Number of hours for both presses together to print the 4,500 fliers

$$\begin{pmatrix} \text{Number of} \\ \text{fliers printed} \\ \text{by first press} \\ \text{in } T \text{ minutes} \end{pmatrix} + \begin{pmatrix} \text{Number of} \\ \text{fliers printed} \\ \text{by second press} \\ \text{in } T \text{ minutes} \end{pmatrix} = (\text{Total printing})$$

$$45T \quad + \quad 80T \quad = 4{,}500 \qquad Q = RT$$
$$125T = 4{,}500$$
$$T = 36 \text{ minutes} \quad \text{Total time}$$

Old press: $Q = RT = (45)(36) = 1{,}620$ fliers in 36 minutes

20. Let x = Time for second car to catch up to first car

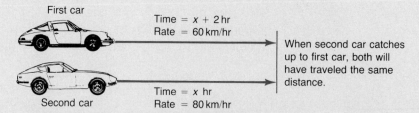

First car
Time = $x + 2$ hr
Rate = 60 km/hr

When second car catches up to first car, both will have traveled the same distance.

Second car
Time = x hr
Rate = 80 km/hr

(Distance first car travels) = (Distance second car travels)

$$60(x + 2) = 80x \qquad D = RT$$
$$60x + 120 = 80x$$
$$-20x = -120$$
$$x = 6 \text{ hours}$$

21. Let
$$x = \text{Rate of current in miles per hour}$$
$$9 - x = \text{Rate of boat upstream}$$
$$9 + x = \text{Rate of boat downstream}$$
$$(\text{Time upstream}) = 2(\text{Time downstream})$$
$$\frac{25}{9 - x} = 2\left(\frac{25}{9 + x}\right) \qquad T = \frac{D}{R}$$
$$\frac{25}{9 - x} = \frac{50}{9 + x}$$
$$25(9 + x) = 50(9 - x)$$
$$225 + 25x = 450 - 50x$$
$$75x = 225$$
$$x = 3 \text{ miles per hour} \qquad \text{Rate of current}$$

22. Let $x = $ Time for both pumps together to fill the tank
$$\text{Rate of faster pump} = \tfrac{1}{6} \text{ tank per hour}$$
$$\text{Rate of slower pump} = \tfrac{1}{9} \text{ tank per hour}$$
$$\text{Amount of tank filled} = \text{Rate} \times \text{Time}$$
$$\begin{pmatrix} \text{Amount filled} \\ \text{by faster pump} \\ \text{in x hours} \end{pmatrix} + \begin{pmatrix} \text{Amount filled} \\ \text{by slower pump} \\ \text{in x hours} \end{pmatrix} = 1 \text{ full tank}$$
$$\frac{1}{6}x \qquad + \qquad \frac{1}{9}x \qquad = 1 \qquad Q = RT$$
$$\frac{x}{6} + \frac{x}{9} = 1$$
$$\boxed{18 \cdot \frac{x}{6} + 18 \cdot \frac{x}{9} = 18 \cdot 1}$$
$$3x + 2x = 18$$
$$5x = 18$$
$$x = 3.6 \text{ hours} \qquad \text{Time for both to fill the tank}$$

23. Let
$$x = \text{The number of nickels}$$
$$40 - x = \text{The number of quarters}$$
$$\text{Value of coins before mixing} = \text{Value of coins after mixing}$$
$$\begin{pmatrix} \text{Value of} \\ \text{nickels} \\ \text{in cents} \end{pmatrix} + \begin{pmatrix} \text{Value of} \\ \text{quarters} \\ \text{in cents} \end{pmatrix} = \begin{pmatrix} \text{Total value} \\ \text{of all coins} \\ \text{in cents} \end{pmatrix}$$
$$5x \quad + 25(40 - x) = \quad 400 \qquad \text{Not 4}$$
$$5x + 1{,}000 - 25x = 400$$
$$-20x = -600$$
$$x = 30 \qquad \text{Nickels}$$
$$40 - x = 10 \qquad \text{Quarters}$$

24. Let x = Amount of pure alcohol to be added

$$\left(\begin{array}{c}\text{Amount of alcohol}\\\text{before mixing}\end{array}\right) = \left(\begin{array}{c}\text{Amount of alcohol}\\\text{after mixing}\end{array}\right)$$

$$\left(\begin{array}{c}\text{Amount of}\\\text{alcohol in}\\\text{35 centiliters}\\\text{of 20\% solution}\end{array}\right) + \left(\begin{array}{c}\text{Amount of}\\\text{pure alcohol}\\\text{added}\end{array}\right) = \left(\begin{array}{c}\text{Amount of}\\\text{alcohol in new}\\\text{30\% solution}\end{array}\right)$$

$$(0.2)(35) \;+\; x \;=\; 0.3(35 + x)$$
$$7 + x = 10.5 + 0.3x$$
$$0.7x = 3.5$$
$$x = 5 \text{ centiliters} \quad \text{Pure alcohol}$$
must be added

25. Let   x = Amount of 10% solution used
    24 − x = Amount of 40% solution used

$$\left(\begin{array}{c}\text{Amount of alcohol}\\\text{before mixing}\end{array}\right) = \left(\begin{array}{c}\text{Amount of alcohol}\\\text{after mixing}\end{array}\right)$$

$$\left(\begin{array}{c}\text{Amount of}\\\text{alcohol in}\\\text{x centiliters}\\\text{of 10\% solution}\end{array}\right) + \left(\begin{array}{c}\text{Amount of}\\\text{alcohol in}\\\text{(24 − x) centiliters}\\\text{of 40\% solution}\end{array}\right) = \left(\begin{array}{c}\text{Amount of}\\\text{alcohol in}\\\text{24 centiliters}\\\text{of 30\% solution}\end{array}\right)$$

$$0.1x \;+\; 0.4(24 - x) \;=\; 0.3(24)$$
$$x + 4(24 - x) = 3(24) \quad \text{Multiply both}$$
$$x + 96 - 4x = 72 \quad \text{sides by 10.}$$
$$-3x = -24$$
$$x = 8 \text{ centiliters} \quad \text{10\% solution}$$
$$24 - x = 16 \text{ centiliters} \quad \text{40\% solution}$$

26. Let   x = Amount of $2.75-per-pound coffee used
    50 − x = Amount of $4-per-pound coffee used

$$\text{Value before mixing} = \text{Value after mixing}$$

$$\left(\begin{array}{c}\text{Value of}\\\$2.75\text{-per-pound}\\\text{coffee used}\end{array}\right) + \left(\begin{array}{c}\text{Value of}\\\$4\text{-per-pound}\\\text{coffee used}\end{array}\right) = \left(\begin{array}{c}\text{Total value of}\\\$3.50\text{-per-pound}\\\text{blend}\end{array}\right)$$

$$2.75x \;+\; 4(50 - x) \;=\; 3.50(50)$$
$$2.75x + 200 - 4x = 175$$
$$-1.25x = -25$$
$$x = 20 \text{ pounds} \quad \$2.75 \text{ coffee}$$
$$50 - x = 30 \text{ pounds} \quad \$4 \text{ coffee}$$

Practice Exercise 4-3 ■

Work odd-numbered problems first, check answers, and then work even-numbered problems in areas of weakness. Answers to all problems are in the back of the book. Make every effort to work a problem before you look at an answer.

A *Set up appropriate equations and solve.*

Rate–Time Problems

1. Two cars leave Chicago at the same time and travel in opposite directions. If one travels at 62 kilometers per hour and the other at 88 kilometers per hour, how long will it take them to be 750 kilometers apart? _____

2. Two airplanes leave Miami at the same time and fly in opposite directions. If one flies at 840 kilometers per hour and the other at 510 kilometers per hour, how long will it take them to be 3,510 kilometers apart? _____

3. The distance between towns *A* and *B* is 750 kilometers. If a passenger train leaves town *A* and travels toward town *B* at 90 kilometers per hour at the same time a freight train leaves town *B* and travels toward *A* at 35 kilometers per hour, how long will it take the two trains to meet?

_____

4. Repeat Problem 3 using 630 kilometers for the distance between the two towns, 100 kilometers per hour as the rate for the passenger train, and 40 kilometers per hour as the rate for the freight train. _____

5. An office worker can fold and stuff 14 envelopes per minute. If another office worker can do 10, how long will it take them working together to fold and stuff 1,560 envelopes? _____

6. One file clerk can file 12 folders per minute and a second clerk 9. How long will it take them working together to file 672 folders? _____

Mixture Problems

Set up appropriate equations and solve.

7. A vending machine takes only dimes and quarters. If it contains 100 coins with a total value of $14.50, how many of each type of coin are in the machine? _____

8. A parking meter contains only nickels and dimes. If it contains 50 coins at a total value of $3.50, how many of each type of coin are in the meter?

_____

9. A school musical production brought in $12,600 on the sale of 3,500 tickets. If the tickets sold for $2 and $4, how many of each type were sold?

_____

10. A concert brought in $60,000 on the sale of 8,000 tickets. If tickets sold for $6 and $10, how many of each type of ticket were sold? _____

11. How many deciliters of alcohol must be added to 100 deciliters of a 40% alcohol solution to obtain a 50% solution? _____

12. How many milliliters of hydrochloric acid must be added to 12 milliliters of a 30% solution to obtain a 40% solution? _____

B *The following problems represent a mixture of rate–time problems and mixture problems. Set up appropriate equations and solve.*

13. A car leaves a town traveling at 50 kilometers per hour. How long will it take a second car traveling at 60 kilometers per hour to catch up to the first car if it leaves 1 hour later? _____

14. Repeat Problem 13 if the first car travels at 45 kilometers per hour and the second car leaves 2 hours later traveling at 75 kilometers per hour.

_____

15. How many liters of distilled water must be added to 140 liters of an 80% alcohol solution to obtain a 70% solution? _____

16. How many centiliters of distilled water must be added to 500 centiliters of a 60% acid solution to obtain a 50% solution? _____

17. Find the total time to complete the job in Problem 5 if the second (slower) office worker is brought on the job 15 minutes after the first person has started. _____

18. Find the total time to complete the job in Problem 6 if the faster file clerk is brought on the job 14 minutes after the slower clerk has started.

_____

19. A chemical stockroom has a 20% alcohol solution and a 50% solution. How many deciliters of each should be used to obtain 90 deciliters of a 30% solution? _____

20. A chemical supply company has a 30% sulfuric acid solution and a 70% sulfuric acid solution. How many liters of each should be used to obtain 100 liters of a 40% solution? _____

21. Pipe *A* can fill a tank in 8 hours and pipe *B* can fill the same tank in 6 hours. How long will it take both pipes together to fill the tank? _____

22. A typist can complete a mailing in 5 hours. If another typist requires 7 hours, how long will it take both working together to complete the mailing?

_____

23. A tea shop wishes to blend a $5-per-kilogram tea with a $6.50-per-kilogram tea to produce a blend selling for $6 per kilogram. How much of each should be used to obtain 75 kilograms of the new blend? _____

24. A gourmet food store wishes to blend a $7-per-kilogram coffee with a $9.50-per-kilogram coffee to produce a blend selling for $8 per kilogram. How much of each should be used to obtain 100 kilograms of the new blend? _____

25. A painter can paint a house in 5 days. With the help of another painter, the house can be painted in 3 days. How long would it take the second painter to paint the house alone? _____

26. You are at a river resort and rent a motorboat for 5 hours at 7 A.M. You are told that the boat will travel at 8 kilometers per hour upstream and 12 kilometers per hour returning. You decide that you would like to go as far up the river as you can and still be back at noon. At what time should you turn back, and how far from the resort will you be at that time? _____

27. You have inherited $20,000 and wish to invest it. If part is invested in a low-risk investment at 10% and the rest in a higher-risk investment at 15%, how much should you invest at each rate to produce the same yield as if all had been invested at 13%? _____

28. An investor has $10,000 to invest. If part is invested in a low-risk investment at 11% and the rest in a higher-risk investment at 16%, how much should be invested at each rate to produce the same yield as if all had been invested at 12%? _____

C **29.** A 9-liter radiator contains a 50% solution of antifreeze in distilled water. How much should be drained and replaced with pure antifreeze to obtain a 70% solution? _____

30. A 12-liter radiator contains a 60% solution of antifreeze and distilled water. How much should be drained and replaced with pure antifreeze to obtain an 80% solution? _____

31. Three seconds after a person fires a rifle at a target, she hears the sound of impact. If sound travels at 335 meters per second and the bullet at 670 meters per second, how far away is the target? _____

The Check Exercise for
this section is on
page 257.

32. An explosion is set off on the surface of the water 11,000 feet from a ship. If the sound reaches the ship through the water 7.77 seconds before it arrives through the air and if sound travels through water 4.5 times faster than through air, how fast (to the nearest foot per second) does sound travel in air and in water? _____

Section 4-4 Applications: Miscellaneous Problems (Optional)

In the preceding two sections you were often told what type of problem you were dealing with (number, mixture, rate–time, and so on). In a sense, having this information at the beginning provides a partial solution to the problem. This is not the case in this section, however; applications are arranged by subject and not by method of solution. Thus, it is important that you reread the suggested strategy for solving word problems in Section 4-2 before you start the exercises. If you have difficulty, don't despair. Keep working on the easier problems, then gradually work up to the more difficult ones. After a certain amount of practice and perseverance many of the problems will not seem as difficult.

Practice Exercise 4-4 ■

This supplemental set of exercises contains a variety of applications arranged according to subject area. The more difficult problems are marked with two asterisks (**), the moderately difficult problems with one asterisk (*), and the easier problems are not marked.

Life Sciences

1. A good approximation for the normal weight w (in kilograms) of a person over 150 centimeters tall is given by the formula $w = 0.98h - 100$, where height h is in centimeters. What would be the normal height of a person with a normal weight of 76 kilograms? _____

2. Find the normal height of a person in Problem 1 with a normal weight of 55 kilograms. _____

3. A scuba diver knows that 1 atmosphere of pressure is the weight of a column of air 1 square inch extending straight up from the surface of the earth without end (14.7 pounds per square inch). Also, the water pressure below the surface increases 1 atmosphere for each 33 feet of depth. In terms of a formula,

$$P = 1 + \frac{D}{33}$$

where P is pressure in atmospheres and D is depth of the water in feet. At what depth will the pressure be 3.6 atmospheres? _____

4. A company selling water-resistant watches advertises that they are water-proof to 3 atmospheres of pressure. How deep could a diver go (see Problem 3) and safely use the watch? _____

*5. A wildlife management group approximated the number of chipmunks in a wildlife preserve by using the popular capture-mark-recapture technique. Using live traps, they captured and marked 600 chipmunks and then released them. After a period for mixing, they captured another 500 and found 60 marked ones among them. Assuming that the ratio of the total chipmunk population to the chipmunks marked in the first sample is the same as the ratio of all chipmunks in the second sample to those found marked, estimate the chipmunk population in the preserve.

*6. A naturalist for a fish-and-game department estimated the total number of rainbow trout in a lake by using the method described in Problem 5. The naturalist netted, marked, and released 200 rainbow trout. A week later, after thorough mixing, 200 more were netted and 8 marked trout were found among them. Estimate the total rainbow trout population in the lake. _____

Domestic　**7.** A student needs at least 80% of all points on the tests in a class to get a B. There are three 100-point tests and a 250-point final. The student's test scores are 72, 85, and 78 for the 100-point tests. What is the least score the student can make on the final and still get a B? _____

8. Repeat Problem 7 but this time suppose the student scores 95, 87, and 66 on the three 100-point tests. _____

*9. Five men form a glider club and decide to share the cost of a glider equally. They find, however, that if they let three more join the club, the share for each of the original five will be reduced by $480. What is the total cost of the glider? _____

*10. Three women bought a sailboat together. If they had taken in a fourth person, the cost for each would have been reduced by $400. What was the total cost of the boat? _____

*11. The cruising speed of an airplane is 150 miles per hour (relative to ground). You wish to hire the plane for a 3-hour sightseeing trip. You instruct the pilot to fly north as far as possible and still return to the airport at the end of the allotted time.

(A) How far north should the pilot fly if there is a 30-mile-an-hour wind blowing from the north? _____

(B) How far north should the pilot fly if there is no wind blowing?

*12. Repeat Problem 11 for an airplane with a cruising speed of 350 kilometers per hour (in still air) and a wind blowing at 70 kilometers per hour from the north. _____

Music **13.** Starting with a string tuned to a given note, one can move up and down the scale simply by decreasing or increasing its length (while maintaining the same tension) according to simple whole-number ratios (see the figure). For example, $\frac{8}{9}$ of the C string gives the next higher note D, $\frac{2}{3}$ of the C string gives G, and $\frac{1}{2}$ of the C string gives C 1 octave higher. (The reciprocals of these fractions, $\frac{9}{8}$, $\frac{3}{2}$, and 2, respectively, are proportional to the frequencies of these notes.) Find the lengths of 7 strings (each less than 30 inches) that will produce the following seven chords when paired with a 30-inch string:

(A)	Octave	1:2	**(B)**	Fifth	2:3
(C)	Fourth	3:4	**(D)**	Major third	4:5
(E)	Minor third	5:6	**(F)**	Major sixth	3:5
(G)	Minor sixth	5:8			

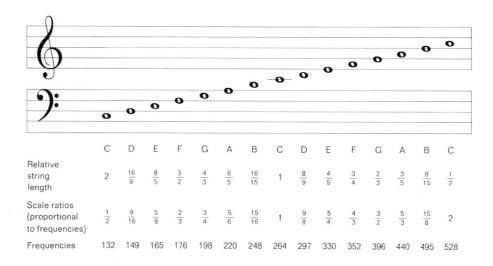

	C	D	E	F	G	A	B	C	D	E	F	G	A	B	C
Relative string length	2	$\frac{16}{9}$	$\frac{8}{5}$	$\frac{3}{2}$	$\frac{4}{3}$	$\frac{6}{5}$	$\frac{16}{15}$	1	$\frac{8}{9}$	$\frac{4}{5}$	$\frac{3}{4}$	$\frac{2}{3}$	$\frac{3}{5}$	$\frac{8}{15}$	$\frac{1}{2}$
Scale ratios (proportional to frequencies)	$\frac{1}{2}$	$\frac{9}{16}$	$\frac{5}{8}$	$\frac{2}{3}$	$\frac{3}{4}$	$\frac{5}{6}$	$\frac{15}{16}$	1	$\frac{9}{8}$	$\frac{5}{4}$	$\frac{4}{3}$	$\frac{3}{2}$	$\frac{5}{3}$	$\frac{15}{8}$	2
Frequencies	132	149	165	176	198	220	248	264	297	330	352	396	440	495	528

14. The three major chords in music are composed of notes whose frequencies are in the ratio 4:5:6. If the first note of a chord has a frequency of 264 hertz (middle C on the piano), find the frequencies of the other two notes. [*Hint:* Set up two proportions using 4:5 and 4:6, respectively.] _____

Business **15.** If you paid $160 for a camera after receiving a discount of 20%, what was the price of the camera before the discount? _____

16. A car rental company charges $21 per day and 10 cents per mile. If a car was rented for 2 days, how far was it driven if the total bill came to $53.20? _____

17. It costs a book publisher $74,200 to prepare a book for publication (type-setting, art, editing, and so on); printing and binding costs are $5.50 per book. If the book is sold to bookstores for $19.50 per copy, how many copies must be sold for the publisher to break even? _____

18. A woman borrowed a sum of money from a bank at 18% simple interest. At the end of 10 months she repaid the bank $1,380. How much was borrowed from the bank? [*Hint:* $A = P + Prt$, where A is the amount repaid, P is the amount borrowed, r is the interest rate expressed as a decimal, and t is time in years.] _____

**19. If a stock that you bought on Monday went up 10% on Tuesday and fell 10% on Wednesday, how much did you pay for the stock on Monday if you sold it on Wednesday for $99? _____

**20. A company's sales decreased 5% in 1984 and increased 5% in 1985. What were the sales in 1983 if sales for 1985 were $9,975,000? _____

Physics and Engineering   **21.** If a small steel ball is thrown downward from a tower with an initial velocity of 15 meters per second, its velocity in meters per second after t seconds is given approximately by

$$v = 15 + 9.75t$$

How many seconds are required for the object to attain a velocity of 93 meters per second? _____

22. How long would it take the ball in Problem 21 to reach a velocity of 120 meters per second? _____

23. If the large cross-sectional area in a hydraulic lift (see the figure) is approximately 630 square centimeters and a person wants to lift 2,250 kilograms with a 25-kilogram force, how large should the small cross-sectional area be? _____

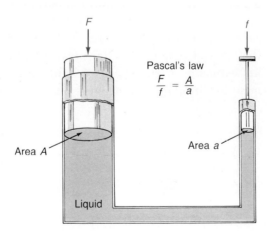

24. If the large cross-sectional area of the hydraulic lift shown in the figure accompanying Problem 23 is 560 square centimeters and the small cross-sectional area is 8 square centimeters, how much force f will be required to lift 2,100 kilograms? _____

25. A type of physics problem with wide applications is the *lever problem*. For a lever, relative to a fulcrum, to be in static equilibrium (balanced), the sum of the downward forces times their respective distances on one side of the fulcrum must equal the sum of the downward forces times their respective distances on the other side of the fulcrum (see the figure). If a person has a 200-centimeter steel wrecking bar and places a fulcrum 20 centimeters from one end, how much can be lifted with a force of 50 kilograms on the long end? _____

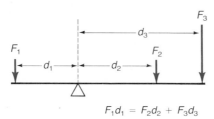

$$F_1 d_1 = F_2 d_2 + F_3 d_3$$

26. Two people decided to move a 1,920-pound rock by use of a 9-foot steel bar (see the figure). If they place the fulcrum 1 foot from the rock and one of them applies a force of 150 pounds on the other end, how much force will the second person have to apply 2 feet from that end to lift the rock?

*27. In 1849, during a celebrated experiment, the French mathematician Fizeau made the first accurate approximation of the speed of light. By using a rotating disk with notches equally spaced on the circumference and a reflecting mirror 5 miles away (see the figure), he was able to measure the elapsed time for the light traveling to the mirror and back. Calculate his estimate for the speed of light (in miles per second) if his measurement for the elapsed time was $\frac{1}{20,000}$ second ($d = rt$). _____

**28. An earthquake emits a primary wave and a secondary wave. Near the surface of the earth the primary wave travels at about 5 miles per second, and the secondary wave at about 3 miles per second. From the time lag between the two waves arriving at a given seismic station, it is possible to estimate the distance to the quake. (The *epicenter* can be located by getting distance bearings at three or more stations.) Suppose a station measured a time difference of 12 seconds between the arrival of the two waves. How far would the earthquake be from the station? _____

Puzzles   **29.** A pole is located in a pond. One-fifth of the length of the pole is in the sand, 4 meters is in water, and two-thirds of the length is in the air. How long is the pole? _____

*30. Diophantus, an early Greek algebraist (A.D. 280), was the subject for a famous ancient puzzle. See if you can find Diophantus' age at death from the following information: Diophantus was a boy for one-sixth of his life; after one-twelfth more he grew a beard; after one-seventh more he married, and after 5 years of marriage he was granted a son; the son lived one-half as long as his father; and Diophantus died 4 years after his son's death.

**31. A classic problem is the courier problem. If a column of soldiers 3 miles long is marching at 5 miles per hour, how long will it take a courier on a motorcycle traveling at 25 miles per hour to deliver a message from the end of the column to the front and then return? _____

The Check Exercise for this section is on page 259.   ****32.** After 12:00 noon exactly, what time will the hands of a clock be together again? _____

Section 4-5   Solving for a Particular Variable

One of the immediate applications you will have for algebra is the changing of formulas or equations to equivalent forms. In the process we will make frequent use of the symmetric property of equality introduced in Section 1-2. Recall:

Symmetric Property of Equality

If $a = b$, then $b = a$.   An equation can be reversed without changing any signs.

Thus, if we are given the formula

$$c = \frac{wrt}{1,000}$$

then we may reverse it if we wish to obtain

$$\frac{wrt}{1,000} = c$$

We will do exactly this in the next example, where we solve the formula for t in terms of the other variables.

EXAMPLE 27   Solve the formula $c = wrt/1{,}000$ for t. (The formula gives the cost of using an electrical appliance; $w = $ Power in watts, $r = $ Rate per kilowatt-hour, $t = $ Time in hours.)

Solution

$$c = \frac{wrt}{1{,}000} \qquad \text{Start with the given formula.}$$

$$\frac{wrt}{1{,}000} = c \qquad \text{Reverse the equation to get } t \text{ on the left side.}$$

$$1{,}000 \cdot \frac{wrt}{1{,}000} = 1{,}000c \qquad \text{Multiply both sides by 1,000.}$$

$$wrt = 1{,}000c \qquad \text{Divide both sides by } wr.$$

$$\frac{wrt}{wr} = \frac{1{,}000c}{wr}$$

$$t = \frac{1{,}000c}{wr} \qquad \text{We have solved for } t.$$

PROBLEM 27   Solve the formula in Example 27 for w.

Solution

EXAMPLE 28   Solve the formula $A = P + Prt$ for r (simple interest formula).

Solution

$$A = P + Prt$$

$$P + Prt = A \qquad \text{Reverse the equation; then perform operations to isolate } r \text{ on the left side.}$$

$$Prt = A - P$$

$$\frac{Prt}{Pt} = \frac{A - P}{Pt}$$

$$r = \frac{A - P}{Pt}$$

PROBLEM 28   Solve the formula $A = P + Prt$ for t.

Solution

EXAMPLE 29   Solve the formula $A = P + Prt$ for P.

Solution $A = P + Prt$

                $P + Prt = A$

Since P is a common factor to both terms on the left, we factor P out and complete the problem:

$P(1 + rt) = A$

$$\frac{P(1 + rt)}{1 + rt} = \frac{A}{1 + rt}$$

$$P = \frac{A}{1 + rt}$$ Note that P appears only on the left side.

COMMON ERROR   If we write $P = A - Prt$, we have not solved for P. To solve for P is to isolate P on the left side with a coefficient of 1. In general, if the variable for which we are solving appears on both sides of the equation, we have not solved for it!

PROBLEM 29   Solve $A = xy + xz$ for x.

Solution

Solutions to Matched
Problems

27.

$$c = \frac{wrt}{1,000}$$

$$\frac{wrt}{1,000} = c$$

$$\frac{1,000}{rt}\left(\frac{wrt}{1,000}\right) = \frac{1,000}{rt}\,c$$

$$w = \frac{1,000c}{rt}$$

28.

$$A = P + Prt$$
$$P + Prt = A$$
$$Prt = A - P$$
$$t = \frac{A - P}{Pr}$$

29.

$$A = xy + xz$$
$$xy + xz = A$$
$$x(y + z) = A$$
$$x = \frac{A}{y + z}$$

Practice Exercise 4-5 ∎

Work odd-numbered problems first, check answers, and then work even-numbered problems in areas of weakness. Answers to all problems are in the back of the book. Make every effort to work a problem yourself before you look at an answer.

A **1.** Solve $d = rt$ for r.   *Distance–rate–time* _____

  2. Solve $d = 1,100t$ for t.   *Sound distance in air* _____

  3. Solve $C = 2\pi r$ for r.   *Circumference of a circle* _____

  4. Solve $I = Prt$ for t.   *Simple interest* _____

  5. Solve $C = \pi D$ for π.   *Circumference of a circle* _____

  6. Solve $e = mc^2$ for m.   *Mass–energy equation* _____

  7. Solve $ax + b = 0$ for x.   *First-degree polynomial equation* _____

  8. Solve $p = 2a + 2b$ for a.   *Perimeter of a rectangle* _____

  9. Solve $y = 2x - 5$ for x.   *Slope–intercept equation for a line* _____

  10. Solve $y = mx + b$ for m.   *Slope–intercept equation for a line* _____

B **11.** Solve $3x - 4y - 12 = 0$ for y.   *Linear equation in two variables* _____

  12. Solve $Ax + By + C = 0$ for y.   *Linear equation in two variables* _____

13.   Solve $I = \dfrac{E}{R}$ for R.   *Electrical circuits—Ohm's law* _____

14.   Solve $m = \dfrac{b}{a}$ for a.   *Optics—magnification* _____

15.   Solve $C = \dfrac{100B}{L}$ for B.   *Anthropology—cephalic index* _____

16.   Solve $(IQ) = \dfrac{100(MA)}{(CA)}$ for (CA).   *Psychology—intelligence quotient*

_____

17.   Solve $F = G\dfrac{m_1 m_2}{d^2}$ for G.   *Gravitational force between two masses*

_____

18.   Solve $F = G\dfrac{m_1 m_2}{d^2}$ for m_1.   *Gravitational force between two masses*

_____

19.   Solve $F = \frac{9}{5}C + 32$ for C.   *Celsius–Fahrenheit* _____

20.   Solve $C = \frac{5}{9}(F - 32)$ for F.   *Celsius–Fahrenheit* _____

C   21.   Solve $\dfrac{1}{f} = \dfrac{1}{a} + \dfrac{1}{b}$ for f.   *Optics—focal length* _____

22.   Solve $\dfrac{1}{R} = \dfrac{1}{R_1} + \dfrac{1}{R_2}$ for R.   *Electrical circuits* _____

23.   Solve $a_n = a_1 + (n - 1)d$ for n.   *Arithmetic progression* _____

24.   Solve $a_n = a_1 + (n - 1)d$ for d.   *Arithmetic progression* _____

25.   Solve $\dfrac{P_1 V_1}{T_1} = \dfrac{P_2 V_2}{T_2}$ for T_2.   *Gas law* _____

26.   Solve $\dfrac{P_1 V_1}{T_1} = \dfrac{P_2 V_2}{T_2}$ for V_1.   *Gas law* _____

27.   Solve $y = \dfrac{2x - 3}{3x - 5}$ for x.   *Rational equation* _____

**The Check Exercise for
this section is on
page 261.**   28.   Solve $y = \dfrac{3x + 2}{2x - 4}$ for x.   *Rational equation* _____

Section 4-6   Solving Inequalities

- Interval Notation
- Solving Linear Inequalities
- Applications

In Section 1-2 we introduced simple inequality statements of the form

$$x > 2 \qquad -4 < x \leq 3 \qquad x \leq -1$$

which have obvious solutions. In this section we will consider inequality statements that do not have obvious solutions. Try to guess the real number solutions for

$$3(x - 2) + 1 < 3x - (x + 7)$$

By the end of this section you will be able to solve this type of inequality almost as easily as you solved first-degree equations. First, however, we digress for a moment to introduce interval notation.

■ Interval Notation

In Section 1-2 we used parentheses ( ) and brackets [ ] in the graphic representation of certain inequality statements. For example,

$$-3 < x \leq 4 \qquad \text{and}$$

are two ways of indicating that x is between -3 and 4, including 4 but excluding -3. Another convenient way of representing this fact is in terms of the interval notation

$$(-3, 4]$$

Table 1 shows the use of **interval notation** in its most common forms.

TABLE 1	INTERVAL NOTATION	INEQUALITY NOTATION	LINE GRAPH
	$[a, b]$	$a \leq x \leq b$	
	$[a, b)$	$a \leq x < b$	
	$(a, b]$	$a < x \leq b$	
	(a, b)	$a < x < b$	
	$[b, \infty)^{\dagger}$	$x \geq b$	
	(b, ∞)	$x > b$	
	$(-\infty, a]$	$x \leq a$	
	$(-\infty, a)$	$x < a$	

† The symbol ∞ (read "infinity") is not a number. When we write $[b, \infty)$, we are simply referring to the interval starting at b and continuing indefinitely to the right. We would never write $[b, \infty]$.

EXAMPLE 30   Write each of the following in inequality notation and graph on a real number line:

(A) $[-2, 3)$ **(B)** $(-4, 2)$ **(C)** $[-2, \infty)$ **(D)** $(-\infty, 3)$

Solution   **(A)** $-2 \le x < 3$ 

(B) $-4 < x < 2$ 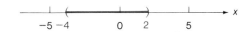

(C) $x \ge -2$ 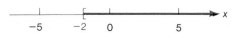

(D) $x < 3$ 

PROBLEM 30   Write each of the following in interval notation and graph on a real number line:

(A) $-3 < x \le 3$ **(B)** $-1 \le x \le 2$ **(C)** $x > 1$ **(D)** $x \le 2$

Solution   **(A)**

(B)

(C)

(D)

■ Solving Linear Inequalities

We now turn to the problem of solving linear inequalities in one variable, such as

$$2(2x + 3) < 6(x - 2) + 10 \qquad \text{and} \qquad -3 < 2x + 3 \le 9$$

The **solution set** for an inequality is the set of elements from its replacement set that make the inequality a true statement. Any element of the solution set is called a **solution** of the inequality. To **solve an inequality** is to find its solution set. Two inequalities are **equivalent** if they have the same solution set. Just as with equations, we try to perform operations on inequalities that produce simpler equivalent inequalities. We continue the process until an inequality is reached whose solution is obvious. The properties of inequalities given in Theorem 2 produce equivalent inequalities when applied.

THEOREM 2 | **Inequality Properties**

For a, b, and c any real numbers:

1. If $a < b$, then $a + c < b + c$.   ADDITION PROPERTY

   $-2 < 4$ $-2 + 3 < 4 + 3$

2. If $a < b$, then $a - c < b - c$.   SUBTRACTION PROPERTY

   $-2 < 4$ $-2 - 3 < 4 - 3$

3. If $a < b$ and c is positive, then $ca < cb$.

   $-2 < 4$ $3(-2) < 3(4)$

MULTIPLICATION PROPERTY

(Note the difference between 3 and 4.)

4. If $a < b$ and c is negative, then $ca > cb$.

   $-2 < 4$ $(-3)(-2) > (-3)(4)$

5. If $a < b$ and c is positive, then $\dfrac{a}{c} < \dfrac{b}{c}$.

   $-2 < 4$ $\dfrac{-2}{2} < \dfrac{4}{2}$

DIVISION PROPERTY

(Note the difference between 5 and 6.)

6. If $a < b$ and c is negative, then $\dfrac{a}{c} > \dfrac{b}{c}$.

   $-2 < 4$ $\dfrac{-2}{-2} > \dfrac{4}{-2}$

Similar properties hold if each inequality sign is reversed or if $<$ is replaced with $\le$ and $>$ is replaced with $\ge$. Thus, we find that we can perform essentially the same operations on inequalities that we perform on equations. When working with inequalities, we have to be particularly careful of the use of the multiplication and division properties.

The sense of the inequality reverses if we multiply or divide both sides of an inequality statement by a negative number.

Let us sketch a proof of the multiplication property: if $a < b$, then by definition of $<$, there exists a positive number p such that $a + p = b$. If we multiply both sides of $a + p = b$ by a positive number c, we obtain $ca + cp = cb$, where cp is positive. (Why?) Thus, by definition of $<$, we see that $ca < cb$. Now if we multiply both sides of $a + p = b$ by a negative number c, we obtain $ca + cp = cb$ or $ca = cb - cp = cb + (-cp)$, where $-cp$ is positive. (Why?) Hence, by definition of $<$, we see that $cb < ca$ or $ca > cb$.

Now let us see how the inequality properties are used to solve linear inequalities. Several examples will illustrate the process.

EXAMPLE 31   Solve and graph: $2(2x + 3) - 10 < 6(x - 2)$

Solution $2(2x + 3) - 10 < 6(x - 2)$ Simplify left and right sides.

         $4x + 6 - 10 < 6x - 12$

            $4x - 4 < 6x - 12$

$$\boxed{4x - 4 + 4 \leq 6x - 12 + 4}$$ Addition property

$$4x < 6x - 8$$

$$\boxed{4x - \mathbf{6x} < 6x - 8 - \mathbf{6x}}$$ Subtraction property

$$-2x < -8$$

$$\boxed{\frac{-2x}{-2} > \frac{-8}{-2}}$$ Division property—note that the sense reverses. (Why?)

$$x > 4 \quad \text{or} \quad (4, \infty)$$

PROBLEM 31   Solve and graph: $3(x - 1) \geq 5(x + 2) - 5$

Solution

EXAMPLE 32   Solve and graph: $\dfrac{2x - 3}{4} + 6 \geq 2 + \dfrac{4x}{3}$

Solution

$$\frac{2x - 3}{4} + 6 \geq 2 + \frac{4x}{3}$$ Multiply both sides by 12, the LCM of 4 and 3.

$$\boxed{12 \cdot \frac{2x - 3}{4} + 12 \cdot 6 \geq 12 \cdot 2 + 12 \cdot \frac{4x}{3}}$$ Sense does not reverse. (Why?)

$$3(2x - 3) + 72 \geq 24 + 4 \cdot 4x$$

$$6x - 9 + 72 \geq 24 + 16x$$

$$6x + 63 \geq 24 + 16x$$

$$-10x \geq -39$$

$$x \leq 3.9 \quad \text{or} \quad (-\infty, 3.9]$$ Sense reverses. (Why?)

PROBLEM 32   Solve and graph: $\dfrac{4x - 3}{3} + 8 < 6 + \dfrac{3x}{2}$

Solution

EXAMPLE 33   Solve and graph: $-3 \le 4 - 7x < 18$

Solution   We proceed as in the preceding examples, except that we try to isolate x in the middle with a coefficient of 1. That is, we try to solve the two inequalities $-3 \le 4 - 7x$ and $4 - 7x < 18$ at the same time:

$$-3 \le 4 - 7x < 18 \qquad \text{Subtract 4 from each member.}$$

$$-3 - 4 \le 4 - 7x - 4 < 18 - 4$$

$$-7 \le -7x < 14 \qquad \text{Divide each member by } -7.$$

$$\frac{-7}{-7} \ge \frac{-7x}{-7} > \frac{14}{-7} \qquad \text{Sense reverses. (Why?)}$$

$$1 \ge x > -2 \quad \text{or} \quad -2 < x \le 1 \quad \text{or} \quad (-2, 1]$$

PROBLEM 33   Solve and graph: $-3 < 7 - 2x \le 7$

Solution

■ Applications

EXAMPLE 34

In a chemistry experiment a solution of hydrochloric acid is to be kept between 30° and 35° Celsius—that is, $30 \leq C \leq 35$. What is the range in temperature in degrees Fahrenheit? $[C = \frac{5}{9}(F - 32)]$

Solution

$$30 \leq C \leq 35 \qquad \text{Replace C with } \tfrac{5}{9}(F - 32).$$

$$30 \leq \frac{5}{9}(F - 32) \leq 35 \qquad \text{Multiply each member by } \tfrac{9}{5}. \text{ (Why?)}$$

$$\frac{9}{5} \cdot 30 \leq \frac{9}{5} \cdot \frac{5}{9}(F - 32) \leq \frac{9}{5} \cdot 35$$

$$54 \leq F - 32 \leq 63 \qquad \text{Add 32 to each member.}$$

$$54 + 32 \leq F - 32 + 32 \leq 63 + 32$$

$$86 \leq F \leq 95 \quad \text{or} \quad [86, 95]$$

PROBLEM 34

A film developer is to be kept between 68° and 77° Fahrenheit—that is, $68 \leq F \leq 77$. What is the range in temperature in degrees Celsius? $(F = \frac{9}{5}C + 32)$

Solution

Solutions to Matched Problems

30. (A) $(-3, 3]$

(B) $[-1, 2]$

(C) $(1, \infty)$

(D) $(-\infty, 2]$

31. $3(x - 1) \geq 5(x + 2) - 5$

$ 3x - 3 \geq 5x + 10 - 5$

$ 3x \geq 5x + 8$

$ -2x \geq 8$

$ x \leq -4 \quad \text{or} \quad (-\infty, -4]$

32.
$$\frac{4x - 3}{3} + 8 < 6 + \frac{3x}{2}$$

$$6 \cdot \frac{4x - 3}{3} + 6 \cdot 8 < 6 \cdot 6 + 6 \cdot \frac{3x}{2}$$

$$2(4x - 3) + 48 < 36 + 3 \cdot 3x$$
$$8x - 6 + 48 < 36 + 9x$$
$$8x + 42 < 36 + 9x$$
$$-x < -6$$
$$x > 6 \quad \text{or} \quad (6, \infty)$$

33.
$$-3 < 7 - 2x \le 7$$

$$-3 - 7 < 7 - 2x - 7 \le 7 - 7$$

$$-10 < -2x \le 0$$

$$\frac{-10}{-2} > \frac{-2x}{-2} \ge \frac{0}{-2}$$

$$5 > x \ge 0$$
$$\text{or} \quad 0 \le x < 5 \quad \text{or} \quad [0, 5)$$

34.
$$68 \le \tfrac{9}{5}C + 32 \le 77$$

$$68 - 32 \le \tfrac{9}{5}C + 32 - 32 \le 77 - 32$$

$$36 \le \tfrac{9}{5}C \le 45$$

$$\tfrac{5}{9} \cdot 36 \le \tfrac{5}{9} \cdot \tfrac{9}{5}C \le \tfrac{5}{9} \cdot 45$$

$$20 \le C \le 25 \quad \text{or} \quad [20, 25]$$

Practice Exercise 4-6 ■

Work odd-numbered problems first, check answers, and then work even-numbered problems in areas of weakness. Answers to all problems are in the back of the book. Make every effort to work a problem yourself before you look at an answer.

The replacement set for all variables is the set of real numbers.

A   Write in inequality notation and graph on a real number line.

1. $[-8, 7]$ _____

2. $(-4, 8)$ _____

3. $[-6, 6)$ _____

4. $(-3, 3]$ _____

5. $[-6, \infty)$ _____   **6.** $(-\infty, 7)$ _____

Write in interval notation and graph on a real number line.

7. $-2 < x \le 6$ _____   **8.** $-5 \le x \le 5$ _____

9. $-7 < x < 8$ _____   **10.** $-4 \le x < 5$ _____

11. $x \le -2$ _____   **12.** $x > 3$ _____

Write in interval and inequality notation.

13.   _____

14.   _____

15.   _____

16.   _____

Solve and graph.

17. $7x - 8 < 4x + 7$ _____   **18.** $4x + 8 \ge x - 1$ _____

19. $3 - x \ge 5(3 - x)$ _____   **20.** $2(x - 3) + 5 < 5 - x$ _____

21. $\dfrac{N}{-2} > 4$ _____   **22.** $\dfrac{M}{-3} \le -2$ _____

23. $-5t < -10$ _____   **24.** $-7n \ge 21$ _____

25. $3 - m < 4(m - 3)$ _____   **26.** $2(1 - u) \ge 5u$ _____

27. $-2 - \dfrac{B}{4} \le \dfrac{1 + B}{3}$ _____   **28.** $\dfrac{y - 3}{4} - 1 > \dfrac{y}{2}$ _____

29. $-4 < 5t + 6 \le 21$ _____   **30.** $2 \le 3m - 7 < 14$ _____

B *What numbers satisfy the given conditions? Solve using inequality methods.*

31. 3 less than twice the number is greater than or equal to -6. _____

32. 5 more than twice the number is less than or equal to 7. _____

33. 15 reduced by 3 times the number is less than 6. _____

34. 5 less than 3 times the number is less than or equal to 4 times the number. _____

Solve and graph.

35. $\dfrac{q}{7} - 3 > \dfrac{q-4}{3} + 1$ _____   **36.** $\dfrac{p}{3} - \dfrac{p-2}{2} \leq \dfrac{p}{4} - 4$ _____

37. $\dfrac{2x}{5} - \dfrac{1}{2}(x-3) \leq \dfrac{2x}{3} - \dfrac{3}{10}(x+2)$ _____

38. $\dfrac{2}{3}(x+7) - \dfrac{x}{4} > \dfrac{1}{2}(3-x) + \dfrac{x}{6}$ _____

39. $-4 \leq \frac{9}{5}x + 32 \leq 68$ _____   **40.** $-1 \leq \frac{2}{3}A + 5 \leq 11$ _____

41. $-12 < \frac{3}{4}(2-x) \leq 24$ _____   **42.** $24 \leq \frac{2}{3}(x-5) < 36$ _____

43. $16 < 7 - 3x \leq 31$ _____   **44.** $-1 \leq 9 - 2x < 5$ _____

45. $-6 < -\frac{2}{5}(1-x) \leq 4$ _____   **46.** $15 \leq 7 - \frac{2}{5}x \leq 21$ _____

C **47.** If both a and b are negative numbers and b/a is greater than 1, then is $a - b$ positive or negative? _____

48. If both a and b are positive numbers and b/a is greater than 1, then is $a - b$ positive or negative? _____

49. Indicate true (T) or false (F):

(A) If $p > q$ and $m > 0$, then $mp < mq$. _____

(B) If $p < q$ and $m < 0$, then $mp > mq$. _____

(C) If $p > 0$ and $q < 0$, then $p + q > q$. _____

50. Assume that $m > n > 0$; then

$$mn > n^2$$
$$mn - m^2 > n^2 - m^2$$
$$m(n - m) > (n + m)(n - m)$$
$$m > n + m$$
$$0 > n$$

But it was assumed $n > 0$. Can you find the error? _____

APPLICATIONS   *Set up inequalities and solve.*

51. *Earth science:* As dry air moves upward it expands, and in so doing cools at a rate of about 5.5°F for each 1,000-foot rise up to about 40,000 feet. If the ground temperature is 70°F, then the temperature T at height h is given approximately by $T = 70 - 0.0055h$. For what range in altitude will the temperature be between 26° and −40°F? _____

*52. *Energy:* If the power demands in a 110-volt electric circuit in a home vary between 220 and 2,750 watts, what is the range of current flowing through the circuit? ($W = EI$, where W = Power in watts, E = Pressure in volts, and I = Current in amperes.) _____

53. *Business and economics:* For a business to make a profit it is clear that revenue R must be greater than cost C; in short, a profit will result only if $R > C$. If a company manufactures records and its cost equation for a week is $C = 300 + 1.5x$ and its revenue equation is $R = 2x$, where x is the number of records sold in a week, how many records must be sold for the company to realize a profit? _____

54. *Psychology:* IQ is given by the formula

$$IQ = \frac{MA}{CA} 100$$

where MA is mental age and CA is chronological age. If

$$80 \le IQ \le 140$$

for a group of 12-year-old children, find the range of their mental ages.

**55. *Puzzle:* A railroad worker is walking through a train tunnel (see the figure) when he notices an unscheduled train approaching him. If he is three-quarters of the way through the tunnel and the train is one tunnel length ahead of him, which way should he run to maximize his chances of escaping? _____

The Check Exercise for this section is on page 263.

Section 4-7   Absolute Value in Equations and Inequalities

- ■ Absolute Value and Distance
- ■ Absolute Value in Equations and Inequalities

■ Absolute Value and Distance

We start with a review of the geometric and algebraic definitions of absolute value (Section 1-4). If *a* is the coordinate of a point on a real number line, then the (nondirected) distance from the origin to a, a nonnegative quantity, is represented by $|a|$ and is referred to as the **absolute value** of a (Figure 1). Thus, if $|x| = 5$, then x can be either -5 or 5.

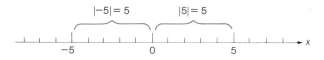

FIGURE 1 Absolute value

Algebraically, recall that we defined absolute value as follows:

Absolute Value

$$|x| = \begin{cases} x & \text{if } x \text{ is positive} \\ 0 & \text{if } x \text{ is 0} \\ -x & \text{if } x \text{ is negative} \end{cases}$$

[*Note:* $-x$ is positive if x is negative.]

Both the geometric and algebraic definitions of absolute value are useful, as will be seen in the material that follows. Remember:

The absolute value of a number is never negative.

EXAMPLE 35 **(A)** $|7| = 7$

(B) $|\pi - 3| = \pi - 3$ Since $\pi - 3$ is nonnegative ($\pi > 3$)

(C) $|-7| = -(-7) = 7$

(D) $|3 - \pi| = -(3 - \pi) = \pi - 3$ Since $3 - \pi$ is negative

PROBLEM 35   Write without the absolute value sign:

(A) $|8|$ **(B)** $\left|\sqrt{5} - 2\right|$ **(C)** $\left|-\sqrt{2}\right|$ **(D)** $\left|2 - \sqrt{5}\right|$

Solution   **(A)**   **(B)**   **(C)**   **(D)**

Following the same reasoning used in Example 35(B) and (D), it can be shown (see Problem 61 in Exercise 4-7) that:

For all real numbers a and b:

$|b - a| = |a - b|$ $|7 - 4| = |3| = 3 = |-3| = |4 - 7|$

We use this result in defining the distance between two points on a real number line.

> **Distance between Points *A* and *B***
>
> Let *A* and *B* be two points on a real number line with coordinates *a* and *b*, respectively. The **distance between *A* and *B*** (also called the **length of the line segment** joining *A* and *B*) is given by
>
> $$d(A, B) = |b - a|$$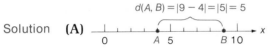

EXAMPLE 36   Find the distance between points *A* and *B* with coordinates *a* and *b*, respectively, as given:

(A) $a = 4, b = 9$ (B) $a = 9, b = 4$
(C) $a = 0, b = 6$ (D) $a = -3, b = 5$

Solution   (A)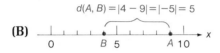

(B)

(C)

(D)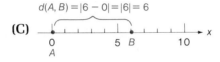

It is clear, since $|b - a| = |a - b|$, that

$$d(A, B) = d(B, A)$$

Hence, in computing the distance between two points on a real number line, it does not matter how the two points are labeled—point *A* can be to the left or to the right of point *B*. Note also that if *A* is at the origin *O*, then

$$d(O, B) = |b - 0| = |b|$$

PROBLEM 36   Find the indicated distances given:

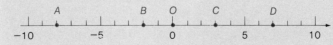

(A) $d(C, D)$ (B) $d(D, C)$ (C) $d(A, B)$
(D) $d(A, C)$ (E) $d(O, A)$ (F) $d(D, A)$

Solution   (A)   (B)   (C)

(D)   (E)   (F)

■ Absolute Value in Equations and Inequalities

Absolute value is frequently encountered in equations and inequalities. Some of these forms have immediate geometric interpretation.

EXAMPLE 37   Solve geometrically and graph:

(A) $|x - 3| = 5$ **(B)** $|x - 3| < 5$ **(C)** $0 < |x - 3| < 5$ **(D)** $|x - 3| > 5$

Solution   **(A)** Geometrically, $|x - 3|$ represents the distance between x and 3; thus, in $|x - 3| = 5$, x is a number whose distance from 3 is 5. That is, x is 5 units to the left of 3 or 5 units to the right of 3:

$$x = 3 - 5 \quad \text{or} \quad x = 3 + 5 \quad \text{More compactly:} \quad x = 3 \pm 5 = -2 \text{ or } 8$$
$$x = -2 \qquad\qquad x = 8$$

(B) Geometrically, in $|x - 3| < 5$, x is a number whose distance from 3 is less than 5; that is, x is within 5 units of 3:

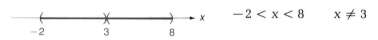

$$-2 < x < 8 \quad \text{or} \quad (-2, 8)$$

(C) The form $0 < |x - 3| < 5$ is encountered in calculus and advanced mathematics. Geometrically, x is a number whose distance from 3 is less than 5, but x cannot equal 3. Thus,

$$-2 < x < 8 \qquad x \neq 3$$

(D) Geometrically, in $|x - 3| > 5$, x is a number whose distance from 3 is greater than 5; that is,

$$x < -2 \quad \text{or} \quad x > 8 \quad \text{Note:} \quad \text{This cannot be written as a double inequality.}$$

We summarize the preceding results in Table 2.

TABLE 2

FORM ($d > 0$)	GEOMETRIC INTERPRETATION	GRAPH
$\|x - c\| = d$	Distance between x and c is equal to d.	
$\|x - c\| < d$	Distance between x and c is less than d.	
$0 < \|x - c\| < d$	Distance between x and c is less than d, but $x \neq c$.	
$\|x - c\| > d$	Distance between x and c is greater than d.	

PROBLEM 37 Solve geometrically and graph:

(A) $|x + 2| = 6$ **(B)** $|x + 2| < 6$
(C) $0 < |x + 2| < 6$ **(D)** $|x + 2| > 6$

[*Hint:* $|x + 2| = |x - (-2)|$]

Solution **(A)**   **(B)**

(C)   **(D)**

Reasoning geometrically as before (noting that $|x| = |x - 0|$), we can establish Theorem 3.

THEOREM 3

For $p > 0$:

1. $|x| = p$ is equivalent to $x = \pm p$
2. $|x| < p$ is equivalent to $-p < x < p$
3. $|x| > p$ is equivalent to $x < -p$ or $x > p$ Not $-p > x > p$. (Why?)

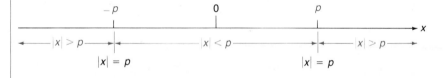

If we replace x in Theorem 3 with $ax + b$, we obtain a more general theorem:

THEOREM 4

For $p > 0$:

1. $|ax + b| = p$ is equivalent to $ax + b = \pm p$
2. $|ax + b| < p$ is equivalent to $-p < ax + b < p$
3. $|ax + b| > p$ is equivalent to $ax + b < -p$ or $ax + b > p$

In fact, if we replace x in Theorem 3 with any algebraic expression, we will obtain other variations of Theorem 3.

EXAMPLE 38   Solve:

(A) $|3x + 5| = 4$ **(B)** $|x| < 5$ **(C)** $|2x - 1| < 3$ **(D)** $|7 - 3x| \leq 2$

Solution   **(A)** $|3x + 5| = 4$ **(B)** $|x| < 5$

$$3x + 5 = \pm 4$$ $$-5 < x < 5$$

$$3x = -5 \pm 4$$

$$x = \frac{-5 \pm 4}{3}$$

$$x = -3, -\tfrac{1}{3}$$

(C) $|2x - 1| < 3$ **(D)** $|7 - 3x| \leq 2$

$$-3 < 2x - 1 < 3$$ $$-2 \leq 7 - 3x \leq 2$$

$$-2 < 2x < 4$$ $$-9 \leq -3x \leq -5$$

$$-1 < x < 2$$ $$3 \geq x \geq \tfrac{5}{3}$$

$$\tfrac{5}{3} \leq x \leq 3$$

PROBLEM 38   Solve:

(A) $|2x - 1| = 8$ **(B)** $|x| \leq 7$ **(C)** $|3x + 3| \leq 9$ **(D)** $|5 - 2x| < 9$

Solution   **(A)**   **(B)**

(C)   **(D)**

EXAMPLE 39   Solve:

(A) $|x| > 3$ **(B)** $|2x - 1| \geq 3$ **(C)** $|7 - 3x| > 2$

Solution   **(A)** $|x| > 3$

$$x < -3 \qquad \text{or} \qquad x > 3$$

(B) $|2x - 1| \geq 3$

$$2x - 1 \leq -3 \qquad \text{or} \qquad 2x - 1 \geq 3$$

$$2x \leq -2 \qquad \text{or} \qquad 2x \geq 4$$

$$x \leq -1 \qquad \text{or} \qquad x \geq 2$$

(C) $|7 - 3x| > 2$

$$7 - 3x < -2 \qquad \text{or} \qquad 7 - 3x > 2$$

$$-3x < -9 \qquad \text{or} \qquad -3x > -5$$

$$x > 3 \qquad \text{or} \qquad x < \tfrac{5}{3}$$

PROBLEM 39

Solve:

(A) $|x| \geq 5$ **(B)** $|4x - 3| > 5$ **(C)** $|6 - 5x| > 16$

Solution

(A)

(B)

(C)

Solutions to Matched Problems

35. (A) 8 **(B)** $\sqrt{5} - 2$ **(C)** $\sqrt{2}$ **(D)** $\sqrt{5} - 2$

36. (A) 4 **(B)** 4 **(C)** 6 **(D)** 11 **(E)** 8 **(F)** 15

37. (A) $x = -8, 4$

(B) $-8 < x < 4$ or $(-8, 4)$

(C) $-8 < x < 4,$ $x \neq -2$

(D) $x < -8$ or $x > 4$

38. (A) $|2x - 1| = 8$ **(B)** $|x| < 7$

$\quad\quad 2x - 1 = \pm 8 \quad\quad\quad\quad\quad -7 < x < 7$

$\quad\quad\quad 2x = 1 \pm 8$

$\quad\quad\quad\quad x = \dfrac{1 \pm 8}{2}$

$\quad\quad\quad\quad x = \dfrac{9}{2}, -\dfrac{7}{2}$

(C) $\quad\quad |3x + 3| \leq 9$ **(D)** $\quad\quad\quad\quad |5 - 2x| < 9$

$\quad -9 \leq 3x + 3 \leq 9 \quad\quad\quad\quad\quad -9 < 5 - 2x < 9$

$\quad -12 \leq 3x \leq 6 \quad\quad\quad\quad\quad\quad -14 < -2x < 4$

$\quad\quad -4 \leq x \leq 2 \quad\quad\quad\quad\quad\quad\quad 7 > x > -2$

$\quad\quad\quad\quad\quad\quad\quad\quad\quad\quad\quad\quad\quad$ or $\;-2 < x < 7$

39. (A) $|x| \geq 5;$ $x \leq -5$ or $x \geq 5$

(B) $|4x - 3| > 5;$ $4x - 3 < -5$ or $4x - 3 > 5$

$\quad\quad\quad\quad\quad\quad\quad\quad\quad 4x < -2$ or $4x > 8$

$\quad\quad\quad\quad\quad\quad\quad\quad\quad\; x < -\frac{1}{2}$ or $x > 2$

(C) $|6 - 5x| > 16;$ $6 - 5x < -16$ or $6 - 5x > 16$

$\quad\quad\quad\quad\quad\quad\quad\quad\quad\quad -5x < -22$ or $-5x > 10$

$\quad\quad\quad\quad\quad\quad\quad\quad\quad\quad\quad x > \frac{22}{5}$ or $x < -2$

Practice Exercise 4-7 ■

Work odd-numbered problems first, check answers, and then work even-numbered problems in areas of weakness. Answers to all problems are in the back of the book. Make every effort to work a problem yourself before you look at an answer.

A   *Simplify, and write without absolute-value signs. Leave radicals in radical form (see Problem 35 following Example 35).*

1. $\left|\sqrt{5}\right|$ _____   **2.** $\left|-\frac{3}{4}\right|$ _____

3. $\left|(-6)-(-2)\right|$ _____   **4.** $\left|(-2)-(-6)\right|$ _____

5. $\left|5-\sqrt{5}\right|$ _____   **6.** $\left|\sqrt{7}-2\right|$ _____

7. $\left|\sqrt{5}-5\right|$ _____   **8.** $\left|2-\sqrt{7}\right|$ _____

Find the distance between points A and B with coordinates a and b, respectively, as given.

9. $a=-7,\quad b=5$ _____   **10.** $a=3,\quad b=12$ _____

11. $a=5,\quad b=-7$ _____   **12.** $a=12,\quad b=3$ _____

13. $a=-16,\quad b=-25$ _____   **14.** $a=-9,\quad b=-17$ _____

Find the indicated distances, given

15. $d(B, O)$ _____   **16.** $d(A, B)$ _____   **17.** $d(O, B)$ _____

18. $d(B, A)$ _____   **19.** $d(B, C)$ _____   **20.** $d(D, C)$ _____

Solve and graph.

21. $\left|x\right|=7$ _____   **22.** $\left|x\right|=5$ _____   **23.** $\left|x\right|\le 7$ _____

24. $\left|t\right|\le 5$ _____   **25.** $\left|x\right|\ge 7$ _____   **26.** $\left|x\right|\ge 5$ _____

27. $\left|y-5\right|=3$ _____   **28.** $\left|t-3\right|=4$ _____

29. $\left|y-5\right|<3$ _____   **30.** $\left|t-3\right|<4$ _____

31. $\left|y-5\right|>3$ _____   **32.** $\left|t-3\right|>4$ _____

33. $\left|u+8\right|=3$ _____   **34.** $\left|x+1\right|=5$ _____

35. $\left|u+8\right|\le 3$ _____   **36.** $\left|x+1\right|\le 5$ _____

37. $\left|u+8\right|\ge 3$ _____   **38.** $\left|x+1\right|\ge 5$ _____

B   *Solve.*

39. $|3x + 4| = 8$ _____   **40.** $|2x - 3| = 5$ _____

41. $|5x - 3| \leq 12$ _____   **42.** $|2x - 3| \leq 5$ _____

43. $|2y - 8| > 2$ _____   **44.** $|3u + 4| > 3$ _____

45. $|5t - 7| = 11$ _____   **46.** $|6m + 9| = 13$ _____

47. $|9 - 7u| < 14$ _____   **48.** $|7 - 9M| < 15$ _____

49. $|1 - \frac{2}{3}x| \geq 5$ _____   **50.** $|\frac{3}{4}x + 3| \geq 9$ _____

51. $|\frac{9}{5}C + 32| < 31$ _____   **52.** $|\frac{5}{9}(F - 32)| < 40$ _____

C   *For what values of x does each of the following hold?*

53. $|x - 5| = x - 5$ _____   **54.** $|x + 7| = x + 7$ _____

55. $|x + 8| = -(x + 8)$ _____   **56.** $|x - 11| = -(x - 11)$ _____

57. $|4x + 3| = 4x + 3$ _____   **58.** $|5x - 9| = (5x - 9)$ _____

The Check Exercise for   **59.** $|5x - 2| = -(5x - 2)$ _____   **60.** $|3x + 7| = -(3x + 7)$ _____
this section is on
page 265.   **61.** Show that $|b - a| = |a - b|$ for all real numbers a and b. _____

Section 4-8   Chapter Review

A **solution** or **root** of an equation in one variable x is a replacement value for x that makes the equality true. To **solve** an equation means to find all solutions—the **solution set**. Two equations are **equivalent** if they have the same solution set. Equations are generally solved by performing operations that produce simpler equivalent equations. The following **properties of equality** yield equivalent equations:

If $a = b$, then $a + c = b + c$ and $a - c = b - c$.

If $a = b$ and $c \neq 0$, then $a \cdot c = b \cdot c$ and $a/c = b/c$.

An **equation-solving strategy** is to:

1. Clear fractions.
2. Simplify both sides.
3. Get all variable terms to one side, constants to the other, and combine like terms.
4. Isolate the variable.

A **linear equation** $ax + b = 0$ has the unique solution $x = -b/a$ if $a \neq 0$ and either no solution or infinitely many if $a = 0$.   *(4-1)*

There is a strategy for solving word problems that can be applied to number problems, geometric problems, and ratio and proportion problems. The **ratio** of a to b is a/b; a **proportion** is a statement that two ratios are equal.   *(4-2)*

The method of Section 4-2 may be applied to rate–time problems and mixture problems. The formula Quantity = Rate × Time involves average rates over time.   *(4-3)*

The **solution set** of a **linear inequality**—an inequality in one variable with no exponents—is the set of all values (**solutions**) that make the inequality true. Inequalities are **equivalent** if they have the same solution set. These **inequality properties** produce equivalent inequalities:

If $a < b$, then $a + c < b + c$ and $a - c < b - c$.

If $a < b$ and $c > 0$, then $a \cdot c < b \cdot c$ and $a/c < b/c$.

If $a < b$ and $c < 0$, then $a \cdot c > b \cdot c$ and $a/c > b/c$.

Solutions to inequalities can be written in **interval notation**, summarized in Table 1.   *(4-6)*

The distance between two points A and B on the real line is given by $d(A, B) = |A - B| = |B - A|$. Equations and inequalities involving absolute value can thus sometimes be solved geometrically.   *(4-7)*

Diagnostic (Review) Exercise 4-8 ■

Work through all the problems in this chapter review and check answers in the back of the book. (Answers to all problems are there, and following each answer is a number in italics indicating the section in which that type of problem is discussed.) Where weaknesses show up, review appropriate sections in the text. When you are satisfied that you know the material, take the practice test following this review.

A   *Solve.*

1. $4x - 9 = x - 15$ _____

2. $2x + 3(x - 1) = 5 - (x - 4)$ _____

3. $4x - 9 < x - 15$ _____   **4.** $-3 < 2x - 5 < 7$ _____

5. $|x| = 6$ _____   **6.** $|x| < 6$ _____

7. $|x| > 6$ _____   **8.** $|y + 9| = 5$ _____

9. $|y + 9| < 5$ _____   **10.** $|y + 9| > 5$ _____

11. $0.4x + 0.3x = 6.3$ _____   **12.** $-\frac{3}{5}y = \frac{2}{3}$ _____

13. $\dfrac{x}{4} - 3 = \dfrac{x}{5}$ _____   **14.** $\dfrac{x}{4} - 1 \geq \dfrac{x}{3}$ _____

15.   Solve $A = bh/2$ for b.   *Area of a triangle* _____

16. If the coordinates of A and B on a real number line are -8 and -2, respectively, find:

   (A) $d(A, B)$ _____   **(B)** $d(B, A)$ _____

B   *Solve each and graph each inequality or absolute-value statement.*

17. $-14 \leq 3x - 2 < 7$ _____   18. $-3 \leq 5 - 2x < 3$ _____

19. $3(2 - x) - 2 \leq 2x - 1$ _____   20. $0.4x - 0.3(x - 3) = 5$ _____

21. $\dfrac{x}{4} - \dfrac{x - 3}{3} = 2$ _____   22. $\dfrac{2}{3m} - \dfrac{1}{4m} = \dfrac{1}{12}$ _____

23. $\dfrac{3x}{x - 5} - 8 = \dfrac{15}{x - 5}$ _____   24. $|4x - 7| = 5$ _____

25. $|4x - 7| \leq 5$ _____   26. $|4x - 7| > 5$ _____

27. $0.05n + 0.1(n - 3) = 1.35$ _____

28. $\dfrac{x + 3}{8} \leq 5 - \dfrac{2 - x}{3}$ _____   29. $-6 < \tfrac{3}{5}(x - 4) \leq -3$ _____

30. $\dfrac{5}{2x + 3} - 5 = \dfrac{-5x}{2x + 3}$ _____   31. $\dfrac{3}{x} - \dfrac{2}{x + 1} = \dfrac{1}{2x}$ _____

32. $\dfrac{11}{9x} - \dfrac{1}{6x^2} = \dfrac{3}{2x}$ _____   33. $\dfrac{u - 3}{2u - 2} = \dfrac{1}{6} - \dfrac{1 - u}{3u - 3}$ _____

34. $\dfrac{x}{x^2 - 6x + 9} - \dfrac{1}{x^2 - 9} = \dfrac{1}{x + 3}$ _____

35.   Solve $S = \dfrac{n(a + L)}{2}$ for L.   *Arithmetic progression* _____

36.   Solve $P = M - Mdt$ for M.   *Mathematics of finance* _____

37.   Indicate true (T) or false (F):

   (A)   If $x < y$ and $a > 0$, then $ax < ay$. _____

   (B)   If $x < y$ and $a < 0$, then $ax > ay$. _____

C   *Solve each and graph each inequality or absolute-value statement.*

38. $|3 - 2x| \leq 5$ _____

39. $\dfrac{x - 3}{12} - \dfrac{x + 2}{9} = \dfrac{1 - x}{6} - 1$ _____

40. $\dfrac{3x}{5} - \dfrac{1}{2}(x - 3) \le \dfrac{1}{3}(x + 2)$ _____

41. $-4 \le \frac{2}{3}(6 - 2x) \le 8$ _____

42. $\dfrac{7}{2 - x} = \dfrac{10 - 4x}{x^2 + 3x - 10}$ _____

43. $|2x - 3| < -2$ _____

44. Solve $y = \dfrac{4x + 3}{2x - 5}$ for x in terms of y. _____

45. Solve $\dfrac{1}{f} = \dfrac{1}{f_1} + \dfrac{1}{f_2}$ for f_1.   *Optics* _____

46. $|2x - 3| = 2x - 3$ for what values of x? _____

47. $|2x - 3| = -(2x - 3)$ for what values of x? _____

APPLICATIONS   *Set up equations or inequalities and solve.*

48. If the width of a rectangle with perimeter 76 centimeters is 2 centimeters less than three-fifths the length, what are the dimensions of the rectangle?

_____

*49. If one car leaves a town traveling at 56 kilometers per hour, how long will it take a second car traveling at 76 kilometers per hour to catch up if the second car leaves 1.5 hours later? _____

50. If you paid \$210 for a stereo that was on sale for 30% off list price, what was the price before the sale? _____

*51. Suppose one printing press can print 45 brochures per minute and a newer press can print 55. How long will it take to print 3,000 brochures if the newer press is brought on the job 10 minutes after the older press has started and both continue until finished? _____

52. If there are 2.54 centimeters in 1 inch, how many inches are in 127 centimeters? _____

53. If 50 milliliters of a solution contains 18 milliliters of alcohol, how many milliliters of alcohol are in 70 milliliters of the same solution? _____

54. A student received grades of 65 and 80 on two tests. What must the student receive on a third test to have an average not lower than 75?

_____

*55. How much distilled water must be added to 60 milliliters of a 30% hydrochloric acid solution to obtain a 25% solution? _____

*56. A 50% alcohol solution and an 80% alcohol solution are in a stockroom. How many deciliters of each should a chemist take to obtain 36 deciliters of a 60% solution? _____

57. If the ratio of all the squirrels in a forest to the ones that were captured, marked, and released is 55/6, and there are 360 marked squirrels, how many squirrels are in the forest? _____

58. A chemical solution is to be kept between 10 and 15°C, inclusive; that is, $10 \le C \le 15$. What is the temperature range in Fahrenheit degrees? $[C = \frac{5}{9}(F - 32)]$ _____

*59. A bottle contains 24 deciliters of a 40% acid solution. How much must be drained off and replaced with pure acid to obtain a 50% solution?

_____

Practice Test Chapter 4 ∎

Take this as if it were a graded test by working the problems within a 50-minute time period. Do not look back in the chapter. Choose one of three levels of difficulty: least difficult, Problems 1–12; more difficult, add Problem 13; most difficult, add Problems 13 and 14. Use the answers in the back of the book to correct your work. The answers are keyed to appropriate text sections so that you can easily locate and review sections where difficulties still persist.

Solve.

1. $3x + 2(x - 1) = 4 - (5x + 6)$ _____

2. $\dfrac{x + 1}{x - 1} + \dfrac{2}{x + 1} = \dfrac{x^2}{x^2 - 1}$ _____

3. $y = \dfrac{2x + 3}{7x - 1}$ for x _____

4. $|3x + 2| = 8$ _____

Solve and graph.

5. $\dfrac{x - 2}{6} \le \dfrac{x - 3}{8} + \dfrac{1}{4}$ _____

6. $|2x - 3| < 5$ _____

7. $-1 \le 2x + 3 < 11$ _____

8. Find a number such that one-third the number is one-half the quantity that is 6 less than the number. _____

9. If a car can travel 260 kilometers on 39 liters of gas, how far will it go on 24 liters? _____

10. Towns A and B are 3,600 kilometers apart. If a truck leaves town A and travels toward town B at 82 kilometers per hour at the same time a truck leaves town B and travels toward A at 38 kilometers per hour, how long will it take the trucks to meeet? _____

11. If one side of a triangle is two-fifths the perimeter P, the second side is 26 meters, and the third side is one-sixth the perimeter, what is the perimeter? _____

12. A gardener needs to mix fertilizer that is 30% nitrogen with some that is 10% nitrogen to obtain 80 pounds of fertilizer that is 18% nitrogen. How much of each component fertilizer should she use? _____

13. A 3-gallon radiator contains a 40% antifreeze solution. How much should be drained and replaced by pure antifreeze to obtain a 60% solution?

_____

14. A pump can drain a pool in 15 hours. If a second pump is added for the job, the time is reduced to 6 hours and 40 minutes. How long would it take for the second pump alone to drain the pool? _____

Check Exercise 4-1 ▪

ANSWER COLUMN

Work the following problems without looking at any text examples. Show your work in the space provided. Write your answer in the answer column.

1. _____

Solve and check Problems 1–5.

2. _____

   1. $-2x + 8 = 2x - 8$

3. _____

4. _____

5. _____

   2. $5(20 - x) + 10x = 360$

6. _____

7. _____

8. _____

   3. $2(3x - 5) - 8x = x - (6 - x)$

9. _____

10. _____

   4. $2[x - (2 - x)] = 3(x - 2) - (2 - x)$

   5. $2x(x - 3) - (x^2 - 16) = x[5 - 3(x + 1)] + 4x^2$

Solve each equation in Problems 6–10.

6. $\dfrac{2x - 8}{6} - \dfrac{x - 1}{3} = \dfrac{x}{2}$

7. $0.25(20 - x) + 0.05x = 3.20$

8. $2 - \dfrac{4x}{3x - 2} = \dfrac{2}{3x - 2}$

9. $\dfrac{y - 4}{10y - 20} = \dfrac{1}{6} - \dfrac{y + 1}{5y - 10}$

10. $3 - \dfrac{2x}{3 - x} = \dfrac{6}{x - 3}$

Check Exercise 4-2 ■

ANSWER COLUMN

Work the following problems without looking at any text examples. Show your work in the space provided. Write your answer in the answer column.

1. _____

For each problem set up an appropriate equation and solve.

2. _____

1. Find a number such that 5 less than one-third the number is 3 less than one-fourth the number.

3. _____

4. _____

5. _____

2. Find three consecutive odd numbers such that twice the sum of the first two is 21 more than 3 times the third.

3. Find the dimensions of a rectangle with perimeter 44 centimeters, if the width is five-sixths the length.

4. If there are 2.2 pounds in 1 kilogram, how many kilograms are in 22 pounds?

5. A 35 by 23-millimeter colored slide is used to make an enlargement whose longest side is 16 inches. How wide (to two decimal places) will the enlargement be if the entire slide is used?

Check Exercise 4-3 ■

Work the following problems without looking at any text examples. Show your work in the space provided. Write your answer in the answer column.

1. _____

Set up an equation and solve. Compute all decimal answers to two decimal places.

2. _____

1. An older capping machine can fill and cap 15 bottles per minute, while a newer machine can fill and cap 20 bottles per minute. How long will it take both machines working together to fill and cap an order of 2,450 bottles?

3. _____

4. _____

5. _____

2. A vending machine contains 30 nickels and dimes worth $2.35. How many of each type of coin is in the machine?

3. You have $1,000 to invest. If you invest a part at 10% and the rest at 18%, how much should you invest at each rate to produce the same yield as if all had been invested at 12%?

4. A speedboat takes 1.25 times longer to go 36 kilometers up a river than to return. If the boat cruises at 30 kilometers per hour in still water, what is the rate of the current?

5. A chemical storeroom has a 20% hydrochloric acid solution and a 60% hydrochloric acid solution. How many milliliters of each should be used to obtain 60 milliliters of a 50% solution?

Check Exercise 4-4 ■

ANSWER COLUMN

Work the following problems without looking at any text examples. Show your work in the space provided. Write your answer in the answer column.

1. _____

2. _____

Set up an equation and solve. Compute all decimal answers to two decimal places.

1. If there is 0.62 mile in 1 kilometer, how many kilometers are in 124 miles?

3. _____

4. _____

5. _____

2. How much pure acid must be added to 60 centiliters of a 60% solution to obtain a 70% solution?

3. A person starts jogging around a large lake at 7 A.M., moving at 200 meters per minute. How long will it take a friend to catch up if the friend starts at 7:10 A.M. and jogs at 300 meters per minute?

4. A candy store wishes to mix $2-per-pound candy with $3.50-per-pound candy to obtain a mix selling for $3 per pound. How much of each should be used to obtain 60 pounds of the mix?

5. If an automated typewriter can type all of the "personalized letters" for an advertising campaign in 15 hours, while a later model can do the job in 10 hours, how long will it take both models together to do the whole job?

Check Exercise 4-5 ∎

Work the following problems without looking at any text examples. Show your work in the space provided. Write your answer in the answer column.

1. _____

Solve each formula or equation for the indicated letter.

2. _____

1. $r = \dfrac{d}{t}$ for t

3. _____

4. _____

5. _____

2. $Q = \dfrac{100M}{C}$ for M

6. _____

7. _____

8. _____

3. $y = 3x - 4$ for x

9. _____

10. _____

4. $3x + 2y = 6$ for y

5. $y = mx + b$ for m

6. $A = \frac{4}{3}\pi R^3$ for π

7. $A = P(1 + rt)$ for r

8. $U = R + RST$ for R

9. $\dfrac{1}{a} = \dfrac{1}{b} + \dfrac{1}{c}$ for b

10. $y = \dfrac{5x - 3}{3x - 1}$ for x

Check Exercise 4-6 ■

ANSWER COLUMN

Work the following problems without looking at any text examples. Show your work in the space provided. Write your answer and/or draw your final graph in the answer column.

1. _____

⊢⊣⊢⊣⊢⊣⊢⊣⊢⊣⊢⊣ → x
-10   -5   0   5   10

2. _____

3. _____

4. _____

5. _____

6. _____

⊢⊣⊢⊣⊢⊣⊢⊣⊢⊣⊢⊣ → x
-10   -5   0   5   10

7. _____

⊢⊣⊢⊣⊢⊣⊢⊣⊢⊣⊢⊣ → x
-10   -5   0   5   10

8. _____

⊢⊣⊢⊣⊢⊣⊢⊣⊢⊣⊢⊣ → m
-10   -5   0   5   10

9. _____

⊢⊣⊢⊣⊢⊣⊢⊣⊢⊣⊢⊣ → u
-10   -5   0   5   10

10. _____

1. Write in interval and inequality notation:

⊢⊣⊢⊣⊢⊣(⊢⊣⊢⊣⊢⊣⊢⊣]⊢⊣ → x
-10   -5   0   5   10

2. Write in interval notation and graph on a real number line:

$$-3 \leq x < 2$$

⊢⊣⊢⊣⊢⊣⊢⊣⊢⊣⊢⊣ → x
-10   -5   0   5   10

Solve. Express your answer in interval notation.

3. $-6x < 24$

4. $\dfrac{y}{-3} \geq -2$

5. $x + 4 > -2$

6. $0 \le 7 - x \le 9$

Solve and graph.

7. $5 - (3 + x) \le x + 6$

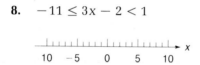

8. $-11 \le 3x - 2 < 1$

9. $\dfrac{3m}{5} - \dfrac{2m - 2}{15} > \dfrac{2}{3} + m$

10. $-3 < \dfrac{3}{4}u - 6 < 3$

Check Exercise 4-7 ■

ANSWER COLUMN

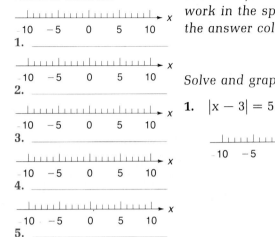

1. _____

-10 -5   0   5   10

2. _____

-10 -5   0   5   10

3. _____

-10 -5   0   5   10

4. _____

-10 -5   0   5   10

5. _____

Work the following problems without looking at any text examples. Show your work in the space provided. Write your answer and draw your final graph in the answer column.

Solve and graph.

1. $|x - 3| = 5$

-10 -5   0   5   10

2. $|x| \leq 4$

-10 -5   0   5   10

3. $|x| > 4$

-10 -5   0   5   10

4. $|2x - 5| \leq 7$

```
  |ˌııⁱıⁱıˌⁱıⁱıⁱˌⁱıⁱıⁱˌⁱıⁱıⁱˌⁱıⁱıⁱˌ→ x
 -10    -5     0     5    10
```

5. $|2x - 5| > 7$

```
  |ˌııⁱıⁱıˌⁱıⁱıⁱˌⁱıⁱıⁱˌⁱıⁱıⁱˌⁱıⁱıⁱˌ→ x
 -10    -5     0     5    10
```

Exponents, Radicals, and Complex Numbers

■ 5

INSTRUCTIONS FOR STUDENTS IN A SELF-PACED CLASS OR LAB

(YES) ── | HAVE YOU HAD INTERMEDIATE ALGEBRA BEFORE THIS COURSE? | ── (NO)

1. Work Diagnostic (Review) Exercise 5-8 on page 317. Check answers in back of book; then work through text sections corresponding to problems missed. (Section numbers are in italics following each answer.)
2. When finished with step 1, take Practice Test Chapter 5 on page 320 as a final check of your understanding of the chapter. Check answers in the back of the book; then review sections where weakness still prevails. (Corresponding section numbers are in italics following each answer.)
3. When you think you are ready, ask your instructor for a graded test for Chapter 5.
4. If your instructor approves, after the test is corrected, go to the next chapter.

1. Work through each section in the chapter as follows:
   (A) Read discussion.
   (B) Read each example and work the corresponding matched problem. Check your solutions to the matched problem in Solutions to Matched Problems on the indicated page.
   (C) At the end of a section work the odd-numbered problems in the Practice Exercise and check answers; then work even-numbered problems in areas of weakness. (Answers to *all* Practice Exercise sets are in the back of the book.)
   (D) Work Check Exercise as instructed. Tear out and turn in as directed by your instructor. (Answers are not in the text.)
2. Repeat each step in item 1 for each section in the chapter.
3. After the instructional part of the chapter is completed, proceed with steps 1 to 4 in the box above this one.

Chapter 5 ■ Exponents, Radicals, and Complex Numbers

Section 5-1　Integer Exponents

- ■ Positive-Integer Exponents
- ■ Zero Exponents
- ■ Negative-Integer Exponents
- ■ Common Errors

■ **Positive-Integer Exponents**

In Section 2-1 we introduced the concept of natural number (positive-integer) exponents and five basic properties that control their use. Recall that

$$a^5 = a \cdot a \cdot a \cdot a \cdot a \quad \text{Five factors of } a$$

and in general:

Definition of Positive-Integer Exponent
For n a positive integer: $a^n = a \cdot a \cdot \cdots \cdot a \quad n \text{ factors of } a$

As a consequence of this definition we stated the following five important properties of exponents:

Exponent Properties
For m and n positive integers: **1.** $a^m a^n = a^{m+n}$ **2.** $(a^n)^m = a^{mn}$ **3.** $(ab)^m = a^m b^m$ **4.** $\left(\dfrac{a}{b}\right)^m = \dfrac{a^m}{b^m}$ **5.** $\dfrac{a^m}{a^n} = \begin{cases} a^{m-n} & \text{if } m > n \\ 1 & \text{if } m = n \\ \dfrac{1}{a^{n-m}} & \text{if } n > m \end{cases}$

Now let us turn to other types of exponents. For example, how should symbols such as

$$8^0 \quad \text{and} \quad 7^{-3}$$

be defined? In this section we will extend the meaning of exponent to include 0 and negative integers. Thus, typical scientific expressions such as the following will then make sense:

The diameter of a red corpuscle is approximately 8×10^{-5} centimeters.

The amount of water found in the air as vapor is about 9×10^{-6} times that found in the sea.

The focal length of a thin lens is given by $f^{-1} = a^{-1} + b^{-1}$.

In extending the concept of exponent beyond the natural numbers, we will require that any new exponent symbol be defined in such a way that all five laws of exponents for natural numbers continue to hold. Thus, we will need only one set of laws for all types of exponents rather than a new set for each new exponent.

■ **Zero Exponents**

We will start by defining the 0 exponent. If all the exponent laws must hold even if some of the exponents are 0, then a^0 ($a \neq 0$) should be defind so that when the first law of exponents is applied,

$$a^0 \cdot a^2 = a^{0+2} = a^2$$

This suggests that a^0 should be defined as 1 for all nonzero real numbers a, since 1 is the only real number that gives a^2 when multiplied by a^2. If we let $a = 0$ and follow the same reasoning, we find that we require

$$0^0 \cdot 0^2 = 0^{0+2} = 0^2 = 0$$

However, since $0^0 \cdot 0^2 = 0^0 \cdot 0 = 0$ is true whatever value is assigned to 0^0, 0^0 could be any real number and is not uniquely determined. For this reason and others, we choose not to define 0^0.

Definition of 0 Exponent
For all real numbers $a \neq 0$: $a^0 = 1$ 0^0 is not defined.

EXAMPLE 1 **(A)** $5^0 = 1$ **(B)** $325^0 = 1$ **(C)** $\left(\frac{1}{3}\right)^0 = 1$
(D) $t^0 = 1, \quad t \neq 0$ **(E)** $(x^2 y^3)^0 = 1, \quad x \neq 0, \quad y \neq 0$

PROBLEM 1

Simplify:

(A) 12^0 **(B)** 999^0 **(C)** $(\frac{2}{7})^0$ **(D)** x^0, $x \neq 0$
(E) $(m^3n^3)^0$, $m, n \neq 0$

Solution

(A)   **(B)**   **(C)**   **(D)**   **(E)**

■ **Negative-Integer Exponents**

To get an idea of how a negative-integer exponent should be defined, we can proceed as above. If the first law of exponents is to hold, then a^{-2} ($a \neq 0$) must be defined so that

$$a^{-2} \cdot a^2 = a^{-2+2} = a^0 = 1$$

Thus a^{-2} must be the reciprocal of a^2; that is,

$$a^{-2} = \frac{1}{a^2}$$

This kind of reasoning leads us to the following general definition.

Definition of Negative-Integer Exponents

If n is a positive integer and a is a nonzero real number, then

$$a^{-n} = \frac{1}{a^n}$$

Of course, it follows, using equality properties,[†] that

$$a^n = \frac{1}{a^{-n}}$$

EXAMPLE 2   **(A)** $a^{-7} = \frac{1}{a^7}$ **(B)** $\frac{1}{x^{-8}} = x^8$ **(C)** $10^{-3} = \frac{1}{10^3}$ or $\frac{1}{1,000}$ or 0.001

(D) $\frac{x^{-3}}{y^{-5}} \left| = \frac{x^{-3}}{1} \cdot \frac{1}{y^{-5}} = \frac{1}{x^3} \cdot \frac{y^5}{1} \right| = \frac{y^5}{x^3}$

[†] Multiply both sides of $a^{-n} = 1/a^n$ by a^n/a^{-n} to obtain

$$\frac{a^n}{a^{-n}} \cdot a^{-n} = \frac{a^n}{a^{-n}} \cdot \frac{1}{a^n}$$

$$a^n = \frac{1}{a^{-n}}$$

PROBLEM 2   Write using positive exponents or no exponents:

(A) x^{-5} (B) $\dfrac{1}{y^{-4}}$ (C) 10^{-2} (D) $\dfrac{m^{-2}}{n^{-3}}$

Solution   (A)   (B)

(C)   (D)

With the definitions of negative exponent and 0 exponent behind us, we can now replace the fifth law of exponents with a simpler form that does not have any restrictions on the relative size of the exponents. Thus,

$$\frac{a^m}{a^n} = a^{m-n} = \frac{1}{a^{n-m}}$$

EXAMPLE 3   Simplify, leaving answers with negative exponents:

(A) $\dfrac{2^5}{2^8} = 2^{5-8} = 2^{-3}$ (B) $\dfrac{10^{-3}}{10^6} = 10^{-3-6} = 10^{-9}$

Simplify, leaving answers with positive exponents:

(C) $\dfrac{2^5}{2^8} = \dfrac{1}{2^{8-5}} = \dfrac{1}{2^3}$ (D) $\dfrac{10^{-3}}{10^6} = \dfrac{1}{10^{6-(-3)}} = \dfrac{1}{10^9}$

PROBLEM 3   Simplify, leaving answers with negative exponents:

(A) $\dfrac{3^4}{3^9}$ (B) $\dfrac{x^{-2}}{x^3}$

Simplify, leaving answers with positive exponents:

(C) $\dfrac{3^4}{3^9}$ (D) $\dfrac{x^{-2}}{x^3}$

Solution   (A)   (B)

(C)   (D)

Table 1 provides a summary of all of our work on exponents to this point.

TABLE 1
INTEGER EXPONENTS AND
THEIR LAWS (SUMMARY)

DEFINITION OF a^p: p AN INTEGER, a A REAL NUMBER	LAWS OF EXPONENTS: n AND m INTEGERS, a AND b REAL NUMBERS
1. If p is a positive integer, then $\qquad a^p = a \cdot a \cdot \cdots \cdot a \qquad p$ factors of a $\qquad$ Example: $3^5 = 3 \cdot 3 \cdot 3 \cdot 3 \cdot 3$ **2.** If $p = 0$, then $\qquad a^p = 1 \qquad a \neq 0$ $\qquad$ Example: $3^0 = 1$ **3.** If p is a negative integer, then $\qquad a^p = \dfrac{1}{a^{-p}} \qquad a \neq 0$ $\qquad$ Example: $3^{-4} \left[= \dfrac{1}{3^{-(-4)}} \right] = \dfrac{1}{3^4}$	**1.** $a^m a^n = a^{m+n}$ **2.** $(a^n)^m = a^{mn}$ **3.** $(ab)^m = a^m b^m$ **4.** $\left(\dfrac{a}{b}\right)^m = \dfrac{a^m}{b^m}$ **5.** $\dfrac{a^m}{a^n} = a^{m-n} = \dfrac{1}{a^{n-m}}$

EXAMPLE 4   Simplify and express answers using positive exponents only:[†]

(A) $a^5 a^{-2} \left[= a^{5-2} \right] = a^3$

(B) $(a^{-3} b^2)^{-2} \left[= (a^{-3})^{-2}(b^2)^{-2} \right] = a^6 b^{-4} = \dfrac{a^6}{b^4}$

(C) $\left(\dfrac{a^{-5}}{a^{-2}}\right)^{-1} \left[= \dfrac{(a^{-5})^{-1}}{(a^{-2})^{-1}} \right] = \dfrac{a^5}{a^2} = a^3$

(D) $\dfrac{4x^{-3} y^{-5}}{6x^{-4} y^3} \left[= \dfrac{2x^{-3-(-4)}}{3y^{3-(-5)}} = \dfrac{2x^{-3+4}}{3y^{3+5}} \right] = \dfrac{2x}{3y^8}$

or, changing to positive exponents first,

$\qquad \dfrac{4x^{-3} y^{-5}}{6x^{-4} y^3} = \dfrac{2x^4}{3x^3 y^3 y^5} = \dfrac{2x}{3y^8}$

(E) $\dfrac{10^{-4} \cdot 10^2}{10^{-3} \cdot 10^5} \left[= \dfrac{10^{-4+2}}{10^{-3+5}} \right] = \dfrac{10^{-2}}{10^2} = \dfrac{1}{10^4} = \dfrac{1}{10,000} = 0.0001$

(F) $\left(\dfrac{m^{-3} m^3}{n^{-2}}\right)^{-2} \left[= \left(\dfrac{m^{-3+3}}{n^{-2}}\right)^{-2} = \left(\dfrac{m^0}{n^{-2}}\right)^{-2} \right]$

$\qquad\qquad\qquad = \left(\dfrac{1}{n^{-2}}\right)^{-2} = \dfrac{1^{-2}}{(n^{-2})^{-2}} = \dfrac{1}{n^4}$

[†] It is important to realize that there are situations where it is desirable to allow negative exponents in an answer (see Section 5-2, for example). In this section we ask you to write answers using positive exponents only so that problems in the exercise set will have unique answer forms.

PROBLEM 4

Simplify and express answers using positive exponents only:

(A) $x^{-2}x^6$ **(B)** $(x^3y^{-2})^{-2}$ **(C)** $\left(\dfrac{x^{-6}}{x^{-2}}\right)^{-1}$

(D) $\dfrac{8m^{-2}n^{-4}}{6m^{-5}n^2}$ **(E)** $\dfrac{10^{-3}\cdot 10^5}{10^{-2}\cdot 10^6}$

Solution

(A)   **(B)**

(C)   **(D)**

(E)

■ Common Errors

As stated earlier, laws of exponents involve products and quotients, not sums and differences. Consider:

CORRECT   COMMON ERROR

$$\frac{a^{-2}y}{b} = \frac{y}{a^2 b} \qquad \frac{a^{-2}+y}{b} = \frac{y}{a^2 b}$$

The plus sign in the numerator of the second illustration makes a big difference. Actually, $\dfrac{a^{-2}+y}{b}$ represents a compact way of writing a complex fraction. To simplify, we replace a^{-2} with $1/a^2$, then proceed as in Section 3-6:

$$\frac{a^{-2}+y}{b} = \frac{\dfrac{1}{a^2}+y}{b} = \frac{a^2\left(\dfrac{1}{a^2}+y\right)}{a^2\cdot b}$$

$$= \frac{1+a^2 y}{a^2 b}$$

Also, consider the following:

CORRECT   COMMON ERROR

$$(a^{-1}b^{-1})^2 = a^{-2}b^{-2} \qquad (a^{-1}+b^{-1})^2 = a^{-2}+b^{-2}$$

$$= \frac{1}{a^2 b^2} \qquad\qquad\qquad = \frac{1}{a^2+b^2}$$

The second illustration contains two errors:

$$(a^{-1} + b^{-1})^2 \qquad \text{is not equal to} \qquad a^{-2} + b^{-2}$$

and

$$a^{-2} + b^{-2} \qquad \text{is not equal to} \qquad \frac{1}{a^2 + b^2}$$

The problem is worked correctly in Example 5(B).

EXAMPLE 5   Simplify and express answers using positive exponents only:

(A) $\dfrac{3^{-2} + 2^{-1}}{11} = \dfrac{\frac{1}{3^2} + \frac{1}{2}}{11} = \dfrac{\frac{2}{18} + \frac{9}{18}}{11} = \dfrac{\frac{11}{18}}{11} = \dfrac{11}{18} \div 11 = \dfrac{11}{18} \cdot \dfrac{1}{11} = \dfrac{1}{18}$

(B) $(a^{-1} + b^{-1})^2 = \left(\dfrac{1}{a} + \dfrac{1}{b}\right)^2 = \left(\dfrac{b + a}{ab}\right)^2 = \dfrac{b^2 + 2ab + a^2}{a^2 b^2}$

PROBLEM 5   Simplify and express answers using positive exponents only:

(A) $\dfrac{2^{-2} + 3^{-1}}{5}$ **(B)** $(x^{-1} + y^{-1})^2$

Solution   **(A)**   **(B)**

Solutions to Matched Problems

1. All are equal to 1

2. (A) $x^{-5} = \dfrac{1}{x^5}$ **(B)** $\dfrac{1}{y^{-4}} = y^4$ **(C)** $10^{-2} = \dfrac{1}{10^2} = \dfrac{1}{100} = 0.01$

    (D) $\dfrac{m^{-2}}{n^{-3}} = \dfrac{n^3}{m^2}$

3. (A) $\dfrac{3^4}{3^9} = 3^{4-9} = 3^{-5}$ **(B)** $\dfrac{x^{-2}}{x^3} = x^{-2-3} = x^{-5}$

    (C) $\dfrac{3^4}{3^9} = \dfrac{1}{3^{9-4}} = \dfrac{1}{3^5}$

    (D) $\dfrac{x^{-2}}{x^3} = \dfrac{1}{x^{3-(-2)}} = \dfrac{1}{x^5}$

4. (A) $x^{-2}x^6 = x^4$ **(B)** $(x^3y^{-2})^{-2} = x^{-6}y^4 = \dfrac{y^4}{x^6}$

(C) $\left(\dfrac{x^{-6}}{x^{-2}}\right) = \dfrac{x^6}{x^2} = x^4$

(D) $\dfrac{8m^{-2}n^{-4}}{6m^{-5}n^2} = \dfrac{4m^{-2-(-5)}}{3n^{2-(-4)}} = \dfrac{4m^3}{3n^6}$

(E) $\dfrac{10^{-3} \cdot 10^5}{10^{-2} \cdot 10^6} = \dfrac{10^2}{10^4} = \dfrac{1}{10^2} = \dfrac{1}{100} = 0.01$

5. (A) $\dfrac{2^{-2} + 3^{-1}}{5} = \dfrac{\frac{1}{2^2} + \frac{1}{3}}{5} = \dfrac{\frac{1}{4} + \frac{1}{3}}{5} = \boxed{\dfrac{12 \cdot \frac{1}{4} + 12 \cdot \frac{1}{3}}{12 \cdot 5}} = \dfrac{3 + 4}{60} = \dfrac{7}{60}$

(B) $(x^{-1} + y^{-1})^2 = \left(\dfrac{1}{x} + \dfrac{1}{y}\right)^2 = \left(\dfrac{y + x}{xy}\right)^2$ or $\dfrac{x^2 + 2xy + y^2}{x^2y^2}$

Practice Exercise 5-1 ■

Work odd-numbered problems first, check answers, and then work even-numbered problems in areas of weakness. Answers to all problems are in the back of the book. Make every effort to work a problem yourself before you look at an answer.

Simplify and write answers using positive exponents only.

A **1.** 23^0 _____   **2.** 10^0 _____   **3.** y^0 _____

4. x^0 _____   **5.** 3^{-3} _____   **6.** 2^{-2} _____

7. m^{-7} _____   **8.** x^{-4} _____   **9.** $\dfrac{1}{4^{-3}}$ _____

10. $\dfrac{1}{3^{-2}}$ _____   **11.** $\dfrac{1}{y^{-5}}$ _____   **12.** $\dfrac{1}{x^{-3}}$ _____

13. $10^7 \cdot 10^{-5}$ _____   **14.** $10^{-4} \cdot 10^6$ _____   **15.** $y^{-3}y^4$ _____

16. x^6x^{-2} _____   **17.** u^5u^{-5} _____   **18.** $m^{-3}m^3$ _____

19. $\dfrac{10^3}{10^{-7}}$ _____   **20.** $\dfrac{10^8}{10^{-3}}$ _____   **21.** $\dfrac{x^9}{x^{-2}}$ _____

22. $\dfrac{a^8}{a^{-4}}$ _____   **23.** $\dfrac{z^{-2}}{z^3}$ _____   **24.** $\dfrac{b^{-3}}{b^5}$ _____

25. $\dfrac{10^{-1}}{10^6}$ _____   **26.** $\dfrac{10^{-4}}{10^2}$ _____   **27.** $(10^{-4})^{-3}$ _____

28. $(2^{-3})^{-2}$ _____   **29.** $(y^{-2})^{-4}$ _____   **30.** $(x^{-5})^{-2}$ _____

31. $(u^{-5}v^{-3})^{-2}$ _____   **32.** $(x^{-3}y^{-2})^{-1}$ _____

33. $(x^2y^{-3})^2$ _____   **34.** $(x^{-2}y^3)^2$ _____

35. $(x^{-2}y^3)^{-1}$ _____   **36.** $(x^2y^{-3})^{-1}$ _____

B **37.** $(m^2)^0$ _____   **38.** $1,231^0$ _____   **39.** $\dfrac{10^{-3}}{10^{-5}}$ _____

40. $\dfrac{10^{-2}}{10^{-4}}$ _____   **41.** $\dfrac{y^{-2}}{y^{-3}}$ _____   **42.** $\dfrac{x^{-3}}{x^{-2}}$ _____

43. $\dfrac{10^{-13} \cdot 10^{-4}}{10^{-21} \cdot 10^3}$ _____   **44.** $\dfrac{10^{23} \cdot 10^{-11}}{10^{-3} \cdot 10^{-2}}$ _____

45. $\dfrac{18 \times 10^{12}}{6 \times 10^{-4}}$ _____   **46.** $\dfrac{8 \times 10^{-3}}{2 \times 10^{-5}}$ _____

47. $\left(\dfrac{y}{y^{-2}}\right)^3$ _____   **48.** $\left(\dfrac{x^2}{x^{-1}}\right)^2$ _____   **49.** $\dfrac{1}{(3mn)^{-2}}$ _____

50. $(2cd^2)^{-3}$ _____   **51.** $(2mn^{-3})^3$ _____   **52.** $(3x^3y^{-2})^2$ _____

53. $(m^4n^{-5})^{-3}$ _____   **54.** $(x^{-3}y^2)^{-2}$ _____

55. $(2^2 3^{-3})^{-1}$ _____   **56.** $(2^{-3}3^2)^{-2}$ _____

57. $(10^{12} \cdot 10^{-12})^{-1}$ _____   **58.** $(10^2 \cdot 3^0)^{-2}$ _____

59. $\dfrac{8x^{-3}y^{-1}}{6x^2y^{-4}}$ _____   **60.** $\dfrac{9m^{-4}n^3}{12m^{-1}n^{-1}}$ _____

61. $\dfrac{2a^6b^{-2}}{16a^{-3}b^2}$ _____   **62.** $\dfrac{4x^{-2}y^{-3}}{2x^{-3}y^{-1}}$ _____

63. $\left(\dfrac{x^{-1}}{x^{-8}}\right)^{-1}$ _____   **64.** $\left(\dfrac{n^{-3}}{n^{-2}}\right)^{-2}$ _____

65. $\left(\dfrac{m^{-2}n^3}{m^4n^{-1}}\right)^2$ _____   **66.** $\left(\dfrac{x^4y^{-1}}{x^{-2}y^3}\right)^2$ _____

67. $\left(\dfrac{6nm^{-2}}{3m^{-1}n^2}\right)^{-3}$ _____   **68.** $\left(\dfrac{2x^{-3}y^2}{4xy^{-1}}\right)^{-2}$ _____

69. $\left[\left(\dfrac{x^{-2}y^3t}{x^{-3}y^{-2}t^2}\right)^2\right]^{-1}$ _____   **70.** $\left[\left(\dfrac{u^3v^{-1}w^{-2}}{u^{-2}v^{-2}w}\right)^{-2}\right]^2$ _____

71. $\left(\dfrac{2^2x^2y^0}{8x^{-1}}\right)^{-2}\left(\dfrac{x^{-3}}{x^{-5}}\right)^3$ _____

72. $\left(\dfrac{3^3x^0y^{-2}}{2^3x^3y^{-5}}\right)^{-1}\left(\dfrac{3^3x^{-1}y}{2^2x^2y^{-2}}\right)^2$ _____

C **73.** $(a^2-b^2)^{-1}$ _____ **74.** $(x+2)^{-2}$ _____

75. $\dfrac{x^{-1}+y^{-1}}{x+y}$ _____ **76.** $\dfrac{2^{-1}+3^{-1}}{25}$ _____

77. $\dfrac{c-d}{c^{-1}-d^{-1}}$ _____ **78.** $\dfrac{12}{2^{-2}+3^{-1}}$ _____

79. $(x^{-1}+y^{-1})^{-1}$ _____ **80.** $(2^{-2}+3^{-2})^{-1}$ _____

81. $(x^{-1}-y^{-1})^2$ _____ **82.** $(10^{-2}+10^{-3})^{-1}$ _____

The Check Exercise for this section is on page 321.

83. $\left(\dfrac{x^{-1}}{x^{-1}-y^{-1}}\right)^{-1}$ _____ **84.** $\left[\dfrac{u^{-2}-v^{-2}}{(u^{-1}-v^{-1})^2}\right]^{-1}$ _____

Section 5-2 Scientific Notation and Applications

Work in science and engineering often involves the use of very, very large numbers:

The estimated free oxygen of the earth weighs approximately 1,500,000,000,000,000,000,000 grams.

Also involved is the use of very, very small numbers:

The probable mass of a hydrogen atom is 0.000 000 000 000 000 000 000 001 7 gram.

Writing and working with numbers of this type in standard decimal notation is generally awkward. It is often convenient to represent numbers of this type in **scientific notation**—that is, as the product of a number in the interval [1, 10) and a power of 10. Any decimal fraction, however large or small, can be represented in scientific notation.

EXAMPLE 6 Here are some examples of decimal fractions and scientific notation:

$$100 = 1 \times 10^2 \qquad\qquad 0.001 = 1 \times 10^{-3}$$
$$1{,}000{,}000 = 1 \times 10^6 \qquad\qquad 0.000\,01 = 1 \times 10^{-5}$$
$$5 = 5 \times 10^0 \qquad\qquad 0.7 = 7 \times 10^{-1}$$
$$35 = 3.5 \times 10 \qquad\qquad 0.083 = 8.3 \times 10^{-2}$$
$$430 = 4.3 \times 10^2 \qquad\qquad 0.004\,3 = 4.3 \times 10^{-3}$$
$$5{,}870 = 5.87 \times 10^3 \qquad\qquad 0.000\,687 = 6.87 \times 10^{-4}$$
$$8{,}910{,}000 = 8.91 \times 10^6 \qquad\qquad 0.000\,000\,36 = 3.6 \times 10^{-7}$$

Can you discover a simple mechanical rule that relates the number of decimal places the decimal is moved with the power of 10 that is used?

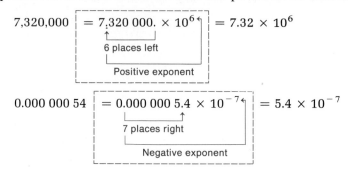

$$7{,}320{,}000 \quad = 7.320\,000. \times 10^6 \quad = 7.32 \times 10^6$$

6 places left

Positive exponent

$$0.000\,000\,54 \quad = 0.000\,000\,5.4 \times 10^{-7} \quad = 5.4 \times 10^{-7}$$

7 places right

Negative exponent

PROBLEM 6   Write in scientific notation:

(A) 450   (B) 27,000   (C) 0.05
(D) 0.000 006 3 (E) 0.000 1   (F) 10,000

Solution   (A)   (B)   (C)

(D)   (E)   (F)

EXAMPLE 7   Evaluation of a complicated arithmetic problem:

$$\frac{(0.000\,000\,000\,000\,026)(720)}{(48{,}000{,}000{,}000)(0.001\,3)}$$

Write each number in scientific notation.

$$= \frac{(2.6 \times 10^{-14})(7.2 \times 10^2)}{(4.8 \times 10^{10})(1.3 \times 10^{-3})}$$

Collect all powers of 10 together.

$$= \frac{\overset{2}{(2.6)}\overset{6}{(7.2)}}{\underset{4}{(4.8)}\underset{1}{(1.3)}} \cdot \frac{(10^{-14})(10^2)}{(10^{10})(10^{-3})}$$

Simplify each group and write the result in scientific notation.

$$= 3 \times 10^{-19}$$

Note: If you try to work this problem directly using a hand calculator, you will find that some of the numbers will not fit unless they are first converted to scientific notation. If you have a calculator, try it. Some calculators can compute directly in scientific notation and read out in scientific notation.

PROBLEM 7   Convert to scientific notation and evaluate:

$$\frac{(42{,}000)(0.000\,000\,000\,09)}{(600{,}000{,}000{,}000)(0.000\,21)}$$

Solution

Figure 1 shows the relative size of a number of familiar objects on a power-of-10 scale. Note that 10^{10} is not just double 10^5.

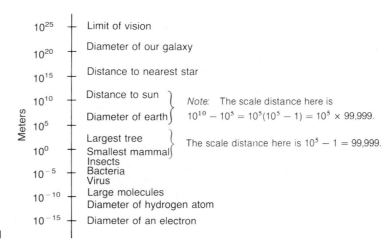

FIGURE 1

We are able to look back into time by looking out into space. Since light travels at a fast but finite rate, we see heavenly bodies not as they exist now, but as they existed sometime in the past. If the distance between the sun and the earth is approximately 9.3×10^7 miles and if light travels at the rate of approximately 1.86×10^5 miles per second, we see the sun as it was how many minutes ago?

$$t = \frac{d}{r} \approx \frac{9.3 \times 10^7}{1.86 \times 10^5} = 5 \times 10^2 = 500 \text{ seconds} \qquad \text{or} \qquad \frac{500}{60} \approx 8.3 \text{ minutes}$$

Hence, we always see the sun as it was approximately 8.3 minutes ago.

Solutions to Matched
Problems

6. **(A)** 4.5×10^2 **(B)** 2.7×10^4 **(C)** 5×10^{-2}
    (D) 6.3×10^{-6} **(E)** 1.0×10^{-4} **(F)** 1.0×10^4

7. $\dfrac{(42,000)(0.000\,000\,000\,09)}{(600,000,000,000)(0.000\,21)} = \dfrac{(4.2 \times 10^4)(9 \times 10^{-11})}{(6 \times 10^{11})(2.1 \times 10^{-4})}$

$$= \frac{(4.2)(9)}{(6)(2.1)} \cdot \frac{(10^4)(10^{-11})}{(10^{11})(10^{-4})} = 3 \times 10^{-14}$$

Practice Exercise 5-2 ∎

Work odd-numbered problems first, check answers, and then work even-numbered problems in areas of weakness. Answers to all problems are in the back of the book. Make every effort to work a problem yourself before you look at an answer.

A   *Write in scientific notation.*

1. 70 _____   **2.** 50 _____   **3.** 800 _____

4. 600 _____   **5.** 80,000 _____   **6.** 600,000 _____

7. 0.008 _____   **8.** 0.06 _____

9. 0.000 000 08 _____   **10.** 0.000 06 _____

11. 52 _____   **12.** 35 _____   **13.** 0.63 _____

14. 0.72 _____   **15.** 340 _____   **16.** 270 _____

17. 0.085 _____   **18.** 0.032 _____   **19.** 6,300 _____

20. 5,200 _____   **21.** 0.000 006 8 _____

22. 0.000 72 _____

Write as a decimal fraction.

23. 8×10^2 _____   **24.** 5×10^2 _____   **25.** 4×10^{-2} _____

26. 8×10^{-2} _____   **27.** 3×10^5 _____   **28.** 6×10^6 _____

29. 9×10^{-4} _____   **30.** 2×10^{-5} _____   **31.** 5.6×10^4 _____

32. 7.1×10^3 _____   **33.** 9.7×10^{-3} _____

34. 8.6×10^{-4} _____   **35.** 4.3×10^5 _____

36. 8.8×10^6 _____   **37.** 3.8×10^{-7} _____

38. 6.1×10^{-6} _____

B   *Write in scientific notation.*

39. 5,460,000,000 _____   **40.** 42,700,000 _____

41. 0.000 000 072 9 _____   **42.** 0.000 072 3 _____

43. The energy of a laser beam can go as high as 10,000,000,000,000 watts.

_____

44. The distance that light travels in 1 year is called a light-year. It is approximately 5,870,000,000,000 miles. _____

45. The nucleus of an atom has a diameter of a little more than 1/100,000 that of the whole atom. _____

46. The mass of one water molecule is 0.000 000 000 000 000 000 000 03 gram.

_____

Write as a decimal fraction.

47. 8.35×10^{10} _____   **48.** 3.46×10^9 _____

49. 6.14×10^{-12} _____   **50.** 6.23×10^{-7} _____

51. The diameter of the sun is approximately 8.65×10^5 miles. _____

52. The distance from the earth to the sun is approximately 9.3×10^7 miles.

_____

53. The probable mass of a hydrogen atom is 1.7×10^{-24} gram. _____

54. The diameter of a red corpuscle is approximately 7.5×10^{-5} centimeter.

_____

Simplify and express answer in scientific notation.

55. $(3 \times 10^{-6})(3 \times 10^{10})$ _____   **56.** $(4 \times 10^5)(2 \times 10^{-3})$ _____

57. $(2 \times 10^3)(3 \times 10^{-7})$ _____   **58.** $(4 \times 10^{-8})(2 \times 10^5)$ _____

59. $\dfrac{6 \times 10^{12}}{2 \times 10^7}$ _____   **60.** $\dfrac{9 \times 10^8}{3 \times 10^5}$ _____

61. $\dfrac{15 \times 10^{-2}}{3 \times 10^{-6}}$ _____   **62.** $\dfrac{12 \times 10^3}{4 \times 10^{-4}}$ _____

Convert each numeral to scientific notation and simplify. Express your answer in scientific notation and as a decimal fraction.

63. $\dfrac{(90{,}000)(0.000\ 002)}{0.006}$ _____   **64.** $\dfrac{(0.000\ 6)(4{,}000)}{0.000\ 12}$ _____

65. $\dfrac{(60{,}000)(0.000\ 003)}{(0.000\ 4)(1{,}500{,}000)}$ _____   **66.** $\dfrac{(0.000\ 039)(140)}{(130{,}000)(0.000\ 21)}$ _____

C **67.** If the mass of the earth is 6×10^{27} grams and each gram is 1.1×10^{-6} ton, find the mass of the earth in tons. _____

68. In 1929 Vernadsky, a biologist, estimated that all the free oxygen of the earth weighs 1.5×10^{21} grams and is produced by life alone. If one gram is approximately 2.2×10^{-3} pound, what is the amount of free oxygen in pounds? _____

69. Designers of high-speed computers are currently thinking of single-addition times of 10^{-7} second (100 nanoseconds). How many additions would such a computer be able to perform in 1 second? In 1 minute? _____

70. If electricity travels in a computer circuit at the speed of light (1.86×10^5 miles per second), how far will it travel in the time it takes the computer in the preceding problem to complete a single addition? (Size of circuits is becoming a critical problem in computer design.) Give the answer in miles and in feet. _____

71. India has a population of 713,000,000 people and a land area of 1,269,000 square miles. What is the population density (people per square mile)?

_____

72. The United States has a population (1980) of 227,000,000 and a land area of 3,539,000 square miles. What is the population density? _____

OPTIONAL CALCULATOR
EXERCISE

The Check Exercise for
this section is on
page 323.

Use a scientific calculator and rework Problems 55–72 using scientific notation.

Section 5-3   Rational Exponents

- Roots of Real Numbers
- Rational Exponents

Roots of Real Numbers

What do we mean by a root of a number? Perhaps you recall that a square root of a number b is a number a such that $a^2 = b$ and a cube root of a number b is a number a such that $a^3 = b$.

What are the square roots of 4?

  2 is a square root of 4, since $2^2 = 4$

  -2 is a square root of 4, since $(-2)^2 = 4$

Thus, 4 has two real square roots, one the negative of the other.

What are the cube roots of 8?

  2 is a cube root of 8, since $2^3 = 8$

and 2 is the only real number with this property. In general:

Definition of an _n_th Root

For n a natural number:

  a is an nth root of b if $a^n = b$

  2 is a fourth root of 16, since $2^4 = 16$

How many real square roots of 9 exist? Of 7? Of -4? How many real fourth roots of 7 exist? Of -7? How many real cube roots of 27 are there? The following important theorem (which we state without proof) answers these questions completely.

THEOREM 1 **NUMBER OF REAL *n*th ROOTS OF A REAL NUMBER *b*[†]**

	n EVEN	*n* ODD
b positive	Two real *n*th roots −2 and 2 are both fourth roots of 16	One real *n*th root 2 is the only real cube root of 8
b negative	No real *n*th root −4 has no real square roots	One real *n*th root −2 is the only real cube root of −8

Thus:

5 has two real square roots, two real fourth roots, and so on.

7 has one real cube root, one real fifth root, and so on.

What symbols do we use to represent these roots? We turn to this question now.

■ **Rational Exponents**

If all exponent laws are to continue to hold even if some of the exponents are not integers, then

$$(5^{1/2})^2 = 5^{2/2} = 5 \qquad \text{and} \qquad (7^{1/3})^3 = 7^{3/3} = 7$$

Hence, $5^{1/2}$ must name a square root of 5, since $(5^{1/2})^2 = 5$. Similarly $7^{1/3}$ must name the cube root of 7, since $(7^{1/3})^3 = 7$.

In general, for n a natural number and b not negative when n is even,

$$(b^{1/n})^n = b^{n/n} = b$$

Thus, $b^{1/n}$ must name an nth root of b. Which real nth root of b does $b^{1/n}$ represent if n is even and b is positive? (According to Theorem 1 there are two real nth roots.) We answer this question in the following definition:

Definition of $b^{1/n}$

For n a natural number:

$b^{1/n}$ is an nth root of b

If n is even and b is positive, then $b^{1/n}$ represents the *positive* real nth root of b (sometimes called the **principal nth root of b**), $-b^{1/n}$ represents the negative real nth root of b, and $(-b)^{1/n}$ does not represent a real number.

If n is odd, then $b^{1/n}$ represents the real nth root of b.

$0^{1/n} = 0$.

$16^{1/2} = 4$	$-16^{1/2} = -4$	$(-16)^{1/2}$ is not real
(not −4 and 4)	($-16^{1/2}$ and $(-16)^{1/2}$ are not the same)	
$32^{1/5} = 2$	$(-32)^{1/5} = -2$	$0^{1/9} = 0$

[†] In this section we limit our discussion to real roots of real numbers. After the real numbers are extended to the complex numbers (see Section 5-7), then additional roots can be considered. For example, it turns out that 8 has three cube roots; in addition to the real number 2, there are two other cube roots in the complex number system. A thorough discussion of roots in the complex number system is reserved for advanced courses on the subject.

EXAMPLE 8   **(A)** $4^{1/2} = 2$

(B) $-4^{1/2} = -2$ Note carefully the difference between parts (B) and (C).

(C) $(-4)^{1/2}$ is not a real number   **(D)** $8^{1/3} = 2$

(E) $(-8)^{1/3} = -2$ **(F)** $0^{1/5} = 0$

PROBLEM 8   Find each of the following:

(A) $9^{1/2}$ **(B)** $-9^{1/2}$ **(C)** $(-9)^{1/2}$

(D) $27^{1/3}$ **(E)** $(-27)^{1/3}$ **(F)** $0^{1/4}$

Solution   **(A)**   **(B)**   **(C)**

(D)   **(E)**   **(F)**

How should an expression such as $5^{2/3}$ be defined? If the properties of exponents are to continue to hold for all rational exponents, then $5^{2/3} = (5^{1/3})^2$; that is, $5^{2/3}$ must represent the square of the cube root of 5. Thus, we are led to the following general definition:

Definition of $b^{m/n}$ and $b^{-m/n}$

For m and n natural numbers and b any real number (except b cannot be negative when n is even):

$$b^{m/n} = (b^{1/n})^m \qquad \text{and} \qquad b^{-m/n} = \frac{1}{b^{m/n}}$$

$4^{3/2} = (4^{1/2})^3 = 2^3 = 8$ $4^{-3/2} = \frac{1}{4^{3/2}} = \frac{1}{8}$ $(-4)^{3/2}$ is not real

$(-32)^{3/5} = [(-32)^{1/5}]^3 = (-2)^3 = -8$

We have now discussed $b^{m/n}$ for all rational numbers m/n and real numbers b. It can be shown, though we will not do so, that all five properties of exponents discussed in Section 5-1 continue to hold for rational exponents so long as we avoid even roots of negative numbers. With the latter restriction in effect, the following useful relationship is an immediate consequence of the exponent properties:

$$b^{m/n} = (b^{1/n})^m = (b^m)^{1/n}$$

EXAMPLE 9   **(A)** $8^{2/3} = (8^{1/3})^2 = 2^2 = 4$ or $8^{2/3} = (8^2)^{1/3} = 64^{1/3} = 4$

(B) $(-8)^{5/3} = [(-8)^{1/3}]^5 = (-2)^5 = -32$ Easier than computing $[(-8)^5]^{1/3}$

(C) $(3x^{1/3})(2x^{1/2}) = 6x^{1/3 + 1/2} = 6x^{5/6}$

(D) $(2x^{1/3}y^{-2/3})^3 = 8xy^{-2}$ or $\dfrac{8x}{y^2}$

(E) $\left(\dfrac{4x^{1/3}}{x^{1/2}}\right)^{1/2} = \dfrac{4^{1/2}x^{1/6}}{x^{1/4}} = \dfrac{2}{x^{1/4-1/6}} = \dfrac{2}{x^{1/12}}$ or $2x^{-1/12}$

(F) $(2a^{1/2} + b^{1/2})(a^{1/2} + 3b^{1/2}) = 2a + 7a^{1/2}b^{1/2} + 3b$

PROBLEM 9

Simplify, and express the answers using positive exponents only:

(A) $9^{3/2}$ **(B)** $(-27)^{4/3}$ **(C)** $(5y^{3/4})(2y^{1/3})$

(D) $(2x^{-3/4}y^{1/4})^4$ **(E)** $\left(\dfrac{8x^{1/2}}{x^{2/3}}\right)^{1/3}$ **(F)** $(x^{1/2} - 2y^{1/2})(3x^{1/2} + y^{1/2})$

Solution

(A) **(B)**

(C) **(D)**

(E) **(F)**

The properties of exponents can be used as long as we are dealing with symbols that name real numbers. Can you resolve the following contradiction?

$$-1 = (-1)^{2/2} = [(-1)^2]^{1/2} = 1^{1/2} = 1$$

The second member of the equality chain, $(-1)^{2/2}$, involves the even root of a negative number, which is not real. Thus we see that the properties of exponents do not necessarily hold when we are dealing with nonreal quantities unless further restrictions are imposed. One such restriction is to require all rational exponents to be reduced to lowest terms.

Solutions to Matched Problems

8. (A) $9^{1/2} = 3$ **(B)** $-9^{1/2} = -3$ **(C)** $(-9)^{1/2}$ is not a real number.
(D) $27^{1/3} = 3$ **(E)** $(-27)^{1/3} = -3$ **(F)** $0^{1/4} = 0$

9. (A) $9^{3/2} = (9^{1/2})^3 = 3^3 = 27$

(B) $(-27)^{4/3} = [(-27)^{1/3}]^4 = (-3)^4 = 81$

(C) $(5y^{3/4})(2y^{1/3}) = 10y^{3/4+1/3} = 10y^{13/12}$

(D) $(2x^{-3/4}y^{1/4})^4 = 2^4x^{-3}y = \dfrac{16y}{x^3}$

(E) $\left(\dfrac{8x^{1/2}}{x^{2/3}}\right)^{1/3} = \dfrac{8^{1/3}x^{1/6}}{x^{2/9}} = \dfrac{2}{x^{2/9-1/6}} = \dfrac{2}{x^{1/18}}$

(F) $(x^{1/2} - 2y^{1/2})(3x^{1/2} + y^{1/2}) = 3x - 5x^{1/2}y^{1/2} - 2y$

Practice Exercise 5-3 ■

Work odd-numbered problems first, check answers, and then work even-numbered problems in areas of weakness. Answers to all problems are in the back of the book. Make every effort to work a problem yourself before you look at an answer.

In Problems 1–72 all variables represent positive real numbers.

A   *Most of the following are integers. Find them.*

1. $25^{1/2}$ _____

2. $36^{1/2}$ _____

3. $(-25)^{1/2}$ _____

4. $(-36)^{1/2}$ _____

5. $8^{1/3}$ _____

6. $27^{1/3}$ _____

7. $(-8)^{1/3}$ _____

8. $(-27)^{1/3}$ _____

9. $-8^{1/3}$ _____

10. $-27^{1/3}$ _____

11. $16^{3/2}$ _____

12. $25^{3/2}$ _____

13. $8^{2/3}$ _____

14. $27^{2/3}$ _____

Simplify, and express the answer using positive exponents only.

15. $x^{1/4}x^{3/4}$ _____

16. $y^{1/5}y^{2/5}$ _____

17. $\dfrac{x^{2/5}}{x^{3/5}}$ _____

18. $\dfrac{a^{2/3}}{a^{1/3}}$ _____

19. $(x^4)^{1/2}$ _____

20. $(y^{1/2})^4$ _____

21. $(a^3b^9)^{1/3}$ _____

22. $(x^4y^2)^{1/2}$ _____

23. $\left(\dfrac{x^9}{y^{12}}\right)^{1/3}$ _____

24. $\left(\dfrac{m^{12}}{n^{16}}\right)^{1/4}$ _____

25. $(x^{1/3}y^{1/2})^6$ _____

26. $\left(\dfrac{u^{1/2}}{v^{1/3}}\right)^{12}$ _____

B   *Most of the following are rational numbers. Find them.*

27. $(\frac{4}{25})^{1/2}$ _____

28. $(\frac{9}{4})^{1/2}$ _____

29. $(\frac{4}{25})^{3/2}$ _____

30. $(\frac{9}{4})^{3/2}$ _____

31. $(\frac{1}{8})^{2/3}$ _____

32. $(\frac{1}{27})^{2/3}$ _____

33. $36^{-1/2}$ _____

34. $25^{-1/2}$ _____

35. $25^{-3/2}$ _____

36. $16^{-3/2}$ _____

37. $5^{3/2}\cdot 5^{1/2}$ _____

38. $7^{2/3}\cdot 7^{4/3}$ _____

39. $(3^6)^{-1/3}$ _____

40. $(4^{-8})^{3/16}$ _____

Simplify, and express the answer using positive exponents only.

41. $x^{1/4}x^{-3/4}$ _____

42. $\dfrac{d^{2/3}}{d^{-1/3}}$ _____

43. $n^{3/4}n^{-2/3}$ _____

44. $m^{1/2}m^{-1/3}$ _____

45. $(x^{-2/3})^{-6}$ _____

46. $(y^{-8})^{1/16}$ _____

47. $(4u^{-2}v^4)^{1/2}$ _____

48. $(8x^3y^{-6})^{1/3}$ _____

49. $(x^4y^6)^{-1/2}$ _____

50. $(4x^{1/2}y^{3/2})^2$ _____

51. $\left(\dfrac{x^{-2/3}}{y^{-1/2}}\right)^{-6}$ _____

52. $\left(\dfrac{m^{-3}}{n^2}\right)^{-1/6}$ _____

53. $\left(\dfrac{25x^5y^{-1}}{16x^{-3}y^{-5}}\right)^{1/2}$ _____

54. $\left(\dfrac{8a^{-4}b^3}{27a^2b^{-3}}\right)^{1/3}$ _____

55. $\left(\dfrac{8y^{1/3}y^{-1/4}}{y^{-1/12}}\right)^2$ _____

56. $\left(\dfrac{9x^{1/3}x^{1/2}}{x^{-1/6}}\right)^{1/2}$ _____

Multiply, and express the answer using positive exponents only.

57. $3m^{3/4}(4m^{1/4} - 2m^8)$ _____

58. $2x^{1/3}(3x^{2/3} - x^6)$ _____

59. $(2x^{1/2} + y^{1/2})(x^{1/2} + y^{1/2})$ _____

60. $(x^{1/2} + y^{1/2})(x^{1/2} - y^{1/2})$ _____

61. $(x^{1/2} + y^{1/2})^2$ _____

62. $(x^{1/2} - y^{1/2})^2$ _____

C Simplify, and express the answer using positive exponents only.

63. $(-16)^{-3/2}$ _____

64. $-16^{-3/2}$ _____

65. $(a^{-1/2} + 3b^{-1/2})(2a^{-1/2} - b^{-1/2})$ _____

66. $(x^{-1/2} - y^{-1/2})^2$ _____

67. $(a^{n/2}b^{n/3})^{1/n}, n > 0$ _____

68. $(a^{3/n}b^{3/m})^{1/3}, n > 0, m > 0$ _____

69. $\left(\dfrac{x^{m+2}}{x^m}\right)^{1/2}, m > 0$ _____

70. $\left(\dfrac{a^m}{a^{m-2}}\right)^{1/2}, m > 0$ _____

71. $(x^{m/4}x^{m/4})^{-2}, m > 0$ _____

72. $(y^{m^2+1}y^{2m})^{1/(m+1)}, m > 0$ _____

73. **(A)** Find a value of x such that $(x^2)^{1/2} \neq x$. _____

(B) Find a real number x and a natural number n such that $(x^n)^{1/n} \neq x$.

_____

The Check Exercise for this section is on page 325.

74. For which real numbers does $(x^2)^{1/2} = |x|$? (More will be said about this form in Section 5-5.) _____

Section 5-4 Radical Forms and Rational Exponents

In the preceding section we introduced the symbol $b^{1/n}$ to represent an nth root of b and found that the symbol could be combined with other exponent forms by using the properties of exponents. Another symbol is also used to represent an nth root: the radical sign. Both symbols are widely used, and you should become familiar with them and their respective properties.

---

nth-Root Radical

For n a natural number greater than 1 and b any real number:

$$\sqrt[n]{b} = b^{1/n}$$

Thus, $\sqrt[n]{b}$ represents an nth root of b.

The symbol $\sqrt{}$ is called a **radical**, n is called the **index**, and b is called the **radicand**. If $n = 2$, we write

$$\sqrt{b} \qquad \text{and not} \qquad \sqrt[2]{b}$$

and refer to $\sqrt{b}$ as "the positive square root of b." Thus, it follows that

$$\sqrt{3} = 3^{1/2} \qquad \sqrt[8]{5} = 5^{1/8}$$
$$\sqrt[3]{x^2} = (x^2)^{1/3} = x^{2/3} \qquad (\sqrt[3]{x})^2 = (x^{1/3})^2 = x^{2/3}$$

There are occasions when it is more convenient to work with radicals than with rational exponents and vice versa. It is often an advantage to be able to shift back and forth between the two forms. The following relationships, suggested by the preceding examples, are useful in this regard.

For b nonnegative when n is even:

$$b^{m/n} = (b^m)^{1/n} = \sqrt[n]{b^m} \qquad \text{and} \qquad b^{m/n} = (b^{1/n})^m = (\sqrt[n]{b})^m$$

The following examples should make the process of changing from one form to the other clear. All variables represent _positive_ real numbers.

EXAMPLE 10   Convert from rational exponent form to radical form:

(A) $5^{1/2} = \sqrt{5}$ Positive square root of 5.

(B) $x^{1/7} = \sqrt[7]{x}$

(C) $7m^{2/3} = 7\sqrt[3]{m^2}$ or $7(\sqrt[3]{m})^2$ First form is usually more useful.

(D) $(3u^2v^3)^{3/5} = \sqrt[5]{(3u^2v^3)^3}$ or $(\sqrt[5]{3u^2v^3})^3$ First form is usually more useful.

(E) $y^{-2/3} = \dfrac{1}{y^{2/3}} = \dfrac{1}{\sqrt[3]{y^2}}$ or $\dfrac{1}{(\sqrt[3]{y})^2}$ Index of radical cannot be negative.

(F) $(x^2 + y^2)^{1/2} = \sqrt{x^2 + y^2}$ $\neq x + y$ (Why?)

PROBLEM 10   Convert to radical form:

(A) $7^{1/2}$ **(B)** $u^{1/5}$ **(C)** $3x^{3/5}$
(D) $(2x^3y^2)^{2/3}$ **(E)** $x^{-3/4}$ **(F)** $(x^3 + y^3)^{1/3}$

Solution   **(A)**   **(B)**   **(C)**

              (D)   **(E)**   **(F)**

EXAMPLE 11   Convert from radical form to rational exponent form:

(A) $\sqrt{13} = 13^{1/2}$ Positive square root of 13.

(B) $\sqrt[5]{x} = x^{1/5}$

(C) $\sqrt[4]{w^3} = w^{3/4}$

(D) $\sqrt[5]{(3x^2y^2)^4} = (3x^2y^2)^{4/5}$

(E) $\dfrac{1}{\sqrt[3]{x^2}} = \dfrac{1}{x^{2/3}} = x^{-2/3}$

(F) $\sqrt[4]{x^4 + y^4} = (x^4 + y^4)^{1/4}$ $\neq x + y$ (Why?)

PROBLEM 11   Convert to rational exponent form:

(A) $\sqrt{17}$ **(B)** $\sqrt[7]{m}$ **(C)** $\sqrt[5]{x^2}$

(D) $(\sqrt[7]{5m^3n^4})^3$ **(E)** $\dfrac{1}{\sqrt[6]{u^5}}$ **(F)** $\sqrt[5]{x^5 - y^5}$

Solution   **(A)**   **(B)**   **(C)**

              (D)   **(E)**   **(F)**

Solutions to Matched
Problems

10. **(A)** $7^{1/2} = \sqrt{7}$ **(B)** $u^{1/5} = \sqrt[5]{u}$ **(C)** $3x^{3/5} = 3\sqrt[5]{x^3}$ or $3(\sqrt[5]{x})^3$

(D) $(2x^3y^2)^{2/3} = \sqrt[3]{(2x^3y^2)^2}$ or $(\sqrt[3]{2x^3y^2})^2$

(E) $x^{-3/4} = \dfrac{1}{\sqrt[4]{x^3}}$ or $\dfrac{1}{(\sqrt[4]{x})^3}$ **(F)** $(x^3 + y^3)^{1/3} = \sqrt[3]{x^3 + y^3}$

11. **(A)** $\sqrt{17} = 17^{1/2}$ **(B)** $\sqrt[7]{m} = m^{1/7}$ **(C)** $\sqrt[5]{x^2} = x^{2/5}$

(D) $(\sqrt[7]{5m^3n^4})^3 = (5m^3n^4)^{3/7}$ **(E)** $\dfrac{1}{\sqrt[6]{u^5}} = u^{-5/6}$

(F) $\sqrt[5]{x^5 - y^5} = (x^5 - y^5)^{1/5}$

Practice Exercise 5-4 ∎

Work odd-numbered problems first, check answers, and then work even-numbered problems in areas of weakness. Answers to all problems are in the back of the book. Make every effort to work a problem yourself before you look at an answer.

All variables are restricted to avoid even roots of negative numbers.

A   Change to radical form. (Do not simplify.)

1. $11^{1/2}$ _____   **2.** $7^{1/2}$ _____   **3.** $5^{1/3}$ _____

4. $6^{1/4}$ _____   **5.** $u^{3/5}$ _____   **6.** $x^{3/4}$ _____

7. $4y^{3/7}$ _____   **8.** $5m^{2/3}$ _____   **9.** $(4y)^{3/7}$ _____

10. $(5m)^{2/3}$ _____   **11.** $(4ab^3)^{2/5}$ _____   **12.** $(7x^2y)^{2/3}$ _____

13. $(a + b)^{1/2}$ _____   **14.** $(a^2 + b^2)^{1/2}$ _____

Change to rational exponent form. (Do not simplify.)

15. $\sqrt{6}$ _____   **16.** $\sqrt{3}$ _____   **17.** $\sqrt[4]{m}$ _____

18. $\sqrt[7]{m}$ _____   **19.** $\sqrt[5]{y^3}$ _____   **20.** $\sqrt[3]{a^2}$ _____

21. $\sqrt[4]{(xy)^3}$ _____   **22.** $\sqrt[5]{(7m^3n^3)^4}$ _____

23. $\sqrt{x^2 - y^2}$ _____   **24.** $\sqrt{1 + y^2}$ _____

B   Change to radical form. (Do not simplify.)

25. $-5y^{2/5}$ _____   **26.** $-3x^{1/2}$ _____

27. $(1 + m^2n^2)^{3/7}$ _____   **28.** $(x^2y^2 - w^3)^{4/5}$ _____

29. $w^{-2/3}$ _____   **30.** $y^{-3/5}$ _____

31. $(3m^2n^3)^{-3/5}$ _____

32. $(2xy)^{-2/3}$ _____

33. $a^{1/2} + b^{1/2}$ _____

34. $x^{-1/2} + y^{-1/2}$ _____

35. $(a^3 + b^3)^{2/3}$ _____

36. $(x^{1/2} + y^{-1/2})^{1/3}$ _____

Change to rational exponent form. (Do not simplify.)

37. $\sqrt[3]{(a + b)^2}$ _____

38. $\sqrt[5]{(x - y)^2}$ _____

39. $-3x\sqrt[4]{a^3b}$ _____

40. $-5\sqrt[3]{2x^2y^2}$ _____

41. $\sqrt[9]{-2x^3y^7}$ _____

42. $\sqrt[5]{-4m^2n^3}$ _____

43. $\dfrac{3}{\sqrt[3]{y}}$ _____

44. $\dfrac{2x}{\sqrt{y}}$ _____

45. $\dfrac{-2x}{\sqrt{x^2 + y^2}}$ _____

46. $\dfrac{2}{\sqrt{x}} + \dfrac{3}{\sqrt{y}}$ _____

47. $\sqrt[3]{m^2} - \sqrt{n}$ _____

48. $\dfrac{-5u^2}{\sqrt{u} + \sqrt[5]{v^3}}$ _____

C **49.** Show that $(x^2 + y^2)^{1/2} \neq x + y$. _____

50. Show that $\sqrt{a^2 + b^2} \neq a + b$. _____

51. **(A)** Find a value of x such that $\sqrt{x^2} \neq x$. _____

(B) Find a positive integer n and a real number x such that $\sqrt[n]{x^n} \neq x$.

_____

52. Which of the following statements is true for all real x (n is a positive integer)? _____

integer)?

(A) $\sqrt{x^2} = |x|$ **(B)** $\sqrt[2n]{x^{2n}} = |x|$

The Check Exercise for this section is on page 327.

Section 5-5   Changing and Simplifying Radical Expressions

- Properties of Radicals
- Simplest Radical Form
- Simplifying $\sqrt[n]{x^n}$ for All Real x

■ Properties of Radicals

Changing and simplifying radical expressions is aided by the introduction of several properties of radicals that follow directly from the exponent properties considered earlier. To start, consider the following examples:

1. $\sqrt[5]{2^5} = (2^5)^{1/5} = 2^{5/5} = 2^1 = 2$

2. $\sqrt{4 \cdot 9} = \sqrt{36} = 6$ and $\sqrt{4}\sqrt{9} = 2 \cdot 3 = 6$

3. $\sqrt{\dfrac{36}{4}} = \sqrt{9} = 3$ and $\dfrac{\sqrt{36}}{\sqrt{4}} = \dfrac{6}{2} = 3$

4. $\sqrt[6]{2^4} = (2^4)^{1/6} = 2^{4/6} = 2^{2/3} = (2^2)^{1/3} = \sqrt[3]{2^2}$

These examples suggest the following general properties of radicals:

Properties of Radicals

In the following n, m, and k are natural numbers ≥ 2; x and y are positive real numbers:

1. $\sqrt[n]{x^n} = x$ $\left(\sqrt[n]{x}\right)^n = x$ $\sqrt[3]{x^3} = x, \left(\sqrt[3]{x}\right)^3 = x$

2. $\sqrt[n]{xy} = \sqrt[n]{x}\,\sqrt[n]{y}$ $\sqrt[4]{xy} = \sqrt[4]{x}\,\sqrt[4]{y}$

3. $\sqrt[n]{\dfrac{x}{y}} = \dfrac{\sqrt[n]{x}}{\sqrt[n]{y}}$ $\sqrt[5]{\dfrac{x}{y}} = \dfrac{\sqrt[5]{x}}{\sqrt[5]{y}}$

4. $\sqrt[kn]{x^{km}} = \sqrt[n]{x^m}$ $\sqrt[12]{x^8} = \sqrt[4 \cdot 3]{x^{4 \cdot 2}} = \sqrt[3]{x^2}$

It is important to remember that these four properties hold in general only if x and y are restricted to positive numbers. **In this section, unless otherwise stated, all variables are assumed to represent positive numbers.** Near the end of the section we will discuss what happens if we relax this restriction.

The properties of radicals presented above are readily established using exponent properties:

1. $\sqrt[n]{x^n} = (x^n)^{1/n} = x^{n/n} = x$

2. $\sqrt[n]{xy} = (xy)^{1/n} = x^{1/n}y^{1/n} = \sqrt[n]{x}\,\sqrt[n]{y}$

3. $\sqrt[n]{\dfrac{x}{y}} = \left(\dfrac{x}{y}\right)^{1/n} = \dfrac{x^{1/n}}{y^{1/n}} = \dfrac{\sqrt[n]{x}}{\sqrt[n]{y}}$

4. $\sqrt[kn]{x^{km}} = (x^{km})^{1/kn} = x^{km/kn} = x^{m/n} = \sqrt[n]{x^m}$

The following example illustrates how these properties are used. Properties 2 and 3 are used from right to left as well as from left to right.

EXAMPLE 12   **(A)** $\sqrt[5]{(3x^2y)^5} = 3x^2y$ Property 1

(B) $\sqrt{10}\sqrt{5} = \sqrt{50} = \sqrt{25 \cdot 2} = \sqrt{25}\sqrt{2} = 5\sqrt{2}$ Property 2

(C) $\sqrt[3]{\dfrac{x}{27}} = \dfrac{\sqrt[3]{x}}{\sqrt[3]{27}} = \dfrac{\sqrt[3]{x}}{3}$ or $\tfrac{1}{3}\sqrt[3]{x}$ Property 3

(D) $\sqrt[6]{x^4} = \sqrt[2 \cdot 3]{x^{2 \cdot 2}} = \sqrt[3]{x^2}$ Property 4

PROBLEM 12   Simplify as in Example 12:

(A) $\sqrt[7]{(u^2 + v^2)^7}$ (B) $\sqrt{6}\sqrt{2}$ (C) $\sqrt[3]{\dfrac{x^2}{8}}$ (D) $\dfrac{\sqrt[3]{54x^8}}{\sqrt[3]{2x^2}}$ (E) $\sqrt[8]{y^6}$

Solution   (A)   (B)

(C)   (D)

(E)

■ Simplest Radical Form

The laws of radicals provide us with the means for changing algebraic expressions containing radicals to a variety of equivalent forms. One form often useful is the simplest radical form. An algebraic expression that contains radicals is said to be in the **simplest radical form** if all four of the following conditions are satisfied.

Simplest Radical Form
1. A radicand (the expression within the radical sign) contains no polynomial factor to a power greater than or equal to the index of the radical. $\sqrt{x^3}$ violates this condition.
2. The power of the radicand and the index of the radical have no common factor other than 1. $\sqrt[6]{x^4}$ violates this condition.
3. No radical appears in a denominator. $3/\sqrt{5}$ violates this condition.
4. No fraction appears within a radical. $\sqrt{\frac{2}{3}}$ violates this condition.

It should be understood that forms other than the simplest radical form may be more useful on occasion. The choice depends on the situation.

EXAMPLE 13   Change to simplest radical form:

(A) $\sqrt{8x^3} = \sqrt{(4x^2)(2x)}$ Violates Condition 1. Factor $8x^3$ into a perfect-square
    $= \sqrt{4x^2}\sqrt{2x} = 2x\sqrt{2x}$ part, $4x^2$, and what is left, $2x$; then use multiplication
                                              property 2.

(B) $\sqrt[3]{54x^5} = \sqrt[3]{(27x^3)(2x^2)}$ Violates Condition 1. Factor $54x^5$ into a perfect-cube part,
    $= \sqrt[3]{27x^3}\sqrt[3]{2x^2}$ $27x^3$, and what is left, $2x^2$; then use multiplication
    $= 3x\sqrt[3]{2x^2}$ property 2.

(C) $\sqrt[9]{x^6} = \sqrt[3]{x^2}$ Violates Condition 2. Index 9 and power of radicand 6
                                          have the common factor 3.

(D) $\dfrac{4x}{\sqrt{8x}} = \dfrac{4x}{\sqrt{8x}} \cdot \dfrac{\sqrt{2x}}{\sqrt{2x}}$

$\qquad = \dfrac{4x\sqrt{2x}}{\sqrt{16x^2}}$

$\qquad = \dfrac{4x\sqrt{2x}}{4x} = \sqrt{2x}$

Violates Condition 3 (has a radical in the denominator). Multiply numerator and denominator by "smallest" or "simplest" expression that will make denominator the square root of a perfect square. Using $\sqrt{2x}$ rather than $\sqrt{8x}$ results in less work—try the latter to see why.

(E) $\sqrt[3]{\dfrac{y}{4x}} = \sqrt[3]{\dfrac{y}{4x} \cdot \dfrac{2x^2}{2x^2}}$

$\qquad = \sqrt[3]{\dfrac{2x^2y}{8x^3}} = \dfrac{\sqrt[3]{2x^2y}}{\sqrt[3]{8x^3}}$

$\qquad = \dfrac{\sqrt[3]{2x^2y}}{2x}$

Violates Condition 4. (There is a fraction within the radical.) Multiply numerator and denominator inside the radical by the "smallest" or "simplest" expression that will make the denominator a perfect cube. Using $2x^2$ rather than 4^2x^2 results in less work—try the latter to see why.

As mentioned above, the simplest radical form may not be the most useful form. In a product such as

$$\frac{4x}{\sqrt{8x}} \cdot \frac{\sqrt{8x}}{2}$$

it is clearly easier to multiply directly rather than to change either factor to simplest form. On the other hand, if $4x/\sqrt{8x}$ is to be evaluated for several values of x, the work will be much easier if the expression is first changed to $\sqrt{2x}$ as in Example 13(D).

The process of removing radicals from a denominator as in Example 13(D) is called **rationalizing the denominator**. In general the process is to multiply both numerator and denominator by an expression so that the product in the denominator is free of radicals. For example, if the denominator is $\sqrt{8x}$, multiply by $\sqrt{2x}$ as in Example 13(D):

$$\sqrt{8x} \cdot \sqrt{2x} = \sqrt{16x^2} = 4x$$

If the denominator is $\sqrt[3]{9x}$, multiply by $\sqrt[3]{3x^2}$:

$$\sqrt[3]{9x} \cdot \sqrt[3]{3x^2} = \sqrt[3]{27x^3} = 3x$$

PROBLEM 13

Change to simplest radical form:

(A) $\sqrt{18y^5}$ **(B)** $\sqrt[3]{32m^8}$ **(C)** $\sqrt[12]{y^8}$ **(D)** $\dfrac{6u}{\sqrt[3]{4x}}$ **(E)** $\sqrt{\dfrac{3y}{8x}}$

Solution

(A)   **(B)**

(C)   **(D)**

(E)

EXAMPLE 14 Change to simplest radical form:

(A) $\sqrt{12x^3y^5z^2} = \sqrt{(2^2x^2y^4z^2)(3xy)} = \sqrt{2^2x^2y^4z^2}\sqrt{3xy} = 2xy^2z\sqrt{3xy}$

(B) $\sqrt[6]{16x^4y^2} = \sqrt[6]{(2^2x^2y)^2} = \sqrt[3]{4x^2y}$

(C) $\dfrac{3}{\sqrt{12}} = \dfrac{3}{\sqrt{12}} \cdot \dfrac{\sqrt{3}}{\sqrt{3}} = \dfrac{3\sqrt{3}}{\sqrt{36}} = \dfrac{3\sqrt{3}}{6} = \dfrac{\sqrt{3}}{2}$ or $\dfrac{1}{2}\sqrt{3}$

(D) $\dfrac{6x^2}{\sqrt[3]{9x}} = \dfrac{6x^2}{\sqrt[3]{9x}} \cdot \dfrac{\sqrt[3]{3x^2}}{\sqrt[3]{3x^2}} = \dfrac{6x^2\sqrt[3]{3x^2}}{\sqrt[3]{3^3x^3}} = \dfrac{6x^2\sqrt[3]{3x^2}}{3x} = 2x\sqrt[3]{3x^2}$

(E) $\sqrt[3]{\dfrac{2a^2}{3b^2}} = \sqrt[3]{\dfrac{(2a^2)(3^2b)}{(3b^2)(3^2b)}} = \sqrt[3]{\dfrac{18a^2b}{3^3b^3}} = \dfrac{\sqrt[3]{18a^2b}}{\sqrt[3]{3^3b^3}} = \dfrac{\sqrt[3]{18a^2b}}{3b}$

PROBLEM 14 Change to simplest radical form:

(A) $\sqrt[3]{16}$ **(B)** $\sqrt[3]{16x^7y^4z^3}$ **(C)** $\sqrt[9]{8x^6y^3}$

(D) $\dfrac{6}{\sqrt{2x}}$ **(E)** $\dfrac{10x^3}{\sqrt[3]{4x^2}}$ **(F)** $\sqrt[3]{\dfrac{3y^2}{2x^4}}$

Solution **(A)**   **(B)**

(C)   **(D)**

(E)   **(F)**

■ **Simplifying $\sqrt[n]{x^n}$ for All Real x**

In the preceding discussion we restricted variables to nonnegative quantities. If we lift this restriction, then

$$\sqrt{x^2} = x$$

is true only for certain values of x and is not true for others. If x is positive or 0, then the equation is true; if x is negative, then the equation is false. For example, test the equation for $x = 2$ and for $x = -2$:

$x = 2$	$x = -2$
$\sqrt{x^2} = x$	$\sqrt{x^2} = x$
$\sqrt{2^2} \overset{?}{=} 2$	$\sqrt{(-2)^2} \overset{?}{=} (-2)$
$\sqrt{4} \overset{?}{=} 2$	$\sqrt{4} \overset{?}{=} -2$
$2 \overset{\checkmark}{=} 2$	$2 \neq -2$

Thus, we see that if x is negative, then we must write

$$\sqrt{x^2} = -x$$

Now both sides represent positive numbers. In summary, for x any real number

$$\sqrt{x^2} = \begin{cases} x & \text{if } x \text{ is positive} \\ 0 & \text{if } x \text{ is } 0 \\ -x & \text{if } x \text{ is negative} \end{cases}$$

Also, recall the definition of absolute value from Chapter 1:

$$|x| = \begin{cases} x & \text{if } x \text{ is positive} \\ 0 & \text{if } x \text{ is } 0 \\ -x & \text{if } x \text{ is negative} \end{cases}$$

We see that $\sqrt{x^2}$ and $|x|$ actually are the same, and we can write:

For x *any* real number:

$$\sqrt{x^2} = |x|$$

Thus, only if x is restricted to nonnegative real numbers can we drop the absolute-value sign.

Now let us consider $\sqrt[3]{x^3}$. Here we do not have the same kind of problem that we had above. It turns out that for *all* real numbers,

$$\sqrt[3]{x^3} = x$$

and we do not need the absolute-value sign on the right. As before, let us evaluate the equation for $x = 2$ and $x = -2$:

$x = 2$	$x = -2$
$\sqrt[3]{x^3} = x$	$\sqrt[3]{x^3} = x$
$\sqrt[3]{2^3} \stackrel{?}{=} 2$	$\sqrt[3]{(-2)^3} \stackrel{?}{=} (-2)$
$\sqrt[3]{8} \stackrel{?}{=} 2$	$\sqrt[3]{-8} \stackrel{?}{=} -2$
$2 \stackrel{\vee}{=} 2$	$-2 \stackrel{\vee}{=} -2$

COMMON ERROR   If asked to simplify $\sqrt[3]{x^3} + \sqrt{x^2}$, many students would write

$$\sqrt[3]{x^3} + \sqrt{x^2} = x + x = 2x$$

and not think any more about it. But if we evaluate each side for $x = -2$, we find that

$$\sqrt[3]{(-2)^3} + \sqrt{(-2)^2} = \sqrt[3]{-8} + \sqrt{4} = -2 + 2 = 0 \quad \text{Left side}$$

and

$$2(-2) = -4 \quad \text{Right side}$$

Both sides are not equal! What is wrong? When x is not restricted to positive values or 0, we should write

$$\sqrt[3]{x^3} + \sqrt{x^2} = x + |x|$$

Then the right side will equal the left side for *all* real numbers. Consider the following example and related problem.

EXAMPLE 15   For x a positive number:

$$\sqrt[3]{x^3} + \sqrt{x^2} = x + |x| = x + x = 2x$$

For x a negative number:

$$\sqrt[3]{x^3} + \sqrt{x^2} = x + |x| = x - x = 0$$

PROBLEM 15

Simplify $2\sqrt[3]{x^3} - \sqrt{x^2}$:

(A) For x a positive number   **(B)** For x a negative number

Solution

(A)

(B)

Following the same reasoning as above, we can obtain the more general result:

In general, for x *any* real number and n a positive integer greater than 1:

$$\sqrt[n]{x^n} = \begin{cases} |x| & \text{if } n \text{ is even} \quad \sqrt[4]{x^4} = |x| \\ x & \text{if } n \text{ is odd} \quad \sqrt[5]{x^5} = x \end{cases}$$

Solutions to Matched Problems

12. **(A)** $\sqrt[7]{(u^2 + v^2)^7} = u^2 + v^2$

(B) $\sqrt{6}\sqrt{2} = \sqrt{12} = \sqrt{4 \cdot 3} = \sqrt{4}\sqrt{3} = 2\sqrt{3}$

(C) $\sqrt[3]{\dfrac{x^2}{8}} = \dfrac{\sqrt[3]{x^2}}{\sqrt[3]{8}} = \dfrac{\sqrt[3]{x^2}}{2}$ or $\dfrac{1}{2}\sqrt[3]{x^2}$

(D) $\dfrac{\sqrt[3]{54x^8}}{\sqrt[3]{2x^2}} = \sqrt[3]{\dfrac{54x^8}{2x^2}} = \sqrt[3]{27x^6} = 3x^2$ **(E)** $\sqrt[8]{y^6} = \sqrt[4]{y^3}$

13. **(A)** $\sqrt{18y^5} = \sqrt{(9y^4)(2y)} = \sqrt{9y^4}\sqrt{2y} = 3y^2\sqrt{2y}$

(B) $\sqrt[3]{32m^8} = \sqrt[3]{(8m^6)(4m^2)} = \sqrt[3]{8m^6}\sqrt[3]{4m^2} = 2m^2\sqrt[3]{4m^2}$

(C) $\sqrt[12]{y^8} = \sqrt[3]{y^2}$

(D) $\dfrac{6u}{\sqrt[3]{4x}} = \dfrac{6u}{\sqrt[3]{4x}} \cdot \dfrac{\sqrt[3]{2x^6}}{\sqrt[3]{2x^2}} = \dfrac{6u\sqrt[3]{2x^2}}{\sqrt[3]{2^3x^3}} = \dfrac{6u\sqrt[3]{2x^2}}{2x} = \dfrac{3u\sqrt[3]{2x^2}}{x}$

(E) $\sqrt{\dfrac{3y}{8x}} = \sqrt{\dfrac{3y \cdot 2x}{8x \cdot 2x}} = \sqrt{\dfrac{6xy}{4^2x^2}} = \dfrac{\sqrt{6xy}}{\sqrt{4^2x^2}} = \dfrac{\sqrt{6xy}}{4x}$

14. **(A)** $\sqrt[3]{16} = \sqrt[3]{8 \cdot 2} = \sqrt[3]{8} \cdot \sqrt[3]{2} = 2\sqrt[3]{2}$

(B) $\sqrt[3]{16x^7y^4z^3} = \sqrt[3]{(8x^6y^3z^3)(2xy)} = \sqrt[3]{8x^6y^3z^3}\sqrt[3]{2xy} = 2x^2yz\sqrt[3]{2xy}$

(C) $\sqrt[9]{8x^6y^3} = \sqrt[9]{(2x^2y)^3} = \sqrt[3]{2x^2y}$

(D) $\dfrac{6}{\sqrt{2x}} = \dfrac{6\sqrt{2x}}{\sqrt{2x}\sqrt{2x}} = \dfrac{6\sqrt{2x}}{2x} = \dfrac{3\sqrt{2x}}{x}$

(E) $\dfrac{10x^3}{\sqrt[3]{4x^2}} = \dfrac{10x^3\sqrt[3]{2x}}{\sqrt[3]{4x^2}\sqrt[3]{2x}} = \dfrac{10x^3\sqrt[3]{2x}}{\sqrt[3]{8x^3}} = \dfrac{10x^3\sqrt[3]{2x}}{2x} = 5x^2\sqrt[3]{2x}$

(F) $\sqrt[3]{\dfrac{3y^2}{2x^4}} = \sqrt[3]{\dfrac{3y^2 \cdot 2^2x^2}{2x^4 \cdot 2^2x^2}} = \dfrac{\sqrt[3]{12x^2y^2}}{\sqrt[3]{2^3x^6}} = \dfrac{\sqrt[3]{12x^2y^2}}{2x^2}$

15. **(A)** $2\sqrt[3]{x^3} - \sqrt{x^2} = 2x - x = x$

(B) $2\sqrt[3]{x^3} - \sqrt{x^2} = 2x - (-x) = 2x + x = 3x$

Practice Exercise 5-5 ■

Work odd-numbered problems first, check answers, and then work even-numbered problems in areas of weakness. Answers to all problems are in the back of the book. Make every effort to work a problem yourself before you look at an answer.

Simplify, and write in simplest radical form. All variables represent positive real numbers.

A   **1.** $\sqrt{y^2}$ _____   **2.** $\sqrt{x^2}$ _____   **3.** $\sqrt{4u^2}$ _____

4. $\sqrt{9m^2}$ _____   **5.** $\sqrt{49x^4y^2}$ _____   **6.** $\sqrt{25x^2y^4}$ _____

7. $\sqrt{18}$ _____   **8.** $\sqrt{8}$ _____   **9.** $\sqrt{m^3}$ _____

10. $\sqrt{x^3}$ _____   **11.** $\sqrt{8x^3}$ _____   **12.** $\sqrt{18y^3}$ _____

13. $\sqrt{\frac{1}{9}}$ _____   **14.** $\sqrt{\frac{1}{4}}$ _____   **15.** $\frac{1}{\sqrt{y^2}}$ _____

16. $\frac{1}{\sqrt{x^2}}$ _____   **17.** $\frac{1}{\sqrt{5}}$ _____   **18.** $\frac{1}{\sqrt{3}}$ _____

19. $\sqrt{\frac{1}{5}}$ _____   **20.** $\sqrt{\frac{1}{3}}$ _____   **21.** $\frac{1}{\sqrt{y}}$ _____

22. $\frac{1}{\sqrt{x}}$ _____   **23.** $\sqrt{\frac{1}{y}}$ _____   **24.** $\sqrt{\frac{1}{x}}$ _____

25. $\sqrt{9x^3y^5}$ _____   **26.** $\sqrt{4x^5y^3}$ _____   **27.** $\sqrt{18x^8y^5}$ _____

28. $\sqrt{8x^7y^6}$ _____   **29.** $\frac{1}{\sqrt{2x}}$ _____   **30.** $\frac{1}{\sqrt{3y}}$ _____

31. $\frac{6x^2}{\sqrt{3x}}$ _____   **32.** $\frac{4xy}{\sqrt{2y}}$ _____   **33.** $\frac{3a}{\sqrt{2ab}}$ _____

34. $\frac{2x^2y}{\sqrt{3xy}}$ _____   **35.** $\sqrt{\frac{6x}{7y}}$ _____   **36.** $\sqrt{\frac{3m}{2n}}$ _____

B   **37.** $\sqrt{\frac{9m^5}{2n}}$ _____   **38.** $\sqrt{\frac{4a^3}{3b}}$ _____

39. $\sqrt[4]{16x^8y^4}$ _____   **40.** $\sqrt[5]{32m^5n^{15}}$ _____

41. $\sqrt[3]{2^4x^4y^7}$ _____   **42.** $\sqrt[4]{2^4a^5b^8}$ _____

43. $\sqrt[4]{x^2}$ _____   **44.** $\sqrt[10]{x^6}$ _____

45. $\sqrt{2}\sqrt{8}$ _____

46. $\sqrt[3]{3}\sqrt[3]{9}$ _____

47. $\sqrt{18m^3n^4}\sqrt{2m^3n^2}$ _____

48. $\sqrt[3]{9x^2y}\sqrt[3]{3xy^2}$ _____

49. $\dfrac{6}{\sqrt[3]{3}}$ _____

50. $\dfrac{2}{\sqrt[3]{2}}$ _____

51. $\dfrac{\sqrt{4a^3}}{\sqrt{3b}}$ _____

52. $\dfrac{\sqrt{9m^5}}{\sqrt{2n}}$ _____

53. $\sqrt{a^2+b^2}$ _____

54. $\sqrt[3]{x^3+y^3}$ _____

55. $\sqrt[3]{\dfrac{8x^3}{27y^6}}$ _____

56. $\sqrt[4]{\dfrac{a^8b^4}{16c^{12}}}$ _____

57. $-m\sqrt[5]{3^6m^7n^{11}}$ _____

58. $-2x\sqrt[3]{8x^8y^{13}}$ _____

59. $\sqrt[6]{x^4(x-y)^2}$ _____

60. $\sqrt[8]{2^6(x+y)^6}$ _____

61. $\sqrt[3]{2x^2y^3}\sqrt[3]{3x^5y}$ _____

62. $\sqrt[4]{6u^3v^4}\sqrt[4]{4u^5v}$ _____

63. $\dfrac{4x^3y^2}{\sqrt[3]{2xy^2}}$ _____

64. $\dfrac{8u^3v^5}{\sqrt[3]{4u^2v^2}}$ _____

65. $-2x\sqrt[3]{\dfrac{3y^2}{4x}}$ _____

66. $6c\sqrt[3]{\dfrac{2ab}{9c^2}}$ _____

C **67.** $\dfrac{x-y}{\sqrt[3]{x-y}}$ _____

68. $\dfrac{1}{\sqrt[3]{(x-y)^2}}$ _____

69. $\sqrt[4]{\dfrac{3y^3}{4x}}$ _____

70. $\sqrt[5]{\dfrac{4n^2}{16m^3}}$ _____

71. $-\sqrt{x^4+2x^2}$ _____

72. $\sqrt[4]{m^4+4m^6}$ _____

73. $\sqrt[4]{16x^4}\sqrt[3]{16x^{24}y^4}$ _____

74. $\sqrt[3]{8\sqrt{16x^6y^4}}$ _____

75. $\sqrt[3]{3m^2n^2}\sqrt[4]{3m^3n^2}$ _____

76. $\sqrt{2x^5y^3}\sqrt[3]{16x^7y^7}$ _____

77. $\sqrt[3]{x^{3n}(x+y)^{3n+6}}$ _____

78. $\sqrt[n]{x^{2n}y^{n^2+n}}$ _____

Simplify each of the following for **(A)** x *a positive number and* **(B)** x *a negative number.*

79. $2\sqrt[3]{x^3}+4\sqrt{x^2}$ _____

80. $\sqrt[3]{8x^3}-\sqrt{16x^2}$ _____

The Check Exercise for
this section is on
page 329.

81. $\sqrt[5]{x^5}+\sqrt[4]{x^4}$ _____

82. $\sqrt[6]{x^6}+\sqrt[3]{x^3}$ _____

83. $3\sqrt[4]{x^4}-2\sqrt[5]{x^5}$ _____

84. $5\sqrt[7]{x^7}-3\sqrt[6]{x^6}$ _____

Section 5-6 Basic Operations on Radicals

- Sums and Differences
- Products
- Quotients—Rationalizing Denominators

■ Sums and Differences

Algebraic expressions involving radicals can often be simplified by adding and subtracting terms that contain exactly the same radical expressions. We proceed in essentially the same way as we do when we combine like terms in polynomials. You will recall that the distributive property of real numbers plays a central role in this process. All variables represent positive real numbers.

EXAMPLE 16   Combine as many terms as possible:

(A) $5\sqrt{3} + 4\sqrt{3} \; = (5 + 4)\sqrt{3} \; = 9\sqrt{3}$

(B) $2\sqrt[3]{xy^2} - 7\sqrt[3]{xy^2} \; = (2 - 7)\sqrt[3]{xy^2} \; = -5\sqrt[3]{xy^2}$

(C) $3\sqrt{xy} - 2\sqrt[3]{xy} + 4\sqrt{xy} - 7\sqrt[3]{xy} = 3\sqrt{xy} + 4\sqrt{xy} - 2\sqrt[3]{xy} - 7\sqrt[3]{xy}$
$$= 7\sqrt{xy} - 9\sqrt[3]{xy}$$

PROBLEM 16

Combine as many terms as possible:

(A) $6\sqrt{2} + 2\sqrt{2}$

(B) $3\sqrt[5]{2x^2y^3} - 8\sqrt[5]{2x^2y^3}$

(C) $5\sqrt[3]{mn^2} - 3\sqrt{mn} - 2\sqrt[3]{mn^2} + 7\sqrt{mn}$

Solution   **(A)**

(B)

(C)

Thus, we see that if two terms contain exactly the same radical—having the same index and the same radicand—they can be combined into a single term. Occasionally, terms containing radicals can be combined after they have been expressed in simplest radical form.

EXAMPLE 17   Express terms in simplest radical form and combine where possible:

(A) $4\sqrt{8} - 2\sqrt{18} = 4\sqrt{4 \cdot 2} - 2\sqrt{9 \cdot 2}$
$$= 8\sqrt{2} - 6\sqrt{2}$$
$$= 2\sqrt{2}$$

(B) $2\sqrt{12} - \sqrt{\tfrac{1}{3}} = 2 \cdot \sqrt{4} \cdot \sqrt{3} - \dfrac{1 \cdot \sqrt{3}}{\sqrt{3} \cdot \sqrt{3}}$

$$= 4\sqrt{3} - \frac{\sqrt{3}}{3}$$

$$= (4 - \tfrac{1}{3})\sqrt{3}$$

$$= \frac{11}{3}\sqrt{3} \quad \text{or} \quad \frac{11\sqrt{3}}{3}$$

(C) $\sqrt[3]{81} - \sqrt[3]{\dfrac{1}{9}} = \sqrt[3]{3^3 \cdot 3} - \sqrt[3]{\dfrac{3}{3^3}} = 3\sqrt[3]{3} - \tfrac{1}{3}\sqrt[3]{3}$

$$= (3 - \tfrac{1}{3})\sqrt[3]{3} = \tfrac{8}{3}\sqrt[3]{3}$$

PROBLEM 17   Express terms in simplest radical form and combine where possible:

(A) $\sqrt{12} - \sqrt{48}$ **(B)** $3\sqrt{8} - \sqrt{\tfrac{1}{2}}$ **(C)** $\sqrt[3]{\tfrac{1}{4}} - \sqrt[3]{16}$

Solution   **(A)**

(B)

(C)

■ **Products**

We will now consider several types of special products and quotients that involve radicals. The distributive property of real numbers plays a central role in our approach to these problems. In the discussion that follows, all variables represent positive real numbers.

EXAMPLE 18   Multiply and simplify:

(A) $\sqrt{2}(\sqrt{10} - 3) = \sqrt{2}\sqrt{10} - \sqrt{2} \cdot 3 = \sqrt{20} - 3\sqrt{2} = 2\sqrt{5} - 3\sqrt{2}$

(B) $(\sqrt{2} - 3)(\sqrt{2} + 5) = \sqrt{2}\sqrt{2} - 3\sqrt{2} + 5\sqrt{2} - 15$

$$= 2 + 2\sqrt{2} - 15$$

$$= 2\sqrt{2} - 13$$

(C) $(\sqrt{x} - 3)(\sqrt{x} + 5) = \sqrt{x}\sqrt{x} - 3\sqrt{x} + 5\sqrt{x} - 15$

$$= x + 2\sqrt{x} - 15$$

(D) $(\sqrt[3]{m} + \sqrt[3]{n^2})(\sqrt[3]{m^2} - \sqrt[3]{n}) = \sqrt[3]{m^3} + \sqrt[3]{m^2 n^2} - \sqrt[3]{mn} - \sqrt[3]{n^3}$

$$= m - \sqrt[3]{mn} + \sqrt[3]{m^2 n^2} - n$$

PROBLEM 18   Multiply and simplify:

(A) $\sqrt{3}(\sqrt{6} - 4)$ (B) $(\sqrt{3} - 2)(\sqrt{3} + 4)$

(C) $(\sqrt{y} - 2)(\sqrt{y} + 4)$ (D) $(\sqrt[3]{x^2} - \sqrt[3]{y^2})(\sqrt[3]{x} + \sqrt[3]{y})$

Solution   (A)

(B)

(C)

(D)

EXAMPLE 19   Show that $(2 - \sqrt{3})$ is a solution of the equation $x^2 - 4x + 1 = 0$.

Solution
$$x^2 - 4x + 1 = 0$$
$$(2 - \sqrt{3})^2 - 4(2 - \sqrt{3}) + 1 \stackrel{?}{=} 0$$
$$4 - 4\sqrt{3} + 3 - 8 + 4\sqrt{3} + 1 \stackrel{?}{=} 0$$
$$0 \stackrel{\vee}{=} 0$$

PROBLEM 19   Show that $(2 + \sqrt{3})$ is a solution of $x^2 - 4x + 1 = 0$.

Solution

■ **Quotients—Rationalizing Denominators**

Recall that to express $\sqrt{2}/\sqrt{3}$ in simplest radical form, we multiplied the numerator and denominator by $\sqrt{3}$ to clear the denominator of the radical:

$$\frac{\sqrt{2}}{\sqrt{3}} = \frac{\sqrt{2} \cdot \sqrt{3}}{\sqrt{3} \cdot \sqrt{3}} = \frac{\sqrt{6}}{3}$$

The denominator is thus converted to a rational number. Also recall that the process of converting an irrational denominator to a rational form is called **rationalizing the denominator**.

How can we rationalize the binomial denominator in

$$\frac{1}{\sqrt{3} - \sqrt{2}}$$

Multiplying the numerator and denominator by $\sqrt{3}$ or $\sqrt{2}$ does not help. Try it! Recalling the special product

$$(a - b)(a + b) = a^2 - b^2$$

suggests that we multiply the numerator and denominator by the denominator, only with the middle sign changed. Thus,

$$\frac{1}{\sqrt{3}-\sqrt{2}} = \frac{1(\sqrt{3}+\sqrt{2})}{(\sqrt{3}-\sqrt{2})(\sqrt{3}+\sqrt{2})} = \frac{\sqrt{3}+\sqrt{2}}{3-2} = \sqrt{3}+\sqrt{2}$$

EXAMPLE 20 Rationalize denominators and simplify:

(A) $\dfrac{\sqrt{2}}{\sqrt{6}-2} = \dfrac{\sqrt{2}(\sqrt{6}+2)}{(\sqrt{6}-2)(\sqrt{6}+2)} = \dfrac{\sqrt{12}+2\sqrt{2}}{6-4}$

$$= \frac{2\sqrt{3}+2\sqrt{2}}{2} = \frac{2(\sqrt{3}+\sqrt{2})}{2} = \sqrt{3}+\sqrt{2}$$

(B) $\dfrac{\sqrt{x}-\sqrt{y}}{\sqrt{x}+\sqrt{y}} = \dfrac{(\sqrt{x}-\sqrt{y})(\sqrt{x}-\sqrt{y})}{(\sqrt{x}+\sqrt{y})(\sqrt{x}-\sqrt{y})} = \dfrac{x-2\sqrt{xy}+y}{x-y}$

PROBLEM 20 Rationalize denominators and simplify: (A) $\dfrac{\sqrt{2}}{\sqrt{2}+3}$ (B) $\dfrac{\sqrt{x}+\sqrt{y}}{\sqrt{x}-\sqrt{y}}$

Solution (A)

(B)

Solutions to Matched Problems

16. (A) $6\sqrt{2}+2\sqrt{2}=8\sqrt{2}$ (B) $3\sqrt[5]{2x^2y^3}-8\sqrt[5]{2x^2y^3}=-5\sqrt[5]{2x^2y^3}$

(C) $5\sqrt[3]{mn^2}-3\sqrt{mn}-2\sqrt[3]{mn^2}+7\sqrt{mn}=3\sqrt[3]{mn^2}+4\sqrt{mn}$

17. (A) $\sqrt{12}-\sqrt{48}=\sqrt{4\cdot3}-\sqrt{16\cdot3}=2\sqrt{3}-4\sqrt{3}=-2\sqrt{3}$

(B) $3\sqrt{8}-\sqrt{\frac{1}{2}}=3\sqrt{8}-\dfrac{1}{\sqrt{2}}=3\sqrt{4\cdot2}-\dfrac{1\sqrt{2}}{\sqrt{2}\sqrt{2}}=6\sqrt{2}-\dfrac{\sqrt{2}}{2}$

$$= (6-\tfrac{1}{2})(\sqrt{2})=\tfrac{11}{2}\sqrt{2}\quad\text{or}\quad\dfrac{11\sqrt{2}}{2}$$

(C) $\sqrt[3]{\frac{1}{4}}-\sqrt[3]{16}=3\sqrt{\dfrac{1\cdot2}{4\cdot2}}-\sqrt[3]{8\cdot2}=\dfrac{\sqrt[3]{2}}{\sqrt[3]{8}}-\sqrt[3]{8}\sqrt[3]{2}=\dfrac{\sqrt[3]{2}}{2}-2\sqrt[3]{2}$

$$= (\tfrac{1}{2}-2)(\sqrt[3]{2})=-\tfrac{3}{2}\sqrt[3]{2}$$

18. (A) $\sqrt{3}(\sqrt{6}-4)=\sqrt{18}-4\sqrt{3}=\sqrt{9\cdot2}-4\sqrt{3}=3\sqrt{2}-4\sqrt{3}$

(B) $(\sqrt{3}-2)(\sqrt{3}+4)=3+2\sqrt{3}-8=-5+2\sqrt{3}$

(C) $(\sqrt{y}-2)(\sqrt{y}+4)=y+2\sqrt{y}-8$

(D) $(\sqrt[3]{x^2}-\sqrt[3]{y^2})(\sqrt[3]{x}+\sqrt[3]{y})=x+\sqrt[3]{x^2y}-\sqrt[3]{xy^2}-y$

19. $(2 + \sqrt{3})^2 - 4(2 + \sqrt{3}) + 1 = 4 + 4\sqrt{3} + 3 - 8 - 4\sqrt{3} + 1 = 0$

20. (A) $\dfrac{\sqrt{2}}{\sqrt{2} + 3} = \dfrac{\sqrt{2}(\sqrt{2} - 3)}{(\sqrt{2} + 3)(\sqrt{2} - 3)} = \dfrac{2 - 3\sqrt{2}}{2 - 9}$

$= \dfrac{2 - 3\sqrt{2}}{-7}$ or $\dfrac{3\sqrt{2} - 2}{7}$

(B) $\dfrac{\sqrt{x} + \sqrt{y}}{\sqrt{x} - \sqrt{y}} = \dfrac{(\sqrt{x} + \sqrt{y})(\sqrt{x} + \sqrt{y})}{(\sqrt{x} - \sqrt{y})(\sqrt{x} + \sqrt{y})} = \dfrac{x + 2\sqrt{xy} + y}{x - y}$

Practice Exercise 5-6 ■

Work odd-numbered problems first, check answers, and then work even-numbered problems in areas of weakness. Answers to all problems are in the back of the book. Make every effort to work a problem yourself before you look at an answer.

Express in simplest radical form and combine where possible.

A 1. $7\sqrt{3} + 2\sqrt{3}$ _____

2. $5\sqrt{2} + 3\sqrt{2}$ _____

3. $2\sqrt{a} - 7\sqrt{a}$ _____

4. $\sqrt{y} - 4\sqrt{y}$ _____

5. $\sqrt{n} - 4\sqrt{n} - 2\sqrt{n}$ _____

6. $2\sqrt{x} - \sqrt{x} + 3\sqrt{x}$ _____

7. $\sqrt{5} - 2\sqrt{3} + 3\sqrt{5}$ _____

8. $3\sqrt{2} - 2\sqrt{3} - \sqrt{2}$ _____

9. $\sqrt{m} - \sqrt{n} - 2\sqrt{n}$ _____

10. $2\sqrt{x} - \sqrt{y} + 3\sqrt{y}$ _____

11. $\sqrt{18} + \sqrt{2}$ _____

12. $\sqrt{8} - \sqrt{2}$ _____

13. $\sqrt{8} - 2\sqrt{32}$ _____

14. $\sqrt{27} - 3\sqrt{12}$ _____

Multiply and simplify where possible.

15. $\sqrt{7}(\sqrt{7} - 2)$ _____

16. $\sqrt{5}(\sqrt{5} - 2)$ _____

17. $\sqrt{2}(3 - \sqrt{2})$ _____

18. $\sqrt{3}(2 - \sqrt{3})$ _____

19. $\sqrt{y}(\sqrt{y} - 8)$ _____

20. $\sqrt{x}(\sqrt{x} - 3)$ _____

21. $\sqrt{n}(4 - \sqrt{n})$ _____

22. $\sqrt{m}(3 - \sqrt{m})$ _____

23. $\sqrt{3}(\sqrt{3} + \sqrt{6})$ _____

24. $\sqrt{5}(\sqrt{10} + \sqrt{5})$ _____

25. $(2 - \sqrt{3})(3 + \sqrt{3})$ _____

26. $(\sqrt{2} - 1)(\sqrt{2} + 3)$ _____

27. $(\sqrt{5} + 2)^2$ _____

28. $(\sqrt{3} - 3)^2$ _____

29. $(\sqrt{m} - 3)(\sqrt{m} - 4)$ _____

30. $(\sqrt{x} + 2)(\sqrt{x} - 3)$ _____

Rationalize denominators and simplify.

31. $\dfrac{1}{\sqrt{5}+2}$ _____

32. $\dfrac{1}{\sqrt{11}-3}$ _____

33. $\dfrac{2}{\sqrt{5}+1}$ _____

34. $\dfrac{4}{\sqrt{6}-2}$ _____

35. $\dfrac{\sqrt{2}}{\sqrt{10}-2}$ _____

36. $\dfrac{\sqrt{2}}{\sqrt{6}+2}$ _____

37. $\dfrac{\sqrt{y}}{\sqrt{y}+3}$ _____

38. $\dfrac{\sqrt{x}}{\sqrt{x}-2}$ _____

B *Express in simplest radical form and combine where possible.*

39. $\sqrt{8mn}+2\sqrt{18mn}$ _____

40. $\sqrt{4x}-\sqrt{9x}$ _____

41. $\sqrt{8}-\sqrt{20}+4\sqrt{2}$ _____

42. $\sqrt{24}-\sqrt{12}+3\sqrt{3}$ _____

43. $\sqrt[5]{a}-4\sqrt[5]{a}+2\sqrt[5]{a}$ _____

44. $3\sqrt[3]{u}-2\sqrt[3]{u}-2\sqrt[3]{u}$ _____

45. $2\sqrt[3]{x}+3\sqrt[3]{x}-\sqrt{x}$ _____

46. $5\sqrt[5]{y}-2\sqrt[5]{y}+3\sqrt[4]{y}$ _____

47. $\sqrt{\dfrac{1}{8}}+\sqrt{8}$ _____

48. $\sqrt{\dfrac{2}{3}}-\sqrt{\dfrac{3}{2}}$ _____

49. $\sqrt{\dfrac{3uv}{2}}-\sqrt{24uv}$ _____

50. $\sqrt{\dfrac{xy}{2}}+\sqrt{8xy}$ _____

Multiply and simplify where possible.

51. $(4\sqrt{3}-1)(3\sqrt{3}-2)$ _____

52. $(2\sqrt{7}-\sqrt{3})(2\sqrt{7}+\sqrt{3})$ _____

53. $(\sqrt{x}-\sqrt{y})(\sqrt{x}+\sqrt{y})$ _____

54. $(2\sqrt{x}+3)(2\sqrt{x}-3)$ _____

55. $(5\sqrt{m}+2)(2\sqrt{m}-3)$ _____

56. $(3\sqrt{u}-2)(2\sqrt{u}+4)$ _____

57. $(\sqrt[3]{4}+\sqrt[3]{9})(\sqrt[3]{2}+\sqrt[3]{3})$ _____

58. $\sqrt[3]{4}(\sqrt[3]{2}-\sqrt[3]{16})$ _____

59. Show that $3-\sqrt{2}$ is a solution to $x^2-6x+7=0$. _____

60. Show that $3+\sqrt{2}$ is a solution to $x^2-6x+7=0$. _____

Rationalize denominators and simplify.

61. $\dfrac{\sqrt{3}+2}{\sqrt{3}-2}$ _____

62. $\dfrac{\sqrt{2}-1}{\sqrt{2}+1}$ _____

63. $\dfrac{\sqrt{2}+\sqrt{3}}{\sqrt{3}-\sqrt{2}}$ _____

64. $\dfrac{3-\sqrt{a}}{\sqrt{a}-2}$ _____

65. $\dfrac{2+\sqrt{x}}{\sqrt{x}-3}$ _____

66. $\dfrac{\sqrt{5}-\sqrt{2}}{\sqrt{5}+\sqrt{2}}$ _____

67. $\dfrac{3\sqrt{x}}{2\sqrt{x}-3}$ _____

68. $\dfrac{5\sqrt{a}}{3-2\sqrt{a}}$ _____

C *Express in simplest radical form and combine where possible.*

69. $\dfrac{\sqrt{3}}{3} + 2\sqrt{\dfrac{1}{3}} + \sqrt{12}$ _____   **70.** $\sqrt{\dfrac{1}{2}} + \dfrac{\sqrt{2}}{2} + \sqrt{8}$ _____

71. $\sqrt[3]{\dfrac{1}{3}} + \sqrt[3]{3^5}$ _____   **72.** $\sqrt[4]{32} - \sqrt[4]{\dfrac{1}{8}}$ _____

Multiply and simplify where possible.

73. $(\sqrt[3]{x} - \sqrt[3]{y^2})(\sqrt[3]{x^2} + 2\sqrt[3]{y})$ _____

74. $(\sqrt[5]{u^2} - \sqrt[5]{v^3})(\sqrt[5]{u^3} + \sqrt[5]{v^2})$ _____

75. $(\sqrt[3]{x} + \sqrt[3]{y})(\sqrt[3]{x^2} - \sqrt[3]{x}\sqrt[3]{y} + \sqrt[3]{y^2})$ _____

76. $(\sqrt[3]{x} - \sqrt[3]{y})(\sqrt[3]{x^2} + \sqrt[3]{x}\sqrt[3]{y} + \sqrt[3]{y^2})$ _____

Rationalize denominators and simplify (see Problems 75 and 76 for Problems 79 and 80).

77. $\dfrac{2\sqrt{x} + 3\sqrt{y}}{4\sqrt{x} + 5\sqrt{y}}$ _____   **78.** $\dfrac{3\sqrt{x} + 2\sqrt{y}}{2\sqrt{x} - 5\sqrt{y}}$ _____

79. $\dfrac{1}{\sqrt[3]{x} + \sqrt[3]{y}}$ _____   **80.** $\dfrac{1}{\sqrt[3]{x} - \sqrt[3]{y}}$ _____

81. $\dfrac{1}{\sqrt{x} + \sqrt{y} - \sqrt{z}}$ _____   **82.** $\dfrac{1}{\sqrt{x} - \sqrt{y} + \sqrt{z}}$ _____

The Check Exercise for this section is on page 331.

[*Hint:* Start by multiplying numerator and denominator by $(\sqrt{x} + \sqrt{y}) + \sqrt{z}.$]

Section 5-7 Complex Numbers

- Introductory Remarks
- The Complex Number System
- Complex Numbers and Radicals
- Concluding Remarks

■ Introductory Remarks

The Pythagoreans (500–275 B.C.) found that the simple equation

$$x^2 = 2 \tag{1}$$

had no rational number solutions. If Equation (1) were to have a solution, then a new kind of number had to be invented—the irrational numbers. The irrational numbers $\sqrt{2}$ and $-\sqrt{2}$ are both solutions to (1). Irrational numbers were not put on a firm mathematical foundation until the last century. The rational and irrational numbers together constitute the real number system.

Is there any need to extend the real number system further? Yes, since we find that another simple equation

$$x^2 = -1$$

has no real solutions. (What real number squared is negative?) Once again, we are forced to invent a new kind of number—a number that has the possibility of being negative when it is squared. This new system of numbers is called the **complex numbers**. The complex numbers evolved over a long period of time,[†] but, like the real numbers, it was not until the last century that they were placed on a firm mathematical basis.

The Complex Number System

A **complex number** is any number of the form

a + bi

where a and b are real numbers; i is called the **imaginary unit**. Thus

$$5 + 2i \qquad \tfrac{1}{4} + 2i \qquad \sqrt{2} - \tfrac{1}{3}i \qquad 0 + 5i \qquad 6 + 0i \qquad 0 + 0i$$

are all complex numbers. Particular kinds of complex numbers are given special names:

$a + 0i = a$	REAL NUMBER
$0 + bi = bi$	PURE IMAGINARY NUMBER
$0 + 0i = 0$	ZERO
$1i = i$	IMAGINARY UNIT
$a - bi$	**CONJUGATE** OF $a + bi$

[†] BRIEF HISTORY OF COMPLEX NUMBERS

APPROXIMATE DATE	PERSON	EVENT
50	Heron of Alexandria	First recorded encounter of a square root of a negative number
850	Mahavira of India	Said that a negative has no square root, since it is not a square
1545	Cardano of Italy	Found that solutions to cubic equations involved square roots of negative numbers
1637	Descartes of France	Introduced the terms "real" and "imaginary"
1748	Euler of Switzerland	Used i for $\sqrt{-1}$
1832	Gauss of Germany	Introduced the term "complex number"

Thus, we see that just as every integer is a rational number, every real number is a complex number; that is, the real numbers form a subset of the set of complex numbers. The complex number system is related to the other number systems that we have studied as shown in Figure 2.

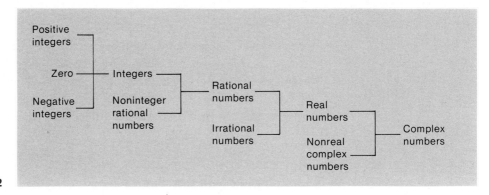

FIGURE 2

To use complex numbers we must know how to add, subtract, multiply, and divide them. We start by defining equality, addition, and multiplication.

EQUALITY	$a + bi = c + di$ if and only if $a = c$ and $b = d$
ADDITION	$(a + bi) + (c + di) = (a + c) + (b + d)i$
MULTIPLICATION	$(a + bi)(c + di) = (ac - bd) + (ad + bc)i$

These definitions, particularly the one for multiplication, may seem a little strange to you. But it turns out that if we want the real number properties discussed in Chapter 1 to continue to hold for complex numbers, and if we also want the possibility of having the square of a number negative, then we must define addition and multiplication as above. Let us use the definition of multiplication to see what happens to i when it is squared:

$$i^2 = \overset{a\quad b}{(0 + 1i)}\overset{c\quad d}{(0 + 1i)}$$

$$= \overset{a\ c\quad b\ d}{(0 \cdot 0 - 1 \cdot 1)} + \overset{a\ d\quad b\ c}{(0 \cdot 1 + 1 \cdot 0)}i$$

$$= -1 + 0i$$

$$= -1$$

Thus,

$$i^2 = -1$$

and we have a number whose square is negative (and a solution to $x^2 = -1$). We choose to let

$$i = \sqrt{-1} \quad \text{and} \quad -i = -\sqrt{-1}$$

Fortunately, you do not have to memorize the definitions of addition and multiplication presented above. We can show that the complex numbers under these definitions are associative and commutative and that multiplication distributes over addition. As a consequence, we can manipulate complex numbers as if they were binomial forms in real number algebra, with the exception that i^2 is to be replaced with -1. The following example illustrates the mechanics of carrying out addition, subtraction, multiplication, and division.

EXAMPLE 21 Carry out the indicated operations and write each answer in the form $a + bi$:

(A) $(3 + 2i) + (2 - i)$ **(B)** $(3 + 2i) - (2 - i)$

(C) $(3 + 2i)(2 - i)$ **(D)** $\dfrac{3 + 2i}{2 - i}$

Solution **(A)** $(3 + 2i) + (2 - i) = 3 + 2i + 2 - i$ Remove parentheses and combine like terms.

$= 5 + i$

(B) $(3 + 2i) - (2 - i) = 3 + 2i - 2 + i$ Remove parentheses and combine like terms.

$= 1 + 3i$

(C) $(3 + 2i)(2 - i) = 6 + i - 2i^2$ Multiply and replace i^2 with -1.

$= 6 + i - 2(-1)$

$= 6 + i + 2$

$= 8 + i$

(D) In order to eliminate i from the denominator, we multiply the numerator and denominator by the conjugate of $2 - i$, that is, by $2 + i$:

$$\frac{3 + 2i}{2 - i} \cdot \frac{2 + i}{2 + i} = \frac{6 + 7i + 2i^2}{4 - i^2} = \frac{6 + 7i + 2(-1)}{4 - (-1)}$$

$$= \frac{4 + 7i}{5} = \frac{4}{5} + \frac{7}{5}i$$

Recall that subtraction and division are defined, in general, as follows:

$A - B = C$ if and only if $A = B + C$

$A \div B = C$ if and only if $A = BC$, $B \neq 0$, and C is unique

The results obtained by the procedures illustrated in Examples 21(B) and (D) are consistent with these definitions, as can easily be checked. And with a little extra work, these procedures can be shown to hold in general (see the C-level problems in Exercise 5-7).

PROBLEM 21   Carry out the indicated operations and write each answer in the form $a + bi$:

(A) $(3 + 2i) + (6 - 4i)$ (B) $(3 - 5i) - (1 - 3i)$

(C) $(2 - 4i)(3 + 2i)$ (D) $\dfrac{2 + 4i}{3 + 2i}$

Solution   (A)

(B)

(C)

(D)

EXAMPLE 22   Carry out the indicated operations and write each answer in the form $a + bi$:

(A) $(2 - 3i)^2 - (4i)^2$ (B) $\dfrac{2 + i}{3i}$

Solution   (A) $(2 - 3i)^2 - (4i)^2 = 4 - 12i + 9i^2 - 16i^2$

$$= 4 - 12i + 9(-1) - 16(-1)$$
$$= 4 - 12i - 9 + 16$$
$$= 11 - 12i$$

(B) $\dfrac{2 + i}{3i} = \dfrac{2 + i}{3i} \cdot \dfrac{i}{i} = \dfrac{2i + i^2}{3i^2} = \dfrac{2i + (-1)}{3(-1)}$

$$= \dfrac{2i - 1}{-3} = \dfrac{-1}{-3} + \dfrac{2}{-3}i = \dfrac{1}{3} - \dfrac{2}{3}i$$

PROBLEM 22   Carry out the indicated operations and write each answer in the form $a + bi$:

(A) $(3i)^2 - (3 - 2i)^2$ (B) $\dfrac{3 + i}{2i}$

Solution   (A)

(B)

■ Complex Numbers and Radicals

Recall that we say y is a square root of x if $y^2 = x$. It can be shown that if x is a positive real number, then x has two real square roots, one the negative of the other; if x is negative, then x has two complex square roots, one also the negative of the other. In particular, if we let $x = -a$, $a > 0$, then one of the square roots of x is given by[†]

$$\sqrt{-a} = i\sqrt{a} \qquad a > 0 \qquad \sqrt{-9} = i\sqrt{9} = 3i$$

To check this, we square $i\sqrt{a}$ and obtain $-a$:

$$(i\sqrt{a})^2 = i^2(\sqrt{a})^2 = (-1)a = -a$$

EXAMPLE 23   Write in the form $a + bi$:

(A) $\sqrt{-4}$ (B) $4 + \sqrt{-40}$ (C) $\dfrac{-3 - \sqrt{-7}}{2}$

Solution   (A) $\sqrt{-4} = i\sqrt{4} = 2i$

(B) $4 + \sqrt{-40} = 4 + i\sqrt{40} = 4 + i\sqrt{4 \cdot 10} = 4 + 2i\sqrt{10}$

(C) $\dfrac{-3 - \sqrt{-7}}{2} = \dfrac{-3 - i\sqrt{7}}{2} = -\dfrac{3}{2} - \dfrac{\sqrt{7}}{2}i$

PROBLEM 23   Write in the form $a + bi$:

(A) $\sqrt{-16}$ (B) $5 - \sqrt{-36}$ (C) $\dfrac{-5 - \sqrt{-2}}{2}$

Solution   (A)   (B)   (C)

EXAMPLE 24   Convert square roots of negative numbers to complex form, perform the indicated operations, and express your answers in the form $a + bi$:

(A) $(3 + \sqrt{-4})(2 - \sqrt{-9})$ (B) $\dfrac{1}{3 - \sqrt{-4}}$

[†] Note that if in $a + bi$, $b = \sqrt{k}$, then we often write $a + i\sqrt{k}$ instead of $a + \sqrt{k}\,i$ so that i will not accidentally end up under the radical sign.

Solution **(A)** $(3 + \sqrt{-4})(2 - \sqrt{-9})$

$\qquad = (3 + i\sqrt{4})(2 - i\sqrt{9})$

$\qquad = (3 + 2i)(2 - 3i)$

$\qquad = 6 - 5i - 6i^2$

$\qquad = 6 - 5i - 6(-1)$

$\qquad = 6 - 5i + 6$

$\qquad = 12 - 5i$

Note that

$\sqrt{-4}\sqrt{-9} \neq \sqrt{(-4)(-9)}$

since

$\sqrt{-4}\sqrt{-9} = (2i)(3i) = 6i^2$

$\qquad\qquad = -6$

while

$\sqrt{(-4)(-9)} = \sqrt{36} = 6$

(B) $\dfrac{1}{3 - \sqrt{-4}} = \dfrac{1}{3 - i\sqrt{4}} = \dfrac{1}{3 - 2i}$

$\qquad\qquad = \dfrac{1}{3 - 2i} \cdot \dfrac{3 + 2i}{3 + 2i} = \dfrac{3 + 2i}{9 - 4i^2}$

$\qquad\qquad = \dfrac{3 + 2i}{9 - 4(-1)} = \dfrac{3 + 2i}{9 + 4}$

$\qquad\qquad = \dfrac{3 + 2i}{13} = \dfrac{3}{13} + \dfrac{2}{13}i$

PROBLEM 24

Convert square roots of negative numbers to complex form, perform the indicated operations, and express your answers in the form $a + bi$:

(A) $(4 - \sqrt{-25})(3 + \sqrt{-49})$ **(B)** $\dfrac{1}{2 + \sqrt{-9}}$

Solution   **(A)**

(B)

■ Concluding Remarks

Complex numbers are used extensively by electrical, aeronautical, and space scientists, as well as chemists and physicists. The interpretation of complex numbers $a + bi$ relative to the real world is not readily seen until you have had more experience in some of these fields. We state only that a and b in the complex number $a + bi$ often represent real-world quantities. Our use of the complex numbers will be in connection with solutions to second-degree equations such as

$\qquad x^2 - 4x + 5 = 0$

which we will study in the next chapter.

Solutions to Matched Problems

21. **(A)** $(3 + 2i) + (6 - 4i) = 3 + 2i + 6 - 4i = 9 - 2i$

(B) $(3 - 5i) - (1 - 3i) = 3 - 5i - 1 + 3i = 2 - 2i$

(C) $(2 - 4i)(3 + 2i) = 6 - 8i - 8i^2 = 6 - 8i + 8 = 14 - 8i$

(D) $\dfrac{2 + 4i}{3 + 2i} = \dfrac{(2 + 4i)}{(3 + 2i)} \cdot \dfrac{(3 - 2i)}{(3 - 2i)} = \dfrac{6 + 8i - 8i^2}{9 - 4i^2} = \dfrac{6 + 8i + 8}{9 + 4}$

$= \dfrac{14 + 8i}{13} = \dfrac{14}{13} + \dfrac{8}{13}i$

22. **(A)** $(3i)^2 - (3 - 2i)^2 = 9i^2 - (9 - 12i + 4i^2) = 9i^2 - 9 + 12i - 4i^2$

$= -9 + 12i + 5i^2 = -9 + 12i - 5 = -14 + 12i$

(B) $\dfrac{3 + i}{2i} = \dfrac{3 + i}{2i} \cdot \dfrac{i}{i} = \dfrac{3i + i^2}{2i^2} = \dfrac{3i - 1}{-2} = \dfrac{-1}{-2} + \dfrac{3}{-2}i = \dfrac{1}{2} - \dfrac{3}{2}i$

23. **(A)** $\sqrt{-16} = i\sqrt{16} = 4i \quad \text{or} \quad 0 + 4i$

(B) $5 - \sqrt{-36} = 5 - i\sqrt{36} = 5 - 6i$

(C) $\dfrac{-5 - \sqrt{-2}}{2} = \dfrac{-5 - i\sqrt{2}}{2} = \dfrac{-5}{2} - \dfrac{\sqrt{2}}{2}i \quad \text{or} \quad -\dfrac{5}{2} - \dfrac{\sqrt{2}}{2}i$

24. **(A)** $\left(4 - \sqrt{-25}\right)\left(3 + \sqrt{-49}\right) = \left(4 - i\sqrt{25}\right)\left(3 + i\sqrt{49}\right)$

$= (4 - 5i)(3 + 7i) = 12 + 13i - 35i^2$

$= 12 + 13i + 35 = 47 + 13i$

(B) $\dfrac{1}{2 + \sqrt{-9}} = \dfrac{1}{2 + i\sqrt{9}} = \dfrac{1}{2 + 3i} = \dfrac{1}{(2 + 3i)} \cdot \dfrac{(2 - 3i)}{(2 - 3i)} = \dfrac{2 - 3i}{4 - 9i^2}$

$= \dfrac{2 - 3i}{4 + 9} = \dfrac{2 - 3i}{13} = \dfrac{2}{13} - \dfrac{3}{13}i$

Practice Exercise 5-7 ■

Work odd-numbered problems first, check answers, and then work even-numbered problems in areas of weakness. Answers to all problems are in the back of the book. Make every effort to work a problem yourself before you look at an answer.

A *Perform the indicated operations and write each answer in the form a + bi.*

1. $(5 + 2i) + (3 + i)$ _____

2. $(6 + i) + (2 + 3i)$ _____

3. $(-8 + 5i) + (3 - 2i)$ _____

4. $(2 - 3i) + (5 - 2i)$ _____

5. $(8 + 5i) - (3 + 2i)$ _____

6. $(9 + 7i) - (2 + 5i)$ _____

7. $(4 + 7i) - (-2 - 6i)$ _____

8. $(9 - 3i) - (12 - 5i)$ _____

9. $(3 - 7i) + 5i$ _____

10. $12 + (5 - 2i)$ _____

11. $(5i)(3i)$ _____

12. $(2i)(4i)$ _____

13. $-2i(5 - 3i)$ _____

14. $-3i(2 - 4i)$ _____

15. $(2 - 3i)(3 + 3i)$ _____

16. $(3 - 5i)(-2 - 3i)$ _____

17. $(7 - 6i)(2 - 3i)$ _____

18. $(2 - i)(3 + 2i)$ _____

19. $(7 + 4i)(7 - 4i)$ _____

20. $(5 - 3i)(5 + 3i)$ _____

21. $\dfrac{1}{2 + i}$ _____

22. $\dfrac{1}{3 - i}$ _____

23. $\dfrac{3 + i}{2 - 3i}$ _____

24. $\dfrac{2 - i}{3 + 2i}$ _____

25. $\dfrac{13 + i}{2 - i}$ _____

26. $\dfrac{15 - 3i}{2 - 3i}$ _____

B *Convert square roots of negative numbers to complex form; perform the indicated operations; and express your answers in the form a + bi.*

27. $\left(5 - \sqrt{-9}\right) + \left(2 - \sqrt{-4}\right)$ _____

28. $\left(-8 + \sqrt{-25}\right) + \left(3 - \sqrt{-4}\right)$ _____

29. $\left(9 - \sqrt{-9}\right) - \left(12 - \sqrt{-25}\right)$ _____

30. $\left(4 + \sqrt{-49}\right) - \left(-2 - \sqrt{-36}\right)$ _____

31. $\left(-2 + \sqrt{-49}\right)\left(3 - \sqrt{-4}\right)$ _____

32. $\left(5 + \sqrt{-9}\right)\left(2 - \sqrt{-1}\right)$ _____

33. $\dfrac{5 - \sqrt{-4}}{3}$ _____

34. $\dfrac{6 - \sqrt{-64}}{2}$ _____

35. $\dfrac{1}{2 - \sqrt{-9}}$ _____

36. $\dfrac{1}{3 - \sqrt{-16}}$ _____

37. $\dfrac{2}{5i}$ _____

38. $\dfrac{1}{3i}$ _____

39. $\dfrac{1 + 3i}{2i}$ _____

40. $\dfrac{2 - i}{3i}$ _____

41. $(2 - i)^2 + 3(2 - i) - 5$ _____

42. $(2 - 3i)^2 - 2(2 - 3i) + 9$ _____

43. Evaluate $x^2 - 2x + 2$ for $x = 1 - i$. _____

44. Evaluate $x^2 - 2x + 2$ for $x = 1 + i$. _____

45. Evaluate $x^2 - 4x + 5$ for $x = 2 + i$. _____

46. Evaluate $x^2 - 4x + 5$ for $x = 2 - i$. _____

47. Simplify: i^2, i^3, i^4, i^5, i^6, i^7, and i^8 _____

48. Simplify: i^{12}, i^{13}, i^{14}, i^{15}, and i^{16} _____

C *Perform the indicated operations and write each answer in the form a + bi.*

49. $(a + bi) + (c + di)$ _____   50. $(a + bi) - (c + di)$ _____

51. $(a + bi)(a - bi)$ _____   52. $(u - vi)(u + vi)$ _____

53. $(a + bi)(c + di)$ _____   54. $\dfrac{a + bi}{c + di}$ _____

55. $\left(-\dfrac{1}{2} - \dfrac{\sqrt{3}}{2}i\right)^3$ _____   56. $\left(-\dfrac{1}{2} + \dfrac{\sqrt{3}}{2}i\right)^3$ _____

Solve each equation.

57. $y^2 = -36$ _____   58. $x^2 = -25$ _____

59. $(x - 9)^2 = -9$ _____   60. $(x - 3)^2 = -4$ _____

61. For what values of x will $\sqrt{x - 10}$ be real? _____

62. When will
$$\frac{-b \pm \sqrt{b^2 - 4ac}}{2a}$$
represent a real number, assuming a, b, and c are all real numbers $(a \neq 0)$?
When will it represent a nonreal complex number? _____

63. Evaluate
$$(a + bi)\left(\frac{a}{a^2 + b^2} - \frac{b}{a^2 + b^2}i\right) \qquad a \neq 0 \text{ and } b \neq 0$$

The Check Exercise for
this section is on
page 333.

thus showing that each nonzero complex number $a + bi$ has an inverse relative to multiplication. _____

Section 5-8   Chapter Review

The concept of exponent is extended to **0 and negative number exponents** by

$$a^0 = 1 \qquad a \neq 0, \ 0^0 \text{ not defined}$$

$$a^{-n} = \frac{1}{a^n} \qquad \text{for } n \text{ a positive integer}$$

The **five basic properties of exponents** continue to hold:

1. $a^m a^n = a^{m+n}$ 4. $\left(\dfrac{a}{b}\right)^m = \dfrac{a^m}{b^m}$

2. $(a^n)^m = a^{mn}$

3. $(ab)^m = a^m b^m$ 5. $\dfrac{a^m}{a^n} = a^{m-n}$ *(5-1)*

Any real number can be written in **scientific notation**—that is, in the form $a \cdot 10^n$, where a is in the interval $[1, 10)$ and n is an integer.   *(5-2)*

For a natural number n, a is an **nth root** of b if $a^n = b$. For even n, b has two real nth roots when b is positive, none when b is negative; for odd n, b always has one real nth root. The concept of exponents is extended to **rational number exponents** by

$$b^{1/n} = \begin{cases} \text{The positive nth root of } b \text{ if } b \text{ is positive} \\ \text{The nth root of } b \text{ if } b \text{ is negative, } n \text{ odd} \\ \text{Not a real number if } b \text{ is negative, } n \text{ even} \end{cases}$$

and

$$b^{m/n} = (b^{1/n})^m = (b^m)^{1/n}$$

The five basic properties of exponents continue to hold for rational exponents as long as undefined roots are avoided.   *(5-3)*

The nth root of b is also denoted by the **radical** form $\sqrt[n]{b}$; here n is called the **index** and b the **radicand**. In this notation $b^{m/n} = (\sqrt[n]{b})^m = \sqrt[n]{b^m}$.   *(5-4)*

For positive a and b, these properties of radicals hold:

1. $\sqrt[n]{b^n} = b$

2. $\sqrt[n]{ab} = \sqrt[n]{a} \cdot \sqrt[n]{b}$

3. $\sqrt[n]{\dfrac{a}{b}} = \dfrac{\sqrt[n]{a}}{\sqrt[n]{b}}$

4. $\sqrt[kn]{b^{km}} = \sqrt[n]{b^m}$

These properties can be used to convert an expression involving radicals to **simplest radical form** wherein the index of the radical is less than the power of any factor of the radicand, the index has no common factor with the power of the radicand, no radical occurs in a denominator, and no fraction appears in a radical. Property 1 can be extended to any real b:

$$\sqrt[n]{b^n} = \begin{cases} b & \text{if } n \text{ is odd} \\ |b| & \text{if } n \text{ is even} \end{cases} \quad \text{(5-5)}$$

Expressions involving radicals can often be added, subtracted, or multiplied by making use of the distributive property for real numbers. Radicals can sometimes be removed from denominators by **rationalizing the denominator**—multiplying the fraction numerator and denominator by a factor that makes the denominator a rational number.   *(5-6)*

A **complex number** is a number of the form $a + bi$, where i is called the **imaginary unit** and a and b are real numbers. If $b = 0$, the number is real; if $a = 0$, the number is **pure imaginary**. The **conjugate** of $a + bi$ is $a - bi$. Two complex numbers $a + bi$ and $c + di$ are **equal** when $a = c$ and $b = d$. Basic operations are defined by

$$(a + bi) + (c + di) = (a + c) + (b + d)i$$

$$(a + bi) - (c + di) = (a - c) + (b - d)i$$

$$(a + bi) \cdot (c + di) = (ac - bd) + (ad + bc)i$$

Division of complex numbers makes use of rationalizing the denominator by

$$\frac{1}{a + bi} = \frac{1}{a + bi} \cdot \frac{a - bi}{a - bi} = \frac{a}{a^2 + b^2} - \frac{b}{a^2 + b^2}i$$

Since $i^2 = -1$, i is also denoted $\sqrt{-1}$. Similarly, if b is positive then $\sqrt{-b} = i\sqrt{b}$.   *(5-7)*

Diagnostic (Review) Exercise 5-8 ■

Work through all the problems in this chapter review and check answers in the back of the book. (Answers to all problems are there, and following each answer is a number in italics indicating the section in which that type of problem is discussed.) Where weaknesses show up, review appropriate sections in the text. When you are satisfied that you know the material, take the practice test following this review.

Unless otherwise stated, all variables represent positive real numbers.

A *Evaluate, if possible, using only real numbers.*

1. $\left(\dfrac{1}{3}\right)^0$ _____

2. 3^{-2} _____

3. $\dfrac{1}{2^{-3}}$ _____

4. $4^{-1/2}$ _____

5. $(-9)^{3/2}$ _____

6. $(-8)^{2/3}$ _____

7. Write in scientific notation:

   (A) 4,280,000,000 _____

   (B) 0.000 031 8 _____

8. Write as a decimal fraction:

   (A) 7.29×10^5 _____

   (B) 6.03×10^{-4} _____

Simplify, and write answers using positive exponents only.

9. $(3x^3y^2)(2xy^5)$ _____

10. $\dfrac{9u^8v^6}{3u^4v^8}$ _____

11. $6(xy^3)^5$ _____

12. $\left(\dfrac{c^2}{d^5}\right)^3$ _____

13. $\left(\dfrac{2x^2}{3y^3}\right)^2$ _____

14. $(x^{-3})^{-4}$ _____

15. $\dfrac{y^{-3}}{y^{-5}}$ _____

16. $(x^2y^{-3})^{-1}$ _____

17. $(x^9)^{1/3}$ _____

18. $(x^4)^{-1/2}$ _____

19. $x^{1/3}x^{-2/3}$ _____

20. $\dfrac{u^{5/3}}{u^{2/3}}$ _____

21. Change to radical form:

   (A) $(3m)^{1/2}$ _____

   (B) $3m^{1/2}$ _____

22. Change to rational exponent form:

   (A) $\sqrt{2x}$ _____

   (B) $\sqrt{a+b}$ _____

Simplify, and write in simplest radical form.

23. $\sqrt{4x^2y^4}$ _____

24. $\sqrt{\dfrac{25}{y^2}}$ _____

25. $\sqrt{36x^4y^7}$ _____

26. $\dfrac{1}{\sqrt{2y}}$ _____

27. $\dfrac{6ab}{\sqrt{3a}}$ _____

28. $\sqrt{2x^2y^5}\sqrt{18x^3y^2}$ _____

29. $\sqrt{\dfrac{y}{2x}}$ _____

30. $4\sqrt{x} - 7\sqrt{x}$ _____

31. $\sqrt{7} + 2\sqrt{3} - 4\sqrt{3}$ _____

32. $\sqrt{5}(\sqrt{5} + 2)$ _____

33. $(\sqrt{3} - 1)(\sqrt{3} + 2)$ _____

34. $\dfrac{\sqrt{5}}{3 - \sqrt{5}}$ _____

Perform the indicated operations and write the answer in the form a + bi.

35. $(-3 + 2i) + (6 - 8i)$ _____

36. $(3 - 3i)(2 + 3i)$ _____

37. $\dfrac{13 - i}{5 - 3i}$ _____

38. $\dfrac{2 - i}{2i}$ _____

B **39.** Convert each number to scientific notation, simplify, and write your answer in scientific notation and as a decimal fraction:

$$\dfrac{0.000\,052}{130(0.000\,2)}$$ _____

Simplify, and write answers using positive exponents only.

40. $\dfrac{3m^4n^{-7}}{6m^2n^{-2}}$ _____

41. $(x^{-3}y^2)^{-2}$ _____

42. $\dfrac{1}{(2x^2y^{-3})^{-2}}$ _____

43. $\left(-\dfrac{a^2b}{c}\right)^2 \left(\dfrac{c}{b^2}\right)^3 \left(\dfrac{1}{a^3}\right)^2$ _____

44. $\left(\dfrac{8u^{-1}}{2^2u^2v^0}\right)^{-2} \left(\dfrac{u^{-5}}{u^{-3}}\right)^3$ _____

45. $\left(\dfrac{9m^3n^{-3}}{3m^{-2}n^2}\right)^{-2}$ _____

46. $(x - y)^{-2}$ _____

47. $(9a^4b^{-2})^{1/2}$ _____

48. $\left(\dfrac{27x^2y^{-3}}{8x^{-4}y^3}\right)^{1/3}$ _____

49. $\dfrac{m^{-1/4}}{m^{3/4}}$ _____

50. $(2x^{1/2})(3x^{-1/3})$ _____

51. $\dfrac{3x^{-1/4}}{6x^{-1/3}}$ _____

52. $\dfrac{5^0}{3^2} + \dfrac{3^{-2}}{2^{-2}}$ _____

53. $(x^{1/2} + y^{1/2})^2$ _____

54. If a is a square root of b, then does $a^2 = b$ or does $b^2 = a$?

_____

55. Change to radical form:

(A) $(2mn)^{2/3}$ _____ **(B)** $3x^{2/5}$ _____

56. Change to rational exponent form:

(A) $\sqrt[7]{x^5}$ _____ **(B)** $-3\sqrt[3]{(xy)^2}$ _____

Simplify, and write in simplest radical form.

57. $\sqrt[3]{(2x^2y)^3}$ _____ **58.** $3x\sqrt[3]{x^5y^4}$ _____

59. $\dfrac{\sqrt{8m^3n^4}}{\sqrt{12m^2}}$ _____ **60.** $\sqrt[8]{y^6}$ _____

61. $-2x\sqrt[5]{3^6x^7y^{11}}$ _____ **62.** $\dfrac{2x^2}{\sqrt[3]{4x}}$ _____

63. $\sqrt[5]{\dfrac{3y^2}{8x^2}}$ _____

64. $(2\sqrt{x} - 5\sqrt{y})(\sqrt{x} + \sqrt{y})$ _____

65. $\dfrac{\sqrt{x} - 2}{\sqrt{x} + 2}$ _____ **66.** $\dfrac{3\sqrt{x}}{2\sqrt{x} - \sqrt{y}}$ _____

67. $\sqrt{\tfrac{2}{3}} + \sqrt{\tfrac{3}{2}}$ _____

Perform the indicated operations and write the answer in the form $a + bi$.

68. $(2 - 2\sqrt{-4}) - (3 - \sqrt{-9})$ _____

69. $\dfrac{2 - \sqrt{-1}}{3 + \sqrt{-4}}$ _____ **70.** $(3 + i)^2 - 2(3 + i) + 3$ _____

Simplify, and write answers using positive exponents only.

71. $(x^{-1} + y^{-1})^{-1}$ _____ **72.** $\left(\dfrac{a^{-2}}{b^{-1}} + \dfrac{b^{-2}}{a^{-1}}\right)^{-1}$ _____

C _Simplify, and write in simplest radical form._

73. $\sqrt[9]{8x^6y^{12}}$ _____ **74.** $\sqrt[3]{3} - \dfrac{6}{\sqrt[3]{9}} + 3\sqrt[3]{\dfrac{1}{9}}$ _____

75. Simplify $3\sqrt[3]{x^3} - 2\sqrt{x^2}$:

(A) For x a positive number _____

(B) For x a negative number _____

Practice Test Chapter 5 ∎

Take this as if it were a graded test by working the problems within a 50-minute time period. Do not look back in the chapter. Choose one of three levels of difficulty: least difficult, Problems 1–12; more difficult, add Problem 13; most difficult, add Problems 13 and 14. Use the answers in the back of the book to correct your work. The answers are keyed to appropriate text sections so that you can easily locate and review sections where difficulties still persist.

Simplify Problems 1–4, and write answers using positive exponents only. All variables represent positive real numbers.

1. $\left(\dfrac{3u^2v^5}{9u^3v^2}\right)^{-2}$ _____

2. $\left(\dfrac{4x^{-2}y^8}{x^2y^2}\right)^{-1/2}$ _____

3. $\left(\dfrac{x^{-1/2}y^{1/3}}{x^{-1/3}y^{1/2}}\right)^{-6}$ _____

4. $(x^{-1/2} + y^{-1/2})^2$ _____

5. Convert each number to scientific notation, simplify, and write the answer in scientific notation:

   $$\dfrac{(0.000\ 075)(4,000)}{(2,000,000)(0.05)}$$ _____

6. Simplify for x a negative number: $4\sqrt[3]{x^3} + 5\sqrt{x^2}$ _____

Perform any indicated operations in Problems 7–10, and express answers in simplest radical form. All variables represent positive real numbers.

7. $\sqrt[3]{2} - \dfrac{6}{\sqrt[3]{4}}$ _____

8. $\sqrt[6]{(2xy)^4}$ _____

9. $(3 - \sqrt{2})^2 - 6(3 - \sqrt{2}) + 7$ _____

10. $\dfrac{\sqrt{x} + \sqrt{y}}{\sqrt{x} - \sqrt{y}}$ _____

Perform the indicated operations in Problems 11 and 12 (after converting square roots of negative numbers to complex form), and write answers in the form $a + bi$.

11. $(2 - 5i)(1 + 3i) - (4 - 2i)$ _____

12. $(1 + \sqrt{-36})(6 + \sqrt{-1})$ _____

13. $\dfrac{1 - i}{1 + \sqrt{-1}}$ _____

14. Express in simplest radical form; all variables represent positive real numbers:

    $$\dfrac{\sqrt{x^2y^3z^4}}{\sqrt[3]{x^3y^4z^5}}$$ _____

Check Exercise 5-1 ■

ANSWER COLUMN

Work the following problems without looking at any text examples. Show your work in the space provided. Write your answer in the answer column.

1. _____

2. _____

3. _____

4. _____

5. _____

6. _____

7. _____

8. _____

9. _____

10. _____

Simplify and write answers using positive exponents only.

1. $\dfrac{x^2 x^{-5}}{x^{-2}}$

2. $(2^3 \cdot 3^0)^{-2}$

3. $(2x^{-1})^{-2}$

4. $(3x^{-3}y)^{-2}$

5. $\dfrac{8 \times 10^{-1}}{2 \times 10^{-5}}$

6. $\dfrac{9x^3y^{-2}}{12x^{-3}y}$

7. $\left(\dfrac{3^2x^3}{9x^{-3}y^0}\right)^{-2}$

8. $\left|\left(\dfrac{3x^2y^{-1}}{x^{-1}y}\right)^{-2}\right|^{-1}$

9. $(2^{-1} + 3^{-1})^0$

10. $(a^{-2} + b^{-2})^{-1}$

Check Exercise 5-2 ■

Work the following problems without looking at any text examples. Show your work in the space provided. Write your answer in the answer column.

1. _____

2. _____

3. _____

4. _____

5. _____

6. _____

7. _____

8. _____

9. _____

10. _____

1. Write in scientific notation:

    43,200,000

2. Write in scientific notation:

    0.000 000 081

3. Write in scientific notation:

    0.6435

4. Write as a decimal fraction:

    5.03×10^5

5. Write as a decimal fraction:

    5.07×10^{-2}

6. Write as a decimal fraction:

6.17×10^{-6}

7. Multiply and express the answer in scientific notation:

$(3 \times 10^{-9})(2 \times 10^4)$

8. Divide and express answer in scientific notation:

$$\frac{8 \times 10^{-3}}{2 \times 10^3}$$

9. Convert each numeral to scientific notation and simplify. Express the answer in scientific notation:

$$\frac{(0.000\ 000\ 032)(210)}{(0.000\ 24)(14)}$$

10. If the mass of the earth is 6×10^{27} grams and each gram is 3.53×10^{-2} ounce, find the mass of the earth in ounces. Express the answer in scientific notation.

Check Exercise 5-3 ■

Work the following problems without looking at any text examples. Show your work in the space provided. Write your answer in the answer column.

1. _____

2. _____

Evaluate each expression in Problems 1–5, if possible.

1. $4^{3/2}$

3. _____

4. _____

5. _____

2. $(-8)^{4/3}$

6. _____

7. _____

8. _____

3. $(-16)^{3/4}$

9. _____

10. _____

4. $64^{-1/2}$

5. $\left(-\dfrac{8}{27}\right)^{2/3}$

Simplify Problems 6–10, and express answers using positive exponents only. All variables represent positive real numbers.

6. $(4x^4y^{16})^{3/2}$

7. $(2m^{3/4})(3m^{-1/3})$

8. $\left(\dfrac{27x^2y^{-3}}{8x^{-4}y^3}\right)^{1/3}$

9. $\left(\dfrac{3x^{-1/2}}{x^{1/3}}\right)^{-2}$

10. $(2x^{1/2} - y^{1/2})(x^{1/2} + 2y^{1/2})$

Check Exercise 5-4 ∎

Work the following problems without looking at any text examples. Show your work in the space provided. Write your answer in the answer column.

1. _____

Change Problems 1–5 to radical form (do not simplify).

2. _____

   1. $5m^{3/4}$

3. _____

4. _____

5. _____

   2. $(x^4 - y^4)^{1/4}$

6. _____

7. _____

8. _____

   3. $x^{-2/3}$

9. _____

10. _____

   4. $\dfrac{3}{x^{1/2} + y^{1/2}}$

   5. $(x^3 - 2y^6)^{2/3}$

Change Problems 6–10 to rational exponent form (do not simplify).

6. $\sqrt[3]{(2xy^2)^2}$

7. $\dfrac{5}{\sqrt[7]{x^2}}$

8. $7\sqrt[3]{x^2}$

9. $\sqrt[5]{x^5 + y^5}$

10. $\sqrt{x} - \sqrt[3]{x^4}$

Check Exercise 5-5 ■

Work the following problems without looking at any text examples. Show your work in the space provided. Write your answer in the answer column.

1. _____

In Problems 1–8, simplify and write in simplest radical form. All variables represent positive real numbers.

2. _____

1. $\sqrt[3]{54x^{10}y^5z^3}$

3. _____

4. _____

5. _____

2. $\dfrac{12x^2}{\sqrt{3x}}$

6. _____

7. _____

8. _____

3. $\sqrt{\dfrac{2u}{3v}}$

9. _____

10. _____

4. $\sqrt[12]{(a + b)^8}$

5. $\dfrac{x}{\sqrt[4]{x^3}}$

6. $\sqrt[12]{16x^4y^8}$

7. $\dfrac{6x^3y}{\sqrt[3]{9x^2y}}$

8. $\sqrt[5]{\dfrac{3y}{16x^2}}$

Simplify Problems 9 and 10.

9. For x a positive number: $\quad 3\sqrt[3]{x^3} - 5\sqrt[4]{x^4}$

10. For x a negative number: $\quad \sqrt[7]{x^7} + \sqrt{x^2}$

Check Exercise 5-6 ■

Work the following problems without looking at any text examples. Show your work in the space provided. Write your answer in the answer column.

1. _____

In Problems 1–4 combine terms where possible after expressing all terms in simplest radical form.

2. _____

    1. $3\sqrt{m} - 2\sqrt{n} - \sqrt{m} + \sqrt{n}$

3. _____

4. _____

5. _____

    2. $\sqrt[3]{8x} + \sqrt[3]{27x}$

6. _____

7. _____

8. _____

    3. $\sqrt{18} - \dfrac{4}{\sqrt{2}}$

9. _____

10. _____

    4. $\sqrt{\dfrac{5}{3}} - \sqrt{\dfrac{3}{5}}$

Multiply in Problems 5–8 and simplify where possible. Represent answers in simplest radical form.

    5. $\sqrt{3}(\sqrt{6} - \sqrt{3})$

6. $(\sqrt{m} - \sqrt{n})(\sqrt{m} + \sqrt{n})$

7. $(2\sqrt{x} - 3)(3\sqrt{x} + 2)$

8. $(\sqrt[3]{m^2} - 2)(\sqrt[3]{m} + 2)$

Rationalize denominators in Problems 9 and 10 and represent answers in simplest radical form.

9. $\dfrac{\sqrt{5}}{\sqrt{5} - 2}$

10. $\dfrac{\sqrt{c} + \sqrt{d}}{\sqrt{c} - \sqrt{d}}$

Check Exercise 5-7 ■

Work the following problems without looking at any text examples. Show your work in the space provided. Write your answer in the answer column.

1. _____

2. _____

Perform the indicated operations in Problems 1–6 and write each answer in the form $a + bi$.

1. $(-3 + 5i) - (2 - 4i)$

3. _____

4. _____

5. _____

2. $(7 + 3i) + (2 - i) - (1 - 2i)$

6. _____

7. _____

8. _____

3. $(2 + 3i)(1 - 4i)$

9. _____

10. _____

4. $\dfrac{3}{4 - 3i}$

5. $\dfrac{3 - i}{4i}$

6. $(3 - 2i)^2 - 2(3 - 2i) + 4$

7. Evaluate $x^2 - 2x + 5$ for $1 + 2i$.

In Problems 8–10 convert square roots of negative numbers to complex form, perform the indicated operations, and express the answer in the form $a + bi$.

8. $(3 - \sqrt{-4})(2 + \sqrt{-25})$

9. $\dfrac{-4 + \sqrt{-4}}{2}$

10. $\dfrac{2 + \sqrt{-1}}{1 - \sqrt{-4}}$

Second-Degree Equations and Inequalities

■ 6

INSTRUCTIONS FOR STUDENTS IN A SELF-PACED CLASS OR LAB

YES — HAVE YOU HAD INTERMEDIATE ALGEBRA BEFORE THIS COURSE? — NO

1. Work Diagnostic (Review) Exercise 6-7 on page 375. Check answers in back of book; then work through text sections corresponding to problems missed. (Section numbers are in italics following each answer.)
2. When finished with step 1, take Practice Test Chapter 6 on page 377 as a final check of your understanding of the chapter. Check answers in the back of the book; then review sections where weakness still prevails. (Corresponding section numbers are in italics following each answer.)
3. When you think you are ready, ask your instructor for a graded test for Chapter 6.
4. If your instructor approves, after the test is corrected, go to the next chapter.

1. Work through each section in the chapter as follows:
   (A) Read discussion.
   (B) Read each example and work the corresponding matched problem. Check your solutions to the matched problem in Solutions to Matched Problems on the indicated page.
   (C) At the end of a section work the odd-numbered problems in the Practice Exercise and check answers; then work even-numbered problems in areas of weakness. (Answers to *all* Practice Exercise sets are in the back of the book.)
   (D) Work Check Exercise as instructed. Tear out and turn in as directed by your instructor. (Answers are not in the text.)
2. Repeat each step in item 1 for each section in the chapter.
3. After the instructional part of the chapter is completed, proceed with steps 1 to 4 in the box above this one.

Chapter 6 ■ Second-Degree Equations and Inequalities

Section 6-1 Solving Quadratic Equations by Square Roots and Factoring

- Quadratic Equations
- Solution by Square Roots
- Solution by Factoring

■ Quadratic Equations

The equation

$$\tfrac{1}{2}x - \tfrac{1}{3}(x + 3) = 2 - x$$

though complicated-looking, is actually a first-degree equation in one variable, since it can be transformed into the equivalent equation

$$7x - 18 = 0$$

which is a special case of

$$ax + b = 0 \qquad a \neq 0$$

We have solved many equations of this type and found that they always have a single solution. From a mathematical point of view we have essentially taken care of the problem of solving first-degree equations in one variable.

In this chapter we will consider the next class of polynomial equations, called second-degree equations or quadratic equations. A **quadratic equation** in one variable is any equation that can be written in the form

Quadratic Equation (Standard Form)	
$ax^2 + bx + c = 0 \qquad a \neq 0$	(1)

where x is a variable and a, b, and c are constants. We will refer to this form as the **standard form** for the quadratic equation. The equations

$$2x^2 - 3x + 5 = 0 \qquad \text{and} \qquad 15 = 180t - 16t^2$$

are both quadratic equations since they are either in the standard form or can be converted into this form.

Applications that give rise to quadratic equations are many and varied. A brief glance at Section 6-4 will give you some indication of the variety.

■ Solution by Square Roots

The easiest type of quadratic equation to solve is the special form where the first-degree term is missing; that is, when Equation (1) is of the form

$$ax^2 + c = 0 \qquad a \neq 0$$

The method of solution makes direct use of the definition of square root. The process is illustrated in the following example.

EXAMPLE 1   Solve by the square root method:

(A) $x^2 - 8 = 0$ (B) $2x^2 - 3 = 0$ (C) $3x^2 + 27 = 0$ (D) $(x + \tfrac{1}{2})^2 = \tfrac{5}{4}$

Solution   (A) $x^2 - 8 = 0$

$$x^2 = 8 \qquad \text{What number squared is 8?}$$

$$x = \pm\sqrt{8} \quad \text{or} \quad \pm 2\sqrt{2} \quad \text{$\pm 2\sqrt{2}$ is a short way of writing $-2\sqrt{2}$ or $+2\sqrt{2}$.}$$

(B) $2x^2 - 3 = 0$

$$2x^2 = 3 \quad \text{Do not write } 2x = \pm\sqrt{3} \text{ next. (Why?)}$$

$$x^2 = \tfrac{3}{2} \quad \text{What number squared is } \tfrac{3}{2}?$$

$$x = \pm\sqrt{\tfrac{3}{2}} \quad \text{or} \quad \pm\frac{\sqrt{6}}{2}$$

(C) $3x^2 + 27 = 0$

$$3x^2 = -27 \quad \text{Do not write } 3x = \pm\sqrt{-27} \text{ next. (Why?)}$$

$$x^2 = -9 \quad \text{What number squared is } -9?$$

$$x = \pm\sqrt{-9} = \pm 3i$$

(D) $(x + \tfrac{1}{2})^2 = \tfrac{5}{4}$ Solve for $x + \tfrac{1}{2}$; then solve for x.

$$x + \tfrac{1}{2} = \pm\sqrt{\tfrac{5}{4}}$$

$$x = -\frac{1}{2} \pm \frac{\sqrt{5}}{2}$$

$$= \frac{-1 \pm \sqrt{5}}{2} \quad \text{Short for } \frac{-1 + \sqrt{5}}{2} \text{ or } \frac{-1 - \sqrt{5}}{2}$$

PROBLEM 1   Solve by the square root method:

(A) $x^2 - 12 = 0$ (B) $3x^2 - 5 = 0$
(C) $2x^2 + 8 = 0$ (D) $(x + \tfrac{1}{3})^2 = \tfrac{2}{9}$

Solution   (A)   (B)

(C)   (D)

■ Solution by Factoring

If the coefficients a, b, and c in the quadratic equation

$$ax^2 + bx + c = 0$$

are such that $ax^2 + bx + c$ can be written as the product of two first-degree factors with integer coefficients, then the quadratic equation can be quickly and easily solved. The method of solution by factoring rests on the following property of the real numbers:

If a and b are real numbers, then
$a \cdot b = 0$ if and only if $a = 0$ or $b = 0$ (or both)

EXAMPLE 2   Solve by factoring, if possible:

(A) $x^2 + 2x - 15 = 0$ **(B)** $4x^2 = 6x$

(C) $2x^2 - 8x + 3 = 0$ **(D)** $3 + \dfrac{5}{x} = \dfrac{2}{x^2}$

Solution   **(A)** $x^2 + 2x - 15 = 0$

$(x - 3)(x + 5) = 0$ $(x - 3)(x + 5) = 0$ if and only if $(x - 3) = 0$ or $(x + 5) = 0$.

$\qquad x - 3 = 0 \qquad$ or $\qquad x + 5 = 0$

$\qquad\qquad x = 3 \qquad\qquad\qquad x = -5$

(B) $\qquad\qquad 4x^2 = 6x$ If both sides are divided by x, we lose one solution ($x = 0$). But we can

$\qquad\qquad 2x^2 = 3x$ simplify the equation by dividing both sides by 2, a common factor of both coefficients.

$2x^2 - 3x = 0$

$x(2x - 3) = 0$ $x(2x - 3) = 0$ if and only if $x = 0$ or $2x - 3 = 0$.

$x = 0 \qquad$ or $\qquad 2x - 3 = 0$

$x = 0 \qquad\qquad\qquad x = \frac{3}{2}$

(C) The polynomial cannot be factored using integer coefficients; hence, another method must be used to find the solution. This will be discussed later.

(D) $\qquad\qquad 3 + \dfrac{5}{x} = \dfrac{2}{x^2}$ Multiply both sides by x^2, the LCM of the denominators ($x \neq 0$).

$\qquad\qquad 3x^2 + 5x = 2$ Write in standard form: $ax^2 + bx + c$.

$\qquad 3x^2 + 5x - 2 = 0$ Factor the left side, if possible.

$(3x - 1)(x + 2) = 0$

$3x - 1 = 0 \qquad$ or $\qquad x + 2 = 0$

$\qquad 3x = 1 \qquad\qquad\qquad x = -2$

$\qquad\quad x = \frac{1}{3}$

PROBLEM 2   Solve by factoring, if possible:

(A) $x^2 - 2x - 8 = 0$ (B) $9t^2 = 6t$ (C) $x^2 - 3x = 3$

(D) $x = \dfrac{3}{2x - 5}$

Solution   (A)   (B)

         (C)   (D)

EXAMPLE 3   The length of a rectangle is 1 inch more than twice its width. If the area is 21 square inches, find its dimensions.

Solution   Draw a figure and label the sides, as shown.

2x + 1

$$x(2x + 1) = 21$$
$$2x^2 + x - 21 = 0$$
$$(2x + 7)(x - 3) = 0$$
$$2x + 7 = 0 \qquad \text{or} \qquad x - 3 = 0$$
$$2x = -7 \qquad\qquad\quad x = 3 \text{ inches} \quad \text{Width}$$
$$x = -\tfrac{7}{2} \qquad\qquad 2x + 1 = 7 \text{ inches} \quad \text{Length}$$

Not possible, so
must be discarded.

Note: In practical problems involving quadratic equations, one of two solutions must often be discarded because it will not make sense in the problem.

PROBLEM 3   The base of a triangle is 6 meters longer than its height. If the area is 20 square meters, find its dimensions ($A = \tfrac{1}{2}bh$).

Solution

Solutions to Matched
Problems

1. (A) $x^2 - 12 = 0$
$$x^2 = 12$$
$$x = \pm\sqrt{12} \quad \text{or} \quad \pm 2\sqrt{3}$$

(B) $3x^2 - 5 = 0$
$$3x^2 = 5$$
$$x^2 = \tfrac{5}{3}$$
$$x = \pm\sqrt{\tfrac{5}{3}}$$
$$\text{or} \quad \pm\frac{\sqrt{15}}{3}$$

(C) $2x^2 + 8 = 0$
$$2x^2 = -8$$
$$x^2 = -4$$
$$x = \pm\sqrt{-4}$$
$$x = \pm 2i$$

(D) $(x + \tfrac{1}{3})^2 = \tfrac{2}{9}$
$$x + \tfrac{1}{3} = \pm\sqrt{\tfrac{2}{9}}$$
$$x = -\frac{1}{3} \pm \frac{\sqrt{2}}{3}$$
$$= \frac{-1 \pm \sqrt{2}}{3}$$

2. (A) $x^2 - 2x - 8 = 0$
$$(x - 4)(x + 2) = 0$$
$$x - 4 = 0 \quad \text{or} \quad x + 2 = 0$$
$$x = 4 \qquad\qquad x = -2$$

(B) $9t^2 = 6t$
$$3t^2 = 2t$$
$$3t^2 - 2t = 0$$
$$t(3t - 2) = 0$$
$$t = 0 \quad \text{or} \quad 3t - 2 = 0$$
$$3t = 2$$
$$t = \tfrac{2}{3}$$

(C) $x^2 - 3x = 3$
$$x^2 - 3x - 3 = 0$$
Left side not factorable
using integer coefficients.

(D) $x = \dfrac{3}{2x - 5}$
$$x(2x - 5) = 3$$
$$2x^2 - 5x = 3$$
$$2x^2 - 5x - 3 = 0$$
$$(2x + 1)(x - 3) = 0$$
$$2x + 1 = 0 \quad \text{or} \quad x - 3 = 0$$
$$2x = -1 \qquad\qquad x = 3$$
$$x = -\tfrac{1}{2}$$

3.

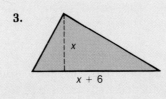

$$\tfrac{1}{2}bh = 20$$
$$\tfrac{1}{2}(x + 6)x = 20$$
$$(x + 6)x = 40$$
$$x^2 + 6x = 40$$
$$x^2 + 6x - 40 = 0$$
$$(x + 10)(x - 4) = 0$$
$$x + 10 = 0 \quad \text{or} \quad x - 4 = 0$$
$$\cancel{x = -10} \qquad\qquad x = 4 \text{ meters} \quad \text{Height}$$
$$x + 6 = 10 \text{ meters} \quad \text{Base}$$

Practice Exercise 6-1 ■

Work odd-numbered problems first, check answers, and then work even-numbered problems in areas of weakness. Answers to all problems are in the back of the book. Make every effort to work a problem yourself before you look at an answer.

A Solve by the square root method.

1. $x^2 - 16 = 0$ _____

2. $x^2 - 25 = 0$ _____

3. $x^2 + 16 = 0$ _____   **4.** $x^2 + 25 = 0$ _____

5. $y^2 - 45 = 0$ _____   **6.** $m^2 - 12 = 0$ _____

7. $4x^2 - 9 = 0$ _____   **8.** $9y^2 - 16 = 0$ _____

9. $16y^2 = 9$ _____   **10.** $9x^2 = 4$ _____

Solve by factoring.

11. $u^2 + 5u = 0$ _____   **12.** $v^2 - 3v = 0$ _____

13. $3A^2 = -12A$ _____   **14.** $4u^2 = 8u$ _____

15. $x^2 - 11x - 12 = 0$ _____   **16.** $y^2 - 6y + 5 = 0$ _____

17. $x^2 + 4x - 5 = 0$ _____   **18.** $x^2 - 4x - 12 = 0$ _____

19. $3Q^2 - 10Q - 8 = 0$ _____   **20.** $2d^2 + 15d - 8 = 0$ _____

B *Solve by the square root method.*

21. $y^2 = 2$ _____   **22.** $x^2 = 3$ _____

23. $16a^2 + 9 = 0$ _____   **24.** $4x^2 + 25 = 0$ _____

25. $9x^2 - 7 = 0$ _____   **26.** $4t^2 - 3 = 0$ _____

27. $(m - 3)^2 = 25$ _____   **28.** $(n + 5)^2 = 9$ _____

29. $(t + 1)^2 = -9$ _____   **30.** $(d - 3)^2 = -4$ _____

31. $(x - \frac{1}{3})^2 = \frac{4}{9}$ _____   **32.** $(x - \frac{1}{2})^2 = \frac{9}{4}$ _____

Solve by factoring. (Write the equations in standard form first.)

33. $u^2 = 2u + 3$ _____   **34.** $m^2 + 2m = 15$ _____

35. $3x^2 = x + 2$ _____   **36.** $2x^2 = 3 - 5x$ _____

37. $y^2 = 5y - 2$ _____   **38.** $3 = t^2 + 7t$ _____

39. $2x(x - 1) = 3(x + 1)$ _____   **40.** $3x(x - 2) = 2(x - 2)$ _____

41. $\dfrac{t}{2} = \dfrac{2}{t}$ _____   **42.** $y = \dfrac{9}{y}$ _____

43. $\dfrac{m}{4}(m + 1) = 3$ _____   **44.** $\dfrac{A^2}{2} = A + 4$ _____

45. $2y = \dfrac{2}{y} + 3$ _____   **46.** $L = \dfrac{15}{L - 2}$ _____

47. $2 + \dfrac{2}{x^2} = \dfrac{5}{x}$ _____ **48.** $1 - \dfrac{3}{x} = \dfrac{10}{x^2}$ _____

49. $\dfrac{x}{6} = \dfrac{1}{x+1}$ _____ **50.** $x + 1 = \dfrac{2}{x}$ _____

51. The width of a rectangle is 8 inches less than its length. If its area is 33 square inches, find its dimensions. _____

52. Find the base and height of a triangle with area 2 square feet if its base is 3 feet longer than its height $(A = \frac{1}{2}bh)$. _____

C *Solve by the square root method.*

53. $(y + \frac{5}{2})^2 = \frac{5}{2}$ _____ **54.** $(x - \frac{3}{2})^2 = \frac{3}{2}$ _____

55. $(x - 2)^2 = -1$ _____ **56.** $(x + \frac{1}{2})^2 = -\frac{3}{4}$ _____

Solve by factoring. (Write the equations in standard form first.)

57. $x = \dfrac{1 + 3x}{x + 3}$ _____ **58.** $\dfrac{1}{x} - \dfrac{1}{x^2} = \dfrac{1}{x+1}$ _____

59. $\dfrac{x + 2}{x - 1} - \dfrac{6x}{x^2 - 1} = \dfrac{2x - 1}{x + 1}$ _____

60. $\dfrac{2(x - 1)}{x - 2} = \dfrac{1}{x} + \dfrac{1}{x - 2}$ _____

Solve for the indicated letters in terms of the other letters. Use positive square roots only.

61. $a^2 + b^2 = c^2$ Solve for a. _____

62. $s = \frac{1}{2}gt^2$ Solve for t. _____

63. In a given city on a given day, the demand equation for gasoline is $d = 900/p$ and the supply equation is $s = p - 80$, where d and s denote the number of gallons demanded and supplied (in thousands), respectively, at a price of p cents per gallon. Find the price at which supply is equal to demand.

_____

64. To find the critical velocity at the top of the loop necessary to keep a steel ball on the track (see the figure), the centripetal force mv^2/r is equated to the force due to gravity mg. The mass m cancels out of the equation, and we are left with $v^2 = gr$. For a loop of radius 0.25 foot, find the critical velocity (in feet per second) at the top of the loop that is required to keep the ball on the track. Use $g = 32$ and compute your answer to two decimal places using a square root table or a calculator. _____

The Check Exercise for this section is on page 379.

Section 6-2 Solution by Completing the Square

- Introduction
- Completing the Square
- Solution of Quadratic Equations by Completing the Square

■ Introduction

The factoring and square root methods discussed in the last two sections are fast and easy to use when they apply. Unfortunately, many quadratic equations will not yield directly to either method. For example, the very simple-looking polynomial in the equation

$$x^2 + 6x - 2 = 0$$

cannot be factored in the integers. The equation requires a new approach if it can be solved at all.

In this section we will discuss a method, called "solution by completing the square," that will work for all quadratic equations. In the next section we will use this method to develop a general formula that will be used in the future whenever the square root or factoring method fails.

The method of completing the square is based on the process of transforming the standard quadratic equation

$$ax^2 + bx + c = 0$$

into the form

$$(x + A)^2 = B$$

where A and B are constants. This last equation can easily be solved by the square root method discussed in the preceding section. Thus,

$$(x + A)^2 = B$$
$$x + A = \pm\sqrt{B}$$
$$x = -A \pm \sqrt{B}$$

■ Completing the Square

Before considering how the first part is accomplished, let's pause for a moment and consider a related problem: what number must be added to $x^2 + 6x$ so that the result is the square of a linear expression? There is an easy mechanical rule for finding this number based on the squares of the following binomials:

$$(x + m)^2 = x^2 + 2mx + m^2 \qquad (x - m)^2 = x^2 - 2mx + m^2$$

In either case, we see that the third term on the right is the square of one-half of the coefficient of x in the second term on the right. This observation leads directly to the rule:

To **complete the square** of a quadratic of the form

$$x^2 + bx$$

add the square of one-half of the coefficient of x, that is

$$\left(\frac{b}{2}\right)^2 \qquad \text{or} \qquad \frac{b^2}{4}$$

Thus,

$$x^2 + bx + \left(\frac{b}{2}\right)^2 = \left(x + \frac{b}{2}\right)^2$$

EXAMPLE 4 **(A)** To complete the square of $x^2 + 6x$, add $(\frac{6}{2})^2$, that is, 9. Thus,

$$x^2 + 6x + 9 = (x + 3)^2$$

(B) To complete the square of $x^2 - 3x$, add $(-\frac{3}{2})^2$, that is, $\frac{9}{4}$. Thus,

$$x^2 - 3x + \tfrac{9}{4} = (x - \tfrac{3}{2})^2$$

PROBLEM 4   Complete the square and factor:

(A) $x^2 + 10x$ **(B)** $x^2 - 5x$

Solution   **(A)**   **(B)**

Note: The rule stated above applies only to quadratic forms where the coefficient of the second-degree term is 1. When solving equations, we will divide through by the leading coefficient so that the rule may be applied; see Example 7 on page 345.

We now use the method of completing the square to solve quadratic equations. In the next section we will use the method to develop a formula that will work for *all* quadratic equations.

■ **Solution of Quadratic Equations by Completing the Square**

Solving quadratic equations by the method of completing the square is best illustrated by examples.

EXAMPLE 5   Solve $x^2 + 6x - 2 = 0$ by the method of completing the square.

Solution $x^2 + 6x - 2 = 0$ Add 2 to both sides of the equation to remove −2 from the left side.

$$x^2 + 6x = 2$$ To complete the square of the left side, add the square of one-half of the coefficient of x to each side of the equation.

$$x^2 + 6x + 9 = 2 + 9 \quad \text{Factor the left side.}$$

$$(x + 3)^2 = 11 \quad \text{Solve by square root method.}$$

$$x + 3 = \pm\sqrt{11}$$

$$x = -3 \pm \sqrt{11}$$

PROBLEM 5

Solve $x^2 - 8x + 10 = 0$ by completing the square.

Solution

EXAMPLE 6 Solve $x^2 - 4x + 13 = 0$ by completing the square.

Solution

$$x^2 - 4x + 13 = 0$$

$$x^2 - 4x = -13 \quad \text{Add 4 to each side to complete the square on the left side.}$$

$$x^2 - 4x + 4 = 4 - 13$$

$$(x - 2)^2 = -9$$

$$x - 2 = \pm\sqrt{-9}$$

$$x - 2 = \pm 3i$$

$$x = 2 \pm 3i$$

PROBLEM 6

Solve $x^2 - 2x + 3 = 0$ by completing the square.

Solution

EXAMPLE 7 Solve $2x^2 - 4x - 3 = 0$ by completing the square.

Solution

$$2x^2 - 4x - 3 = 0 \quad \text{Note that the coefficient of } x^2 \text{ is not 1. Divide through by the leading coefficient and proceed as in the last example.}$$

$$x^2 - 2x - \tfrac{3}{2} = 0$$

$$x^2 - 2x = \tfrac{3}{2}$$

$$x^2 - 2x + 1 = \tfrac{3}{2} + 1$$

$$(x - 1)^2 = \tfrac{5}{2}$$

$$x - 1 = \pm\sqrt{\tfrac{5}{2}}$$

$$x = 1 \pm \frac{\sqrt{10}}{2}$$

$$x = \frac{2 \pm \sqrt{10}}{2}$$

PROBLEM 7   Solve $2x^2 + 8x + 3 = 0$ by completing the square.

Solution

Solutions to Matched
Problems

4. (A) $x^2 + 10x + 25 = (x + 5)^2$ **(B)** $x^2 - 5x + (\frac{5}{2})^2 = (x - \frac{5}{2})^2$

5. $x^2 - 8x + 10 = 0$ **6.** $x^2 - 2x + 3 = 0$
$\qquad x^2 - 8x = -10$ $\qquad x^2 - 2x = -3$
$\quad x^2 - 8x + 16 = 16 - 10$ $\quad x^2 - 2x + 1 = 1 - 3$
$\qquad (x - 4)^2 = 6$ $\qquad (x - 1)^2 = -2$
$\qquad x - 4 = \pm\sqrt{6}$ $\qquad x - 1 = \pm\sqrt{-2}$
$\qquad x = 4 \pm \sqrt{6}$ $\qquad x = 1 \pm i\sqrt{2}$

7. $2x^2 + 8x + 3 = 0$
$\quad 2x^2 + 8x = -3$
$\quad x^2 + 4x = -\frac{3}{2}$
$\quad x^2 + 4x + 4 = 4 - \frac{3}{2}$
$\quad (x + 2)^2 = \frac{5}{2}$
$\quad x + 2 = \pm\sqrt{\frac{5}{2}}$
$\quad x = -2 \pm \sqrt{\frac{5}{2}} \quad \text{or} \quad \dfrac{-4 \pm \sqrt{10}}{2}$

Practice Exercise 6-2 ■

Work odd-numbered problems first, check answers, and then work even-numbered problems in areas of weakness. Answers to all problems are in the back of the book. Make every effort to work a problem yourself before you look at an answer.

A   *Complete the square and factor.*

1. $x^2 + 4x$ _____   **2.** $x^2 + 8x$ _____   **3.** $x^2 - 6x$ _____

4. $x^2 - 10x$ _____   **5.** $x^2 + 12x$ _____   **6.** $x^2 + 2x$ _____

Solve by completing the square.

7. $x^2 + 4x + 2 = 0$ _____   **8.** $x^2 + 8x + 3 = 0$ _____

9. $x^2 - 6x - 3 = 0$ _____   **10.** $x^2 - 10x - 3 = 0$ _____

B   *Complete the square and factor.*

11. $x^2 + 3x$ _____   **12.** $x^2 + x$ _____

13. $u^2 - 5u$ _____ 14. $m^2 - 7m$ _____

Solve by completing the square.

15. $x^2 + x - 1 = 0$ _____ 16. $x^2 + 3x - 1 = 0$ _____

17. $u^2 - 5u + 2 = 0$ _____ 18. $n^2 - 3n - 1 = 0$ _____

19. $m^2 - 4m + 8 = 0$ _____ 20. $x^2 - 2x + 3 = 0$ _____

21. $2y^2 - 4y + 1 = 0$ _____ 22. $2x^2 - 6x + 3 = 0$ _____

23. $2u^2 + 3u - 1 = 0$ _____ 24. $3x^2 + x - 1 = 0$ _____

C 25. $2u^2 - 3u + 2 = 0$ _____ 26. $3x^2 - 5x + 3 = 0$ _____

27. $x^2 + x + 1 = 0$ _____ 28. $2x^2 - 3x + 4 = 0$ _____

29. $x^2 + 2\sqrt{2}x - 2 = 0$ _____ 30. $x^2 - 2\sqrt{5}x + 5 = 0$ _____

31. $x^2 - 4\sqrt{3}x + 13 = 0$ _____ 32. $x^2 + 2\sqrt{2}x + 3 = 0$ _____

33. $x^2 - 2ix - 4 = 0$ _____ 34. $x^2 + 2ix + 2 = 0$ _____

The Check Exercise for this section is on page 381.

35. Solve for x: $x^2 + mx + n = 0$ _____

36. Solve for x: $ax^2 + bx + c = 0, a \neq 0$ _____

Section 6-3   The Quadratic Formula

■ Quadratic Formula
■ The Discriminant
■ Which Method?

■ Quadratic Formula

The method of completing the square can be used to solve any quadratic equation, but the process is often tedious. If you had a very large number of quadratic equations to solve by completing the square, before you finished you would probably ask yourself if the process could not be made more efficient. Why not take the general equation

$$ax^2 + bx + c = 0 \qquad a \neq 0$$

and solve it once and for all for x in terms of the coefficients a, b, and c by the method of completing the square—thus obtaining a formula that could be memorized and used whenever a, b, and c are known?

We start by making the leading coefficient 1. How? Multiply both sides of the equation by $1/a$. Thus,

$$x^2 + \frac{b}{a}x + \frac{c}{a} = 0$$

Add $-c/a$ to both sides to clear c/a from the left side:

$$x^2 + \frac{b}{a}x = -\frac{c}{a}$$

Complete the square on the left side by adding the square of one-half the coefficient of x to each side:

$$x^2 + \frac{b}{a}x + \frac{b^2}{4a^2} = \frac{b^2}{4a^2} - \frac{c}{a}$$

We now factor the left side and solve by the square root method:

$$\left(x + \frac{b}{2a}\right)^2 = \frac{b^2 - 4ac}{4a^2}$$

$$x + \frac{b}{2a} = \pm\sqrt{\frac{b^2 - 4ac}{4a^2}}$$

$$x = -\frac{b}{2a} \pm \frac{\sqrt{b^2 - 4ac}}{2a}$$

Thus:

Quadratic Formula

$$x = \frac{-b \pm \sqrt{b^2 - 4ac}}{2a} \qquad a \neq 0$$

This equation is called the **quadratic formula**. It should be memorized and used to solve quadratic equations when simpler methods fail.

EXAMPLE 8 Solve $2x^2 - 4x - 3 = 0$ by use of the quadratic formula.

Solution

$$2x^2 - 4x - 3 = 0$$

$$x = \frac{-b \pm \sqrt{b^2 - 4ac}}{2a}$$

Write down the quadratic formula and identify a, b, and c. Here, $a = 2$, $b = -4$, $c = -3$.

$$= \frac{-(-4) \pm \sqrt{(-4)^2 - 4(2)(-3)}}{2(2)}$$

Substitute into the formula and simplify. Be careful of sign errors here.

$$= \frac{4 \pm \sqrt{40}}{4} = \frac{4 \pm 2\sqrt{10}}{4}$$

$$= \frac{2 \pm \sqrt{10}}{2}$$

PROBLEM 8 Solve $x^2 - 2x - 1 = 0$ using the quadratic formula.

Solution

EXAMPLE 9 Solve $x^2 + 11 = 6x$ using the quadratic formula.

Solution

$$x^2 + 11 = 6x$$ Write in standard form.

$$x^2 - 6x + 11 = 0$$

$$x = \frac{-b \pm \sqrt{b^2 - 4ac}}{2a}$$ $a = 1, b = -6, c = 11$

$$= \frac{-(-6) \pm \sqrt{(-6)^2 - 4(1)(11)}}{2(1)}$$ Be careful of sign errors here.

$$= \frac{6 \pm \sqrt{-8}}{2}$$

$$= \frac{6 \pm 2i\sqrt{2}}{2} = 3 \pm i\sqrt{2}$$

PROBLEM 9 Solve $2x^2 + 3 = 4x$ using the quadratic formula.

Solution

■ **The Discriminant**

The expression $b^2 - 4ac$ that occurs under the radical in the quadratic formula is called the **discriminant**. It provides useful information about the number and nature of the roots.

Discriminant Test	
$ax^2 + bx + c = 0$ a, b, c real numbers; $a \neq 0$	
$b^2 - 4ac$	ROOTS
Positive	Two real roots
0	One real root
Negative	Two nonreal roots (the roots will be complex conjugates)

EXAMPLE 10 Apply the discriminant test to determine the number and nature of the roots:

(A) $3x^2 - 4x + 1 = 0$ **(B)** $9x^2 - 6x + 1 = 0$ **(C)** $x^2 + 5x + 7 = 0$

Solution **(A)** The discriminant $b^2 - 4ac = 4^2 - 4 \cdot 3 \cdot 1 = 4$ is positive. The equation has two real roots. (Check that they are 1 and $\frac{2}{3}$.)

(B) The discriminant $b^2 - 4ac = 6^2 - 4 \cdot 9 \cdot 1 = 0$, so there is one real root. (Check that it is $\frac{1}{3}$.)

(C) The discriminant $b^2 - 4ac = 25 - 4 \cdot 1 \cdot 7 = -3$ is negative, so there are two nonreal, complex conjugate roots. (Check that they are $-\frac{5}{2} \pm i\sqrt{3}/2$.)

PROBLEM 10   Apply the discriminant test to determine the number and nature of the roots:

(A) $4x^2 - 20x + 25 = 0$ **(B)** $2x^2 + x - 1 = 0$
(C) $x^2 - 6x + 10 = 0$

Solution   **(A)**

(B)

(C)

■ **Which Method?**

In normal practice the quadratic formula is used whenever the square root method or the factoring method does not produce results easily. These latter methods are generally faster when they apply and should be used when possible.[†]

Note that any equation of the form

$$ax^2 + c = 0$$

can always be solved by the square root method. And any equation of the form

$$ax^2 + bx = 0$$

can always be solved by factoring since $ax^2 + bx = x(ax + b)$.

It is important to realize, however, that the quadratic formula can always be used and will produce the same results as any other method.

EXAMPLE 11   Solve $\dfrac{30}{8 + x} + 2 = \dfrac{30}{8 - x}$ by the most efficient method.

Solution

$$\frac{30}{8 + x} + 2 = \frac{30}{8 - x}$$

Multiply both sides by the LCM of the denominators $(8 + x)(8 - x)$
Note that $x \neq -8, 8$.

$$30(8 - x) + 2(8 + x)(8 - x) = 30(8 + x)$$
$$240 - 30x + 128 - 2x^2 = 240 + 30x$$
$$-2x^2 - 60x + 128 = 0$$ Divide both sides by -2.
$$x^2 + 30x - 64 = 0$$ Factor the left side, if possible.
$$(x + 32)(x - 2) = 0$$

$x + 32 = 0$ or $x - 2 = 0$
$x = -32$ $x = 2$

We could also have solved $x^2 + 30x - 64 = 0$ by using the quadratic formula:

$$x^2 + 30x - 64 = 0$$

[†] The process of completing the square, in addition to producing the quadratic formula, is used in many other places in mathematics. See Section 7-4, for example.

$$x = \frac{-b \pm \sqrt{b^2 - 4ac}}{2a} \qquad a = 1, b = 30, c = -64$$

$$= \frac{-30 \pm \sqrt{30^2 - 4(1)(-64)}}{2(1)}$$

$$= \frac{-30 \pm \sqrt{1,156}}{2}$$

$$= \frac{-30 \pm 34}{2}$$

Thus, $x = -32$ or $x = 2$.

It is clear that the factoring method was much easier in Example 11. Nevertheless, we got the same result, as expected.

PROBLEM 11

Solve $\dfrac{6}{x - 2} + 2 = \dfrac{4}{x}$ by the most efficient method.

Solution

Solutions to Matched Problems

8. $x^2 - 2x - 1 = 0$

$$x = \frac{-b \pm \sqrt{b^2 - 4ac}}{2a} \qquad a = 1, b = -2, c = -1$$

$$= \frac{-(-2) \pm \sqrt{(-2)^2 - 4(1)(-1)}}{2(1)}$$

$$= \frac{2 \pm \sqrt{8}}{2} = \frac{2 \pm 2\sqrt{2}}{2} = 1 \pm \sqrt{2}$$

9. $\qquad 2x^2 + 3 = 4x$

$\qquad 2x^2 - 4x + 3 = 0$

$$x = \frac{-b \pm \sqrt{b^2 - 4ac}}{2a} \qquad a = 2, b = -4, c = 3$$

$$= \frac{-(-4) \pm \sqrt{(-4)^2 - 4(2)(3)}}{2(2)}$$

$$= \frac{4 \pm \sqrt{-8}}{4} = \frac{4 \pm 2i\sqrt{2}}{4} = 1 \pm \frac{\sqrt{2}}{2} i$$

10. (A) $b^2 - 4ac = 400 - 400 = 0$, one real root
(B) $b^2 - 4ac = 1 + 8 = 9$, two real roots
(C) $b^2 - 4ac = 36 - 40 = -4$, two nonreal, complex conjugate roots

11. $\dfrac{6}{x - 2} + 2 = \dfrac{4}{x}$ $x \neq 0, 2$

$6x + 2x(x - 2) = 4(x - 2)$
$6x + 2x^2 - 4x = 4x - 8$
$2x^2 - 2x + 8 = 0$
$x^2 - x + 4 = 0$

Left side does not factor in the integers, so we go directly to the quadratic formula.

$$x = \frac{-b \pm \sqrt{b^2 - 4ac}}{2a}   a = 1, b = -1, c = 4$$

$$= \frac{-(-1) \pm \sqrt{(-1)^2 - 4(1)(4)}}{2(1)}$$

$$= \frac{1 \pm \sqrt{-15}}{2} = \frac{1 \pm i\sqrt{15}}{2} = \frac{1}{2} \pm \frac{\sqrt{15}}{2} i$$

Practice Exercise 6-3 ■

Work odd-numbered problems first, check answers, and then work even-numbered problems in areas of weakness. Answers to all problems are in the back of the book. Make every effort to work a problem yourself before you look at an answer.

A *Specify the constants a, b, and c for each quadratic equation when written in the standard form $ax^2 + bx + c = 0$.*

1. $2x^2 - 5x + 3 = 0$ _____

2. $3x^2 - 2x + 1 = 0$ _____

3. $m = 1 - 3m^2$ _____

4. $2u^2 = 1 - 3u$ _____

5. $3y^2 - 5 = 0$ _____

6. $2x^2 - 5x = 0$ _____

Solve by use of the quadratic formula.

7. $x^2 + 8x + 3 = 0$ _____

8. $x^2 + 4x + 2 = 0$ _____

9. $y^2 - 10y - 3 = 0$ _____

10. $y^2 - 6y - 3 = 0$ _____

B **11.** $u^2 = 1 - 3u$ _____

12. $t^2 = 1 - t$ _____

13. $y^2 + 3 = 2y$ _____

14. $x^2 + 8 = 4x$ _____

15. $2m^2 + 3 = 6m$ _____

16. $2x^2 + 1 = 4x$ _____

17. $p = 1 - 3p^2$ _____

18. $3q + 2q^2 = 1$ _____

Apply the discriminant test to determine the number and nature of the roots.

19. $4x^2 + 5x - 6 = 0$ _____

20. $3x^2 + 2x + 1 = 0$ _____

21. $9x^2 - 24x + 16 = 0$ _____

22. $25x^2 + 10x + 1 = 0$ _____

23. $x^2 - 8x + 17 = 0$ _____

24. $x^2 + 11x + 30 = 0$ _____

Solve each of the following equations by any method, excluding completing the square.

25. $(x - 5)^2 = 7$ _____

26. $(y + 4)^2 = 11$ _____

27. $x^2 + 2x = 2$ _____

28. $x^2 - 1 = 3x$ _____

29. $2u^2 + 3u = 0$ _____

30. $2n^2 = 4n$ _____

31. $x^2 - 2x + 9 = 2x - 4$ _____

32. $x^2 + 15 = 2 - 6x$ _____

33. $y^2 = 10y + 3$ _____

34. $3(2x + 1) = x^2$ _____

35. $2d^2 + 1 = 4d$ _____

36. $2y(3 - y) = 3$ _____

37. $\dfrac{2}{u} = \dfrac{3}{u^2} + 1$ _____

38. $1 + \dfrac{8}{x^2} = \dfrac{4}{x}$ _____

39. $\dfrac{1.2}{y - 1} + \dfrac{1.2}{y} = 1$ _____

40. $\dfrac{24}{10 + m} + 1 = \dfrac{24}{10 - m}$ _____

Solve for the indicated letter in terms of the other letters.

41. $d = \frac{1}{2}gt^2$ for t (positive) _____

42. $a^2 + b^2 = c^2$ for a (positive) _____

43. $A = P(1 + r)^2$ for r (positive) _____

44. $P = EI - RI^2$ for I _____

C *Solve by use of the quadratic formula.*

45. $x^2 - \sqrt{7}x + 2 = 0$ _____

46. $3x^2 - 2\sqrt{15}x + 5 = 0$ _____

47. $\sqrt{3}x^2 + 4x + \sqrt{3} = 0$ _____

48. $\sqrt{2}x^2 + 2\sqrt{3}x + \sqrt{2} = 0$ _____

49. $2x^2 + 3ix + 2 = 0$ _____

50. $x^2 - ix + 6 = 0$ _____

51. $x^2 + ix - 1 = 0$ _____

52. $3x^2 - 5ix + 2 = 0$ _____

Solve for x in terms of the remaining variables.

53. $y^2 + xy - x^2 = 0$ _____

54. $x^2 + y^2 = x + y$ _____

55. $\dfrac{x + y}{x - y} = \dfrac{x}{y}$ _____

56. $x^2 + 3xy + y^2 - 2x + y + 1 = 0$ _____

57. For what values of c does $2x^2 - 3x + c = 0$ have exactly one solution?

_____

58. For what values of a does $ax^2 + 6x + 5 = 0$ have two real solutions?

_____

59. Show that if r_1 and r_2 are the two roots of $ax^2 + bx + c = 0$, then $r_1 r_2 = c/a$. _____

60. For r_1 and r_2 in Problem 59, show that $r_1 + r_2 = -b/a$. _____

CALCULATOR PROBLEMS   *Solve to two decimal places using a hand calculator.*

61. $2.07x^2 - 3.79x + 1.34 = 0$ _____

62. $0.61x^2 - 4.28x + 2.93 = 0$ _____

63. $4.83x^2 + 2.04x - 3.18 = 0$ _____

64. $5.13x^2 + 7.27x - 4.32 = 0$ _____

Use the discriminant to determine which equations have real solutions.

65. $0.013\,4x^2 + 0.041\,4x + 0.030\,4 = 0$ _____

66. $0.543x^2 - 0.182x + 0.003\,12 = 0$ _____

The Check Exercise for   **67.** $0.013\,4x^2 + 0.021\,4x + 0.030\,4 = 0$ _____
this section is on
page 383.   **68.** $0.543x^2 - 0.182x + 0.031\,2 = 0$ _____

Section 6-4   Applications

We will now consider a number of applications from several fields. Since quadratic equations often have two solutions, it is important to check both solutions in the original problem to see if one or the other must be rejected. Also, a review of the strategy of solving word problems in Section 4-2 should prove helpful.

EXAMPLE 12   The sum of a number and its reciprocal is $\frac{5}{2}$. Find the number.

Solution   Let

$\quad x =$ The number

Then

$$x + \frac{1}{x} = \frac{5}{2} \qquad \text{Clear fractions.}$$

$$2x^2 + 2 = 5x \qquad \text{Write in standard form.}$$

$$2x^2 - 5x + 2 = 0 \qquad \text{Solve by factoring.}$$
$$(2x - 1)(x - 2) = 0$$
$$x = \tfrac{1}{2} \qquad \text{or} \qquad 2 \qquad \text{Both answers are solutions to the problem,}$$
as can easily be checked.

PROBLEM 12   If the reciprocal of a number is subtracted from the original number, the difference is $\tfrac{8}{3}$. Find the number.

Solution

EXAMPLE 13   A tank can be filled in 4 hours by two pipes when both are used. How many hours are required for each pipe to fill the tank alone if the smaller pipe requires 3 hours more than the larger one?

Solution   Let

$4 = $ Time for both pipes to fill the tank together

$x = $ Time for the larger pipe to fill the tank alone

$x + 3 = $ Time for the smaller pipe to fill the tank alone

Then

$\dfrac{1}{4} = $ Rate for both pipes together $\tfrac{1}{4}$ tank per hour

$\dfrac{1}{x} = $ Rate for larger pipe $\dfrac{1}{x}$ tank per hour

$\dfrac{1}{x + 3} = $ Rate for smaller pipe $\dfrac{1}{x + 3}$ tank per hour

Sum of individual rates $=$ Rate together

$$\frac{1}{x} + \frac{1}{x + 3} = \frac{1}{4}$$

$$\boxed{4x(x + 3) \cdot \frac{1}{x} + 4x(x + 3) \cdot \frac{1}{x + 3} = 4x(x + 3) \cdot \frac{1}{4}} \qquad \text{Clear fractions.}$$

$$4(x + 3) + 4x = x(x + 3)$$
$$4x + 12 + 4x = x^2 + 3x$$
$$x^2 - 5x - 12 = 0 \qquad \text{Use the quadratic formula.}$$
$$x = \frac{5 \pm \sqrt{73}}{2} \qquad \begin{array}{l}\text{Why should we discard} \\ \text{the negative answer?}\end{array}$$
$$x = \frac{5 + \sqrt{73}}{2}$$
$$\approx 6.77 \text{ hours} \qquad \text{Larger pipe}$$
$$x + 3 \approx 9.77 \text{ hours} \qquad \text{Smaller pipe}$$

Note: Example 13 is typical of most significant real-world problems in that decimal quantities rather than convenient small numbers are involved.

PROBLEM 13   Two pipes can fill a tank in 3 hours when used together. Alone, one can fill the tank 2 hours faster than the other. How long will it take each pipe to fill the tank alone? Compute the answers to two decimal places, using a calculator.

Solution

EXAMPLE 14   For a car traveling at a speed of v miles per hour, the least number of feet d under the best possible conditions that is necessary to stop a car (including a reaction time) is given approximately by the formula $d = 0.044v^2 + 1.1v$. Estimate the speed of a car requiring 200 feet to stop after danger is realized. A hand calculator will be useful for this problem. Compute the answer to two decimal places.

Solution

$$0.044v^2 + 1.1v = 200 \qquad \text{Write in standard form.}$$

$$0.044v^2 + 1.1v - 200 = 0 \qquad \text{Use the quadratic formula.}$$

$$v = \frac{-b \pm \sqrt{b^2 - 4ac}}{2a} \qquad a = 0.044,\ b = 1.1,\ c = -200$$

$$= \frac{-1.1 \pm \sqrt{1.1^2 - 4(0.044)(-200)}}{2(0.044)}$$

$$= \frac{-1.1 \pm \sqrt{36.41}}{0.088} \qquad \begin{array}{l}\text{Disregard the negative answer, since}\\ \text{we are only interested in positive } v.\end{array}$$

$$= \frac{-1.1 + 6.03}{0.088} = 56.02 \text{ miles per hour} \qquad \text{Complete to two decimal places.}$$

Note: Example 14 is typical of most significant real-world problems in that decimal quantities rather than convenient small numbers are involved.

PROBLEM 14   Repeat Example 14 for a car requiring 300 feet to stop after danger is realized.

Solution

12. Let x = The number

$$x - \frac{1}{x} = \frac{8}{3} \qquad x \neq 0$$

$$3x^2 - 3 = 8x$$

$$3x^2 - 8x - 3 = 0$$

$$(3x + 1)(x - 3) = 0$$

$$3x + 1 = 0 \qquad \text{or} \quad x - 3 = 0$$

$$3x = -1 \qquad\qquad x = 3$$

$$x = -\tfrac{1}{3}$$

13. Let

3 = Time for both pipes to fill tank together

x = Time for faster pipe to fill tank alone

$x + 2$ = Time for slower pipe to fill tank alone

Then

$\frac{1}{3}$ = Rate for both pipes together ($\frac{1}{3}$ tank per hour)

$\dfrac{1}{x}$ = Rate for faster pipe

$\dfrac{1}{x + 2}$ = Rate for slower pipe

$$\frac{1}{x} + \frac{1}{x + 2} = \frac{1}{3} \qquad x \neq -2, 0$$

$$3(x + 2) + 3x = x(x + 2)$$

$$3x + 6 + 3x = x^2 + 2x$$

$$-x^2 + 4x + 6 = 0$$

$$x^2 - 4x - 6 = 0$$

$$x = \frac{-(-4) \pm \sqrt{(-4)^2 - 4(1)(-6)}}{2(1)}$$

$$= \frac{4 \pm \sqrt{40}}{2}$$

$$= \frac{4 \pm 6.32}{2} \qquad \text{Keep positive answer only.}$$

$$x = 5.16 \text{ hours} \qquad \text{Faster pipe}$$

$$x + 2 = 7.16 \text{ hours} \qquad \text{Slower pipe}$$

14.

$$0.044v^2 + 1.1v = 300$$

$$0.044v^2 + 1.1v - 300 = 0$$

$$v = \frac{-1.1 \pm \sqrt{1.1^2 - 4(0.044)(-300)}}{2(0.044)}$$

$$= \frac{-1.1 \pm \sqrt{54.01}}{0.088}$$

$$= \frac{-1.1 + 7.35}{0.088}$$

$$= 71.02 \text{ miles per hour}$$

Thus, comparing this result with that obtained in Example 13, we see that a 27% increase in speed causes the car to go 50% farther before stopping.

Practice Exercise 6-4 ■

These problems are not grouped from easy (A) to difficult or theoretical (C). They are grouped somewhat according to type. The most difficult problems are marked with two asterisks (**), those of moderate difficulty are marked with one asterisk (*), and the easier problems are not marked.

Number Problems

1. Find two consecutive positive even integers whose product is 168.

_____

2. Find two positive numbers having a sum of 21 and a product of 104.

_____

3. Find all numbers with the property that when the number is added to itself the sum is the same as when the number is multiplied by itself.

_____

4. The sum of a number and its reciprocal is $\frac{10}{3}$. Find the number. _____

Geometry   *The following theorem may be used where needed:*

Pythagorean theorem: A triangle is a right triangle if and only if the square of the longest side is equal to the sum of the squares of the two shorter sides.

$$c^2 = a^2 + b^2$$

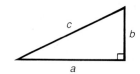

*5. Approximately how far would the horizon be from an airplane 2 miles high? Assume that the radius of the earth is 4,000 miles and use a calculator to estimate the answer to the nearest mile (see the figure).

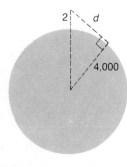

_____

6. Find the base and height of a triangle with area 2 square meters if its base is 3 meters longer than its height ($A = \frac{1}{2}bh$). _____

*7. If the length and width of a 4 by 2-centimeter rectangle are each increased by the same amount, the area of the new rectangle will be twice the old. What are the dimensions to two decimal places of the new rectangle?

_____

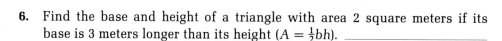

8. The width of a rectangle is 2 meters less than its length. Find its dimensions to two decimal places if its area is 12 square meters. _____

*9. A flag has a white cross of uniform width on a red background. Find the width of the cross so that it takes up exactly one-half the total area of a 4 by 3-foot flag. _____

Physics and Engineering

10. The pressure p in pounds per square foot of wind blowing at v miles per hour is given by $p = 0.003v^2$. If a pressure gauge on a bridge registers a wind pressure of 14.7 pounds per square foot, what is the velocity of the wind? _____

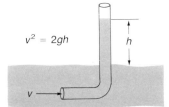

$v^2 = 2gh$

h

$v \rightarrow$

11. One method of measuring the velocity of water in a stream or river is to use an L-shaped tube as indicated in the figure. Torricelli's law in physics tells us that the height (in feet) that the water is pushed up into the tube above the surface is related to the water's velocity (in feet per second) by the formula $v^2 = 2gh$, where g is approximately 32 feet per second per second. [Note: The device can also be used as a simple speedometer for a boat.] How fast is a stream flowing if $h = 0.5$ foot? Find the answer to two decimal points. _____

12. At 20 miles per hour a car collides with a stationary object with the same force it would have if it had been dropped $13\frac{1}{2}$ feet—that is, if it had been pushed off the roof of an average one-story house. In general, a car moving at r miles per hour hits a stationary object with a force of impact that is equivalent to that force with which it would hit the ground when falling from a certain height h given by the formula $h = 0.0336r^2$. Approximately how fast would a car have to be moving if it crashed as hard as if it had been pushed off the top of a 12-story building 121 feet high? _____

*13. For a car traveling at a speed of v miles per hour, the least number of feet d under the best possible conditions that is necessary to stop a car (including a reaction time) is given by the empirical formula $d = 0.044v^2 + 1.1v$. Estimate the speed of a car requiring 165 feet to stop after danger is realized. (See Example 14.) _____

*14. If an arrow is shot vertically in the air (from the ground) with an initial velocity of 176 feet per second, its distance y above the ground t seconds after it is released (neglecting air resistance) is given by $y = 176t - 16t^2$.

(A) Find the time when y is 0, and interpret physically. _____

(B) Find the times when the arrow is 16 feet off the ground. Compute answers to two decimal places. _____

*15. A barrel 2 feet in diameter and 4 feet in height has a 1-inch diameter drainpipe in the bottom. It can be shown that the height h of the surface of the water above the bottom of the barrel at time t minutes after the drain has been opened is given by the formula $h = \left(\sqrt{h_0} - \frac{5}{12}t\right)^2$, where h_0 is the water level above the drain at time $t = 0$. If the barrel is full and the drain opened, how long will it take to empty one-half of the contents?

_____

Rate–Time Problems ****16.** One pipe can fill a tank in 5 hours less than another; together they fill the tank in 5 hours. How long would it take each alone to fill the tank? Compute the answer to two decimal places. _____

**17. A new printing press can do a job in 1 hour less than an older press. Together they can do the same job in 1.2 hours. How long would it take each alone to do the job? _____

18. Two boats travel at right angles to each other after leaving the same dock at the same time; 1 hour later they are 13 kilometers apart. If one travels 7 kilometers per hour faster than the other, what is the rate of each? [*Hint:* Use the Pythagorean theorem stated on page 358.] _____

*19. A speedboat takes 1 hour longer to go 24 kilometers up a river than to return. If the boat cruises at 10 kilometers per hour in still water, what is the rate of the current? _____

Economics and Business

The Check Exercise for this section is on page 385.

20. If P dollars is invested at r percent compounded annually, at the end of 2 years it will grow to $A = P(1 + r)^2$. At what interest rate will \$100 grow to \$144 in 2 years? [*Note:* If $A = 144$ and $P = 100$, find r.] _____

*21. In a certain city the demand equation for popular records is $d = 3,000/p$, where d would be the quantity of records demanded on a given day if the selling price were p dollars per record. (Notice that as the price goes up, the number of records the people are willing to buy goes down and vice versa.) On the other hand, the supply equation is $s = 200p - 700$, where s is the quantity of records a supplier is willing to supply at p dollars per record. (Notice that as the price goes up, the number of records a supplier is willing to sell goes up and vice versa.) At what price will supply equal demand; that is, at what price will $d = s$? In economic theory the price at which supply equals demand is called the **equilibrium point**, the point at which the price ceases to change. _____

Section 6-5   Equations Reducible to Quadratic Form

- Radical Equations
- Other Forms Reducible to Quadratic Form

■ Radical Equations

Consider the equation

$$x - 1 = \sqrt{x + 11}$$

Such an equation is called a **radical equation** since it contains a variable in the radicand. What can we do to solve this equation? Perhaps doing something to the equation to eliminate the radical will help. What? Let us square both members to see what happens—certainly if $a = b$, then $a^2 = b^2$. (Why?) Thus,

$$(x - 1)^2 = (\sqrt{x + 11})^2$$
$$x^2 - 2x + 1 = x + 11$$
$$x^2 - 3x - 10 = 0$$
$$(x + 2)(x - 5) = 0$$
$$x = -2, 5$$

Check $\quad$ $x = -2$: $\qquad -2 - 1 \overset{?}{=} \sqrt{-2 + 11}$

$\qquad\qquad\qquad\qquad -3 \overset{?}{=} \sqrt{9}$ $\quad$ Recall that $\sqrt{9}$ names the positive square root of 9.

$\qquad\qquad\qquad\qquad -3 \neq 3$

Hence, $x = -2$ is not a solution.

$\quad x = 5$: $\qquad 5 - 1 \overset{?}{=} \sqrt{5 + 11}$

$\qquad\qquad\qquad 4 \overset{?}{=} \sqrt{16}$

$\qquad\qquad\qquad 4 \overset{\checkmark}{=} 4$

Hence, $x = 5$ is a solution.

Therefore, 5 is a solution and -2 is not. The process of squaring introduced an "extraneous" solution. In general, one can prove the following important theorem.

THEOREM 1

> If both members of an equation are raised to a natural number power, then the solution set of the original equation is a subset of the solution set of the new equation.

Thus, any new equation obtained by raising both members of an equation to the same natural number power may have solutions (called **extraneous solutions**) that are not solutions of the original equation. On the other hand, any solution of the original equation must be among those of the new equation. We need only check all of the solutions at the end of the process to eliminate the so-called extraneous ones.

EXAMPLE 15 $\quad$ Solve: $\quad x + \sqrt{x - 4} = 4$

Solution $\qquad\qquad x + \sqrt{x - 4} = 4$ $\qquad\qquad$ Isolate the radical on one side.

$\qquad\qquad\qquad \sqrt{x - 4} = 4 - x$ $\qquad\qquad$ Square both sides.

$\qquad\qquad\qquad\quad x - 4 = 16 - 8x + x^2$ $\qquad$ Write in standard form.

$\qquad\qquad x^2 - 9x + 20 = 0$

$\qquad\qquad (x - 5)(x - 4) = 0$

$\qquad\qquad\qquad\qquad x = 4, 5$

Check $\quad$ $x = 4$: $\qquad 4 + \sqrt{4 - 4} = 4$, so 4 is a solution

$\quad x = 5$: $\qquad 5 + \sqrt{5 - 4} = 5 + 1 \neq 4$, so 5 is extraneous

PROBLEM 15

Solution

Solve: $x = 5 + \sqrt{x - 3}$

Radical equations involving two radicals are usually easiest to solve if the radicals are on opposite sides of the equation.

EXAMPLE 16

Solution

Solve: $\sqrt{2x + 3} - \sqrt{x - 2} = 2$

$\sqrt{2x + 3} - \sqrt{x - 2} = 2$ Easier to solve with a radical on each side.

$\sqrt{2x + 3} = \sqrt{x - 2} + 2$ Square both sides.

$2x + 3 = x - 2 + 4\sqrt{x - 2} + 4$ Isolate the radical on one side.

$x + 1 = 4\sqrt{x - 2}$ Square both sides again.

$x^2 + 2x + 1 = 16(x - 2)$

$x^2 - 14x + 33 = 0$

$(x - 11)(x - 3) = 0$

$x = 3, 11$

Check

$x = 3$: $\sqrt{2 \cdot 3 + 3} - \sqrt{3 - 2} = \sqrt{9} - \sqrt{1} = 3 - 1 = 2$,
so 3 is a solution

$x = 11$: $\sqrt{2 \cdot 11 + 3} - \sqrt{11 - 2} = \sqrt{25} - \sqrt{9} = 5 - 3 = 2$,
so 11 is a solution

PROBLEM 16

Solution

Solve: $\sqrt{2x + 7} - \sqrt{x + 3} = 1$

■ Other Forms Reducible to Quadratic Form

Many equations that are not immediately recognizable as quadratic can be transformed into a quadratic form and then solved. Let's look at some examples.

EXAMPLE 17   If asked to solve the equation

$$x^4 - x^2 - 12 = 0$$

you might at first have trouble. But if you recognize that the equation is quadratic in x^2, you can solve for x^2 first, then solve for x. You might find it convenient to make the substitution $u = x^2$ and then solve the equation

$$u^2 - u - 12 = 0$$
$$(u - 4)(u + 3) = 0$$
$$u = 4, -3$$

Replacing u with x^2, we obtain

$$x^2 = 4 \qquad\qquad x^2 = -3$$
$$x = \pm 2 \qquad\qquad x = \pm i\sqrt{3}$$

You can check that each of these four values is a solution of the original equation. Remember, however, that the only operation that might introduce extraneous roots is raising to powers. Since this solution did not involve raising both members of the equation to a higher power, extraneous roots will not occur.

PROBLEM 17

Solution

Solve: $x^6 + 6x^3 - 16 = 0$ for real solutions only

In general, if an equation that is not quadratic can be transformed into the form

$$au^2 + bu + c = 0$$

where u is an expression in some other variable, then the equation is said to be in **quadratic form**. Once recognized as a quadratic form, an equation can often be solved using one of the quadratic methods.

EXAMPLE 18   Solve: $x^{2/3} - x^{1/3} - 6 = 0$

Solution   Let $u = x^{1/3}$; then $u^2 = x^{2/3}$. After substitution, the original equation becomes

$$u^2 - u - 6 = 0$$
$$(u - 3)(u + 2) = 0$$
$$u = 3, -2$$

Replacing u with $x^{1/3}$, we obtain

$$x^{1/3} = 3 \qquad\qquad x^{1/3} = -2$$
$$x = 27 \qquad\qquad x = -8 \quad \textit{Note:}\ \ x \neq \sqrt[3]{3} \text{ and } x \neq \sqrt[3]{-2} \text{ (common errors)}.$$

PROBLEM 18

Solution

Solve: $x^{2/3} - x^{1/3} - 12 = 0$

Solutions to Matched
Problems

15.
$$x = 5 + \sqrt{x - 3}$$
$$x - 5 = \sqrt{x - 3}$$
$$(x - 5)^2 = (\sqrt{x - 3})^2$$
$$x^2 - 10x + 25 = x - 3$$
$$x^2 - 11x + 28 = 0$$
$$(x - 4)(x - 7) = 0$$
$$x - 4 = 0 \quad \text{or} \quad x - 7 = 0$$
$$\cancel{x = 4} \qquad\qquad x = 7$$

16. $\sqrt{2x + 7} - \sqrt{x + 3} = 1$
$$\sqrt{2x + 7} = \sqrt{x + 3} + 1$$
$$(\sqrt{2x + 7})^2 = (\sqrt{x + 3} + 1)^2$$
$$2x + 7 = x + 3 + 2\sqrt{x + 3} + 1$$
$$x + 3 = 2\sqrt{x + 3}$$
$$(x + 3)^2 = (2\sqrt{x + 3})^2$$
$$x^2 + 6x + 9 = 4(x + 3)$$
$$x^2 + 2x - 3 = 0$$
$$(x + 3)(x - 1) = 0$$
$$x + 3 = 0 \quad\quad \text{or} \quad x - 1 = 0$$
$$x = -3 \qquad\qquad x = 1$$
Checking, we find both are solutions.

17. $x^6 + 6x^3 - 16 = 0$
Let $u = x^3$; then
$$u^2 + 6u - 16 = 0$$
$$(u + 8)(u - 2) = 0$$
$$u + 8 = 0 \quad\quad \text{or} \quad u - 2 = 0$$
$$u = -8 \qquad\qquad u = 2$$
Replace u with x^3, and solve for x.
$$x^3 = -8 \qquad\qquad x^3 = 2$$
$$x = \sqrt[3]{-8} \qquad\qquad x = \sqrt[3]{2}$$
$$= -2$$

18. $x^{2/3} - x^{1/3} - 12 = 0$
Let $u = x^{1/3}$; then
$$u^2 - u - 12 = 0$$
$$(u - 4)(u + 3) = 0$$
$$u - 4 = 0 \quad \text{or} \quad u + 3 = 0$$
$$u = 4 \qquad\qquad u = -3$$
Replace u with $x^{1/3}$, and solve for x.
$$x^{1/3} = 4 \qquad\qquad x^{1/3} = -3$$
$$(x^{1/3})^3 = 4^3 \qquad (x^{1/3})^3 = (-3)^3$$
$$x = 64 \qquad\qquad x = -27$$

Practice Exercise 6-5 ■

Work odd-numbered problems first, check answers, and then work even-numbered problems in areas of weakness. Answers to all problems are in the back of the book. Make every effort to work a problem yourself before you look at an answer.

Solve.

A   **1.** $x - 2 = \sqrt{x}$ _____   **2.** $\sqrt{x} = x - 6$ _____

3. $m - 13 = \sqrt{m + 7}$ _____   **4.** $\sqrt{5n + 9} = n - 1$ _____

5. $x^4 - 10x^2 + 9 = 0$ _____   **6.** $x^4 - 13x^2 + 36 = 0$ _____

7. $x^4 - 7x^2 - 18 = 0$ _____   **8.** $y^4 - 2y^2 - 8 = 0$ _____

9. $\sqrt{x^2 - 3x} = 2$ _____   **10.** $\sqrt{x^2 + 8x} = 3$ _____

B   **11.** $m - 7\sqrt{m} + 12 = 0$ _____   **12.** $t - 11\sqrt{t} + 18 = 0$ _____

13. $1 + \sqrt{x + 5} = x$ _____   **14.** $x - \sqrt{x + 10} = 2$ _____

15. $\sqrt{3x + 1} = \sqrt{x} - 1$ _____   **16.** $\sqrt{3x + 4} = 2 + \sqrt{x}$ _____

17. $\sqrt{3t + 4} + \sqrt{t} = -3$ _____   **18.** $\sqrt{3w - 2} - \sqrt{w} = 2$ _____

19. $\sqrt{u - 2} = 2 + \sqrt{2u + 3}$ _____

20. $\sqrt{3y - 2} = 3 - \sqrt{3y + 1}$ _____

21. $\sqrt{2x - 1} - \sqrt{x - 4} = 2$ _____

22. $\sqrt{y - 2} - \sqrt{5y + 1} = -3$ _____

23. $x^6 - 7x^3 - 8 = 0$ (Find real solutions only.) _____

24. $x^6 + 3x^3 - 10 = 0$ (Find real solutions only.) _____

25. $y^8 - 17y^4 + 16 = 0$ _____   **26.** $3m^4 - 4m^2 - 7 = 0$ _____

27. $x^{2/3} - 3x^{1/3} - 10 = 0$ _____   **28.** $2x^{2/3} + 3x^{1/3} - 2 = 0$ _____

29. $y^{1/2} - 3y^{1/4} + 2 = 0$ _____   **30.** $y^{1/2} - 5y^{1/4} + 6 = 0$ _____

31. $6x^{-2} - 5x^{-1} - 6 = 0$ _____

32. $3n^{-2} - 11n^{-1} - 20 = 0$ _____

C **33.** $4x^{-4} - 17x^{-2} + 4 = 0$ _____   **34.** $9y^{-4} - 10y^{-2} + 1 = 0$ _____

35. $(m^2 - m)^2 - 4(m^2 - m) = 12$ _____

36. $(x^2 + 2x)^2 - (x^2 + 2x) = 6$ _____

37. $(x - 3)^4 + 3(x - 3)^2 = 4$ _____

38. $(m - 5)^4 + 36 = 13(m - 5)^2$ _____

39. $\sqrt{3x + 6} - \sqrt{x + 4} = \sqrt{2}$ _____

40. $\sqrt{7x - 2} - \sqrt{x + 1} = \sqrt{3}$ _____

41. $\dfrac{1}{\sqrt{x - 2}} + \dfrac{2}{3} = 1$ _____   **42.** $\dfrac{1}{\sqrt{x + 5}} = \dfrac{\sqrt{x}}{6}$ _____

43. $\dfrac{x}{3} + \dfrac{2}{x} = \dfrac{6x + 1}{3x}$ _____   **44.** $x + \dfrac{1}{x} = \dfrac{x + 3}{2}$ _____

The Check Exercise for
this section is on
page 387.

45. $\dfrac{1}{x - 1} + \dfrac{1}{x - 2} = \dfrac{5}{6}$ _____   **46.** $\sqrt{x} = 3 - \dfrac{2}{\sqrt{x}}$ _____

Section 6-6 Nonlinear Inequalities

- Quadratic Inequalities
- Set Union and Intersection
- Other Inequalities

- **Quadratic Inequalities**

You were introduced to solving first-degree linear inequalities such as

$$2x - 3 \le 4(x - 4)$$

in Section 4-6. But how do we solve inequalities such as

$$x^2 + 2x < 8$$

Using the quadratic formula directly doesn't work. However, if we move all terms to the left and factor, then we will be able to observe something that will lead to a solution. Thus,

$$x^2 + 2x - 8 < 0$$

$$(x + 4)(x - 2) < 0$$

We are looking for values of x that will make the left side less than 0—that is, negative. What will the signs of (x + 4) and (x − 2) have to be so that their product is negative? They must have opposite signs!

Let us see where each of the factors is positive, negative, and 0. The point at which either factor is 0 is called a **critical point**. We will see why shortly.

Sign analysis for (x + 4):

Critical point	(x + 4) is positive when	(x + 4) is negative when
x + 4 = 0	x + 4 > 0	x + 4 < 0
x = −4	x > −4	x < −4

Geometrically:

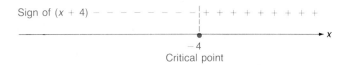

Thus, (x + 4) is negative for values of x to the left of the critical point and is positive for values of x to the right of the critical point.

Sign analysis for (x − 2):

Critical point	(x − 2) is positive when	(x − 2) is negative when
x − 2 = 0	x − 2 > 0	x − 2 < 0
x = 2	x > 2	x < 2

Geometrically:

Thus, (x − 2) is negative for values of x to the left of the critical point and is positive for values of x to the right of the critical point.

Combining these results in a single geometric representation leads to a simple solution of the original problem:

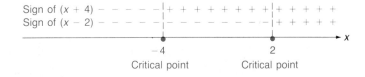

Now it is very easy to see that the factors have opposite signs for x between -4 and 2. Thus, the solution of $x^2 + 2x < 8$ is

$$-4 < x < 2$$

This discussion leads to the general result stated in Theorem 2, which is behind the sign-analysis method of solving quadratic inequalities.

THEOREM 2

The value of x at which the linear expression $(ax + b)$ is 0 is called a **critical point**. To the left of the critical point, on the real number line, $(ax + b)$ has one sign and to the right of the critical point the opposite sign.

EXAMPLE 19   Solve and graph: $x^2 \geq x + 6$.

Solution

$$x^2 \geq x + 6$$
$$x^2 - x - 6 \geq 0$$
$$(x - 3)(x + 2) \geq 0$$

Critical points: -2 and 3

Locate these points on a real number line and indicate the sign of each factor to the left and to the right of its critical point:

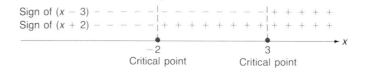

The inequality statement is satisfied when both factors have the same sign or when one or the other factor is 0. The former occurs when x is to the left of -2 or to the right of 3; the latter occurs at the critical points. Thus,

Solution: $x \leq -2$ or $x \geq 3$

Graph:

PROBLEM 19

Solve and graph:

(A) $x^2 < x + 12$ **(B)** $x^2 \geq x + 12$

Solution   **(A)**   **(B)**

The solution $x \le -2$ or $x \ge 3$ to Example 19 can also be described as the interval $(-\infty, -2]$ combined with the interval $[3, \infty)$. There is a convenient notation used to describe such a combined set.

■ Set Union and Intersection

The **union** of sets A and B, denoted by $A \cup B$, is the set of all elements formed by combining all the elements of A and all the elements of B into one set. The **intersection** of sets A and B, denoted by $A \cap B$, is the set of elements in A that are also in B. Symbolically:

UNION:	$A \cup B = \{x \mid x \in A \text{ or } x \in B\}$
	$\{2, 3\} \cup \{3, 4\} = \{2, 3, 4\}$
INTERSECTION:	$A \cap B = \{x \mid x \in A \text{ and } x \in B\}$
	$\{2, 3\} \cap \{3, 4\} = \{3\}$

The word "or" is used in the way it is generally used in mathematics; that is, x may be an element of set A or set B or both. If $A \cap B = \varnothing$ (that is, if the set of all elements common to both A and B is empty), then sets A and B are said to be **disjoint**.

Venn diagrams are useful aids in visualizing set relationships. Union and intersection of sets are illustrated in Figure 1.

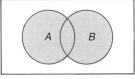 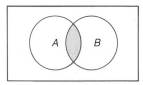 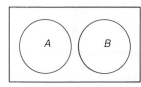

FIGURE 1 Venn diagrams (a) Union of two sets (b) Intersection of two sets (c) Two disjoint sets

EXAMPLE 20 If $A = \{1, 2, 3, 4\}$, $B = \{1, 3, 5, 7\}$, and $C = \{2, 4, 6\}$, then

$A \cup B = \{1, 2, 3, 4, 5, 7\}$ Elements in A combined with elements in B.

$A \cap B = \{1, 3\}$ Elements in A that are also in B.

$B \cap C = \varnothing$ Sets B and C are disjoint.

PROBLEM 20   If $A = \{3, 6, 9\}$, $B = \{3, 4, 5, 6, 7\}$, and $C = \{4, 5, 7, 8\}$, find:

(A) $A \cup B$ **(B)** $A \cap B$ **(C)** $A \cap C$

Solution   **(A)**   **(B)**   **(C)**

EXAMPLE 21   If $P = [-4, 2)$, $Q = (-1, 6]$, and $R = [3, \infty)$, find:

(A) $P \cup Q$ and $P \cap Q$ (B) $P \cup R$ and $P \cap R$

Solution   (A) Graphically:

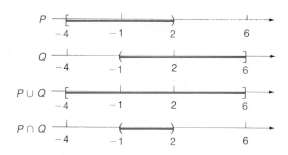

Symbolically:

$P \cup Q = [-4, 6]$ Combine elements in P with those in Q.

$P \cap Q = (-1, 2)$ Elements in P that are also in Q.

(B) Graphically:

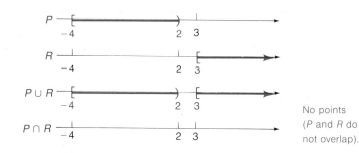

No points
(P and R do
not overlap).

Symbolically:

$P \cup R = [-4, 2) \cup [3, \infty)$

$P \cap R = \varnothing$ Null or empty set (P and R are disjoint).

PROBLEM 21   If $A = [0, 3]$, $B = (-2, 2)$, and $C = (-1, \infty)$, find:

(A) $A \cup B$ (B) $A \cap B$ (C) $B \cup C$ (D) $B \cap C$

Solution   (A)   (B)   (C)   (D)

■   **Other Inequalities**

The procedures discussed above can also be used on some inequalities that are not quadratic.

EXAMPLE 22   Solve and graph: $\dfrac{x^2 - x + 1}{2 - x} \geq 1$

Solution $\dfrac{x^2 - x + 1}{2 - x} \geq 1$ Since we do not know the sign of $2 - x$, we do not multiply both sides by it; instead, we subtract 1 from each side.

$\dfrac{x^2 - x + 1}{2 - x} - 1 \geq 0$ Combine terms on the left side into a single fraction.

$\dfrac{x^2 - x + 1}{2 - x} - \dfrac{2 - x}{2 - x} \geq 0$

$\dfrac{x^2 - 1}{2 - x} \geq 0$ Factor the numerator.

$\dfrac{(x - 1)(x + 1)}{2 - x} \geq 0$ Proceed as in Example 19.

Critical points: $-1, 1, 2$

Locate these points on a real number line and indicate the sign of each first-degree form to the left and to the right of its critical point:

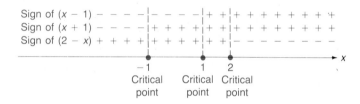

The inequality statement is satisfied when $(x - 1)$, $(x + 1)$, and $(2 - x)$ are all positive, two are negative and one is positive, or the numerator is 0. Two are negative to the left of -1 and all are positive between 1 and 2. The equality part of the inequality holds when x is 1 or -1, but not when $x = 2$. Thus,

Solution: $x \leq -1$ or $1 \leq x < 2$ That is, $(-\infty, -1] \cup [1, 2)$.

Graph:

PROBLEM 22   Solve and graph: $\dfrac{3}{2 - x} \leq \dfrac{1}{x + 4}$

Solution

19. (A) $x^2 < x + 12$

$x^2 - x - 12 < 0$

$(x - 4)(x + 3) < 0$

Critical points: $-3, 4$

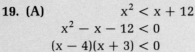

Solution and graph: $-3 < x < 4$

(B) $x^2 \geq x + 12$

$x^2 - x - 12 \geq 0$

$(x - 4)(x + 3) \geq 0$

Referring to part (A) we see that the solution and graph are as follows:

$x \leq -3$ or $x \geq 4$

20. (A) $\{3, 6, 9\} \cup \{3, 4, 5, 6, 7\} = \{3, 4, 5, 6, 7, 9\}$

(B) $\{3, 6, 9\} \cap \{3, 4, 5, 6, 7\} = \{3, 6\}$

(C) $\{3, 6, 9\} \cap \{4, 5, 7, 8\} = \varnothing$ Null set

21.

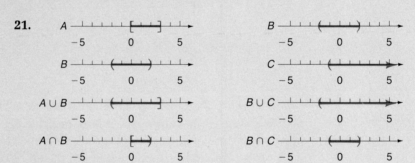

(A) $A \cup B = (-2, 3]$ **(B)** $A \cap B = [0, 2)$

(C) $B \cup C = (-2, \infty)$ **(D)** $B \cap C = (-1, 2)$

22. $\dfrac{3}{2 - x} \leq \dfrac{1}{x + 4}$ Move all terms to left side.

$\dfrac{3}{2 - x} - \dfrac{1}{x + 4} \leq 0$ Combine terms on left side.

$\dfrac{3(x + 4) - (2 - x)}{(2 - x)(x + 4)} \leq 0$ Simplify left side.

$\dfrac{4x + 10}{(2 - x)(x + 4)} \leq 0$ Find critical points.

Critical points: $-4, -\frac{5}{2}, 2$

The inequality would be satisfied if $(4x + 10)$, $(2 - x)$, and $(x + 4)$ were all negative (which does not happen in this problem), or if one were negative and two were positive (which happens in two regions in this problem). The equality is satisfied only at $x = -\frac{5}{2}$. Thus, the solution and graph are as follows:

$$-4 < x \le -\tfrac{5}{2} \quad \text{or} \quad x > 2$$

Practice Exercise 6-6 ∎

Work odd-numbered problems first, check answers, and then work even-numbered problems in areas of weakness. Answers to all problems are in the back of the book. Make every effort to work a problem yourself before you look at an answer.

Solve and graph.

A **1.** $(x - 3)(x + 4) < 0$ _____

    2. $(x + 2)(x - 4) < 0$ _____

  3. $(x - 3)(x + 4) \ge 0$ _____

    4. $(x + 2)(x - 4) > 0$ _____

  5. $x^2 + x < 12$ _____

    6. $x^2 < 10 - 3x$ _____

  7. $x^2 + 21 > 10x$ _____

    8. $x^2 + 7x + 10 > 0$ _____

Write each set in Problems 9–14, using the listing method; that is, list the elements between braces. If the set is empty, write $\varnothing$.

  9. $\{1, 3, 5\} \cup \{2, 3, 4\}$ _____

  10. $\{3, 4, 6, 7\} \cup \{3, 4, 5\}$ _____

  11. $\{1, 3, 5\} \cap \{2, 3, 4\}$ _____

  12. $\{3, 4, 6, 7\} \cap \{3, 4, 5\}$ _____

  13. $\{1, 5, 9\} \cap \{3, 4, 6, 8\}$ _____

  14. $\{6, 8, 9\} \cap \{4, 5, 7\}$ _____

Write each set in Problems 15–20 as a single interval, if possible.

  15. $(-3, 0] \cap [-2, 1)$ _____

  16. $(-3, 4) \cup [0, 8]$ _____

  17. $(-1, 1) \cup (0, \infty)$ _____

  18. $(-\infty, -1) \cap [-3, 2]$ _____

  19. $(-\infty, 3) \cup (2, 5]$ _____

  20. $[0, 2) \cup [2, 5)$ _____

Solve and graph.

B **21.** $x(x + 6) \ge 0$ _____

    22. $x(x - 8) \le 0$ _____

  23. $x^2 \ge 9$ _____

    24. $x^2 > 4$ _____

25. $\dfrac{x-5}{x+2} \le 0$ _____

26. $\dfrac{x+2}{x-3} < 0$ _____

27. $\dfrac{x-5}{x+2} > 0$ _____

28. $\dfrac{x+2}{x-3} \ge 0$ _____

29. $\dfrac{x-4}{x(x+2)} \le 0$ _____

30. $\dfrac{x(x+5)}{x-3} \ge 0$ _____

31. $\dfrac{1}{x} < 4$ _____

32. $\dfrac{5}{x} > 3$ _____

33. $x^2 + 4 \ge 4x$ _____

34. $6x \le x^2 + 9$ _____

35. $x^2 + 9 < 6x$ _____

36. $x^2 + 4 < 4x$ _____

37. $x^2 \ge 3$ _____

38. $x^2 < 2$ _____

39. $\dfrac{2}{x-3} \le -2$ _____

40. $\dfrac{2x}{x+3} \ge 1$ _____

C **41.** $\dfrac{2}{x-3} \le \dfrac{2}{x+2}$ _____

42. $\dfrac{2}{x+1} \ge \dfrac{1}{x-2}$ _____

43. $\dfrac{(x-1)(x+3)}{x} > 0$ _____

44. $\dfrac{x(x-3)}{x+2} < 0$ _____

45. $\dfrac{1}{x+1} + \dfrac{1}{x-2} \ge 0$ _____

46. $\dfrac{x^2 - 6x + 8}{x+2} > 0$ _____

47. $\dfrac{(x+1)^2}{x^2 + 2x - 3} \le 0$ _____

48. $\dfrac{(x-1)(x+3)}{x(x+2)} > 0$ _____

49. For what values of x will $\sqrt{x^2 - 3x + 2}$ be a real number? _____

50. For what values of x will $\sqrt{\dfrac{x-3}{x+5}}$ be a real number? _____

51. If an object is shot straight up from the ground with an initial velocity of 160 feet per second, its distance d in feet above the ground at the end of t seconds (neglecting air resistance) is given by $d = 160t - 16t^2$. Find the duration of time for which $d \ge 256$. _____

52. Repeat Problem 51 for $d \ge 0$. _____

The Check Exercise for this section is on page 389. **53.** Prove Theorem 2 for the case $a > 0$.

54. Prove Theorem 2 for the case $a < 0$.

Section 6-7 Chapter Review

The **standard form** of a **quadratic equation** is $ax^2 + bx + c = 0$ with $a \neq 0$. When $b = 0$, a quadratic equation can be solved by taking square roots. When $ax^2 + bx + c$ can be factored into two first-degree factors, the equation can be solved by setting each factor equal to 0. *(6-1)*

Any quadratic equation can be solved by **completing the square**; to complete the square in $x^2 + bx$, add the square of one-half the coefficient of x to obtain a perfect square:

$$x^2 + bx + \left(\frac{b}{2}\right)^2 = \left(x + \frac{b}{2}\right)^2 \quad \text{(6-2)}$$

This method leads to a general solution, the **quadratic formula**:

$$x = \frac{-b \pm \sqrt{b^2 - 4ac}}{2a}$$

The expression $b^2 - 4ac$ is called the **discriminant** and provides information about the roots: if positive, two real solutions; if negative, two nonreal, complex solutions; if 0, one real solution. *(6-3)*

Many applications lead to quadratic equations. *(6-4)*. Other equations may sometimes be reduced to quadratic form and solved. Radicals can be removed from equations by raising both sides to a power, but this procedure may introduce **extraneous solutions**. *(6-5)*

A **quadratic inequality** comparing $ax^2 + bx + c$ to 0 may be solved by factoring the quadratic and checking the signs of the factors. The points where the factors are 0 are called **critical points**. Sign analysis can also be applied to other inequalities comparing a product or quotient of linear factors to 0. Solutions of such inequalities can be written compactly in interval notation using set unions and intersections. The **union** $A \cup B$ of two sets A and B is the set of all elements in A or B (or both); the **intersection** $A \cap B$ is the set of all elements common to both A and B. *(6-6)*

Diagnostic (Review) Exercise 6-7 ■

Work through all the problems in this chapter review and check answers in the back of the book. (Answers to all problems are there, and following each answer is a number in italics indicating the section in which that type of problem is discussed.) Where weaknesses show up, review appropriate sections in the text. When you are satisfied that you know the material, take the practice test following this review.

A *Find all solutions by factoring or square root methods.*

1. $x^2 - 3x = 0$ _____   **2.** $x^2 = 25$ _____

3. $x^2 - 5x + 6 = 0$ _____   **4.** $x^2 - 2x - 15 = 0$ _____

5. $x^2 - 7 = 0$ _____

6. Write $4x = 2 - 3x^2$ in standard form, $ax^2 + bx + c = 0$, and identify a, b, and c. _____

7. Write down the quadratic formula associated with the standard form $ax^2 + bx + c = 0$. _____

8. Use the quadratic formula to solve $x^2 + 3x + 1 = 0$. _____

9. Find two positive numbers whose product is 27 if one is 6 more than the other. _____

10. Find the discriminant of $7x^2 - 11x + 5 = 0$ and use it to determine the nature and number of roots. _____

In Problems 11–14, let $A = \{1, 2, 3, 4, 5\}$, $B = \{1, 3, 5, 7, 9\}$, $C = [0, 5)$, and $D = (-1, 3]$. Find the set indicated.

11. $A \cup B$ _____   12. $A \cap B$ _____

13. $C \cup D$ _____   14. $C \cap D$ _____

Solve and graph.

15. $x^2 + x < 20$ _____   16. $x^2 + x \geq 20$ _____

B   Find all solutions by factoring or square root methods.

17. $10x^2 = 20x$ _____   18. $3x^2 = 36$ _____

19. $3x^2 + 27 = 0$ _____   20. $(x - 2)^2 = 16$ _____

21. $3t^2 - 8t - 3 = 0$ _____   22. $2x = \dfrac{3}{x} - 5$ _____

Solve using the quadratic formula.

23. $3x^2 = 2(x + 1)$ _____   24. $2x(x - 1) = 3$ _____

Solve using any method.

25. $2x^2 - 2x = 40$ _____   26. $\dfrac{8m^2 + 15}{2m} = 13$ _____

27. $m^3 + m - 1 = 0$ _____   28. $u + \dfrac{3}{u} = 2$ _____

29. $\sqrt{5x - 6} - x = 0$ _____   30. $8\sqrt{x} = x + 15$ _____

31. $m^4 + 5m^2 - 36 = 0$ _____   32. $2m^{2/3} - 5x^{1/3} - 12 = 0$ _____

Solve and graph.

33. $x^2 \geq 4x + 21$ _____   34. $\dfrac{1}{x} < 2$ _____

35. $10x > x^2 + 25$ _____   36. $x^2 + 16 \geq 8x$ _____

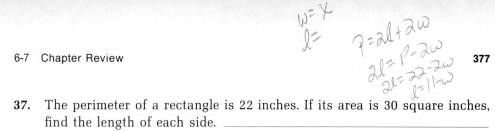

37. The perimeter of a rectangle is 22 inches. If its area is 30 square inches, find the length of each side. _____

C **38.** Solve $x^2 - 6x - 3 = 0$ by completing the square. _____

Solve using any method.

39. $\left(t - \dfrac{3}{2}\right)^2 = -\dfrac{3}{2}$ _____   **40.** $3x - 1 = \dfrac{2(x + 1)}{x + 2}$ _____

41. $y^8 - 17y^4 + 16 = 0$ _____

42. $\sqrt{y - 2} - \sqrt{5y + 1} = -3$ _____

43. $\dfrac{3}{x - 4} \leq \dfrac{2}{x - 3}$ Graph the solution. _____

44. If the hypotenuse of a right triangle is 15 centimeters and its area is 54 square centimeters, what are the lengths of the two sides? [*Hint:* If x represents one side, use the Pythagorean theorem to express the other side in terms of x; then use the formula for the area of a triangle, $A = \frac{1}{2}bh$.]

45. Cost equations for manufacturing companies are often quadratic in nature. (At very high or very low outputs the costs are more per unit because of inefficiency of plant operation at these extremes.) If the cost equation for producing paint is $C = x^2 - 10x + 31$, where C is the cost of producing x gallons per week (both in thousands), find:

(A)   The output for a \$15,000 weekly cost _____

(B)   The output for a \$6,000 weekly cost _____

Practice Test Chapter 6 ■

Take this as if it were a graded test by working the problems within a 50-minute time period. Do not look back in the chapter. Choose one of three levels of difficulty: least difficult, Problems 1–12; more difficult, add Problem 13; most difficult, add Problems 13 and 14. Use the answers in the back of the book to correct your work. The answers are keyed to appropriate text sections so that you can easily locate and review sections where difficulties still persist.

Solve by factoring or square root methods. Write answers in simplest radical form.

1. $5x^2 = 25x$ _____   **2.** $4x^2 + 36 = 0$ _____

3. $(3x + 1)^2 - 5 = 0$ _____   **4.** $x = 5 - \dfrac{6}{x}$ _____

Solve by any method. Write answers in simplest radical form.

5. $x^2 + 5x + 3 = 0$ _____

6. $\sqrt{10x - 21} = x$ _____

7. $x^4 + x^2 - 6 = 0$ _____

8. $\sqrt{x} = 3\sqrt[4]{x} - 2$ _____

9. $1 = \sqrt{x + 6} - \sqrt{x - 1}$ _____

Solve and graph Problems 10 and 11.

10. $x^2 \leq 11x - 30$ _____

11. $\dfrac{2}{x + 5} < \dfrac{1}{x + 3}$ _____

12. An object thrown down from a height of 576 feet with an initial velocity of 80 feet per second will fall to a height of $576 - 80t - 16t^2$ feet after t seconds. How long will it take for the object to hit the ground?

_____

13. For what values of k will the roots of $3x^2 + 6x + k = 0$ be nonreal complex numbers? _____

14. Solve $Px + Q + \dfrac{R}{x} = 0$ for x. _____

Check Exercise 6-1 ■

Work the following problems without looking at any text examples. Show your work in the space provided. Write your answer in the answer column.

1. _____

2. _____

Solve Problems 1–5 by the square root method. By solve we mean to find all solutions.

3. _____

    1. $2x^2 - 32 = 0$

4. _____

5. _____

    2. $u^2 = 72$

6. _____

7. _____

8. _____

    3. $x^2 + 81 = 0$

9. _____

10. _____

    4. $\left(x - \dfrac{3}{2}\right)^2 = \dfrac{3}{4}$

    5. $(x + 3)^2 = -4$

Solve Problems 6–9 by factoring, if possible.

6. $m^2 = 2m + 15$

7. $4x^2 + 23x = 6$

8. $2x^2 = 4x + 3$

9. $6 = \dfrac{1}{x} + \dfrac{2}{x^2}$

10. The height of a triangle is 2 centimeters less than its base. Find its base and height if its area is 12 square centimeters. ($A = \frac{1}{2}bh$)

Check Exercise 6-2 ■

Work the following problems without looking at any text examples. Show your work in the space provided. Write your answer in the answer column.

1. _____

2. _____

Solve by the method of completing the square.

1. $x^2 + 2x - 1 = 0$

3. _____

4. _____

5. _____

2. $x^2 - 2x - 8 = 0$

3. $u^2 - 6u + 10 = 0$

4. $m^2 + 3m - 1 = 0$

5. $2x^2 - 8x - 3 = 0$

Check Exercise 6-3 ∎

Work the following problems without looking at any text examples. Show your work in the space provided. Write your answer in the answer column.

1. _____

2. _____

Solve Problems 1 and 2, using the quadratic formula.

1. $x^2 = 6x - 7$

3. _____

4. _____

5. _____

2. $4x^2 + 3 = 4x$

3. Use the discriminant test to determine which of the following have two real roots.

(A) $x^2 + 6x + 7 = 0$
(B) $3x^2 - 8x + 5 = 0$
(C) $x^2 + 10x + 28 = 0$
(D) $-x^2 + 3x - 3 = 0$

Solve Problems 4 and 5, using the most efficient method.

4. $(x - 3)^2 = -4$

5. $\dfrac{24}{7 + x} + 1 = \dfrac{24}{7 - x}$

Check Exercise 6-4 ■

ANSWER COLUMN

Work the following problems without looking at any text examples. Show your work in the space provided. Write your answer in the answer column.

1. _____

2. _____

Set up an equation and solve. Use a hand calculator to express answers to two decimal places.

1. Find all consecutive odd number pairs whose product is 143.

3. _____

4. _____

5. _____

2. The length of a rectangle is 3 centimeters less than twice the width. What are the rectangle's dimensions if its area is 10 square centimeters?

$$w = x$$
$$l = 2x - 3$$
$$(2x-3)(x) = 10$$
$$2x^2 - 3x = 10$$

3. If P dollars is invested at r percent compounded annually, at the end of 2 years it will grow to $A = P(1 + r)^2$. At what rate of interest will \$10 grow to \$12.10?

4. If an arrow is shot vertically into the air (from the ground) with an initial velocity of 192 feet per second, its distance y above the ground t seconds after it is released (neglecting air resistance) is given by $y = 192t - 16t^2$. Find the time at which y is 0.

5. One electronic typewriter can complete a personalized mailing in 1 hour less than an older model. Together they can complete the job in 3 hours. How long would it take each alone to complete the job?

Check Exercise 6-5 ■

Work the following problems without looking at any text examples. Show your work in the space provided. Write your answer in the answer column.

1. _____

Solve completely.

2. _____

1. $\sqrt{x} = 2 - x$

3. _____

4. _____

5. _____

2. $x = \sqrt{x + 5} + 1$

3. $\sqrt{m - 1} + \sqrt{3m + 3} = 4$

4. $x^{-4} - 3x^{-2} - 4 = 0$

5. $x^{2/3} + 7x^{1/3} - 8 = 0$

Check Exercise 6-6 ■

ANSWER COLUMN

1. _____

2. _____

3. _____

4. _____

5. _____

Work the following problems without looking at any text examples. Show your work in the space provided. Write your answer in the answer column.

Solve. Express answers in interval notation, and graph.

1. $(x - 3)(x + 1) < 0$

2. $x^2 + 2x \geq 15$

3. $\dfrac{x - 5}{x + 2} \leq 0$

4. $\dfrac{2}{x} < 1$

5. $\dfrac{3}{x + 3} \geq \dfrac{1}{x - 1}$

Graphing Involving Two Variables ■ 7

**INSTRUCTIONS FOR STUDENTS IN A
SELF-PACED CLASS OR LAB**

YES — HAVE YOU HAD INTERMEDIATE ALGEBRA BEFORE THIS COURSE? — NO

1. Work Diagnostic (Review) Exercise 7-6 on page 456. Check answers in back of book; then work through text sections corresponding to problems missed. (Section numbers are in italics following each answer.)
2. When finished with step 1, take Practice Test Chapter 7 on page 459 as a final check of your understanding of the chapter. Check answers in the back of the book; then review sections where weakness still prevails. (Corresponding section numbers are in italics following each answer.)
3. When you think you are ready, ask your instructor for a graded test for Chapter 7.
4. If your instructor approves, after the test is corrected, go to the next chapter.

1. Work through each section in the chapter as follows:
   (A) Read discussion.
   (B) Read each example and work the corresponding matched problem. Check your solutions to the matched problem in Solutions to Matched Problems on the indicated page.
   (C) At the end of a section work the odd-numbered problems in the Practice Exercise and check answers; then work even-numbered problems in areas of weakness. (Answers to *all* Practice Exercise sets are in the back of the book.)
   (D) Work Check Exercise as instructed. Tear out and turn in as directed by your instructor. (Answers are not in the text.)
2. Repeat each step in item 1 for each section in the chapter.
3. After the instructional part of the chapter is completed, proceed with steps 1 to 4 in the box above this one.

Chapter 7 ■ Graphing Involving Two Variables

Section 7-1　Graphing Linear Equations

- Cartesian Coordinate System
- Graphing a First-Degree Equation in Two Variables
- Vertical and Horizontal Lines
- Use of Different Scales for Each Coordinate Axis

In Chapters 4 and 6 we graphed equations and inequalities in one variable on a real number line. Recall:

STATEMENT　　GRAPH

$-3 < x \leq 5$

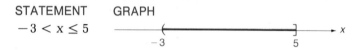

Every real number can be associated with a unique point on a line, and, conversely, every point on a line can be associated with a unique real number. We now develop a system that will enable us to graph equations and inequalities in two variables such as

$$3x - 2y = 5 \qquad \text{and} \qquad y \leq 4x - 1$$

■ Cartesian Coordinate System

To form a cartesian coordinate system we select two real number lines, one vertical and one horizontal, and let them cross through their origins (0's) as indicated in Figure 1a.[†] Up and to the right are the usual choices for the positive directions. These two number lines are called the **vertical axis** and the **horizontal axis** or (together) the **coordinate axes**. The coordinate axes divide the plane into

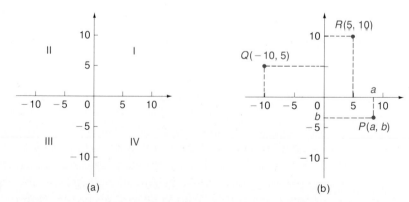

FIGURE 1　　　　　　(a)　　　　　　　　　　　(b)

[†] Here we use the same scale on each real number line. Later in this section we will consider a particular type of problem where it is useful to use a different scale on each real number line.

four parts called **quadrants**. The quadrants are numbered counterclockwise from I to IV. All points in the plane lie in one of the four quadrants except for points on the coordinate axes.

Pick a point P in the plane at random (see Figure 1b). Pass horizontal and vertical lines through the point. The vertical line will intersect the horizontal axis at a point with coordinate a, and the horizontal line will intersect the vertical axis at a point with coordinate b. These two numbers form the **coordinates**

$$(a, b)$$

of the point P in the plane. In particular, the coordinates of the point Q are $(-10, 5)$ and those of the point R are $(5, 10)$.

The first coordinate a of the coordinates of point P is also called the **abscissa** of P; the second coordinate b of the coordinates of point P is also called the **ordinate** of P. The abscissa for Q in Figure 1b is -10 and the ordinate for Q is 5. The point with coordinates $(0, 0)$ is called the **origin**.

We know that coordinates (a, b) exist for each point in the plane since every point on each axis has a real number associated with it. Hence, by the procedure described, each point located in the plane can be labeled with a unique pair of real numbers. Conversely, by reversing the process, each pair of real numbers can be associated with a unique point in the plane.

The system that we have just defined is called a **cartesian coordinate system** (sometimes referred to as a **rectangular coordinate system**).

EXAMPLE 1   Find the coordinates of each of the points A, B, C, and D.

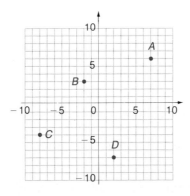

Solution $A(7, 6)$ $B(-2, 3)$ $C(-8, -4)$ $D(2, -7)$

PROBLEM 1   Find the coordinates, using the figure in Example 1, for each of the following points:

(A) 2 units to the right and 1 unit up from A
(B) 2 units to the left and 2 units down from C
(C) 1 unit up and 1 unit to the left of D
(D) 2 units to the right of B

Solution   (A)   (B)   (C)   (D)

EXAMPLE 2   Plot (associate each ordered pair of numbers with a point in the cartesian co-ordinate system):

$(2, 7)$ $(7, 2)$ $(-8, 4)$ $(4, -8)$ $(-8, -4)$ $(-4, -8)$

$(3, 0)$ $(0, 3)$

Solution

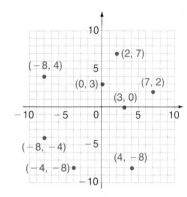

To plot $(-8, 4)$, for example, start at the origin and count 8 units to the left; then go straight up 4 units.

It is very important to note that the ordered pair $(2, 7)$ and the set $\{2, 7\}$ are not the same thing; $\{2, 7\} = \{7, 2\}$, but $(2, 7) \neq (7, 2)$.

PROBLEM 2   Plot: $(3, 4), (-3, 2), (-2, -2), (4, -2), (0, 1), (-4, 0)$

Solution

The development of the cartesian coordinate system represented a very important advance in mathematics. It was through the use of this system that René Descartes (1596–1650), a French philosopher-mathematician, was able to transform geometric problems requiring long tedious reasoning into algebraic problems that could be solved almost mechanically. This joining of algebra and geometry has now become known as **analytic geometry**.

Two fundamental problems of analytic geometry are the following:

1. Given an equation, find its graph.
2. Given a geometric figure, such as a straight line, circle, or ellipse, find its equation.

In this section we will be mainly interested in the first problem. In other parts of this chapter we will consider the second problem. Before we take up the first problem, however, let us refresh your memory on what is meant by "the graph of an equation." In general:

The Graph of an Equation

The graph of an equation in two variables in a rectangular coordinate system must meet the following two conditions:

1. If an ordered pair of numbers is a solution to the equation, the corresponding point must be on the graph of the equation.
2. If a point is on the graph of an equation, its coordinates must satisfy the equation.

■ Graphing a First-Degree Equation in Two Variables

Suppose we are interested in graphing

$$y = 2x - 4$$

We start by finding some of its solutions. A **solution** of an equation in two variables is an ordered pair of real numbers that satisfies the equation. If we agree that the first element in the ordered pair will replace x and the second y, then

$$(0, -4)$$

is a solution of $y = 2x - 4$, as can easily be checked. How do we find other solutions? The answer is easy: we simply assign to x in $y = 2x - 4$ any convenient value and solve for y. For example, if $x = 3$, then

$$y = 2(3) - 4 = 2$$

Hence,

$$(3, 2)$$

is another solution of $y = 2x - 4$. It is clear that by proceeding in this manner we can get solutions to this equation without end. Thus, the solution set is infinite. Table 1 lists some solutions, and we have graphed these solutions in a cartesian coordinate system in Figure 2, identifying the horizontal axis with x and the vertical axis with y.

TABLE 1	CHOOSE x	COMPUTE $2x - 4 = y$	WRITE ORDERED PAIR (x, y)
	-4	$2(-4) - 4 = -12$	$(-4, -12)$
	-2	$2(-2) - 4 = -8$	$(-2, -8)$
	0	$2(0) - 4 = -4$	$(0, -4)$
	2	$2(2) - 4 = 0$	$(2, 0)$
	4	$2(4) - 4 = 4$	$(4, 4)$
	6	$2(6) - 4 = 8$	$(6, 8)$
	8	$2(8) - 4 = 12$	$(8, 12)$

FIGURE 2

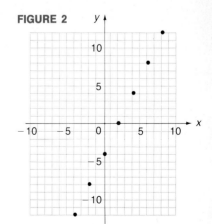

It appears in Figure 2 that the graph of the equation is a straight line. If we knew this for a fact, then graphing $y = 2x - 4$ would be easy. We would simply find two solutions of the equation, plot them, then graph as much of $y = 2x - 4$ as we like by drawing a line through the two points using a straightedge. It turns out that it is true that the graph of $y = 2x - 4$ is a straight line. In fact, we have the following general theorem, which we state without proof.

THEOREM 1

The graph of any equation of the form

$$y = mx + b \qquad \text{or} \qquad Ax + By = C$$

where m, b, A, B, and C are constants (A and B both not 0) and x and y are variables is a straight line.

Thus, the graphs of

$$y = \tfrac{2}{3}x - 5 \qquad \text{and} \qquad 2x - 3y = 12$$

are straight lines, since the first is of the form $y = mx + b$ and the second is of the form $Ax + By = C$.

Graphing Equations of the Form $y = mx + b$ or $Ax + By = C$

Step 1:   Find two solutions of the equation. (A third solution is sometimes useful as a check point.)

Step 2:   Plot the solutions in a coordinate system.

Step 3:   Using a straightedge, draw a line through the points plotted in step 2.

Note: The third solution provides a check point, since if the line does not pass through all three points, a mistake has been made in finding the solutions.

EXAMPLE 3   **(A)** Graph $y = 2x - 4$.   **(B)** Graph $x + 3y = 6$.

Solution   **(A)** Make up a table of at least two solutions (ordered pairs of numbers that satisfy the equation), plot them, then draw a line through these points with a straightedge.

x	$2x - 4 = y$	(x, y)
0	$2(0) - 4 = -4$	$(0, -4)$
2	$2(2) - 4 = 0$	$(2, 0)$
4	$2(4) - 4 = 4$	$(4, 4)$

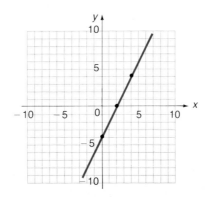

(B) To graph x + 3y = 6, assign to either x or y any convenient value and solve for the other variable. If we let x = 0, a convenient value, then

$$0 + 3y = 6$$
$$3y = 6$$
$$y = 2$$

Thus, (0, 2) is a solution.
 If we let y = 0, another convenient choice, then

$$x + 3(0) = 6$$
$$x + 0 = 6$$
$$x = 6$$

Thus, (6, 0) is a solution.
 To find a check point, choose another value for x or y, say x = −6. Then

$$-6 + 3y = 6$$
$$3y = 12$$
$$y = 4$$

Thus, (−6, 4) is also a solution.
 We summarize these results in a table and then draw the graph. The first two solutions indicate where the graph crosses the coordinate axes and are called the **y and x intercepts**, respectively. The intercepts are often the easiest points to find. To find the y intercept we let x = 0 and solve for y; to find the x intercept let y = 0 and solve for x. This is called the **intercept method** of graphing a straight line.

(x, y)	
(0, 2)	y intercept
(6, 0)	x intercept
(−6, 4)	Check point

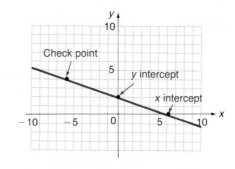

If a straight line does not pass through all three points, then we know we have made a mistake and must go back and check our work.

PROBLEM 3   Graph: **(A)** $y = 2x - 6$ **(B)** $3x + y = 6$

Solution

(A)

(B)

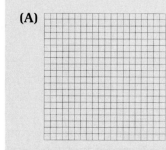

■ Vertical and Horizontal Lines

Vertical and horizontal lines in rectangular coordinate systems have particularly simple equations.

EXAMPLE 4   Graph the equations $y = 4$ and $x = 3$ in a rectangular coordinate system.

Solution   To graph $y = 4$ or $x = 3$ in a rectangular coordinate system, each equation must be provided with the missing variable (usually done mentally) as follows:

> $y = 4$ is equivalent to $0x + y = 4$
> $x = 3$ is equivalent to $x + 0y = 3$

In the first case, we see that no matter what value is assigned to x, $0x = 0$; thus, as long as $y = 4$, x can assume any value, and the graph of $y = 4$ is a horizontal line crossing the y axis at 4. Similarly, in the second case y can assume any value as long as $x = 3$, and the graph of $x = 3$ is a vertical line crossing the x axis at 3. Thus:

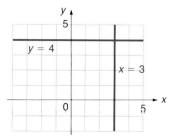

PROBLEM 4   Graph $y = -3$ and $x = -4$ in a rectangular coordinate system.

Solution

■ Use of Different Scales for Each Coordinate Axis

Given the problem of graphing

$$A = 50 + 5t \qquad 0 \le t \le 10$$

we first note that t is restricted to values from 0 to 10. Let us find A for three values in this interval. We choose each end value and the middle value:

t	0	5	10
A	50	75	100

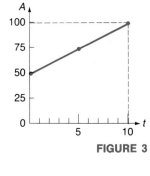

FIGURE 3

We see that as t varies from 0 to 10, A varies from 50 to 100. To keep the graph on the paper, we choose a different scale on the vertical axis (representing A) than that used on the horizontal axis (representing t). In addition, we note that the whole graph (under the restrictions for t) lies in the first quadrant. Thus, we obtain Figure 3. The graph is a line segment joining the two points $(0, 50)$ and $(10, 100)$. The dotted lines are guidelines used to guide one's eyes to the endpoint and are not part of the graph.

It should now be clear why equations of the form $Ax + By = C$ and $y = mx + b$ are called **linear equations**: their graphs are straight lines.

Solutions to Matched Problems

1. (A) $(9, 7)$ **(B)** $(-10, -6)$ **(C)** $(1, -6)$ **(D)** $(0, 3)$

2.

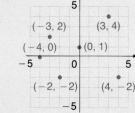

3. (A)

(x, y)
$(0, -6)$
$(3, 0)$
$(2, -2)$

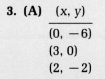

(B)

(x, y)
$(0, 6)$
$(2, 0)$
$(1, 3)$

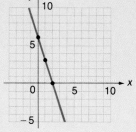

4.

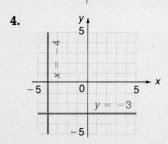

Practice Exercise 7-1 ■

Work odd-numbered problems first, check answers, and then work even-numbered problems in areas of weakness. Answers to all problems are in the back of the book. Make every effort to work a problem yourself before you look at an answer.

A   Write down the coordinates of each labeled point.

1.

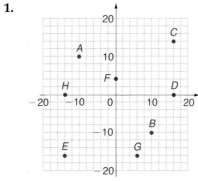

2.

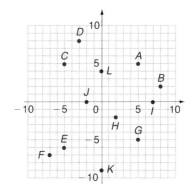

Plot each set of ordered pairs of numbers on the same coordinate system.

3. $(4, 4), (-4, 1), (-3, -3), (5, -1), (0, 2), (-2, 0)$

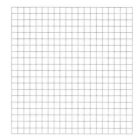

4. $(3, 1), (-2, 3), (-5, -1), (2, -1), (4, 0), (0, -5)$

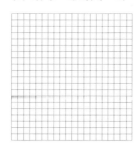

5. $(2, 7), (7, 2), (-6, 3), (-4, -7), (2, 3), (0, -8), (9, 0)$

6. $(-9, 8)$, $(8, -9)$, $(0, 5)$, $(4, -8)$, $(-3, 0)$, $(7, 7)$, $(-6, -6)$

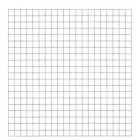

Write down the coordinates of each labeled point to the nearest quarter of a unit.

7.

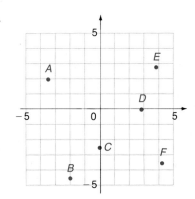

8.

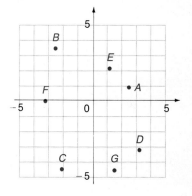

9. Plot the following ordered pairs of numbers on the same coordinate system: $A(3\frac{1}{2}, 2\frac{1}{2})$, $B(-4\frac{1}{2}, 3)$, $C(0, -3\frac{3}{4})$, $D(-2\frac{3}{4}, -3\frac{3}{4})$, $E(4\frac{1}{4}, -3\frac{3}{4})$

10. Plot the following ordered pairs of numbers on the same coordinate system: $A(1\frac{1}{2}, 3\frac{1}{2})$, $B(-3\frac{1}{4}, 0)$, $C(3, -2\frac{1}{2})$, $D(-4\frac{1}{2}, 1\frac{3}{4})$, $E(-2\frac{1}{2}, -4\frac{1}{4})$

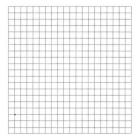

Graph in a rectangular coordinate system.

11. $y = 2x$

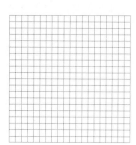

12. $y = x$

13. $y = 2x - 3$

14. $y = x - 1$

15. $y = \dfrac{x}{3}$

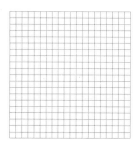

16. $y = \dfrac{x}{2}$

17. $y = \dfrac{x}{3} + 2$

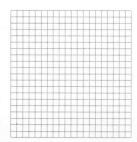

18. $y = \dfrac{x}{2} + 1$

19. $x + y = -4$

20. $x + y = 6$

21. $x - y = 3$

22. $x - y = 5$

23. $3x + 4y = 12$

24. $2x + 3y = 12$

25. $8x - 3y = 24$

26. $3x - 5y = 15$

27. $y = 3$

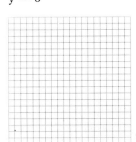

28. $x = 2$

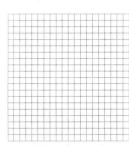

29. $x = -4$

30. $y = -3$

B **31.** $y = \frac{1}{2}x$

32. $y = \frac{1}{4}x$

33. $y = \frac{1}{2}x - 1$

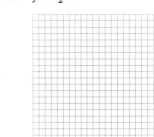

34. $y = \frac{1}{4}x + 1$

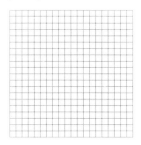

35. $y = -x + 2$

36. $y = -2x + 6$

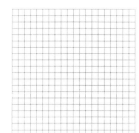

37. $y = \frac{1}{3}x - 1$

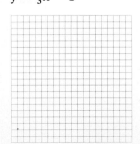

38. $y = -\frac{1}{2}x + 2$

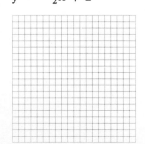

39. $3x + 2y = 10$

40. $2x + y = 7$

41. $5x - 6y = 15$

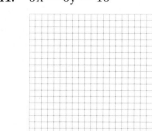

42. $7x - 4y = 21$

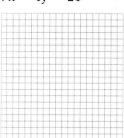

43. $y = 0$

44. $x = 0$

Write in the form $y = mx + b$ and graph.

45. $x + 6 = 3x + 2 - y$

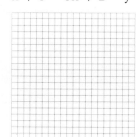

46. $y - x - 2 = x + 1$

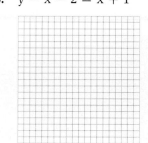

Write in the form $Ax + By = C$, $A > 0$, and graph.

47. $y + 8 = 2 - x - y$

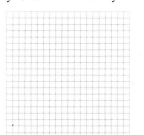

48. $6x - 3 + y = 2y + 4x + 5$

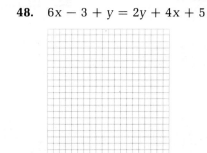

Graph each of the following using a different scale on the vertical axis to keep the size of the graph within reason.

49. $I = 6t, 0 \leq t \leq 10$

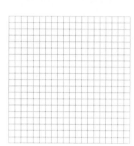

50. $d = 60t, 0 \leq t \leq 10$

51. $v = 10 + 32t, 0 \leq t \leq 5$

52. $A = 100 + 10t, 0 \leq t \leq 10$

53.   Graph $x + y = 3$ and $2x - y = 0$ on the same coordinate system. Determine by inspection the coordinates of the point where the two graphs cross. Show that the coordinates of the point of intersection satisfy both equations.

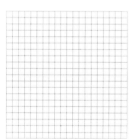

54.   Repeat Problem 53 with the equations $2x - 3y = -6$ and $x + 2y = 11$.

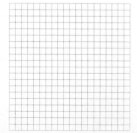

C 55. Graph $y = mx - 2$ for $m = 2$, $m = \frac{1}{2}$, $m = 0$, $m = -\frac{1}{2}$, and $m = -2$, all on the same coordinate system.

56. Graph $y = -\frac{1}{2}x + b$ for $b = -6$, $b = 0$, and $b = 6$, all on the same coordinate system.

57. Graph $y = |x|$. [*Hint:* Graph $y = x$ for $x \geq 0$, and graph $y = -x$ for $x < 0$.]

58. Graph $y = |2x|$ and $y = |\frac{1}{2}x|$ on the same coordinate system (see Problem 57).

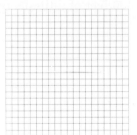

APPLICATIONS 59. *Psychology:* In 1948 Professor Brown, a psychologist, trained a group of rats (in an experiment on motivation) to run down a narrow passage in a cage to receive food in a box. A harness was put on each rat and the harness was then connected to an overhead wire that was attached to a scale. In this way the rat could be placed at different distances (in centimeters) from the food and Professor Brown could then measure the pull

(in grams) of the rat toward the food. It was found that a relation between motivation (pull) and position was given approximately by the equation $p = -\frac{1}{5}d + 70, 30 \le d \le 175$. Graph this equation for the indicated values of d.

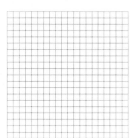

60.   *Electronics:*   In a simple electric circuit, such as found in a flashlight, if the resistance is 30 ohms, the current in the circuit I (in amperes) and the electromotive force E (in volts) are related by the equation $E = 30I$. Graph this equation for $0 \le I \le 1$.

61.   *Biology:*   In biology there is an approximate rule, called the bioclimatic rule, for temperate climates. This rule states that in spring and early summer, periodic phenomena such as blossoming for a given species, appearance of certain insects, and ripening of fruit usually come about 4 days later for each 500 feet of altitude. Stated as a formula,

$$d = 4\left(\frac{h}{500}\right)$$

where d = Change in days and h = Change in altitude in feet. Graph the equation for $0 \le h \le 4,000$.

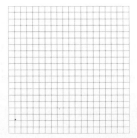

The Check Exercise for this section is on page 463.

Section 7-2 Slope and Equations of a Line

- ■ Slope of a Line
- ■ Slope–Intercept Form
- ■ Point–Slope Form
- ■ Vertical and Horizontal Lines
- ■ Parallel and Perpendicular Lines

In the preceding section we considered this problem: given a linear equation of the form

$$Ax + By = C \quad \text{or} \quad y = mx + b$$

find its graph. Now we will consider the reverse problem: given certain information about a straight line in a rectangular coordinate system, find its equation. We start by introducing a measure of the steepness of a line called slope.

■ Slope of a Line

If we take two points (x_1, y_1) and (x_2, y_2) on a line, then the ratio of the change in y to the change in x as we move from point P_1 to P_2 is called the **slope** of the line.

Slope Formula

If a line passes through $P_1(x_1, y_1)$ and $P_2(x_2, y_2)$, then its slope is given by the formula

$$m = \frac{y_2 - y_1}{x_2 - x_1} \quad x_1 \neq x_2$$

$$= \frac{\text{Vertical change (rise)}}{\text{Horizontal change (run)}}$$

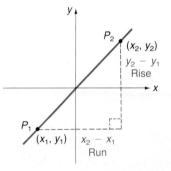

EXAMPLE 5   Find the slope of the line passing through $(-3, -2)$ and $(3, 4)$.

Solution   Let $(x_1, y_1) = (-3, -2)$ and $(x_2, y_2) = (3, 4)$; then

$$m = \frac{y_2 - y_1}{x_2 - x_1} = \frac{4 - (-2)}{3 - (-3)} = \frac{6}{6} = 1$$

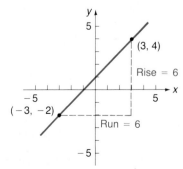

Note: It does not matter which point we call P_1 or P_2 as long as we stick to the choice once it is made. If we reverse our first choice, we obtain the same value for the slope, since the sign of the numerator and the denominator both change:

$$m = \frac{(-2) - 4}{(-3) - 3} = \frac{-6}{-6} = 1$$

PROBLEM 5

Solution

Graph the line passing through $(-2, 7)$ and $(3, -3)$; then compute its slope.

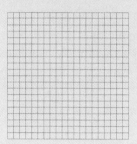

For a horizontal line, y does not change as x changes; hence, its slope is 0. On the other hand, for a vertical line, x does not change as y changes; hence, $x_1 = x_2$ and

$$m = \frac{y_2 - y_1}{x_2 - x_1} = \frac{y_2 - y_1}{0} \qquad \text{Vertical-line slope is not defined.}$$

In general, the slope of a line may be positive, negative, 0, or not defined. Each of these cases is interpreted geometrically as shown in Table 2.

TABLE 2
GOING FROM LEFT
TO RIGHT

LINE	SLOPE	EXAMPLE
Rising	Positive	
Falling	Negative	
Horizontal	0	
Vertical	Not defined	

■ **Slope–Intercept Form**

Any equation of the form $Ax + By = C$, $B \neq 0$, can always be written in the form

$$y = mx + b$$

where m and b are constants. For example, starting with

$$2x + 3y = 6 \qquad \text{Form } Ax + By = C.$$

we solve for y to obtain

$$3y = -2x + 6$$

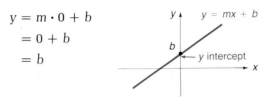

$$\frac{1}{3}(3y) = \frac{1}{3}(-2x + 6)$$

$$y = -\tfrac{2}{3}x + 2 \quad \text{Form } y = mx + b.$$

The constants m and b in $y = mx + b$ have special geometric meaning. If we let $x = 0$, then

$$y = m \cdot 0 + b$$
$$= 0 + b$$
$$= b$$

Thus, b is the y coordinate of the y intercept, the point where the graph crosses the y axis. For brevity, the value b is also called the **y intercept**. In the example given above, $y = -\tfrac{2}{3}x + 2$, the y intercept is 2.

Now let us determine the geometric significance of m in $y = mx + b$. We choose two points (x_1, y_1) and (x_2, y_2) on the graph of $y = mx + b$ (Figure 4). Since the two points are on the graph, they are solutions to the equation $y = mx + b$. Thus,

$$y_1 = mx_1 + b \qquad \text{and} \qquad y_2 = mx_2 + b$$

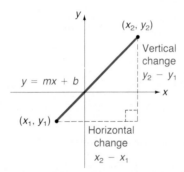

FIGURE 4

Solving both equations for b, we obtain

$$b = y_1 - mx_1 \qquad \text{and} \qquad b = y_2 - mx_2$$

Since $y_1 - mx_1$ and $y_2 - mx_2$ are both equal to b, they are equal to each other. Thus,

$$y_1 - mx_1 = y_2 - mx_2 \quad \text{Now solve for } m.$$
$$mx_2 - mx_1 = y_2 - y_1$$
$$(x_2 - x_1)m = y_2 - y_1$$
$$m = \frac{y_2 - y_1}{x_2 - x_1} \qquad x_1 \neq x_2$$

Thus, **m is the slope** of the graph of $y = mx + b$.

In summary, if an equation of a line is written in the form $y = mx + b$, then b is the y intercept and m is the slope. Conversely, if we know the slope and y intercept of a line, we can write its equation in the form $y = mx + b$.

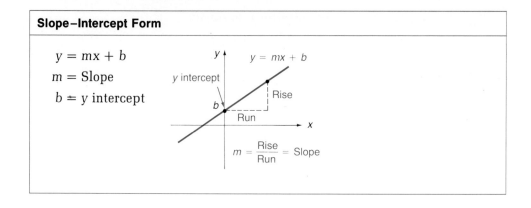

EXAMPLE 6 **(A)** Find the slope and y intercept of the line $y = \frac{1}{3}x + 2$.
(B) Find the equation of a line with slope -2 and y intercept 3.

Solution **(A)** $y = \frac{1}{3}x + 2$
    (Slope) (y intercept)

(B) Since $m = -2$ and $b = 3$, then $y = mx + b = -2x + 3$ is the equation.

PROBLEM 6 **(A)** Find the slope and y intercept of the line $y = \frac{x}{2} - 7$.

(B) Find the equation of a line with slope $-\frac{1}{3}$ and y intercept 6.

Solution **(A)**   **(B)**

The slope–intercept form allows us to graph a linear equation very efficiently.

EXAMPLE 7 Graph $y = -\frac{1}{2}x + 3$.

Solution The y intercept is 3, so the point $(0, 3)$ is on the graph. The slope of the line is $-\frac{1}{2}$, so if the x coordinate is increased (run) by 2 units, the y coordinate changes (rises) by -1. The resulting point is easily graphed and the two points yield the graph of the line.

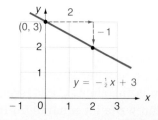

PROBLEM 7

Graph $y = \frac{2}{3}x - 2$.

Solution

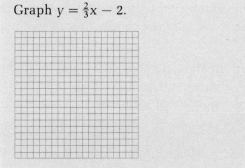

■ **Point–Slope Form**

In Example 6(B) we found the equation of a line given its slope and y intercept. Often it is necessary to find the equation of a line given its slope and the coordinates of a point through which it passes, or to find the equation of a line given the coordinates of two points through which it passes.

Let a line have slope m and pass through the fixed point (x_1, y_1). If the variable point (x, y) is to be a point on the line, the slope of the line passing through (x, y) and (x_1, y_1) must be m (see Figure 5).

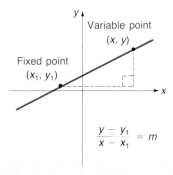

FIGURE 5

Thus, the equation

$$\frac{y - y_1}{x - x_1} = m$$

restricts the variable point (x, y), so that only those points in the plane lying on the line will have coordinates that satisfy the equation and vice versa. This equation is usually written in the following form:

Point–Slope Form
The equation of a line passing through (x_1, y_1) with slope m is given by $$y - y_1 = m(x - x_1)$$

It is referred to as the **point–slope form of the equation of a line**. Using this equation in conjunction with the slope formula, we can also find the equation of a line knowing only the coordinates of two points through which it passes.

EXAMPLE 8   **(A)** Find an equation of a line with slope $-\frac{1}{3}$ that passes through $(6, -3)$. Write the resulting equation in the form $y = mx + b$.

(B) Find an equation of a line that passes through the two points $(-2, -6)$ and $(2, 2)$.

Solution   **(A)** $y - y_1 = m(x - x_1)$

$y - (-3) = -\frac{1}{3}(x - 6)$

$y + 3 = -\frac{1}{3}(x - 6)$

$y + 3 = -\dfrac{x}{3} + 2$

$y = -\frac{1}{3}x - 1$

(B) First find the slope of the line using the slope formula

$$m = \frac{y_2 - y_1}{x_2 - x_1} = \frac{2 - (-6)}{2 - (-2)} = 2$$

Now proceed as in part (A), using the coordinates of either point for (x_1, y_1).

Use $(x_1, y_1) = (-2, -6)$	or	Use $(x_1, y_1) = (2, 2)$
$y - y_1 = m(x - x_1)$		$y - y_1 = m(x - x_1)$
$y - (-6) = 2[x - (-2)]$		$y - 2 = 2(x - 2)$
$y + 6 = 2(x + 2)$		$y - 2 = 2x - 4$
$y + 6 = 2x + 4$		$y = 2x - 2$
$y = 2x - 2$		

PROBLEM 8   **(A)** Find the equation of a line with slope $\frac{2}{3}$ that passes through $(-3, 4)$.

(B) Find the equation of a line that passes through the two points $(6, -1)$ and $(-2, 3)$. Transform the equation into the form $y = mx + b$.

Solution   **(A)**   **(B)**

The slope–intercept form can also be used to find the equation of the line given the slope and one point. In Example 8(A), for instance, given the slope $-\frac{1}{3}$, we know the equation has the form

$$y = -\tfrac{1}{3}x + b$$

Since $(6, -3)$ must satisfy the equation, we can solve for b:

$$-3 = -\tfrac{1}{3}(6) + b$$
$$-3 = -2 + b$$
$$-1 = b$$

Thus, the equation of the line is $y = -\tfrac{1}{3}x - 1$.

■ Vertical and Horizontal Lines

If a line is vertical, its slope is not defined. Since points on a vertical line have constant abscissas and arbitrary ordinates, the equation of a vertical line is of the form

$$x + 0y = c$$

or simply

$$x = c \quad \text{Vertical line}$$

where c is the abscissa of each point on the line. Similarly, if a line is horizontal (slope 0), then every point on the line has constant ordinate and arbitrary abscissa. Thus, the equation of a horizontal line is of the form

$$0x + y = c$$

or simply

$$y = c \quad \text{Horizontal line}$$

where c is the ordinate of each point on the line. Also, since a horizontal line has slope 0 ($m = 0$), then, using the slope–intercept form, we obtain

$$y = mx + b$$
$$y = 0x + c$$
$$y = c$$

EXAMPLE 9   The equation of a vertical line through $(-2, -4)$ is $x = -2$, and the equation of a horizontal line through the same point is $y = -4$.

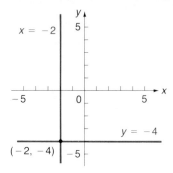

PROBLEM 9   What are the equations of vertical and horizontal lines through $(3, -8)$?

Solution

■ Parallel and Perpendicular Lines

It can be shown that if two nonvertical lines are parallel, then they have the same slope. And if two lines have the same slope, they are parallel. It can also be shown that if two nonvertical lines are perpendicular, then their slopes are the negative reciprocals of each other (that is, $m_2 = -1/m_1$, or, equivalently, $m_1 m_2 = -1$). And if the slopes of two lines are the negative reciprocals of each other, the lines are perpendicular. Symbolically:

Parallel and Perpendicular Lines

Given nonvertical lines L_1 and L_2 with slopes m_1 and m_2, respectively, then

$L_1 \| L_2$ if and only if $m_1 = m_2$

$L_1 \perp L_2$ if and only if $m_1 m_2 = -1$ or $m_2 = -\dfrac{1}{m_1}$

Note: $\|$ means "is parallel to" and $\perp$ means "is perpendicular to."

The lines determined by

$$y = \tfrac{2}{3}x - 5 \qquad \text{and} \qquad y = \tfrac{2}{3}x + 8$$

are parallel since both have the same slope, $\tfrac{2}{3}$. And the lines that are determined by

$$y = \frac{x}{3} + 5 \qquad \text{and} \qquad y = -3x - 7$$

are perpendicular, since the product of their slopes is -1; that is,

$$(\tfrac{1}{3})(-3) = -1$$

EXAMPLE 10   Given the line $x - 2y = 4$, find the equation of a line that passes through $(2, -3)$ and is:

(A) Parallel to the given line   (B) Perpendicular to the given line

Write final equations in the form $y = mx + b$.

Solution   First find the slope of the given line by writing $x - 2y = 4$ in the form $y = mx + b$:

$$x - 2y = 4$$
$$-2y = 4 - x$$
$$\boxed{-\tfrac{1}{2}(-2y) = -\tfrac{1}{2}(4 - x)}$$
$$y = -2 + \tfrac{1}{2}x$$
$$\overset{\text{Slope}}{y = \tfrac{1}{2}x - 2}$$

The slope of the given line is $\tfrac{1}{2}$.

(A) The slope of a line parallel to the given line is also $\tfrac{1}{2}$. We have only to find the equation of a line through $(2, -3)$ with slope $\tfrac{1}{2}$ to solve part (A):

$$y - y_1 = m(x - x_1) \quad m = \tfrac{1}{2} \text{ and } (x_1, y_1) = (2, -3)$$
$$y - (-3) = \tfrac{1}{2}(x - 2)$$
$$y + 3 = \tfrac{1}{2}x - 1$$
$$y = \tfrac{1}{2}x - 4$$

(B) The slope of the line perpendicular to the given line is the negative reciprocal of $\tfrac{1}{2}$, that is, -2. We have only to find the equation of a line through $(2, -3)$ with slope -2 to solve part (B):

$$y - y_1 = m(x - x_1) \quad m = -2 \text{ and } (x_1, y_1) = (2, -3)$$
$$y - (-3) = -2(x - 2)$$
$$y + 3 = -2x + 4$$
$$y = -2x + 1$$

PROBLEM 10   Given the line $2x = 6 - 3y$, find the equation of a line that passes through $(-3, 9)$ and is:

(A) Parallel to the given line   (B) Perpendicular to the given line

Write final equations in the form $y = mx + b$.

Solution   (A)   (B)

Solutions to Matched
Problems

5.

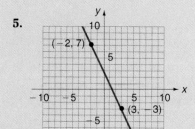

$$m = \frac{y_1 - y_2}{x_1 - x_2} = \frac{(-3) - 7}{3 - (-2)} = \frac{-10}{5} = -2$$

6. **(A)** $m = \frac{1}{2}$, $b = -7$ **(B)** $y = -\frac{1}{3}x + 6$

7.

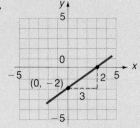

8. **(A)** $y - y_1 = m(x - x_1)$
      $y - 4 = \frac{2}{3}(x + 3)$
      $y - 4 = \frac{2}{3}x + 2$
         $y = \frac{2}{3}x + 6$

     (B) First find the slope; then use the
         point–slope formula.

$$m = \frac{y_1 - y_2}{x_1 - x_2} = \frac{3 - (-1)}{(-2) - 6}$$

$$= \frac{4}{-8} = -\frac{1}{2}$$

$$y - y_1 = m(x - x_1)$$
$$y + 1 = -\tfrac{1}{2}(x - 6)$$
$$y + 1 = -\tfrac{1}{2}x + 3$$
$$y = -\tfrac{1}{2}x + 2$$

9. Vertical line through $(3, -8)$: $x = 3$
     Horizontal line through $(3, -8)$: $y = -8$

10. First find the slope of the graph of $2x = 6 - 3y$:
     $2x = 6 - 3y$
     $3y = -2x + 6$
       $y = -\frac{2}{3}x + 2$
     Slope of the given line is $-\frac{2}{3}$

     (A) Find the equation of a line with slope $-\frac{2}{3}$ that passes through
         $(-3, 9)$:
$$y - y_1 = m(x - x_1)$$
$$y - 9 = -\tfrac{2}{3}(x + 3)$$
$$y - 9 = -\tfrac{2}{3}x - 2$$
$$y = -\tfrac{2}{3}x + 7$$

     (B) Find the equation of a line with slope $\frac{3}{2}$ that passes through $(-3, 9)$:
$$y - y_1 = m(x - x_1)$$
$$y - 9 = \tfrac{3}{2}(x + 3)$$
$$y - 9 = \tfrac{3}{2}x + \tfrac{9}{2}$$
$$y = \tfrac{3}{2}x + \tfrac{27}{2}$$

Practice Exercise 7-2 ■

Work odd-numbered problems first, check answers, and then work even-numbered problems in areas of weakness. Answers to all problems are in the back of the book. Make every effort to work a problem yourself before you look at an answer.

A   *Find the slope and y intercept, and graph each equation.*

1. $y = 2x - 3$

2. $y = x + 1$

 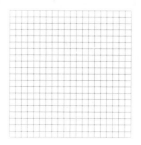

3. $y = -x + 2$

4. $y = -2x + 1$

 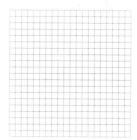

Write the equation of the line with slope and y intercept as indicated.

5. Slope $= 5$
       y intercept $= -2$ _____

6. Slope $= 3$
       y intercept $= -5$ _____

7. Slope $= -2$
       y intercept $= 4$ _____

8. Slope $= -1$
       y intercept $= 2$ _____

Write the equation of the line that passes through the given point with the indicated slope.

9. $m = 2$; $(5, 4)$ _____

10. $m = 3$; $(2, 5)$ _____

11. $m = -2$; $(2, 1)$ _____

12. $m = -3$; $(1, 3)$ _____

Find the slope of the line that passes through the given points.

13. $(3, 2)$ and $(5, 6)$ _____

14. $(1, 3)$ and $(2, 4)$ _____

15. $(2, 1)$ and $(10, 5)$ _____

16. $(1, 3)$ and $(7, 5)$ _____

Write the equation of the line through each indicated pair of points.

17. $(3, 2)$ and $(5, 6)$ _____

18. $(1, 3)$ and $(2, 4)$ _____

19. (2, 1) and (10, 5) _____   **20.** (1, 3) and (7, 5) _____

B   *Find the slope and y intercept, and graph each equation.*

21. $y = -\dfrac{x}{3} + 2$ **22.** $y = -\dfrac{x}{4} - 1$

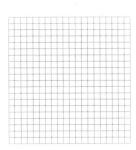   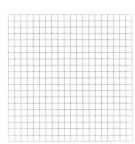

23. $x + 2y = 4$ **24.** $x - 3y = -6$

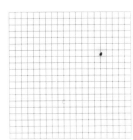   

25. $2x + 3y = 6$ **26.** $3x + 4y = 12$

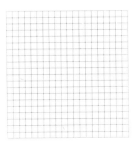   

Write the equation of the line with slope and y intercept as given.

27. Slope $= -\frac{1}{2}$
y intercept $- -2$ _____   **28.** Slope $= -\frac{1}{3}$
y intercept $= 5$ _____

29. Slope $= \frac{2}{3}$
y intercept $= \frac{3}{2}$ _____   **30.** Slope $= -\frac{3}{2}$
y intercept $= \frac{5}{2}$ _____

Write the equation of the line that passes through the given point with the indicated slope. Transform the equation into the form y = mx + b.

31. $m = -2$; $(-3, 2)$ _____   **32.** $m = -3$; $(4, -1)$ _____

33. $m = \frac{1}{2}$; $(-4, 3)$ _____   **34.** $m = \frac{2}{3}$; $(-6, -5)$ _____

Find the slope of the line that passes through the given points.

35. (3, 7) and $(-6, 4)$ _____   **36.** $(-5, -2)$ and $(-5, -4)$ _____

37. $(4, -2)$ and $(-4, 0)$ _____ **38.** $(-3, 0)$ and $(3, -2)$ _____

Write the equation of the line through each of the indicated pairs of points. Transform the equation into the form $y = mx + b$.

39. $(3, 7)$ and $(-6, 4)$ _____ **40.** $(-5, -2)$ and $(5, -4)$ _____

41. $(4, -2)$ and $(-4, 0)$ _____ **42.** $(-3, 0)$ and $(3, -4)$ _____

Write the equations of the vertical and horizontal lines through each point.

43. $(-3, 5)$ _____ **44.** $(6, -2)$ _____

45. $(-1, 22)$ _____ **46.** $(5, 0)$ _____

C *Given the indicated equation of a line and the indicated point, find the equation of the line through the given point that is **(A)** parallel to the given line and **(B)** perpendicular to the given line. Write the answers in the form $y = mx + b$.*

47. $y = \frac{3}{5}x + 1$; $(1, 4)$ _____

48. $y = -\frac{2}{3}x + 3$; $(-1, -2)$ _____

49. $y = -3x + \frac{1}{3}$; $(0, 2)$ _____

50. $y = 5x - \frac{1}{4}$; $(-2, 0)$ _____

51. $x + y = 3$; $(1, 1)$ _____

52. $x - y = \frac{1}{7}$; $(2, 3)$ _____

53. $y = 3$; $(-2, 5)$ _____

54. $x = 4$; $(3, -2)$ _____

APPLICATIONS

55. *Business:* A sporting goods store sells a pair of cross-country ski boots costing $20 for $33 and a pair of cross-country skis costing $60 for $93.

    (A) If the markup policy of the store for items costing over $10 is assumed to be linear and is reflected in the pricing of these two items, write an equation that relates retail price R with cost C. _____

    (B) Graph this equation for $10 \le C \le 300$.

    (C) Use the equation to find the cost of a surfboard retailing for $240.

_____

56. *Business:*   The management of a company manufacturing ballpoint pens estimates costs for running the company to be $200 per day at zero output and $700 per day at an output of 1,000 pens.

    (A)   Assuming total cost per day C is linearly related to total output per day x, write an equation relating these two quantities. _____

    (B)   Graph the equation for $0 \leq x \leq 2{,}000$.

57. *Physics:*   Water freezes at 32° Fahrenheit and 0° Celsius and boils at 212° Fahrenheit and 100° Celsius. Find the linear relationship between the two scales. _____

58. *Physics:*   It is known from physics (Hooke's law) that the relationship between the stretch s of a spring and the weight w causing the stretch is linear (a principle upon which most spring scales are constructed). A 10-pound weight stretches a spring 1 inch and with no weight the stretch of the spring is 0.

    (A)   Find a linear equation, $s = mw + b$, that represents this relationship. [*Hint:*   Both points (10, 1) and (0, 0) are on its graph.] _____

    (B)   Find the stretch of the spring for 15-pound and 30-pound weights.

    _____

    (C)   What is the slope of the graph? (The slope indicates the increase in stretch for each pound increase in weight.) _____

    (D)   Graph the equation for $0 \leq w \leq 40$.

59. *Business:*   An electronic computer was purchased by a company for $20,000 and is assumed to have a salvage value of $2,000 after 10 years (for tax purposes). Its value is depreciated linearly from $20,000 to $2,000.

    (A)   Find the linear equation, $V = mt + b$, that relates value V in dollars to time t in years. _____

    (B)   Find the values of the computer after 4 and 8 years, respectively.

    _____

(C) Find the slope of the graph. (The slope indicates the decrease in value per year.) _____

(D) Graph the equation for $0 \le t \le 10$.

60. *Biology:* A biologist needs to prepare a special diet for a group of experimental animals. Two food mixes, M and N, are available. If mix M contains 20% protein and mix N contains 10% protein, what combinations of each mix will provide exactly 20 grams of protein? Let x be the amount of M used and y the amount of N used. Then write a linear equation relating x, y, and 20. Graph this equation for $x \ge 0$ and $y \ge 0$.

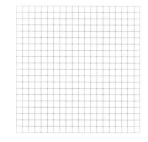

The Check Exercise for this section is on page 465.

Section 7-3 Graphing Linear Inequalities

We know how to graph first-degree equations such as

$$y = 2x - 3 \qquad \text{or} \qquad 2x - 3y = 5$$

but how do we graph first-degree inequalities such as

$$y \le 2x - 3 \qquad \text{or} \qquad 2x - 3y > 5$$

We will find that graphing inequalities is almost as easy as graphing equations. The following discussion leads to a simple solution to the problem.

A line in a cartesian coordinate system divides the plane into two **half-planes**. A vertical line divides the plane into left and right half-planes; a nonvertical line divides the plane into upper and lower half-planes (Figure 6).

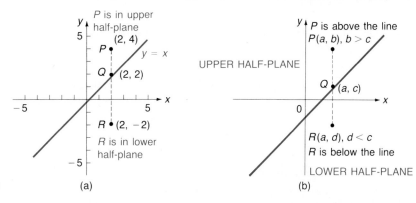

FIGURE 6   (a)   (b)

Now let us compare the graphs of

$$y < 2x - 3, \qquad y = 2x - 3, \qquad \text{and} \qquad y > 2x - 3$$

Consider the vertical line $x = 3$, and ask what the relationship of y is to $2 \cdot 3 - 3$ as we move $(3, y)$ up and down this vertical line (see Figure 7a). If we are at point Q, a point on the graph of $y = 2x - 3$, then $y = 2 \cdot 3 - 3$; if we move up the vertical line to P, the ordinate of $(3, y)$ increases and $y > 2 \cdot 3 - 3$; if we move down the line to R, the ordinate of $(3, y)$ decreases and $y < 2 \cdot 3 - 3$. Since the same results are obtained for each point x_0 on the x axis (see Figure 7b), we conclude that the graph of $y > 2x - 3$ is the upper half-plane determined by $y = 2x - 3$, and $y < 2x - 3$ is the lower half-plane.

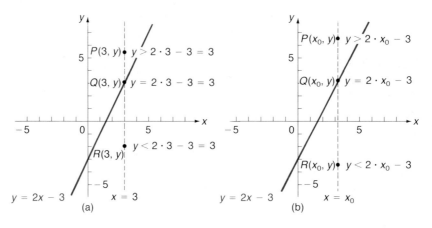

FIGURE 7

In graphing $y > 2x - 3$, we show the line $y = 2x - 3$ as a broken line, indicating that it is not part of the graph; in graphing $y \geq 2x - 3$, we show the line $y = 2x - 3$ as a solid line, indicating that it is part of the graph. Figure 8 illustrates four typical cases.

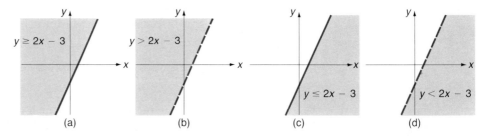

FIGURE 8   (a)   (b)   (c)   (d)

The preceding discussion suggests the following important theorem, which we state without proof.

THEOREM 2

The graph of a linear inequality

$$Ax + By < C \qquad \text{or} \qquad Ax + By > C$$

with $B \neq 0$, is either the upper half-plane or the lower half-plane (but not both) determined by the line $Ax + By = C$. If $B = 0$, the graph of

$$Ax < C \qquad \text{or} \qquad Ax > C$$

is either the left half-plane or the right half-plane (but not both) determined by the line $Ax = C$.

This theorem leads to a simple, fast procedure for graphing linear inequalities in two variables:

Steps in Graphing Linear Inequalities

1. Graph the corresponding equation $Ax + By = C$—as a broken line if equality is not included in original statement, as a solid line if equality is included in original statement.
2. Choose a test point in the plane not on the line—the origin is the best choice if it is not on the line—and substitute the coordinates into the inequality.
3. The graph of the original inequality includes:
   (A) The half-plane containing the test point if the inequality is satisfied by that point
   (B) The half-plane not containing the test point if the inequality is not satisfied by that point
4. Shade the half-plane to show the graph of the solution set.

EXAMPLE 11   Graph $3x - 4y \leq 12$.

Solution   *Step 1:*   First graph the line $3x - 4y = 12$ as a solid line, since equality is included in $3x - 4y \leq 12$.

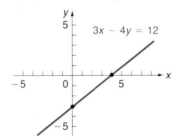

Step 2:   Choose a convenient test point—any point not on the line will do. In this case the origin results in the simplest computation.

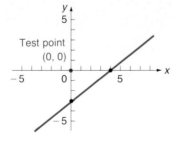

We see that the origin (0, 0) satisfies the original inequality:

$$3x - 4y \leq 12$$
$$3 \cdot 0 - 4 \cdot 0 \leq 12$$
$$0 \leq 12$$

Hence, all other points on the same side as the origin are also part of the graph. Thus, the graph is the upper half-plane.

Step 3:   The final graph is the upper half-plane and the line:

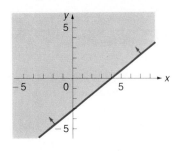

PROBLEM 11   Graph $2x + 2y \leq 5$.

Solution

EXAMPLE 12   Graph in a rectangular coordinate system:

(A) $y > -3$ **(B)** $2x < 5$

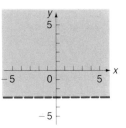 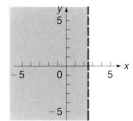

(C) $-2 \leq x \leq 4$

REMARK   If we are to graph an inequality involving only one variable, it is important to
know the dimension of the coordinate system in which the graph is to occur.
In Example 12 the graphs are to occur in a rectangular coordinate system;
hence, in each case the presence of a second variable is assumed, but with a

0 coefficient. For example, $y > -3$ is assumed to mean $0x + y > -3$. Thus, x can take on any value as long as y is greater than -3. The graph is the upper half-plane above the line $y = -3$. If we were to graph $y > -3$ on a single number line, however, then we would obtain

PROBLEM 12   Graph in a rectangular coordinate system:

(A) $y < 2$ **(B)** $3x > -8$ **(C)** $1 \leq y \leq 5$

Solution   **(A)**   **(B)**

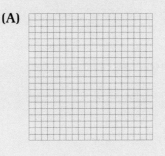

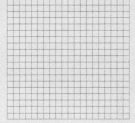

(C)

EXAMPLE 13   Graph $\{(x, y) \mid x \geq 0, y \geq 0, 2x + 3y \leq 18\}$.

Solution   This is the set of all ordered pairs of real numbers (x, y) such that x and y are both nonnegative and satisfy $2x + 3y \leq 18$. We graph each inequality in the same coordinate system and take the intersection (see Section 6-6) of the graphs of these solution sets.

PROBLEM 13   Graph $\{(x, y)\,|\,1 \leq x \leq 4,\ -2 \leq y \leq 2\}$.

Solution

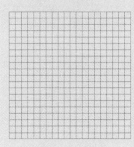

Solutions to Matched   **11.** $2x + 2y \leq 5$
Problems

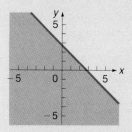

12. (A) $y < 2$ **(B)** $3x > -8$

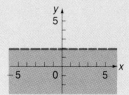   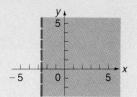

(C) $1 \leq y \leq 5$

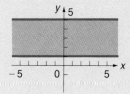

13. $\{(x, y)\,|\,1 \leq x \leq 4,\ -2 \leq y \leq 2\}$

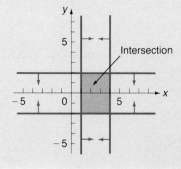

Practice Exercise 7-3 ∎

Work odd-numbered problems first, check answers, and then work even-numbered problems in areas of weakness. Answers to all problems are in the back of the book. Make every effort to work a problem yourself before you look at an answer.

Graph each inequality in a rectangular coordinate system.

A   **1.** $x + y \le 6$ **2.** $x + y \ge 4$

3. $x - y > 3$ **4.** $x - y < 5$

 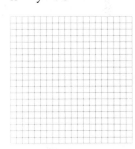

5. $y \ge x - 2$ **6.** $y \le x + 1$

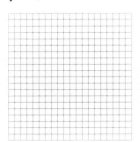 

B   **7.** $2x - 3y < 6$ **8.** $3x + 4y < 12$

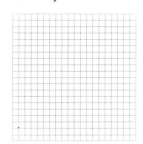 

9. $3y - 2x \geq 24$

10. $3x + 2y \geq 18$

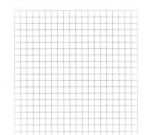

11. $y \geq \dfrac{x}{3} - 2$

12. $y \leq \dfrac{x}{2} - 4$

13. $y \leq \dfrac{2}{3}x + 5$

14. $y > \dfrac{x}{3} + 2$

15. $x \geq -5$

16. $y \leq 8$

17. $y < 0$

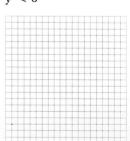

18. $x \geq 0$

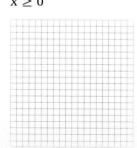

19. $-1 < x \le 3$

20. $-3 \le y < 2$

21. $-2 \le y \le 2$

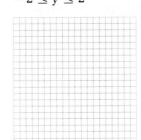

22. $-1 \le x \le 4$

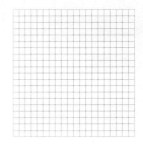

C   *Graph each set.*

23. $\{(x, y) \mid -3 \le x \le 3 \text{ and } -1 \le y \le 5\}$

24. $\{(x, y) \mid -1 \le x \le 5 \text{ and } -2 \le y \le 2\}$

25. $\{(x, y) \mid x \ge 0, \ y \ge 0, \text{ and } 3x + 4y \le 12\}$

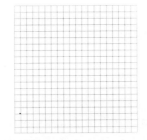

26. $\{(x, y) \mid x \geq 0, y \geq 0, \text{ and } 3x + 2y \leq 18\}$

APPLICATIONS

27. *Business:*   A manufacturer of sailboards makes a standard model and a competition model. The pertinent manufacturing data are summarized in the following table where the total number of fabricating hours for both boards cannot exceed 120 and the total number of finishing hours for both boards cannot exceed 30. If x is the number of standard models and y is the number of competition models produced per week, write a system of inequalities that indicates the restrictions on x and y. Graph this system showing the region of permissible values for x and y.

	STANDARD MODEL (WORKHOURS PER BOARD)	COMPETITION MODEL (WORKHOURS PER BOARD)	MAXIMUM WORKHOURS AVAILABLE PER WEEK
FABRICATING	6	8	120
FINISHING	1	3	30

The Check Exercise for
this section is on
page 467.

Section 7-4   Conic Sections; Circles

■ Introduction
■ Distance-Between-Two-Points Formula
■ Circles

■ **Introduction**

In the first part of this chapter we discussed equations of a straight line—that is, equations such as

$$2x - 3y = 5 \qquad \text{and} \qquad y = -\tfrac{1}{3}x + 2$$

These are first-degree equations in two variables. If we increase the degree of the equations by 1, what kind of graphs will we get? That is, what kind of graphs will second-degree equations such as

$$y = x^2 \qquad x^2 + y^2 = 25 \qquad \frac{x^2}{4} - \frac{y^2}{16} = 1 \qquad x^2 + 4y^2 - 3x + 7y = 4$$

produce? It can be shown that the graphs will be one of the plane curves you would get by intersecting a plane and a general cone—thus the name **conic sections**. Some typical curves are illustrated in Figure 9. The principal conic sections are circles, parabolas, ellipses, and hyperbolas.

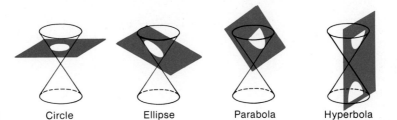

FIGURE 9   Conic sections   Circle   Ellipse   Parabola   Hyperbola

In this book we will consider only a few simple, interesting, and useful second-degree equations in two variables. The subject is treated more thoroughly in a course in analytic geometry.

■ Distance-Between-Two-Points Formula

A basic tool in determining equations of conics is the distance-between-two-points formula. The derivation of the formula is easy, making direct use of the Pythagorean theorem. Let P_1 and P_2 have coordinates as indicated in Figure 10.

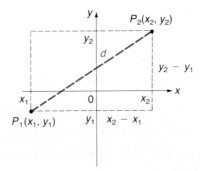

FIGURE 10

Using the Pythagorean theorem, we can write

$$d^2 = |x_2 - x_1|^2 + |y_2 - y_1|^2$$

or, since $|N|^2 = N^2$, the following:

Distance-Between-Two-Points Formula
$d = \sqrt{(x_2 - x_1)^2 + (y_2 - y_1)^2}$

Note that it doesn't make any difference which point you call P_1 or P_2, since $(a - b)^2 = (b - a)^2$.

EXAMPLE 14   Find the distance between $(-3, 6)$ and $(4, -2)$.

Solution   Let $P_1 = (4, -2)$ and $P_2 = (-3, 6)$; then

$$d = \sqrt{(x_2 - x_1)^2 + (y_2 - y_1)^2}$$
$$= \sqrt{[(-3) - 4]^2 + [6 - (-2)]^2}$$
$$= \sqrt{(-7)^2 + (8)^2}$$
$$= \sqrt{113}$$

Or let $P_1 = (-3, 6)$ and $P_2 = (4, -2)$; then

$$d = \sqrt{[4 - (-3)]^2 + [(-2) - 6]^2}$$
$$= \sqrt{(7)^2 + (-8)^2}$$
$$= \sqrt{113}$$

PROBLEM 14   Find the distance between $(4, -2)$ and $(3, 1)$.

Solution

■ **Circles**

We start with the definition of a circle and then use the distance formula to find the standard equation of a circle in a plane.

Definition of a Circle

A **circle** is the set of points equidistant from a fixed point. The fixed distance is called the **radius**, and the fixed point is called the **center**.

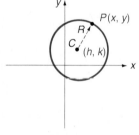

FIGURE 11   Circle

Let a circle have radius R and center at (h, k). Referring to Figure 11, we see that an arbitrary point $P(x, y)$ is on the circle if and only if

$$R = \sqrt{(x - h)^2 + (y - k)^2} \qquad \text{Distance from } C \text{ to } P \text{ is constant.}$$

or, equivalently,

$$(x - h)^2 + (y - k)^2 = R^2$$

Thus, we can state the following:

Equations of a Circle

1. Radius R and center (h, k):

$$(x - h)^2 + (y - k)^2 = R^2$$

2. Radius R and center at the origin:

$$x^2 + y^2 = R^2 \qquad \text{since } (h, k) = (0, 0)$$

EXAMPLE 15   Graph: **(A)** $x^2 + y^2 = 25$ **(B)** $(x - 3)^2 + (y + 2)^2 = 9$

Solution   **(A)** This is a circle with radius 5 and center at the origin:

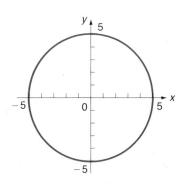

(B) $(x - 3)^2 + (y + 2)^2 = 9$ is the same as $(x - 3)^2 + [y - (-2)]^2 = 3^2$, which is the equation of a circle with radius 3 and center at $(h, k) = (3, -2)$. Thus:

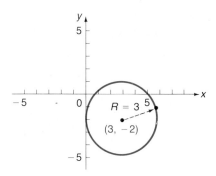

PROBLEM 15   Graph: **(A)** $x^2 + y^2 = 4$ **(B)** $(x + 2)^2 + (y - 3)^2 = 16$

Solution   **(A)** 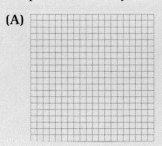   **(B)**

EXAMPLE 16   What is the equation of a circle with radius 8 and center at the origin? Center at $(-5, 4)$?

Solution   If the center is at the origin, then $(h, k) = (0, 0)$; thus, the equation is

$$x^2 + y^2 = 8^2 \quad \text{or} \quad x^2 + y^2 = 64$$

If the center is at $(-5, 4)$, then $h = -5$ and $k = 4$, and the equation is

$$[x - (-5)]^2 + (y - 4)^2 = 8^2 \quad \text{or} \quad (x + 5)^2 + (y - 4)^2 = 64$$

PROBLEM 16   What is the equation of a circle with radius $\sqrt{7}$ and center at the origin? Center at $(6, -4)$?

Solution   Center at $(0, 0)$:

Center at $(6, -4)$:

EXAMPLE 17   Graph $x^2 + y^2 - 6x + 8y + 9 = 0$.

Solution   We transform the equation into the form

$$(x - h)^2 + (y - k)^2 = R^2$$

by completing the square relative to x and relative to y (see Section 6-2).

$$x^2 - 6x + \,? + y^2 + 8y + \,? = -9 + \,? + \,?$$
$$x^2 - 6x + 9 + y^2 + 8y + 16 = -9 + 9 + 16$$
$$(x - 3)^2 + (y + 4)^2 = 16$$

or

$$(x - 3)^2 + [y - (-4)]^2 = 16 \quad \text{Thus, } (h, k) = (3, -4).$$

This is the equation of a circle with radius 4 and center at $(h, k) = (3, -4)$.

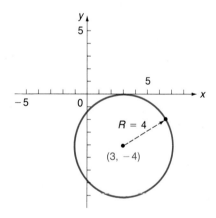

PROBLEM 17   Graph $x^2 + y^2 + 10x - 4y + 20 = 0$.

Solution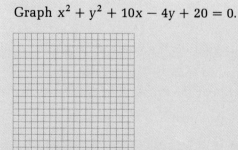

Solutions to Matched
Problems

14. $d = \sqrt{(x_2 - x_1)^2 + (y_2 - y_1)^2} = \sqrt{(3 - 4)^2 + [1 - (-2)]^2} = \sqrt{10}$

15. (A)   **(B)**

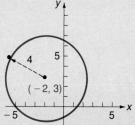 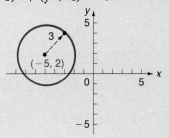

16. Center at (0, 0): $x^2 + y^2 = (\sqrt{7})^2$
$$x^2 + y^2 = 7$$
Center at (6, −4): $(x - 6)^2 + [y - (-4)]^2 = (\sqrt{7})^2$
$$(x - 6)^2 + (y + 4)^2 = 7$$

17. $(x + 5)^2 + (y - 2)^2 = 9$

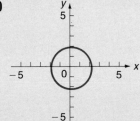

Practice Exercise 7-4 ■

Work odd-numbered problems first, check answers, and then work even-numbered problems in areas of weakness. Answers to all problems are in the back of the book. Make every effort to work a problem yourself before you look at an answer.

A   Find the distance between each pair of points. Leave your answer in exact radical form.

1.   (2, 4), (4, 5) _____   **2.**   (7, 3), (8, 6) _____

3.   (3, −7), (−2, 1) _____   **4.**   (−5, 3), (−1, −2) _____

5.   (−8, −2), (−5, 1) _____   **6.**   (2, −7), (−4, −3) _____

Graph each equation in a cartesian coordinate system.

7. $x^2 + y^2 = 16$ **8.** $x^2 + y^2 = 36$

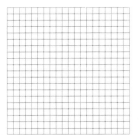   

9. $x^2 + y^2 = 6$

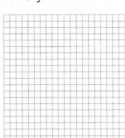

10. $x^2 + y^2 = 10$

Write the equation of a circle with center at the origin and radius as given.

11. 7 _____

12. 8 _____

13. $\sqrt{5}$ _____

14. $\sqrt{10}$ _____

B **15.** Is the triangle with vertices $(-1, 2)$, $(2, -1)$, $(3, 3)$ an isosceles triangle?

_____

16. Is the triangle in the preceding problem an equilateral triangle? _____

Graph each equation in a cartesian coordinate system.

17. $(x - 3)^2 + (y - 4)^2 = 16$

18. $(x - 4)^2 + (y - 2)^2 = 9$

19. $(x - 4)^2 + (y + 3)^2 = 9$

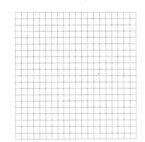

20. $(x + 4)^2 + (y - 2)^2 = 25$

21. $(x + 3)^2 + (y + 3)^2 = 16$

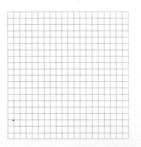

22. $(x + 2)^2 + (y + 4)^2 = 25$

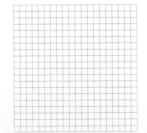

Write the equation of a circle in the form

$$(x - h)^2 + (y - k)^2 = R^2$$

with radius and center as given.

23. 7, (3, 5) _____

24. 2, (4, 1) _____

25. 8, (−3, 3) _____

26. 6, (5, −2) _____

27. $\sqrt{3}$, (−4, −1) _____

28. $\sqrt{14}$, (−7, −5) _____

C   *Use the method of completing the square to graph each of the following circles. Indicate the center and radius of each.*

29. $x^2 + y^2 - 4x - 6y + 4 = 0$

30. $x^2 + y^2 - 6x - 4y + 4 = 0$

31. $x^2 + y^2 - 6x + 6y + 2 = 0$

32. $x^2 + y^2 + 6x - 4y - 3 = 0$

33. $x^2 + y^2 + 6x + 4y + 4 = 0$

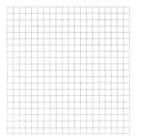

34. $x^2 + y^2 + 4x + 4y - 8 = 0$

35.   Find x so that (x, 8) is 13 units from (2, −4). _____

36.   Find an equation of the set of points equidistant from (3, 3) and (6, 0).

37. Find the equation of the circle centered at the origin and passing through the point $(4, -3)$. _____

38. Find the equation of the circle centered at $(4, -3)$ and passing through the origin. _____

APPLICATIONS **39.** An ancient stone bridge in the form of a circular arc has a span of 80 feet (see the figure). If the height of the arch above its ends is 20 feet, find an equation of the circle containing the arch if its center is at the origin as indicated. [*Hint:* $(40, R - 20)$ must satisfy $x^2 + y^2 = R^2$.] _____

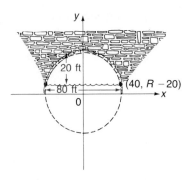

40. A cylindrical oil drum is cut to make a watering trough (see the figure). The end of the resulting trough is 20 centimeters high and 80 centimeters wide at the top. What was the radius R of the original drum? _____

The Check Exercise for this section is on page 469.

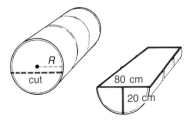

Section 7-5   Parabolas, Ellipses, and Hyperbolas

- Parabolas
- Ellipses
- Hyperbolas

This section provides a very brief glimpse into a subject that is treated in detail in a course in analytic geometry. Many results are simply stated without development or motivation. Nevertheless, a brief exposure to these important curves at this time should help to increase your understanding of a more detailed development in a future course. In addition, you will have gained some concrete experience with graphs other than straight lines and circles.

Parabolas

We start with a definition of a parabola:

Definition of a Parabola

A **parabola** is the set of all points equidistant from a fixed point and a fixed line. The fixed point is called the **focus**, and the fixed line is the **directrix**.

We will begin by finding an equation of a parabola with focus $(a, 0)$, $a > 0$, and directrix $x = -a$ (see Figure 12).

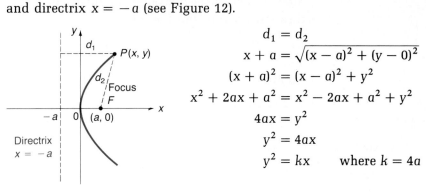

$$d_1 = d_2$$
$$x + a = \sqrt{(x - a)^2 + (y - 0)^2}$$
$$(x + a)^2 = (x - a)^2 + y^2$$
$$x^2 + 2ax + a^2 = x^2 - 2ax + a^2 + y^2$$
$$4ax = y^2$$
$$y^2 = 4ax$$
$$y^2 = kx \qquad \text{where } k = 4a$$

FIGURE 12

Proceeding in the same way, we can obtain similar equations for parabolas opening upward, to the left, and downward. We summarize four cases as follows:

Equations of a Parabola (Standard Forms)

$y^2 = kx \qquad k > 0$	$y^2 = kx \qquad k < 0$
Opens right	Opens left

$x^2 = ky \qquad k > 0$	$x^2 = ky \qquad k < 0$
Opens up	Opens down

We now show how examples of these four cases can be sketched rather quickly.

Rapid Sketching of Parabolas

To graph $y^2 = kx$ and $x^2 = ky$:

Step 1:   The origin $(0, 0)$ is always part of the graph.

Step 2:   Two other points can be found easily by assigning x in $y^2 = kx$ (or y in $x^2 = ky$) a value that will make the right side a positive perfect square.

Step 3:   Locate the two points found in step 2; then sketch a parabola through these two points and the origin.

Note: If more accuracy is desired, assign x in $y^2 = kx$ (or y in $x^2 = ky$) additional values (making the right side positive) and use a calculator or a square root table.

EXAMPLE 18   Graph $y^2 = -8x$.

Solution   *Step 1:*   The origin $(0, 0)$ is part of the graph.

Step 2:   What can we assign x to make the right side a positive perfect square? We let $x = -2$; then

$$y^2 = -8(-2)$$
$$y^2 = 16$$
$$y = \pm\sqrt{16} = \pm 4$$

Thus, the points $(-2, -4)$ and $(-2, 4)$ are also on the graph.

Step 3:   We plot $(0, 0)$, $(-2, -4)$, and $(-2, 4)$; then we sketch a parabola through these three points:

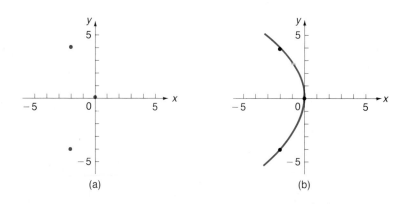

(a)   (b)

Note: With a little practice most of the computation can be done mentally and the graph sketched with little effort. A very common error is to draw the graph opening the wrong way. If you find the three points as directed, you will generally avoid this type of error.

PROBLEM 18

Solution

Graph $x^2 = -3y$.

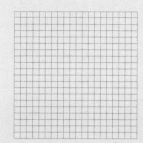

Parabolas are encountered frequently in the physical world. Suspension bridges, arch bridges, reflecting telescopes, radiotelescopes, radar equipment, solar furnaces, and searchlights all utilize parabolic forms in their design.

■ Ellipses

You are no doubt aware of many uses or occurrences of elliptical forms; orbits of satellites, orbits of planets and comets, gears and cams, and domes in buildings are but a few examples. Formally, we define an ellipse as follows:

Definition of an Ellipse
An **ellipse** is the set of all points such that the sum of the distances of each to two fixed points is constant. The fixed points are called **foci**, and each separately is a **focus**.

An ellipse is easy to draw. Place two pins in a piece of cardboard at the foci and tie a piece of loose string (representing the constant sum) to the pins; then move a pencil within the string, keeping it taut.

With regard to an equation of an ellipse, we will limit ourselves to the cases in which the foci are symmetrically located on either coordinate axis. Thus, if $(-c, 0)$ and $(c, 0)$, $c > 0$, are the foci, and $2a$ is the constant sum of the distances [note from Figure 13 that $2a > 2c$ (why?), hence $a > c$], then

$$d_1 + d_2 = 2a$$

$$\sqrt{(x + c)^2 + y^2} + \sqrt{(x - c)^2 + y^2} = 2a$$

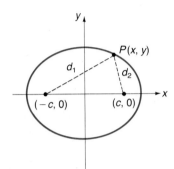

FIGURE 13

After eliminating radicals and simplifying—a good exercise for the reader—we eventually obtain

$$(a^2 - c^2)(x^2) + a^2 y^2 = a^2(a^2 - c^2)$$

or

$$\frac{x^2}{a^2} + \frac{y^2}{a^2 - c^2} = 1$$

Since $a > c$, then $a^2 - c^2 > 0$. To simplify the equation further, we choose to let $b^2 = a^2 - c^2$, $b > 0$. Thus,

$$\frac{x^2}{a^2} + \frac{y^2}{b^2} = 1$$

Proceeding similarly with the foci on the vertical axis, we arrive at

$$\frac{x^2}{b^2} + \frac{y^2}{a^2} = 1$$

Combining these results, we can write

$$\frac{x^2}{m^2} + \frac{y^2}{n^2} = 1$$

as a standard form for an equation of an ellipse located as described above. (We shift over to m and n to simplify our approach and because a and b have special significance in a more advanced treatment of the subject.) It can be shown that if $m > n > 0$, then the foci are on the x axis; and if $n > m > 0$, then the foci are on the y axis. The two cases are summarized as follows:

Equations of Ellipses (Standard Forms)

$$\frac{x^2}{m^2} + \frac{y^2}{n^2} = 1$$

Case 1: $m > n > 0$ 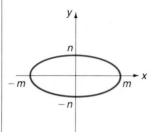   *Case 2:* $n > m > 0$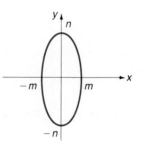

We now show how particular examples of these two cases can be sketched rather quickly.

Rapid Sketching of Ellipses

To graph $\dfrac{x^2}{m^2} + \dfrac{y^2}{n^2} = 1$:

Step 1: Find the x intercepts by letting $y = 0$ and solving for x.

Step 2: Find the y intercepts by letting $x = 0$ and solving for y.

Step 3: Sketch an ellipse passing through these intercepts.

EXAMPLE 19   Graph $\dfrac{x^2}{16} + \dfrac{y^2}{9} = 1$.

Solution   *Step 1:*   Find x intercepts:

$$\frac{x^2}{16} + \frac{0}{9} = 1$$
$$x^2 = 16$$
$$x = \pm\sqrt{16} = \pm 4$$

Step 2:   Find y intercepts:

$$\frac{0}{16} + \frac{y^2}{9} = 1$$
$$y^2 = 9$$
$$y = \pm\sqrt{9} = \pm 3$$

Step 3:   Plot the intercepts and draw in the ellipse.

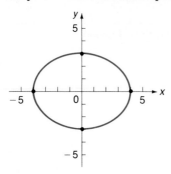

PROBLEM 19   Graph $\dfrac{x^2}{9} + \dfrac{y^2}{16} = 1$.

Solution

Hyperbolas

Definition of a Hyperbola

A **hyperbola** is the set of all points such that the absolute value of the difference of the distances of each to two fixed points is constant. The two fixed points are called **foci**.

As with the ellipse, we will limit our investigation to cases in which the foci are symmetrically located on either coordinate axis. Thus, if $(-c, 0)$ and $(c, 0)$ are the foci and $2a$ is the constant difference (Figure 14), then

$$|d_1 - d_2| = 2a$$

$$\left|\sqrt{(x + c)^2 + y^2} - \sqrt{(x - c)^2 + y^2}\right| = 2a$$

After eliminating radicals and absolute-value signs (by appropriate use of squaring) and simplifying—another good exercise for the reader—we eventually obtain

$$\frac{x^2}{a^2} + \frac{y^2}{a^2 - c^2} = 1$$

which looks exactly like the equation we obtained for the ellipse. However, from Figure 14 we see that $2a < 2c$; hence, $a^2 - c^2 < 0$. To simplify the equation

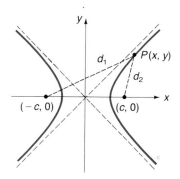

FIGURE 14

further, we let $-b^2 = a^2 - c^2$. Thus,

$$\frac{x^2}{a^2} - \frac{y^2}{b^2} = 1$$

Proceeding similarly with the foci on the vertical axis, we obtain

$$\frac{y^2}{a^2} - \frac{x^2}{b^2} = 1$$

Combining these results, we can write

$$\frac{x^2}{m^2} - \frac{y^2}{n^2} = 1 \qquad \frac{y^2}{n^2} - \frac{x^2}{m^2} = 1$$

as standard forms for equations of hyperbolas located as described above. (Again, we shift over to m and n to simplify our approach and because a and b have special significance in a more advanced treatment of the subject.) In summary:

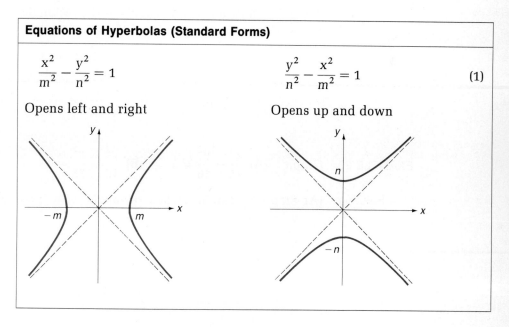

Equations of Hyperbolas (Standard Forms)

$$\frac{x^2}{m^2} - \frac{y^2}{n^2} = 1 \qquad\qquad \frac{y^2}{n^2} - \frac{x^2}{m^2} = 1 \qquad (1)$$

Opens left and right   Opens up and down

As an aid to graphing Equations (1), we solve each equation for y in terms of x. From the first equation, we obtain

$$y = \pm\frac{nx}{m}\sqrt{1 - \frac{m^2}{x^2}}$$

and from the second we obtain

$$y = \pm \frac{nx}{m}\sqrt{1 + \frac{m^2}{x^2}}$$

As x gets large, the radicals approach 1; hence, the equations (for large x) behave very much as if:

Asymptotes

$$y = \pm \frac{n}{m}x \qquad\qquad (2)$$

The two straight lines in Equation (2) are guidelines called **asymptotes**. The graphs of Equations (1) will approach these guidelines, but never touch them, as the graph moves farther and farther away from the origin. Quick sketches of hyperbolas can be made rather easily by following these steps:

Rapid Sketching of Hyperbolas

To graph $\dfrac{x^2}{m^2} - \dfrac{y^2}{n^2} = 1$ and $\dfrac{y^2}{n^2} - \dfrac{x^2}{m^2} = 1$:

Step 1: Draw a dashed rectangle with intercepts $x = \pm m$ and $y = \pm n$.

Step 2: Draw dashed diagonals of the rectangle and extend to form asymptotes. (These are the graphs of Equation 2.)

Step 3: Determine the true intercepts of the hyperbola (be particularly careful in this step); then sketch in the hyperbola (both branches).

EXAMPLE 20   Graph: **(A)** $\dfrac{x^2}{25} - \dfrac{y^2}{16} = 1$ **(B)** $\dfrac{y^2}{16} - \dfrac{x^2}{25} = 1$

Solution   **(A)** *Step 1:* Draw a dashed rectangle with intercepts $x = \pm\sqrt{25} = \pm 5$ and $y = \pm\sqrt{16} = \pm 4$.

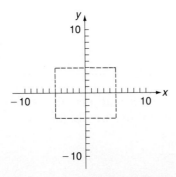

Step 2:   Draw in asymptotes (extended diagonals of the rectangle).

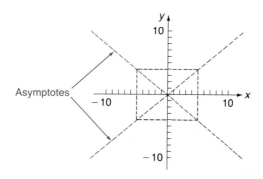

Step 3:   Determine the true intercepts for the hyperbola; then sketch it in. Let y = 0; then

$$\frac{x^2}{25} - \frac{0}{16} = 1$$

$$x^2 = 25$$

$$x = \pm\sqrt{25} = \pm 5$$

If we let x = 0, then

$$\frac{0}{25} - \frac{y^2}{16} = 1$$

$$y^2 = -16$$

$$y = \pm\sqrt{-16} = \pm 4i$$

These are complex numbers and do not represent real intercepts. We conclude that the only real intercepts are x = ±5, and the hyperbola opens left and right.

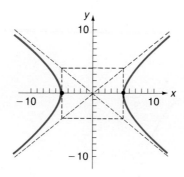

(B)

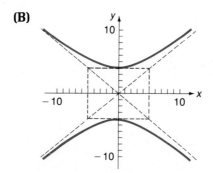

PROBLEM 20   Graph: (A) $\dfrac{x^2}{4} - \dfrac{y^2}{9} = 1$ (B) $\dfrac{y^2}{9} - \dfrac{x^2}{4} = 1$

Solution   **(A)**   **(B)**

Hyperbolic forms are encountered in the study of comets, the loran system of navigation for ships and aircraft, some modern architectural structures, and optics, to name but a few examples among many.

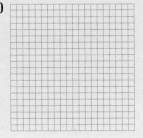

This roof structure is called a hyperbolic paraboloid

Solutions to Matched Problems

18.

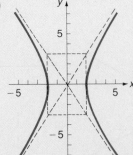

19.

20. (A)

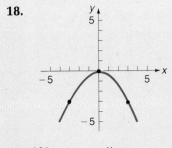

(B)

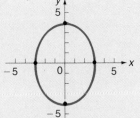
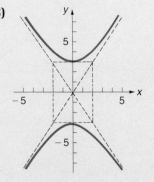
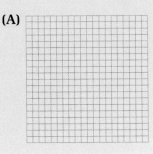

Practice Exercise 7-5 ∎

Work odd-numbered problems first, check answers, and then work even-numbered problems in areas of weakness. Answers to all problems are in the back of the book. Make every effort to work a problem yourself before you look at an answer.

A   *Graph each of the following parabolas.*

1. $y^2 = 4x$

2. $y^2 = x$

3. $y^2 = -12x$

4. $y^2 = -16x$

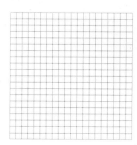

5. $x^2 = y$

6. $x^2 = 12y$

7. $x^2 = -16y$

8. $x^2 = -4y$

Graph each of the following ellipses.

9. $\dfrac{x^2}{25} + \dfrac{y^2}{4} = 1$

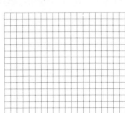

10. $\dfrac{x^2}{9} + \dfrac{y^2}{4} = 1$

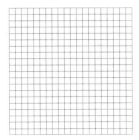

11. $\dfrac{x^2}{4} + \dfrac{y^2}{25} = 1$

12. $\dfrac{x^2}{4} + \dfrac{y^2}{9} = 1$

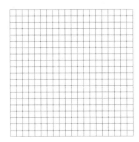

B   *Graph each of the following hyperbolas.*

13. $\dfrac{x^2}{4} - \dfrac{y^2}{25} = 1$

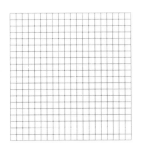

14. $\dfrac{x^2}{16} - \dfrac{y^2}{9} = 1$

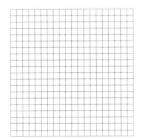

15. $\dfrac{y^2}{25} - \dfrac{x^2}{4} = 1$

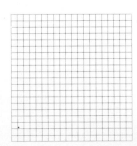

16. $\dfrac{y^2}{9} - \dfrac{x^2}{16} = 1$

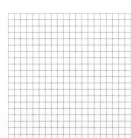

Graph each of the following equations after first writing the equation in one of the standard forms discussed in this section. For example,

$$9x^2 + 4y^2 = 36$$

can be written in the form

$$\frac{x^2}{4} + \frac{y^2}{9} = 1$$

by dividing through by 36.

17. $4y^2 - 8x = 0$

18. $3x^2 + 9y = 0$

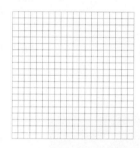

19. $4x^2 + 9y^2 = 36$

20. $4x^2 + 25y^2 = 100$

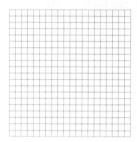

21. $4x^2 - 9y^2 = 36$

22. $25y^2 - 4x^2 = 100$

C **23.** Find the coordinates of the focus and the equation of the directrix for the parabola in Problem 1. _____

24. Find the coordinates of the focus and the equation of the directrix for the parabola in Problem 3. _____

25. Find the coordinates of the foci for the ellipse in Problem 11. _____

26. Find the coordinates of the foci for the ellipse in Problem 10. _____

27. Find the coordinates of the foci for the hyperbola in Problem 14.

_____

28. Find the coordinates of the foci for the hyperbola in Problem 15.

_____

29. Use the definition of a parabola and the distance formula to find the equation of a parabola with directrix $y = 4$ and focus $(2, 2)$. _____

30. Use the definition of a parabola and the distance formula to find the equation of a parabola with directrix $x = 2$ and focus at $(6, 4)$. _____

APPLICATIONS

The Check Exercise for
this section is on
page 471.

31. A parabolic concrete bridge is to span 100 meters. If the arch rises 25 meters above its ends, find the equation of the parabola, assuming it passes through the origin of a coordinate system and has its focus on the y axis. [*Hint:* $(50, -25)$ must satisfy $x^2 = -4ay$.] _____

32. A radar bowl, in the form of a rotated parabola, is 20 meters in diameter and 5 meters deep. Find the equation of the parabola, assuming it passes through the origin of a coordinate system and has its focus on the positive y axis. (See Problem 31.) _____

Section 7-6   Chapter Review

A **cartesian coordinate system** is formed with two real number lines—one as a **horizontal axis**, the other as a **vertical axis**—intersecting at their origins. These axes divide the plane into four **quadrants**. Every point in the plane corresponds to its **coordinates**—a pair (a, b) where the **abscissa** a is the coordinate of the point projected to the horizontal axis and the **ordinate** b is the coordinate of the point projected to the vertical axis. The point $(0, 0)$ is the **origin**.   *(7-1)*

The **graph of an equation** in two variables x and y is the set of all points whose coordinates (x, y) are solutions of the equation. The graph of a first-degree equation in x and y is a line. The point or value where a graph crosses the y axis is called the **y intercept**; where it crosses the x axis, it is called the **x intercept**.   *(7-1)*

The **slope** of a nonvertical line is given by

$$\frac{y_1 - y_2}{x_1 - x_2}$$

where (x_1, y_1) and (x_2, y_2) are any two points on the line. The equation $y = mx + b$ represents a line in **slope–intercept form**; m is the slope of the line and

b is the y intercept. The equation $y - y_1 = m(x - x_1)$ represents a line in **point–slope form**; m is the slope and (x_1, y_1) is any point on the line. A **vertical line** has equation of the form $x = c$ and has no slope. A **horizontal line** has equation of the form $y = c$ and slope 0. Two nonvertical lines are parallel when their slopes are equal, perpendicular when the product of their slopes is -1.   *(7-2)*

A nonvertical line divides the plane into two **half-planes**: the **upper half-plane** above the line and the **lower half-plane** below. The graph of a linear inequality in two variables is a half-plane, possibly including the boundary line.   *(7-3)*

The distance between two points (x_1, y_1) and (x_2, y_2) in the plane is given by

$$d = \sqrt{(x_2 - x_1)^2 + (y_2 - y_1)^2}$$

A **circle** is the set of all points a fixed distance (**radius**) from a fixed point (**center**). The equation of a circle of radius R with center (h, k) is $(x - h)^2 + (y - k)^2 = R^2$; if the center is the origin, this becomes $x^2 + y^2 = R^2$.   *(7-4)*

A **parabola** is the set of all points equidistant from a fixed point (**focus**) and a fixed line (**directrix**). A parabola in standard form has equation

$$y^2 = kx \qquad \text{Opens horizontally}$$

or

$$x^2 = ky \qquad \text{Opens vertically}$$

An **ellipse** is the set of all points such that the sum of the distances of each to two fixed points (**foci**; each separately is a **focus**) is constant. An ellipse in standard form has equation

$$\frac{x^2}{m^2} + \frac{y^2}{n^2} = 1$$

A **hyperbola** is the set of all points such that the absolute value of the difference of the distances of each to two fixed points (**foci**) is constant. A hyperbola in standard form has equation

$$\frac{x^2}{m^2} - \frac{y^2}{n^2} = 1 \qquad \text{Opens horizontally}$$

or

$$\frac{y^2}{n^2} - \frac{x^2}{m^2} = 1 \qquad \text{Opens vertically}$$

The lines

$$y = \pm\frac{n}{m}x$$

are **asymptotes** for the hyperbola. Circles, parabolas, ellipses, and hyperbolas are the principal **conic sections**—geometric figures obtained by intersecting a plane and a cone.   *(7-5)*

Diagnostic (Review) Exercise 7-6 ■

Work through all the problems in this chapter review and check answers in the back of the book. (Answers to all problems are there, and following each answer is a number in italics indicating the section in which that type of problem is discussed.) Where weaknesses show up, review appropriate sections in the text. When you are satisfied that you know the material, take the practice test following this review.

A   *Graph each in a rectangular coordinate system.*

1. $y = 2x - 3$

2. $2x + y = 6$

3. $x - y \geq 6$

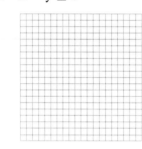

4. $y > x - 1$

5. $x^2 + y^2 = 36$

6.   What is the slope and y intercept for the graph of $y = -2x - 3$?

_____

7.   Write an equation of a line that passes through (2, 4) with slope -2. Write the answer in the form $Ax + By = C, A > 0$. _____

8. What is the slope of the line that passes through (1, 3) and (3, 7)?

_____

9. What is an equation of a line that passes through (1, 3) and (3, 7)? Write the answer in the form $y = mx + b$. _____

10. What is the distance between the two points (1, 3) and (3, 7)?

_____

11. What is an equation of a circle with radius 5 and center at the origin?

_____

B *Graph each in a rectangular coordinate system.*

12. $3x - 2y = 9$

13. $y = \frac{1}{3}x - 2$

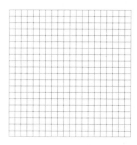

14. $x = -3$

15. $4x - 5y \le 20$

16. $y < \dfrac{x}{2} + 1$

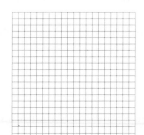

17. $x \ge -3$

18. $-4 \le y < 3$

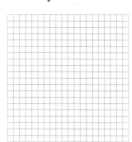

19. $x^2 + y^2 = 49$

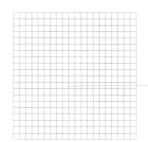

20. $(x - 2)^2 + (y + 3)^2 = 16$

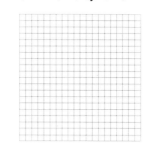

21. $y^2 = -2x$

22. $\dfrac{x^2}{9} + \dfrac{y^2}{16} = 1$

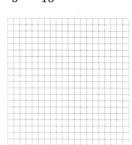

23. $\dfrac{y^2}{16} - \dfrac{x^2}{9} = 1$

24. What is the slope and y intercept for the graph of $x + 2y = -6$?

_____

25. Write an equation of a line that passes through $(-3, 2)$ with slope $-\frac{1}{3}$. Write the final answer in the form $y = mx + b$. _____

26. Write the equation of a line that passes through $(-3, 2)$ and $(3, -2)$. Write the final answer in the form $Ax + By = C$, $A > 0$. _____

27. Find the equation of a line that passes through $(3, -4)$ and is perpendicular to $x + 2y = -6$. Write the final answer in the form $y = mx + b$.

_____

28. Write the equations of the vertical and horizontal lines that pass through $(5, -2)$. _____

29. Write an equation of a circle in the form $(x - h)^2 + (y - k)^2 = R^2$ if its center is at $(-3, 4)$ and it has a radius of 7. _____

C **30.** Find an equation of a circle with center at the origin that passes through $(12, -5)$. _____

31. Write the equation of a line that passes through $(-6, 2)$ and is parallel to $3x - 2y = 5$. Write your final answer in the form $y = mx + b$.

_____

32. Transform the equation

$$x^2 + y^2 + 6x - 8y = 0$$

into the form

$$(x - h)^2 + (y - k)^2 = R^2$$

Since the graph is a circle, what is its radius and what are the coordinates of its center? _____

33. Graph the set $\{(x, y)\,|\,y \geq 0,\ 1 \leq x \leq 5,\ 2x + 3y \leq 18\}$.

Practice Test Chapter 7 ■

Take this as if it were a graded test by working the problems within a 50-minute time period. Do not look back in the chapter. Choose one of three levels of difficulty: least difficult, Problems 1–12; more difficult, add Problem 13; most difficult, add Problems 13 and 14. Use the answers in the back of the book to correct your work. The answers are keyed to appropriate text sections so that you can easily locate and review sections where difficulties still persist.

Graph Problems 1–6 in a rectangular coordinate system.

1. $4x - 2y = 10$

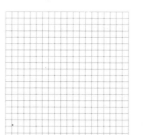

2. $4x - 3y \geq 12$

3. $(x + 3)^2 + (y - 2)^2 = 9$

4. $x^2 = -12y$

5. $\dfrac{x^2}{9} - \dfrac{y^2}{16} = 1$

6. $y = -\dfrac{1}{2}x + 3$

7. Write an equation of a line with y intercept 5 and slope -3. Write the final answer in the form $y = mx + b$. _____

8. What is the equation of the line that passes through $(-2, -3)$ and $(0, -2)$? Write the final answer in the form $y = mx + b$. _____

9. Write the equation of the horizontal line that passes through $(-3, -2)$.

_____

10. Write an equation of a line that passes through $(-3, 4)$ and is perpendicular to $3x - 2y = 4$. Write the final equation in the form $y = mx + b$.

_____

11. Transform $x^2 + y^2 - 8x + 6y + 9 = 0$ into the form $(x - h)^2 + (y - k)^2 = R^2$ and identify the radius and coordinates of the center of the circle.

_____

12. Find the equation of the circle with center at the origin and passing through $(-2, 8)$. _____

13.   Graph $\dfrac{x^2}{9} + \dfrac{y^2}{16} = 1$.

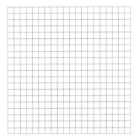

14.   Graph $\{(x, y) \,|\, y \geq 0, 1 \leq x \leq 6, y + x \leq 9\}$.

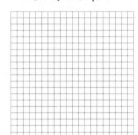

Check Exercise 7-1 ■

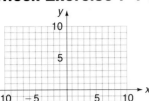

1. _____

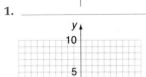

2. _____

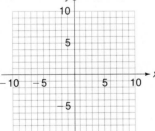

3. _____

4. _____

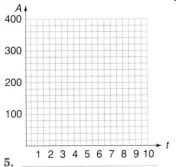

5. _____

Work the following problems without looking at any text examples. Show your work in the space provided. Draw the final graph in the answer column.

Graph.

1. $y = \frac{3}{4}x$

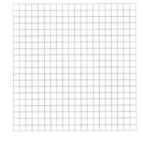

2. $6x - 8y = 24$

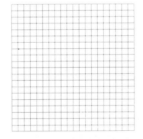

3. $y = -\frac{2}{3}x + 2$

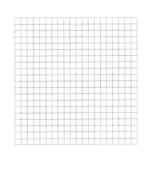

4. $3x + 4y = 10$

5. $A = 100 + 25t, 0 \leq t \leq 6$

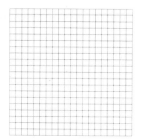

Check Exercise 7-2 ■

ANSWER COLUMN

Work the following problems without looking at any text examples. Show your work in the space provided. Write your answer in the answer column.

1. _____

2. _____

3. _____

4. _____

5. _____

1. What are the slope and y intercept of the graph for $3x - 4y = 8$?

2. Write the equation of a line with slope $-\frac{3}{4}$ that passes through $(4, -1)$. Transform the equation into the form $y = mx + b$.

3. Write the equations of the horizontal and vertical lines that pass through $(-5, 3)$.

4. Write the equation of a line that passes through $(-4, 3)$ and $(5, -3)$. Transform the equation into the form $y = mx + b$.

5. Write the equation of the line that is perpendicular to $3x + 4y = 9$ and passes through $(6, -5)$. Transform the equation into the form $y = mx + b$.

Check Exercise 7-3 ▪

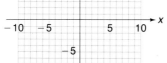

Work the following problems without looking at any text examples. Show your work in the space provided. Draw the final graph in the answer column.

Graph.

1. $y \geq \frac{1}{2}x - 2$

1. _____

2. $4x - 3y > 12$

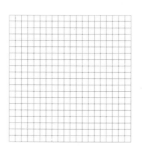

2. _____

3. _____

3. $5x + 6y \leq 30$

4. _____

5. _____

4. $-2 \leq y < 4$

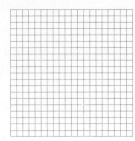

5. Graph the set $\{(x, y) \mid x \geq 0, \ y \geq 0, \ 6x + 8y \leq 48\}$.

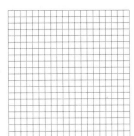

Check Exercise 7-4 ■

Work the following problems without looking at any text examples. Show your work in the space provided. Write your answer or draw the final graph in the answer column.

1. _____

2. _____

3. _____

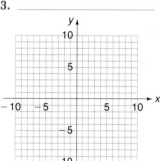

4. _____

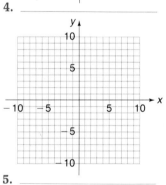

5. _____

1. Find the distance between $(-2, 5)$ and $(3, 2)$. Leave answer in exact radical form.

2. Write the equation of a circle with radius $\sqrt{3}$ and center at the origin.

3. Write the equation of a circle with radius 7 and center at $(-2, 3)$.

4. Graph $(x - 3)^2 + (y + 2)^2 = 9$.

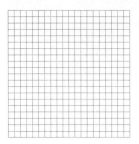

5. Use the method of completing the square as an aid to graphing $x^2 + y^2 + 6x - 4y + 9 = 0$.

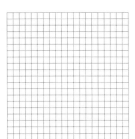

Check Exercise 7-5 ■

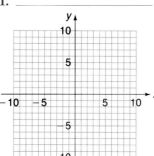

1. _____

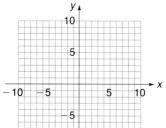

2. _____

Work the following problems without looking at any text examples. Show your work in the space provided. Draw the final graph in the answer column.

1. Graph $y^2 = 5x$.

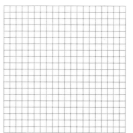

2. Graph $x^2 = -16y$.

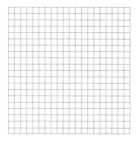

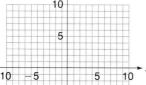

3. _____

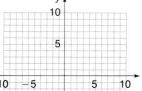

4. _____

3. Graph $\dfrac{x^2}{36} + \dfrac{y^2}{9} = 1$.

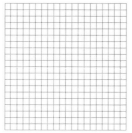

5. _____

4. Graph $\dfrac{y^2}{9} - \dfrac{x^2}{25} = 1.$

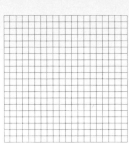

5. Graph $25x^2 + 4y^2 = 100.$

Relations and Functions

■ 8

**INSTRUCTIONS FOR STUDENTS IN A
SELF-PACED CLASS OR LAB**

(YES) ── **HAVE YOU HAD INTERMEDIATE ALGEBRA BEFORE THIS COURSE?** ── (NO)

1. Work Diagnostic (Review) Exercise 8-6 on page 529. Check answers in back of book; then work through text sections corresponding to problems missed. (Section numbers are in italics following each answer.)
2. When finished with step 1, take Practice Test Chapter 8 on page 534 as a final check of your understanding of the chapter. Check answers in the back of the book; then review sections where weakness still prevails. (Corresponding section numbers are in italics following each answer.)
3. When you think you are ready, ask your instructor for a graded test for Chapter 8.
4. If your instructor approves, after the test is corrected, go to the next chapter.

---

1. Work through each section in the chapter as follows:
   (A) Read discussion.
   (B) Read each example and work the corresponding matched problem. Check your solutions to the matched problem in Solutions to Matched Problems on the indicated page.
   (C) At the end of a section work the odd-numbered problems in the Practice Exercise and check answers; then work even-numbered problems in areas of weakness. (Answers to *all* Practice Exercise sets are in the back of the book.)
   (D) Work Check Exercise as instructed. Tear out and turn in as directed by your instructor. (Answers are not in the text.)
2. Repeat each step in item 1 for each section in the chapter.
3. After the instructional part of the chapter is completed, proceed with steps 1 to 4 in the box above this one.

Chapter 8 ■ Relations and Functions

Section 8-1 Relations and Functions

■ Introduction
■ Relations and Functions
■ Common Ways of Specifying Relations and Functions
■ Relations Specified by Equations
■ A Brief History of Function

■ Introduction

Relations among various sets of objects abound in our daily activities. For example:

To each person there corresponds an age.

To each item in a drugstore there corresponds a price.

To each automobile there corresponds a license number.

To each circle there corresponds an area.

To each number there corresponds its cube.

To each nonzero real number there corresponds two square roots.

One of the most important aspects of science is establishing relations among various types of phenomena. Once a relation is known, predictions can be made. A chemist can use a gas law to predict the pressure of an enclosed gas given its temperature; an engineer can use a formula to predict the deflections of a beam subject to different loads; an economist would like to be able to predict interest rates given the rate of change of the money supply; and so on.

Establishing and working with relations are so fundamental to both pure and applied science that people have found it necessary to describe them in the precise language of mathematics. Special relations called "functions" represent one of the most important concepts in all of mathematics. Your efforts to understand and use this concept correctly right from the beginning will be rewarded many times.

■ Relations and Functions

What do all the examples of relations given above have in common? Each deals with the matching of elements from a first set, called the **domain** of the relation, with elements in a second set, called the **range** of the relation.

Consider Table 1 showing three relations involving the cube, square, and square root. (The choice of small domains enables us to introduce two important concepts in a relatively simple setting. Shortly, we will consider relations with

TABLE 1

RELATION 1		RELATION 2		RELATION 3	
Domain (Number)	Range (Cube)	Domain (Number)	Range (Square)	Domain (Number)	Range (Square Root)

RELATION 1

0 ⟶ 0
1 ⟶ 1
2 ⟶ 8

RELATION 2

−2, −1, 0, 1, 2 ⟶ 4, 1, 0

RELATION 3

0 ⟶ 0
1 ⟶ 1, −1
4 ⟶ 2, −2
9 ⟶ 3, −3

infinite domains.) The first two relations are examples of functions. The third is not a function. These two very important terms, "relation" and "function," are defined in the box below.

Definition of a Relation and of a Function: Rule Form

A **relation** is a rule (process or method) that produces a correspondence between a first set of elements called the **domain** and a second set of elements called the **range** such that to each element in the domain there corresponds *one or more* elements in the range.

A **function** is a relation with the added restriction that to each domain element there corresponds *one and only one* range element.

(All functions are relations, but some relations are not functions.)

In the cube, square, and square root examples, we see that all three are relations according to the definition.[†] Relations 1 and 2 are also functions, since to each domain value there corresponds exactly one range value. (For example, the square of −2 is 4 and no other number.) On the other hand, relation 3 is not a function, since to at least one domain value there corresponds more than one range value. (For example, to the domain value 9 there corresponds −3 and 3, both square roots of 9.)

EXAMPLE 1   Out of the following three relations, two are functions:

(A) DOMAIN   RANGE

1 ⟶ 5
2 ⟶ 7
3 ⟶ 9

(B) DOMAIN   RANGE

−2 ⟶ −1
0 ⟶ 0
2 ⟶ 0
4 ⟶ 1

(C) DOMAIN   RANGE

3 ⟶ 1
3 ⟶ 3
7 ⟶ 8
9 ⟶ 9

Function
(Exactly one range value corresponds to each domain value.)

Function
(Exactly one range value corresponds to each domain value.)

Not a function
(Two range values correspond to the domain value 3.)

[†] We have used the word "relation" earlier as a word from our ordinary language. After the formal definition, the word "relation" becomes part of our technical mathematical vocabulary. From now on when we use this word in a mathematical context, it will have the meaning as specified.

PROBLEM 1   Indicate which relations are functions:

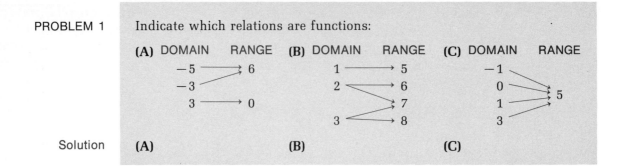

Solution   **(A)**   **(B)**   **(C)**

Since in a relation (or function) elements in the range are paired with elements in the domain by some rule or process, this correspondence (pairing) can be illustrated using ordered pairs of elements where the first component represents a domain element and the second component a corresponding range element. Thus, we can write Relations 1–3 in Table 1 as

Relation 1 = $\{(0, 0), (1, 1), (2, 8)\}$

Relation 2 = $\{(-2, 4), (-1, 1), (0, 0), (1, 1), (2, 4)\}$

Relation 3 = $\{(0, 0), (1, 1), (1, -1), (4, 2), (4, -2), (9, 3), (9, -3)\}$

This suggests an alternative but equivalent way of defining relations and functions that provides additional insight into these concepts:

Definition of a Relation and of a Function: Set Form

A **relation** is any set of ordered pairs of elements.

A **function** is a relation with the added restriction that no two distinct ordered pairs can have the same first component.

The set of first components in a relation (or function) is called the **domain** of the relation, and the set of second components is called the **range**.

According to this definition, we see (as before) that Relation 3 is not a function, since there are two distinct ordered pairs [(1, 1) and (1, −1), for example] that have the same first component. (More than one range element is associated with a given domain element.)

The rule form of the definition of a relation and a function suggests a formula or a "machine" operating on domain values to produce range values—a dynamic process. On the other hand, the set definition of these concepts is closely related to graphs in a cartesian coordinate system—a static form. Each approach has its advantages in certain situations.

■ Common Ways of Specifying Relations and Functions

One of the main objectives of this section is to expose you to the more common ways in which relations and functions are specified and to provide you with experience in determining whether a given relation is or is not a function.

As a consequence of the definitions, we find that a relation (or function) can be specified in many different ways: by an equation, by a table, by a set of ordered pairs of elements, and by a graph, to name a few of the more common ways (see Table 2). All that matters is that we are given a set of elements called the domain and a rule (method or process) for obtaining corresponding range values for each domain value.

TABLE 2
COMMON WAYS OF
SPECIFYING RELATIONS
AND FUNCTIONS

METHOD	ILLUSTRATION	EXAMPLE
Equation	$y = x^2 - x,\ x \in R^\dagger$	$x = -1$ corresponds to $y = 2$
Table	$\begin{array}{c\|c\|c\|c} m & 1 & 2 & 3 \\ \hline n & 1 & 8 & 27 \end{array}$	$m = 2$ corresponds to $n = 8$
Sets of ordered pairs of elements	**(A)** $\{(1, 1), (2, 8), (3, 27)\}$ **(B)** $\{(x, y) \mid y = x^3, x \in R\}$	3 corresponds to 27 $x = -2$ corresponds to $y = -8$
Graph	 	$u = 0$ corresponds to $v = \pm 2$ Domain is understood to be $(-\infty, 4]$ on the u axis

† Recall that R is the set of real numbers.

Which relation in Table 2 is not a function? The relation specified by the graph is not a function, since it is possible for a domain value to correspond to more than one range value. (What does $u = -5$ correspond to?)

For a given function or relation it is often convenient to be able to shift from one representation to another or to use more than one representation. Consider Example 2.

EXAMPLE 2

A relation can be specified by a table and a graph. For example, a laboratory experiment may yield the table and graph shown.

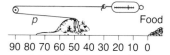

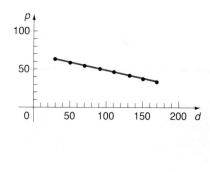

DISTANCE d (in centimeters)	PULL TOWARD FOOD p (in grams)
30	64
50	60
70	56
90	52
110	48
130	44
150	40
170	36

Both the table and the graph in Example 2 establish the same correspondence between domain values d and range values p; hence, both specify the same relation and we call the graph the graph of the relation. More generally, for any relation between real numbers, the graph of all ordered pairs (x, y) in the relation is called the **graph of the relation**.

Note: It is the usual practice to associate domain values with the horizontal axis and range values with the vertical axis. Thus, the first coordinate (abscissa) of the coordinates of a point on the graph is a domain value and the second coordinate (ordinate) is a range value.

PROBLEM 2

Solution

Is the relation in Example 2 a function? Explain.

It is very easy to determine whether a relation is a function if you have its graph.

Vertical Line Test for a Function

A relation between real numbers is a function if each vertical line in the coordinate system passes through *at most* one point on the graph of the relation. (If a vertical line passes through two or more points on the graph of a relation, then the relation is not a function.)

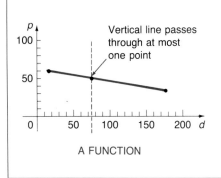

A FUNCTION

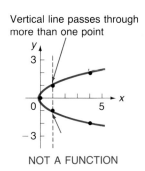

NOT A FUNCTION

■ **Relations Specified by Equations**

Most of the domains and ranges included in this text will be sets of numbers, and the rules associating range values with domain values will be equations in two variables.

Consider the equation

$$y = x^2 - x \qquad x \in R$$

For each **input** x we obtain one **output** y. For example:

If $x = 3$, then $y = 3^2 - 3 = 6$.

If $x = -\frac{1}{2}$, then $y = (-\frac{1}{2})^2 - (-\frac{1}{2}) = \frac{1}{4} + \frac{1}{2} = \frac{3}{4}$.

The input values are domain values and the output values are range values. The equation (a rule) assigns each domain value x a range value y. The variable x is called an independent variable (since values are independently assigned to x from the domain), and y is called a dependent variable (since y's value depends on the value assigned to x). In general, any variable used as a place-holder for domain values is called an **independent variable**; any variable that is used as a placeholder for range values is called a **dependent variable**.

Unless stated to the contrary, we will adhere to the following convention regarding domains and ranges for relations and functions specified by equations:

Agreement on Domains and Ranges
If a relation or function is specified by an equation and the domain is not indicated, then we will assume that the domain is the set of all real number replacements of the independent variable (inputs) that produce real values for the dependent variable (outputs). The range is the set of all outputs corresponding to input values.

EXAMPLE 3   Assuming x is an independent variable, find the domain of the relation specified by the indicated equation:

(A) $y = 2x - 1$ **(B)** $y = \sqrt{x - 2}$ **(C)** $y = \sqrt{x^2 - 4}$

Solution   **(A)** For each real x, y is defined and is real. Thus,

Domain: R

(B) For y to be real, $x - 2$ cannot be negative; that is,

$$x - 2 \geq 0$$
$$x \geq 2$$

Thus,

Domain: $x \geq 2$ or $[2, \infty)$

(C) For y to be real, $x^2 - 4$ cannot be negative; that is,

$$x^2 - 4 \geq 0$$
$$(x - 2)(x + 2) \geq 0$$

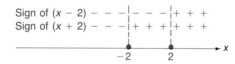

Thus,

Domain: $x \leq -2$ or $x \geq 2$ or $(-\infty, -2] \cup [2, \infty)$

PROBLEM 3   Assuming x is an independent variable, find the domain of the relation specified by the indicated equation:

(A) $y = \dfrac{1}{x^2 - 4}$ **(B)** $y = \sqrt{2 - x}$ **(C)** $y = \sqrt{4 - x^2}$

Solution   **(A)**   **(B)**

(C)

Most equations in two variables specify relations, but when does an equation specify a function?

Equations and Functions

If, in an equation in two variables, there corresponds exactly one value of the dependent variable (output) for each value of the independent variable (input), then the equation specifies a function. If there is more than one output for at least one input, then the equation does not specify a function.

EXAMPLE 4   **(A)** Is the relation specified by the equation $y^2 = x + 1$ a function, assuming x is the independent variable?
(B) What is the domain of the relation?

Solution   **(A)** The relation is not a function, since, for example, if $x = 3$, then $y = \pm 2$.
(B) The domain of the relation (since it is not explicitly given) is the set of all real x that produce real y. Solving for y in terms of x, we obtain

$$y = \pm \sqrt{x + 1}$$

For y to be real, $x + 1$ must be greater than or equal to 0; that is,

$$x + 1 \geq 0$$
$$x \geq -1$$

Thus,

Domain: $x \geq -1$ or $[-1, \infty)$

PROBLEM 4   **(A)** Is the relation specified by the equation $x^2 + y^2 = 25$ a function, assuming x is the independent variable?
(B) What is the domain of the relation?

Solution   **(A)**   **(B)**

■ A Brief History of Function

In reviewing the history of function, we are made aware of the tendency of mathematicians to extend and generalize a concept. The word "function" appears to have been first used by Leibniz in 1694 to stand for any quantity associated with a curve. By 1718, Johann Bernoulli considered a function any expression made up of constants and a variable. Later in the same century, Euler came to regard a function as any equation made up of constants and variables. Euler made extensive use of the extremely important notation $f(x)$, which we will consider in the next section, although its origin is generally attributed to Clairaut (1734).

The form of the definition of function that was used until well into this century (many texts still contain this definition) was formulated by Dirichlet (1805–1859). He stated that if two variables x and y are so related that for each value of x there corresponds exactly one value of y, then y is said to be a (single-valued) function of x. He called x, the variable to which values are assigned at will, the independent variable, and y, the variable whose values depend on the values assigned to x, the dependent variable. He called the values assumed by x the domain of the function and the corresponding values assumed by y he termed the range of the function.

Now, since set concepts permeate almost all mathematics, we have the more general definitions of function presented in this section in terms of sets of ordered pairs of elements. The function is one of the most important concepts in mathematics, and as such it plays a central role as a guide for the selection and development of material in many mathematics courses. (Look at the section titles in this chapter and the next.)

Solutions to Matched Problems

1. **(A)** Function   **(B)** Not a function   **(C)** Function
2. Yes, each domain value corresponds to exactly one range value.
3. **(A)** $x^2 - 4$ cannot be 0; that is, $x \neq 2, -2$. The domain is all real numbers except ± 2.
   (B) $2 - x$ must be nonnegative; that is, $2 - x \geq 0$, so $2 \geq x$. The domain is $(-\infty, 2]$ or $x \leq 2$.
   (C) $4 - x^2$ must be nonnegative; that is, $4 - x^2 \geq 0$. Since $4 - x^2 = (2 - x)(2 + x) \geq 0$ when $-2 \leq x \leq 2$, the domain is $[-2, 2]$ or $-2 \leq x \leq 2$.
4. **(A)** No; for example, when $x = 0$, $y = \pm 5$.
   (B) $y^2 = 25 - x^2$ or $y = \pm\sqrt{25 - x^2}$; therefore, $25 - x^2$ must be nonnegative; that is, $25 - x^2 = (5 - x)(5 + x) \geq 0$. The domain is $-5 \leq x \leq 5$ or $[-5, 5]$.

Practice Exercise 8-1 ■

Work odd-numbered problems first, check answers, and then work even-numbered problems in areas of weakness. Answers to all problems are in the back of the book. Make every effort to work a problem yourself before you look at an answer.

A   *Indicate whether each relation in Problems 1–6 is or is not a function.*

1. DOMAIN	RANGE		2. DOMAIN	RANGE
3 $\longrightarrow$	0		$-1 \longrightarrow$	5
5 $\longrightarrow$	1		$-2 \longrightarrow$	7
7 $\longrightarrow$	2	_____	$-3 \longrightarrow$	9 _____

3. DOMAIN	RANGE		4. DOMAIN	RANGE
3	5		8 $\longrightarrow$	0
	6		9	1
4 $\longrightarrow$	7			2
5 $\longrightarrow$	8	_____	10 $\longrightarrow$	3 _____

5. DOMAIN   RANGE

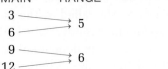

_____

6. DOMAIN   RANGE

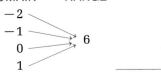

_____

Each relation in Problems 7–12 is specified by a graph. Indicate whether the relation is a function.

7.

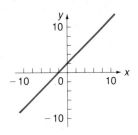

_____

8.

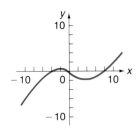

_____

9.

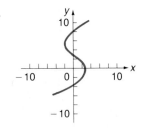

_____

10.

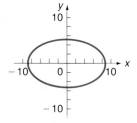

_____

11.

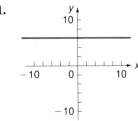

_____

12.

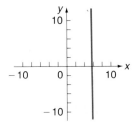

_____

B   _Each equation specifies a relation. Which specify functions given that x is an independent variable?_

13. $y = 3x - 1$ _____

14. $y = \dfrac{x}{2} - 1$ _____

15. $y = x^2 - 3x + 1$ _____

16. $y = x^3$ _____

17. $y^2 = x$ _____

18. $x^2 + y^2 = 25$ _____

19. $x = y^2 - y$ _____

20. $x = (y - 1)(y + 2)$ _____

21. $y = x^4 - 3x^2$ _____

22. $2x - 3y = 5$ _____

23. $y = \dfrac{x + 1}{x - 1}$ _____

24. $y = \dfrac{x^2}{1 - x}$ _____

Graph each relation in Problems 25–36. State its domain and range and indicate which are functions. The variable x is independent.

25. $F = \{(1, 1), (2, 1), (3, 2), (3, 3)\}$

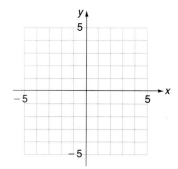

Domain: _____

Range: _____

Function? _____

26. $f = \{(2, 4), (4, 2), (2, 0), (4, -2)\}$

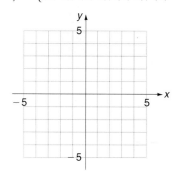

Domain: _____

Range: _____

Function? _____

27. $G = \{(-1, -2), (0, -1), (1, 0), (2, 1), (3, 2), (4, 1)\}$

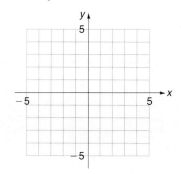

Domain: _____

Range: _____

Function? _____

28. $g = \{(-2, 0), (0, 2), (2, 0)\}$

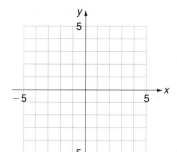

Domain: _____

Range: _____

Function? _____

29. $y = 6 - 2x, \ x \in \{0, 1, 2, 3, 4\}$

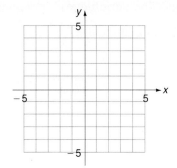

Domain: _____

Range: _____

Function? _____

30. $y = \dfrac{x}{2} - 4, \ x \in \{0, 1, 2, 3, 4\}$

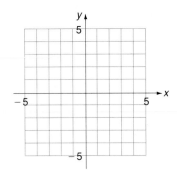

Domain: _____

Range: _____

Function? _____

31. $y^2 = x, \ x \in \{0, 1, 4\}$

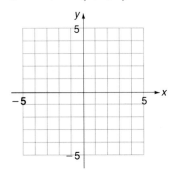

Domain: _____

Range: _____

Function? _____

32. $y = x^2, \ x \in \{-2, 0, 2\}$

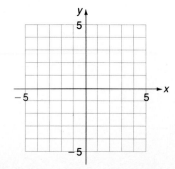

Domain: _____

Range: _____

Function? _____

33. $x^2 + y^2 = 4$, $x \in \{-2, 0, 2\}$

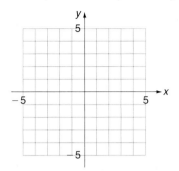

Domain: _____
Range: _____
Function? _____

34. $x^2 + y^2 = 9$, $x \in \{-3, 0, 3\}$

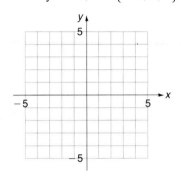

Domain: _____
Range: _____
Function? _____

35. $y = |x|$, $x \in \{-2, 0, 2\}$

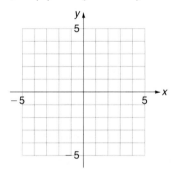

Domain: _____
Range: _____
Function? _____

36. $|y| = x$, $x \in \{0, 1, 4\}$

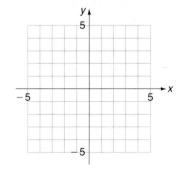

Domain: _____
Range: _____
Function? _____

Determine the domains of the relations in Problems 37–50.

37. $y = 5 - x$ _____

38. $y = 5x + 2$ _____

39. $y = 3x^2 - 2x + 1$ _____

40. $y = (5 - x)^2$ _____

41. $|y| = x + 3$ _____

42. $y = |x - 3|$ _____

43. $x^2 + y^2 = 64$ _____

44. $25x^2 + y^2 = 25$ _____

45. $y = \dfrac{x - 4}{2x^2 + 5x - 3}$ _____

46. $y = \dfrac{x + 2}{x^2 + x - 12}$ _____

47. $y^2 = 4 - x$ _____

48. $y^2 = x - 5$ _____

49. $y = \sqrt{\dfrac{x - 1}{x + 3}}$ _____

50. $y = \sqrt{x^2 + 3x - 10}$ _____

C *Graph each relation in Problems 51–54. State domain and range and indicate which relations are functions. The variable x is independent.*

51. $H = \left\{(x, y)\,\middle|\, y = \dfrac{x}{2},\ x \in \{-4, -2, 0, 2, 4\}\right\}$

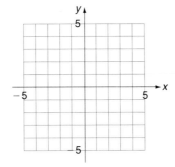

Domain: _____
Range: _____
Function? _____

52. $h = \left\{(x, y)\,\middle|\, y = x + 3,\ x \in \{-3, -1, 0, 2\}\right\}$

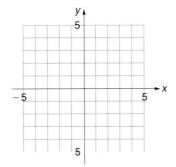

Domain: _____
Range: _____
Function? _____

53. $F = \left\{(x, y)\,\middle|\, 0 \le y \le x,\ 0 \le x \le 3;\ x, y \in J\right\}$

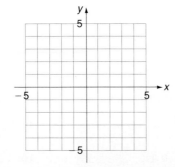

Domain: _____
Range: _____
Function? _____

54. $G = \{(x, y) \mid 0 \le y < |x|, \ -2 \le x \le 2; \ x, y \in J\}$

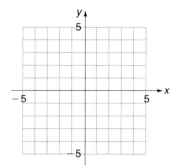

Domain: _____

Range: _____

Function? _____

APPLICATIONS

55. If an arrow is shot straight upward from the ground with an initial velocity of 160 feet per second, its distance d in feet above the ground at the end of t seconds (neglecting air resistance) is given by

$$d = 160t - 16t^2 \qquad 0 \le t \le 10$$

(A) Graph this relation (t is the independent variable).

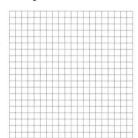

(B) What are its domain and range? _____

(C) Is the relation a function? _____

56. The distance s that an object falls (neglecting air resistance) in t seconds is given by

$$s = 16t^2 \qquad t \ge 0$$

(A) Graph this relation (t is the independent variable).

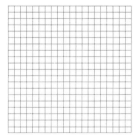

The Check Exercise for
this section is on
page 537.

(B) What are its domain and range? _____

(C) Is the relation a function? _____

Section 8-2   Function Notation

- ■   The Function Symbol $f(x)$
- ■   Use of the Function Symbol $f(x)$

■   The Function Symbol $f(x)$

We have just seen that a function involves two sets of elements, a domain and a range, and a rule of correspondence that enables us to assign each element in the domain to exactly one element in the range. We use different letters to denote names for numbers; in essentially the same way, we will now use different letters to denote names for functions. For example, f and g may be used to name the two functions

$$f: \quad y = 2x + 1$$
$$g: \quad y = x^2 + 2x - 3$$

If x represents an element in the domain of a function f, then we will often use the symbol

$$f(x)$$

in place of y to designate the number in the range of the function f to which x is paired (Figure 1).

The function f "maps" the domain value x into the range value $f(x)$

FIGURE 1   Domain   Range

Do not confuse this new function symbol and think of it as the product of f and x. The symbol $f(x)$ is read "f of x" or "the value of f at x." The variable x is an independent variable; both y and $f(x)$ are dependent variables.

■   Use of the Function Symbol $f(x)$

This new function notation is extremely important, and its correct use should be mastered as early as possible. For example, in place of the more formal representation of the functions f and g given above, we can now write

$$f(x) = 2x + 1 \quad \text{and} \quad g(x) = x^2 + 2x - 3$$

The function symbols $f(x)$ and $g(x)$ have certain advantages over the variable y in certain situations. For example, if we write $f(3)$ and $g(5)$, then each symbol indicates in a concise way that these are range values of particular functions associated with particular domain values. Let us find $f(3)$ and $g(5)$.

To find $f(3)$, we replace x by 3 wherever x occurs in

$$f(x) = 2x + 1$$

and evaluate the right side:

$$f(3) = 2 \cdot 3 + 1$$
$$= 6 + 1$$
$$= 7$$

Thus,

$f(3) = 7$ The function f assigns the range value 7 to the domain value 3; the ordered pair (3, 7) belongs to f.

To find $g(5)$, we replace x by 5 wherever x occurs in

$$g(x) = x^2 + 2x - 3$$

and evaluate the right side:

$$g(5) = 5^2 + 2 \cdot 5 - 3$$
$$= 25 + 10 - 3$$
$$= 32$$

Thus,

$g(5) = 32$ The function g assigns the range value 32 to the domain value 5; the ordered pair (5, 32) belongs to g.

It is very important to understand and remember the definition of $f(x)$:

The Function Symbol $f(x)$

For any element x in the domain of the function f, the function symbol

$f(x)$

represents the element in the range of f corresponding to x in the domain of f. [If x is an input value, then $f(x)$ is an output value; or, symbolically, $f: \quad x \to f(x)$.] The ordered pair $(x, f(x))$ belongs to the function f.

Figure 2, illustrating a "function machine," may give you additional insight into the nature of function and the new function symbol $f(x)$. We can think of a function machine as a device that produces exactly one output (range) value for each input (domain) value. (If more than one output value is produced for an input value, then the machine would be a "relation machine" and not a function machine.)

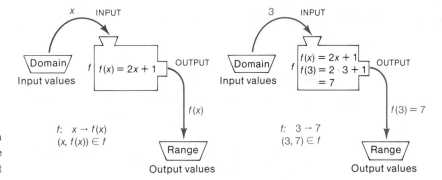

FIGURE 2 "Function machine"—exactly one output for each input

For the function $f(x) = 2x + 1$, the machine takes each domain value (input), multiplies it by 2, then adds 1 to the result to produce the range value (output). Different rules inside the machine result in different functions.

EXAMPLE 5   If $f(x) = \dfrac{x}{2} + 1$ and $g(x) = 1 - x^2$, find:

(A) $f(6)$ (B) $g(-2)$ (C) $f(4) + g(0)$ (D) $\dfrac{2g(-3) + 6}{f(8)}$

Solution   (A) $f(6) = \frac{6}{2} + 1 = 3 + 1 = 4$

(B) $g(-2) = 1 - (-2)^2 = 1 - 4 = -3$

(C) $f(4) + g(0) = (\overset{f(4)}{\frac{4}{2} + 1}) + (\overset{g(0)}{1 - 0^2}) = 3 + 1 = 4$

(D) $\dfrac{2g(-3) + 6}{f(8)} = \dfrac{2[\overset{g(-3)}{1 - (-3)^2}] + 6}{\underset{f(8)}{\frac{8}{2} + 1}} = \dfrac{2(-8) + 6}{5} = \dfrac{-10}{5} = -2$

PROBLEM 5   If $f(x) = \dfrac{x}{3} - 2$ and $g(x) = 4 - x^2$, find:

(A) $f(9)$ (B) $g(-2)$ (C) $f(0) + g(2)$ (D) $\dfrac{4g(-1) - 4}{f(12)}$

Solution   (A)   (B)

(C)   (D)

EXAMPLE 6   For $f(x) = 5x + 3$, find: (A) $f(z)$ (B) $f(z + 1)$ (C) $f(2q)$

Solution   (A) $f(z) = 5z + 3$ Replace x in $f(x) = 5x + 3$ with z.

(B) $f(z + 1) = 5(z + 1) + 3$ Replace x in $f(x) = 5x + 3$ with $z + 1$ and simplify.
        $= 5z + 5 + 3$
        $= 5z + 8$

(C) $f(2q) = 5(2q) + 3$ Replace x in $f(x) = 5x + 3$ with $2q$ and simplify.
        $= 10q + 3$

PROBLEM 6   For $f(x) = 2 - 3x$, find: **(A)** $f(m)$ **(B)** $f(m - 1)$ **(C)** $f(5p)$

Solution   **(A)**   **(B)**

(C)

EXAMPLE 7   For $f(x) = \dfrac{x}{2} + 1$ and $g(x) = 1 - x^2$, find:

(A) $g(2 + h)$ **(B)** $\dfrac{g(2 + h) - g(2)}{h}$ **(C)** $f[g(3)]$

Solution   **(A)** $g(2 + h) = 1 - (2 + h)^2 = 1 - (4 + 4h + h^2)$ Replace x in $g(x) = 1 - x^2$
$$= -3 - 4h - h^2$$ with $(2 + h)$.

(B) $\dfrac{g(2 + h) - g(2)}{h} = \dfrac{[1 - (2 + h)^2] - (1 - 2^2)}{h}$ Be careful here! The brackets and parentheses are important.

$$= \dfrac{-3 - 4h - h^2 + 3}{h}$$

$$= \dfrac{-4h - h^2}{h} = -4 - h$$

(C) $f[g(3)] = f(1 - 3^2)$ Evaluate $g(3)$ first; then evaluate f for this value.

$$= f(-8) = \dfrac{-8}{2} + 1 = -3$$

PROBLEM 7   For $f(x) = \dfrac{x}{3} - 2$ and $g(x) = 4 - x^2$, find:

(A) $g(3 + h)$ **(B)** $\dfrac{g(3 + h) - g(3)}{h}$ **(C)** $g[f(3)]$

Solution   **(A)**

(B)

(C)

Solutions to Matched
Problems

5. (A) $f(9) = \frac{9}{3} - 2 = 3 - 2 = 1$ **(B)** $g(-2) = 4 - (-2)^2 = 4 - 4 = 0$

(C) $f(0) + g(2) = (\frac{0}{3} - 2) + (4 - 2^2) = (-2) + (0) = -2$

(D) $\dfrac{4g(-1) - 4}{f(12)} = \dfrac{4[4 - (-1)^2] - 4}{\frac{12}{3} - 2} = \dfrac{4(3) - 4}{2} = \dfrac{8}{2} = 4$

6. (A) $f(m) = 2 - 3m$

(B) $f(m - 1) = 2 - 3(m - 1) = 2 - 3m + 3 = 5 - 3m$

(C) $f(5p) = 2 - 3(5p) = 2 - 15p$

7. (A) $g(3 + h) = 4 - (3 + h)^2 = 4 - (9 + 6h + h^2) = -5 - 6h - h^2$

(B) $\dfrac{g(3 + h) - g(3)}{h} = \dfrac{-5 - 6h - h^2 - (4 - 3^2)}{h}$

$= \dfrac{-5 - 6h - h^2 - (-5)}{h} = \dfrac{-6h - h^2}{h} = -6 - h$

(C) $g[f(3)] = g(\frac{3}{3} - 2) = g(-1) = 4 - (-1)^2 = 4 - 1 = 3$

Practice Exercise 8-2 ∎

Work odd-numbered problems first, check answers, and then work even-numbered problems in areas of weakness. Answers to all problems are in the back of the book. Make every effort to work a problem yourself before you look at an answer.

A   If $f(x) = 3x - 2$, find the following.

1. $f(2)$ _____   **2.** $f(1)$ _____   **3.** $f(-2)$ _____

4. $f(-1)$ _____   **5.** $f(0)$ _____   **6.** $f(4)$ _____

If $g(x) = x - x^2$, find the following.

7. $g(2)$ _____   **8.** $g(1)$ _____   **9.** $g(4)$ _____

10. $g(5)$ _____   **11.** $g(-2)$ _____   **12.** $g(-1)$ _____

B   For $f(x) = 10x - 7$, $g(t) = 6 - 2t$, $F(u) = 3u^2$, and $G(v) = v - v^2$, find the following.

13. $f(-2)$ _____   **14.** $F(-1)$ _____   **15.** $g(2)$ _____

16. $G(-3)$ _____   **17.** $g(0)$ _____   **18.** $G(0)$ _____

19. $f(3) + g(2)$ _____   **20.** $F(2) + G(3)$ _____

21. $2g(-1) - 3G(-1)$ _____   **22.** $4G(-2) - g(-3)$ _____

23. $\dfrac{f(2) \cdot g(-4)}{G(-1)}$ _____

24. $\dfrac{F(-1) \cdot G(2)}{g(-1)}$ _____

25. $g(u - 2)$ _____

26. $f(v + 1)$ _____

27. $G(3a)$ _____

28. $F(2c)$ _____

29. $g(2 + h)$ _____

30. $F(2 + h)$ _____

31. $\dfrac{g(2 + h) - g(2)}{h}$ _____

32. $\dfrac{F(2 + h) - F(2)}{h}$ _____

33. $\dfrac{f(3 + h) - f(3)}{h}$ _____

34. $\dfrac{G(2 + h) - G(2)}{h}$ _____

35. $F[g(1)]$ _____

36. $G[F(1)]$ _____

37. $g[f(1)]$ _____

38. $g[G(0)]$ _____

39. $f[G(1)]$ _____

40. $G[g(2)]$ _____

41. If $A(w) = \dfrac{w - 3}{w + 5}$, find $A(5)$, $A(0)$, and $A(-5)$. _____

42. If $h(s) = \dfrac{s}{s - 2}$, find $h(3)$, $h(0)$, and $h(2)$. _____

C For $f(x) = 10x - 7$ and $g(t) = 6 - 2t$, find the following.

43. $\dfrac{f(x + h) - f(x)}{h}$ _____

44. $\dfrac{g(t + h) - g(t)}{h}$ _____

45. For $f(x) = 5x$:

(A) Does $f(at) = af(t)$? _____

(B) Does $f(a + b) = f(a) + f(b)$? _____

(C) Does $f(ab) = f(a) \cdot f(b)$? _____

46. For $g(x) = x^2$:

(A) Does $g(at) = ag(t)$? _____

(B) Does $g(a + b) = g(a) + g(b)$? _____

(C) Does $g(ab) = g(a) \cdot g(b)$? _____

APPLICATIONS *Each of the statements in Problems 47–50 can be described by a function. Write an equation that specifies the function.*

47. *Cost function:* The cost $C(x)$ of x records at \$5 per record. (The cost depends on the number of records purchased.) _____

48. *Cost function:* The cost $C(x)$ of manufacturing x pairs of skis if fixed costs are \$800 per day and the variable costs are \$60 per pair of skis. (The cost per day depends on the number of skis manufactured per day.)

_____

49. *Temperature conversion:* The temperature in Celsius degrees C(F) can be found from the temperature in Fahrenheit degrees F by subtracting 32 from the Fahrenheit temperature and multiplying the difference by $\frac{5}{9}$.

_____

50. *Earth science:* The pressure $P(d)$ in the ocean in pounds per square inch depends on the depth d. To find the pressure, divide the depth by 33, add 1 to the quotient, then multiply the result by 15. _____

51. *Distance–rate–time:* Let the distance that a car travels at 30 miles per hour in t hours be given by $d(t) = 30t$. Find:

(A) $d(1)$, $d(10)$ _____   **(B)** $\dfrac{d(2 + h) - d(2)}{h}$ _____

*52. *Physics:* The distance in feet that an object falls in t seconds in a vacuum is given by $s(t) = 16t^2$. Find:

(A) $s(0)$, $s(1)$, $s(2)$, and $s(3)$ _____

(B) $\dfrac{s(2 + h) - s(2)}{h}$ _____

The Check Exercise for
this section is on
page 539.

What happens as h tends to 0? Interpret physically. _____

Section 8-3 Graphing Polynomial Functions

- Linear Functions
- Quadratic Functions
- Higher-Degree Polynomial Functions
- Application

In Chapters 6 and 7 we studied linear and quadratic forms in some detail. These forms with certain restrictions define **linear** and **quadratic functions**. Linear and quadratic functions are important special cases of a larger class of functions called **polynomial functions**. In this section we will look at the graphs of linear and quadratic functions (and some of their properties) and at an effective technique for graphing polynomial functions in general. The **graph of a function** is the graph of all ordered pairs of numbers that constitute the function.

■ Linear Functions

Any nonvertical line in a rectangular coordinate system defines a linear function. (A vertical line does not define a function. Why?) Thus, any function defined by an equation of the form

$$f(x) = ax + b \quad \text{Linear function}$$

where a and b are constants and x is a variable, is called a **linear function**. We know from Section 7-2 that the graph of this equation is a straight line (nonvertical) with slope a and y intercept b.

EXAMPLE 8   Graph the linear function defined by $f(x) = x/3 + 1$, and indicate its slope and y intercept.

Solution

x	$f(x)$
-3	0
0	1
3	2

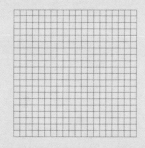

$y = \frac{1}{3}x + 1$

y intercept: 1

Slope: $\frac{1}{3}$

PROBLEM 8   Graph the linear function defined by $f(x) = -x/2 + 3$, and indicate its slope and y intercept.

Solution

■ Quadratic Functions

Any function defined by an equation of the form

$$f(x) = ax^2 + bx + c \qquad a \neq 0 \qquad \text{Quadratic function}$$

where a, b, and c are constants and x is a variable, is called a **quadratic function**.

Let us first graph the two simple quadratic functions:

$$f(x) = x^2 \qquad \text{and} \qquad g(x) = -x^2$$

We evaluate these functions for integer values from their domains, find corresponding range values, then plot the resulting ordered pairs and join these points with a smooth curve. The work in the dashed boxes in the following calculations is usually done mentally or on scratch paper.

Graphing $f(x) = x^2$:

x	$y = f(x)$	$(x, f(x))$
-2	$y = f(-2) = (-2)^2 = 4$	$(-2, 4)$
-1	$y = f(-1) = (-1)^2 = 1$	$(-1, 1)$
0	$y = f(0) = 0^2 = 0$	$(0, 0)$
1	$y = f(1) = 1^2 = 1$	$(1, 1)$
2	$y = f(2) = 2^2 = 4$	$(2, 4)$
Domain values	Range values	Elements of f

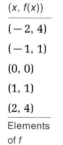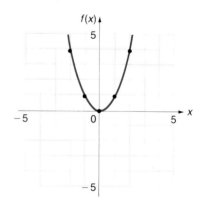

Graphing $g(x) = -x^2$:

x	$y = g(x)$	$(x, g(x))$
-2	$y = g(-2) = -(-2)^2 = -4$	$(-2, -4)$
-1	$y = g(-1) = -(-1)^2 = -1$	$(-1, -1)$
0	$y = g(0) = -0^2 = 0$	$(0, 0)$
1	$y = g(1) = -1^2 = -1$	$(1, -1)$
2	$y = g(2) = -2^2 = -4$	$(2, -4)$
Domain values	Range values	Elements of g

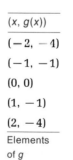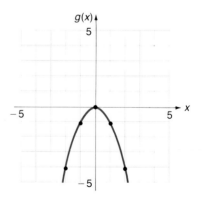

From our work on conic sections in Section 7-5, we recognize these graphs (graphs of $y = x^2$ and $y = -x^2$) to be parabolas. It can be shown that the graph of *any* quadratic function is also a parabola. In general:

Graph of $f(x) = ax^2 + bx + c$, $a \neq 0$

The graph of a quadratic function f is a parabola that has its **axis** (line of symmetry) parallel to the vertical axis. It opens upward if $a > 0$ and downward if $a < 0$. The intersection point of the axis and parabola is called the **vertex**.

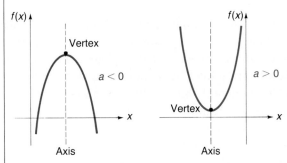

Note: $f(x)$ has a maximum or minimum value at a vertex.[†]

Given an arbitrary quadratic function, we can speed up the graphing process by factoring x out (on the right) of the first two terms to obtain a "nested factored" form:

$$f(x) = ax^2 + bx + c$$
$$= (ax + b)x + c \quad \text{Nested factored form}$$

The nested factored form (the term will become clear before we finish this section) is particularly convenient for evaluating $f(x)$ for various values of x mentally—and even more convenient for use with a hand calculator when x is not a small integer. When using a hand calculator, store the chosen value of x and recall it as necessary as you proceed from left to right.

EXAMPLE 9 Graph $f(x) = -2x^2 + 10x + 5$, $-2 \leq x \leq 7$. From the graph of the parabola, estimate the coordinates of the vertex and the equation of the axis. Estimate the maximum and minimum values of $f(x)$ if they exist.

Solution First write the equation in a nested factored form:

$$f(x) = -2x^2 + 10x + 5$$
$$= (-2x + 10)x + 5$$

All of the calculations involving the integers from -2 to 7 can be done mentally. If decimal values between the integers are desired (for increased graph clarity), a hand calculator will be of help.

Mental calculations proceed as follows:

$$f(-2) = [-2(-2) + 10](-2) + 5 = -23$$
$$f(-1) = [-2(-1) + 10](-1) + 5 = -7$$

and so on

[†] $f(c)$ is the **maximum value of $f(x)$** if $f(c) \geq f(x)$ for all x in the domain of f; $f(d)$ is the **minimum value of $f(x)$** if $f(d) \leq f(x)$ for all x in the domain of f.

Calculator calculations (if desired) proceed as follows:

$f(-2)$: [2] [+/−] [STO] [1] [CLR] [(] [2] [+/−] [×] [RCL] [1] [+] [10] [)] [×] [RCL] [1] [+] [5]

$f(-1)$: [1] [+/−] [STO] [1] [CLR] [(] [2] [+/−] [×] [RCL] [1] [+] [10] [)] [×] [RCL] [1] [+] [5]

and so on. The specific calculator steps are included here only as an aid when you are stuck; it is better to proceed using only the equation $f(x) = (-2x + 10)x + 5$ to guide your calculations.

We construct a table of ordered pairs of numbers belonging to the function f. We then plot these points and join them with a smooth curve. (Be sure to plot enough points so that it is clear what happens between the points when the points are joined by a smooth curve.) If no restrictions on the independent variable are given, start at the origin and work in both directions until enough points are determined to give a clear picture of the graph near the origin and a reasonable indication of what happens to the graph as it moves away from the origin. Scales on the two axes can be different.

x	$f(x)$	
−2	−23	← Minimum
−1	−7	
0	5	
1	13	
2	17	
2.5	17.5	← Maximum (vertex)
3	17	
4	13	
5	5	
6	−7	
7	−23	← Minimum

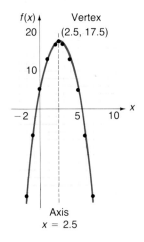

Note: To clarify the graph between $x = 2$ and $x = 3$, we use $x = 2.5$.

PROBLEM 9

Graph $f(x) = 3x^2 - 9x + 8$. Estimate the coordinates of the vertex, the equation of the axis, and maximum or minimum values of $f(x)$ if either exists.

Solution

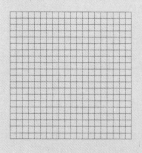

■ Higher-Degree Polynomial Functions

As we indicated at the start of this section, linear and quadratic functions are special cases of a general class of functions called **polynomial functions** (functions whose range values are determined by use of a polynomial). In general, a function f defined by an equation of the form

$$f(x) = a_n x^n + a_{n-1} x^{n-1} + \cdots + a_1 x + a_0 \qquad a_n \neq 0$$

where the coefficients a_i are constants and n is a nonnegative integer, is called an **nth-degree polynomial function**. The following equations define polynomial functions of various degrees:

$$f(x) = 2x - 3 \qquad \text{First-degree (linear)}$$
$$g(x) = 2x^2 - 3x + 2 \qquad \text{Second-degree (quadratic)}$$
$$P(x) = x^3 - 2x^2 + x - 1 \qquad \text{Third-degree}$$
$$Q(x) = x^4 - 5 \qquad \text{Fourth-degree}$$

Unless otherwise stated, the domain of a polynomial function is the set of all real numbers. In advanced courses it can be shown that graphs of polynomial functions have no holes or breaks in them—the graphs are continuous smooth curves.

We will now illustrate the use of the nested factoring technique in graphing higher-degree polynomial functions. A hand calculator will prove useful for values of the independent variable that are not small integers.

EXAMPLE 10   Graph: $P(x) = x^3 + 3x^2 - x - 3, \ -4 \leq x \leq 2$

Solution   We first write $P(x)$ in a nested factored form as follows:

$$P(x) = x^3 + 3x^2 - x - 3 \qquad \text{Factor the first two terms, and repeat factoring each}$$
$$= (x + 3)x^2 - x - 3 \qquad \text{resulting first two terms until you cannot go any further.}$$
$$= [(x + 3)x - 1]x - 3$$

Mental calculations proceed as follows:

$$P(-4) = \{[(-4) + 3](-4) - 1\}(-4) - 3 = -15$$
$$P(-3) = \{[(-3) + 3](-3) - 1\}(-3) - 3 = 0$$
and so on

Calculator calculations (if desired) proceed as follows:

$P(-4)$: $\boxed{4}\ \boxed{+/-}\ \boxed{\text{STO}}\ \boxed{1}\ \boxed{\text{CLR}}\ \boxed{(}\ \boxed{(}\ \boxed{\text{RCL}}\ \boxed{1}\ \boxed{+}\ \boxed{3}\ \boxed{)}\ \boxed{\times}\ \boxed{\text{RCL}}\ \boxed{1}\ \boxed{-}\ \boxed{1}\ \boxed{)}\ \boxed{\times}\ \boxed{\text{RCL}}\ \boxed{1}\ \boxed{-}\ \boxed{3}\ \boxed{=}$

$P(-3)$: $\boxed{3}\ \boxed{+/-}\ \boxed{\text{STO}}\ \boxed{1}\ \boxed{\text{CLR}}\ \boxed{(}\ \boxed{(}\ \boxed{\text{RCL}}\ \boxed{1}\ \boxed{+}\ \boxed{3}\ \boxed{)}\ \boxed{\times}\ \boxed{\text{RCL}}\ \boxed{1}\ \boxed{-}\ \boxed{1}\ \boxed{)}\ \boxed{\times}\ \boxed{\text{RCL}}\ \boxed{1}\ \boxed{-}\ \boxed{3}\ \boxed{=}$

and so on. As in Example 9, these specific calculator steps are included only as an aid when you get stuck; it is better to proceed using only the equation $P(x) = [(x + 3)x - 1]x - 3$ to guide your calculations.

Note: A programmable calculator is even more efficient; if you have one, try it.

Continuing, either mentally or with a calculator, we construct a table of ordered pairs of numbers belonging to the function P. We then plot these points and join them with a smooth curve. The more points we compute between two given points, the greater the accuracy of the graph. Here, integer values for x give us a reasonable picture:

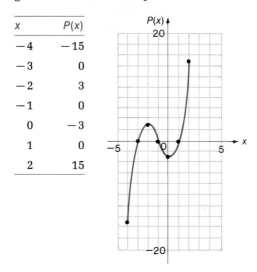

x	P(x)
−4	−15
−3	0
−2	3
−1	0
0	−3
1	0
2	15

PROBLEM 10   Graph $P(x) = x^3 - 4x^2 - 4x + 16$, $-3 \le x \le 5$, using the nested factoring method.

Solution

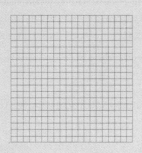

■ **Application**

EXAMPLE 11   A rectangular dog pen is to be made with 160 feet of fencing.

(A) If x represents the width of the pen, express its area $A(x)$ in terms of x.
(B) What is the domain of the function A (determined by the physical restrictions)?

Solution   **(A)** Draw a figure and label the sides:

x (Width)

80 − x (Length)

Perimeter = 160
Half the perimeter = 80
If x = Width, then 80 − x = Length.

$$A(x) = (\text{Width})(\text{Length}) = (80 - x)x \quad \text{Area depends on width } x.$$

(B) The area cannot be negative; hence, x cannot be negative and x cannot be greater than 80. [Look at $A(x) = (80 - x)x$.] Thus,

Domain: $0 \le x \le 80$ Inequality notation

$[0, 80]$ Interval notation

PROBLEM 11   Work Example 11 with the added assumption that a large barn is to be used as one side of the pen.

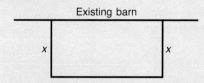

Solution

Solutions to Matched
Problems

8.

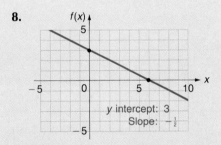

y intercept: 3
Slope: $-\frac{1}{2}$

9. Minimum $f(x) = f(1.5) = 1.25$

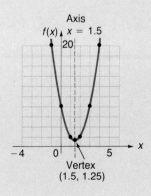

Axis
$x = 1.5$

Vertex
(1.5, 1.25)

10.

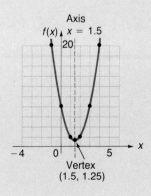

11.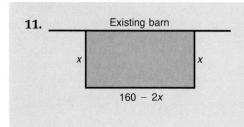

(A) Area $= A(x) = $ (Width)(Length) $=$ x(160 − 2x)

(B) Domain: $A(x)$ must be nonnegative; that is, x(160 − 2x) = 2x(80 − x) ≥ 0. Thus, 0 ≤ x ≤ 80 or [0, 80] is the domain.

Practice Exercise 8-3 ■

Work odd-numbered problems first, check answers, and then work even-numbered problems in areas of weakness. Answers to all problems are in the back of the book. Make every effort to work a problem yourself before you look at an answer.

A   *Graph the following linear functions. Indicate the slope and y intercept for each.*

1. $f(x) = 2x - 4$

2. $g(x) = \dfrac{x}{2}$

3. $h(x) = 4 - 2x$

4. $f(x) = -\dfrac{x}{2} + 3$

5. $g(x) = -\frac{2}{3}x + 4$

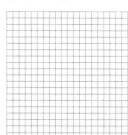

6. $f(x) = 3$

Graph each of the following quadratic functions. Estimate axis of symmetry, vertex, and maximum or minimum value for each.

7. $f(x) = x^2 + 8x + 16$

8. $h(x) = x^2 - 2x - 3$

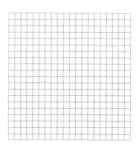

9. $f(u) = u^2 - 2u + 4$

10. $f(x) = x^2 - 10x + 25$

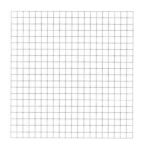

11. $h(x) = 2 + 4x - x^2$

12. $g(x) = -x^2 - 6x - 4$

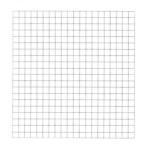

13. $f(x) = 6x - x^2$

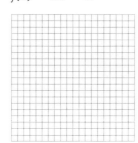

14. $G(x) = 16x - 2x^2$

15. $F(s) = s^2 - 4$

16. $g(t) = t^2 + 4$

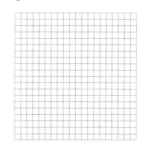

17. $F(x) = 4 - x^2$

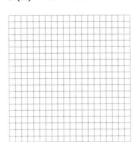

18. $G(x) = 9 - x^2$

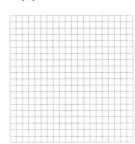

B   *Graph each of the following quadratic functions. Estimate axis of symmetry, vertex, and maximum or minimum value for each.*

19. $f(x) = x^2 - 7x + 10$

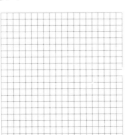

20. $g(t) = t^2 - 5t + 2$

21. $g(t) = 4 + 3t - t^2$

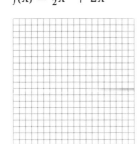

22. $h(x) = 2 - 5x - x^2$

23. $f(x) = \frac{1}{2}x^2 + 2x$

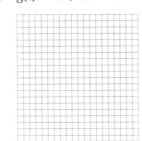

24. $f(x) = 2x^2 - 12x + 14$

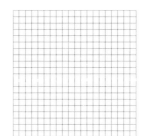

25. $f(x) = -2x^2 - 8x - 2$

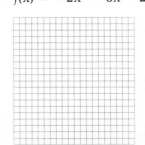

26. $f(x) = -\frac{1}{2}x^2 + 4x - 4$

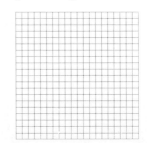

Graph each polynomial function using the nested factoring method.

27. $P(x) = x^3 - 5x^2 + 2x + 8;$ $\quad -2 \leq x \leq 5$

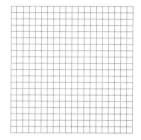

28. $P(x) = x^3 + 2x^2 - 5x - 6;$ $\quad -4 \leq x \leq 3$

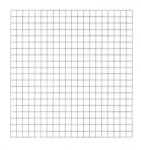

29. $P(x) = x^3 + 4x^2 - x - 4;$ $\quad -5 \leq x \leq 2$

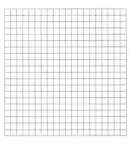

30. $P(x) = x^3 - 2x^2 - 5x + 6;$ $\quad -3 \leq x \leq 4$

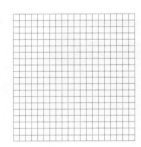

C Graph each polynomial function using the nested factoring method.

31. $P(x) = x^4 - 2x^3 - 2x^2 + 8x - 8$

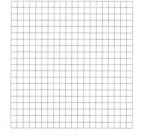

32. $P(x) = x^4 - 2x^2 + 16x - 15$

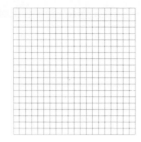

33. $P(x) = x^4 + 4x^3 - x^2 - 20x - 20$

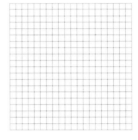

34. $P(x) = x^4 - 4x^2 - 4x - 1$

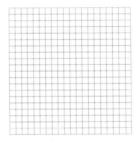

APPLICATIONS

35.   *Cost equation:*   The cost equation for a particular company to produce stereos is found to be

$$C = g(n) = 96,000 + 80n$$

where \$96,000 is fixed costs (tooling and overhead) and \$80 is the variable cost per unit (material, labor, and so on). Graph this function for $0 \leq n \leq 1,000$.

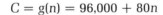

36.   *Demand equation:*   After extensive surveys the research department in a stereo company produced the demand equation

$$n = f(p) = 8,000 - 40p \quad 100 \leq p \leq 200$$

where n is the number of units that retailers are likely to purchase per

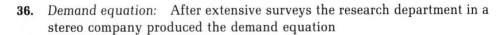

week at a price of p dollars per unit. Graph the function for the indicated domain.

37. *Construction:* A rectangular feeding pen for cattle is to be made with 100 meters of fencing.

   (A) If x represents the width of the pen, express its area $A(x)$ in terms of x. _____

   (B) What is the domain of the function A (determined by the physical restrictions)? _____

   (C) Graph the function for this domain.

   (D) What dimension pen will produce the largest area? What is the maximum area? _____

38. *Construction:* Work the preceding problem with the added assumption that a large straight river is to be used as one side of the pen.

*39. *Packaging:* A candy box is to be made out of a rectangular piece of cardboard that measures 8 by 12 inches. Equal-sized squares (x by x inches) will be cut out of each corner, and then the ends and sides will be folded up to form a rectangular box.

   (A) Write the volume of the box $V(x)$ in terms of x. _____

   (B) Considering the physical limitations, what is the domain of the function V? _____

(C)   Graph the function for this domain.

(D)   From the graph, estimate to the nearest half-inch the size square that must be cut from each corner to yield a box with the maximum volume. What is the maximum volume? _____

**40.   *Packaging:*   A parcel delivery service will deliver only packages with length plus girth (distance around) not exceeding 108 inches. A packaging company wishes to design a box with a square base (x by x inches) that will have a maximum volume but meet the delivery service's restrictions.

(A)   Write the volume of the box V(x) in terms of x. _____

(B)   Considering the physical limitations imposed by the delivery service, what is the domain of the function V? _____

(C)   Graph the function for this domain.

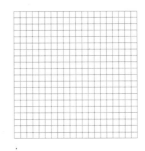

The Check Exercise for
this section is on
page 541.

(D)   From the graph, estimate to the nearest inch the dimensions of the box with the maximum volume. What is the maximum volume?

_____

Section 8-4   Inverse Relations and Functions

■ Inverses
■ One-to-One Correspondence and Inverses

In this section we are going to discuss an important method for obtaining new relations and functions from old relations and functions. We will use this method in Chapter 9 to obtain the logarithmic functions from the exponential functions.

■   Inverses

Given a relation G, if we interchange the order of the components in each ordered pair belonging to G, we obtain a new relation G^{-1} called the **inverse**

of G. [*Note:* G^{-1} is a relation–function symbol; it does not mean $1/G$.] For example, if

$$G = \{(2, 4), (-1, 3), (0, 4)\}$$

then by reversing the components in each ordered pair in G we obtain

$$G^{-1} = \{(4, 2), (3, -1), (4, 0)\}$$

It follows from the definition (and is evident from the example) that the domains and ranges of G and G^{-1} are interchanged.

Inverse of G

If G is a relation, the inverse of G, denoted by G^{-1}, is given by

$$G^{-1} = \{(b, a) \,|\, (a, b) \in G\}$$
$$\text{Domain of } G^{-1} = \text{Range of } G$$
$$\text{Range of } G^{-1} = \text{Domain of } G$$

If a relation G is specified by an equation, say

$$G: \quad y = 2x - 1 \tag{1}$$

then how do we find G^{-1}? The answer is easy: we interchange the variables in Equation (1). Thus,

$$G^{-1}: \quad x = 2y - 1 \tag{2}$$

or, solving for y,

$$G^{-1}: \quad y = \frac{x + 1}{2} \tag{3}$$

Any ordered pair of numbers that satisfies Equation (1), when reversed in order, will satisfy Equations (2) and (3). For example, (3, 5) satisfies Equation (1) and (5, 3) satisfies Equations (2) and (3), as can easily be checked.

If we sketch a graph of G, G^{-1}, and $y = x$ on the same coordinate system (Figure 3), we will observe something interesting. If we fold the paper along the line $y = x$, the graphs of G and G^{-1} match. Actually, we can graph G^{-1} by drawing G with wet ink and folding the paper along $y = x$ before the ink dries; G will then print G^{-1}. [To prove this in general, one has to show that the line $y = x$ is the perpendicular bisector of the line joining (a, b) and (b, a).] Knowing that the graphs of G and G^{-1} are symmetric relative to the line $y = x$ makes it easy to graph G^{-1} if G is known and vice versa.

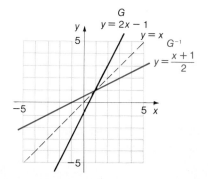

FIGURE 3

In Figure 3 observe also that G and G^{-1} are both functions. This is not always the case, however. Inverses of some functions may not be functions. Consider the following example.

EXAMPLE 12   The relation f is given by $y = x^2$.

    (A) Find f^{-1}.
    (B) Graph f, f^{-1}, and $y = x$. Is either f or f^{-1} a function?
    (C) Indicate the domain and range of f and f^{-1}.

Solution   **(A)** f^{-1}: $x = y^2$ or $y = \pm\sqrt{x}$
    (B)

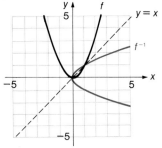

        f is a function

        f^{-1} is not a function

    (C) Domain of $f = R = $ Range of f^{-1}
        Range of $f = [0, \infty) = $ Domain of f^{-1}

PROBLEM 12   The relation f is given by $y = |x| + 2$.

    (A) Find f^{-1}.
    (B) Graph f, f^{-1}, and $y = x$. Is either f or f^{-1} a function?
    (C) Indicate the domain and range of f and f^{-1}.

Solution   **(A)**   **(B)**

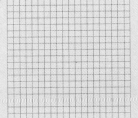

    (C)

■ One-to-One Correspondence and Inverses

If we are given a function f, how can we tell in advance whether its inverse f^{-1} will be a function? The answer is contained in the concept of one-to-one correspondence. A **one-to-one correspondence** exists between two sets if each element in the first set corresponds to exactly one element in the second set, and each element in the second set corresponds to exactly one element in the first set. Consider the two functions f and g and their inverses:

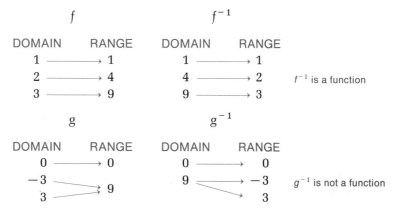

Function f has a one-to-one correspondence between domain and range values. (Notice that f^{-1} is also a function.) Function g does not have a one-to-one correspondence between domain and range values. (Notice that g^{-1} is not a function.)

THEOREM 1

> **Inverses**
>
> A function f has an inverse that is a function if and only if there exists a one-to-one correspondence between domain and range values of f. In this case, we say that f is a **one-to-one function** and note that
>
> $$f[f^{-1}(y)] = y \quad \text{and} \quad f^{-1}[f(x)] = x$$

Theorem 1 is interpreted schematically in Figure 4.

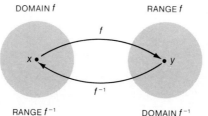

If the function f is one-to-one and if f maps x into y, then f^{-1} will map y back into x.

FIGURE 4

Figure 5 illustrates some functions that are one-to-one, and Figure 6 (page 512) illustrates some that are not. In Figure 5, each domain value corresponds to exactly one range value, and each range value corresponds to exactly one domain

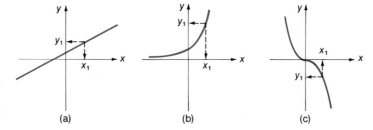

FIGURE 5 Functions that are one-to-one (each has an inverse that is a function)

value. In Figure 6, each domain value corresponds to exactly one range value, but some range values correspond to more than one domain value. (In Figure 6a, y_1 corresponds to x_1 and x_2; in Figure 6b, y_1 corresponds to x_1, x_2, and x_3.)

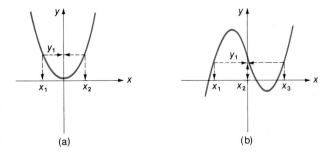

FIGURE 6 Functions that are not one-to-one (neither has an inverse that is a function)

(a)   (b)

When a function is defined by an equation and the function is one-to-one, we can often find its inverse in terms of an equation.

EXAMPLE 13 Given $f(x) = 3x + 2$, find:

(A) $f^{-1}(x)$ **(B)** $f^{-1}(5)$ **(C)** $f^{-1}[f(5)]$ **(D)** $f^{-1}[f(x)]$

Solution Since f is linear and has slope 3, it is one-to-one; hence, f has an inverse that is a function.

(A) f: $y = 3x + 2$ Replace $f(x)$ with y in $f(x) = 3x + 2$.

　　　f^{-1}: $x = 3y + 2$ Interchange variables x and y to obtain f^{-1}.

　　　　　$y = \dfrac{x - 2}{3}$ Solve for y in terms of x.

Thus,

$$f^{-1}(x) = \frac{x - 2}{3}$$ Replace y with $f^{-1}(x)$.

(B) $f^{-1}(5) = \dfrac{5 - 2}{3} = \dfrac{3}{3} = 1$

(C) $f^{-1}[f(5)] = \dfrac{f(5) - 2}{3}$ We are just verifying the results of Theorem 1.

$$= \frac{17 - 2}{3}$$

$$= \tfrac{15}{3} = 5$$

(D) $f^{-1}[f(x)] = \dfrac{f(x) - 2}{3}$ See comment in part (C).

$$= \frac{(3x + 2) - 2}{3} = x$$

PROBLEM 13

Given $g(x) = \dfrac{x}{3} - 2$, find:

(A) $g^{-1}(x)$　　**(B)** $g^{-1}(-2)$　　**(C)** $g^{-1}[g(3)]$　　**(D)** $g^{-1}[g(x)]$

Solution

(A)

(B)

(C)

(D)

Solutions to Matched Problems

12. (A) f^{-1}:　$x = |y| + 2$ or $|y| = x - 2$ or $y = \pm(x - 2)$

(B) 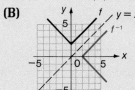　f is a function
　　　　f^{-1} is not a function

(C) Domain of $f = R =$ Range of f^{-1}
　　Range of $f = [2, \infty) =$ Domain of f^{-1}

13. (A)　　g:　$y = \dfrac{x}{3} - 2$

$$g^{-1}: \quad x = \frac{y}{3} - 2$$

$$3x = y - 6$$
$$y = 3x + 6$$
$$\text{or} \quad g^{-1}(x) = 3x + 6$$

(B) $g^{-1}(-2) = 3(-2) + 6 = 0$

(C) $g^{-1}(g(3)) = g^{-1}(\tfrac{3}{3} - 2) = g^{-1}(-1)$
　　　　$= 3(-1) + 6 = 3$

(D) $g^{-1}[g(x)] = g^{-1}\left(\dfrac{x}{3} - 2\right)$

$$= 3\left(\frac{x}{3} - 2\right) + 6 = x - 6 + 6 = x$$

Practice Exercise 8-4 ■

Work odd-numbered problems first, check answers, and then work even-numbered problems in areas of weakness. Answers to all problems are in the back of the book. Make every effort to work a problem yourself before you look at an answer.

A   *Find the inverse for each of the following relations.*

1. $R = \{(-2, 1), (0, 3), (2, 2)\}$ _____

2. $F = \{(-3, -1), (0, 1), (3, 2)\}$ _____

3. $G = \{(-2, 4), (-1, 1), (0, 0), (1, 1), (2, 4)\}$ _____

4. $H = \{(-5, 0), (-2, 1), (0, 0), (2, 1), (5, 0)\}$ _____

Graph on the same coordinate system along with y = x.

5.   R and R^{-1} in Problem 1

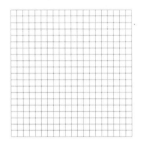

6.   F and F^{-1} in Problem 2

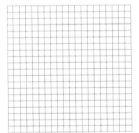

7.   G and G^{-1} in Problem 3

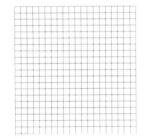

8.   H and H^{-1} in Problem 4

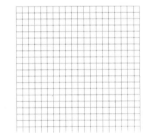

Indicate which are functions.

9.   R or R^{-1} in Problem 1 _____

10.   F or F^{-1} in Problem 2 _____

11.   G or G^{-1} in Problem 3 _____

12.   H or H^{-1} in Problem 4 _____

B   *Find the inverse for each of the following relations in the form of an equation.*

13.   f: $y = 3x - 2$ _____

14.   g: $y = 2x + 3$ _____

15.   F: $y = \dfrac{x}{3} - 2$ _____

16.   G: $y = \dfrac{x}{2} + 5$ _____

17. h: $y = \dfrac{x^2}{2}$ _____

18. H: $y = |2x|$ _____

In Problems 19–22 graph each pair of relations in the same coordinate system along with $y = x$. Indicate which relations are functions and which are one-to-one.

19. f and f^{-1} in Problem 13

20. g and g^{-1} in Problem 14

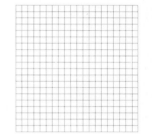

21. h and h^{-1} in Problem 17

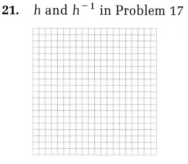

22. H and H^{-1} in Problem 18

23. For $f(x) = 3x - 2$, find:

(A) $f^{-1}(x)$ _____

(B) $f^{-1}(2)$ _____

(C) $f[f^{-1}(3)]$ _____

24. For $g(x) = 2x + 3$, find:

(A) $g^{-1}(x)$ _____

(B) $g^{-1}(5)$ _____

(C) $g[g^{-1}(4)]$ _____

25. For $F(x) = \dfrac{x}{3} - 2$, find:

(A) $F^{-1}(x)$ _____

(B) $F^{-1}(-1)$ _____

(C) $F^{-1}[F(4)]$ _____

26. For $G(x) = \dfrac{x}{2} + 5$, find:

(A) $G^{-1}(x)$ _____

(B) $G^{-1}(8)$ _____

(C) $G^{-1}[G(-4)]$ _____

C **27.** For $f(x) = \dfrac{x}{3} + 2$, find:

   (A) $f^{-1}(x)$ _____   **(B)** $f[f^{-1}(a)]$ _____

28. For $g(x) = 4x + 2$, find:

   (A) $g^{-1}(x)$ _____   **(B)** $g^{-1}[g(a)]$ _____

29. Find $F[F^{-1}(x)]$ for Problem 15. _____

30. Find $G^{-1}[G(x)]$ for Problem 16. _____

31. Let $G(x) = 3^x$, $x \in \{-2, -1, 0, 1, 2\}$.

   (A) Graph G, G^{-1}, and $y = x$ using the same coordinate axes.

   (B) Indicate whether G and G^{-1} are functions. _____

32. Let $H(x) = \sqrt{x}$, $x \geq 0$.

   (A) Find $H^{-1}(x)$. _____

   (B) Graph H, H^{-1}, and $y = x$ using the same coordinate axes.

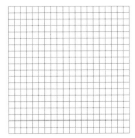

The Check Exercise for
this section is on
page 543.

   (C) Find $H^{-1}(3)$ and $H^{-1}[H(x)]$. _____

Section 8-5   Variation

■ Direct Variation
■ Inverse Variation
■ Joint Variation
■ Combined Variation

In reading scientific material, one is likely to come across statements such as "the pressure of an enclosed gas varies directly as the absolute temperature," or "the frequency of vibration of air in an organ pipe varies inversely as the length of the pipe," or even more complicated statements such as "the force of attraction between two bodies varies jointly as their masses and inversely as the square of the distance between the two bodies." These statements have precise mathematical meaning in that they represent particular types of functions. The purpose of this section is to investigate these special functions.

■ Direct Variation

The statement **y varies directly as x** means[†]

$$y = kx \qquad k \neq 0$$

where k is a constant called the **constant of variation**. Similarly, the statement "y varies directly as the square of x" means

$$y = kx^2 \qquad k \neq 0$$

and so on. The first equation defines a linear function, and the second a quadratic function.

Direct variation is illustrated by the familiar formulas

$$C = \pi D \qquad \text{and} \qquad A = \pi r^2$$

where the first formula asserts that the circumference of a circle varies directly as the diameter and the second states that the area of a circle varies directly as the square of the radius. In both cases, π is the constant of variation.

EXAMPLE 14   Translate each statement into an appropriate equation, and find the constant of variation if $y = 16$ when $x = 4$.

(A) y varies directly as x.
(B) y varies directly as the cube of x.

Solution   **(A)** y varies directly as x means

$$y = kx \quad \text{Do not forget } k.$$

To find the constant of variation k, substitute $x = 4$ and $y = 16$ and solve for k:

$$16 = k \cdot 4$$
$$k = \tfrac{16}{4} = 4$$

Thus, $k = 4$ and the equation of variation is

$$y = 4x$$

(B) y varies directly as the cube of x means

$$y = kx^3 \quad \text{Do not forget } k.$$

[†] You will sometimes hear "y is proportional to x" in place of "y varies directly as x." Both mean the same thing.

To find k, substitute $x = 4$ and $y = 16$:

$$16 = k \cdot 4^3$$
$$k = \frac{16}{64} = \frac{1}{4}$$

Thus, the equation of variation is

$$y = \frac{1}{4}x^3$$

PROBLEM 14   If $y = 4$ when $x = 8$, find the equation of variation for each statement:

(A) y varies directly as x.
(B) y varies directly as the cube root of x.

Solution   **(A)**   **(B)**

■ **Inverse Variation**

The statement **y varies inversely as x** means[†]

$$y = \frac{k}{x} \qquad k \neq 0$$

where k is a constant (the constant of variation). As in the case of direct variation, we also discuss y varying inversely as the square of x and so on.

An illustration of inverse variation is given in the distance–rate–time formula $d = rt$ in the form $t = d/r$ for a fixed distance d. In driving a fixed distance, say $d = 400$ miles, time varies inversely as the rate; that is,

$$t = \frac{400}{r}$$

where 400 is the constant of variation—as the rate increases, the time decreases and vice versa.

EXAMPLE 15   Translate each statement into an appropriate equation, and find the constant of variation if $y = 16$ when $x = 4$.

(A) y varies inversely as x.
(B) y varies inversely as the square root of x.

[†] You will sometimes hear "y is inversely proportional to x" in place of "y varies inversely as x." Both mean the same thing.

Solution   **(A)** y varies inversely as x means

$$y = \frac{k}{x} \quad \text{Do not forget } k.$$

To find k, substitute $x = 4$ and $y = 16$:

$$16 = \frac{k}{4}$$

$$k = 64$$

Thus, the equation of variation is

$$y = \frac{64}{x}$$

(B) y varies inversely as the square root of x means

$$y = \frac{k}{\sqrt{x}}$$

To find k, substitute $x = 4$ and $y = 16$:

$$16 = \frac{k}{\sqrt{4}}$$

$$k = 32$$

Thus, the equation of variation is

$$y = \frac{32}{\sqrt{x}}$$

PROBLEM 15   If $y = 4$ when $x = 8$, find the equation of variation for each statement:

(A) y varies inversely as x.
(B) y varies inversely as the square of x.

Solution   **(A)**   **(B)**

■ **Joint Variation**

The statement **w varies jointly as x and y** means

$$w = kxy \qquad k \neq 0$$

where k is a constant (the constant of variation). Similarly, if

$$w = kxyz^2 \qquad k \neq 0$$

we would say that "w varies jointly as x, y, and the square of z," and so on. For example, the area of a rectangle varies jointly as its length and width (recall $A = lw$), and the volume of a right circular cylinder varies jointly as the square of its radius and its height (recall $V = \pi r^2 h$). What is the constant of variation in each case?

■ Combined Variation

The basic types of variation introduced above are often combined. For example, the statement "w varies jointly as x and y and inversely as the square of z" means

$$w = k\frac{xy}{z^2} \qquad k \neq 0 \quad \text{We do not write:} \quad w = \frac{kxy}{kz^2}.$$

This is wrong because it eliminates the proportionality constant k.

Thus, the statement "the force of attraction F between two bodies varies jointly as their masses m_1 and m_2 and inversely as the square of the distance d between the two bodies" means

$$F = k\frac{m_1 m_2}{d^2} \qquad k \neq 0$$

If (assuming k is positive) either of the two masses is increased, the force of attraction increases; on the other hand, if the distance is increased, the force of attraction decreases.

EXAMPLE 16   The pressure P of an enclosed gas varies directly as the absolute temperature T and inversely as the volume V. If 500 cubic feet of gas yields a pressure of 10 pounds per square foot at a temperature of 300 K (absolute temperature[†]), what will be the pressure of the same gas if the volume is decreased to 300 cubic feet and the temperature increased to 360 K?

Solution   *Method 1:*   Write the equation of variation $P = k(T/V)$, and find k using the first set of values:

$$10 = k(\tfrac{300}{500})$$
$$k = \tfrac{50}{3}$$

Hence, the equation of variation for this particular gas is $P = \tfrac{50}{3}(T/V)$.
    Now find the new pressure P using the second set of values:

$$P = \tfrac{50}{3}(\tfrac{360}{300}) = 20 \text{ pounds per square foot}$$

Method 2 (generally faster than Method 1):   Write the equation of variation $P = k(T/V)$; then convert to the equivalent form:

$$\frac{PV}{T} = k$$

If P_1, V_1, and T_1 are the first set of values for the gas and P_2, V_2, and T_2 are

[†] A Kelvin (absolute) and a Celsius degree are the same size, but 0 on the Kelvin scale is $-273°$ on the Celsius scale. This is the point at which molecular action is supposed to stop and is called "absolute zero."

the second set, then

$$\frac{P_1 V_1}{T_1} = k \qquad \text{and} \qquad \frac{P_2 V_2}{T_2} = k$$

Hence,

$$\frac{P_1 V_1}{T_1} = \frac{P_2 V_2}{T_2}$$

Since all values are known except P_2, substitute and solve. Thus,

$$\frac{(10)(500)}{300} = \frac{P_2(300)}{360}$$

$$P_2 = 20 \text{ pounds per square foot}$$

PROBLEM 16   The length L of skid marks of a car's tires (when brakes are applied) varies directly as the square of the speed v of the car. If skid marks of 20 feet are produced at 30 miles per hour, how fast would the same car be going if it produced skid marks of 80 feet? Solve in two ways (see Example 16).

Solution   *Method 1:*   *Method 2:*

EXAMPLE 17   The frequency of pitch f of a given musical string varies directly as the square root of the tension T and inversely as the length L. What is the effect on the frequency if the tension is increased by a factor of 4 and the length is cut in half?

Solution   Write the equation of variation:

$$f = \frac{k\sqrt{T}}{L} \qquad \text{or equivalently} \qquad \frac{f_2 L_2}{\sqrt{T_2}} = \frac{f_1 L_1}{\sqrt{T_1}}$$

We are given that $T_2 = 4T_1$ and $L_2 = 0.5L_1$. Substituting in the second equation, we have

$$\frac{f_2 0.5 L_1}{\sqrt{4T_1}} = \frac{f_1 L_1}{\sqrt{T_1}} \qquad \text{Solve for } f_2.$$

$$\frac{f_2 0.5 L_1}{2\sqrt{T_1}} = \frac{f_1 L_1}{\sqrt{T_1}}$$

$$f_2 = \frac{2\sqrt{T_1} f_1 L_1}{0.5 L_1 \sqrt{T_1}} = 4f_1$$

Thus, the frequency of pitch is increased by a factor of 4.

PROBLEM 17   The weight w of an object on or above the surface of the earth varies inversely as the square of the distance d between the object and the center of the earth. If an object on the surface of the earth is moved into space so as to double its distance from the earth's center, what effect will this move have on its weight?

Solution

Solutions to Matched Problems

14. (A) $y = kx$
$\quad 4 = k \cdot 8$
$\quad k = \frac{1}{2}$
$\quad y = \frac{1}{2}x$

(B) $y = k\sqrt[3]{x}$
$\quad 4 = k\sqrt[3]{8}$
$\quad k = 2$
$\quad y = 2\sqrt[3]{x}$

15. (A) $y = \dfrac{k}{x}$
$\quad 4 = \dfrac{k}{8}$
$\quad k = 32$
$\quad y = \dfrac{32}{x}$

(B) $y = \dfrac{k}{x^2}$
$\quad 4 = \dfrac{k}{8^2}$
$\quad k = 256$
$\quad y = \dfrac{256}{x^2}$

16. *Method 1:*

$L = kv^2 \qquad\qquad 80 = \dfrac{2}{90}v^2$
$20 = k(30)^2$
$\quad k = \dfrac{2}{90} \qquad\qquad v^2 = \dfrac{(80)(90)}{2}$
$\qquad\qquad\qquad\qquad\quad = 3{,}600$
$\quad L = \dfrac{2}{90}v^2 \qquad\quad v = \sqrt{3{,}600} = 60$ miles per hour

Method 2:

$L = kv^2 \qquad$ Now find v_2, given $L_1 = 20$,
$\dfrac{L}{v^2} = k \qquad\qquad \dfrac{20}{30^2} = \dfrac{80}{v_2^2}$
$\dfrac{L_1}{v_1^2} = \dfrac{L_2}{v_2^2} \qquad\quad v_2^2 = \dfrac{(900)(80)}{20}$
$\qquad\qquad\qquad\quad v_2 = \sqrt{3{,}600} = 60$ miles per hour

17. $w = \dfrac{k}{d^2}$; $d_2 = 2d_1$; $w_2 = \dfrac{k}{d_2^2} = \dfrac{k}{(2d_1)^2} = \dfrac{k}{4d_1^2} = \dfrac{1}{4} \cdot w_1$

Thus, it will be one-fourth as heavy.

Practice Exercise 8-5 ■

Work odd-numbered problems first, check answers, and then work even-numbered problems in areas of weakness. Answers to all problems are in the back of the book. Make every effort to work a problem yourself before you look at an answer.

A   *Translate each problem into an equation using k as the constant of variation.*

1. F varies directly as the square of v. _____

2. u varies directly as v. _____

3.   The pitch or frequency f of a guitar string of a given length varies directly as the square root of the tension T of the string. _____

4.   Geologists have found in studies of earth erosion that the erosive force (sediment-carrying power) P of a swiftly flowing stream varies directly as the sixth power of the velocity v of the water. _____

5. y varies inversely as the square root of x. _____

6. I varies inversely as t. _____

7.   The biologist Reaumur suggested in 1735 that the length of time t that it takes fruit to ripen during the growing season varies inversely as the sum T of the average daily temperatures during the growing season. _____

8.   In a study on urban concentration, F. Auerbach discovered an interesting law. After arranging all the cities of a given country according to their population size, starting with the largest, he found that the population P of a city varied inversely as the number n indicating its position in the ordering. _____

9. R varies jointly as S, T, and V. _____

10. g varies jointly as x and the square of y. _____

11.   The volume of a cone V varies jointly as its height h and the square of the radius r of its base. _____

12.   The amount of heat put out by an electrical appliance (in calories) varies jointly as time t, resistance R in the circuit, and the square of the current I. _____

Solve using either of the two methods illustrated in Example 16.

13. u varies directly as the square root of v. If $u = 2$ when $v = 2$, find u when $v = 8$. _____

14. y varies directly as the square of x. If $y = 20$ when $x = 2$, find y when $x = 5$. _____

15. *L* varies inversely as the square root of *M*. If $L = 9$ when $M = 9$, find *L* when $M = 3$. _____

16. *I* varies inversely as the cube of *t*. If $I = 4$ when $t = 2$, find *I* when $t = 4$.

_____

B *Translate each problem into an equation using k as the constant of variation.*

17. *U* varies jointly as *a* and *b* and inversely as the cube of *c*. _____

18. *w* varies directly as the square of *x* and inversely as the square root of *y*. _____

19. The maximum safe load *L* for a horizontal beam varies jointly as its width *w* and the square of its height *h* and varies inversely as its length *l*.

_____

20. Joseph Cavanaugh, a sociologist, found that the number of long-distance phone calls *n* between two cities in a given time period varied (approximately) jointly as the populations P_1 and P_2 of the two cities and inversely as the distance *d* between the two cities. _____

Solve using either of the two methods illustrated in Example 16.

21. *Q* varies jointly as *m* and the square of *n* and inversely as *P*. If $Q = -4$ when $m = 6$, $n = 2$, and $P = 12$, find *Q* when $m = 4$, $n = 3$, and $P = 6$.

_____

22. *w* varies jointly as *x*, *y*, and *z* and inversely as the square of *t*. If $w = 2$ when $x = 2$, $y = 3$, $z = 6$, and $t = 3$, find *w* when $x = 3$, $y = 4$, $z = 2$, and $t = 2$. _____

23. The weight *w* of an object on or above the surface of the earth varies inversely as the square of the distance *d* between the object and the center of the earth. If a girl weighs 100 pounds on the surface of the earth, how much would she weigh (to the nearest pound) 400 miles above the earth's surface? (Assume that the radius of the earth is 4,000 miles.) _____

24. A child was struck by a car in a crosswalk. The driver of the car had slammed on his brakes and left skid marks 160 feet long. He told the police he had been driving at 30 miles per hour. The police know that the length of skid marks *L* (when brakes are applied) varies directly as the square of the speed of the car *v* and that at 30 miles per hour (under ideal conditions) skid marks would be 40 feet long. How fast was the driver actually going before he applied his brakes? _____

25. Ohm's law states that the current *I* in a wire varies directly as the electromotive force *E* and inversely as the resistance *R*. If $I = 22$ amperes when $E = 110$ volts and $R = 5$ ohms, find *I* if $E = 220$ volts and $R = 11$ ohms.

_____

26. Anthropologists, in their study of race and human genetic groupings, often use an index called the "cephalic index." The cephalic index C varies directly as the width w of the head and inversely as the length l of the head (both when viewed from the top). If an Indian in Baja California (Mexico) has measurements of $C = 75$, $w = 6$ inches, and $l = 8$ inches, what is C for an Indian in northern California with $w = 8.1$ inches and $l = 9$ inches? _____

C 27. If the horsepower P required to drive a speedboat through water varies directly as the cube of the speed v of the boat, what change in horsepower is required to double the speed of the boat? _____

28. The intensity of illumination E on a surface varies inversely as the square of its distance d from a light source. What is the effect on the total illumination on a book if the distance between the light source and the book is doubled? _____

29. The frequency of vibration f of a musical string varies directly as the square root of the tension T and inversely as the length L of the string. If the tension of the string is increased by a factor of 4 and the length of the string is doubled, what is the effect on the frequency? _____

30. In an automobile accident the destructive force F of a car varies (approximately) jointly as the weight w of the car and the square of the speed v of the car. (This is why accidents at high speed are generally so serious.) What would be the effect on the destructive force of a car if its weight were doubled and its speed were doubled? _____

ADDITIONAL APPLICATIONS

The following problems include significant applications from many different areas and are arranged according to subject area. The more difficult problems are marked with two asterisks (**), the moderately difficult problems are marked with one asterisk (*), and the easier problems are not marked.

Astronomy 31. The square of the time t required for a planet to make one orbit around the sun varies directly as the cube of its mean (average) distance d from the sun. Write the equation of variation, using k as the constant of variation.

_____

*32. The centripetal force F of a body moving in a circular path at constant speed varies inversely as the radius r of the path. What happens to F if r is doubled? _____

33. The length of time t a satellite takes to complete a circular orbit of the earth varies directly as the radius r of the orbit and inversely as the orbital velocity v of the satellite. If $t = 1.42$ hours when $r = 4,050$ miles and $v = 18,000$ miles per hour (Sputnik I), find t for $r = 4,300$ miles and $v = 18,500$ miles per hour. _____

Life Science 34. The number N of gene mutations resulting from x-ray exposure varies directly as the size of the x-ray dose r. What is the effect on N if r is quadrupled? _____

35. In biology there is an approximate rule, called the "bioclimatic rule" for temperate climates, which states that the difference d in time for fruit to ripen (or insects to appear) varies directly as the change in altitude h. If $d = 4$ days when $h = 500$ feet, find d when $h = 2,500$ feet.

Physics and Engineering

36. Over a fixed distance d, speed r varies inversely as time t. Police use this relationship to set up speed traps. (The graph of the resulting function is a hyperbola.) If in a given speed trap $r = 30$ miles per hour when $t = 6$ seconds, what would be the speed of a car if $t = 4$ seconds? _____

*37. The length L of skid marks of a car's tires (when the brakes are applied) varies directly as the square of the speed v of the car. How is the length of skid marks affected by doubling the speed? _____

38: The time t required for an elevator to lift a weight varies jointly as the weight w and the distance d through which it is lifted and inversely as the power P of the motor. Write the equation of variation, using k as the constant of variation. _____

39. The total pressure P of the wind on a wall varies jointly as the area of the wall A and the square of the velocity of the wind v. If $P = 120$ pounds when $A = 100$ square feet and $v = 20$ miles per hour, find P if $A = 200$ square feet and $v = 30$ miles per hour. _____

**40. The thrust T of a given type of propeller varies jointly as the fourth power of its diameter d and the square of the number of revolutions per minute n it is turning. What happens to the thrust if the diameter is doubled and the number of revolutions per minute is cut in half? _____

Psychology

41. In early psychological studies on sensory perception (hearing, seeing, feeling, and so on), the question was asked: "Given a certain level of stimulation S, what is the minimum amount of added stimulation ΔS that can be detected?" A German physiologist, E. H. Weber (1795–1878) formulated, after many experiments, the famous law that now bears his name: "The amount of change ΔS that will be just noticed varies directly as the magnitude S of the stimulus."

(A) Write the law as an equation of variation. _____

(B) If a person lifting weights can just notice a difference of 1 ounce at the 50-ounce level, what will be the least difference she will be able to notice at the 500-ounce level? _____

(C) Determine the just noticeable difference in illumination a person is able to perceive at 480 candlepower if he is just able to perceive a difference of 1 candlepower at the 60-candlepower level.

---

42. Psychologists in their study of intelligence often use an index called IQ. IQ varies directly as mental age MA and inversely as chronological age CA (up to the age of 15). If a 12-year-old boy with a mental age of 14.4 has an IQ of 120, what will be the IQ of an 11-year-old girl with a mental age of 15.4? _____

Music **43.** The frequency of vibration of air in an open organ pipe varies inversely as the length of the pipe. If the air column in an open 32-foot pipe vibrates 16 times per second (low C), how fast would the air vibrate in a 16-foot pipe? _____

44. The frequency of pitch f of a musical string varies directly as the square root of the tension T and inversely as the length l and the diameter d. Write the equation of variation using k as the constant of variation. (It is interesting to note that if pitch depended on only length, then pianos would have to have strings varying from 3 inches to 38 feet.)

_____

Photography **45.** The f-stop numbers N on a camera, known as focal ratios, vary directly as the focal length F of the lens and inversely as the diameter d of the diaphragm opening (effective lens opening). Write the equation of variation using k as the constant of variation. _____

*46. In taking pictures using flashbulbs, the lens opening (f-stop number) N varies inversely as the distance d from the object being photographed. What adjustment should you make on the f-stop number if the distance between the camera and the object is doubled? _____

Chemistry ***47.** Atoms and molecules that make up the air constantly fly about like microscopic missiles. The velocity v of a particular particle at a fixed temperature varies inversely as the square root of its molecular weight w. If an oxygen molecule in air at room temperature has an average velocity of 0.3 mile per second, what will be the average velocity of a hydrogen molecule, given that the hydrogen molecule is one-sixteenth as heavy as the oxygen molecule? _____

48. The Maxwell–Boltzmann equation says that the average velocity v of a molecule varies directly as the square root of the absolute temperature T and inversely as the square root of its molecular weight w. Write the equation of variation using k as the constant of variation. _____

Business **49.** The amount of work A completed varies jointly as the number of workers W used and the time t they spend. If 10 workers can finish a job in 8 days, how long will it take 4 workers to do the same job? _____

50. The simple interest I earned in a given time varies jointly as the principal p and the interest rate r. If \$100 at 4% interest earns \$8, how much will \$150 at 3% interest earn in the same period? _____

Geometry   ***51.**   The volume of a sphere varies directly as the cube of its radius r. What happens to the volume if the radius is doubled? _____

The Check Exercise for
this section is on
page 545.

*52.   The surface area S of a sphere varies directly as the square of its radius r. What happens to the area if the radius is cut in half? _____

Section 8-6   Chapter Review

A **relation** is a rule that produces a correspondence between a first set (the **domain**) and a second set (the **range**) such that for each element in the domain there is at least one corresponding element in the range. A **function** is a relation such that for each element in the domain there is exactly one corresponding element in the range. A relation may thus be considered a set of ordered pairs; a function is therefore a set of ordered pairs with no two distinct pairs having the same first component. Relations and functions may be specified by diagrams, tables, equations, sets of ordered pairs, or graphs. The values in the domain can be thought of as **input** values, those in the range as **output**. A relation between real numbers can be represented by a set of points in the plane, the **graph of the relation**; such a relation is a function if no vertical line crosses its graph more than once (**vertical-line test**). If a relation is specified by an equation, the variable for values in the domain is called the **independent variable**; for values in the range it is called the **dependent variable**.   *(8-1)*

If a function is denoted by f, the notation $f(x)$ denotes the value in the range corresponding to x in the domain.   *(8-2)*

A function $f(x) = ax + b$ is called a **linear function**; $f(x) = ax^2 + bx + c$ is a **quadratic function** for $a \neq 0$. The graph of a linear function is a nonvertical straight line; that of a quadratic function is a parabola. If $f(x)$ is specified by a polynomial in x, the function is called a **polynomial function**. The **nested factored form** of the polynomial is useful in evaluating the function. Polynomial functions are graphed by plotting a sufficient number of points. Graphing a quadratic function allows us to estimate the **axis of symmetry** and the **maximum** or **minimum** value of the function.   *(8-3)*

If G is a relation, the relation $G^{-1} = \{(b, a) \mid (a, b) \in G\}$ is the **inverse** of G; the domain of G^{-1} is the range of G, and the range of G^{-1} is the domain of G. If G is a relation between real numbers, the graph of G^{-1} is the graph of G reflected about the line $y = x$. A function f is **one-to-one** if every range element corresponds to exactly one domain element. The inverse f^{-1} of a function f is itself a function if and only if f is one-to-one.   *(8-4)*

Direct, inverse, joint, and **combined variations** are relations that occur often in applications:

Direct:	y varies directly as x	$y = kx$
Inverse:	y varies inversely as x	$y = \dfrac{k}{x}$
Joint:	w varies jointly as x and y	$w = kxy$
Combined:	w varies directly as x and inversely as y	$w = \dfrac{kx}{y}$

In each case, k is called the **constant of variation**.   *(8-5)*

Diagnostic (Review) Exercise 8-6 ■

Work through all the problems in this chapter review and check answers in the back of the book. (Answers to all problems are there, and following each answer is a number in italics indicating the section in which that type of problem is discussed.) Where weaknesses show up, review appropriate sections in the text. When you are satisfied that you know the material, take the practice test following this review.

A Which of the relations in Problems 1–12 are functions? The variable x is independent.

1.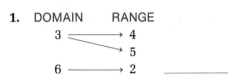

2.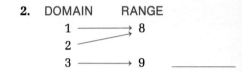

3.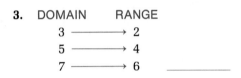

4.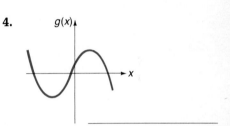

5.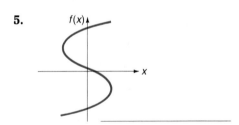

6.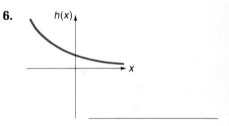

7. $y = x^3 - 2x$ _____

8. $y^2 = x$ _____

9. $x^2 + y^2 = 25$ _____

10. $\{(1, 2), (1, -2), (0, 3)\}$ _____

11. $\{(-1, 2), (1, 3), (2, 4)\}$ _____

12. $\{(-2, 3), (0, 3), (2, 3)\}$ _____

Write the domain and range of the relation in the problem indicated.

13. Problem 10 _____

14. Problem 11 _____

15. Problem 12 _____

16. If $f(x) = 6 - x$, find:

    (A) $f(6)$ _____

    (B) $f(0)$ _____

    (C) $f(-3)$ _____

    (D) $f(m)$ _____

17.   If $G(z) = z - 2z^2$, find:

(A) $G(2)$ _____

(B) $G(0)$ _____

(C) $G(-1)$ _____

(D) $G(c)$ _____

18.   Graph $f(x) = 2x - 4$. Indicate its slope and y intercept.

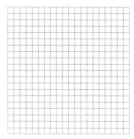

19.   Graph $g(x) = \dfrac{x^2}{2}$.

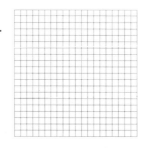

20.   Graph the relation $M = \{(0, 5), (2, 7), (2, 3)\}$, its inverse M^{-1}, and $y = x$, all on the same coordinate system. Indicate whether M or M^{-1} is a function.

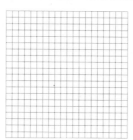

21.   What are the domain and range of M^{-1} in the preceding problem?

_____

Translate each statement into an equation using k as the constant of variation.

22. m varies directly as the square of n. _____

23. P varies inversely as the cube of Q. _____

24. A varies jointly as a and b. _____

25. y varies directly as the cube of x and inversely as the square root of z.

_____

B **26.** Is every relation a function? Explain. _____

27. If $f(t) = 4 - t^2$ and $g(t) = t - 3$, find:

(A) $f(0) - g(0)$ _____ (B) $\dfrac{g(6)}{f(-1)}$ _____

(C) $g(x) - f(x)$ _____ (D) $f[g(2)]$ _____

28. If $f(x) = 2x - 3$, find:

(A) $f(3 + h)$ _____ (B) $\dfrac{f(3 + h) - f(3)}{h}$ _____

29. Graph $g(t) = -\frac{3}{2}t + 6$ and indicate its slope and y intercept.

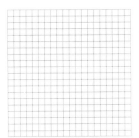

30. Graph $f(x) = x^2 - 4x + 5$. Estimate its vertex, axis, and the maximum or minimum value of $f(x)$.

31. Write $P(x) = x^3 - 2x^2 - 5x + 6$ in a nested factored form and graph for $-3 \leq x \leq 4$.

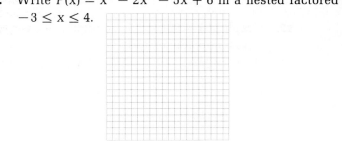

Indicate which relations are one-to-one correspondences in the problems indicated.

32. Problems 1–3 _____ **33.** Problems 4–6 _____

34. Problems 7–9 _____ **35.** Problems 10–12 _____

36. Which of the functions in Problems 4 and 6 have inverses that are functions? _____

37. Which of the functions in Problems 2 and 3 have inverses that are functions? _____

38. Determine the domain of the relation specified by the equation

$$y = \sqrt{\frac{x + 2}{x - 5}}$$ _____

39. Let $M(x) = \dfrac{x + 3}{2}$.

  (A) Find $M^{-1}(x)$. _____

  (B) Are both M and M^{-1} functions? _____

  (C) Find $M^{-1}(2)$. _____

  (D) Find $M^{-1}[M(3)]$. _____

40. If y varies directly as x and inversely as z:

  (A) Write the equation of variation. _____

  (B) If $y = 4$ when $x = 6$ and $z = 2$, find y when $x = 4$ and $z = 4$.

  _____

C **41.** If $g(t) = 1 - t^2$, find:

  (A) $g(2 + h)$ _____   **(B)** $\dfrac{g(2 + h) - g(2)}{h}$ _____

42. Graph $g(t) = 96t - 16t^2$. Indicate its vertex, axis, and the maximum or minimum value of $f(x)$.

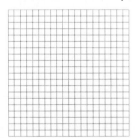

43. Let $f(x) = x^2$, $x \geq 0$.

  (A) Find $f^{-1}(x)$. _____

(B) Graph f and f^{-1} on the same coordinate system along with $y = x$.

(C) Find $f^{-1}(9)$ and $f^{-1}[f(x)]$. _____

44. Let $E(x) = 2^x$, $x \in \{-2, -1, 0, 1, 2\}$.

(A) Graph E, E^{-1}, and $y = x$ on the same coordinate system.

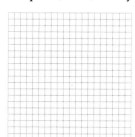

(B) Indicate whether E and E^{-1} are functions. _____

APPLICATIONS

45. *Cost function:* The cost $C(x)$ for renting a business copying machine is $200 for 1 month plus 5 cents a copy for x copies. Express this functional relationship in terms of an equation and graph it for $0 \leq x \leq 3,000$.

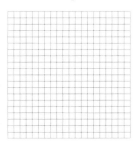

46. *Revenue function:* The revenue function for a company producing stereo radios (for a particular model) is

$$R = f(p) = 6,000p - 30p^2 \qquad 0 \leq p \leq 200$$

where p is the price per unit.

(A) Graph f.

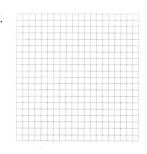

(B)   At what price p will the revenue be maximum? What is the maximum? _____

47. *Variation:*   The time t required for an elevator to lift a weight varies jointly as the weight w and the distance d through which it is lifted and inversely as the power P of the motor. Write the equation of variation using k as the constant of variation. If it takes a 400-horsepower motor 4 seconds to lift an 800-kilogram elevator 8 meters, how long will it take the same motor to lift a 1,600-kilogram elevator 24 meters? _____

*48. *Variation:*   The total force F of a wind on a wall varies jointly as the area of the wall A and the square of the velocity of the wind v. How is the total force on a wall affected if the area is cut in half and the wind velocity is doubled? _____

Practice Test Chapter 8 ■

Take this as if it were a graded test by working the problems within a 50-minute time period. Do not look back in the chapter. Choose one of three levels of difficulty: least difficult, Problems 1–12; more difficult, add Problem 13; most difficult, add Problems 13 and 14. Use the answers in the back of the book to correct your work. The answers are keyed to appropriate text sections so that you can easily locate and review sections where difficulties still persist.

1.   Which of the following relations are functions (x is independent):

(A)   Domain   Range   (B)
    2 $\longrightarrow$ 7
    4
    6 $\longrightarrow$ 11

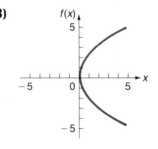

(C) $y = x^2 - 3x + 1$ (D) $\{(-1, 2), (0, 0), (-1, 3)\}$

_____

2.   What is the range of the relation given in Problem 1(C) if the domain is $\{-1, 0, 1, 2\}$? _____

3.   For $f(t) = t - 2$ and $g(t) = 6 - t^2$, find $f(4) + g(2) - 3g(0)$. _____

4.   For $g(x) = 2x - 1$, find $\dfrac{g(a + h) - g(a)}{h}$. _____

5.   For $f(t) = t - 2$ and $g(t) = 6 - t^2$, find $f(g(m))$. _____

6. Find the domain of the function $f(x) = \dfrac{1}{\sqrt{x^2 + x - 2}}$. _____

7. Given $f(x) = \dfrac{x - 2}{4}$, find $f^{-1}(x)$ and $f^{-1}(1)$. _____

8. Write the equation of variation equivalent to: m varies directly as n to the two-thirds' power and inversely as p to the fourth power. _____

9. If y varies directly as the square of x and inversely as z, and if $y = 8$ when $x = 2$ and $z = 3$, find y when $x = 6$ and $z = 4$. _____

10. Graph $f(x) = -\dfrac{x}{2} + 4$ and indicate the slope and y intercept of the graph.

_____

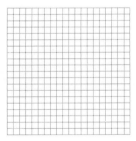

11. Graph $f(x) = x^2 - 4x + 3$ and estimate the maximum or minimum value of $f(x)$, whichever exists. _____

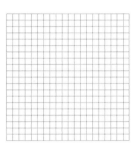

12. For $f(x) = |3x|$, $x \in [-3, 3]$, graph f, f^{-1}, and $y = x$, all on the same coordinate system; indicate whether f or f^{-1} is a function. _____

13.   Graph $f(x) = x^3 + x^2 - 9x - 9$, $-3 \le x \le 3$, using a nested factored form.

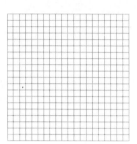

14.   Find the domain for f^{-1} if $f(x)$ is the function $f(x) = \dfrac{x^2 + 1}{1 - x^2}$. _____

Check Exercise 8-1 ■

Work the following problems without looking at any text examples. Show your work in the space provided. Write your answer in the answer column.

1. _____

2. _____

3. _____

4. _____

5. _____

---

1. Which of the relations below are not functions?

    (A) $1 \longrightarrow 0$ **(B)** $-2 \longrightarrow 4$
           $3 \longrightarrow 7$ 2
           5 $-1 \longrightarrow 1$
                                1

    (C) $4 \longrightarrow 2$ **(D)** -2
                -2 $0 \longrightarrow 0$
           $9 \longrightarrow 3$ 2
                -3

2. Which of the relations below are not functions?

    (A) 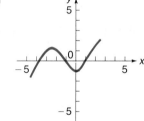   **(B)**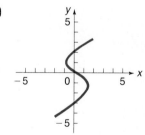

    (C) 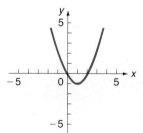   **(D)**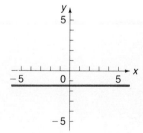

3. Which of the equations below do not specify a function, given that x is the independent variable?

    (A) $y = 2x^2 + 1$ **(B)** $y^2 = 4x$ **(C)** $y = x^4$ **(D)** $y - x = 1$

4. Given the relation $F = \{(-2, 1), (0, 1), (-2, 3)\}$, write the domain and indicate whether F is a function.

5. Given the relation $y = x^2 - x$, $x \in \{1, 2, 3\}$, write the range and indicate whether the relation is a function.

Check Exercise 8-2 ■

ANSWER COLUMN

Work the following problems without looking at any text examples. Show your work in the space provided. Write your answer in the answer column.

1. _____

Problems 1–9 refer to the functions

2. _____

$$f(x) = 2x - 4 \qquad g(t) = 3 - t \qquad F(u) = 2u^2 \qquad G(v) = 2v - v^2$$

Evaluate as indicated.

3. _____

1. $g(-3)$

4. _____

5. _____

6. _____

2. $G(-2)$

7. _____

8. _____

3. $f(3) - F(2)$

9. _____

10. _____

4. $3f(0) - g(0)$

5. $\dfrac{G(-1)}{2F(2) - 13}$

6. $\dfrac{g(2 + h) - g(2)}{h}$

7. $f[g(4)]$

8. $F[G(-1)]$

9. $\dfrac{G(v + h) - G(v)}{h}$

10. Write an equation that specifies the function C: the total cost per week $C(x)$ of manufacturing x electronic games if fixed costs are \$4,000 per week and variable costs are \$9 per game.

Name _____   Class _____   Score _____

Check Exercise 8-3 ■

ANSWER COLUMN

Work the following problems without looking at any text examples. Show your work in the space provided. Draw the final graph or write your answer in the answer column.

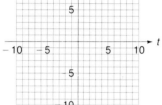

Graph Problems 1–4.

1. _____

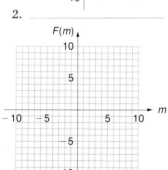

1. $f(x) = 3 - \dfrac{x}{2}$

2. _____

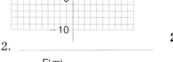

2. $g(t) = t^2 - 4t$

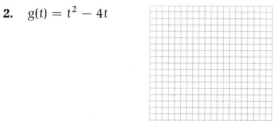

3. _____

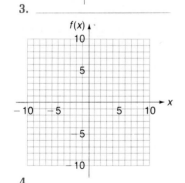

3. $F(m) = -8 - 8m - m^2$

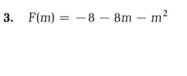

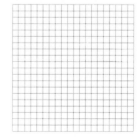

4. _____

5. _____

4. $f(x) = x^3 - 7x^2 + 14x - 8$

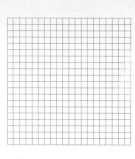

5. Estimate the maximum or minimum value (indicate which) of $f(x) = 4 - 16x - 4x^2$ using its graph.

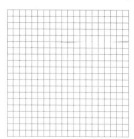

Name _____ Class _____ Score _____

Check Exercise 8-4 ■

ANSWER COLUMN

Work the following problems without looking at any text examples. Show your work in the space provided. Write your answer or draw the final graph in the answer column.

1. _____

2. _____   **1.** For $f(x) = 3 - \dfrac{x}{2}$, find $f^{-1}(x)$ and $f^{-1}(2)$.

3. _____

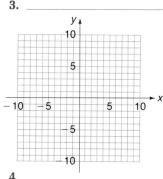

4. _____

                    2. For f and f^{-1} in Problem 1, find $f^{-1}[f(6)]$ and $f^{-1}[f(a)]$.

5. _____

                    3. For F: $y = x^2 + 1$, find F^{-1} in the form of an equation.

4. Graph F and F^{-1} from Problem 3 and $y = x$, all on the same coordinate system.

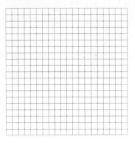

5. Indicate which of F and F^{-1} from Problems 3 and 4 are functions.

Check Exercise 8-5 ■

Work the following problems without looking at any text examples. Show your work in the space provided. Write your answer in the answer column.

1. _____

2. _____

In Problems 1–6 translate each statement into an equation, using k as the constant of variation.

   1. M varies directly as the square root of N.

3. _____

4. _____

5. _____

   2. x varies inversely as the cube of *t*.

6. _____

7. _____

8. _____

   3. P varies jointly as u and v.

9. _____

10. _____

   4. Q varies jointly as m, n, and the cube of s.

   5. F varies directly as the square of c and inversely as the cube root of d.

6. G varies jointly as u and the cube of v, and inversely as the square of w.

Solve Problems 7–10.

7. If y varies directly as x^2, and $y = 60$ when $x = 6$, find y when $x = 3$.

8. If w varies jointly as x and y, and inversely as z, and if $w = 60$ when $x = 3$, $y = 50$, and $z = 2$, find w when $x = 2$, $y = 30$, and $z = 3$.

9. The horsepower P required to drive a speedboat through water varies directly as the cube of the speed v of the boat. If 10 horsepower is required to drive the boat at 20 miles per hour, how much horsepower is required to drive it at 30 miles per hour?

10. The frequency of vibration f of a musical string varies directly with the square root of the tension T and inversely as the length L of the string. If the frequency is 440 hertz when $T = 36$ kilograms and $L = 60$ centimeters, find the frequency when $T = 49$ kilograms and $L = 50$ centimeters.

Exponential and Logarithmic Functions ■ 9

INSTRUCTIONS FOR STUDENTS IN A SELF-PACED CLASS OR LAB

YES ← **HAVE YOU HAD INTERMEDIATE ALGEBRA BEFORE THIS COURSE?** → NO

1. Work Diagnostic (Review) Exercise 9-6 on page 590. Check answers in back of book; then work through text sections corresponding to problems missed. (Section numbers are in italics following each answer.)
2. When finished with step 1, take Practice Test Chapter 9 on page 593 as a final check of your understanding of the chapter. Check answers in the back of the book; then review sections where weakness still prevails. (Corresponding section numbers are in italics following each answer.)
3. When you think you are ready, ask your instructor for a graded test for Chapter 9.
4. If your instructor approves, after the test is corrected, go to the next chapter.

1. Work through each section in the chapter as follows:
   (A) Read discussion.
   (B) Read each example and work the corresponding matched problem. Check your solutions to the matched problem in Solutions to Matched Problems on the indicated page.
   (C) At the end of a section work the odd-numbered problems in the Practice Exercise and check answers; then work even-numbered problems in areas of weakness. (Answers to *all* Practice Exercise sets are in the back of the book.)
   (D) Work Check Exercise as instructed. Tear out and turn in as directed by your instructor. (Answers are not in the text.)
2. Repeat each step in item 1 for each section in the chapter.
3. After the instructional part of the chapter is completed, proceed with steps 1 to 4 in the box above this one.

Chapter 9 ■ Exponential and Logarithmic Functions

Section 9-1 Exponential Functions

- Exponential Functions
- Graphing an Exponential Function
- Typical Exponential Graphs
- Base *e*
- Basic Exponential Properties

■ Exponential Functions

In this section and the next we will consider two new kinds of functions that use variable exponents in their definitions. To start, note that

$$f(x) = 2^x \qquad \text{and} \qquad g(x) = x^2$$

are not the same function. The function g is a quadratic function, which we have already discussed; the function f is a new function called an exponential function. An **exponential function** is a function defined by an equation of the form:

Exponential Function
$f(x) = b^x \qquad b > 0, \quad b \neq 1$

where b is a constant, called the **base**, and the exponent x is a variable. The replacement set for the exponent, the **domain of f**, is the set of real numbers R. The **range of f** is the set of positive real numbers. We require b to be positive to avoid complex numbers such as $(-2)^{1/2}$.

■ Graphing an Exponential Function

Many students, if asked to graph an exponential function such as $f(x) = 2^x$, would not hesitate at all. They would likely make up a table by assigning integers to x, plot the resulting points, and then join these points with a smooth curve (Figure 1).

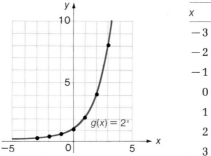

FIGURE 1

The only catch is that 2^x has not been defined at this point for all real numbers. We know what 2^5, 2^{-3}, $2^{2/3}$, $2^{-3/5}$, $2^{1.4}$, and $2^{-3.15}$ all mean (that is, 2^p, where p is a rational number), but what does

$$2^{\sqrt{2}}$$

mean? The question is not easy to answer at this time. In fact, a precise definition of $2^{\sqrt{2}}$ must wait for more advanced courses, where we can show that

$$b^x$$

names a real number for b a positive real number and x any real number and that the graph of $g(x) = 2^x$ is as indicated in Figure 1.

■ Typical Exponential Graphs

It is useful to compare the graphs of $y = 2^x$ and $y = (\frac{1}{2})^x = 2^{-x}$ by plotting both on the same coordinate system (Figure 2a).

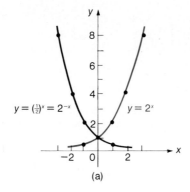

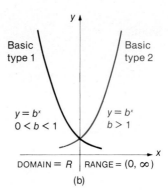

FIGURE 2

(a)

(b)

The graph of

$$f(x) = b^x \qquad b > 1 \text{ (Figure 2b)}$$

will look very much like the graph of $y = 2^x$, and the graph of

$$f(x) = b^x \qquad 0 < b < 1 \text{ (Figure 2b)}$$

will look very much like the graph of $y = (\frac{1}{2})^x$. Note in both cases that the x axis is a horizontal asymptote[†] and the graphs will never touch it.

Note: An exponential function is either increasing or decreasing and hence is one-to-one and has an inverse that is a function. This fact will be important to us in the next section when we define a logarithmic function as an inverse of an exponential function.

EXAMPLE 1   Graph $y = \frac{1}{2}4^x$ for $-3 \le x \le 3$.

Solution

x	y
−3	0.01
−2	0.03
−1	0.13
0	0.50
1	2.00
2	8.00
3	32.00

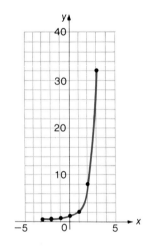

$$\boxed{4}\ \boxed{y^x}\ \boxed{3}\ \boxed{+/-}\ \boxed{=}\ \boxed{÷}\ \boxed{2}\ \boxed{=}$$

PROBLEM 1   Graph $y = \frac{1}{2}4^{-x}$ for $-3 \le x \le 3$.

Solution

x	y

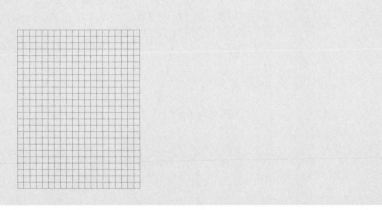

[†] Asymptotes were discussed in Section 7-5.

Exponential functions are often referred to as "growth functions" because of their widespread use in describing different kinds of growth. These functions are used to describe population growth of people, animals, and bacteria; radioactive decay (negative growth); growth of a new chemical substance in a chemical reaction; increase or decline in the temperature of a substance being heated or cooled; growth of money at compound interest; light absorption (negative growth) as it passes through air, water, or glass; decline of atmospheric pressure as altitude is increased; and growth of learning a skill such as swimming or typing relative to practice.

■ Base e

For introductory purposes, the bases 2 and $\frac{1}{2}$ were convenient choices; however, a certain irrational number, denoted by e, is by far the most frequently used exponential base for both theoretical and practical purposes. In fact,

$$f(x) = e^x$$

is often referred to as *the* exponential function because of its widespread use. The reasons for the preference for e as a base are made clear in advanced courses. And at that time, it is shown that e is approximated by $(1 + 1/n)^n$ to any decimal accuracy desired by making n (an integer) sufficiently large. The irrational number e to eight decimal places is

$$e \approx 2.718\ 281\ 83$$

Similarly, e^x can be approximated by using $(1 + 1/n)^{nx}$ for sufficiently large n. Because of the importance of e^x and e^{-x}, tables for their evaluation are readily available and many hand calculators can evaluate these functions directly. We will rely on hand calculators to obtain values for e^x. The important constant e along with two other important constants $\sqrt{2}$ and π are shown on the number line in Figure 3a, and the graph of $y = e^x$ is shown in Figure 3b.

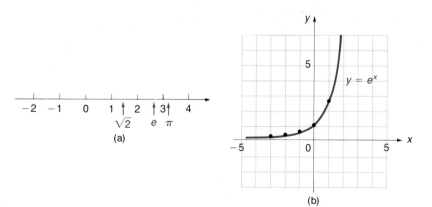

FIGURE 3

EXAMPLE 2   Graph $y = 10e^{-0.5x}$, $-3 \le x \le 3$, using a hand calculator.

Solution

x	y
−3	44.82
−2	27.18
−1	16.49
0	10.00
1	6.07
2	3.68
3	2.23

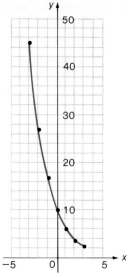

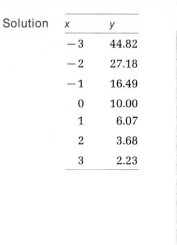

PROBLEM 2   Graph $y = 10e^{0.5x}$, $-3 \le x \le 3$, using a hand calculator.

Solution

x	y

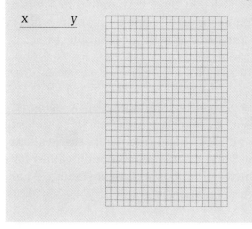

EXAMPLE 3   If $P is invested at $100r\%$ compounded continuously, then the amount A in the account at the end of t years is given by (from the mathematics of finance):

$$A = Pe^{rt}$$

If $100 is invested at 12% compounded continuously, graph the amount in the account relative to time for a period of 10 years.

Solution   We wish to graph

$$A = 100e^{0.12t} \qquad 0 \le t \le 10$$

We make up a table of values using a calculator, graph the points from the table, and then join the points with a smooth curve, as shown at the top of the next page.

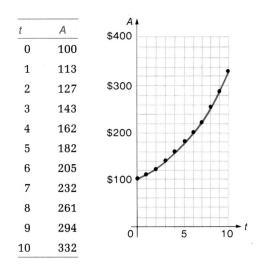

t	A
0	100
1	113
2	127
3	143
4	162
5	182
6	205
7	232
8	261
9	294
10	332

PROBLEM 3   Repeat Example 3 with $5,000 being invested at 20% compounded continuously.

Solution

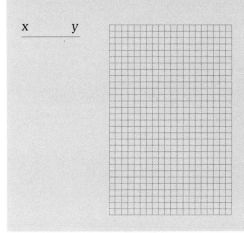

x	y

■ Basic Exponential Properties

Earlier (Sections 2-1, 5-1, 5-3) we discussed five laws of exponents for rational exponents. It can be shown that these same laws hold for irrational exponents. Thus, we now assume that all five laws of exponents hold for any real exponents as long as the bases involved are positive.

As a consequence of exponential functions being either increasing or decreasing and thus one-to-one, we have:

$$b^m = b^n \quad \text{if and only if} \quad m = n, b > 0, b \neq 1$$

Thus, if $2^{15} = 2^{3x}$, then $3x = 15$ and $x = 5$.

Solutions to Matched
Problems

1. $y = \frac{1}{2}4^{-x}$

x	y
−3	32.00
−2	8.00
−1	2.00
0	0.50
1	0.13
2	0.03
3	0.01

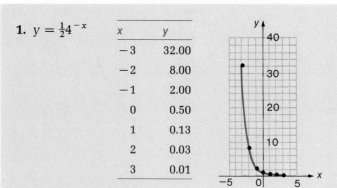

2. $y = 10e^{0.5x}$

x	y
−3	2.23
−2	3.68
−1	6.07
0	10.00
1	16.49
2	27.18
3	44.82

3. $A = 5,000e^{0.2t}$

t	A
0	5,000
1	6,107
2	7,459
3	9,111
4	11,128
5	13,591
6	16,601
7	20,276
8	24,765
9	30,248
10	36,945

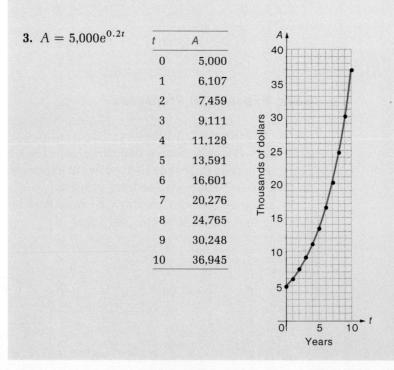

Practice Exercise 9-1 ■

■ *Work odd-numbered problems first, check answers, and then work even-numbered problems in areas of weakness. Answers to all problems are in the back of the book. Make every effort to work a problem yourself before you look at an answer.*

Graph each exponential function for $-3 \leq x \leq 3$ using a calculator. (Plot points using integers for x; then join the points with a smooth curve.)

A   **1.** $y = 3^x$ **2.** $y = 2^x$

3. $y = (\tfrac{1}{3})^x = 3^{-x}$    **4.** $y = (\tfrac{1}{2})^x = 2^{-x}$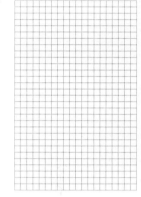

5. $y = 4 \cdot 3^x$ [Note: $4 \cdot 3^x \neq 12^x$]      **6.** $y = 5 \cdot 2^x$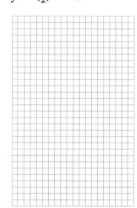

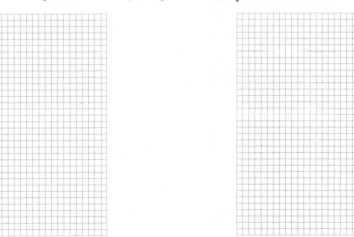

B   **7.** $y = 2^{x+3}$

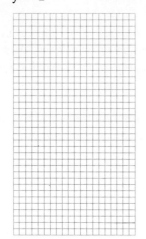

8. $y = 3^{x+1}$

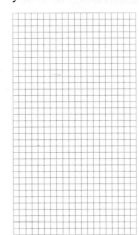

9. $y = 7(\frac{1}{2})^{2x} = 7 \cdot 2^{-2x}$

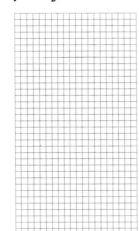

10. $y = 11 \cdot 2^{-2x}$

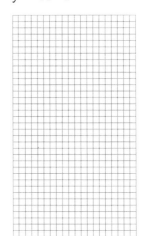

Graph Problems 11–14 for $-3 \leq x \leq 3$. *Use a calculator.*

11. $y = e^x$

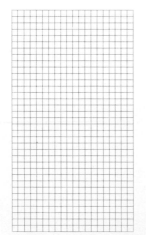

12. $y = e^{-x}$

13. $y = 10e^{-0.12x}$

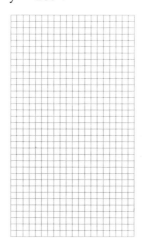

14. $y = 100e^{0.25x}$

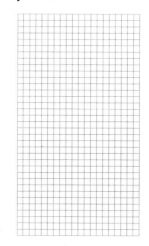

C   **15.**   Graph $y = 10 \cdot 2^{-x^2}$ for $-2 \le x \le 2$.

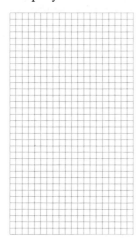

16.   Graph $y = e^{-x^2}$ for x = −1.5, −1.0, −0.5, 0, 0.5, 1.0, 1.5, and join these points with a smooth curve.

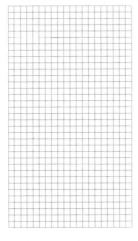

17.   Graph $y = y_0 2^x$, where y_0 is the value of y when $x = 0$. (Express the vertical scale in terms of y_0.)

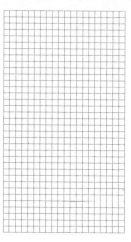

18.   Graph $y = y_0 e^{-0.22x}$, where y_0 is the value of y when $x = 0$. (Express the vertical scale in terms of y_0.)

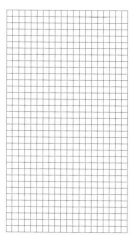

19.   Graph $y = 2^x$ and $x = 2^y$ on the same coordinate system.

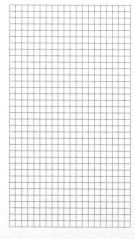

20. Graph $f(x) = 10^x$ and $y = f^{-1}(x)$ on the same coordinate system.

21. If we start with 2 cents and double the amount each day, at the end of n days we will have 2^n cents. Graph $f(n) = 2^n$ for $1 \le n \le 10$. (Pick the scale on the vertical axis so that the graph will not go off the paper.)

22. *Compound interest:* If a certain amount of money P, called the principal, is invested at $100r\%$ interest compounded annually, the amount of money A after t years is given by

$$A = P(1 + r)^t$$

Graph this equation for $P = \$10$, $r = 0.10$, and $0 \le t \le 10$.

*23. *Earth science:*   The atmospheric pressure P, in pounds per square inch, can be calculated approximately using the formula

$$P = 14.7e^{-0.21x}$$

where x is altitude relative to sea level in miles. Graph the equation for $-1 \leq x \leq 5$.

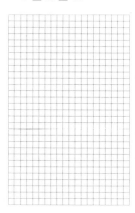

*24. *Bacterial growth:*   If bacteria in a certain culture double every hour, write a formula that gives the number of bacteria N in the culture after n hours, assuming the culture has N_0 bacteria to start with. _____

25. *Radioactive decay:*   Radioactive strontium-90 has a half-life of 28 years; that is, in 28 years one-half of any amount of strontium-90 will change to another substance because of radioactive decay. If we place a bar containing 100 milligrams of strontium-90 in a nuclear reactor, the amount of strontium-90 that will be left after t years is given by $A = 100(\frac{1}{2})^{t/28}$. Graph this exponential function for $t = 0$, 28, 2(28), 3(28), 4(28), 5(28), and 6(28), and join these points with a smooth curve.

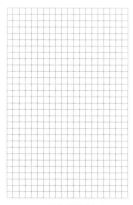

26. *Radioactive decay:*   Radioactive argon-39 has a half-life of 4 minutes; that is, in 4 minutes one-half of any amount of argon-39 will change to another substance because of radioactive decay. If we start with A_0 milligrams of argon-39, the amount left after t minutes is given by $A = A_0(\frac{1}{2})^{t/4}$. Graph this exponential function for $A_0 = 100$ and $t = 0, 4, 8, 12, 16,$ and $20,$ and join these points with a smooth curve.

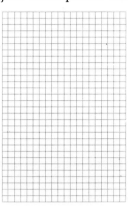

*27. *Sociology—small-group analysis:* Sociologists Stephan and Mischler found that when the members of a discussion group of 10 were ranked according to the number of times each participated, the number of times $N(i)$ the ith-ranked person participated was given approximately by the exponential function

$$N(i) = N_1 e^{-0.11(i-1)} \qquad 1 \le i \le 10$$

where N_1 is the number of times the top-ranked person participated in the discussion. Graph the exponential function, using $N_1 = 100.$

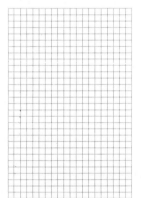

The Check Exercise for this section is on page 595.

Section 9-2 Logarithmic Functions

■ Logarithmic Functions
■ From Logarithmic to Exponential and Vice Versa
■ Finding x, b, or y in $y = \log_b x$
■ Logarithmic–Exponential Identities

■ Logarithmic Functions

We now define a new class of functions, called **logarithmic functions**, as inverses of exponential functions. (Since exponential functions are one-to-one, their inverses are functions.) Here you will see why we placed special emphasis on the general concept of inverse functions in Section 8-4. If you know quite a bit about a function, then (knowing about inverses in general) you will automatically know quite a bit about its inverse. For example, the graph of f^{-1} is the graph of f reflected across the line $y = x$, and the domain and range of f^{-1} are, respectively, the range and domain of f.

If we start with the exponential function

$$f: \quad y = 2^x$$

and interchange the variables x and y, we obtain the inverse of f:

$$f^{-1}: \quad x = 2^y$$

The graphs of f and f^{-1} (along with $y = x$) are shown in Figure 4. This new function is given the name **logarithmic function with base 2** and is symbolized as follows (since we cannot "algebraically" solve $x = 2^y$ for y):

$$y = \log_2 x$$

Thus,

$$y = \log_2 x \quad \text{is equivalent to} \quad x = 2^y$$

That is, $\log_2 x$ is the power to which 2 must be raised to obtain x. (Symbolically, $x = 2^y = 2^{\log_2 x}$.)

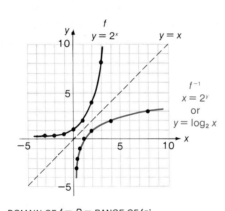

	f		f^{-1}	
x	$y = 2^x$		$x = 2^y$	y
-3	$\frac{1}{8}$		$\frac{1}{8}$	-3
-2	$\frac{1}{4}$		$\frac{1}{4}$	-2
-1	$\frac{1}{2}$		$\frac{1}{2}$	-1
0	1		1	0
1	2		2	1
2	4		4	2
3	8		8	3

Ordered pairs reversed

FIGURE 4   DOMAIN OF $f = R =$ RANGE OF f^{-1}
              RANGE OF $f = (0, \infty) =$ DOMAIN OF f^{-1}

In general, we define the **logarithmic function with base b** to be the inverse of the exponential function with base b ($b > 0$, $b \neq 1$).

Definition of Logarithmic Function

For $b > 0$ and $b \neq 1$:

$$y = \log_b x \quad \text{is equivalent to} \quad x = b^y$$

(The log to the base b of x is the power to which b must be raised to obtain x.)

$$y = \log_{10} x \quad \text{is equivalent to} \quad x = 10^y$$
$$y = \log_e x \quad \text{is equivalent to} \quad x = e^y$$

Remember that $y = \log_b x$ and $x = b^y$ define the same function, and as such can be used interchangeably.

Since the domain of an exponential function includes all real numbers and its range is the set of positive real numbers, the **domain** of a logarithmic function is the set of all positive real numbers and its **range** is the set of all real numbers. Thus, $\log_{10} 3$ is defined, but $\log_{10} 0$ and $\log_{10}(-5)$ are not defined (3 is a logarithmic domain value, but 0 and -5 are not). Typical logarithmic curves are shown in Figure 5.

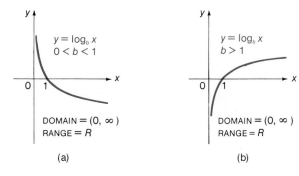

FIGURE 5 Typical logarithmic graphs

■ From Logarithmic to Exponential and Vice Versa

We now look into the matter of converting logarithmic forms to equivalent exponential forms and vice versa.

EXAMPLE 4   From logarithmic form to exponential form:

(A) $\log_2 8 = 3$ is equivalent to $8 = 2^3$
(B) $\log_{25} 5 = \frac{1}{2}$ is equivalent to $5 = 25^{1/2}$
(C) $\log_2(\frac{1}{4}) = -2$ is equivalent to $\frac{1}{4} = 2^{-2}$

PROBLEM 4

Change to equivalent exponential form:

(A) $\log_3 27 = 3$ **(B)** $\log_{36} 6 = \frac{1}{2}$ **(C)** $\log_3(\frac{1}{9}) = -2$

Solution   **(A)**   **(B)**   **(C)**

EXAMPLE 5   From exponential form to logarithmic form:

(A) $49 = 7^2$ is equivalent to $\log_7 49 = 2$
(B) $3 = \sqrt{9}$ is equivalent to $\log_9 3 = \frac{1}{2}$ $\sqrt{9} = 9^{1/2}$
(C) $\frac{1}{5} = 5^{-1}$ is equivalent to $\log_5(\frac{1}{5}) = -1$

PROBLEM 5

Change to equivalent logarithmic form:

(A) $64 = 4^3$ **(B)** $2 = \sqrt[3]{8}$ **(C)** $\frac{1}{16} = 4^{-2}$

Solution   **(A)**   **(B)**   **(C)**

■ Finding x, b, or y in $y = \log_b x$

To gain a little deeper understanding of logarithmic functions and their relationship to the exponential functions, we will look at a few problems where we are to find x, b, or y in $y = \log_b x$, given the other two values. All values were chosen so that the problems can be solved without tables or a calculator.

EXAMPLE 6   Find x, b, or y as indicated.

(A) Find y: $y = \log_4 8$ (B) Find x: $\log_3 x = -2$
(C) Find b: $\log_b 1,000 = 3$

Solution   (A) Write $y = \log_4 8$ in equivalent exponential form:

$$8 = 4^y \qquad \text{Write each number to the same base 2.}$$
$$2^3 = 2^{2y} \qquad \text{Recall that } b^m = b^n \text{ if and only if } m = n.$$
$$2y = 3$$
$$y = \tfrac{3}{2}$$

Thus, $\tfrac{3}{2} = \log_4 8$.

(B) Write $\log_3 x = -2$ in equivalent exponential form:

$$x = 3^{-2}$$
$$x = \frac{1}{3^2} = \frac{1}{9}$$

Thus, $\log_3(\tfrac{1}{9}) = -2$.

(C) Write $\log_b 1,000 = 3$ in equivalent exponential form:

$$1,000 = b^3 \qquad \text{Write 1,000 as a third power.}$$
$$10^3 = b^3$$
$$b = 10$$

Thus, $\log_{10} 1,000 = 3$.

PROBLEM 6   Find x, b, or y as indicated:

(A) Find y: $y = \log_9 27$ (B) Find x: $\log_2 x = -3$
(C) Find b: $\log_b 100 = 2$

Solution   (A)   (B)

(C)

■ Logarithmic–Exponential Identities

Recall from Section 8-4 that if f and f^{-1} are both functions (that is, if f is one-to-one), then

$$f^{-1}[f(x)] = x \qquad \text{and} \qquad f[f^{-1}(x)] = x$$

Applying these general properties to $f(x) = b^x$ and $f^{-1}(x) = \log_b x$, we see that

$$f^{-1}[f(x)] = x \qquad f[f^{-1}(x)] = x$$
$$\log_b[f(x)] = x \qquad b^{f^{-1}(x)} = x$$
$$\log_b b^x = x \qquad b^{\log_b x} = x$$

Thus, we have the useful logarithmic–exponential identities:

Logarithmic–Exponential Identities

For $b > 0$, $b \neq 1$:

1. $\log_b b^x = x$

2. $b^{\log_b x} = x \qquad x > 0$

EXAMPLE 7
(A) $\log_{10} 10^5 = 5$
(B) $\log_{10} 0.01 = \log_{10} 10^{-2} = -2$
(C) $\log_e e^{2x+1} = 2x + 1$
(D) $\log_4 1 = \log_4 4^0 = 0$
(E) $10^{\log_{10} 7} = 7$
(F) $e^{\log_e x^2} = x^2$

PROBLEM 7
Find each of the following:
(A) $\log_{10} 10^{-5}$ (B) $\log_5 25$ (C) $\log_{10} 1$
(D) $\log_e e^{m+n}$ (E) $10^{\log_{10} 4}$ (F) $e^{\log_e(x^4+1)}$

Solution
(A)   (B)   (C)

(D)   (E)   (F)

Solutions to Matched Problems

4. (A) $\log_3 27$ is equivalent to $27 = 3^3$
   (B) $\log_{36} 6 = \frac{1}{2}$ is equivalent to $6 = 36^{1/2}$
   (C) $\log_3 \frac{1}{9} = -2$ is equivalent to $\frac{1}{9} = 3^{-2}$

5. (A) $64 = 4^3$ is equivalent to $\log_4 64 = 3$
   (B) $2 = \sqrt[3]{8}$ is equivalent to $\log_8 2 = \frac{1}{3}$
   (C) $\frac{1}{16} = 4^{-2}$ is equivalent to $\log_4 \frac{1}{16} = -2$

6. (A) $y = \log_9 27$ **(B)** $\log_2 x = -3$ **(C)** $\log_b 100 = 2$
       $27 = 9^y$ $x = 2^{-3}$ $100 = b^2$
       $3^3 = 3^{2y}$ $x = \dfrac{1}{2^3} = \dfrac{1}{8}$ $10^2 = b^2$
       $2y = 3$ $b = 10$
       $y = \frac{3}{2}$

7. (A) $\log_{10} 10^{-5} = -5$ **(B)** $\log_5 25 = \log_5 5^2 = 2$
   (C) $\log_{10} 1 = 0$ **(D)** $\log_e e^{m+n} = m + n$
   (E) $10^{\log_{10} 4} = 4$ **(F)** $e^{\log_e(x^4+1)} = x^4 + 1$

Practice Exercise 9-2 ■

Work odd-numbered problems first, check answers, and then work even-numbered problems in areas of weakness. Answers to all problems are in the back of the book. Make every effort to work a problem yourself before you look at an answer.

A   *Rewrite in exponential form.*

1. $\log_3 9 = 2$ _____   **2.** $\log_2 4 = 2$ _____

3. $\log_3 81 = 4$ _____   **4.** $\log_5 125 = 3$ _____

5. $\log_{10} 1,000 = 3$ _____   **6.** $\log_{10} 100 = 2$ _____

7. $\log_e 1 = 0$ _____   **8.** $\log_8 1 = 0$ _____

Rewrite in logarithmic form.

9. $64 = 8^2$ _____   **10.** $25 = 5^2$ _____

11. $10,000 = 10^4$ _____   **12.** $1,000 = 10^3$ _____

13. $u = v^x$ _____   **14.** $a = b^c$ _____

15. $9 = 27^{2/3}$ _____   **16.** $8 = 4^{3/2}$ _____

Find each of the following.

17. $\log_{10} 10^5$ _____   **18.** $\log_5 5^3$ _____

19. $\log_2 2^{-4}$ _____   **20.** $\log_{10} 10^{-7}$ _____

21. $\log_6 36$ _____   **22.** $\log_3 9$ _____

23. $\log_{10} 1,000$ _____   **24.** $\log_{10} 0.001$ _____

Find x, y, or b as indicated.

25. $\log_2 x = 2$ _____   **26.** $\log_3 x = 2$ _____

27. $\log_4 16 = y$ _____   **28.** $\log_8 64 = y$ _____

29. $\log_b 16 = 2$ _____   **30.** $\log_b 10^{-3} = -3$ _____

B   *Rewrite in exponential form.*

31. $\log_{10} 0.001 = -3$ _____   **32.** $\log_{10} 0.01 = -2$ _____

33. $\log_{81} 3 = \frac{1}{4}$ _____   **34.** $\log_4 2 = \frac{1}{2}$ _____

35. $\log_{1/2} 16 = -4$ _____   **36.** $\log_{1/3} 27 = -3$ _____

37. $\log_a N = e$ _____   **38.** $\log_k u = v$ _____

Rewrite in logarithmic form.

39. $0.01 = 10^{-2}$ _____

40. $0.001 = 10^{-3}$ _____

41. $1 = e^0$ _____

42. $1 = (\frac{1}{2})^0$ _____

43. $\frac{1}{8} = 2^{-3}$ _____

44. $\frac{1}{8} = (\frac{1}{2})^3$ _____

45. $\frac{1}{3} = 81^{-1/4}$ _____

46. $\frac{1}{2} = 32^{-1/5}$ _____

47. $7 = \sqrt{49}$ _____

48. $11 = \sqrt{121}$ _____

Find each of the following.

49. $\log_b b^u$ _____

50. $\log_b b^{uv}$ _____

51. $\log_e e^{1/2}$ _____

52. $\log_e e^{-3}$ _____

53. $\log_2 \sqrt{8}$ _____

54. $\log_5 \sqrt[3]{5}$ _____

55. $\log_{23} 1$ _____

56. $\log_{17} 1$ _____

57. $\log_4 8$ _____

58. $\log_4(\frac{1}{4})$ _____

Find x, y, or b as indicated.

59. $\log_4 x = \frac{1}{2}$ _____

60. $\log_{25} x = \frac{1}{2}$ _____

61. $\log_{1/3} 9 = y$ _____

62. $\log_{49}(\frac{1}{7}) = y$ _____

63. $\log_b 1{,}000 = \frac{3}{2}$ _____

64. $\log_b 4 = \frac{2}{3}$ _____

C **65.** $\log_b 1 = 0$ _____

66. $\log_b b = 1$ _____

67. For $f = \{(x, y) \mid y = 1^x\}$ discuss the domain and range for f and f^{-1}. Are both relations functions? _____

68. Why is 1 not a suitable logarithmic base? [*Hint:* Try to find $\log_1 5$.]

_____

69. **(A)** For $f = \{(x, y) \mid y = 10^x\}$, graph f and f^{-1} using the same coordinate axes.

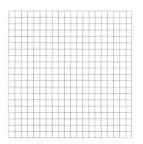

(B)   Discuss the domain and range of f and f^{-1}. _____

(C)   What other name could you use for the inverse of f? _____

The Check Exercise for
this section is on
page 597.

70.   Prove that $\log_b(1/x) = -\log_b x$.

71.   If $\log_b x = 3$, find $\log_b(1/x)$. _____

Section 9-3   Properties of Logarithmic Functions

- Basic Logarithmic Properties
- Use of the Logarithmic Properties

■ Basic Logarithmic Properties

Logarithmic functions have several very useful properties that follow directly from the fact that they are inverses of exponential functions. These properties will enable us to convert multiplication problems into addition problems, division problems into subtraction problems, and power and root problems into multiplication problems. Moreover, we will be able to solve exponential equations such as $2 = 10^x$.

THEOREM 1

Properties of Logarithmic Functions
If b, M, and N are positive real numbers, $b \neq 1$, and p is a real number, then: **1.** $\log_b b^u = u$ **2.** $\log_b MN = \log_b M + \log_b N$ **3.** $\log_b \dfrac{M}{N} = \log_b M - \log_b N$ **4.** $\log_b M^p = p \log_b M$ **5.** $\log_b 1 = 0$

The first property in Theorem 1 follows directly from the definition of a logarithmic function. The proof of the second property is based on the laws of exponents. To bring exponents into the proof, we let

$$u = \log_b M \qquad \text{and} \qquad v = \log_b N$$

and convert these to the equivalent exponential forms

$$M = b^u \qquad \text{and} \qquad N = b^v$$

Now see if you can provide the reasons for each of the following steps:

$$\log_b MN = \log_b b^u b^v = \log_b b^{u+v} = u + v = \log_b M + \log_b N$$

The other properties are established in a similar manner.

■ Use of the Logarithmic Properties

We now see how logarithmic properties can be used to convert multiplication problems into addition problems, division problems into subtraction problems, and power and root problems into multiplication problems.

EXAMPLE 8   **(A)** $\log_{10} 10^5 = 5$ $\log_b b^u = u$

(B) $\log_b 3x = \log_b 3 + \log_b x$ $\log_b MN = \log_b M + \log_b N$

(C) $\log_b \dfrac{x}{5} = \log_b x - \log_b 5$ $\log_b \dfrac{M}{N} = \log_b M - \log_b N$

(D) $\log_b x^7 = 7 \log_b x$ $\log_b M^p = p \log_b M$

(E) $\log_b \dfrac{mn}{pq} = \log_b mn - \log_b pq$ $\log_b \dfrac{M}{N} = \log_b M - \log_b N$

$\qquad = \log_b m + \log_b n - (\log_b p + \log_b q)$ $\log_b MN = \log_b M + \log_b N$

$\qquad = \log_b m + \log_b n - \log_b p - \log_b q$

(F) $\log_b (mn)^{2/3} = \tfrac{2}{3} \log_b mn$ $\log_b M^p = p \log_b M$

$\qquad = \tfrac{2}{3}(\log_b m + \log_b n)$ $\log_b MN = \log_b M + \log_b N$

(G) $\log_b \dfrac{x^8}{y^{1/5}} = \log_b x^8 - \log_b y^{1/5}$ $\log_b \dfrac{M}{N} = \log_b M - \log_b N$

$\qquad = 8 \log_b x - \tfrac{1}{5} \log_b y$ $\log_b M^p = p \log_b M$

PROBLEM 8   Write in terms of simpler logarithmic forms as in Example 8:

(A) $\log_b \left(\dfrac{r}{uv} \right)$ **(B)** $\log_b \left(\dfrac{m}{n} \right)^{3/5}$ **(C)** $\log_b \left(\dfrac{u^{1/3}}{v^5} \right)$

Solution   **(A)**

(B)

(C)

EXAMPLE 9   If $\log_e 3 = 1.10$ and $\log_e 7 = 1.95$, find:

(A) $\log_e(\tfrac{7}{3})$ **(B)** $\log_e \sqrt[3]{21}$

Solution   **(A)** $\log_e(\tfrac{7}{3}) = \log_e 7 - \log_e 3 = 1.95 - 1.10 = 0.85$

(B) $\log_e \sqrt[3]{21} = \log_e(21)^{1/3} = \tfrac{1}{3} \log_e(3 \cdot 7) = \tfrac{1}{3}(\log_e 3 + \log_e 7)$

$\qquad\qquad\qquad = \tfrac{1}{3}(1.10 + 1.95) = 1.02$

PROBLEM 9   If $\log_e 5 = 1.609$ and $\log_e 8 = 2.079$, find:

(A) $\log_e\left(\dfrac{5^{10}}{8}\right)$ (B) $\log_e \sqrt[4]{\dfrac{8}{5}}$

Solution   (A)

(B)

Finally, we note that since logarithmic functions are one-to-one:

$\log_b m = \log_b n$ if and only if $m = n$

Thus, if $\log_{10} x = \log_{10} 32.15$, then $x = 32.15$.

The following example and problem, though somewhat artificial, will give you additional practice in using the properties in Theorem 1.

EXAMPLE 10   Find x so that $\log_b x = \frac{2}{3}\log_b 27 + 2\log_b 2 - \log_b 3$ without using a calculator or table.

Solution $\log_b x = \frac{2}{3}\log_b 27 + 2\log_b 2 - \log_b 3$ Express the right side in terms of a single log.

$\qquad\quad = \log_b 27^{2/3} + \log_b 2^2 - \log_b 3$ Property 4

$\qquad\quad = \log_b 9 + \log_b 4 - \log_b 3$ $27^{2/3} = 9,\ 2^2 = 4$

$\qquad\quad = \log_b \dfrac{9\cdot 4}{3} = \log_b 12$ Properties 2 and 3

Thus,

$\qquad \log_b x = \log_b 12$

Hence,

$\qquad x = 12$

PROBLEM 10   Find x so that $\log_b x = \frac{2}{3}\log_b 8 + \frac{1}{2}\log_b 9 - \log_b 6$ without using a calculator or table.

Solution

Solutions to Matched
Problems

8. (A) $\log_b \dfrac{r}{uv} = \log_b r - \log_b(uv)$

$$= \log_b r - (\log_b u + \log_b v)$$
$$= \log_b r - \log_b u - \log_b v$$

(B) $\log_b \left(\dfrac{m}{n}\right)^{3/5} = \dfrac{3}{5} \log_b \dfrac{m}{n}$

$$= \dfrac{3}{5} (\log_b m - \log_b n)$$

(C) $\log_b \dfrac{u^{1/3}}{v^5} = \log_b u^{1/3} - \log_b v^5$

$$= \tfrac{1}{3} \log_b u - 5 \log_b v$$

9. (A) $\log_e \dfrac{5^{10}}{8} = \log_e 5^{10} - \log_e 8$

$$= 10 \log_e 5 - \log_e 8$$
$$= 10(1.609) - (2.079)$$
$$= 14.01 \quad \text{To four significant digits}^{\dagger}$$

(B) $\log_e \sqrt[4]{\tfrac{8}{5}} = \log_e (\tfrac{8}{5})^{1/4}$
$$= \tfrac{1}{4} \log_e \tfrac{8}{5}$$
$$= \tfrac{1}{4}(\log_e 8 - \log_e 5)$$
$$= \tfrac{1}{4}(2.079 - 1.609)$$
$$= 0.1175 \quad \text{To four significant digits}$$

10. $\log_b x = \tfrac{2}{3} \log_b 8 + \tfrac{1}{2} \log_b 9 - \log_b 6 = \log_b 8^{2/3} + \log_b 9^{1/2} - \log_b 6$

$$= \log_b 4 + \log_b 3 - \log_b 6 = \log_b \dfrac{4 \cdot 3}{6} = \log_b 2$$

$\log_b x = \log_b 2$
$\quad x = 2$

Practice Exercise 9-3 ∎

Work odd-numbered problems first, check answers, and then work even-numbered problems in areas of weakness. Answers to all problems are in the back of the book. Make every effort to work a problem yourself before you look at an answer.

A *Write in terms of simpler logarithmic forms (going as far as you can with logarithmic properties—see Example 8).*

1. $\log_b uv$ _____

2. $\log_b rt$ _____

3. $\log_b(A/B)$ _____

4. $\log_b(p/q)$ _____

5. $\log_b u^5$ _____

6. $\log_b w^{25}$ _____

7. $\log_b N^{3/5}$ _____

8. $\log_b u^{-2/3}$ _____

9. $\log_b \sqrt{Q}$ _____

10. $\log_b \sqrt[5]{M}$ _____

11. $\log_b uvw$ _____

12. $\log_b(u/vw)$ _____

† Significant digits are discussed in Appendix A.

Write each expression in terms of a single logarithm with a coefficient of 1.
[Example: $\log_b u^2 - \log_b v = \log_b(u^2/v)$]

13. $\log_b A + \log_b B$ _____

14. $\log_b P + \log_b Q + \log_b R$ _____

15. $\log_b X - \log_b Y$ _____

16. $\log_b x^2 - \log_b y^3$ _____

17. $\log_b w + \log_b x - \log_b y$

18. $\log_b w - \log_b x - \log_b y$

_____

_____

If $\log_b 2 = 0.69$, $\log_b 3 = 1.10$, and $\log_b 5 = 1.61$, find the logarithm to the base b of each of the following numbers.

19. $\log_b 30$ _____

20. $\log_b 6$ _____

21. $\log_b(\frac{2}{5})$ _____

22. $\log_b(\frac{5}{3})$ _____

23. $\log_b 27$ _____

24. $\log_b 16$ _____

B   Write in terms of simpler logarithmic forms (going as far as you can with logarithmic properties—see Example 8).

25. $\log_b u^2 v^7$ _____

26. $\log_b u^{1/2} v^{1/3}$ _____

27. $\log_b\left(\dfrac{1}{a}\right)$ _____

28. $\log_b\left(\dfrac{1}{M^3}\right)$ _____

29. $\log_b\left(\dfrac{\sqrt[3]{N}}{p^2 q^3}\right)$ _____

30. $\log_b\left(\dfrac{m^5 n^3}{\sqrt{p}}\right)$ _____

31. $\log_b \sqrt[4]{\dfrac{x^2 y^3}{\sqrt{z}}}$ _____

32. $\log_b \sqrt[5]{\left(\dfrac{x}{y^4 z^9}\right)^3}$ _____

Write each expression in terms of a single logarithm with a coefficient of 1.

33. $2 \log_b x - \log_b y$ _____

34. $\log_b m - \frac{1}{2} \log_b n$ _____

35. $3 \log_b x + 2 \log_b y - 4 \log_b z$ _____

36. $\frac{1}{3} \log_b w - 3 \log_b x - 5 \log_b y$ _____

37. $\frac{1}{5}(2 \log_b x + 3 \log_b y)$ _____

38. $\frac{1}{3}(\log_b x - \log_b y)$ _____

If $\log_b 2 = 0.69$, $\log_b 3 = 1.10$, and $\log_b 5 = 1.61$, find the logarithm to the base b of the following numbers.

39. $\log_b 7.5$ _____

40. $\log_b 1.5$ _____

41. $\log_b \sqrt[3]{2}$ _____

42. $\log_b \sqrt{3}$ _____

43. $\log_b \sqrt{0.9}$ _____

44. $\log_b \sqrt[3]{\frac{3}{2}}$ _____

C   **45.** Find x so that $\frac{3}{2} \log_b 4 - \frac{2}{3} \log_b 8 + 2 \log_b 2 = \log_b x$. _____

46. Find x so that $3 \log_b 2 + \frac{1}{2} \log_b 25 - \log_b 20 = \log_b x$. _____

47. Write $\log_b y - \log_b c + kt = 0$ in exponential form free of logarithms.

---

48. Write $\log_e x - \log_e 100 = -0.08t$ in exponential form free of logarithms.

---

49. Prove that $\log_b(M/N) = \log_b M - \log_b N$ under the hypotheses of Theorem 1.

50. Prove that $\log_b M^p = p \log_b M$ under the hypotheses of Theorem 1.

51. Prove that $\log_b MN = \log_b M + \log_b N$ by starting with $M = b^{\log_b M}$ and $N = b^{\log_b N}$.

The Check Exercise for this section is on page 599.

52. Prove that $\log_b(M/N) = \log_b M - \log_b N$ by starting with $M = b^{\log_b M}$ and $N = b^{\log_b N}$.

Section 9-4   Logarithms to Various Bases

- Common and Natural Logarithms—Calculator Evaluation
- Change-of-Base Formula
- Graphing Logarithmic Functions

John Napier (1550–1617) is credited with the invention of logarithms. They evolved out of an interest in reducing the computational strain in astronomy research. This new computational tool was immediately accepted by the scientific world. Now, with the availability of inexpensive hand calculators, logarithms have lost most of their importance as a computational device. However, the logarithmic concept has been greatly generalized since its conception, and logarithmic functions are used widely in both theoretical and applied sciences. For example, even with a very good scientific hand calculator, we still need logarithmic functions to solve the simple-looking exponential equation from population growth studies and the mathematics of finance:

$$2 = 1.08^x$$

Of all possible logarithmic bases, the base e and the base 10 are used almost exclusively. Before we can use logarithms in certain practical problems, we need to be able to approximate the logarithm of any positive number to either base 10 or base e. And conversely, if we are given the logarithm of a number to base 10 or base e, we need to be able to approximate the number. Historically, tables such as Table II and Table III in the back of the book were used for this purpose, but now with inexpensive scientific hand calculators readily available, most people will choose a calculator, since it is faster and far more accurate than almost any table you might use.

■ Common and Natural Logarithms—Calculator Evaluation

Common logarithms (also called **Briggsian logarithms**) are logarithms with base 10. **Natural logarithms** (also called **Napierian logarithms**) are logarithms with base e. Most scientific calculators have a button labeled "log" (or "LOG") and a button labeled "ln" (or "LN"). The former represents a common (base 10) logarithm and the latter a natural (base e) logarithm. In fact, "log" and "ln" are both used extensively in mathematical literature, and whenever you see either used in this book without a base indicated, they will be interpreted as follows:

Logarithmic Notation

$$\log x = \log_{10} x \qquad \ln x = \log_e x$$

To find the common or natural logarithm using a scientific calculator is very easy. You simply enter a number from the domain of the function and push the log or ln button. Tables for evaluating logarithms are also readily available. Their use is discussed in Appendixes B and C. We will rely on the calculator to obtain values for logarithms.

EXAMPLE 11   Use a scientific calculator to find each to six decimal places:

(A) log 3,184   **(B)** ln 0.000 349   **(C)** log(-3.24)

Solution

	ENTER	PRESS	DISPLAY
(A)	3,184	log	3.502973
(B)	0.000 349	ln	-7.960439
(C)	-3.24	log	Error

Why is an error indicated in part (C)? Because -3.24 is not in the domain of the log function.

PROBLEM 11   Use a scientific calculator to find each to six decimal places:

(A) log 0.013 529   **(B)** ln 28.693 28   **(C)** ln(-0.438)

Solution   **(A)**   **(B)**   **(C)**

EXAMPLE 12   Use a scientific calculator to evaluate each to three decimal places:

(A) $n = \dfrac{\log 2}{\log 1.1}$ **(B)** $n = \dfrac{\ln 3}{\ln 1.08}$

Solution   **(A)** First note that $(\log 2)/(\log 1.1) \neq \log 2 - \log 1.1$. Recall (see Section 9-3) that $\log_b(M/N) = \log_b M - \log_b N$, which is, of course, not the same as $(\log_b M)/(\log_b N)$.

$$n = \frac{\log 2}{\log 1.1} = 7.273 \qquad \boxed{2}\ \boxed{\log}\ \boxed{\div}\ \boxed{1.1}\ \boxed{\log}\ \boxed{=}$$

(B) $n = \dfrac{\ln 3}{\ln 1.08} = 14.275 \qquad \boxed{3}\ \boxed{\ln}\ \boxed{\div}\ \boxed{1.08}\ \boxed{\ln}\ \boxed{=}$

PROBLEM 12

Use a scientific calculator to evaluate each to two decimal places:

(A) $n = \dfrac{\ln 2}{\ln 1.1}$ \qquad **(B)** $n = \dfrac{\log 3}{\log 1.08}$

Solution

(A) \qquad\qquad\qquad\qquad\qquad\qquad **(B)**

We now turn to the second problem: given the logarithm of a number, find the number. We make direct use of the logarithmic–exponential relationships we discussed in Section 9-2.

Logarithmic–Exponential Relationships

$\log x = y$	is equivalent to	$x = 10^y$
$\ln x = y$	is equivalent to	$x = e^y$

EXAMPLE 13

Find x to three significant digits, given the indicated logarithms:

(A) $\log x = -9.315$ \qquad **(B)** $\ln x = 2.386$

Solution

(A) $\log x = -9.315$ \qquad\qquad Change to equivalent exponential form.

$\qquad x = 10^{-9.315}$ \qquad $\boxed{9.315}\ \boxed{+/-}\ \boxed{10^x}$

$\qquad x = 4.84 \times 10^{-10}$ \qquad Notice the answer is displayed in scientific notation in the calculator.

(B) $\ln x = 2.386$ \qquad\qquad Change to equivalent exponential form.

$\qquad x = e^{2.386}$ \qquad $\boxed{2.386}\ \boxed{e^x}$

$\qquad x = 10.9$

PROBLEM 13

Find x to four significant figures, given the indicated logarithms:

(A) $\ln x = -5.062$ \qquad **(B)** $\log x = 12.0821$

Solution

(A) \qquad\qquad\qquad\qquad\qquad\qquad **(B)**

■ Change-of-Base Formula

If we have a means (through either a calculator or table) of finding logarithms of numbers to one base, then by means of the change-of-base formula we can find the logarithm of a number to any other base.

Change-of-Base Formula

$$\log_b N = \frac{\log_a N}{\log_a b} = \frac{\log N}{\log b} = \frac{\ln N}{\ln b}$$

Can you supply the reasons for each step in the following derivation of the change-of-base formula?

$$y = \log_b N$$
$$N = b^y$$
$$\log_a N = \log_a b^y$$
$$\log_a N = y \log_a b$$
$$y = \frac{\log_a N}{\log_a b}$$
$$\log_b N = \frac{\log_a N}{\log_a b} \quad \text{Since } y = \log_b N$$

EXAMPLE 14   Find $\log_5 14$ to three decimal places using common logarithms.

Solution $$\log_5 14 = \frac{\log_{10} 14}{\log_{10} 5} = 1.640 \quad \boxed{14}\;\boxed{\log}\;\boxed{\div}\;\boxed{5}\;\boxed{\log}\;\boxed{=}$$

PROBLEM 14   Find $\log_7 729$ to four decimal places.

Solution

■ Graphing Logarithmic Functions

With the aid of a scientific calculator to find values, logarithmic functions can be graphed by plotting points.

EXAMPLE 15   Graph $y = 4 \log x$.

Solution

x	y
0.01	-8
0.1	-4
1	0
2	1.20
5	2.80
10	4

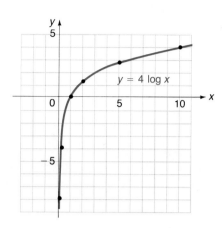

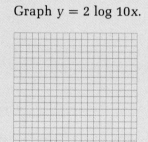

PROBLEM 15   Graph $y = 2 \log 10x$.

Solution

EXAMPLE 16   Graph $y = 3 \ln 2x$.

Solution

x	y
0.1	-4.8
0.5	0
1	2.08
2	4.16
3	5.37
4	6.23
6	7.45
8	8.32

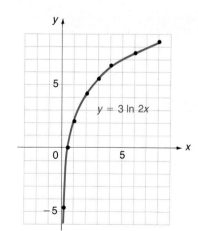

PROBLEM 16   Graph $y = \frac{1}{2} \ln 6x$.

Solution

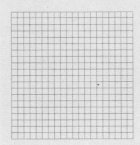

EXAMPLE 17   Graph $y = \log_{0.5} x$.

Solution   To find values for y, use $\log_{0.5} x = \dfrac{\ln x}{\ln 0.5}$.

x	y
0.1	3.32
0.5	1
1	0
2	−1
3	−1.58
4	−2
6	−2.58

$\boxed{.1}\ \boxed{\ln}\ \boxed{\div}\ \boxed{.5}\ \boxed{\ln}\ \boxed{=}$

or

$\boxed{.5}\ \boxed{\ln}\ \boxed{\text{STO}}\ \boxed{.1}\ \boxed{\ln}\ \boxed{\div}\ \boxed{\text{RCL}}\ \boxed{=}$

PROBLEM 17   Graph $y = 2 \log_{1/4} x$.

Solution

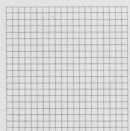

Solutions to Matched Problems

11. (A) $-1.868\ 734$ **(B)** $3.356\ 663$ **(C)** Not possible
12. (A) 7.27 **(B)** 14.27
13. (A) $\ln x = -5.062$ **(B)** $\log x = 12.0821$
     $x = e^{-5.062}$ $x = 10^{12.0821}$
     $x = 0.006\ 333$ $x = 1.21 \times 10^{12}$

14. $\log_7 729 = \dfrac{\log_{10} 729}{\log_{10} 7} = 3.3875$

15.

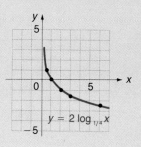

$y = 2 \log 10x$

16.

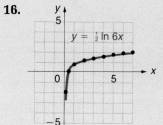

$y = \frac{1}{2} \ln 6x$

17.

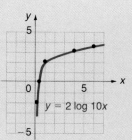

$y = 2 \log_{1/4} x$

Practice Exercise 9-4 ■

Work odd-numbered problems first, check answers, and then work even-numbered problems in areas of weakness. Answers to all problems are in the back of the book. Make every effort to work a problem yourself before you look at an answer.

 Hand calculators are used extensively in this exercise set.

A   Use a calculator to find each to four decimal places.

1. $\log 82{,}734$ _____   **2.** $\log 843{,}250$ _____

3. $\log 0.001\ 439$ _____   **4.** $\log 0.035\ 604$ _____

5. $\ln 43.046$ _____   **6.** $\ln 2{,}843{,}100$ _____

7. $\ln 0.081\ 043$ _____   **8.** $\ln 0.000\ 032\ 4$ _____

B   Use a calculator to find x to four significant digits given the following.

9. $\log x = 5.3027$ _____   **10.** $\log x = 1.9168$ _____

11. $\log x = -3.1773$ _____   **12.** $\log x = -2.0411$ _____

13. $\ln x = 3.8655$ _____   **14.** $\ln x = 5.0884$ _____

15. $\ln x = -0.3916$ _____   **16.** $\ln x = -4.1083$ _____

Evaluate each of the following to three decimal places using a calculator.

17. $n = \dfrac{\log 2}{\log 1.15}$ _____

18. $n = \dfrac{\log 2}{\log 1.12}$ _____

19. $n = \dfrac{\ln 3}{\ln 1.15}$ _____

20. $n = \dfrac{\ln 4}{\ln 1.2}$ _____

21. $x = \dfrac{\ln 0.5}{-0.21}$ _____

22. $t = \dfrac{\log 200}{2 \log 2}$ _____

Use the change-of-base formula and a calculator with either log or ln to find each to four decimal places.

23. $\log_5 372$ _____

24. $\log_4 23$ _____

25. $\log_8 0.0352$ _____

26. $\log_2 0.005\,439$ _____

27. $\log_3 0.1483$ _____

28. $\log_{12} 435.62$ _____

Graph.

29. $y = \frac{1}{2} \ln 5x$

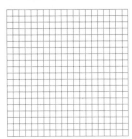

30. $y = 3 \ln \dfrac{x}{4}$

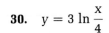

31. $y = 2 \log x^2$

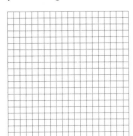

32. $y = 4 \log(\frac{1}{2}x)$

33. $y = 4 \log_5 x$

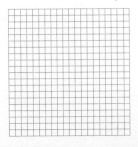

34. $y = 10 \log_{0.10} x$

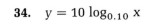
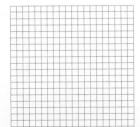
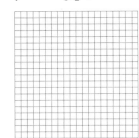

C *Find x to five significant digits using a calculator.*

35. $x = \log(5.3147 \times 10^{12})$ _____   **36.** $x = \log(2.0991 \times 10^{17})$ _____

37. $x = \ln(6.7917 \times 10^{-12})$ _____   **38.** $x = \ln(4.0304 \times 10^{-8})$ _____

The Check Exercise for **39.** $\log x = 32.068\,523$ _____   **40.** $\log x = -12.731\,64$ _____

this section is on

page 601. **41.** $\ln x = -14.667\,13$ _____   **42.** $\ln x = 18.891\,143$ _____

Section 9-5 Exponential and Logarithmic Equations

■ Exponential Equations
■ Logarithmic Equations

Equations involving exponential and logarithmic functions such as

$$2^{3x-2} = 5 \quad \text{and} \quad \log(x + 3) + \log x = 1$$

are called **exponential** and **logarithmic equations**, respectively. Logarithmic properties play a central role in their solution.

■ Exponential Equations

The following examples illustrate the use of logarithmic properties in solving exponential equations.

EXAMPLE 18   Solve $2^{3x-2} = 5$ for x to four decimal places.

Solution

$$2^{3x-2} = 5$$

How can we get *x* out of the exponent? Use logs! If two positive quantities are equal, their logs are equal.

$$\log 2^{3x-2} = \log 5$$

Use $\log_b = N^p = p \log_b N$ to get $3x - 2$ out of the exponent position.

$$(3x - 2)\log 2 = \log 5$$

$$3x - 2 = \frac{\log 5}{\log 2}$$

Remember: $(\log 5)/(\log 2) \neq \log 5 - \log 2$

$$x = \frac{1}{3}\left(2 + \frac{\log 5}{\log 2}\right)$$

$$\boxed{5}\ \boxed{\texttt{log}}\ \boxed{\div}\ \boxed{2}\ \boxed{\texttt{log}}\ \boxed{=}\ \boxed{+}\ \boxed{2}\ \boxed{=}\ \boxed{\div}\ \boxed{3}\ \boxed{=}$$

$$x = 1.4406$$

To four decimal places

PROBLEM 18   Solve $35^{1-2x} = 7$ for x to four decimal places.

Solution

EXAMPLE 19   If a certain amount of money P (principal) is invested at $100r\%$ interest compounded annually, then the amount of money A in the account after n years,
assuming no withdrawals, is given by

$$A = P(1 + r)^n$$

How long will it take the money to double if it is invested at 6% compounded
annually?

Solution   To find the doubling time, we replace A in $A = P(1.06)^n$ with $2P$ and solve for n:

$2P = P(1.06)^n$ Divide both sides by P.

$2 = 1.06^n$ Take the common or natural log of both sides.

$\log 2 = \log 1.06^n$

$\log 2 = n \log 1.06$ Note how log properties are used to get n out of the exponent position.

$$n = \frac{\log 2}{\log 1.06}$$

$= 12$ years   To the nearest year

PROBLEM 19   Repeat Example 19 changing the interest rate from 6% compounded
annually to 9% compounded annually.

Solution

EXAMPLE 20   The atmospheric pressure P (in pounds per square inch) at x miles above sea
level is given approximately by

$$P = 14.7e^{-0.21x}$$

At what height will the atmospheric pressure be half of the sea-level pressure?
Compute the answer to two significant digits.

Solution   Sea-level pressure is the pressure at $x = 0$. Thus,

$$P = 14.7e^0 = 14.7$$

One-half of sea-level pressure is $\frac{14.7}{2} = 7.35$. Now our problem is to find x so that $P = 7.35$; that is, we solve $7.35 = 14.7e^{-0.21x}$ for x:

$$7.35 = 14.7e^{-0.21x} \qquad \text{Divide both sides by 14.7 to simplify.}$$

$$0.5 = e^{-0.21x} \qquad \text{Take the natural log of both sides.}$$

$$\ln 0.5 = \ln e^{-0.21x} \qquad \text{Why use natural logs? Compare with the common log to see why.}$$

$$\ln 0.5 = -0.21x$$

$$x = \frac{\ln 0.5}{-0.21} \qquad \text{Use a hand calculator.}$$

$$= 3.3 \text{ miles} \qquad \text{To two significant digits}$$

PROBLEM 20

Using the formula in Example 20, find the altitude in miles to two significant digits so that the atmospheric pressure will be one-eighth that at sea level.

Solution

■ Logarithmic Equations

The next two examples illustrate approaches to solving some types of logarithmic equations.

EXAMPLE 21 Solve $\log(x + 3) + \log x = 1$ and check.

Solution
$$\log(x + 3) + \log x = 1 \qquad \text{Combine the left side using log } M + \log N = \log MN.$$

$$\log[x(x + 3)] = 1 \qquad \text{Change to the equivalent exponential form.}$$

$$x(x + 3) = 10^1 \qquad \text{Write in } ax^2 + bx + c = 0 \text{ form.}$$

$$x^2 + 3x - 10 = 0 \qquad \text{Solve.}$$

$$(x + 5)(x - 2) = 0$$

$$x = -5, 2$$

Check $x = -5$: $\log(-5 + 3) + \log(-5)$ is not defined   Why?

Thus, -5 must be discarded.

$x = 2$: $\log(2 + 3) + \log 2 = \log 5 + \log 2$

$$= \log(5 \cdot 2) = \log 10 \overset{\vee}{=} 1$$

Thus, 2 is the only solution.

Remember, answers should be checked in the original equation to see whether any should be discarded. An extraneous solution can be introduced in shifting from logarithmic to exponential form because the logarithm has a restricted domain.

PROBLEM 21

Solve $\log(x - 15) = 2 - \log x$ and check.

Solution

EXAMPLE 22   Solve $(\ln x)^2 = \ln x^2$.

Solution

$$(\ln x)^2 = \ln x^2$$

$(\ln x)^2 = 2 \ln x$ This is a quadratic equation in ln x. Move all nonzero terms to the left and factor.

$$(\ln x)^2 - 2 \ln x = 0$$

$$(\ln x)(\ln x - 2) = 0$$

$\ln x = 0$ or $\ln x - 2 = 0$

$x = e^0$ $\ln x = 2$

$x = 1$ $x = e^2$ Check these results.

PROBLEM 22

Solve $\log x^2 = (\log x)^2$.

Solution

Solutions to Matched Problems

18.
$$35^{1-2x} = 7$$
$$\log_{10} 35^{1-2x} = \log_{10} 7$$
$$(1 - 2x)(\log_{10} 35) = \log_{10} 7$$
$$1 - 2x = \frac{\log_{10} 7}{\log_{10} 35}$$
$$-2x = \frac{\log_{10} 7}{\log_{10} 35} - 1$$
$$x = -\frac{1}{2}\left(\frac{\log_{10} 7}{\log_{10} 35} - 1\right) \approx -\frac{1}{2}\left(\frac{0.8451}{1.5441} - 1\right) = 0.2263$$

19.
$$A = P(1 + r)^n$$
$$2P = P(1.09)^n$$
$$2 = 1.09^n$$
$$\log_{10} 2 = \log_{10} 1.09^n$$
$$\log_{10} 2 = n \log_{10} 1.09$$
$$n = \frac{\log_{10} 2}{\log_{10} 1.09} \approx \frac{0.3010}{0.0374} = 8.05$$

It will more than double in 9 years, but not quite double in 8 years.

20.
$$P = 14.7e^{-0.21x}$$
$$\frac{14.7}{8} = 14.7e^{-0.21x}$$
$$0.125 = e^{-0.21x}$$
$$\ln 0.125 = \ln e^{-0.21x}$$
$$\ln 0.125 = -0.21x$$
$$x = \frac{\ln 0.125}{-0.21} \approx \frac{-2.0794}{-0.21} = 9.9 \text{ miles}$$

21.
$$\log_{10}(x - 15) = 2 - \log_{10} x$$
$$\log_{10}(x - 15) + \log_{10} x = 2$$
$$\log_{10} x(x - 15) = 2$$
$$x(x - 15) = 10^2$$
$$x^2 - 15x - 100 = 0$$
$$(x - 20)(x + 5) = 0$$
$$x = -5, 20$$

Check: $x = -5$: Neither side of the equation is defined for $x = -5$.

$x = 20$:
$$\log_x(20 - 15) \overset{?}{=} 2 - \log_{10} 20$$
$$0.6990 \overset{\checkmark}{=} 0.6990$$

Thus, $x = 20$ is the only solution.

22.
$$\log x^2 = (\log x)^2$$
$$2 \log x = (\log x)^2$$
$$(\log x)^2 - 2 \log x = 0$$
$$(\log x)(\log x - 2) = 0$$
$$\log x = 0 \qquad \text{or} \quad \log x - 2 = 0$$
$$x = 10^0 = 1 \qquad\qquad \log x = 2$$
$$x = 10^2 = 100$$

Practice Exercise 9-5 ∎

Work odd-numbered problems first, check answers, and then work even-numbered problems in areas of weakness. Answers to all problems are in the back of the book. Make every effort to work a problem yourself before you look at an answer.

▦ *A scientific hand calculator will prove useful in many of the problems in this exercise.*

A *Solve to three significant digits.*

1. $10^{-x} = 0.0347$ _____   **2.** $10^x = 14.3$ _____

3. $10^{3x+1} = 92$ _____   **4.** $10^{5x-2} = 348$ _____

5. $e^x = 3.65$ _____   **6.** $e^{-x} = 0.0142$ _____

7. $e^{2x-1} = 405$ _____   **8.** $e^{3x+5} = 23.8$ _____

9. $5^x = 18$ _____   **10.** $3^x = 4$ _____

11. $2^{-x} = 0.238$ _____   **12.** $3^{-x} = 0.074$ _____

Solve exactly.

13. $\log 5 + \log x = 2$ _____

14. $\log x - \log 8 = 1$ _____

15. $\log x + \log(x - 3) = 1$ _____

16. $\log(x - 9) + \log 100x = 3$ _____

B *Solve to three significant digits.*

17. $2 = 1.05^x$ _____   **18.** $3 = 1.06^x$ _____

19. $e^{-1.4x} = 13$ _____   **20.** $e^{0.32x} = 632$ _____

21. $123 = 500e^{-0.12x}$ _____   **22.** $438 = 200e^{0.25x}$ _____

Solve exactly.

23. $\log x - \log 5 = \log 2 - \log(x - 3)$ _____

24. $\log(6x + 5) - \log 3 = \log 2 - \log x$ _____

25. $(\ln x)^3 = \ln x^4$ _____   **26.** $(\log x)^3 = \log x^4$ _____

27. $\ln(\ln x) = 1$ _____   **28.** $\log(\log x) = 1$ _____

29. $x^{\log x} = 100x$ _____   **30.** $3^{\log x} = 3x$ _____

C   In Problems 31–36 solve for the indicated letter in terms of all others using common or natural logs, whichever produces the simplest results.

31. $I = I_0 e^{-kx}$ for x X-ray absorption _____

32. $A = P(1 + i)^n$ for n Compound interest _____

33. $N = 10 \log \left(\dfrac{I}{I_0} \right)$ for I Sound intensity—decibels _____

34. $t = \dfrac{-1}{k} (\ln A - \ln A_0)$ for A Radioactive decay _____

35. $I = \dfrac{E}{R} (1 - e^{-Rt/L})$ for t Electric circuits _____

36. $S = R \dfrac{(1 + i)^n - 1}{i}$ for n Future value of an annuity _____

37. Find the fallacy:

$$3 > 2$$
$$(\log \tfrac{1}{2})3 > (\log \tfrac{1}{2})2$$
$$3 \log \tfrac{1}{2} > 2 \log \tfrac{1}{2}$$
$$\log(\tfrac{1}{2})^3 > \log(\tfrac{1}{2})^2$$
$$(\tfrac{1}{2})^3 > (\tfrac{1}{2})^2$$
$$\tfrac{1}{8} > \tfrac{1}{4}$$

38. Find the fallacy:

$$-2 < -1$$
$$\ln e^{-2} < \ln e^{-1}$$
$$2 \ln e^{-1} < \ln e^{-1}$$
$$2 < 1$$

APPLICATIONS

39. *Compound interest:*   How long will it take a sum of money to double if it is invested at 15% interest compounded annually (see Example 19)?

_____

40. *Compound interest:*   How long will it take money to quadruple if it is invested at 20% interest compounded annually (see Example 19)? _____

41. *Bacterial growth:*   A single cholera bacterium divides every $\tfrac{1}{2}$ hour to produce two complete cholera bacteria. If we start with a colony of 5,000 bacteria, then after t hours we will have

$$A = 5,000 \cdot 2^{2t}$$

bacteria. How long will it take for A to equal 1,000,000? _____

42. *Astronomy:*   An optical instrument is required to observe stars beyond the sixth magnitude, the limit of ordinary vision. However, even optical instruments have their limitations. The limiting magnitude L of any optical telescope with lens diameter D in inches is given by

$$L = 8.8 + 5.1 \log D$$

(A) Find the limiting magnitude for a homemade 6-inch reflecting telescope. _____

(B) Find the diameter of a lens that would have a limiting magnitude of 20.6. _____

*43. *World populations:*   A mathematical model for world population growth over short periods of time is given by

$$P = P_0 e^{rt}$$

where P_0 = Population at $t = 0$, r = Rate compounded continuously, t = Time in years, and P = Population at time t. How long will it take the earth's population to double if it continues to grow at its current rate of 2% per year (compounded continuously)? [*Hint:*   Given $r = 0.02$, find t so that $P = 2P_0$.] _____

**44. *World population:*   If the world population is now 4 billion people and if it continues to grow at 2% per year (compounded continuously), how long will it be before there is only 1 square yard of land per person? Use the formula in Problem 43 and the fact that there are 1.7×10^{14} square yards of land on earth. _____

*45. *Nuclear reactors—strontium-90:*   Radioactive strontium-90 is used in nuclear reactors and decays according to

$$A = Pe^{-0.0248t}$$

where P is the amount present at $t = 0$ and A is the amount remaining after t years. Find the half-life of strontium-90; that is, find t so that $A = 0.5P$. _____

*46. *Archaeology—carbon-14 dating:*   Cosmic-ray bombardment of the atmosphere produces neutrons, which in turn react with nitrogen to produce radioactive carbon-14. Radioactive carbon-14 enters all living tissues through carbon dioxide, which is first absorbed by plants. As long as a plant or animal is alive, carbon-14 is maintained in a constant amount in its tissues. Once dead, however, it ceases taking in carbon and, to the slow beat of time, the carbon-14 diminishes by radioactive decay according to the equation

$$A = A_0 e^{-0.000\ 124t}$$

where t is time in years. Estimate the age of a skull uncovered in an archaeological site if 10% of the original amount of carbon-14 is still present. [*Hint:*   Find t such that $A = 0.1A_0$.] _____

*47. *Sound intensity—decibels:*   Because of the extraordinary range of sensitivity of the human ear (a range of over 1,000 million million to 1), it is helpful to use a logarithmic scale to measure sound intensity over this range rather than an absolute scale. The unit of measure is called the *decibel*, after the inventor of the telephone, Alexander Graham Bell. If we let N be the number of decibels, I the power of the sound in question in watts per cubic centimeter, and I_0 the power of sound just below the threshold of hearing (approximately 10^{-16} watt per square centimeter), then

$$I = I_0 10^{N/10}$$

Show that this formula can be written in the form

$$N = 10 \log\left(\frac{I}{I_0}\right)$$

---

48. *Sound intensity—decibels:* Use the formula in Problem 47 (with $I_0 = 10^{-16}$ watt per square centimeter) to find the decibel ratings of the following sounds:

(A) Whisper (10^{-13} watt per square centimeter) _____

(B) Normal conversation (3.16×10^{-10} watt per square centimeter)

_____

(C) Heavy traffic (10^{-8} watt per square centimeter) _____

(D) Jet plane with afterburner (10^{-1} watt per square centimeter)

_____

*49. *Earth science:* For relatively clear bodies of fresh water or salt water, light intensity is reduced according to the exponential function

$$I = I_0 e^{-kd}$$

where I is the intensity at d feet below the surface and I_0 is the intensity at the surface; k is called the coefficient of extinction. Two of the clearest bodies of water in the world are the freshwater Crystal Lake in Wisconsin ($k = 0.0485$) and the saltwater Sargasso Sea off the West Indies ($k = 0.009\,42$). Find the depths (to the nearest foot) in these two bodies of water at which the light is reduced to 1% of that at the surface.

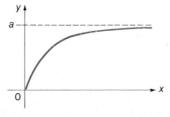

50. *Psychology—learning:* In learning a particular task, such as typing or swimming, one progresses faster at the beginning and then levels off. If you plot the level of performance against time, you will obtain a curve of the type shown in the figure. This is called a learning curve and can be very closely approximated by an exponential equation of the form $y = a(1 - e^{-cx})$, where a and c are positive constants. Curves of this type have applications in psychology, education, and industry. Suppose a particular person's history of learning to type is given by the exponential equation $N = 80(1 - e^{-0.08n})$, where N is the number of words per minute typed after n weeks of instruction. Approximately how many weeks did it take the person to learn to type 60 words per minute?

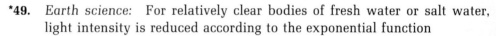

The Check Exercise for this section is on page 603.

Section 9-6   Chapter Review

An **exponential function** is a function defined by $f(x) = b^x$, where the **base** $b > 0$ and $b \neq 1$. The domain of an exponential function is R; the range is the set of positive real numbers. The number $e \approx 2.718$ is the base used for _the_ ex-_ponential function_. The basic properties of exponents continue to hold for b^x, and $b^m = b^n$ if and only if $m = n$ ($b > 0$, $b \neq 1$).   _(9-1)_

The inverse of an exponential function b^x is a **logarithmic function**, $y = \log_b x$; thus, $y = \log_b x$ means $b^y = x$. The domain of a logarithmic function is the set of positive real numbers; the range is the set of all real numbers. The logarithmic–exponential identities are

$$\log_b b^x = x \qquad \text{and} \qquad b^{\log_b x} = x \quad (9\text{-}2)$$

Logarithms satisfy these basic properties:

1. $\log_b b^M = M$

2. $\log_b MN = \log_b M + \log_b N$

3. $\log_b \dfrac{M}{N} = \log_b M - \log_b N$

4. $\log_b M^p = p \log_b M$

5. $\log_b 1 = 0$

Also $\log_b M = \log_b N$ if and only if $M = N$.   _(9-3)_

Logarithms using base 10 are called **common logarithms**; those with base e are **natural logarithms**. These two types are denoted by $\log_{10} x = $ **log** x and $\log_e x = $ **ln** x. The **change-of-base** formula

$$\log_b N = \frac{\log_a N}{\log_a b}$$

converts logarithms using base a to logarithms using base b.   _(9-4)_

Equations involving an exponential function are called **exponential equations**; those involving logarithms are called **logarithmic equations**.   _(9-5)_

Diagnostic (Review) Exercise 9-6 ∎

Work through all the problems in this chapter review and check answers in the back of the book. (Answers to all problems are there, and following each answer is a number in italics indicating the section in which that type of problem is discussed.) Where weaknesses show up, review appropriate sections in the text. When you are satisfied that you know the material, take the practice test following this review.

⊞ _You will find a scientific hand calculator useful in many of the problems in this exercise._

A   **1.**   Write $m = 10^n$ in logarithmic form with base 10. _____

2.   Write $\log x = y$ in exponential form. _____

Solve for x exactly. Do not use a calculator or table.

3. $\log_2 x = 3$ _____   **4.** $\log_x 25 = 2$ _____

5. $\log_3 27 = x$ _____

Solve for x to three significant digits.

6. $10^x = 17.5$ _____   **7.** $e^x = 143{,}000$ _____

In Problems 8 and 9, solve for x exactly. Do not use a calculator or table.

8. $\log x - 2 \log 3 = 2$ _____   **9.** $\log x + \log(x - 3) = 1$ _____

B **10.** Write $\ln y = x$ in exponential form. _____

11. Write $x = e^y$ in logarithmic form with base e. _____

Solve for x exactly. Do not use a calculator or table.

12. $\log_{1/4} 16 = x$ _____   **13.** $\log_x 9 = -2$ _____

14. $\log_{16} x = \frac{3}{2}$ _____   **15.** $\log_x e^5 = 5$ _____

16. $10^{\log_{10} x} = 33$ _____   **17.** $\ln x = 0$ _____

Solve for x to three significant digits.

18. $25 = 5(2)^x$ _____   **19.** $4{,}000 = 2{,}500 e^{0.12x}$ _____

20. $0.01 = e^{-0.05x}$ _____

In Problems 21–24, solve for x exactly. Do not use a table or calculator.

21. $\log 3x^2 - \log 9x = 2$ _____

22. $\log x - \log 3 = \log 4 - \log(x + 4)$ _____

23. $(\log x)^3 = \log x^9$ _____   **24.** $\ln(\log x) = 1$ _____

25. Calculate $\log_5 23$ to three significant digits. _____

Graph.

26. $y = 3e^{-2x}$ **27.** $y = 10^{x/2}$

   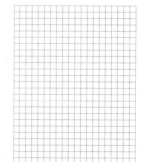

28. $y = 2 \log 3x$

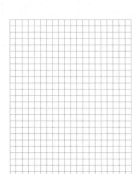

29. $y = 2 \ln \dfrac{x}{3}$

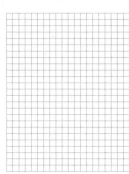

C   **30.**   Write $\ln y = -5t + \ln c$ in an exponential form free of logarithms; then solve for y in terms of the other letters. _____

31.   For $f = \{(x, y) \mid y = \log_2 x\}$, graph f and f^{-1} using the same coordinate system.

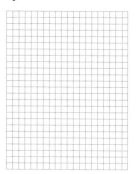

What are the domains and ranges for f and f^{-1}? _____

32.   Explain why 1 cannot be used as a logarithmic base. _____

33.   Prove that $\log_b(M/N) = \log_b M - \log_b N$. _____

APPLICATIONS   **34.**   *Population growth:*   Many countries in the world have a population growth rate of 3% (or more) per year. At this rate how long, to the nearest year, will it take a population to double? Use the population growth model

$$P = P_0(1.03)^t$$

which assumes annual compounding. Compute the answer to three significant digits. _____

35.   *Population growth:*   Repeat Problem 34 using the continuous population growth model

$$P = P_0 e^{0.03t}$$

which assumes continuous compounding. Compute the answer to three significant digits. _____

36. *Carbon-14 dating:* Refer to Problem 46, Exercise 9-5. How long, to three significant digits, will it take for the carbon-14 to diminish to 1% of the original amount after the death of a plant or animal?

$$A = A_0 e^{-0.000\,124t} \qquad \text{where } t \text{ is time in years} \underline{\hspace{3cm}}$$

37. *X-ray absorption:* Solve $x = -(1/k) \ln(I/I_0)$ for I in terms of the other letters. \underline{\hspace{3cm}}

38. *Amortization—time payments:* Solve $r = P\{i/[1 - (1 + i)^{-n}]\}$ for n in terms of the other letters. \underline{\hspace{3cm}}

Practice Test Chapter 9 ■

Take this as if it were a graded test by working the problems within a 50-minute time period. Do not look back in the chapter. Choose one of three levels of difficulty: least difficult, Problems 1–12; more difficult, add Problem 13; most difficult, add Problems 13 and 14. Use the answers in the back of the book to correct your work. The answers are keyed to appropriate text sections so that you can easily locate and review sections where difficulties still persist.

1. Write $\log_{16} 8 = \frac{3}{4}$ in exponential form. \underline{\hspace{3cm}}

2. Write $\log_{10} y = 0.2x$ in exponential form. \underline{\hspace{3cm}}

3. Write $A = e^{0.08t}$ in logarithmic form using base e. \underline{\hspace{3cm}}

In Problems 4–7 solve for x, without using a calculator or table.

4. $\log_2 x = 4$ \underline{\hspace{2cm}}   5. $\log_{16} 4 = x$ \underline{\hspace{2cm}}

6. $\log_x 16 = -2$ \underline{\hspace{3cm}}

7. $\log_{10} 2x + \log_{10} 5x = 3$ \underline{\hspace{3cm}}

Solve Problems 8–10 to three significant figures using a calculator or table.

8. $10^{2x+1} = 252$ \underline{\hspace{2cm}}   9. $24 = 4(3)^x$ \underline{\hspace{2cm}}

10. $0.1 = e^{-0.4x}$ \underline{\hspace{2cm}}

11. If a country has a population growth rate of 4% per year, how long to the nearest year will it take the population to double? [Use $P = P_0(1.04)^t$ and solve for t when $P = 2P_0$.] Use a calculator or table. \underline{\hspace{2cm}}

12. How long, to three significant figures, will it take for the carbon-14 to diminish to one-fourth of the original amount after the death of a plant or animal? ($A = A_0 e^{-0.000\,124t}$, where t is time in years.) Use a calculator or table. \underline{\hspace{3cm}}

13.   Graph $y = 2 \ln(4x)$.

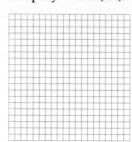

14.   Graph $y = \frac{1}{4}2^{-x}$.

Check Exercise 9-1 ■

ANSWER COLUMN

Work the following problems without looking at any text examples. Show your work in the space provided. Draw the final graph or write your answer in the answer column.

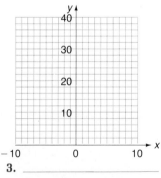

1. _____

Graph Problems 1–4.

1. $y = 4 \cdot 2^x$, $-3 \le x \le 3$

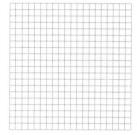

2. _____

3. _____

2. $y = 2 \cdot 3^{-x}$, $-3 \le x \le 3$

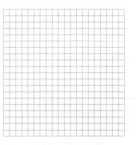

4. _____

5. _____

3. $y = 1.5 \cdot 2^{-2x}, \ -3 \le x \le 3$

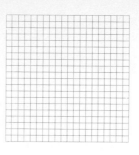

4. $y = 10e^{0.5x}, \ -3 \le x \le 3$

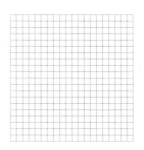

5. The atmospheric pressure P, in pounds per square inch, can be calculated approximately using $P = 14.7e^{-0.21x}$, where x is the altitude in miles relative to sea level. What is the atmospheric pressure in a mine shaft 0.5 mile below sea level?

Check Exercise 9-2 ■

Work the following problems without looking at any text examples. Show your work in the space provided. Write your answer in the answer column.

1. _____

2. _____

    1. Rewrite in exponential form: $\log_3 81 = 4$

3. _____

4. _____

5. _____

    2. Rewrite in exponential form: $\log_{10} 0.0001 = -4$

6. _____

7. _____

8. _____

    3. Rewrite in exponential form: $y = \log_b x,\ b > 0,\ b \neq 1$

9. _____

10. _____

    4. Rewrite in logarithmic form: $\frac{1}{16} = 2^{-4}$

    5. Rewrite in logarithmic form: $1 = 5^0$

6. Rewrite in logarithmic form: $x = b^y$, $b > 0$, $b \neq 1$

7. Find: $\log_3 \sqrt{27}$

8. Find x: $\log_2 x = 3$

9. Find y: $y = \log_{1/2} 4$

10. Find b: $\log_b \frac{1}{8} = -3$

Check Exercise 9-3 ■

Work the following problems without looking at any text examples. Show your work in the space provided. Write your answer in the answer column.

1. _____

2. _____

Write Problems 1–4 in terms of simpler logarithmic forms (going as far as you can using logarithmic properties).

3. _____

1. $\log_b \dfrac{w}{xy}$

4. _____

5. _____

2. $\log_b \sqrt{MN}$

6. _____

7. _____

8. _____

3. $\log_b \sqrt{\dfrac{x^3}{y^5}}$

9. _____

10. _____

4. $\log_b \dfrac{\sqrt[4]{A}}{2B^3}$

Write Problems 5–7 in terms of a single logarithm with a coefficient of 1.

5. $2 \log_b u - 3 \log_b v - 4 \log_b w$

6. $\frac{1}{3}(\log_b A - \log_b B)$

7. $-3 \log_b M - 2 \log_b N$

If $\log_e 2 = 0.69$, $\log_e 3 = 1.10$, and $\log_e 5 = 1.61$, find the indicated logarithms in Problems 8 and 9 to two decimal places.

8. $\log_e 40$

9. $\log_e \sqrt{0.3}$

10. Find x so that $\frac{1}{3} \log_b 8 - 2 \log_b 2 = \log_b$ x.

Check Exercise 9-4 ∎

ANSWER COLUMN

Work the following problems without looking at any text examples. Show your work in the space provided. Write your answer or draw the final graph in the answer column.

1. _____

2. _____

In Problems 1–9, use a calculator to find each to four decimal places.

1. $\log_{10} 0.006\ 77$

3. _____

4. _____

5. _____

2. $\ln 43{,}662$

6. _____

7. _____

8. _____

3. $\dfrac{\log_{10} 55}{\log_{10} 110}$

9. _____

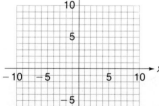

4. x if $\ln x = 0.9250$

10. _____

5. x if $\log_{10} x = 4.2723$

6. x if ln x = −3.764

7. x if $\log_{10}$ x = −2.8021

8. $\log_3$ 100

9. $\log_5$ 0.225

10. Graph y = $\frac{1}{4}$ ln $3x^2$.

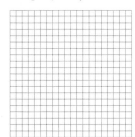

Check Exercise 9-5 ■

Work the following problems without looking at any text examples. Show your work in the space provided. Write your answer in the answer column.

1. _____

Use logarithmic properties and a calculator to solve Problems 1–3 to three significant digits.

2. _____

1. $5^{2x-3} = 3$

3. _____

4. _____

5. _____

2. $\log_{10} x + \log_{10}(x - 9) = 1$

3. $57 = 10e^{-2.3t}$

4. How long to the nearest year will it take money to triple if invested at 12% compounded annually? $[A = P(1 + r)^n]$

5. The atmospheric pressure P (in pounds per square inch) at x miles above sea level is given approximately by

    $$P = 14.7e^{-0.21x}$$

    At what height (to two decimal places) will the pressure be one-tenth that at sea level?

Systems of Equations and Inequalities ■ 10

INSTRUCTIONS FOR STUDENTS IN A SELF-PACED CLASS OR LAB

YES ← HAVE YOU HAD INTERMEDIATE ALGEBRA BEFORE THIS COURSE? → NO

1. Work Diagnostic (Review) Exercise 10-6 on page 655. Check answers in back of book; then work through text sections corresponding to problems missed. (Section numbers are in italics following each answer.)

2. When finished with step 1, take Practice Test Chapter 10 on page 657 as a final check of your understanding of the chapter. Check answers in the back of the book; then review sections where weakness still prevails. (Corresponding section numbers are in italics following each answer.)

3. When you think you are ready, ask your instructor for a graded test for Chapter 10.

4. If your instructor approves, after the test is corrected, go to the next chapter.

1. Work through each section in the chapter as follows:
   (A) Read discussion.
   (B) Read each example and work the corresponding matched problem. Check your solutions to the matched problem in Solutions to Matched Problems on the indicated page.
   (C) At the end of a section work the odd-numbered problems in the Practice Exercise and check answers; then work even-numbered problems in areas of weakness. (Answers to *all* Practice Exercise sets are in the back of the book.)
   (D) Work Check Exercise as instructed. Tear out and turn in as directed by your instructor. (Answers are not in the text.)

2. Repeat each step in item 1 for each section in the chapter.

3. After the instructional part of the chapter is completed, proceed with steps 1 to 4 in the box above this one.

Chapter 10 ■ Systems of Equations and Inequalities

Section 10-1   Systems of Linear Equations in Two Variables

- ■ Solution by Graphing
- ■ Solution by Substitution
- ■ Solution by Elimination Using Addition
- ■ Applications

Many practical problems can be solved conveniently using two-equation–two-unknown methods. For example, if a 12-foot board is cut in two pieces so that one piece is 2 feet longer than the other piece, how long is each piece? We could solve this problem using one-equation–one-unknown methods studied earlier, but we can also proceed as follows, using two variables. Let

x = Length of the longer piece
y = Length of the shorter piece

Then

$x + y = 12$
$x - y = 2$

To **solve** this system is to find all the ordered pairs of real numbers that satisfy both equations at the same time. In general, we are interested in solving linear systems of the type:

Linear System (Standard Form)
$ax + by = m$ $a, b, c, d, m,$ and n are constants $cx + dy = n$ x and y are variables a, b, c, d not all 0

There are several methods of solving systems of this type. We will consider three that are widely used: solution by graphing, solution by substitution, and solution by elimination using addition.

■ **Solution by Graphing**

We proceed by graphing both equations on the same coordinate system. Then the coordinates of any points that the graphs have in common must be solutions to the system since they must satisfy both equations.

EXAMPLE 1   Solve by graphing: $x + y = 12$
$x - y = 2$

Solution   Graph each equation and find coordinates of points of intersection, if they exist.

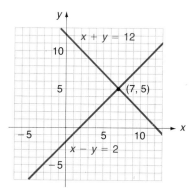

Solution:

$x = 7$ or $(x, y) = (7, 5)$

$y = 5$

Check $x + y = 12$ $x - y = 2$

$7 + 5 \overset{\checkmark}{=} 12$ $7 - 5 \overset{\checkmark}{=} 2$

PROBLEM 1   Solve by graphing: $x + y = 10$
$x - 4y = 0$

Solution

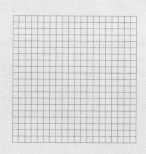

It is clear that the system in Example 1 has exactly one solution since the lines have exactly one point of intersection. In general, two lines in the same rectangular coordinate system must be related to each other in one of three ways: (1) they intersect at one and only one point, (2) they are parallel, or (3) they coincide (see Example 2).

EXAMPLE 2   Solve each of the following systems by graphing:

(A) $2x - 3y = 2$ **(B)** $4x + 6y = 12$ **(C)** $2x - 3y = -6$
        $x + 2y = 8$ $2x + 3y = -6$ $-x + \frac{3}{2}y = 3$

Solution

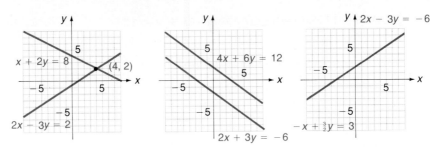

(**A**)   Lines intersect at one point only:
Exactly one solution
$x = 4, y = 2$

(**B**)   Lines are parallel:
No solution

(**C**)   Lines coincide:
Infinite number of solutions

Now we know exactly what to expect when solving a system of two linear equations in two unknowns:

Possible Solutions to a Linear System

1. Exactly one pair of numbers
2. No solution
3. Infinitely many solutions

PROBLEM 2

Solve each of the following systems by graphing:

(**A**) $2x + 3y = 12$
$\quad\;\; x - 3y = -3$

(**B**) $\quad x - 3y = -3$
$\quad -2x + 6y = 12$

(**C**) $2x - 3y = 12$
$\quad -x + \frac{3}{2}y = -6$

Solution

(**A**)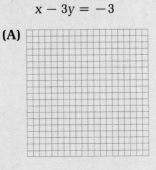

(**B**)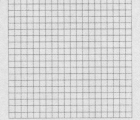

(**C**)

Generally, graphic methods give us only rough approximations of solutions. The methods of substitution and elimination using addition to be considered next will yield results to any decimal accuracy desired—assuming solutions exist.

■ Solution by Substitution

Choose one of the two equations in a system and solve for one variable in terms of the other. (Choose an equation that avoids getting involved with fractions, if possible.) Then substitute the result into the other equation and solve the resulting linear equation in one variable. Now substitute this result back into either of the original equations to find the second variable. An example should make the process clear.

EXAMPLE 3   Solve by substitution:

$$2x - 3y = 7 \tag{1}$$
$$-3x + y = -7 \tag{2}$$

Solution $y = \underbrace{3x - 7}$ Solve Equation (2), the simplest choice, for y in terms of x.

$$2x - 3y = 7$$
$$2x - 3(3x - 7) = 7$$
$$2x - 9x + 21 = 7$$
$$-7x = -14$$

$x = 2$ Substitute $x = 2$ into $y = 3x - 7$ and solve for y.

$$y = 3 \cdot 2 - 7$$
$$y = -1$$

Thus, $(2, -1)$ is a solution to the original system, as we can readily check:

Check $(2, -1)$ must satisfy *both* equations.

$$2x - 3y = 7 \qquad\qquad -3x + y = -7$$
$$2(2) - 3(-1) \overset{?}{=} 7 \qquad -3(2) + (-1) \overset{?}{=} -7$$
$$4 + 3 \overset{\checkmark}{=} 7 \qquad\qquad -6 - 1 \overset{\checkmark}{=} -7$$

PROBLEM 3   Solve by substitution and check: $3x - 4y = 18$
                                                  $2x + y = 1$

Solution

■ Solution by Elimination Using Addition

Now we turn to elimination using addition. This is probably the most important method of solution, since it is readily generalized to higher-order systems. The method involves the replacement of systems of equations with simpler *equivalent systems* (by performing appropriate operations) until we obtain a system with an obvious solution. **Equivalent systems** of equations are, as you would expect, systems that have exactly the same solution set. Theorem 1 lists operations that produce equivalent systems.

THEOREM 1 | Equivalent systems of equations result if:

(A) Two equations are interchanged.
(B) An equation is multiplied by a nonzero constant.
(C) A constant multiple of another equation is added to a given equation.

Solving systems of equations by use of this theorem is best illustrated by examples.

EXAMPLE 4   Solve by elimination using addition:

$$3x - 2y = 8 \tag{3}$$
$$2x + 5y = -1 \tag{4}$$

Solution   We use Theorem 1 to eliminate one of the variables and thus obtain a system with an obvious solution:

$$15x - 10y = 40$$
$$\underline{4x + 10y = -2}$$
$$19x = 38$$
$$x = 2$$

If we multiply Equation (3) by 5 and Equation (4) by 2, and then add, we can eliminate y.

Now substitute $x = 2$ back into either of the original equations, say Equation (4), and solve for y ($x = 2$ paired with either of the two original equations produces an equivalent system):

$$2(2) + 5y = -1$$
$$5y = -5$$
$$y = -1$$

Thus, $(2, -1)$ is a solution to the original system. The check is completed as in Example 3.

PROBLEM 4   Solve by elimination using addition and check your answer. Eliminate the variable x first.

$$2x + 5y = -9$$
$$3x - 4y = -2$$

Solution

EXAMPLE 5   Solve by elimination using addition:

$$x + 3y = 2 \tag{5}$$
$$2x + 6y = -3 \tag{6}$$

Solution

$$\begin{array}{l} -2x - 6y = -4 \quad \text{Multiply Equation (5) by } -2 \text{ and add} \\ \underline{2x + 6y = -3} \\ 0 = -7 \quad \text{A contradiction!} \end{array}$$

Our assumption that there are values for x and y that satisfy Equations (5) and (6) simultaneously must be false. (Otherwise, we have proved that $0 = -7$.) If you check the slope of each line, you will find that they are the same (but the y intercepts are different); hence, the lines are parallel and the system has no solution. Systems of this type are called **inconsistent**—conditions have been placed on the variables x and y that are impossible to meet.

PROBLEM 5   Solve by elimination using addition: $3x - 4y = -2$
$$-6x + 8y = 1$$

Solution

EXAMPLE 6   Solve by elimination using addition:

$$-2x + y = -8 \tag{7}$$
$$x - \tfrac{1}{2}y = 4 \tag{8}$$

Solution

$$\begin{array}{l} -2x + y = -8 \quad \text{Multiply Equation (8) by 2 and add.} \\ \underline{2x - y = 8} \\ 0 = 0 \end{array}$$

Both sides have been eliminated. Actually, if we had multiplied Equation (8) by -2, we would have obtained Equation (7). When one equation is a constant multiple of the other, the system is said to be **dependent**, and their graphs will coincide. There are infinitely many solutions to the system—any solution of one equation will be a solution of the other. One way of expressing all solutions is to solve one of the equations for y in terms of x, say the first equation,

$$y = 2x - 8$$

Then

$$(x, 2x - 8)$$

is a solution to the system for any real number x, as can easily be checked. If x = 1, for example, then y = 2x − 8 = 2 − 8 = −6 and (1, −6) is a solution. We could have chosen to solve for x in terms of y; the solution would then be described as the set of all pairs $(\frac{1}{2}y + 4, y)$.

PROBLEM 6   Solve by elimination using addition: $6x - 3y = -2$
                                                      $-2x + \ \ y = \frac{2}{3}$

Solution

■ Applications

Many of the applications we considered earlier, using one-equation–one-unknown methods, can be set up more naturally using two-equation–two-unknown methods. The following examples will illustrate the process.

EXAMPLE 7   A change machine changes dollar bills into quarters and nickels. If you receive 12 coins after inserting a $1 bill, how many of each type of coin did you receive?

Solution   Let

x = Number of quarters

y = Number of nickels

Then

$x + \ \ y = 12$ Number of coins

$25x + 5y = 100$ Value of coins in cents

$-5x - 5y = -60$

$\underline{25x + 5y = \ \ 100}$

$20x \ \ \ \ \ = \ \ \ 40$

$x = 2$ Quarters

$x + y = 12$

$2 + y = 12$

$y = 10$ Nickels

Check $2 + 10 = 12$ coins in all $25 \cdot 2 + 5 \cdot 10 = 50 + 50$

$= 100$ cents or $1

PROBLEM 7

Solution

Repeat Example 7 with a receipt of 8 coins from a $1 bill.

EXAMPLE 8

Solution

A zoologist wishes to prepare a special diet that contains, among other things, 120 grams of protein and 17 grams of fat. Two available food mixes specify the following percentages of protein and fat:

MIX	PROTEIN (%)	FAT (%)
A	30	1
B	20	5

How many grams of each mix should be used to prepare the diet mix?

Let

$$x = \text{Number of grams of mix } A \text{ used}$$

$$y = \text{Number of grams of mix } B \text{ used}$$

Set up one equation for the protein requirements and one equation for the fat requirements:

$$0.3x + 0.2y = 120 \quad \text{Protein requirements}$$

$$0.01x + 0.05y = 17 \quad \text{Fat requirements}$$

Multiply the top equation by 10 and the bottom equation by 100 to clear decimals (not necessary, but helpful):

$$3x + 2y = 1,200 \quad \text{Multiply bottom equation by } -3; \text{ then add to eliminate } x.$$

$$\underline{x + 5y = 1,700}$$

$$3x + 2y = 1,200$$

$$\underline{-3x - 15y = -5,100}$$

$$-13y = -3,900$$

$$y = 300 \text{ grams} \quad \text{Mix } B$$

$$x + 5y = 1,700$$

$$x + 5(300) = 1,700$$

$$x = 200 \text{ grams} \quad \text{Mix } A$$

The zoologist should use 200 grams of mix A and 300 grams of mix B to meet the diet requirements.

Check   *Protein requirement:*

(Protein from mix *A*) + (Protein from mix *B*) $\overset{?}{=}$ 120 grams

(30% of 200 grams) + (20% of 300 grams) = 60 + 60 $\overset{\checkmark}{=}$ 120 grams

Fat requirement:

(Fat from mix *A*) + (Fat from mix *B*) $\overset{?}{=}$ 17 grams

(1% of 200 grams) + (5% of 300 grams) $\overset{\checkmark}{=}$ 2 + 15 = 17 grams

PROBLEM 8   Repeat Example 8 for a diet mixture that is to contain 110 grams of protein and 8 grams of fat.

Solution

EXAMPLE 9   A jeweler has two bars of gold alloy in stock, one 12-carat and the other 18-carat. (Note that 24-carat gold is pure gold, 12-carat gold is $\frac{12}{24}$ pure, 18-carat gold is $\frac{18}{24}$ pure, and so on.) How many grams of each alloy must be mixed to obtain 10 grams of 14-carat gold?

Solution   Let

x = Number of grams of 12-carat gold used

y = Number of grams of 18-carat gold used

Then

$$x + y = 10 \qquad \text{Amount of new alloy.}$$

$$\tfrac{12}{24}x + \tfrac{18}{24}y = \tfrac{14}{24}(10) \qquad \begin{array}{l}\text{Pure gold present before mixing equals}\\ \text{pure gold present after mixing.}\end{array}$$

$$x + y = 10 \qquad \begin{array}{l}\text{Multiply the second equation by } \tfrac{24}{2} \text{ to simplify;}\\ \text{then solve using the methods described earlier.}\\ \text{(We use elimination here.)}\end{array}$$

$$6x + 9y = 70$$

$$\begin{array}{r} -6x - 6y = -60 \\ 6x + 9y = 70 \\ \hline 3y = 10 \end{array}$$

$$y = 3\tfrac{1}{3} \text{ grams} \qquad \text{18-carat alloy}$$

$$x + 3\tfrac{1}{3} = 10$$

$$x = 6\tfrac{2}{3} \text{ grams} \qquad \text{12-carat alloy}$$

The checking of solutions is left to you.

PROBLEM 9   Repeat Example 9, but suppose that the jeweler has only 10-carat and pure gold in stock.

Solution

Solutions to Matched Problems

1. $x + y = 10$
   $x - 4y = 0$

Solution:
$x = 8$
$y = 2$ or $(x, y) = (8, 2)$

2. (A) $2x + 3y = 12$
        $x - 3y = -3$

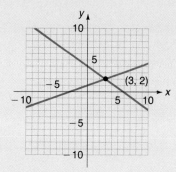

Solution:
$x = 3$
$y = 2$ or $(x, y) = (3, 2)$

(B) $x - 3y = -3$
        $-2x + 6y = 12$

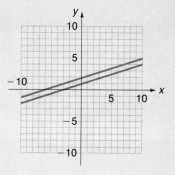

Solution:
Lines are parallel;
no solution.

2. (C) $2x - 3y = 12$
             $-x + \frac{3}{2}y = -6$

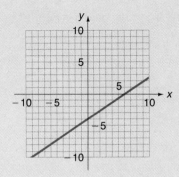

Solution:
Lines coincide;
infinitely many solutions

3. $3x - 4y = 18$ (A)
$2x + \ y = 1$ (B)
Solve (B) for y in terms of x;
then substitute into (A):

$$y = 1 - 2x$$
$$3x - 4(1 - 2x) = 18$$
$$3x - 4 + 8x = 18$$
$$11x = 22$$
$$\boldsymbol{x = 2}$$
$$y = 1 - 2(2)$$
$$\boldsymbol{y = -3}$$

$(2, -3)$ is the solution.

4. $2x + 5y = -9$ (A)
$3x - 4y = -2$ (B)
Multiply (A) by 3 and (B) by
-2; then add to eliminate x:

$$6x + 15y = -27$$
$$\underline{-6x + \ \ 8y = \ \ \ \ 4}$$
$$23y = -23$$
$$\boldsymbol{y = \ -1}$$

Substitute into (A) to find x:
$$2x + 5(-1) = -9$$
$$2x = -4$$
$$\boldsymbol{x = -2}$$

5. $3x - 4y = -2$ (A)
$-6x + 8y = 1$ (B)
Multiply (A) by 2 and add:

$$6x - 8y = -4$$
$$\underline{-6x + 8y = \ \ \ 1}$$
$$0 = -3$$

Contradiction! No solution.

6. $6x - 3y = -2$ (A)
$-2x + \ y = \frac{2}{3}$ (B)
Multiply (B) by 3 and add:

$$6x - 3y = -2$$
$$\underline{-6x + 3y = \ \ \ 2}$$
$$0 = \ \ \ 0$$

Infinitely many solutions;
$(x, 2x + \frac{2}{3})$ is a solution for any
real number x.

7. Let x = Number of quarters
y = Number of nickels

$$x + \ y = 8$$
$$25x + 5y = 100$$

$$-5x - 5y = -40$$
$$\underline{25x + 5y = \ \ 100}$$
$$20x \ \ \ \ \ \ \ = \ \ \ \ 60$$
$$\boldsymbol{x = 3} \quad \text{Quarters}$$
$$x + y = 8$$
$$3 + y = 8$$
$$\boldsymbol{y = 5} \quad \text{Nickels}$$

8. Let x = Number of grams of mix *A* used
  y = Number of grams of mix *B* used

$$0.3x + 0.2y = 110$$
$$0.01x + 0.05y = 8$$

$$3x + 2y = 1{,}100$$
$$x + 5y = 800$$

$$3x + 2y = 1{,}100$$
$$\underline{-3x - 15y = -2{,}400}$$
$$-13y = -1{,}300$$
$$\boldsymbol{y = 100 \text{ grams}} \quad \text{Mix } B$$

$$x + 5y = 800$$
$$x + 5(100) = 800$$
$$\boldsymbol{x = 300 \text{ grams}} \quad \text{Mix } A$$

9. Let x = Number of grams of 10-carat gold used
  y = Number of grams of pure gold used

$$x + y = 10$$
$$\tfrac{10}{24}x + y = \tfrac{14}{24}(10)$$

$$x + y = 10$$
$$5x + 12y = 70$$

$$-5x - 5y = -50$$
$$\underline{5x + 12y = 70}$$
$$\boldsymbol{y = \tfrac{20}{7} \approx 2.86 \text{ grams}} \quad \text{Pure gold}$$

$$x + y = 10$$
$$x + \tfrac{20}{7} = 10$$
$$\boldsymbol{x = \tfrac{50}{7} \approx 7.14 \text{ grams}} \quad \text{10-carat gold}$$

Practice Exercise 10-1 ■

Work odd-numbered problems first, check answers, and then work even-numbered problems in areas of weakness. Answers to all problems are in the back of the book. Make every effort to work a problem yourself before you look at an answer.

A   *Solve by graphing.*

1. $3x - 2y = 12$
  $7x + 2y = 8$

2. $x + 5y = -10$
  $-5x + y = 24$

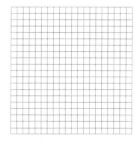

3. $3x +\ 5y = 15$
$6x + 10y = -5$

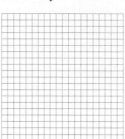

4. $3x - 5y = 15$
$x - \frac{5}{3}y = 5$

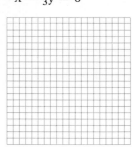

Solve by substitution.

5. $2x + y = 6$
$y = x + 3$ _____

6. $m - 2n = 0$
$-3m + 6n = 8$ _____

7. $3x -\ y = -3$
$5x + 3y = -19$ _____

8. $2m - 3n = 9$
$m + 2n = -13$ _____

Solve by elimination using addition.

9. $3p +\ 8q = 4$
$15p + 10q = -10$ _____

10. $3x -\ y = -3$
$5x + 3y = -19$ _____

11. $6x - 2y = 18$
$-3x +\ y = -9$ _____

12. $4m + 6n = 2$
$6m - 9n = 15$ _____

B *Solve each system by graphing, by substitution, and by elimination using addition.*

13. $x - 3y = -11$
$2x + 5y = 11$

14. $5x +\ y = 4$
$x - 2y = 3$

15. $11x + 2y = 1$
$9x - 3y = 24$

16. $2x + y = 0$
$3x + y = 2$

Use any of the methods discussed in this section to solve each system.

17. $y = 3x - 3$
$6x = 8 + 3y$ _____

18. $3m = 2n$
$n = -7 - 2m$ _____

19. $\frac{1}{2}x - y = -3$
$-x + 2y = 6$ _____

20. $y = 2x - 1$
$6x - 3y = -1$ _____

21. $2x + 3y = 2y - 2$
$3x + 2y = 2x + 2$ _____

22. $2u - 3v = 1 - 3u$
$4v = 7u - 2$ _____

C **23.** $0.2x - 0.5y = 0.07$
$0.8x - 0.3y = 0.79$ _____

24. $0.5m + 0.2n = 0.54$
$0.3m - 0.6n = 0.18$ _____

25. $\frac{1}{4}x - \frac{2}{3}y = -2$
$\frac{1}{2}x - y = -2$ _____

26. $\frac{2}{3}a + \frac{1}{2}b = 2$
$\frac{1}{2}a + \frac{1}{3}b = 1$ _____

APPLICATIONS **27.** *Puzzle:* A bank gave you $1.50 in change consisting of only nickels and dimes. If there were 22 coins in all, how many of each type of coin did you receive? _____

28. *Puzzle:* A friend of yours came out of a post office having spent $1.32 on thirty 4-cent and 5-cent stamps. How many of each type did he buy?

_____

29. *Geometry:* If the sum of two angles in a right triangle is 90° and their difference is 14°, find the two angles. _____

30. *Geometry:* Find the dimensions of a rectangle with perimeter 72 inches if its length is 25% greater than its width. _____

*31. *Business:* A packing carton contains 144 small packages, some weighing $\frac{1}{4}$ pound each and the others $\frac{1}{2}$ pound each. How many of each type are in the carton if the total contents of the carton weigh 51 pounds? _____

*32. *Biology:* A biologist, in a nutrition experiment, wants to prepare a special diet for experimental animals. He requires a food mixture that contains, among other things, 20 ounces of protein and 6 ounces of fat. He is able to purchase food mixes of the following composition:

MIX	PROTEIN (%)	FAT (%)
A	20	2
B	10	6

How many ounces of each mix should he use to prepare the diet mix? Solve graphically and algebraically. _____

*33. *Chemistry:*   A chemist has two concentrations of hydrochloric acid in stock, a 50% solution and an 80% solution. How much of each should she mix to obtain 100 milliliters of a 68% solution? _____

*34. *Business:*   A newspaper printing plant has two folding machines for the final assembling of the evening newspaper—circulation, 29,000. The slower machine can fold papers at the rate of 6,000 per hour, and the faster machine at the rate of 10,000 per hour. If the use of the slower machine is delayed $\frac{1}{2}$ hour because of a minor breakdown, how much total time is required to fold all the papers? How much time does each machine spend on the job? _____

*35. *Earth science:*   A ship using sound-sensing devices above and below water recorded a surface explosion 6 seconds sooner by its underwater device than its abovewater device. Sound travels in air at about 1,100 feet per second and in seawater at about 5,000 feet per second.

(A)   How long did it take each sound wave to reach the ship? _____

(B)   How far was the explosion from the ship? _____

*36. *Earth science:*   An earthquake emits a primary wave and a secondary wave. Near the surface of the earth the primary wave travels at about 5 miles per second, and the secondary wave at about 3 miles per second. From the time lag between the two waves arriving at a given station, it is possible to estimate the distance to the quake. (The epicenter can be located by obtaining distance bearings at three or more stations.) Suppose a station measured a time difference of 16 seconds between the arrival of the two waves. How long did each wave travel, and how far was the earthquake from the station? _____

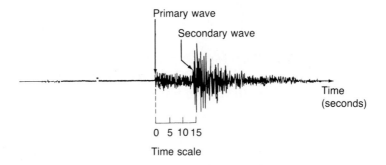

**37. *Business:*   Two companies have offered you a sales position. Both jobs are essentially the same, but one company pays a straight 8% commission and the other pays $51 per week plus a 5% commission. The best sales representatives with either company rarely have sales greater than $4,000 in any one week. Before accepting either offer, it would be helpful to know at what point both companies pay the same, and which of the companies pays more on either side of this point. Solve graphically and algebraically.

**38.   *Business:*   Solve Problem 37 with the straight-commission company paying 7% commission and the salary-plus-commission company paying $75 per week plus 4% commission. _____

**39.   *Economics:*   In a particular city the weekly supply and demand for popular stereo records, relative to average price per record p, are given by the equations

$$d = 5{,}000 - 1{,}000p \qquad \$1 \le p \le \$4$$
$$s = -3{,}000 + 3{,}000p \qquad \$1 \le p \le \$4$$

At what price does the supply equal the demand (equilibrium point)?
(A)   Solve graphically.   **(B)**   Solve algebraically. _____

The Check Exercise for
this section is on
page 659.

Section 10-2   Systems of Linear Equations in Three Variables

■ Solving Systems of Three Equations in Three Variables
■ Application

■ Solving Systems of Three Equations in Three Variables

Having learned how to solve systems of linear equations in two variables, we proceed to higher-order systems. Systems of the form

$$3x - 2y + 4z = 6$$
$$2x + 3y - 5z = -8 \qquad\qquad (1)$$
$$5x - 4y + 3z = 7$$

as well as higher-order systems are encountered frequently and are worth studying. In fact, systems of equations in many variables are so important in solving real-world problems that there are whole courses on this one topic. A triplet of numbers $x = 0$, $y = -1$, and $z = 1$ [also written as an ordered triplet $(0, -1, 1)$] is a **solution** of system (1) since each equation is satisfied by this triplet. The set of all such ordered triplets of numbers is called the **solution set** of the system. Two systems are said to be **equivalent** if they have the same solution set.

We will use an extension of the method of elimination by addition discussed in Section 10-1 to solve systems in the form of (1). Theorem 1 in Section 10-1 is behind the process.

Steps in Solving Systems of Three Equations in Three Variables
Step 1:   Choose two equations from the system and eliminate one of the three variables, using elimination by addition or subtraction. The result is generally one equation in two unknowns.
Step 2:   Now eliminate the same variable from the unused equation and one of those used in step 1. We (generally) obtain another equation in two variables.
Step 3:   The two equations from steps 1 and 2 form a system of two equations and two unknowns. Solve as in the preceding section.
Step 4:   Substitute the solution from step 3 into any of the three original equations and solve for the third variable to complete the solution of the original system.

EXAMPLE 10   Solve:

$$3x - 2y + 4z = 6 \tag{2}$$
$$2x + 3y - 5z = -8 \tag{3}$$
$$5x - 4y + 3z = 7 \tag{4}$$

Solution   *Step 1:*   We look at the coefficients of the various variables and choose to eliminate y from Equations (2) and (4) because of the convenient coefficients -2 and -4. Multiply Equation (2) by -2 and add to Equation (4):

$$
\begin{array}{rl}
-6x + 4y - 8z = -12 & \quad -2[\text{Equation (2)}] \\
\underline{5x - 4y + 3z = 7} & \quad [\text{Equation (4)}] \\
-x - 5z = -5 &
\end{array}
$$

Step 2:   Now let us eliminate y (the same variable) from Equations (2) and (3). Multiply Equation (2) by 3 and Equation (3) by 2 and add:

$$
\begin{array}{rl}
9x - 6y + 12z = 18 & \quad 3[\text{Equation (2)}] \\
\underline{4x + 6y - 10z = -16} & \quad 2[\text{Equation (3)}] \\
13x + 2z = 2 &
\end{array}
$$

Step 3:   From steps 1 and 2 we obtain the system

$$-x - 5z = -5 \tag{5}$$
$$13x + 2z = 2 \tag{6}$$

We solve this system as in the last section. Multiply Equation (5) by 13 and add to Equation (6) to eliminate x:

$$
\begin{array}{rl}
-13x - 65z = -65 & \quad 13[\text{Equation (5)}] \\
\underline{13x + 2z = 2} & \\
-63z = -63 & \\
z = 1 &
\end{array}
$$

Substitute $z = 1$ back into either Equation (5) or (6)—we choose Equation (5)—to find x:

$$-x - 5z = -5$$
$$-x - 5 \cdot 1 = -5$$
$$-x = 0$$
$$x = 0$$

Step 4: Substitute $x = 0$ and $z = 1$ back into any of the three original equations—we choose Equation (2)—to find y:

$$3x - 2y + 4z = 6$$
$$3 \cdot 0 - 2y + 4 \cdot 1 = 6$$
$$-2y + 4 = 6$$
$$-2y = 2$$
$$y = -1$$

Thus, the solution to the original system is $(0, -1, 1)$ or $x = 0$, $y = -1$, $z = 1$.

Check   To check the solution, we must check *each* equation in the original system:

$$3x - 2y + 4z = 6 \qquad\qquad 2x + 3y - 5z = -8$$
$$3 \cdot 0 - 2(-1) + 4 \cdot 1 \overset{?}{=} 6 \qquad\qquad 2 \cdot 0 + 3(-1) - 5 \cdot 1 \overset{?}{=} -8$$
$$6 \overset{\checkmark}{=} 6 \qquad\qquad\qquad\qquad\qquad -8 = -8$$

$$5x - 4y + 3z = 7$$
$$5 \cdot 0 - 4(-1) + 3 \cdot 1 \overset{?}{=} 7$$
$$7 \overset{\checkmark}{=} 7$$

PROBLEM 10   Solve: $2x + 3y - 5z = -12$
$\qquad\qquad\qquad 3x - 2y + 2z = 1$
$\qquad\qquad\qquad 4x - 5y - 4z = -12$

Solution

If we encounter, in the process described above, an equation that states a contradiction, such as $0 = -2$, then we must conclude that the system has no solution (that is, the system is **inconsistent**). If, on the other hand, one of the equations turns out to be $0 = 0$, the system either has infinitely many solutions or it has none. We must proceed further to determine which. Notice how this last result differs from the two-equation–two-unknown case. There, when we obtained $0 = 0$, we *knew* that there were infinitely many solutions. If a system has infinitely many solutions, then it is said to be **dependent**.

For completeness, let us look at a system that turns out to be dependent to see how the solution set can be represented. Consider the system

$$x + y - z = 2 \tag{7}$$
$$3x + 2y - z = 5 \tag{8}$$
$$5x + 2y + z = 7 \tag{9}$$

We choose to eliminate z from two equations by adding Equation (9) to Equation (7) and by adding Equation (9) to Equation (8). Doing this we obtain the system

$$6x + 3y = 9$$
$$8x + 4y = 12$$

By multiplying the top equation by $\frac{1}{3}$ and the bottom equation by $\frac{1}{4}$, we obtain the simpler system

$$2x + y = 3 \tag{10}$$
$$2x + y = 3 \tag{11}$$

Since these two equations are the same, the original system must be dependent. (If we multiply either Equation (10) or (11) by -1 and add the result to the other we will obtain $0 = 0$.) To represent the solution set of the original system, we proceed as follows. Solve Equation (10) for y in terms of x:

$$y = 3 - 2x \tag{12}$$

Now replace y by $3 - 2x$ in any of the original equations and solve for z. We use Equation (9):

$$5x + 2y + z = 7$$
$$5x + 2(3 - 2x) + z = 7$$
$$5x + 6 - 4x + z = 7$$
$$z = 1 - x$$

Thus, for *any* real number x, the ordered triplet

$$(x, 3 - 2x, 1 - x)$$

is a solution of the original system. For example:

If $x = 1$, then $(1, 3 - 2 \cdot 1, 1 - 1) = (1, 1, 0)$ is a solution.

If $x = -3$, then $(-3, 3 - 2(-3), 1 - (-3)) = (-3, 9, 4)$ is a solution.

And so on.

■ Application

We now consider a real-world problem that leads to a system of equations in three variables.

EXAMPLE 11

Production scheduling: A small manufacturing plant makes three types of inflatable boats: one-person, two-person, and four-person models. Each boat requires the services of three departments, as listed in the table. The cutting, assembly, and packaging departments have available a maximum of 380, 330, and 120 workhours per week, respectively. How many boats of each type must be produced each week for the plant to operate at full capacity?

	ONE-PERSON BOAT	TWO-PERSON BOAT	FOUR-PERSON BOAT
CUTTING DEPARTMENT	0.6 hour	1.0 hour	1.5 hours
ASSEMBLY DEPARTMENT	0.6 hour	0.9 hour	1.2 hours
PACKAGING DEPARTMENT	0.2 hour	0.3 hour	0.5 hour

Solution   Let

$$x = \text{Number of one-person boats produced per week}$$
$$y = \text{Number of two-person boats produced per week}$$
$$z = \text{Number of four-person boats produced per week}$$

The 380 workhours available in the cutting department will be used up by 0.6x workhours on the x one-person boats, 1.0y on the y two-person boats, and 1.5z on the z four-person boats, so that $380 = 0.6x + 1.0y + 1.5z$. Similar equations are obtained for the other departments. Thus,

$$0.6x + 1.0y + 1.5z = 380 \quad \text{Cutting department}$$
$$0.6x + 0.9y + 1.2z = 330 \quad \text{Assembly department}$$
$$0.2x + 0.3y + 0.5z = 120 \quad \text{Packaging department}$$

We can clear the system of decimals, if desired, by multiplying each side of each equation by 10. Thus,

$$6x + 10y + 15z = 3{,}800 \tag{13}$$
$$6x + 9y + 12z = 3{,}300 \tag{14}$$
$$2x + 3y + 5z = 1{,}200 \tag{15}$$

Let us start by eliminating x from Equations (13) and (14):

$$\text{Add}\begin{cases} 6x + 10y + 15z = 3{,}800 & \text{Equation (13)} \\ -6x - 9y - 12z = -3{,}300 & -1[\text{Equation (14)}] \\ \hline y + 3z = \phantom{-3{,}}500 \end{cases}$$

Now we eliminate x from Equations (13) and (15):

$$\text{Add}\begin{cases} 6x + 10y + 15z = 3{,}800 & \text{Equation (13)} \\ -6x - 9y - 15z = -3{,}600 & -3[\text{Equation (15)}] \\ \hline y = \phantom{-3{,}}200 \end{cases}$$

Substituting $y = 200$ into $y + 3z = 500$, we can solve for z:

$$200 + 3z = 500$$
$$3z = 300$$
$$z = 100$$

Now use Equation (13), (14), or (15) to find x (we use Equation 15):

$$2x + 3y + 5z = 1,200$$
$$2x + 3(200) + 5(100) = 1,200$$
$$2x = 100$$
$$x = 50$$

Thus, each week the company should produce 50 one-person boats, 200 two-person boats, and 100 four-person boats to operate at full capacity. The check of the solution is left to you.

PROBLEM 11

Repeat Example 11 assuming the cutting, assembly, and packaging departments have available a maximum of 260, 234, and 82 workhours per week, respectively.

Solution

Solutions to Matched Problems

10. $2x + 3y - 5z = -12$ ⠀⠀(A)
$3x - 2y + 2z = 1$ ⠀⠀(B)
$4x - 5y - 4z = -12$ ⠀⠀(C)

We choose to eliminate z from (A) and (B).

$$4x + 6y - 10z = -24 \quad 2\text{(A)}$$
$$\underline{15x - 10y + 10z = 5} \quad 5\text{(B)}$$
$$19x - 4y = -19$$

Now we choose to eliminate z from (B) and (C).

$$6x - 4y + 4z = 2 \quad 2\text{(B)}$$
$$\underline{4x - 5y - 4z = -12} \quad \text{(C)}$$
$$10x - 9y = -10$$

We now have two equations and two unknowns, which we solve.

$$19x - 4y = -19 \quad \text{(D)}$$
$$10x - 9y = -10 \quad \text{(E)}$$

$$
\begin{array}{ll}
171x - 36y = -171 & \text{9(D)} \\
-40x + 36y = 40 & \text{-4(E)} \\
\hline
131x = -131 \\
\end{array}
$$

$$x = -1$$

Substitute $x = -1$ into either (D) or (E) [we choose (E)] and solve for y.

$$10x - 9y = -10$$
$$10(-1) - 9y = -10$$
$$-9y = 0$$
$$y = 0$$

Now substitute $x = -1$ and $y = 0$ into either (A), (B), or (C) [we choose (B)] and solve for z.

$$3x - 2y + 2z = 1$$
$$3(-1) - 2(0) + 2z = 1$$
$$2z = 4$$
$$z = 2$$

Thus, $(-1, 0, 2)$ is the solution to the original system, as the reader can check.

11. Let x, y, z be the number of one-, two-, and four-person boats, respectively. The available workhours are then used as follows:

$$
\begin{array}{ll}
0.6x + 1.0y + 1.5z = 260 & \text{Cutting department} \\
0.6x + 0.9y + 1.2z = 234 & \text{Assembly department} \\
0.2x + 0.3y + 0.5z = 82 & \text{Packaging department}
\end{array}
$$

Clear decimals by multiplying by 10 to obtain

$$
\begin{array}{ll}
6x + 10y + 15z = 2{,}600 & \text{(A)} \\
6x + 9y + 12z = 2{,}340 & \text{(B)} \\
2x + 3y + 5z = 820 & \text{(C)}
\end{array}
$$

Eliminate x from (A) and (B):

$$
\begin{array}{ll}
6x + 10y + 15z = 2{,}600 & \text{(A)} \\
-6x - 9y - 12z = -2{,}340 & \text{-(B)} \\
\hline
y + 3z = 260 & \text{(D)}
\end{array}
$$

Eliminate x from (B) and (C):

$$
\begin{array}{ll}
6x + 9y + 12z = 2{,}340 & \text{(B)} \\
-6x - 9y - 15z = -2{,}460 & \text{-3(C)} \\
\hline
-3z = -120 & \text{(E)}
\end{array}
$$

Solve (D) and (E)

$$
\begin{array}{ll}
y + 3z = 260 & \text{(D)} \\
-3z = -120 & \text{(E)}
\end{array}
$$

to obtain $z = 40$, $y = 140$. Substitute into (C) to find x:

$$2x + 3 \cdot 140 + 5 \cdot 40 = 820$$
$$2x = 200$$
$$x = 100$$

Thus, the company should produce 100 one-person boats, 140 two-person boats, and 40 three-person boats.

Practice Exercise 10-2 ■

Work odd-numbered problems first, check answers, and then work even-numbered problems in areas of weakness. Answers to all problems are in the back of the book. Make every effort to work a problem yourself before you look at an answer.

Solve and check each system.

A 1.
$$-2x = 2$$
$$x - 3y = 2$$
$$-x + 2y + 3z = -7 \underline{\hspace{1cm}}$$

2.
$$2y + z = -4$$
$$x - 3y + 2z = 9$$
$$-y = 3 \underline{\hspace{1cm}}$$

3.
$$4y - z = -13$$
$$3y + 2z = 4$$
$$6x - 5y - 2z = 0 \underline{\hspace{1cm}}$$

4.
$$2x + z = -5$$
$$x - 3z = -6$$
$$4x + 2y - z = -9 \underline{\hspace{1cm}}$$

B 5.
$$2x + y - z = 5$$
$$x - 2y - 2z = 4$$
$$3x + 4y + 3z = 3 \underline{\hspace{1cm}}$$

6.
$$x - 3y + z = 4$$
$$-x + 4y - 4z = 1$$
$$2x - y + 5z = -3 \underline{\hspace{1cm}}$$

7.
$$2a + 4b + 3c = 6$$
$$a - 3b + 2c = -7$$
$$-a + 2b - c = 5 \underline{\hspace{1cm}}$$

8.
$$3u - 2v + 3w = 11$$
$$2u + 3v - 2w = -5$$
$$u + 4v - w = -5 \underline{\hspace{1cm}}$$

9.
$$2x - 3y + 3z = -15$$
$$3x + 2y - 5z = 19$$
$$5x - 4y - 2z = -2 \underline{\hspace{1cm}}$$

10.
$$3x - 2y - 4z = -8$$
$$4x + 3y - 5z = -5$$
$$6x - 5y + 2z = -17 \underline{\hspace{1cm}}$$

11.
$$5x - 3y + 2z = 13$$
$$2x + 4y - 3z = -9$$
$$4x - 2y + 5z = 13 \underline{\hspace{1cm}}$$

12.
$$4x - 2y + 3z = 0$$
$$3x - 5y - 2z = -12$$
$$2x + 4y - 3z = -4 \underline{\hspace{1cm}}$$

C 13.
$$x - 8y + 2z = -1$$
$$x - 3y + z = 1$$
$$2x - 11y + 3z = 2 \underline{\hspace{1cm}}$$

14.
$$-x + 2y - z = -4$$
$$4x + y - 2z = 1$$
$$x + y - z = -1 \underline{\hspace{1cm}}$$

15.
$$4w - x = 5$$
$$-3w + 2x - y = -5$$
$$2w - 5x + 4y + 3z = 13$$
$$2w + 2x - 2y - z = -2$$
$$\underline{\hspace{3cm}}$$

16.
$$2r - s + 2t - u = 5$$
$$r - 2s + t + u = 1$$
$$-r + s - 3t - u = -1$$
$$-r - 2s + t + 2u = -4$$
$$\underline{\hspace{3cm}}$$

APPLICATIONS 17. *Geometry:*   A circle in a rectangular coordinate system can be written in the form $x^2 + y^2 + Dx + Ey + F = 0$. Find D, E, and F so that the circle passes through $(-2, -1)$, $(-1, -2)$, and $(6, -1)$. $\underline{\hspace{3cm}}$

18. *Geometry:*   Repeat Problem 17 with the circle passing through $(6, -8)$, $(6, 0)$, and $(0, -8)$. $\underline{\hspace{3cm}}$

*19. *Production scheduling:*   A garment company manufactures three shirt styles. Each style of shirt requires the services of three departments, as listed in the table. The cutting, sewing, and packaging departments have available a maximum of 1,160, 1,560, and 480 workhours per week, respectively. How many of each style shirt must be produced each week for the plant to operate at full capacity? _____

	STYLE A	STYLE B	STYLE C
CUTTING DEPARTMENT	0.2 hour	0.4 hour	0.3 hour
SEWING DEPARTMENT	0.3 hour	0.5 hour	0.4 hour
PACKAGING DEPARTMENT	0.1 hour	0.2 hour	0.1 hour

*20. *Production scheduling:* Repeat Problem 19 with the cutting, sewing, and packaging departments having available a maximum of 1,180, 1,560, and 510 workhours per week, respectively. _____

*21. *Diet:* In an experiment involving guinea pigs, a zoologist finds she needs a food mix that contains, among other things, 23 grams of protein, 6.2 grams of fat, and 16 grams of moisture. She has on hand mixes with the compositions shown in the table. How many grams of each mix should she use to get the desired diet mix?

MIX	PROTEIN (%)	FAT (%)	MOISTURE (%)
A	20	2	15
B	10	6	10
C	15	5	5

*22. *Diet:* Repeat Problem 21 assuming the diet mix is to contain 18.5 grams of protein, 4.9 grams of fat, and 13 grams of moisture. _____

**23. *Business:* A newspaper firm uses three printing presses, of different ages and capacities, to print the evening paper. With all three presses running, the paper can be printed in 2 hours. If the newest press breaks down, the older two presses can print the paper in 4 hours; if the middle press breaks down, the newest and oldest together can print the paper in 3 hours. How long would it take each press alone to print the paper? [*Hint:* Use $(2/x) + (2/y) + (2/z) = 1$ as one of the equations.] _____

The Check Exercise for this section is on page 661.

Section 10-3   Systems and Augmented Matrices—An Introduction

- Augmented Matrices
- Solving Linear Systems Using Augmented Matrix Methods

■ Augmented Matrices

In solving systems of equations by elimination in the preceding sections, the coefficients of the variables and constant terms played a central role. The process can be made more efficient for generalization and computer work by the introduction of a mathematical form called a matrix. A **matrix** is a rectangular array of numbers written within brackets. Some examples are

$$\begin{bmatrix} 3 & 5 \\ 0 & -2 \end{bmatrix}$$

$$\begin{bmatrix} 2 \\ -3 \\ 0 \end{bmatrix}$$

$$\begin{bmatrix} 1 & -1 & 0 & 5 \end{bmatrix}$$

$$\begin{bmatrix} -1 & 2 & -5 & 0 \\ 0 & 3 & 2 & 1 \end{bmatrix}$$

$$\begin{bmatrix} 1 & 0 & 0 \\ 0 & 1 & 0 \\ 0 & 0 & 1 \end{bmatrix}$$

Each number in a matrix is called an **element** of the matrix.

Associated with each linear system of the form

$$
\begin{aligned}
ax + by &= m \\
cx + dy &= n
\end{aligned}
\tag{1}
$$

where x and y are variables, is a matrix called the **augmented matrix** of the system:

$$\tag{2}$$

This matrix contains the essential parts of system (1). The vertical bar is included only to separate the coefficients of the variables from the constant terms. Our objective is to learn how to manipulate augmented matrices in such a way that a solution to system (1) will result, if a solution exists. The manipulative process is a direct outgrowth of the elimination process discussed in Sections 10-1 and 10-2.

Recall that two linear systems are said to be **equivalent** if they have exactly the same solution set. How did we transform linear systems into equivalent linear systems? We used Theorem 1, which we restate here for convenient reference:

THEOREM 1

Producing Equivalent Systems
A system of linear equations is transformed into an equivalent system if: **1.** Two equations are interchanged.   **2.** An equation is multiplied by a nonzero constant.   **3.** A constant multiple of another equation is added to a given equation.

Paralleling the previous discussion, we say that two augmented matrices are **row-equivalent**, denoted by the symbol $\sim$ between the two matrices, if they are augmented matrices of equivalent systems of equations. (Think about this.) How do we transform augmented matrices into row-equivalent matrices? We use Theorem 2, which is a direct consequence of Theorem 1.

THEOREM 2

> **Producing Row-Equivalent Matrices**
>
> An augmented matrix is transformed into a row-equivalent matrix if:
>
> **1.** Two rows are interchanged ($R_i \leftrightarrow R_j$).
> **2.** A row is multiplied by a nonzero constant ($kR_i \rightarrow R_i$).
> **3.** A constant multiple of another row is added to a given row
>   ($R_i + kR_j \rightarrow R_i$).
>
> [*Note:* The arrow $\rightarrow$ means "replaces."]

■ **Solving Linear Systems Using Augmented Matrix Methods**

The use of Theorem 2 in solving systems in the form of system (1) is best illustrated by examples.

EXAMPLE 12   Solve using augmented matrix methods:

$$3x + 4y = 1$$
$$x - 2y = 7 \tag{3}$$

Solution   We start by writing the augmented matrix corresponding to system (3):

$$\begin{bmatrix} 3 & 4 & | & 1 \\ 1 & -2 & | & 7 \end{bmatrix} \tag{4}$$

Our objective is to use row operations from Theorem 2 to try to transform system (4) into the form

$$\begin{bmatrix} 1 & k & | & m \\ 0 & 1 & | & n \end{bmatrix} \tag{5}$$

where k, m, and n are real numbers. The solution to system (3) will then be easy since matrix (5) will be the augmented matrix of the following system, which can be solved by substitution:

$$x + ky = m$$
$$y = n$$

We now proceed to use row operations to transform (4) into form (5).

Step 1: To get a 1 in the upper left corner, we interchange rows 1 and 2 (Theorem 2, part 1).

$$\begin{bmatrix} 3 & 4 & | & 1 \\ 1 & -2 & | & 7 \end{bmatrix} \overset{R_1 \leftrightarrow R_2}{\sim} \begin{bmatrix} 1 & -2 & | & 7 \\ 3 & 4 & | & 1 \end{bmatrix}$$

Step 2: To get a 0 in the lower left corner, we multiply R_1 by (-3) and add to R_2 (Theorem 2, part 3)—this changes R_2 but not R_1. Some people find it useful to write $(-3)R_1$ outside the matrix to help reduce errors in arithmetic, as shown.

$$\begin{array}{ccc} -3 & 6 & -21 \longleftarrow \end{array}$$
$$\begin{bmatrix} 1 & -2 & | & 7 \\ 3 & 4 & | & 1 \end{bmatrix} \overset{R_2 + (-3)R_1 \rightarrow R_2}{\sim} \begin{bmatrix} 1 & -2 & | & 7 \\ 0 & 10 & | & -20 \end{bmatrix}$$

Step 3: To get a 1 in the second row, second column, we multiply R_2 by $\frac{1}{10}$ (Theorem 2, part 2).

$$\begin{bmatrix} 1 & -2 & | & 7 \\ 0 & 10 & | & -20 \end{bmatrix} \overset{\frac{1}{10}R_2 \rightarrow R_2}{\sim} \begin{bmatrix} 1 & -2 & | & 7 \\ 0 & 1 & | & -2 \end{bmatrix}$$

We have accomplished our objective! The last matrix is the augmented matrix for the system

$$x - 2y = 7$$
$$y = -2$$

(6)

which is easily solved by substitution. Substituting $y = -2$ into the first equation, we obtain $x + 4 = 7$ or $x = 3$. Since system (6) is equivalent to system (3), our starting system, we have solved (3); that is, $x = 3$ and $y = -2$.

Check

$$3x + 4y = 1 \qquad\qquad x - 2y = 7$$
$$3(3) + 4(-2) \overset{?}{=} 1 \qquad\qquad 3 - 2(-2) \overset{?}{=} 7$$
$$9 - 8 \overset{\checkmark}{=} 1 \qquad\qquad 3 + 4 \overset{\checkmark}{=} 7$$

The solution process above is written more compactly as follows:

Step 1:
Need a 1 here
$$\begin{bmatrix} 3 & 4 & | & 1 \\ 1 & -2 & | & 7 \end{bmatrix} \qquad R_1 \leftrightarrow R_2$$

Step 2:
Need a 0 here
$$\sim \begin{bmatrix} 1 & -2 & | & 7 \\ 3 & 4 & | & 1 \end{bmatrix} \qquad R_2 + (-3)R_1 \to R_2$$
$$-3 \qquad 6 \qquad -21$$

Step 3:
Need a 1 here
$$\sim \begin{bmatrix} 1 & -2 & | & 7 \\ 0 & 10 & | & -20 \end{bmatrix} \qquad \tfrac{1}{10}R_2 \to R_2$$

$$\sim \begin{bmatrix} 1 & -2 & | & 7 \\ 0 & 1 & | & -2 \end{bmatrix}$$

PROBLEM 12   Solve using augmented matrix methods: $2x - y = -7$
$$x + 2y = 4$$

Solution

EXAMPLE 13   Solve using augmented matrix methods:

$$2x - 3y = 7$$

$$3x + 4y = 2$$

Solution

Step 1:
Need a 1 here

$$\begin{bmatrix} 2 & -3 & | & 7 \\ 3 & 4 & | & 2 \end{bmatrix} \quad \tfrac{1}{2}R_1 \to R_1$$

Step 2:
Need a 0 here

$$\sim \begin{bmatrix} 1 & -\frac{3}{2} & | & \frac{7}{2} \\ 3 & 4 & | & 2 \end{bmatrix} \quad R_2 + (-3)R_1 \to R_2$$

$$\begin{array}{ccc} -3 & \frac{9}{2} & -\frac{21}{2} \end{array}$$

Step 3:
Need a 1 here

$$\sim \begin{bmatrix} 1 & -\frac{3}{2} & | & \frac{7}{2} \\ 0 & \frac{17}{2} & | & -\frac{17}{2} \end{bmatrix} \quad \tfrac{2}{17}R_2 \to R_2$$

$$\sim \begin{bmatrix} 1 & -\frac{3}{2} & | & \frac{7}{2} \\ 0 & 1 & | & -1 \end{bmatrix}$$

The original system is therefore equivalent to

$$x - \tfrac{3}{2}y = \tfrac{7}{2}$$

$$y = -1$$

Thus, $y = -1$ and $x + \tfrac{3}{2} = \tfrac{7}{2}$ so $x = 2$. You should check the solution $(2, -1)$.

PROBLEM 13   Solve using augmented matrix methods: $5x - 2y = 12$
$2x + 3y = 1$

Solution

EXAMPLE 14   Solve using augmented matrix methods:

$$2x - y = 4$$
$$-6x + 3y = -12$$

Solution

$$\begin{bmatrix} 2 & -1 & | & 4 \\ -6 & 3 & | & -12 \end{bmatrix}$$ $\frac{1}{2}R_1 \to R_1$ (This produces a 1 in the upper left corner.)
$\frac{1}{3}R_2 \to R_2$ (This simplifies R_2.)

$$\sim \begin{bmatrix} 1 & -\frac{1}{2} & | & 2 \\ -2 & 1 & | & -4 \end{bmatrix}$$ $R_2 + 2R_1 \to R_2$ (This produces a 0 in the lower left corner.)

$$\begin{matrix} 2 & -1 & & 4 \end{matrix}$$

$$\sim \begin{bmatrix} 1 & -\frac{1}{2} & | & 2 \\ 0 & 0 & | & 0 \end{bmatrix}$$

The last matrix corresponds to the system

$$x - \tfrac{1}{2}y = 2$$
$$0x + 0y = 0$$

Thus, $x = \frac{1}{2}y + 2$. Hence, for any real number y, $(\frac{1}{2}y + 2, y)$ is a solution. If $y = 6$, for example, then $(5, 6)$ is a solution; if $y = -2$, then $(1, -2)$ is a solution; and so on. Geometrically, the graphs of the two original equations coincide and there are infinitely many solutions. In general, if we end up with a row of 0's in an augmented matrix for a two-equation–two-unknown system, the system is dependent and there are infinitely many solutions.

PROBLEM 14   Solve using augmented matrix methods: $-2x + 6y = 6$
$3x - 9y = -9$

Solution

EXAMPLE 15   Solve using augmented matrix methods:

$$2x + 6y = -3$$
$$x + 3y = 2$$

Solution

$$\begin{bmatrix} 2 & 6 & | & -3 \\ 1 & 3 & | & 2 \end{bmatrix} \quad R_1 \leftrightarrow R_2$$

$$\sim \begin{bmatrix} 1 & 3 & | & 2 \\ 2 & 6 & | & -3 \end{bmatrix} \quad R_2 + (-2)R_1 \rightarrow R_2$$

$$\begin{array}{ccc} -2 & -6 & & -4 \end{array}$$

$$\sim \begin{bmatrix} 1 & 3 & | & 2 \\ 0 & 0 & | & -7 \end{bmatrix} \quad R_2 \text{ implies the contradiction } 0 = -7.$$

The system is inconsistent and has no solution—otherwise, we have proved that $0 = -7$! Thus, if in a row of an augmented matrix we obtain all 0's to the left of the vertical bar and a nonzero number to the right of the bar, the system is inconsistent and there are no solutions.

PROBLEM 15   Solve using augmented matrix methods: $2x - y = 3$
$$4x - 2y = -1$$

Solution

Summary

FORM 1	FORM 2	FORM 3
A UNIQUE SOLUTION	INFINITELY MANY SOLUTIONS (DEPENDENT)	NO SOLUTION (INCONSISTENT)

$$\begin{bmatrix} 1 & 0 & | & m \\ 0 & 1 & | & n \end{bmatrix} \qquad \begin{bmatrix} 1 & m & | & n \\ 0 & 0 & | & 0 \end{bmatrix} \qquad \begin{bmatrix} 1 & m & | & n \\ 0 & 0 & | & p \end{bmatrix}$$

m, n, p real numbers; $p \neq 0$

The augmented matrix method is readily applied to systems with three (or more) variables.

EXAMPLE 16   Solve using augmented matrix methods:

$$2x - 2y + z = 3$$
$$3x + y - z = 7$$
$$x - 3y + 2z = 0$$

Solution

$$\begin{bmatrix} 2 & -2 & 1 & | & 3 \\ 3 & 1 & -1 & | & 7 \\ 1 & -3 & 2 & | & 0 \end{bmatrix} \qquad R_1 \leftrightarrow R_3$$

(Need a 1 here)

$$\sim \begin{bmatrix} 1 & -3 & 2 & | & 0 \\ 3 & 1 & -1 & | & 7 \\ 2 & -2 & 1 & | & 3 \end{bmatrix} \qquad \begin{array}{l} R_2 + (-3)R_1 \to R_2 \\ R_3 + (-2)R_1 \to R_3 \end{array}$$

(Need 0's here)

$$\sim \begin{bmatrix} 1 & -3 & 2 & | & 0 \\ 0 & 10 & -7 & | & 7 \\ 0 & 4 & -3 & | & 3 \end{bmatrix} \qquad \tfrac{1}{10}R_2 \to R_2$$

(Need a 1 here)

$$\sim \begin{bmatrix} 1 & -3 & 2 & | & 0 \\ 0 & 1 & -\tfrac{7}{10} & | & \tfrac{7}{10} \\ 0 & 4 & -3 & | & 3 \end{bmatrix} \qquad R_3 + (-4)R_2 \to R_3$$

(Need a 0 here)

$$\sim \begin{bmatrix} 1 & -3 & 2 & | & 0 \\ 0 & 1 & -\tfrac{7}{10} & | & \tfrac{7}{10} \\ 0 & 0 & -\tfrac{1}{5} & | & \tfrac{1}{5} \end{bmatrix} \qquad (-5)R_3 \to R_3$$

(Need a 1 here)

$$\sim \begin{bmatrix} 1 & -3 & 2 & | & 0 \\ 0 & 1 & -\tfrac{7}{10} & | & \tfrac{7}{10} \\ 0 & 0 & 1 & | & -1 \end{bmatrix}$$

The resulting equivalent system

$$x - 3y + 2z = 0$$
$$y - \tfrac{7}{10}z = \tfrac{7}{10}$$
$$z = -1$$

is solved by substitution:

$$y - \tfrac{7}{10}z = y + \tfrac{7}{10} = \tfrac{7}{10} \qquad so \qquad y = 0$$
$$x - 3y + 2z = x - 0 - 2 = 0 \qquad so \qquad x = 2$$

You should check the solution $(2, 0, -1)$.

PROBLEM 16

Solve using augmented matrix methods:

$$3x + \ y - 2z = 2$$
$$x - 2y + \ z = 3$$
$$2x - \ y - 3z = 3$$

Solution

The process of solving systems of equations described in this section is known as **gaussian elimination**. It is a powerful method that can be used to solve large-scale systems on a computer. The concept of a matrix is also a powerful mathematical tool with applications far beyond solving systems of equations. Matrices are discussed further in Appendixes F and G.

12. $\begin{bmatrix} 2 & -1 & \vline & -7 \\ 1 & 2 & \vline & 4 \end{bmatrix}$ $R_1 \leftrightarrow R_2$

$\sim \begin{bmatrix} 1 & 2 & \vline & 4 \\ 2 & -1 & \vline & -7 \end{bmatrix}$ $-2R_1 + R_2 \rightarrow R_2$

$\sim \begin{bmatrix} 1 & 2 & \vline & 4 \\ 0 & -5 & \vline & -15 \end{bmatrix}$ $-\frac{1}{5}R_2 \rightarrow R_2$

$\sim \begin{bmatrix} 1 & 2 & \vline & 4 \\ 0 & 1 & \vline & 3 \end{bmatrix}$

$x + 2y = 4$
$y = 3$
$x + 6 = 4$
$x = -2$

Solution: $x = -2$, $y = 3$

13. $\begin{bmatrix} 5 & -2 & \vline & 12 \\ 2 & 3 & \vline & 1 \end{bmatrix}$ $-2R_2 + R_1 \rightarrow R_1$

$\sim \begin{bmatrix} 1 & -8 & \vline & 10 \\ 2 & 3 & \vline & 1 \end{bmatrix}$ $-2R_1 + R_2 \rightarrow R_2$

$\sim \begin{bmatrix} 1 & -8 & \vline & 10 \\ 0 & 19 & \vline & -19 \end{bmatrix}$ $\frac{1}{19}R_2 \rightarrow R_2$

$\sim \begin{bmatrix} 1 & -8 & \vline & 10 \\ 0 & 1 & \vline & -1 \end{bmatrix}$

$x - 8y = 10$
$y = -1$
$x + 8 = 10$
$x = 2$

Solution: $x = 2$, $y = -1$

14. $\begin{bmatrix} -2 & 6 & \vline & 6 \\ 3 & -9 & \vline & -9 \end{bmatrix}$ $-\frac{1}{2}R_1 \rightarrow R_1$

$\sim \begin{bmatrix} 1 & -3 & \vline & -3 \\ 3 & -9 & \vline & -9 \end{bmatrix}$ $-3R_1 + R_2 \rightarrow R_2$

$\sim \begin{bmatrix} 1 & -3 & \vline & -3 \\ 0 & 0 & \vline & 0 \end{bmatrix}$

$x - 3y - -3$
$x = 3y - 3$

For any y, $(3y - 3, y)$ is a solution.

15. $\begin{bmatrix} 2 & -1 & \vline & 3 \\ 4 & -2 & \vline & -1 \end{bmatrix}$ $\frac{1}{2}R_1 \rightarrow R_1$

$\sim \begin{bmatrix} 1 & -\frac{1}{2} & \vline & \frac{3}{2} \\ 4 & -2 & \vline & -1 \end{bmatrix}$ $-4R_1 + R_2 \rightarrow R_2$

$\sim \begin{bmatrix} 1 & -\frac{1}{2} & \vline & \frac{3}{2} \\ 0 & 0 & \vline & -7 \end{bmatrix}$

$x - \frac{1}{2}y = \frac{3}{2}$
$0 = -7$

The system is inconsistent and has no solution.

16.
$$\begin{bmatrix} 3 & 1 & -2 & | & 2 \\ 1 & -2 & 1 & | & 3 \\ 2 & -1 & -3 & | & 3 \end{bmatrix} \qquad R_1 \leftrightarrow R_2$$

$$\sim \begin{bmatrix} 1 & -2 & 1 & | & 3 \\ 3 & 1 & -2 & | & 2 \\ 2 & -1 & -3 & | & 3 \end{bmatrix} \qquad \begin{array}{l} -3R_1 + R_2 \to R_2 \\ -2R_1 + R_3 \to R_3 \end{array}$$

$$\sim \begin{bmatrix} 1 & -2 & 1 & | & 3 \\ 0 & 7 & -5 & | & -7 \\ 0 & 3 & -5 & | & -3 \end{bmatrix} \qquad -2R_3 + R_2 \to R_2$$

$$\sim \begin{bmatrix} 1 & -2 & 1 & | & 3 \\ 0 & 1 & 5 & | & -1 \\ 0 & 3 & -5 & | & -3 \end{bmatrix} \qquad -3R_2 + R_3 \to R_3$$

$$\sim \begin{bmatrix} 1 & -2 & 1 & | & 3 \\ 0 & 1 & 5 & | & -1 \\ 0 & 0 & -20 & | & 0 \end{bmatrix} \qquad -\tfrac{1}{20}R_3 \to R_3$$

$$\sim \begin{bmatrix} 1 & -2 & 1 & | & 3 \\ 0 & 1 & 5 & | & -1 \\ 0 & 0 & 1 & | & 0 \end{bmatrix}$$

$$x - 2y + z = 3$$
$$y + 5z = -1$$
$$z = 0$$
$$y + 0 = -1$$
$$y = -1$$
$$x + 2 + 0 = 3$$
$$x = 1$$

Solution: $x = 1$, $y = -1$, $z = 0$

Practice Exercise 10-3 ■

Work odd-numbered problems first, check answers, and then work even-numbered problems in areas of weakness. Answers to all problems are in the back of the book. Make every effort to work a problem yourself before you look at an answer.

A   *Perform each of the indicated row operations on the following matrix:*

$$\begin{bmatrix} 1 & -3 & | & 2 \\ 4 & -6 & | & -8 \end{bmatrix}$$

1. $R_1 \leftrightarrow R_2$ _____

2. $\tfrac{1}{2}R_2 \to R_2$ _____

3. $-4R_1 \to R_1$ _____

4. $-2R_1 \to R_1$ _____

5. $2R_2 \rightarrow R_2$ _____   **6.** $-1R_2 \rightarrow R_2$ _____

7. $R_2 + (-4)R_1 \rightarrow R_2$ _____   **8.** $R_1 + (-\frac{1}{2})R_2 \rightarrow R_1$ _____

9. $R_2 + (-2)R_1 \rightarrow R_2$ _____   **10.** $R_2 + (-3)R_1 \rightarrow R_2$ _____

11. $R_2 + (-1)R_1 \rightarrow R_2$ _____   **12.** $R_2 + (1)R_1 \rightarrow R_2$ _____

Solve using augmented matrix methods.

13. $x + y = 5$
       $x - y = 1$ _____

14. $x - y = 2$
       $x + y = 6$ _____

B   **15.** $x - 2y = 1$
        $2x - y = 5$ _____

16. $x + 3y = 1$
       $3x - 2y = 14$ _____

17. $x - 4y = -2$
       $-2x + y = -3$ _____

18. $x - 3y = -5$
       $-3x - y = 5$ _____

19. $3x - y = 2$
       $x + 2y = 10$ _____

20. $2x + y = 0$
       $x - 2y = -5$ _____

21. $x + 2y = 4$
       $2x + 4y = -8$ _____

22. $2x - 3y = -2$
       $-4x + 6y = 7$ _____

23. $2x + y = 6$
       $x - y = -3$ _____

24. $3x - y = -5$
       $x + 3y = 5$ _____

25. $3x - 6y = -9$
       $-2x + 4y = 6$ _____

26. $2x - 4y = -2$
       $-3x + 6y = 3$ _____

27. $4x - 2y = 2$
       $-6x + 3y = -3$ _____

28. $-6x + 2y = 4$
       $3x - y = -2$ _____

C   **29.** $3x - y = 7$
        $2x + 3y = 1$ _____

30. $2x - 3y = -8$
       $5x + 3y = 1$ _____

31. $3x + 2y = 4$
       $2x - y = 5$ _____

32. $4x + 3y = 26$
       $3x - 11y = -7$ _____

33. $0.2x - 0.5y = 0.07$
       $0.8x - 0.3y = 0.79$ _____

34. $0.3x - 0.6y = 0.18$
       $0.5x - 0.2y = 0.54$ _____

35. $2x + 4y - 10z = -2$
       $3x + 9y - 21z = 0$
       $x + 5y - 12z = 1$ _____

36. $3x + 5y - z = -7$
       $x + y + z = -1$
       $2x + y + 11z = 7$ _____

The Check Exercise for this section is on page 663.

37. $3x + 8y - z = -18$
$2x + y + 5z = 8$
$2x + 4y + 2z = -4$ _____

38. $2x + 7y + 15z = -12$
$4x + 7y + 13z = -10$
$3x + 6y + 12z = -9$ _____

Section 10-4   Systems of Linear Inequalities

■ Systems of Inequality Statements
■ Application

■ **Systems of Inequality Statements**

As in systems of linear equations in two variables, we say that the ordered pair of numbers (x_0, y_0) is a solution of a system of linear inequalities in two variables if the ordered pair satisfies each inequality in the system. Thus, **the graph of a system of linear inequalities** is the intersection of the graphs of each inequality in the system. In this book we will limit our investigation of solutions of systems of inequalities to graphic methods. An example will illustrate the process.

EXAMPLE 17   Solve the following linear system graphically:

$$0 \le x \le 8$$
$$0 \le y \le 4$$

Solution   This system is actually equivalent to the system

$\left. \begin{array}{l} x \ge 0 \\ x \le 8 \\ y \ge 0 \\ y \le 4 \end{array} \right\}$ The solution to the system is the intersection of all four solution sets.

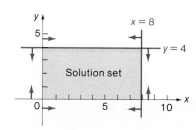

PROBLEM 17   Solve graphically: $2 \le x \le 6$
$1 \le y \le 3$

Solution

EXAMPLE 18   Solve graphically:

$$3x + 5y \leq 60$$
$$4x + 2y \leq 40$$
$$x \geq 0$$
$$y \geq 0$$

Solution

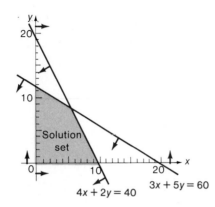

PROBLEM 18   Solve graphically: $x + 2y \geq 12$
$$3x + 2y \geq 24$$
$$x \geq 0$$
$$y \geq 0$$

Solution

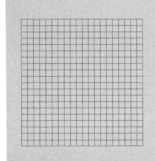

EXAMPLE 19   Solve graphically:

$$x + y \leq 1$$
$$y - x \leq 1$$
$$y \geq -1$$

Solution

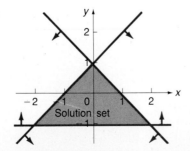

PROBLEM 19

Solve graphically: $x + 3y \leq 6$
$x - 2y \leq 2$
$x \geq -2$

Solution

■ **Application**

EXAMPLE 20

Production scheduling: A manufacturer of sailboards makes a standard model and a competition model. The relevant manufacturing data are shown in the table. What combinations of boards can be produced each week so as not to exceed the number of workhours available in each department per week?

	STANDARD MODEL (WORKHOURS PER BOARD)	COMPETITION MODEL (WORKHOURS PER BOARD)	MAXIMUM WORKHOURS AVAILABLE PER WEEK
FABRICATING	6	8	120
FINISHING	1	3	30

Solution

Let x and y be the respective number of standard and competition boards produced per week. These variables are restricted as follows:

$6x + 8y \leq 120$ Fabricating
$x + 3y \leq 30$ Finishing
$x \geq 0$
$y \geq 0$

The solution set of this system of inequalities is the shaded area in the figure and is referred to as the **feasible region**. Any point within the shaded area would represent a possible production schedule. Any point outside the shaded area would represent an impossible schedule. For example, it would be possible to produce 10 standard boards and 5 competition boards per week, but it would

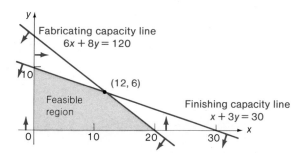

not be possible to produce 13 standard boards and 6 competition boards per week.

PROBLEM 20

Repeat Example 20 using 5 hours for fabricating a standard board in place of 6 hours and a maximum of 27 workhours for the finishing department.

Solution

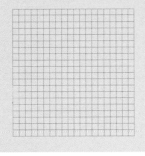

Solutions to Matched Problems

17.

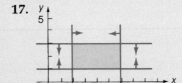

18.

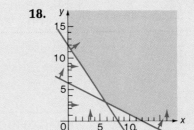

19.

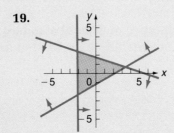

20. The feasible region is determined by

$$5x + 8y \leq 120 \quad \text{Fabricating}$$
$$x + 3y \leq 27 \quad \text{Finishing}$$
$$x \geq 0$$
$$y \geq 0$$

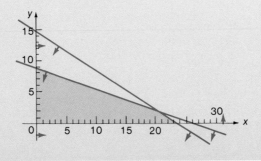

Practice Exercise 10-4 ∎

Work odd-numbered problems first, check answers, and then work even-numbered problems in areas of weakness. Answers to all problems are in the back of the book. Make every effort to work a problem yourself before you look at an answer.

A   *Find the solution set of each system graphically.*

1. $-2 \leq x < 2$
    $-1 < y \leq 6$

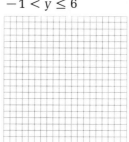

2. $-4 \leq x < -1$
    $-2 < y \leq 5$

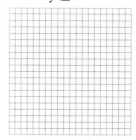

3. $-1 \leq y \leq 2$
    $x \geq -3$
    $x + y \leq 1$

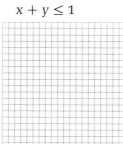

4. $1 \leq x \leq 3$
    $x + y \leq 5$
    $y \geq 0$

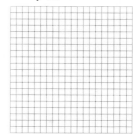

5. $0 \leq y \leq 3$
    $x - y \geq -3$
    $x - y \leq 1$

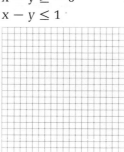

6. $-2 \leq x \leq 1$
    $y - x \leq 1$
    $y - x \geq -2$

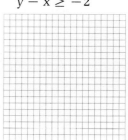

B   7. $2x + y \leq 8$
    $0 \leq x \leq 3$
    $0 \leq y \leq 5$

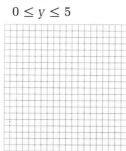

8. $x + 3y \leq 12$
    $0 \leq x \leq 8$
    $0 \leq y \leq 3$

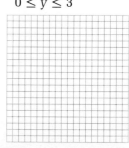

9. $2x + \ y \leq 8$
$x + 3y \leq 12$
$x \geq 0$
$y \geq 0$

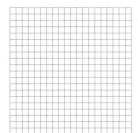

10. $x + 2y \leq 10$
$3x + \ y \leq 15$
$x \geq 0$
$y \geq 0$

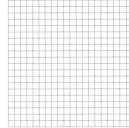

11. $6x + 3y \leq 24$
$3x + 6y \leq 30$
$x \geq 0$
$y \geq 0$

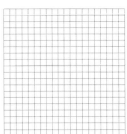

12. $2x + \ y \leq 10$
$x + 2y \leq 8$
$x \geq 0$
$y \geq 0$

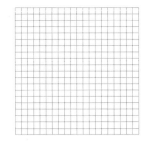

13. $3x + 4y \geq 8$
$4x + 3y \geq 24$
$x \geq 0$
$y \geq 0$

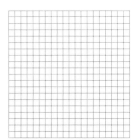

14. $x + 2y \geq 8$
$2x + \ y \geq 10$
$x \geq 0$
$y \geq 0$

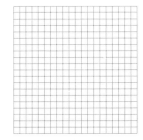

15. $2y - x \geq 0$
$2y + x \leq 4$
$x \geq -2$

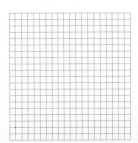

16. $y - 2x \leq 4$
$2y + x \leq 1$
$y \geq -1$

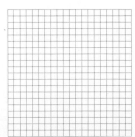

C **17.** $3x + 5y \geq 60$ **18.** $2x + y \geq 8$
         $4x + 2y \geq 40$ $x + 2y \geq 10$
         $2 \leq x \leq 14$ $1 \leq x \leq 7$
         $6 \leq y \leq 18$ $3 \leq y \leq 9$

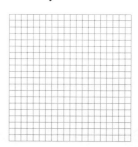

APPLICATIONS **19.** *Manufacturing—resource allocation:*   A manufacturing company makes two types of water skis: a trick ski and a slalom ski. The trick ski requires 6 workhours for fabricating and 1 workhour for finishing. The slalom ski requires 4 workhours for fabricating and 1 workhour for finishing. The maximum workhours available per day for fabricating and finishing are 108 and 24, respectively. If x is the number of trick skis and y is the number of slalom skis produced per day, write a system of inequalities that indicates appropriate restraints on x and y. Find the set of feasible solutions graphically for the number of each type of ski that can be produced.

20. *Nutrition:*   A dietitian in a hospital is to arrange a special diet using two foods. Each ounce of food M contains 30 units of calcium, 10 units of iron, and 10 units of vitamin A. Each ounce of food N contains 10 units of calcium, 10 units of iron, and 30 units of vitamin A. The minimum requirements in the diet are 360 units of calcium, 160 units of iron, and 240 units of vitamin A. If x is the number of ounces of food M used and y is the number of ounces of food N used, write a system of linear inequalities that reflects the conditions indicated. Find the set of feasible solutions graphically for the amount of each kind of food that can be used.

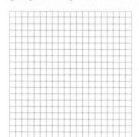

The Check Exercise for this section is on page 665.

Section 10-5   Systems Involving Second-Degree Equations

In this section we will investigate several special types of systems that involve at least one second-degree equation in two variables. The methods used to solve these systems are best illustrated through examples.

EXAMPLE 21   Solve the system:

$$4x^2 + y^2 = 25$$

$$2x + y = 7$$

Solution   In this type of problem the substitution principle is effective. Solve the linear equation for one variable in terms of the other, and then substitute into the nonlinear equation to obtain a quadratic equation in one variable.

$$4x^2 + y^2 = 25$$ Solve $2x + y = 7$ for y in terms of x.

$$2x + y = 7$$

$$y = 7 - 2x$$ Substitute into $4x^2 + y^2 = 25$.

$$4x^2 + y^2 = 25$$

$$4x^2 + (7 - 2x)^2 = 25$$ Simplify, and write in standard quadratic

$$4x^2 + 49 - 28x + 4x^2 = 25$$ form $ax^2 + bx + c = 0$.

$$8x^2 - 28x + 24 = 0$$ Multiply both sides by $\frac{1}{4}$.

$$2x^2 - 7x + 6 = 0$$ Solve—we use the factoring method.

$$(2x - 3)(x - 2) = 0$$

$$2x - 3 = 0 \qquad \text{or} \qquad x - 2 = 0$$

$$x = \tfrac{3}{2} \qquad\qquad\qquad x = 2$$

These values are substituted back into the linear equation $y = 7 - 2x$ to find the corresponding values for y. (Note that if we substitute these values back into the second-degree equation, we may obtain "extraneous" roots; try it and see why. Recall that a solution of a system is an ordered pair of numbers that satisfies *both* equations.)

For $x = \frac{3}{2}$:   For $x = 2$:

$$y = 7 - 2(\tfrac{3}{2}) \qquad\quad y = 7 - 2 \cdot 2$$

$$y = 4 \qquad\qquad\qquad y = 3$$

Thus, $(\frac{3}{2}, 4)$ and $(2, 3)$ are the solutions to the system, as can easily be checked.

The equations in Example 21 can be graphed. The first equation is an ellipse; the second is a straight line. The intersections of the ellipse and the line are the solutions of the system (see Figure 1).

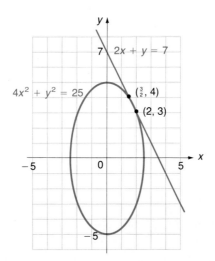

FIGURE 1

Graphing the equations provides another check on our solutions. In situations where the system cannot be solved algebraically, graphing allows us to estimate the solutions.

PROBLEM 21   Solve the system: $2x^2 - y^2 = 1$
$$3x + y = 2$$

Solution

EXAMPLE 22   Solve:

$$x^2 - y^2 = 5$$
$$x^2 + 2y^2 = 17$$

Solution

$\quad x^2 - \;\;y^2 = 5 \qquad$ Proceed as with linear equations—subtract to eliminate x.
$\quad \underline{x^2 + 2y^2 = 17}$
$\qquad\quad -3y^2 = -12$
$\qquad\qquad\;\; y^2 = 4$
$\qquad\qquad\quad y = \pm 2$

For $y = 2$: $\qquad$ Substitute into either original equation; then solve for x.

$\quad x^2 - (2)^2 = 5$
$\qquad\quad\; x^2 = 9$
$\qquad\qquad x = \pm 3$

For y = −2:   Substitute into either original equation; then solve for *x*.

$$x^2 - (-2)^2 = 5$$
$$x^2 = 9$$
$$x = \pm 3$$

Thus $(3, -2)$, $(3, 2)$, $(-3, -2)$, and $(-3, 2)$ are the four solutions to the system. Checking the solutions is left to you. The graph of the system shows the reasonableness of the solutions:

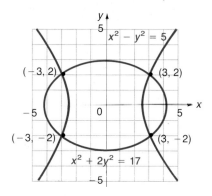

PROBLEM 22   Solve the system: $2x^2 - 3y^2 = 5$
                                    $3x^2 + 4y^2 = 16$

Solution

EXAMPLE 23   Solve:

$$x^2 + 3xy + y^2 = 20 \qquad (1)$$
$$xy - y^2 = 0 \qquad (2)$$

Solution $y(x - y) = 0$ Factor Equation (2).

$y = 0$ or $x - y = 0$ Substitute each of these in turn into Equation (1),
                   $y = x$ and proceed as before.

For y = 0 [replace y with 0 in Equation (1) and solve for x]:

$$x^2 + 3x(0) + (0)^2 = 20$$
$$x^2 = 20$$
$$x = \pm 2\sqrt{5}$$

For y = x [replace y with x in Equation (1) and solve for x]:

$$x^2 + 3x(x) + (x)^2 = 20$$
$$5x^2 = 20$$
$$x^2 = 4$$
$$x = \pm 2 \quad \text{Substitute these values back into } y = x \text{ to find}$$
corresponding values of y.

For $x = 2$, $y = 2$; for $x = -2$, $y = -2$.

Thus, $(2\sqrt{5}, 0)$, $(-2\sqrt{5}, 0)$, $(2, 2)$, and $(-2, -2)$ are the four solutions to the system. Checking the solutions is left to you. These equations require more advanced techniques to graph efficiently; hence, their graphs are omitted.

COMMON ERROR   If we had started by solving $xy - y^2 = 0$ for x as follows,

$$xy - y^2 = 0$$
$$xy = y^2$$
$$x = y$$

we would have lost two solutions: $(2\sqrt{5}, 0)$ and $(-2\sqrt{5}, 0)$. Dividing both sides of $xy = y^2$ by the variable y is behind this loss. **Do not divide both sides of an equation by an expression involving a variable for which you are solving.** The operation often results in the loss of solutions.

PROBLEM 23

Solve the system: $x^2 + xy - y^2 = 4$
$2x^2 - xy = 0$

Solution

Example 23 is somewhat specialized. However, it suggests a procedure that, when used alone or in combination with other procedures, is effective for some problems.

To obtain an idea of how many real solutions you might expect from a system of second-degree equations in two variables, you have only to look at the number of ways two conics can intersect. In general, it turns out that a system of one linear and one quadratic equation can have at most two solutions, and a system of two quadratic equations can have at most four solutions. Of course, some of the solutions may be complex. The graphs of Examples 21 and 22 illustrate this result; see Figure 2 (page 652) for additional examples.

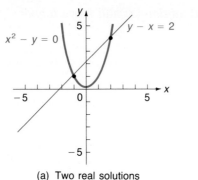

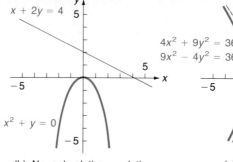

(a) Two real solutions

(b) No real solutions; solutions are nonreal complex

(c) Four real solutions

FIGURE 2

Solutions to Matched Problems

Checks are left to the reader.

21.
$$2x^2 - y^2 = 1$$
$$3x + y = 2$$
$$y = 2 - 3x$$
$$2x^2 - (2 - 3x)^2 = 1$$
$$2x^2 - (4 - 12x + 9x^2) = 1$$
$$2x^2 - 4 + 12x - 9x^2 = 1$$
$$-7x^2 + 12x - 5 = 0$$
$$7x^2 - 12x + 5 = 0$$
$$(7x - 5)(x - 1) = 0$$
$$7x - 5 = 0 \quad \text{or} \quad x - 1 = 0$$
$$x = \tfrac{5}{7} \qquad\qquad x = 1$$

For $x = 1$:
$$y = 2 - 3x$$
$$y = 2 - 3(1)$$
$$= -1$$

For $x = \tfrac{5}{7}$:
$$y = 2 - 3x$$
$$y = 2 - 3(\tfrac{5}{7})$$
$$= -\tfrac{1}{7}$$

Thus, $(1, -1)$ and $(\tfrac{5}{7}, -\tfrac{1}{7})$ are the solutions to the system.

22.
$$2x^2 - 3y^2 = 5 \qquad \text{(A)}$$
$$3x^2 + 4y^2 = 16 \qquad \text{(B)}$$
$$8x^2 - 12y^2 = 20 \qquad 4\text{(A)}$$
$$\underline{9x^2 + 12y^2 = 48} \qquad 3\text{(B)}$$
$$17x^2 \qquad\quad = 68$$
$$x^2 = 4$$
$$x = \pm 2$$

For $x = 2$:
$$2x^2 - 3y^2 = 5$$
$$2(2)^2 - 3y^2 = 5$$
$$-3y^2 = -3$$
$$y^2 = 1$$
$$y = \pm 1$$

For $x = -2$:
$$2x^2 - 3y^2 = 5$$
$$2(-2)^2 - 3y^2 = 5$$
$$-3y^2 = -3$$
$$y^2 = 1$$
$$y = \pm 1$$

Thus, $(2, 1)$, $(2, -1)$, $(-2, 1)$, and $(-2, -1)$ are the solutions to the system.

23.
$$x^2 + xy - y^2 = 4$$
$$2x^2 - xy = 0$$
$$x(2x - y) = 0$$
$$x = 0 \quad \text{or} \quad 2x - y = 0$$
$$y = 2x$$

For $y = 2x$:
$$x^2 + xy - y^2 = 4$$
$$x^2 + x(2x) - (2x)^2 = 4$$
$$-x^2 = 4$$
$$x^2 = -4$$
$$x = \pm\sqrt{-4} = \pm 2i$$

For $x = 0$:
$$x^2 + xy - y^2 = 4$$
$$0^2 + 0y - y^2 = 4$$
$$y^2 = -4$$
$$y = \pm\sqrt{-4} = \pm 2i$$

For $x = 2i$:
$$y = 2x$$
$$y = 2(2i) = 4i$$

For $x = -2i$:
$$y = 2x$$
$$y = 2(-2i) = -4i$$

Thus, $(0, 2i)$, $(0, -2i)$, $(2i, 4i)$, and $(-2i, -4i)$ are the solutions to the system.

Practice Exercise 10-5 ■

Work odd-numbered problems first, check answers, and then work even-numbered problems in areas of weakness. Answers to all problems are in the back of the book. Make every effort to work a problem yourself before you look at an answer.

Solve each system.

A 1. $x^2 + y^2 = 25$
      $y = -4$ _____

2. $x^2 + y^2 = 169$
      $x = -12$ _____

3. $y^2 = 2x$
      $x = y - \frac{1}{2}$ _____

4. $8x^2 - y^2 = 16$
      $y = 2x$ _____

5. $x^2 + 4y^2 = 32$
      $x + 2y = 0$ _____

6. $2x^2 - 3y^2 = 25$
      $x + y = 0$ _____

7. $x^2 = 2y$
      $3x = y + 5$ _____

8. $y^2 = -x$
      $x - 2y = 5$ _____

9. $x^2 - y^2 = 3$
      $x^2 + y^2 = 5$ _____

10. $2x^2 + y^2 = 24$
      $x^2 - y^2 = -12$ _____

11. $x^2 - 2y^2 = 1$
      $x^2 + 4y^2 = 25$ _____

12. $x^2 + y^2 = 10$
      $16x^2 + y^2 = 25$ _____

B 13. $xy - 6 = 0$
       $x - y = 4$ _____

14. $xy = -4$
       $y - x = 2$ _____

15. $x^2 + xy - y^2 = -5$
       $y - x = 3$ _____

16. $x^2 - 2xy + y^2 = 1$
       $x - 2y = 2$ _____

17. $2x^2 - 3y^2 = 10$
       $x^2 + 4y^2 = -17$ _____

18. $2x^2 + 3y^2 = -4$
       $4x^2 + 2y^2 = 8$ _____

19. $x^2 + y^2 = 20$
       $x^2 = y$ _____

20. $x^2 - y^2 = 2$
       $y^2 = x$ _____

21. $x^2 + y^2 = 16$
       $y^2 = 4 - x$ _____

22. $x^2 + y^2 = 5$
       $x^2 = 4(2 - y)$ _____

23. Find the dimensions of a rectangle with area 32 square feet and perimeter 36 square feet. _____

24. Find two numbers such that their sum is 1 and their product is 1.

_____

C 25. $2x^2 + y^2 = 18$
       $xy = 4$ _____

26. $x^2 - y^2 = 3$
       $xy = 2$ _____

27. $x^2 + 2xy + y^2 = 36$
$x^2 - xy = 0$ _____

28. $2x^2 - xy + y^2 = 8$
$(x - y)(x + y) = 0$ _____

The Check Exercise for
this section is on
page 667.

29. $x^2 - 2xy + 2y^2 = 16$
$x^2 - y^2 = 0$ _____

30. $x^2 + xy - 3y^2 = 3$
$x^2 + 4xy + 3y^2 = 0$ _____

Section 10-6 Chapter Review

The standard form of a system of two linear equations in two unknowns is

$ax + by = m$
$cx + dy = n$

The **solutions of the system** are all ordered pairs (x, y) that satisfy both equations. Solutions can be estimated by **solving graphically**—that is, by graphing each equation and estimating the intersection point, if any, from the graph. Solutions can be found by **substitution** by solving one equation for one variable in terms of the other, substituting the result in the other equation, and then solving. **Equivalent systems** are those with exactly the same solution set. These operations produce equivalent systems:

1. Two equations are interchanged.
2. An equation is multiplied by a nonzero constant.
3. A constant multiple of another equation is added to a given equation.

These operations can be used to solve the system by **elimination**—that is, by eliminating one variable. A system that has no solution is called **inconsistent**; one that has an infinite number of solutions is called **dependent**. *(10-1)*

A system of three equations in three unknowns can be solved by elimination. **Solutions**, **equivalent systems**, and **dependent systems** mean the same as for the two-equation–two-unknown case. *(10-2)*

A **matrix** is a rectangular array of numbers (**elements** of the matrix) written within brackets. The **augmented matrix** of the standard system of two equations in two unknowns is

$$\begin{bmatrix} a & b & | & m \\ c & d & | & n \end{bmatrix}$$

The operations that are performed on the system need only be done on the coefficients and constants in the augmented matrix to obtain a **row-equivalent matrix**—that is, an augmented matrix of an equivalent system. These operations are:

1. Two rows are interchanged.
2. A row is multiplied by a nonzero constant.
3. A constant multiple of another row is added to a given row.

The **augmented matrix method** is a simple, concise way to solve systems and can be extended to large-scale systems. *(10-3)*

The **graph of a system of linear inequalities** is the intersection of half-planes representing the graph of each inequality. *(10-4)*

Systems involving second-degree equations can sometimes be solved by substitution. *(10-5)*

Diagnostic (Review) Exercise 10-6 ■

Work through all the problems in this chapter review and check answers in the back of the book. (Answers to all problems are there, and following each answer is a number in italics indicating the section in which that type of problem is discussed.) Where weaknesses show up, review appropriate sections in the text. When you are satisfied that you know the material, take the practice test following this review.

A   **1.**   Solve graphically: $x - y = 5$
                                  $x + y = 7$

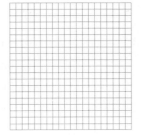

   2.   Solve graphically: $2x + y \le 8$
                                  $2x + 3y \le 12$
                                       $x \ge 0$
                                       $y \ge 0$

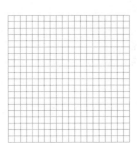

   3.   Solve by substitution: $2x + 3y = 7$
                                   $3x - y = 5$ _____

   4.   Solve by elimination using addition: $2x + 3y = 7$
                                                  $3x - y = 5$ _____

Solve each system in Problems 5–8.

   5. $y + 2z = 4$ **6.** $x^2 + y^2 = 2$
         $x - z = -2$ $2x - y = 3$ _____
         $x + y = 1$ _____

   7. $x^2 - y^2 = 7$ **8.** $x + 2y + z = 3$
         $x^2 + y^2 = 25$ _____ $2x + 3y + 4z = 3$
                                            $x + 2y + 3z = 1$ _____

B   **9.**   Solve graphically: $2x + y \ge 8$
                                  $x + 3y \ge 12$
                                       $x \ge 0$
                                       $y \ge 0$

Solve each system.

10. $3x - 2y = -1$
$-6x + 4y = 3$ _____

11. $3x - 2y - 7z = -6$
$-x + 3y + 2z = -1$
$x + 5y + 3z = 3$ _____

12. $3x^2 - y^2 = -6$
$2x^2 + 3y^2 = 29$ _____

13. $x^2 = y$
$y = 2x - 2$ _____

C **14.** $2x - 6y = -3$
$-\frac{2}{3}x + 2y = 1$ _____

15. $x^2 + 2xy - y^2 = -4$
$x^2 - xy = 0$ _____

Solve using augmented matrix methods.

16. $3x + 2y = 3$
$x + 3y = 8$ _____

17. $x + y = 1$
$x - z = -2$
$y + 2z = 4$ _____

18. $x + 2y + 3z = 1$
$2x + 3y + 4z = 3$
$x + 2y + z = 3$ _____

19. $x + 2y - z = 2$
$2x + 3y + z = -3$
$3x + 5y + z = -1$ _____

APPLICATIONS **20.** *Puzzle:* If you have 30 nickels and dimes worth $2.30 in your pocket, how many of each do you have? _____

21. *Business:* If $6,000 is to be invested, part at 10% and the rest at 6%, how much should be invested at each rate so that the total annual return from both investments is $440? _____

22. *Geometry:* The perimeter of a rectangle is 22 centimeters. If its area is 30 square centimeters, find the length of each side. _____

*23. *Chemistry:* A chemist has one 40% and one 70% solution of acid in stock. How much of each should she take to get 100 grams of a 49% solution? _____

*24. *Business:* A container contains 120 packages. Some of the packages weigh $\frac{1}{2}$ pound each, and the rest weigh $\frac{1}{3}$ pound each. If the total contents of the container weigh 48 pounds, how many are there of each type of package? _____

*25. *Diet:* A lab assistant wishes to obtain a food mix that contains, among other things, 27 grams of protein, 5.4 grams of fat, and 19 grams of moisture. He has available mixes of the compositions as listed in the table. How many grams of each mix should be used to get the desired diet mix? Set up a system of equations and solve using augmented matrix methods.

MIX	PROTEIN (%)	FAT (%)	MOISTURE (%)
A	30	3	10
B	20	5	20
C	10	4	10

Practice Test Chapter 10 ■

Take this as if it were a graded test by working the problems within a 50-minute time period. Do not look back in the chapter. Choose one of three levels of difficulty: least difficult, Problems 1–12; more difficult, add Problem 13; most difficult, add Problems 13 and 14. Use the answers in the back of the book to correct your work. The answers are keyed to appropriate text sections so that you can easily locate and review sections where difficulties still persist.

Solve Problems 1–3 graphically.

1. $x + 2y = 10$
 $4x - 5y = 1$

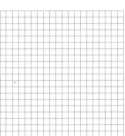

2. $x + 4y \leq 8$
 $0 \leq x \leq 4$
 $y \geq 0$

3. $2x + 3y \leq 12$
 $x - y \leq 2$
 $x \geq 0$
 $y \geq 0$

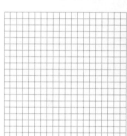

Solve Problems 4–9 algebraically.

4. $-2x + y = 7$
   $x - 2y = -8$ _____

5. $3x - y = -2$
   $-6x + 2y = 4$ _____

6. $x - 5y = -2$
   $-3x + 15y = 8$ _____

7. $x + y + z = -2$
   $3x - y - 2z = 1$
   $-2x + y + z = 1$ _____

8. $x^2 + 4y^2 = 4$
   $x - y = 1$ _____

9. $x^2 + y^2 = 5$
   $2x^2 - 3y^2 = 0$ _____

10. Use row operations to reduce the matrix

$$\begin{bmatrix} 2 & 3 & \bigl| & 4 \\ 5 & 7 & \bigl| & 12 \end{bmatrix} \quad \text{to the form} \quad \begin{bmatrix} 1 & k & \bigl| & m \\ 0 & 1 & \bigl| & n \end{bmatrix}$$

Write the number n as your answer. _____

Solve Problems 11–13 using system-of-equation methods.

11. If a parking meter contains 56 coins in nickels and dimes, worth $4.05, how many of each type of coin are in the meter? _____

12. A farmer placed an order with a chemical company for a chemical fertilizer that would contain, among other things, 120 pounds of nitrogen and 90 pounds of phosphoric acid. The company had two mixtures on hand, with the following compositions:

MIX	NITROGEN (%)	PHOSPHORIC ACID (%)
A	20	10
B	6	6

How many pounds of each mixture should the chemist mix to fill the order?

13. Solve:

$$x^2 - 4y^2 = 0$$
$$4x^2 + y^2 = 4$$

14. A garment industry manufactures three shirt styles. Each style of shirt requires the services of three departments, as listed in the table. The cutting, sewing, and packaging departments have available a maximum of 1,160, 1,560, and 480 workhours per week, respectively. Let x, y, and z represent the number of shirts of styles A, B, and C, respectively, that the plant must produce to operate at full capacity. Set up a system of three equations in x, y, and z to represent this problem. (Do not solve.)

	STYLE A	STYLE B	STYLE C
CUTTING DEPARTMENT	0.2 hour	0.4 hour	0.3 hour
SEWING DEPARTMENT	0.3 hour	0.5 hour	0.4 hour
PACKAGING DEPARTMENT	0.1 hour	0.2 hour	0.1 hour

Check Exercise 10-1 ▪

ANSWER COLUMN

Work the following problems without looking at any text examples. Show your work in the space provided. Write your answer or draw the final graph in the answer column.

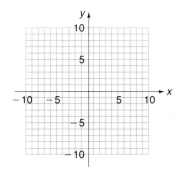

1. Solve by graphing:

$$3x + 2y = 13$$
$$2x - y = 4$$

1. _____

2. _____

3. _____

4. _____

Solve Problems 2–5 by substitution or by elimination using addition.

2. $2x - 4y = 4$
   $x = 10 - 2y$

5. _____

6. _____

7. _____

3. $4x - 6y = -2$
   $-6x + 9y = 6$

8. _____

4. $2x - 3y = 2$
   $3x - 7y = -2$

5. $8x - 4y = 12$
         $-2x + \ y = -3$

6. A shipping carton contains 86 small packages, some weighing $\frac{1}{3}$ pound each and the rest $\frac{3}{4}$ pound each. How many of each type of package are in the carton if the carton weighs 57 pounds total?

7. Find the dimensions of a rectangle with perimeter 320 meters, if its length is 3 times its width.

8. A jeweler has two bars of silver alloy in stock, one 50% silver and the other 100% silver. How much of each should be used to obtain 100 grams of an alloy that is 60% silver?

Check Exercise 10-2 ∎

Work the following problems without looking at any text examples. Show your work in the space provided. Write your answer in the answer column.

1. _____

2. _____

Solve and check.

3. _____

1. $3x - y = 5$
   $x + 2y - z = 5$
   $2x - 3y + z = 0$

4. _____

5. _____

2. $2x + y + z = 3$
   $x - 2y - z = 8$
   $3x - y + 2z = 9$

3. $2x - 3y + 4z = 4$
   $-4x + 2y - 2z = -6$
   $3x - 5y + 3z = 10$

4. $4x + y - 2z = 2$
$3x - y + z = -2$
$x + 2y - 3z = 1$

5. The equation of a circle in a rectangular coordinate system can be written in the form $x^2 + y^2 + Dx + Ey + F = 0$. Find D, E, and F so that the circle passes through $(-2, 5)$, $(5, 6)$, and $(7, 2)$.

Check Exercise 10-3 ■

Work the following problems without looking at any text examples. Show your work in the space provided. Write your answer in the answer column.

1. _____

2. _____

3. _____

4. _____

5. _____

1. Perform the row operation $-3R_2 + R_1 \rightarrow R_1$ on the matrix

$$\begin{bmatrix} 3 & 5 & -2 & | & 11 \\ 1 & 2 & -1 & | & 3 \end{bmatrix}$$

Solve Problems 2–5 using augmented matrix methods.

2. $\quad 4x - y = 1$
$\quad\quadx - 3y = 4$

3. $x - 3y = 5$
$-2x + 6y = 10$

4. $4x - 2y = 8$
$-2x + y = -4$

5. $x + 2y + 3z = -1$
$2x - 3y + z = 1$
$3x - 2y + 4z = -2$

Check Exercise 10-4 ■

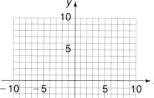

1. _____

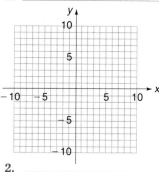

2. _____

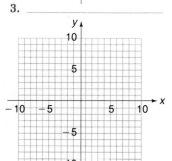

3. _____

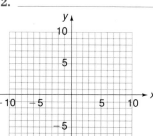

4. _____

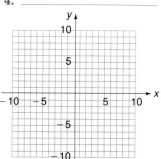

5. _____

Work the following problems without looking at any text examples. Show your work in the space provided. Draw the final graph in the answer column.

Solve graphically.

1. $-1 \leq x \leq 3$
  $2 \leq y \leq 8$

2. $0 \leq x \leq 3$
  $0 \leq y \leq 6$
  $x + y \leq 4$

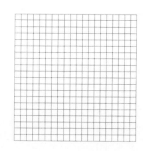

3. $2x + y \le 8$
$x + 2y \le 8$
$x \ge 0$
$y \ge 0$

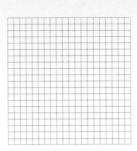

4. $y - x \ge 3$
$2x + y \le 10$
$x \ge 0$
$y \le 8$

5. $5x + 3y \ge 30$
$2x + 3y \ge 18$
$x \le 10$
$y \le 10$

Check Exercise 10-5 ▪

Work the following problems without looking at any text examples. Show your work in the space provided. Write your answer in the answer column.

1. _____

Solve each system.

2. _____

1. $2x^2 - y^2 = 9$
   $y - x = 0$

3. _____

4. _____

5. _____

2. $x^2 - 3y^2 = -3$
   $x - 2y = 1$

3. $2x^2 - 3y^2 = 15$
   $3x^2 + 4y^2 = 31$

4. $x^2 - xy + y^2 = 7$
$x - y = 3$

5. $x^2 + xy - y^2 = -9$
$2xy - y^2 = 0$

Sequences and Series ■ 11

INSTRUCTIONS FOR STUDENTS IN A SELF-PACED CLASS OR LAB

```
  YES  ◀──    HAVE YOU HAD INTERMEDIATE ALGEBRA BEFORE THIS COURSE?    ──▶  NO
```

1. Work Diagnostic (Review) Exercise 11-5 on page 694. Check answers in back of book; then work through text sections corresponding to problems missed. (Section numbers are in italics following each answer.)
2. When finished with step 1, take Practice Test Chapter 11 on page 695 as a final check of your understanding of the chapter. Check answers in the back of the book; then review sections where weakness still prevails. (Corresponding section numbers are in italics following each answer.)
3. When you think you are ready, ask your instructor for a graded test for Chapter 11.
4. If your instructor approves, after the test is corrected, review for a final examination.

1. Work through each section in the chapter as follows:
   (A) Read discussion.
   (B) Read each example and work the corresponding matched problem. Check your solutions to the matched problem in Solutions to Matched Problems on the indicated page.
   (C) At the end of a section work the odd-numbered problems in the Practice Exercise and check answers; then work even-numbered problems in areas of weakness. (Answers to *all* Practice Exercise sets are in the back of the book.)
   (D) Work Check Exercise as instructed. Tear out and turn in as directed by your instructor. (Answers are not in the text.)
2. Repeat each step in item 1 for each section in the chapter.
3. After the instructional part of the chapter is completed, proceed with steps 1 to 4 in the box above this one.

Chapter 11 ■ Sequences and Series

Section 11-1 Sequences and Series

- ■ Sequences
- ■ Series

■ Sequences

In this chapter we are going to consider functions whose domains are the set of natural numbers or particular subsets of the natural numbers. These special functions, called sequences, are encountered with increased frequency as one progresses in mathematics.

An **infinite sequence** is a function a whose domain is the set of all natural numbers $N = \{1, 2, 3, \ldots, n, \ldots\}$. The range of the function is $a(1), a(2), a(3), \ldots, a(n), \ldots$, which is usually written:

Infinite Sequence

$$a_1, a_2, a_3, \ldots, a_n, \ldots \qquad \text{where} \qquad a_n = a(n)$$

The elements in the range are called the **terms of the sequence**; a_1 is the first term, a_2 the second term, and a_n the nth term. For example, if

$$a_n = \frac{n-1}{n} \qquad n \in N \tag{1}$$

the function a is a sequence with terms

$$0, \frac{1}{2}, \frac{2}{3}, \frac{3}{4}, \ldots, \frac{n-1}{n}, \ldots \qquad \text{Replace } n \text{ in Equation (1) with 1, 2, 3, and so on.}$$

If the domain of a function a is the set of positive integers $\{1, 2, 3, \ldots, n\}$ for some fixed n, then a is called a **finite sequence**. Thus, if

$$a_1 = 5$$

and

$$a_n = a_{n-1} + 2 \qquad n \in \{2, 3, 4\}$$

the function a is a finite sequence with terms

$$5, 7, 9, 11 \qquad a_2 = a_{2-1} + 2 = a_1 + 2 = 5 + 2 = 7 \text{ and so on.}$$

The two examples just presented illustrate two common ways in which sequences are specified:

1. The nth term a_n is expressed by means of n.
2. One or more terms are given, and the nth term is expressed by means of preceding terms.

EXAMPLE 1 Find the first four terms of a sequence whose nth term is

$$a_n = \frac{1}{2^n}$$

Solution $\qquad a_1 = \frac{1}{2^1} = \frac{1}{2} \qquad a_2 = \frac{1}{2^2} = \frac{1}{4} \qquad a_3 = \frac{1}{2^3} = \frac{1}{8} \qquad a_4 = \frac{1}{2^4} = \frac{1}{16}$

PROBLEM 1 Find the first four terms of a sequence whose nth term is

$$a_n = \frac{n}{n^2 + 1}$$

Solution

EXAMPLE 2 Find the first five terms of a sequence specified by

$$a_1 = 5 \qquad a_n = \tfrac{1}{2}a_{n-1} \qquad n \geq 2$$

Solution $\quad a_1 = 5$

$a_2 = \tfrac{1}{2}a_{2-1} = \tfrac{1}{2}a_1 = \tfrac{1}{2}(5) = \tfrac{5}{2}$

$a_3 = \tfrac{1}{2}a_{3-1} = \tfrac{1}{2}a_2 = \tfrac{1}{2}\left(\tfrac{5}{2}\right) = \tfrac{5}{4}$

$a_4 = \tfrac{1}{2}a_{4-1} = \tfrac{1}{2}a_3 = \tfrac{1}{2}\left(\tfrac{5}{4}\right) = \tfrac{5}{8}$

$a_5 = \tfrac{1}{2}a_{5-1} = \tfrac{1}{2}a_4 = \tfrac{1}{2}\left(\tfrac{5}{8}\right) = \tfrac{5}{16}$

PROBLEM 2 Find the first five terms of a sequence specified by

$$a_1 = 3 \qquad a_n = a_{n-1} + 4 \qquad n \geq 2$$

Solution

Now let us look at the problem in reverse. That is, given the first few terms of a sequence (and assuming the sequence continues in the indicated pattern), find a_n in terms of n.

EXAMPLE 3   Find a_n in terms of n for the given sequences.

   (A) 5, 6, 7, 8, . . .   **(B)** 2, -4, 8, -16, . . .

Solution   **(A)** $a_n = n + 4$
         (B) $a_n = (-1)^{n+1} 2^n$ Note how $(-1)^{n+1}$ functions as a sign alternator.

PROBLEM 3   Find a_n in terms of n for:

   (A) 3, 5, 7, 9, . . .   **(B)** 1, $-\frac{1}{2}$, $\frac{1}{4}$, $-\frac{1}{8}$, . . .

Solution   **(A)**   **(B)**

■ **Series**

The sum of the terms of a sequence is called a **series**. Thus, if $a_1, a_2, a_3, \ldots, a_n$ are the terms of a sequence, then

$$S_n = a_1 + a_2 + a_3 + \cdots + a_n$$

is called a series. If the sequence is infinite, the corresponding series is called an **infinite series**. We will restrict our attention to finite series in this section.

A series is often represented in a compact form using **summation notation**, as follows:

Summation Notation

$$S_n = \sum_{k=1}^{n} a_k = a_1 + a_2 + a_3 + \cdots + a_n$$

---

where the terms of the series on the right are obtained from the middle expression by successively replacing k in a_k with integers, starting with 1 and ending with n. Thus, if a sequence is given by

$$\frac{1}{2}, \frac{1}{4}, \frac{1}{8}, \ldots, \frac{1}{2^n}$$

the corresponding series is given by

$$S_n = \frac{1}{2} + \frac{1}{4} + \frac{1}{8} + \cdots + \frac{1}{2^n}$$

or

$$S_n = \sum_{k=1}^{n} \frac{1}{2^k}$$

EXAMPLE 4   Write $S_5 = \displaystyle\sum_{k=1}^{5} \frac{k-1}{k}$ without summation notation.

Solution $S_5 = \displaystyle\sum_{k=1}^{5} \frac{k-1}{k}$ Replace k in $\dfrac{k-1}{k}$ successively with 1, 2, 3, 4, and 5; then add.

$$= \frac{1-1}{1} + \frac{2-1}{2} + \frac{3-1}{3} + \frac{4-1}{4} + \frac{5-1}{5}$$

$$= 0 + \tfrac{1}{2} + \tfrac{2}{3} + \tfrac{3}{4} + \tfrac{4}{5}$$

PROBLEM 4   Write $S_6 = \displaystyle\sum_{k=1}^{6} \frac{(-1)^{k+1}}{2k-1}$ without summation notation.

Solution

EXAMPLE 5   Write the following series using summation notation:

$$S_6 = 1 - \tfrac{1}{2} + \tfrac{1}{3} - \tfrac{1}{4} + \tfrac{1}{5} - \tfrac{1}{6}$$

Solution   We first note that the nth term of the series is given by

$$a_n = (-1)^{n+1}\frac{1}{n}$$

Hence,

$$S_6 = \sum_{k=1}^{6} (-1)^{k+1}\frac{1}{k}$$

PROBLEM 5   Write the following series using summation notation:

$$S_5 = 1 - \tfrac{2}{3} + \tfrac{4}{9} - \tfrac{8}{27} + \tfrac{16}{81}$$

Solution

Solutions to Matched
Problems

1. $a_1 = \tfrac{1}{2}$, $a_2 = \tfrac{2}{5}$, $a_3 = \tfrac{3}{10}$, $a_4 = \tfrac{4}{17}$

2. $a_1 = 3$, $a_2 = 7$, $a_3 = 11$, $a_4 = 15$, $a_5 = 19$

3. (A) $a_n = 2n + 1$
   (B) $a_n = (-1)^{n+1}/2^{n-1}$ or $a_n = (-\tfrac{1}{2})^{n-1}$ (two choices of many)

4. $S_6 = 1 - \tfrac{1}{3} + \tfrac{1}{5} - \tfrac{1}{7} + \tfrac{1}{9} - \tfrac{1}{11}$

5. $S_5 = \displaystyle\sum_{k=1}^{5} (-\tfrac{2}{3})^{k-1}$ or $S_5 = \displaystyle\sum_{k=1}^{5} (-1)^{k+1}(\tfrac{2}{3})^{k-1}$ (two choices of many)

Practice Exercise 11-1 ∎

Work odd-numbered problems first, check answers, and then work even-numbered problems in areas of weakness. Answers to all problems are in the back of the book. Make every effort to work a problem yourself before you look at an answer.

A   Write the first four terms for each sequence.

1. $a_n = n - 2$ _____

2. $a_n = n + 3$ _____

3. $a_n = \dfrac{n-1}{n+1}$ _____

4. $a_n = \left(1 + \dfrac{1}{n}\right)^n$ _____

5. $a_n = (-2)^{n+1}$ _____

6. $a_n = \dfrac{(-1)^{n+1}}{n^2}$ _____

7. Write the 8th term of the sequence in Problem 1. _____

8. Write the 10th term of the sequence in Problem 2. _____

9. Write the 100th term of the sequence in Problem 3. _____

10. Write the 200th term of the sequence in Problem 4. _____

Write each series in expanded form without summation notation.

11. $S_5 = \displaystyle\sum_{k=1}^{5} k$ _____

12. $S_4 = \displaystyle\sum_{k=1}^{4} k^2$ _____

13. $S_3 = \displaystyle\sum_{k=1}^{3} \dfrac{1}{10^k}$ _____

14. $S_5 = \displaystyle\sum_{k=1}^{5} (\tfrac{1}{3})^k$ _____

15. $S_4 = \displaystyle\sum_{k=1}^{4} (-1)^k$ _____

16. $S_6 = \displaystyle\sum_{k=1}^{6} (-1)^{k+1} k$ _____

B   Write the first five terms of each sequence.

17. $a_n = (-1)^{n+1} n^2$ _____

18. $a_n = (-1)^{n+1} \dfrac{1}{2^n}$ _____

19. $a_n = \dfrac{1}{3}\left(1 - \dfrac{1}{10^n}\right)$ _____

20. $a_n = n[1 - (-1)^n]$ _____

21. $a_1 = 7;\ a_n = a_{n-1} - 4,\ n \geq 2$ _____

22. $a_1 = a_2 = 1;\ a_n = a_{n-1} + a_{n-2},\ n \geq 3$ (Fibonacci sequence) _____

23. $a_1 = 4;\ a_n = \tfrac{1}{4} a_{n-1},\ n \geq 2$ _____

24. $a_1 = 2;\ a_n = 2a_{n-1},\ n \geq 2$ _____

Find a_n in terms of n.

25. $4, 5, 6, 7, \ldots$ _____

26. $-2, -1, 0, 1, \ldots$ _____

27. $3, 6, 9, 12, \ldots$ _____

28. $-2, -4, -6, -8, \ldots$ _____

29. $\frac{1}{2}, \frac{2}{3}, \frac{3}{4}, \frac{4}{5}, \ldots$ _____

30. $\frac{1}{2}, \frac{3}{4}, \frac{5}{6}, \frac{7}{8}, \ldots$ _____

31. $1, -1, 1, -1, \ldots$ _____

32. $1, -2, 3, -4, \ldots$ _____

33. $-2, 4, -8, 16, \ldots$ _____

34. $1, -3, 5, -7, \ldots$ _____

35. $x, \dfrac{x^2}{2}, \dfrac{x^3}{3}, \dfrac{x^4}{4}, \ldots$ _____

36. $x, -x^3, x^5, -x^7, \ldots$ _____

Write each series in expanded form without summation notation.

37. $\displaystyle\sum_{k=1}^{4} \frac{(-2)^{k+1}}{k}$ _____

38. $\displaystyle\sum_{k=1}^{5} (-1)^{k+1}(2k-1)^2$ _____

39. $S_3 = \displaystyle\sum_{k=1}^{3} \frac{1}{k} x^{k+1}$ _____

40. $S_5 = \displaystyle\sum_{k=1}^{5} x^{k-1}$ _____

41. $\displaystyle\sum_{k=1}^{5} \frac{(-1)^{k+1}}{k} x^k$ _____

42. $\displaystyle\sum_{k=0}^{4} \frac{(-1)^k x^{2k+1}}{2k+1}$ _____

Write each series using summation notation.

43. $S_4 = 1^2 + 2^2 + 3^2 + 4^2$ _____

44. $S_5 = 2 + 3 + 4 + 5 + 6$ _____

45. $S_5 = \dfrac{1}{2} + \dfrac{1}{2^2} + \dfrac{1}{2^3} + \dfrac{1}{2^4} + \dfrac{1}{2^5}$ _____

46. $S_4 = 1 - \frac{1}{2} + \frac{1}{3} - \frac{1}{4}$ _____

47. $S_n = 1 + \dfrac{1}{2^2} + \dfrac{1}{3^2} + \cdots + \dfrac{1}{n^2}$ _____

48. $S_n = 2 + \dfrac{3}{2} + \dfrac{4}{3} + \cdots + \dfrac{n+1}{n}$ _____

49. $S_n = 1 - 4 + 9 + \cdots + (-1)^{n+1} n^2$ _____

50. $S_n = \dfrac{1}{2} - \dfrac{1}{4} + \dfrac{1}{8} + \cdots + \dfrac{(-1)^{n+1}}{2^n}$ _____

C   **51.** Show that: $\displaystyle\sum_{k=1}^{n} ca_k = c \sum_{k=1}^{n} a_k$

52. Show that: $\displaystyle\sum_{k=1}^{n} (a_k + b_k) = \sum_{k=1}^{n} a_k + \sum_{k=1}^{n} b_k$

The sequence

$$a_n = \frac{a_{n-1}^2 + P}{2a_{n-1}} \qquad n \geq 2, P \text{ a positive real number}$$

can be used to find $\sqrt{P}$ to any decimal accuracy desired. To start the sequence, choose a_1 arbitrarily from the positive real numbers.

53. **(A)** Find the first four terms of the sequence to three decimal places:

$$a_1 = 3 \qquad a_n = \frac{a_{n-1}^2 + 2}{2a_{n-1}} \qquad n \geq 2$$

(A small hand calculator is useful, but not necessary.) _____

   (B) Compare terms with decimal approximation of $\sqrt{2}$ from a table or a calculator. _____

   (C) Repeat parts (A) and (B) by letting a_1 be any other positive number, say 1. _____

54. **(A)** Find the first four terms of the sequence to three decimal places:

$$a_1 = 2 \qquad a_n = \frac{a_{n-1}^2 + 5}{2a_{n-1}} \qquad n \geq 2 \text{ _____}$$

   (B) Find $\sqrt{5}$ in a table or by a calculator and compare with part (A).

_____

The Check Exercise for
this section is on
page 697.

   (C) Repeat parts (A) and (B) by letting a_1 be any other positive number, say 3. _____

Section 11-2 Arithmetic Sequences and Series

- ■ Arithmetic Sequences
- ■ Arithmetic Series

■ Arithmetic Sequences

Consider the sequence

  5, 9, 13, 17, ...

Can you guess what the 5th term is? If you guessed 21, you have observed that each term after the first can be obtained from the preceding one by adding 4 to it. This is an example of an arithmetic sequence. In general, a sequence

  $a_1, a_2, a_3, \ldots, a_n, \ldots$

is called an **arithmetic sequence** (or an **arithmetic progression**) if there exists a constant d, called the **common difference**, such that

$$a_n = a_{n-1} + d \qquad \text{for every } n > 1 \quad \text{ARITHMETIC SEQUENCE}$$

EXAMPLE 6   Which sequence is an arithmetic sequence and what is its common difference?

(A) $1, 2, 3, 5, \ldots$ **(B)** $3, 5, 7, 9, \ldots$

Solution   Sequence (B) is an arithmetic sequence with $d = 2$.

PROBLEM 6   Repeat Example 6 with: **(A)** $-4, -1, 2, 5, \ldots$ **(B)** $2, 4, 8, 16, \ldots$

Solution

Arithmetic sequences have several convenient properties. For example, it is easy to derive formulas for the nth term in terms of n and the sum of any number of consecutive terms. To obtain an nth-term formula, we note that if a is an arithmetic sequence, then

$$a_2 = a_1 + d$$
$$a_3 = a_2 + d = a_1 + 2d$$
$$a_4 = a_3 + d = a_1 + 3d$$

which suggests:

$$a_n = a_1 + (n-1)d \qquad \text{for every } n > 1 \quad \text{nTH-TERM FORMULA}$$

EXAMPLE 7   If the 1st and 10th terms of an arithmetic sequence are 3 and 30, respectively, find the 50th term of the sequence.

Solution   First find d:

$$a_n = a_1 + (n-1)d$$
$$a_{10} = a_1 + (10-1)d$$
$$30 = 3 + 9d$$
$$d = 3$$

Now find a_{50}:

$$a_{50} = 3 + (50-1)3 \qquad \text{Use } a_n = a_1 + (n-1)d.$$
$$= 3 + 49 \cdot 3$$
$$= 150$$

PROBLEM 7   If the 1st and 15th terms of an arithmetic sequence are -5 and 23, respectively, find the 73d term of the sequence.

Solution

■ Arithmetic Series

The sum of the terms of an arithmetic sequence is called an **arithmetic series**. We will derive two simple and very useful formulas for finding the **sum of the first n terms of an arithmetic series**. Let

$$S_n = a_1 + (a_1 + d) + \cdots + [a_1 + (n-2)d] + [a_1 + (n-1)d]$$

Reversing the order of the sum, we obtain

$$S_n = [a_1 + (n-1)d] + [a_1 + (n-2)d] + \cdots + (a_1 + d) + a_1$$

Adding left-hand sides and corresponding elements of the right sides of the two equations, we see that

$$2S_n = [2a_1 + (n-1)d] + [2a_1 + (n-1)d] + \cdots + [2a_1 + (n-1)d]$$
$$= n[2a_1 + (n-1)d]$$

or

$$S_n = \frac{n}{2}[2a_1 + (n-1)d] \quad \text{SUM FORMULA—FORM 1}$$

By replacing $a_1 + (n-1)d$ with a_n, we obtain a second useful formula for the sum:

$$S_n = \frac{n}{2}(a_1 + a_n) \quad \text{SUM FORMULA—FORM 2}$$

EXAMPLE 8   Find the sum of the first 26 terms of an arithmetic series if the 1st term is -7 and $d = 3$.

Solution $$S_n = \frac{n}{2}[2a_1 + (n-1)d] \quad \text{Use sum formula—form 1.}$$

$$S_{26} = \frac{26}{2}[2(-7) + (26-1)3]$$

$$= 793$$

PROBLEM 8   Find the sum of the first 52 terms of an arithmetic series if the 1st term
              is 23 and $d = -2$.

Solution

EXAMPLE 9   Find the sum of all the odd numbers from 51 to 99, inclusive.

Solution   First find n:

$$a_n = a_1 + (n - 1)d$$
$$99 = 51 + (n - 1)2$$
$$n = 25$$

Now find S_{25}:

$$S_n = \frac{n}{2}(a_1 + a_n) \quad \text{Use sum formula—form 2.}$$

$$S_{25} = \frac{25}{2}(51 + 99)$$
$$= 1,875$$

PROBLEM 9   Find the sum of all the even numbers from -22 to 52, inclusive.

Solution

Solutions to Matched
Problems

6. Sequence (A) is an arithmetic sequence with $d = 3$.

7. $a_n = a_1 + (n - 1)d$
$\quad a_{15} = a_1 + 14d$
$\quad 23 = -5 + 14d$
$\quad d = 2$
$\quad a_{73} = a_1 + 72d$
$\quad\quad = -5 + 72 \cdot 2$
$\quad\quad = 139$

8. $S_n = \frac{n}{2}[2a_1 + (n - 1)d]$

$\quad S_{52} = \frac{52}{2}[2 \cdot 23 + 51(-2)]$

$\quad\quad = -1,456$

9. $a_n = a_1 + (n - 1)d$
$\quad 52 = -22 + (n - 1)2$
$\quad n = 38$

$\quad S_n = \frac{n}{2}[2a_1 + (n - 1)d]$

$\quad S_{38} = \frac{38}{2}[2(-22) + 37 \cdot 2]$

$\quad\quad = 570$

Practice Exercise 11-2 ∎

Work odd-numbered problems first, check answers, and then work even-numbered problems in areas of weakness. Answers to all problems are in the back of the book. Make every effort to work a problem yourself before you look at an answer.

A   1. Determine which of the following are arithmetic sequences. Find d and the next two terms for those that are.

   (A)   2, 4, 8, . . . _____   **(B)**   7, 6.5, 6, . . . _____

   (C) $-11, -16, -21, \ldots$ _____   **(D)** $\frac{1}{2}, \frac{1}{6}, \frac{1}{18}, \ldots$ _____

   2. Repeat Problem 1 for:

   (A)   5, -1, -7, . . . _____   **(B)**   12, 4, $\frac{4}{3}$, . . . _____

   (C) $\frac{1}{2}, \frac{2}{3}, \frac{3}{4}, \ldots$ _____   **(D)**   16, 48, 80, . . . _____

Let $a_1, a_2, a_3, \ldots, a_n, \ldots$ be an arithmetic sequence. In Problems 3–18 find the indicated quantities.

   3. $a_1 = -5, d = 4, a_2 = ?, a_3 = ?, a_4 = ?$ _____

   4. $a_1 = -18, d = 3, a_2 = ?, a_3 = ?, a_4 = ?$ _____

   5. $a_1 = -3, d = 5, a_{15} = ?, S_{11} = ?$ _____

   6. $a_1 = 3, d = 4, a_{22} = ?, S_{21} = ?$ _____

   7. $a_1 = 1, a_2 = 5, S_{21} = ?$ _____

   8. $a_1 = 5, a_2 = 11, S_{11} = ?$ _____

B   9. $a_1 = 7, a_2 = 5, a_{15} = ?$ _____

   10. $a_1 = -3, d = -4, a_{10} = ?$ _____

   11. $a_1 = 3, a_{20} = 117, d = ?, a_{101} = ?$ _____

   12. $a_1 = 7, a_8 = 28, d = ?, a_{25} = ?$ _____

   13. $a_1 = -12, a_{40} = 22, S_{40} = ?$ _____

   14. $a_1 = 24, a_{24} = -28; S_{24} = ?$ _____

   15. $a_1 = \frac{1}{3}, a_2 = \frac{1}{2}, a_{11} = ?, S_{11} = ?$ _____

16. $a_1 = \frac{1}{6}, a_2 = \frac{1}{4}, a_{19} = ?, S_{19} = ?$ _____

17. $a_3 = 13, a_{10} = 55, a_1 = ?$ _____

18. $a_9 = -12, a_{13} = 3, a_1 = ?$ _____

19. $S_{51} = \sum_{k=1}^{51} (3k + 3) = ?$ _____

20. $S_{40} = \sum_{k=1}^{40} (2k - 3) = ?$ _____

C **21.** Find $g(1) + g(2) + g(3) + \cdots + g(51)$ if $g(t) = 5 - t$. _____

22. Find $f(1) + f(2) + f(3) + \cdots + f(20)$ if $f(x) = 2x - 5$. _____

23. Find the sum of all the even integers between 21 and 135. _____

24. Find the sum of all the odd integers between 100 and 500. _____

25. Show that the sum of the first n odd natural numbers is n^2, using appropriate formulas from this section. _____

26. Show that the sum of the first n even natural numbers is $n + n^2$, using appropriate formulas from this section. _____

27. If in a given sequence $a_1 = -3$ and $a_n = a_{n-1} + 3, n > 1$, find a_n in terms of n. _____

28. For the sequence in Problem 27 find $S_n = \sum_{k=1}^{n} a_k$ in terms of n. _____

APPLICATIONS **29.** *Earth science:* An object falling from rest in a vacuum near the surface of the earth falls 16 feet during the 1st second, 48 feet during the 2d second, and 80 feet during the 3d second, and so on.

(A) How far will the object fall during the 11th second? _____

(B) How far will the object fall in 11 seconds? _____

(C) How far will the object fall in t seconds? _____

30. *Business:* In investigating different job opportunities, you find that firm A will start you at $10,000 per year and guarantee you a raise of $500 each year, while firm B will start you at $13,000 per year but will only guarantee you a raise of $200 each year. Over a 15-year period which firm will pay the greatest total amount? _____

The Check Exercise for this section is on page 699.

Section 11-3   Geometric Sequences and Series

■ Geometric Sequences
■ Geometric Series
■ Infinite Geometric Series

■ Geometric Sequences

Consider the sequence

$$2, -4, 8, -16, \ldots$$

Can you guess what the 5th and 6th terms are? If you guessed 32 and -64, respectively, you have observed that each term after the first can be obtained from the preceding one by multiplying it by -2. This is an example of a geometric sequence. In general, a sequence

$$a_1, a_2, a_3, \ldots, a_n, \ldots$$

is called a **geometric sequence** (or a **geometric progression**) if there exists a nonzero constant r, called the **common ratio**, such that

$$a_n = ra_{n-1} \qquad \text{for every } n > 1 \quad \text{GEOMETRIC SEQUENCE}$$

EXAMPLE 10   Which sequence is a geometric sequence and what is its common ratio?

(A) $2, 6, 8, 10, \ldots$ **(B)** $-1, 3, -9, 27, \ldots$

Solution   Sequence (B) is a geometric sequence with $r = -3$.

PROBLEM 10   Repeat Example 10 with: **(A)** $\frac{1}{4}, \frac{1}{2}, 1, 2, \ldots$ **(B)** $\frac{1}{2}, \frac{1}{4}, \frac{1}{16}, \frac{1}{256}, \ldots$

Solution

Just as with arithmetic sequences, geometric sequences have several convenient properties. It is easy to derive formulas for the nth term in terms of n and the sum of any number of consecutive terms. To obtain an nth-term formula, we note that if a is a geometric sequence, then

$$a_2 = ra_1$$
$$a_3 = ra_2 = r^2 a_1$$
$$a_4 = ra_3 = r^3 a_1$$

which suggests that

$$a_n = a_1 r^{n-1} \qquad \text{for every } n > 1 \quad \text{nTH-TERM FORMULA}$$

EXAMPLE 11 Find the 7th term of the geometric sequence $1, \frac{1}{2}, \frac{1}{4}, \ldots$.

Solution

$$r = \frac{1}{2}$$

$$a_n = a_1 r^{n-1}$$

$$a_7 = 1 \left(\frac{1}{2} \right)^{7-1} = \frac{1}{2^6} = \frac{1}{64}$$

PROBLEM 11 Find the 8th term of the geometric sequence $\frac{1}{64}, -\frac{1}{32}, \frac{1}{16}, \ldots$.

Solution

EXAMPLE 12 If the 1st and 10th terms of a geometric sequence are 1 and 2, respectively, find the common ratio r.

Solution

$$a_n = a_1 r^{n-1}$$

$$2 = 1 r^{10-1}$$

$$r = 2^{1/9} \approx 1.08 \quad \text{Calculation by calculator [or by logarithms (see Appendix B)]}$$

PROBLEM 12 If the 1st and 8th terms of a geometric sequence are 2 and 16, respectively, find the common ratio r.

Solution

■ Geometric Series

A **geometric series** is any series whose terms form a geometric sequence. As was the case with an arithmetic series, we can derive two simple and very useful formulas for finding the **sum of the first n terms of a geometric series**. Let

$$S_n = a_1 + a_1 r + a_1 r^2 + a_1 r^3 + \cdots + a_1 r^{n-1}$$

and multiply both members by r to obtain

$$rS_n = a_1 r + a_1 r^2 + a_1 r^3 + \cdots + a_1 r^{n-1} + a_1 r^n$$

Now subtract the left side of the second equation from the left side of the first, and the right side of the second equation from the right side of the first, to obtain

$$S_n - rS_n = a_1 - a_1r^n \quad \text{Note how many terms dropped out on the right side.}$$
$$S_n(1 - r) = a_1 - a_1r^n$$

Thus:

$$S_n = \frac{a_1 - a_1r^n}{1 - r} = \frac{a_1(1 - r^n)}{1 - r} \qquad r \neq 1 \quad \text{SUM FORMULA—FORM 1}$$

Since $a_n = a_1r^{n-1}$, or $ra_n = a_1r^n$, the sum formula can also be written as

$$S_n = \frac{a_1 - ra_n}{1 - r} \qquad r \neq 1 \quad \text{SUM FORMULA—FORM 2}$$

EXAMPLE 13   Find the sum to three significant figures of the first 20 terms of a geometric series if the 1st term is 1 and $r = 2$.

Solution $$S_n = \frac{a_1 - a_1r^n}{1 - r} \qquad\qquad \text{Use sum formula—form 1.}$$

$$S_n = \frac{1 - 1 \cdot 2^{20}}{1 - 2} \approx 1{,}050{,}000 \qquad \begin{array}{l}\text{Calculation by calculator}\\ \text{[or by logarithms (see Appendix B)]}\end{array}$$

PROBLEM 13   Find the sum to four significant figures of the first 14 terms of a geometric series if the 1st term is $\frac{1}{64}$ and $r = -2$.

Solution

■ **Infinite Geometric Series**

Given a geometric series, what happens to the sum S_n as n increases? To answer this question, we first write the sum formula in the more convenient form

$$S_n = \frac{a_1 - a_1r^n}{1 - r} = \frac{a_1}{1 - r} - \frac{a_1r^n}{1 - r}$$

It is possible to show that if $|r| < 1$, that is, if $-1 < r < 1$, then r^n will tend to 0 as n increases. (See what happens, for example, if $r = \frac{1}{2}$ and n gets large.) Thus, S_n can be made as close to

$$\frac{a_1}{1 - r}$$

as we wish by taking n sufficiently large. Thus, we define

$$S_\infty = \frac{a_1}{1 - r} \qquad |r| < 1$$

and call this the **sum of an infinite geometric series**. If $|r| \geq 1$, an infinite geometric series has no sum, since r^n can (in absolute value) be made as large as we like by making n sufficiently large. (What happens to r^n if $r = 3$ and n gets large?)

EXAMPLE 14   Represent the repeating decimal $0.4545\overline{45}$ as the quotient of two integers. (The bar over the last two digits indicates that these digits repeat indefinitely.)

Solution $0.4545\overline{45} = 0.45 + 0.0045 + 0.000045 + \cdots$

The right side of the equation is an infinite geometric series with $a_1 = 0.45$ and $r = 0.01$. Thus,

$$S_\infty = \frac{a_1}{1 - r} = \frac{0.45}{1 - 0.01} = \frac{0.45}{0.99} = \frac{5}{11}$$

Hence, $0.4545\overline{45}$ and $\frac{5}{11}$ name the same rational number. Check the result by dividing 5 by 11.

PROBLEM 14   Repeat Example 14 for $0.8181\overline{81}$.

Solution

Solutions to Matched Problems

10. Sequence (A) is a geometric sequence with $r = 2$.

11. $r = -2$
$a_n = a_1 r^{n-1}$
$a_8 = \frac{1}{64} \cdot (-2)^7$
$\quad = \frac{-128}{64}$
$\quad = -2$

12. $a_n = a_1 r^{n-1}$
$16 = 2r^7$
$r^7 = 8$
$r = 8^{1/7}$
$r \approx 1.346$

13. $S_n = \frac{a_1 - a_1 r^n}{1 - r}$
$S_{14} = \frac{\frac{1}{64} - \frac{1}{64}(-2)^{14}}{1 - (-2)}$
$\quad \approx -85.33$

14. $a_1 = 0.81, r = 0.01$
$S_\infty = \frac{a_1}{1 - r} = \frac{0.81}{0.99} = \frac{9}{11}$

Practice Exercise 11-3 ∎

Work odd-numbered problems first, check answers, and then work even-numbered problems in areas of weakness. Answers to all problems are in the back of the book. Make every effort to work a problem yourself before you look at an answer.

A   **1.** Determine which of the following are geometric sequences. Find r and the next two terms for those that are.

(A) 2, -4, 8, . . . _____   **(B)** 7, 6.5, 6, . . . _____

(C) -11, -16, -21, . . . _____   **(D)** $\frac{1}{2}$, $\frac{1}{6}$, $\frac{1}{18}$, . . . _____

2. Repeat Problem 1 for:

(A) 5, -1, -7, . . . _____   **(B)** 12, 4, $\frac{4}{3}$, . . . _____

(C) $\frac{1}{2}$, $\frac{2}{3}$, $\frac{3}{4}$, . . . _____   **(D)** 16, 48, 80, . . . _____

Let $a_1, a_2, a_3, \ldots, a_n, \ldots$ be a geometric sequence. Find each of the indicated quantities.

3. $a_1 = -6$, $r = -\frac{1}{2}$, $a_2 = ?$, $a_3 = ?$, $a_4 = ?$ _____

4. $a_1 = 12$, $r = \frac{2}{3}$, $a_2 = ?$, $a_3 = ?$, $a_4 = ?$ _____

5. $a_1 = 81$, $r = \frac{1}{3}$, $a_{10} = ?$ _____

6. $a_1 = 64$, $r = \frac{1}{2}$, $a_{13} = ?$ _____

7. $a_1 = 3$, $a_7 = 2{,}187$, $r = 3$, $S_7 = ?$ _____

8. $a_1 = 1$, $a_7 = 729$, $r = -3$, $S_7 = ?$ _____

B   **9.** $a_1 = 100$, $a_6 = 1$, $r = ?$ _____

10. $a_1 = 10$, $a_{10} = 30$, $r = ?$ _____

11. $a_1 = 5$, $r = -2$, $S_{10} = ?$ _____

12. $a_1 = 3$, $r = 2$, $S_{10} = ?$ _____

13. $a_1 = 9$, $a_4 = \frac{8}{3}$, $a_2 = ?$, $a_3 = ?$ _____

14. $a_1 = 12$, $a_4 = -\frac{4}{9}$, $a_2 = ?$, $a_3 = ?$ _____

15. $S_7 = \displaystyle\sum_{k=1}^{7} (-3)^{k-1} = ?$ _____   **16.** $S_7 = \displaystyle\sum_{k=1}^{7} 3^k = ?$ _____

17. Find $g(1) + g(2) + \cdots + g(10)$ if $g(x) = (\frac{1}{2})^x$. _____

18. Find $f(1) + f(2) + \cdots + f(10)$ if $f(x) = 2^x$. _____

19. Find a positive number x such that $-2 + x - 6$ is a geometric series with three terms. _____

20. Find a positive number x such that $6 + x + 8$ is a geometric series with three terms. _____

Find the sum of each infinite geometric series that has a sum.

21. $3 + 1 + \frac{1}{3} + \cdots$ _____   **22.** $16 + 4 + 1 + \cdots$ _____

23. $2 + 4 + 8 + \cdots$ _____   **24.** $4 + 6 + 9 + \cdots$ _____

25. $2 - \frac{1}{2} + \frac{1}{8} - \cdots$ _____   **26.** $21 - 3 + \frac{3}{7} - \cdots$ _____

C *Represent each repeating decimal fraction as the quotient of two integers.*

27. $0.777\overline{7}$ _____   **28.** $0.555\overline{5}$ _____

29. $0.5454\overline{54}$ _____   **30.** $0.27272\overline{7}$ _____

31. $3.216216\overline{216}$ _____   **32.** $5.6363\overline{63}$ _____

APPLICATIONS ***33.** *Business:* If P dollars is invested at r% compounded annually, the amount A present after n years forms a geometric progression with a constant ratio $(1 + r)$. Write a formula for the amount present after n years. How long will it take for a sum of money P to double if invested at 6% interest compounded annually? _____

*34. *Life science:* If a population of A_0 people grows at the constant rate of r% per year, the population after t years forms a geometric progression with a constant ratio $(1 + r)$. Write a formula for the total population after t years. If the world's population is increasing at the rate of 2% per year, how long will it take to double? _____

35. *Engineering:* A rotating flywheel coming to rest rotates 300 revolutions the first minute. If in each subsequent minute it rotates two-thirds as many times as in the preceding minute, how many revolutions will the wheel make before coming to rest? _____

36. *Physics:* The first swing of a bob on a pendulum is 10 inches. If on each subsequent swing it travels 0.9 as far as on the preceding swing, how far will the bob travel before coming to rest? _____

37. *Economics:* The government, through a subsidy program, distributes $1,000,000. If we assume that each individual or agency spends 0.8 of what is received, and 0.8 of this is spent, and so on, how much total increase in spending results from this government action? (Let $a_1 = \$800,000$.)

*38. *Zeno's paradox:* Visualize a hypothetical 440-yard oval race track that has tapes stretched across the track at the halfway point and at each point that marks the halfway point of each remaining distance thereafter. A runner running around the track has to break the first tape before the second, the second before the third, and so on. From this point of view it appears that the runner will never finish the race. (This famous paradox is attributed to the Greek philosopher Zeno, 495−435 B.C.) If we assume the runner runs at 440 yards per minute, the times between tape breakings form an infinite geometric progression. What is the sum of this progression?

The Check Exercise for this section is on page 701.

Section 11-4 Binomial Formula

■ Factorial
■ Binomial Formula

The binomial form

$(a + b)^n$

with n a natural number, appears more frequently than one might expect. The coefficients in its expansion play an important role in probability studies. In this section we will give an informal derivation of the famous binomial formula, which will enable us to expand $(a + b)^n$ directly for any natural number n, however large. First, we introduce the useful concept of factorial.

■ Factorial

For n a natural number, **n factorial**—denoted by $n!$—is the product of the first n natural numbers. **Zero factorial** is defined to be 1. Symbolically:

n Factorial
$n! = n(n - 1) \cdot \cdots \cdot 2 \cdot 1$
$1! = 1$
$0! = 1$

It is also useful to note that

$n! = n \cdot (n - 1)!$

EXAMPLE 15   Evaluate each:

(A) $4! = 4 \cdot 3 \cdot 2 \cdot 1 = 24$

(B) $5! = 5 \cdot 4 \cdot 3 \cdot 2 \cdot 1 = 120$

(C) $\dfrac{7!}{6!} = \dfrac{7 \cdot 6!}{6!} = 7$

(D) $\dfrac{8!}{5!} = \dfrac{8 \cdot 7 \cdot 6 \cdot 5!}{5!} = 336$

PROBLEM 15   Evaluate: **(A)** $7!$ **(B)** $\dfrac{8!}{7!}$ **(C)** $\dfrac{6!}{3!}$

Solution   **(A)**   **(B)**   **(C)**

A form involving factorials that is very useful is given by the formula

$$\binom{n}{r} = \frac{n!}{(n-r)!\,r!} \qquad 0 \leq r \leq n,\, n \in N$$

EXAMPLE 16   Find: **(A)** $\dbinom{5}{2}$ **(B)** $\dbinom{4}{4}$

Solution   **(A)** $\dbinom{5}{2} = \dfrac{5!}{(5-2)!\,2!} = \dfrac{5!}{3!\,2!} = 10$

(B) $\dbinom{4}{4} = \dfrac{4!}{(4-4)!\,4!} = \dfrac{4!}{0!\,4!} = 1$

PROBLEM 16   Find: **(A)** $\dbinom{9}{2}$ **(B)** $\dbinom{5}{5}$

Solution   **(A)**   **(B)**

■ **Binomial Formula**

Let us try to discover a formula for the expansion of $(a + b)^n$, for n a natural number:

$(a + b)^1 = a + b$

$(a + b)^2 = a^2 + 2ab + b^2$

$(a + b)^3 = a^3 + 3a^2b + 3ab^2 + b^3$

$(a + b)^4 = a^4 + 4a^3b + 6a^2b^2 + 4ab^3 + b^4$

$(a + b)^5 = a^5 + 5a^4b + 10a^3b^2 + 10a^2b^3 + 5ab^4 + b^5$

OBSERVATIONS

1. The expansion of $(a + b)^n$ has $n + 1$ terms.
2. The power of a starts at n and decreases by 1 for each term until it is 0 in the last term.
3. The power of b starts at 0 in the 1st term and increases by 1 for each term until it is n in the last term.
4. The sum of the powers of a and b in each term is the constant n.
5. The coefficient of any term after the first can be obtained from the preceding term as follows. In the preceding term multiply the coefficient by the exponent of a, and then divide this product by the number representing the position of the preceding term in the series.

We now postulate these same properties for the general case:

$$(a + b)^n = a^n + \frac{n}{1} a^{n-1}b + \frac{n(n-1)}{1 \cdot 2} a^{n-2}b^2$$

$$+ \frac{n(n-1)(n-2)}{1 \cdot 2 \cdot 3} a^{n-3}b^3 + \cdots + b^n$$

$$= \frac{n!}{0!(n-0)!} a^n + \frac{n!}{1!(n-1)!} a^{n-1}b + \frac{n!}{2!(n-2)!} a^{n-2}b^2$$

$$+ \frac{n!}{3!(n-3)!} a^{n-3}b^3 + \cdots + \frac{n!}{n!(n-n)!} b^n$$

$$= \binom{n}{0}a^n + \binom{n}{1}a^{n-1}b + \binom{n}{2}a^{n-2}b^2 + \binom{n}{3}a^{n-3}b^3 + \cdots + \binom{n}{n}b^n$$

Thus, it appears that:

Binomial Formula

$$(a + b)^n = \sum_{k=0}^{n} \binom{n}{k} a^{n-k}b^k \qquad n \geq 1$$

This result is known as the **binomial formula**, and its general proof requires a method of proof called mathematical induction, which is considered in advanced courses.

EXAMPLE 17   Use the binomial formula to expand $(x + y)^6$.

Solution $$(x + y)^6 = \sum_{k=0}^{6} \binom{6}{k} x^{6-k}y^k$$

$$= \binom{6}{0}x^6 + \binom{6}{1}x^5y + \binom{6}{2}x^4y^2 + \binom{6}{3}x^3y^3 + \binom{6}{4}x^2y^4$$

$$+ \binom{6}{5}xy^5 + \binom{6}{6}y^6$$

$$= x^6 + 6x^5y + 15x^4y^2 + 20x^3y^3 + 15x^2y^4 + 6xy^5 + y^6$$

PROBLEM 17   Use the binomial formula to expand $(x + 1)^5$.

Solution

EXAMPLE 18   Use the binomial formula to find the 4th term in the expansion of $(x - 2)^{20}$.

Solution   4th term $= \dbinom{20}{3} x^{20-3}(-2)^3$ Since the 1st term is given when $k = 0$, the 4th term is given when $k = 3$.

$$= \frac{20!}{(20-3)!3!} x^{17}(-2)^3$$

$$= \frac{20 \cdot 19 \cdot 18}{3 \cdot 2 \cdot 1} x^{17}(-8)$$

$$= -9{,}120x^{17}$$

PROBLEM 18   Use the binomial formula to find the 5th term in the expansion of $(u - 1)^{18}$.

Solution

Solutions to Matched Problems

15. (A) $7! = 7 \cdot 6 \cdot 5 \cdot 4 \cdot 3 \cdot 2 \cdot 1 = 5{,}040$

    (B) $\dfrac{8!}{7!} = \dfrac{8 \cdot 7!}{7!} = 8$ **(C)** $\dfrac{6!}{3!} = \dfrac{6 \cdot 5 \cdot 4 \cdot 3!}{3!} = 120$

16. (A) $\dbinom{9}{2} = \dfrac{9!}{7!2!} = \dfrac{9 \cdot 8 \cdot 7!}{7!2!} = \dfrac{72}{2} = 36$ **(B)** $\dbinom{5}{5} = \dfrac{5!}{0!5!} = 1$

17. $(x + 1)^5 = \displaystyle\sum_{k=0}^{5} \dbinom{5}{k} x^{5-k} 1^k$

$$= \dbinom{5}{0}x^5 + \dbinom{5}{1}x^4 + \dbinom{5}{2}x^3 + \dbinom{5}{3}x^2 + \dbinom{5}{4}x^1 + \dbinom{5}{5}x^0$$

$$= x^5 + 5x^4 + 10x^3 + 10x^2 + 5x + 1$$

18. 5th term $= \dbinom{18}{4} u^{18-4}(-1)^4 = \dfrac{18 \cdot 17 \cdot 16 \cdot 15 \cdot 14!}{14!4!} u^{14} \cdot 1$

$$= 3{,}060u^{14}$$

Practice Exercise 11-4 ■

Work odd-numbered problems first, check answers, and then work even-numbered problems in areas of weakness. Answers to all problems are in the back of the book. Make every effort to work a problem yourself before you look at an answer.

A *Evaluate.*

1. 6! _____   **2.** 4! _____   **3.** $\dfrac{20!}{19!}$ _____

4. $\dfrac{5!}{4!}$ _____   **5.** $\dfrac{10!}{7!}$ _____   **6.** $\dfrac{9!}{6!}$ _____

7. $\dfrac{6!}{4!2!}$ _____   **8.** $\dfrac{5!}{2!3!}$ _____   **9.** $\dfrac{9!}{0!(9-0)!}$ _____

10. $\dfrac{8!}{8!(8-8)!}$ _____   **11.** $\dfrac{8!}{2!(8-2)!}$ _____   **12.** $\dfrac{7!}{3!(7-3)!}$ _____

Write as the quotient of two factorials.

13. 9 _____   **14.** 12 _____

15. $6 \cdot 7 \cdot 8$ _____   **16.** $9 \cdot 10 \cdot 11 \cdot 12$ _____

B *Evaluate.*

17. $\dbinom{9}{5}$ _____   **18.** $\dbinom{5}{2}$ _____   **19.** $\dbinom{6}{5}$ _____

20. $\dbinom{7}{1}$ _____   **21.** $\dbinom{9}{9}$ _____   **22.** $\dbinom{5}{0}$ _____

23. $\dbinom{17}{13}$ _____   **24.** $\dbinom{20}{16}$ _____

Expand using the binomial formula.

25. $(u + v)^5$ _____   **26.** $(x + y)^4$ _____

27. $(y - 1)^4$ _____   **28.** $(x - 2)^5$ _____

29. $(2x - y)^5$ _____   **30.** $(m + 2n)^6$ _____

Find the indicated term in each expansion.

31. $(u + v)^{15}$; 7th term _____   **32.** $(a + b)^{12}$; 5th term _____

33. $(2m + n)^{12}$; 11th term _____   **34.** $(x + 2y)^{20}$; 3d term _____

35. $[(w/2) - 2]^{12}$; 7th term _____   **36.** $(x - 3)^{10}$; 4th term _____

C **37.** Evaluate $(1.01)^{10}$ to four decimal places, using the binomial formula. [*Hint:* Let $1.01 = 1 + 0.01$.] _____

38. Evaluate $(0.99)^6$ to four decimal places, using the binomial formula.

_____

39. Show that: $\begin{pmatrix} n \\ r \end{pmatrix} = \begin{pmatrix} n \\ n-r \end{pmatrix}$ _____

40. Can you guess what the next two rows in **Pascal's triangle** are? Compare the numbers in the triangle with the binomial coefficients obtained with the binomial formula. _____

The Check Exercise for this section is on page 703.

```
            1
         1     1
      1     2     1
   1     3     3     1
1     4     6     4     1
```

Section 11-5 Chapter Review

An **infinite sequence** is a function whose domain is the set of all natural numbers. A **finite sequence** is a function whose domain is a set $\{1, 2, 3, \ldots, n\}$. The element associated with the natural number n is called the **nth term** of the sequence and is usually denoted with a subscript as a_n. The sum of the terms of a sequence is called a **series**—an **infinite series** if the sequence is infinite. **Summation notation** is useful for representing a series:

$$S_n = \sum_{k=1}^{n} a_k = a_1 + a_2 + \cdots + a_n \quad (11\text{-}1)$$

An **arithmetic sequence** has nth term $a_n = a_{n-1} + d = a_1 + (n-1)d$, where the constant d is called the **common difference**. The sum of the first n terms of an arithmetic sequence is called an **arithmetic series** and is given by the formula

$$S_n = \frac{n}{2}[2a_1 + (n-1)d] = \frac{n}{2}(a_1 + a_n) \quad (11\text{-}2)$$

A **geometric sequence** has nth term $a_n = ra_{n-1} = a_1 r^{n-1}$, where the constant $r \neq 0$ is called the **common ratio**. A **geometric series** is the sum of terms in a geometric sequence. For n terms the series sum is

$$S_n = \frac{a_1 - a_1 r^n}{1-r} = \frac{a_1(1-r^n)}{1-r} = \frac{a_1 - ra_n}{1-r} \qquad r \neq 1$$

For an infinite series with $|r| < 1$, the sum is

$$S_\infty = \frac{a_1}{1-r} \quad (11\text{-}3)$$

The **binomial formula** is

$$(a + b)^n = \sum_{k=0}^{n} \binom{n}{k} a^{n-k} b^k$$

where

$$\binom{n}{k} = \frac{n!}{(n-r)!r!}$$

$n! = n(n-1) \cdot \cdots \cdot 2 \cdot 1$, $1! = 1$, and $0! = 1$. The number $n!$ is called **n factorial**.
(11-4)

Diagnostic (Review) Exercise 11-5 ▣

Work through all the problems in this chapter review and check answers in the back of the book. (Answers to all problems are there, and following each answer is a number in italics indicating the section in which that type of problem is discussed.) Where weaknesses show up, review appropriate sections in the text. When you are satisfied that you know the material, take the practice test following this review.

A   **1.** Identify all arithmetic and all geometric sequences from the following list of sequences:

(A) $16, -8, 4, \ldots$ _____   **(B)** $5, 7, 9, \ldots$ _____

(C) $-8, -5, -2, \ldots$ _____   **(D)** $2, 3, 5, 8, \ldots$ _____

(E) $-1, 2, -4, \ldots$ _____

In Problems 2–5:

(A)   Write the first four terms of each sequence.
(B)   Find a_{10}.
(C)   Find S_{10}.

2. $a_n = 2n + 3$ _____   **3.** $a_n = 32(\tfrac{1}{2})^n$ _____

4. $a_1 = -8; a_n = a_{n-1} + 3, n \geq 2$ _____

5. $a_1 = -1; a_n = (-2)a_{n-1}, n \geq 2$ _____

6. Find S_∞ in Problem 3. _____

Evaluate.

7. $6!$ _____   **8.** $\dfrac{22!}{19!}$ _____   **9.** $\dfrac{7!}{2!(7-2)!}$ _____

B   Write Problems 10 and 11 without summation notation and find the sums.

10. $S_{10} = \sum_{k=1}^{10} (2k - 8)$ _____   **11.** $S_7 = \sum_{k=1}^{7} \dfrac{16}{2^k}$ _____

12. $S_\infty = 27 - 18 + 12 - \cdots = ?$ _____

13. Write $S_n = \dfrac{1}{3} - \dfrac{1}{9} + \dfrac{1}{27} + \cdots + \dfrac{(-1)^{n+1}}{3^n}$ using summation notation and find S_∞. _____

14. If in an arithmetic sequence $a_1 = 13$ and $a_7 = 31$, find the common difference d and the 5th term a_5. _____

Evaluate.

15. $\dfrac{20!}{18!(20-18)!}$ _____

16. $\begin{pmatrix} 16 \\ 12 \end{pmatrix}$ _____ **17.** $\begin{pmatrix} 11 \\ 11 \end{pmatrix}$ _____

18. Expand $(x - y)^5$ using the binomial formula. _____

19. Find the 10th term in the expansion of $(2x - y)^{12}$. _____

C **20.** Write $0.72\overline{72}$ as the quotient of two integers. _____

21. A free-falling body travels $g/2$ feet in the first second, $3g/2$ feet during the next second, $5g/2$ feet the next, and so on, where g is the gravitational constant. Find the distance fallen during the 25th second and the total distance fallen from the start to the end of the 25th second. Express answers in terms of g. _____

22. Expand $(x + i)^6$, i the complex unit, using the binomial formula.

_____

Practice Test Chapter 11 ▪

Take this as if it were a graded test by working the problems within a 50-minute time period. Do not look back in the chapter. Choose one of three levels of difficulty: least difficult, Problems 1–12; more difficult, add Problem 13; most difficult, add Problems 13 and 14. Use the answers in the back of the book to correct your work. The answers are keyed to appropriate text sections so that you can easily locate and review sections where difficulties still persist.

In Problems 1–3, consider the sequence given by $a_n = 5 - 3n$.

1. Find the first four terms. _____

2. Find the 45th term. _____

3. Find the sum of the first 45 terms. _____

In Problems 4–6, consider the sequence given by $a_n = 0.2(0.1)^{n-1}$.

4. Find the 4th term. _____

5. Find the sum of the first five terms. _____

6. Find S_∞. _____

7. Write the sum $\displaystyle\sum_{k=1}^{5} (-1)^{k-1}(k^2 - 1)$ in expanded notation. _____

8. Write the sum $\frac{2}{3} + \frac{4}{5} + \frac{6}{7} + \frac{8}{9} + \frac{10}{11}$ in summation notation. _____

9. Find a positive number x so that 6, x, 12 is a geometric sequence. _____

10. Find a number x so that 6, x, 12 is an arithmetic sequence. _____

11. Evaluate $\dbinom{11}{3}$. _____

12. Find the 9th term in the expansion of $(x - 1)^{11}$ using the binomial formula.

_____

13. Evaluate $\displaystyle\sum_{k=1}^{8} (5k - 2^k)$. _____

14. If \$100 is invested at 6% compounded annually, how long will it take to triple in value? Express your answer in terms of logarithms. _____

Check Exercise 11-1 ■

Work the following problems without looking at any text examples. Show your work in the space provided. Write your answer in the answer column.

1. _____

2. _____

3. _____

4. _____

5. _____

6. _____

7. _____

8. _____

9. _____

10. _____

1. Find the first four terms of the sequence

$$a_n = n^2 - 1$$

2. Find the 13th term of the sequence

$$a_n = n^2 - 1$$

3. Find the first four terms of the sequence

$$a_n = \frac{n}{n + 3}$$

4. Find the 43d term of the sequence

$$a_n = \frac{n}{n + 3}$$

In Problems 5–7, find a_n in terms of n.

5. 8, 11, 14, 17, . . .

6. $\frac{1}{4}, \frac{1}{6}, \frac{1}{8}, \frac{1}{10}, \ldots$

7. $2, -4, 8, -16, \ldots$

In Problems 8 and 9, write each sum in expanded form.

8. $\displaystyle\sum_{k=0}^{3} (-1)^k k^3$

9. $\displaystyle\sum_{k=2}^{5} 2^{k-1} x^k$

10. Write the sum

$$1 - \frac{1}{2^3} + \frac{1}{3^3} - \frac{1}{4^3} + \frac{1}{5^3}$$

in summation notation.

Check Exercise 11-2 ■

Work the following problems without looking at any text examples. Show your work in the space provided. Write your answer in the answer column.

1. _____

2. _____

1. Which of the following are arithmetic sequences?

    (A) 1, 4, 9, 16, 25, . . .

3. _____

    (B) 0, 3, 6, 9, 12, . . .

    (C) 10, 7, 4, 1, -2, . . .

    (D) 1, 3, 6, 10, 15, . . .

4. _____

In Problems 2–8, let $a_1, a_2, a_3, \ldots, a_n, \ldots$ be an arithmetic sequence. Find the quantity indicated.

5. _____

    2. a_{12} when $a_1 = 5$, $d = 3$

6. _____

7. _____

8. _____

    3. d when $a_4 = 7$ and $a_{13} = 10$

9. _____

10. _____

    4. a_1 when $a_9 = 11$ and $a_{19} = 36$

    5. S_8 when $a_1 = 6$ and $d = -3$

6. S_5 when $a_5 = 5$ and $d = 5$

7. a_1 when $S_7 = 0$ and $d = -2$

8. n when $a_1 = -30$, $d = 3$, and $a_{30} = 57$

9. Evaluate $\displaystyle\sum_{k=1}^{31} 2k + 3$.

10. Find the sum of all odd integers from 11 to 121 inclusive.

Check Exercise 11-3 ■

ANSWER COLUMN

Work the following problems without looking at any text examples. Show your work in the space provided. Write your answer in the answer column.

1. _____

2. _____

3. _____

4. _____

5. _____

6. _____

7. _____

8. _____

9. _____

10. _____

1. Which of the following are geometric sequences?

    (A)   1, 3, 6, 10, 15, . . .
    (B)   1, 3, 9, 27, 81, . . .
    (C)   1, -3, -7, -11, -15, . . .
    (D)   2, -2, 2, -2, 2, . . .

In Problems 2–7, let $a_1, a_2, a_3, \ldots, a_n, \ldots$ be a geometric sequence. Find the quantity indicated.

2. a_4 when $a_1 = 160$ and $r = \frac{1}{2}$

3. r when $a_8 = 36$ and $a_{12} = \frac{4}{9}$

4. a_1 when $a_5 = 3$ and $a_8 = -24$

5. S_{10} when $a_1 = 5$ and $d = 2$

6. S_∞ when $a_1 = 5$ and $d = \frac{1}{2}$

7. S_∞ when $a_1 = 0.3$ and $d = 0.1$

8. Evaluate $\displaystyle\sum_{n=1}^{5} (-2)^n$.

9. Represent $0.23\overline{23}$ as the quotient of two integers.

10. If you pay a child a penny the first day and double the payment each day for 2 weeks, what would the last payment be?

Check Exercise 11-4 ■

ANSWER COLUMN

Work the following problems without looking at any text examples. Show your work in the space provided. Write your answer in the answer column.

1. _____

2. _____

3. _____

4. _____

5. _____

1. Evaluate $\dfrac{11!}{8!}$.

2. Evaluate $\begin{pmatrix} 7 \\ 3 \end{pmatrix}$.

3. Write $12 \cdot 13 \cdot 14 \cdot 15$ as the quotient of two factorials.

4. Expand $(a + b)^5$ by the binomial formula.

5. Find the 8th term in $(x - y)^{10}$.

■ Appendix

A Significant Digits

Most calculations involving problems of the real world deal with figures that are only approximate. It would therefore seem reasonable to assume that a final answer could not be any more accurate than the least accurate figure used in the calculation. This is an important point, since calculators tend to give the impression that greater accuracy is achieved than is warranted.

Suppose we wish to compute the length of the diagonal of a rectangular field from measurements of its sides of 237.8 meters and 61.3 meters. Using the Pythagorean theorem and a calculator, we find

$$d = \sqrt{237.8^2 + 61.3^2}$$
$$= 245.573\,878 \ldots$$

61.3 meters
237.8 meters

The calculator answer suggests an accuracy that is not justified. What accuracy is justified? To answer this question, we introduce the idea of **significant digits**.

The measurement 61.3 meters indicates that the measurement was made to the nearest tenth of a meter; that is, the actual width is between 61.25 and 61.35 meters. The number 61.3 has three significant digits. If we had written, instead, 61.30 meters as the width, then the actual width would be between 61.295 and 61.305 meters, and our measurement, 61.30 meters, would have four significant digits.

The number of significant digits in a number is found by counting the digits from left to right, starting with the first nonzero digit and ending with the last digit present.

The significant digits in the following numbers are underlined:

719.37   82,395   5.600   0.000 830   0.000 08

The definition takes care of all cases except one. Consider, for example, the number 7,800. It is not clear whether the number has been rounded to the hundreds place, the tens place, or the units place. This ambiguity can be resolved by writing this type of number in scientific notation. Thus,

$7.8 \ \ \times 10^3$ has two significant digits
$7.80 \ \times 10^3$ has three significant digits
7.800×10^3 has four significant digits

All three are equal to 7,800 when written without powers of 10.

In calculations involving multiplication, division, powers, and roots, we adopt the following convention:

We will round off the answer to match the number of significant digits in the number with the least number of significant digits used in the calculation.

Thus, in computing the length of the diagonal of the field, we would write the answer to three significant digits because the width, the least accurate of the two numbers involved, has three significant digits:

$$d = 246 \text{ meters} \quad \text{Three significant digits}$$

One final note: in rounding a number that is exactly halfway between a larger and a smaller number, we will use the convention of making the final result even.

EXAMPLE 1   Round each number to three significant digits:

(A) 43.0690   **(B)** 48.05   **(C)** 48.15   **(D)** $8.017\,632 \times 10^{-3}$

Solution   **(A)** 43.1

     (B) 48.0 ⎫   Use the convention of making the digit before the
     (C) 48.2 ⎭   5 even if it is odd or leaving it alone if it is even.

     (D) 8.02×10^{-3}

PROBLEM 1   Round each number to three significant digits:

(A) 3.1495   **(B)** 0.004 135   **(C)** 32,450   **(D)** $4.314\,764\,09 \times 10^{12}$

Solution   **(A)**   **(B)**   **(C)**   **(D)**

Solution to Matched Problem

1. **(A)** 3.15   **(B)** 0.004 14   **(C)** 32,400   **(D)** 4.31×10^{12}

B   Table Evaluation of Common and Natural Logarithms

■ Common Logarithms
■ Natural Logarithms

■ Common Logarithms

We now show how Table II in the back of this book can be used to approximate common logarithms. Recalling that any decimal fraction can be written in scientific notation (see Section 5-2), we see that

$$\log_{10} 33,800 = \log_{10}(3.38 \times 10^4)$$
$$= \log_{10} 3.38 + \log_{10} 10^4$$
$$= \log_{10} 3.38 + 4$$

and that

$$\log_{10} 0.003\,51 = \log_{10}(3.51 \times 10^{-3})$$
$$= \log_{10} 3.51 + \log_{10} 10^{-3}$$
$$= \log_{10} 3.51 - 3$$

In general:

If a number N is written in scientific notation

$$N = r \times 10^k \qquad 1 \le r < 10, \quad k \text{ an integer}$$

then

$$\log_{10} N = \log_{10}(r \times 10^k) \quad {\scriptstyle \log = \log_{10}}$$
$$= \log_{10} r + \log_{10} 10^k$$
$$= \underset{\text{Mantissa}}{\log_{10} r} + \underset{\text{Characteristic}}{k}$$

Thus, if common logarithms of r, $1 \le r < 10$, are given in a table, we will be able to approximate the common logarithm of any positive decimal fraction to the accuracy of the table.

Using methods of advanced mathematics, a table of common logarithms of numbers from 1 to 10 can be computed to any decimal accuracy desired. Table II in the back of this book is such a table to four-decimal-place accuracy. It is useful to remember that if x is between 1 and 10, then $\log x$ is between 0 and 1 (Figure 1).

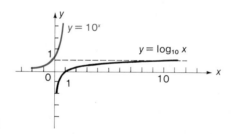

FIGURE 1

To illustrate the use of Table II, a small portion of it is reproduced here. To find log 3.47, for example, we first locate 3.4 under the x heading; then we move across to the column headed 7, where we find .5403. Thus, log 3.47 = 0.5403.[†]

TABLE II   (SMALL PORTION)

x	0	1	2	3	4	5	6	7	8	9
3.2	0.5051	0.5065	0.5079	0.5092	0.5105	0.5119	0.5132	0.5145	0.5159	0.5172
3.3	0.5185	0.5198	0.5211	0.5224	0.5237	0.5250	0.5263	0.5276	0.5289	0.5302
3.4	0.5315	0.5328	0.5340	0.5353	0.5366	0.5378	0.5391	**0.5403**	0.5416	0.5428
3.5	0.5441	0.5453	0.5465	0.5478	0.5490	0.5502	0.5514	0.5527	0.5539	0.5551

[†] Throughout the rest of this Appendix we will use $=$ in place of $\approx$ in many places, realizing that values are only approximately equal. Occasionally, we will use $\approx$ when a special emphasis is desired.

Now let us finish finding the common logarithms of 33,800 and 0.003 51.

EXAMPLE 1 **(A)** $\log 33{,}800 = \log(3.38 \times 10^4)$

$$= \log 3.38 + \log 10^4 \quad \text{Use Table II.}$$
$$= 0.5289 + 4$$
$$= 4.5289$$

(B) $\log 0.003\ 51 = \log(3.51 \times 10^{-3})$

$$= \log 3.51 + \log 10^{-3} \quad \text{Use Table II.}$$
$$= 0.5453 - 3$$
$$= -2.4547$$

PROBLEM 1   Use Table II to find: **(A)** $\log 328{,}000$ **(B)** $\log 0.000\ 342$

Solution   **(A)**   **(B)**

Now let us reverse the problem—that is, given the log of a number, find the number. To find the number, we first write the log of the number in the form

$$m + c$$

where m (the mantissa) is a nonnegative number between 0 and 1, and c (the characteristic) is an integer. Then reverse the process illustrated in Example 1.

EXAMPLE 2 **(A)** If $\log x = 2.5224$, find x.   **(B)** If $\log x = 0.5172 - 4$, find x.
(C) If $\log x = -4.4685$, find x.

Solution   **(A)** $\log x = 2.5224$

Write 2.5224 in the form $m + c$, $0 \le m < 1$ and c an integer. Look for 0.5224 in the body of Table II. Thus, we see that $0.5224 = \log 3.33$.

$$= 0.5224 + 2$$
$$= \log 3.33 + \log 10^2$$
$$= \log(3.33 \times 10^2)$$

Thus,

$$x = 3.33 \times 10^2 \quad \text{or} \quad 333$$

(B) $\log x = 0.5172 - 4$

$$= \log 3.29 + \log 10^{-4} \quad \text{Use Table II.}$$
$$= \log(3.29 \times 10^{-4})$$

Thus,

$$x = 3.29 \times 10^{-4} \quad \text{or} \quad 0.000\ 329$$

(C) $\log x = -4.4685$

Convert -4.4685 to $m + c$ form with $0 \le m < 1$ by adding and subtracting 5:

$$= 0.5315 - 5$$
$$= \log 3.40 + \log 10^{-5}$$
$$= \log(3.40 \times 10^{-5})$$

$$\begin{array}{r} 5.0000 - 5 \\ -4.4685 \\ \hline 0.5315 - 5 \end{array}$$

Thus,

$$x = 3.40 \times 10^{-5} \quad \text{or} \quad 0.000\ 0340$$

PROBLEM 2   Use Table II to find x if:

   (A) $\log x = 5.5378$ **(B)** $\log x = 0.5289 - 3$ **(C)** $\log x = -2.4921$

Solution   **(A)**   **(B)**   **(C)**

What if a number has more significant digits than are included in a table? We then use the nearest table value, or, if more accuracy is desired, we use a calculator. Another alternative, called *linear interpolation*, is discussed in Appendix C.

EXAMPLE 3   **(A)** $\log 32{,}683 \approx \log(3.27 \times 10^4)$ Round 3.2683 to the nearest table value—that is, to 3.27.

$$= \log 3.27 + \log 10^4$$

$$= 0.5145 + 4$$

$$= 4.5145$$

   (B) To find x if $\log x = 0.5241 - 3$, we observe (in Table II) that 0.5241 is between 0.5237 and 0.5250, but it is closer to 0.5237. Thus, we write

$$\log x = 0.5241 - 3$$ Since 0.5241 is not in the body of the table, select the value in the table that is closest; that is, select 0.5237.

$$\approx 0.5237 - 3$$

$$= \log 3.34 + \log 10^{-3}$$

$$= \log(3.34 \times 10^{-3})$$

Thus,

$$x = 3.34 \times 10^{-3} \quad \text{or} \quad 0.003\ 34$$

PROBLEM 3   Find:

   (A) $\log 0.034\ 319$ **(B)** x if $\log x = 6.5473$ **(C)** x if $\log x = -4.4942$

Solution   **(A)**   **(B)**   **(C)**

■ Natural Logarithms

Approximating natural logarithms using Table III in the back of this book proceeds in much the same way as finding common logarithms using Table II, except the arithmetic is a little more complicated. A couple of examples will illustrate the process.

EXAMPLE 4   **(A)** $\ln 52,400 = \ln(5.24 \times 10^4)$ $\ln = \log_e$

$$= \ln 5.24 + 4 \ln 10$$

Note that ln 5.24 is read out of the main part of Table III, and 4 ln 10 is obtained from the list at the top of the table.

$$= 1.6563 + 9.2103$$
$$= 10.8666$$

(B) $\ln 0.002\,78 = \ln(2.78 \times 10^{-3})$

$$= \ln 2.78 - 3 \ln 10$$
$$= 1.0225 - 6.9078$$
$$= -5.8853$$

PROBLEM 4   Use Table III to find:

(A) $\ln 0.000\,683$ **(B)** $\ln 328,000$ **(C)** $\ln 23,582$

Solution   **(A)**   **(B)**   **(C)**

Solutions to Matched Problems

1. (A) $\log_{10} 328,000 = \log_{10}(3.28 \times 10^5)$

$$= \log_{10} 3.28 + \log_{10} 10^5$$

$$= 0.5159 + 5 = 5.5159$$

(B) $\log_{10} 0.000\,342 = \log_{10}(3.42 \times 10^{-4})$

$$= \log_{10} 3.42 + \log_{10} 10^{-4}$$

$$= 0.5340 - 4 = -3.4660$$

2. (A) $\log_{10} x = 5.5378$

$$= 0.5378 + 5$$

$$= \log_{10} 3.45 + \log_{10} 10^5$$
$$= \log_{10}(3.45 \times 10^5)$$

$$x = 3.45 \times 10^5 \quad \text{or} \quad 345,000$$

2. (B) $\log_{10} x = 0.5289 - 3$

$$= \log_{10} 3.38 + \log_{10} 10^{-3}$$
$$= \log_{10}(3.38 \times 10^{-3})$$

$\quad x = 3.38 \times 10^{-3} \quad$ or $\quad 0.003\ 38$

(C) $\log_{10} x = -2.4921 \qquad$ Add and subtract 3: $\quad 3.0000 - 3$
$$= 0.5079 - 3 \qquad\qquad\qquad\qquad \frac{-2.4921}{0.5079 - 3}$$

$$= \log_{10} 3.22 + \log_{10} 10^{-3}$$
$$= \log_{10}(3.22 \times 10^{-3})$$

$\quad x = 3.22 \times 10^{-3} \quad$ or $\quad 0.003\ 22$

3. (A) $\log_{10} 0.034\ 319 \approx \log_{10}(3.43 \times 10^{-2}) \quad$ Nearest table value

$$= \log_{10} 3.43 + \log_{10} 10^{-2}$$

$$= 0.5353 - 2 = 1.4647$$

(B) $\log_{10} x = 6.5473$
$$= 0.5473 + 6$$

$$\approx 0.5478 + 6 \quad \text{Nearest table value}$$
$$= \log_{10} 3.53 + \log_{10} 10^{6}$$
$$= \log_{10}(3.53 \times 10^{6})$$

$\quad x = 3.53 \times 10^{6} \quad$ or $\quad 3{,}530{,}000$

(C) $\qquad\qquad\qquad\quad \mathbf{5.0000 - 5} \quad$ Add and subtract 5.
$$\log_{10} x = \frac{-4.4942}{}$$
$$= \quad 0.5058 - 5$$

$$\approx 0.5051 - 5 \quad \text{Nearest table value}$$
$$= \log_{10} 3.20 + \log_{10} 10^{-5}$$
$$= \log_{10}(3.20 \times 10^{-5})$$

$\quad x = 3.20 \times 10^{-5} \quad$ or $\quad 0.000\ 032\ 0$

4. (A) $\ln 0.000\ 683 = \ln(6.83 \times 10^{-4})$

$$= \ln 6.83 - 4 \ln 10$$

$$= 1.9213 - 9.2103$$
$$= -7.2890$$

(B) $\ln 328{,}000 = \ln(3.28 \times 10^{5})$

$$= \ln 3.28 + 5 \ln 10$$

$$= 1.1878 + 11.5130$$
$$= 12.7008$$

(C) $\ln 23{,}582 \approx \ln(2.36 \times 10^{4}) \quad$ Nearest table value

$$= \ln 2.36 + 4 \ln 10$$

$$= 0.8587 + 9.2103$$
$$= 10.0690$$

Practice Exercise Appendix B ■

Work odd-numbered problems first, check answers, and then work even-numbered problems in areas of weakness. Answers to all problems are in the back of the book. Make every effort to work a problem yourself before you look at an answer.

A   *Use Table II to find each.*

1. log 7.29 _____   **2.** log 6.37 _____

3. log 2,040 _____   **4.** log 327 _____

5. log 0.0413 _____   **6.** log 0.000 927

B   *Use the nearest values in Table II to approximate each of the following.*

7. log 304,918 _____   **8.** log 82,734 _____

9. log 0.004 769 _____   **10.** log 0.061 94 _____

Find x using the nearest values in Table II.

11. log x = 7.437 15 _____   **12.** log x = 9.113 64 _____

13. log x = −4.8013 _____   **14.** log x = −3.4128 _____

C   *Find each of the following using the nearest values in Table III.*

15. ln 2.35 _____   **16.** ln 7.02 _____

The Check Exercise for
this section is on
page 745. **17.** ln 603,517 _____   **18.** ln 5,233 _____

19. ln 0.003 1687 _____   **20.** ln 0.071 33 _____

C   Linear Interpolation for Common Logarithms

A printed table is necessarily limited to a finite number of entries. The logarithmic function

$$y = \log x \qquad 1 \leq x < 10$$

is defined for an infinite number of values—all real numbers between 1 and 10. What, then, do we do about values of x not in a table? How do we find, for example, log 3.276? If a certain amount of accuracy can be sacrificed, we can round 3.276 to the closest entry in the table and proceed as before. Thus,

$$\log 3.276 \approx \log 3.28 = 0.5159$$

We can do better than this, however, without too much additional work, by using a process called **linear interpolation** (see Figure 1).

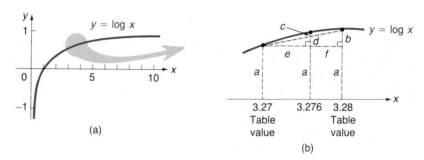

FIGURE 1

In Figure 1, we would like to find $a + d + c$. Using proportional parts of similar triangles, we will be able to find d without difficulty, since a and $a + b$ can be determined from the table. Hence, we will settle for $a + d$ as an approximation for $a + d + c$. This approximation is better than what appears in the figure, since the curve is distorted in the drawing for clarity. The logarithmic curve is actually much flatter.

Now to the process of linear interpolation. We find d using the proportion

$$\frac{d}{b} = \frac{e}{e + f}$$

We organize the work as follows for convenient computation:

$$0.010 \left\{ \begin{matrix} & x & \log x \\ & 3.280 & 0.5159 \\ 0.006 \left\{ \begin{matrix} 3.276 & n \\ 3.270 & 0.5145 \end{matrix} \right\} d \end{matrix} \right\} 0.0014$$

Here $b = 0.0014$, $e = 0.006$, and $e + f = 0.010$. Thus,

$$\frac{d}{0.0014} = \frac{0.006}{0.010}$$
$$d = 0.0008$$

Hence,

$$\log 3.276 = 0.5145 + 0.0008 = 0.5153$$

In practice, the linear interpolation process is carried out by means of a few key operational steps (often done mentally), as indicated in the next example. Notice how decimal points have been dropped to simplify the arithmetic. This is convenient, and no harm is done as long as the decimal points are properly reintroduced at the end of the calculation. If in doubt, proceed with all the decimal points, as in the example above.

EXAMPLE 1   Use linear interpolation to find log 3,514.

Solution $\log 3{,}514 = \log(3.514 \times 10^3) = (\log 3.514) + 3$

$$
10\left\{\begin{array}{c}
\begin{array}{cc}
x & \log x \\
3.520 & 0.5465
\end{array} \\
4\left\{\begin{array}{cc}
3.514 & n \\
3.510 & 0.5453
\end{array}\right\}d
\end{array}\right\}12
$$

$$\frac{4}{10} = \frac{d}{12}$$

$$d = 5$$

Thus,

$$\log 3.514 = 0.5453 + 0.0005 = 0.5458$$

and

$$\log 3{,}514 = 0.5458 + 3 = 3.5458$$

PROBLEM 1   Use linear interpolation to find log 326.6.

Solution

The linear interpolation process just described is also used to find x given log x. The following example illustrates the procedure.

EXAMPLE 2   Use linear interpolation to find x given log x = 2.5333.

Solution $\log x = 2.5333$

                $= 0.5333 + 2$

$$
10\left\{\begin{array}{c}
\begin{array}{cc}
x & \log x \\
3.420 & 0.5340
\end{array} \\
d\left\{\begin{array}{cc}
n & 0.5333 \\
3.410 & 0.5328
\end{array}\right\}5
\end{array}\right\}12
$$

$$\frac{d}{10} = \frac{5}{12}$$

$$d = 4$$

$$n = 3.410 + 0.004 = 3.414$$

Thus,

$$x = 3.414 \times 10^2 \quad \text{or} \quad 341.4$$

PROBLEM 2   Use linear interpolation to find x given log x = 7.5230.

Solution

Solutions to Matched
Problems

1. $\log 326.6 = \log(3.266 \times 10^2)$

$= \log 3.266 + \log 10^2$
$= 0.5140 + 2$

$= 2.5140$

<u>Linear Interpolation</u>

	x	log *x*
	3.270	0.5145
	3.266	*n*
	3.260	0.5132

$10 \left\{ 6 \left\{ \begin{array}{c} 3.270 \\ 3.266 \\ 3.260 \end{array} \right. \right.$ — $\left. \begin{array}{c} 0.5145 \\ n \\ 0.5132 \end{array} \right\} d \right\} 13$

$\dfrac{d}{13} = \dfrac{6}{10}$

$d \approx 8$

$n = 0.5132 + 0.0008 = 0.5140$

2. $\log x = 7.5230$
$= 0.5230 + 7$

$= \log 3.335 + \log 10^7$
$= \log(3.335 \times 10^7)$

$x = 3.335 \times 10^7$

<u>Linear Interpolation</u>

	x	log *x*
	3.340	0.5237
	n	0.5230
	3.330	0.5224

$10 \left\{ d \left\{ \begin{array}{c} 3.340 \\ n \\ 3.330 \end{array} \right. \right.$ — $\left. \begin{array}{c} 0.5237 \\ 0.5230 \\ 0.5224 \end{array} \right\} 6 \right\} 13$

$\dfrac{d}{10} = \dfrac{6}{13}$

$d \approx 5$

$n = 3.330 + 0.005 = 3.335$

Practice Exercise Appendix C ■

Work odd-numbered problems first, check answers, and then work even-numbered problems in areas of weakness. Answers to all problems are in the back of the book. Make every effort to work a problem yourself before you look at an answer.

Use linear interpolation to find each logarithm.

1. log 2.317 _____

2. log 5.143 _____

3. log 703,400 _____

4. log 28,430 _____

5. log 65.03 _____

6. log 20.35 _____

7. log 0.004 006 _____

8. log 0.037 13 _____

9. log 0.9008 _____

10. log 0.6413 _____

11. log 692,300 _____

12. log 84,660 _____

Use linear interpolation to find x.

13. log x = 0.7163 _____

14. log x = 0.4085 _____

15. log x = 5.5458 _____

16. log x = 2.4735 _____

17. log x = 3.4303 _____

18. log x = 1.9141 _____

19. log x = 0.6038 − 3 _____

20. log x = 0.2177 − 1 _____

The Check Exercise for this section is on page 747.

21. log x = 0.8392 − 1 _____

22. log x = 0.8509 − 4 _____

23. log x = −0.8315 _____

24. log x = −2.6651 _____

D Determinants

■ Second-Order Determinants
■ Third-Order Determinants

Determinants arise quite naturally in many areas in mathematics, including the solving of linear systems, vector analysis, calculus, and so on. We will consider a few of their uses in this and the next appendix.

■ Second-Order Determinants

A square array of four real numbers, such as

$$\begin{vmatrix} 2 & -3 \\ 5 & 1 \end{vmatrix}$$

is called a determinant of order 2. (It is important to note that the array of numbers is between parallel lines and not square brackets. If square brackets are used, then the symbol has another meaning.) The determinant shown here has two **rows** and two **columns**—rows are across and columns are up and down. Each number in the determinant is called an **element** of the determinant.

In general, we can symbolize a **second-order determinant** as follows:

$$\begin{vmatrix} a_{11} & a_{12} \\ a_{21} & a_{22} \end{vmatrix}$$

where we use a single letter with a **double subscript** to facilitate generalization to higher-order determinants. The first subscript number indicates the row in which the element lies, and the second subscript number indicates the column. Thus, a_{21} is the element in the second row and first column and a_{12} is the element in the first row and second column. Each second-order determinant represents a real number given by this formula:

$$\begin{vmatrix} a_{11} & a_{12} \\ a_{21} & a_{22} \end{vmatrix} = a_{11}a_{22} - a_{21}a_{12}$$

EXAMPLE 1 $\quad \begin{vmatrix} -1 & 2 \\ -3 & 4 \end{vmatrix} = (-1)(-4) - (-3)(2) = 4 - (-6) = 10$

PROBLEM 1 $\quad$ Find: $\begin{vmatrix} 3 & -5 \\ 4 & -2 \end{vmatrix}$

Solution

■ Third-Order Determinants

A determinant of order 3 is a square array of nine elements and represents a real number given by the formula

$$\begin{vmatrix} a_{11} & a_{12} & a_{13} \\ a_{21} & a_{22} & a_{23} \\ a_{31} & a_{32} & a_{33} \end{vmatrix} = a_{11}a_{22}a_{33} - a_{11}a_{32}a_{23} + a_{21}a_{32}a_{13} - a_{21}a_{12}a_{33} + a_{31}a_{12}a_{23} - a_{31}a_{22}a_{13} \tag{1}$$

Note that each term in the expansion on the right of (1) contains exactly one element from each row and each column. Don't panic! You do not need to memorize Formula (1). After we introduce the ideas of "minor" and "cofactor," we will state a theorem that can be used to obtain the same result with much less memory strain.

The **minor of an element** in a third-order determinant is a second-order determinant obtained by deleting the row and column that contain the element. For example, in the determinant in Formula (1)

$$\text{Minor of } a_{23} = \begin{vmatrix} a_{11} & a_{12} & a_{13} \\ a_{21} & a_{22} & a_{23} \\ a_{31} & a_{32} & a_{33} \end{vmatrix} = \begin{vmatrix} a_{11} & a_{12} \\ a_{31} & a_{32} \end{vmatrix}$$

$$\text{Minor of } a_{32} = \begin{vmatrix} a_{11} & a_{12} & a_{13} \\ a_{21} & a_{22} & a_{23} \\ a_{31} & a_{32} & a_{33} \end{vmatrix} = \begin{vmatrix} a_{11} & a_{13} \\ a_{21} & a_{23} \end{vmatrix}$$

A quantity closely associated with the minor of an element is the cofactor of an element. The **cofactor of an element** a_{ij} (from the ith row and jth column) is the product of the minor of a_{ij} and $(-1)^{i+j}$. That is:

$$\boxed{\text{Cofactor of } a_{ij} = (-1)^{i+j}(\text{Minor of } a_{ij})}$$

Thus, a cofactor of an element is nothing more than a signed minor. The sign is determined by raising -1 to a power that is the sum of the numbers indicating the row and column in which the element lies. Note that $(-1)^{i+j}$ is -1 if $i + j$ is odd and 1 if $i + j$ is even. Referring again to the determinant in Formula (1),

$$\text{Cofactor of } a_{23} = (-1)^{2+3}\begin{vmatrix} a_{11} & a_{12} \\ a_{31} & a_{32} \end{vmatrix} = -\begin{vmatrix} a_{11} & a_{12} \\ a_{31} & a_{32} \end{vmatrix}$$

$$\text{Cofactor of } a_{11} = (-1)^{1+1}\begin{vmatrix} a_{22} & a_{23} \\ a_{32} & a_{33} \end{vmatrix} = \begin{vmatrix} a_{22} & a_{23} \\ a_{32} & a_{33} \end{vmatrix}$$

EXAMPLE 2   Find the cofactor of -2 and 5 in the determinant

$$\begin{vmatrix} -2 & 0 & 3 \\ 1 & -6 & 5 \\ -1 & 2 & 0 \end{vmatrix}$$

Solution $$\text{Cofactor of } -2 = (-1)^{1+1}\begin{vmatrix} -6 & 5 \\ 2 & 0 \end{vmatrix} = \begin{vmatrix} -6 & 5 \\ 2 & 0 \end{vmatrix}$$

$$= (-6)(0) - (2)(5) = -10$$

$$\text{Cofactor of } 5 = (-1)^{2+3}\begin{vmatrix} -2 & 0 \\ -1 & 2 \end{vmatrix} = -\begin{vmatrix} -2 & 0 \\ -1 & 2 \end{vmatrix}$$

$$= -[(-2)(2) - (-1)(0)] = 4$$

PROBLEM 2   Find the cofactors of 2 and 3 in the determinant in Example 2.

Solution

Note: The sign in front of the minor, $(-1)^{i+j}$, can be determined mechanically by using a checkerboard pattern of plus and minus signs over the determinant, starting with $+$ in the upper left-hand corner:

$$\begin{array}{ccc} + & - & + \\ - & + & - \\ + & - & + \end{array}$$

Use either the checkerboard or the exponent method, whichever is easier for you, to determine the sign in front of the minor.

Now we are ready for the central theorem of this section. It will provide us with an efficient means of evaluating third-order determinants. Moreover, it is worth noting that the theorem generalizes completely to include determinants of arbitrary order.

THEOREM 1   The value of a determinant of order 3 is the sum of three products obtained by multiplying each element of any one row (or each element of any one column) by its cofactor.

To prove this theorem we must show that the expansions indicated by the theorem for any row or any column (six cases) produce the expression on the right of Formula (1). Proofs of special cases of this theorem are left to the C-level problems of Exercise D.

EXAMPLE 3   Evaluate by expanding **(A)** by the first row and **(B)** by the second column:

$$\begin{vmatrix} 2 & -2 & 0 \\ -3 & 1 & 2 \\ 1 & -3 & -1 \end{vmatrix}$$

Solution   **(A)** $\begin{vmatrix} 2 & -2 & 0 \\ -3 & 1 & 2 \\ 1 & -3 & -1 \end{vmatrix}$ Expand by the first row.

$$= a_{11}\left(\begin{matrix}\text{Cofactor}\\\text{of } a_{11}\end{matrix}\right) + a_{12}\left(\begin{matrix}\text{Cofactor}\\\text{of } a_{12}\end{matrix}\right) + a_{13}\left(\begin{matrix}\text{Cofactor}\\\text{of } a_{13}\end{matrix}\right)$$

$$= 2\left((-1)^{1+1}\begin{vmatrix} 1 & 2 \\ -3 & -1 \end{vmatrix}\right) + (-2)\left((-1)^{1+2}\begin{vmatrix} -3 & 2 \\ 1 & -1 \end{vmatrix}\right) + 0$$

$$= (2)(1)[(1)(-1) - (-3)(2)] + (-2)(-1)[(-3)(-1) - (1)(2)]$$

$$= (2)(5) + (2)(1) = 12$$

(B) $\begin{vmatrix} 2 & -2 & 0 \\ -3 & 1 & 2 \\ 1 & -3 & -1 \end{vmatrix}$ Expand by the second column.

$$= a_{12} \begin{pmatrix} \text{Cofactor} \\ \text{of } a_{12} \end{pmatrix} + a_{22} \begin{pmatrix} \text{Cofactor} \\ \text{of } a_{22} \end{pmatrix} + a_{32} \begin{pmatrix} \text{Cofactor} \\ \text{of } a_{32} \end{pmatrix}$$

$$= (-2) \left((-1)^{1+2} \begin{vmatrix} -3 & 2 \\ 1 & -1 \end{vmatrix} \right) + (1) \left((-1)^{2+2} \begin{vmatrix} 2 & 0 \\ 1 & -1 \end{vmatrix} \right)$$

$$+ (-3) \left((-1)^{3+2} \begin{vmatrix} 2 & 0 \\ -3 & 2 \end{vmatrix} \right)$$

$$= (-2)(-1)[(-3)(-1) - (1)(2)] + (1)(1)[(2)(-1) - (1)(0)]$$

$$+ (-3)(-1)[(2)(2) - (-3)(0)]$$

$$= (2)(1) + (1)(-2) + (3)(4)$$

$$= 12$$

PROBLEM 3   Evaluate by expanding **(A)** by the first row and **(B)** by the third column:

$$\begin{vmatrix} 2 & 1 & -1 \\ -2 & -3 & 0 \\ -1 & 2 & 1 \end{vmatrix}$$

Solution   **(A)**

(B)

It should now be clear that we can greatly reduce the work involved in evaluating a determinant by choosing to expand by a row or column with the greatest number of 0's.

Where are determinants used? Many equations and formulas have particularly simple and compact representations in determinant form that are easily remembered. See, for example, Problems 39–42 in Exercise D and Cramer's rule in Appendix E.

Solutions to Matched
Problems

1. $\begin{vmatrix} 3 & -5 \\ 4 & -2 \end{vmatrix} = (3)(-2) - (4)(-5) = (-6) - (-20) = 14$

2. Cofactor of $2 = (-1)^{3+2} \begin{vmatrix} -2 & 3 \\ 1 & 5 \end{vmatrix} = (-1)[(-2)(5) - (1)(3)]$

$= (-1)[(-10) - (3)] = 13$

Cofactor of $3 = (-1)^{1+3} \begin{vmatrix} 1 & -6 \\ -1 & 2 \end{vmatrix} = (+1)[(1)(2) - (-1)(-6)]$

$= (2) - (6) = -4$

3. **(A)** $\begin{vmatrix} 2 & 1 & -1 \\ -2 & -3 & 0 \\ -1 & 2 & 1 \end{vmatrix} = (2)\left((-1)^{1+1}\begin{vmatrix} -3 & 0 \\ 2 & 1 \end{vmatrix}\right) + (1)\left((-1)^{1+2}\begin{vmatrix} -2 & 0 \\ -1 & 1 \end{vmatrix}\right)$

$+ (-1)\left((-1)^{1+3}\begin{vmatrix} -2 & -3 \\ -1 & 2 \end{vmatrix}\right)$

$= (2)(1)[(-3)(1) - 0] + (1)(-1)[(-2)(1) - 0]$

$+ (-1)(1)[(-2)(2) - (-1)(-3)]$

$= 2[(-3) - 0] + (-1)[(-2) - 0]$

$+ (-1)[(-4) - (3)]$

$= (-6) + (2) + (7) = 3$

(B) $\begin{vmatrix} 2 & 1 & -1 \\ -2 & -3 & 0 \\ -1 & 2 & 1 \end{vmatrix} = (-1)\left((-1)^{1+3}\begin{vmatrix} -2 & -3 \\ -1 & 2 \end{vmatrix}\right)$

$+ 0 + (1)\left((-1)^{3+3}\begin{vmatrix} 2 & 1 \\ -2 & -3 \end{vmatrix}\right)$

$= (-1)(1)[(-2)(2) - (-1)(-3)]$

$+ (1)(1)[(2)(-3) - (-2)(1)]$

$= (-1)[(-4) - (3)] + (1)[(-6) - (-2)]$

$= 7 + (-4) = 3$

Practice Exercise Appendix D ■

Work odd-numbered problems first, check answers, and then work even-numbered problems in areas of weakness. Answers to all problems are in the back of the book. Make every effort to work a problem yourself before you look at an answer.

A   Evaluate each second-order determinant.

1. $\begin{vmatrix} 2 & 4 \\ 3 & -1 \end{vmatrix}$ _____   2. $\begin{vmatrix} 2 & 2 \\ -3 & 1 \end{vmatrix}$ _____

3. $\begin{vmatrix} 5 & -4 \\ -2 & 2 \end{vmatrix}$ _____   4. $\begin{vmatrix} 6 & -2 \\ -1 & -3 \end{vmatrix}$ _____

5. $\begin{vmatrix} 3 & -3.1 \\ -2 & 1.2 \end{vmatrix}$ _____   6. $\begin{vmatrix} -1.4 & 3 \\ -0.5 & -2 \end{vmatrix}$ _____

Given the determinant

$$\begin{vmatrix} a_{11} & a_{12} & a_{13} \\ a_{21} & a_{22} & a_{23} \\ a_{31} & a_{32} & a_{33} \end{vmatrix}$$

write the minor of each of the following elements.

7. a_{11} _____   8. a_{33} _____

9. a_{23} _____   10. a_{22} _____

Write the cofactor of each of the following elements.

11. a_{11} _____   12. a_{33} _____

13. a_{23} _____   14. a_{22} _____

Given the determinant

$$\begin{vmatrix} -2 & 3 & 0 \\ 5 & 1 & -2 \\ 7 & -4 & 8 \end{vmatrix}$$

write the minor of each of the following elements. (Leave your answer in determinant form.)

15. a_{11} _____   16. a_{22} _____

17. a_{32} _____   18. a_{21} _____

Write the cofactor of each of the following elements and evaluate each.

19. a_{11} _____   20. a_{22} _____

21. a_{32} _____   22. a_{21} _____

B Evaluate each of the following determinants using cofactors.

23. $\begin{vmatrix} 1 & 0 & 0 \\ -2 & 4 & 3 \\ 5 & -2 & 1 \end{vmatrix}$ _____   24. $\begin{vmatrix} 2 & -3 & 5 \\ 0 & -3 & 1 \\ 0 & 6 & 2 \end{vmatrix}$ _____

25. $\begin{vmatrix} 0 & 1 & 5 \\ 3 & -7 & 6 \\ 0 & -2 & -3 \end{vmatrix}$ _____   26. $\begin{vmatrix} 4 & -2 & 0 \\ 9 & 5 & 4 \\ 1 & 2 & 0 \end{vmatrix}$ _____

27. $\begin{vmatrix} 4 & -4 & 6 \\ 2 & 8 & -3 \\ 0 & -5 & 0 \end{vmatrix}$ _____   28. $\begin{vmatrix} 3 & -2 & -8 \\ -2 & 0 & -3 \\ 1 & 0 & -4 \end{vmatrix}$ _____

29. $\begin{vmatrix} -1 & 2 & -3 \\ -2 & 0 & -6 \\ 4 & -3 & 2 \end{vmatrix}$ _____   30. $\begin{vmatrix} 0 & 2 & -1 \\ -6 & 3 & 1 \\ 7 & -9 & -2 \end{vmatrix}$ _____

31. $\begin{vmatrix} 1 & 4 & 1 \\ 1 & 1 & -2 \\ 2 & 1 & -1 \end{vmatrix}$ _____

32. $\begin{vmatrix} 3 & 2 & 1 \\ -1 & 5 & 1 \\ 2 & 3 & 1 \end{vmatrix}$ _____

33. $\begin{vmatrix} 1 & 4 & 3 \\ 2 & 1 & 6 \\ 3 & -2 & 9 \end{vmatrix}$ _____

34. $\begin{vmatrix} 4 & -6 & 3 \\ -1 & 4 & 1 \\ 5 & -6 & 3 \end{vmatrix}$ _____

C *Assuming that Theorem 1 applies to determinants of arbitrary order, use it to evaluate the following fourth- and fifth-order determinants.*

35. $\begin{vmatrix} 0 & 1 & 0 & 1 \\ 2 & 4 & 7 & 6 \\ 0 & 3 & 0 & 1 \\ 0 & 6 & 2 & 5 \end{vmatrix}$ _____

36. $\begin{vmatrix} 2 & 6 & 1 & 7 \\ 0 & 3 & 0 & 0 \\ 3 & 4 & 2 & 5 \\ 0 & 9 & 0 & 2 \end{vmatrix}$ _____

37. $\begin{vmatrix} 2 & 0 & 0 & 0 & 0 \\ 0 & 3 & 0 & 0 & 0 \\ 0 & 0 & 2 & 0 & 0 \\ 0 & 0 & 0 & 1 & 0 \\ 0 & 0 & 0 & 0 & 4 \end{vmatrix}$ _____

38. $\begin{vmatrix} -2 & 0 & 0 & 0 & 0 \\ 9 & -1 & 0 & 0 & 0 \\ 2 & 1 & 3 & 0 & 0 \\ -1 & 4 & 2 & 2 & 0 \\ 7 & -2 & 3 & 5 & 5 \end{vmatrix}$ _____

39. Show that

$$\begin{vmatrix} x & y & 1 \\ 2 & 3 & 1 \\ -1 & 2 & 1 \end{vmatrix} = 0$$

is the equation of a line that passes through $(2, 3)$ and $(-1, 2)$.

40. Show that

$$\begin{vmatrix} x & y & 1 \\ x_1 & y_1 & 1 \\ x_2 & y_2 & 1 \end{vmatrix} = 0$$

is the equation of a line that passes through (x_1, y_1) and (x_2, y_2).

41. In analytic geometry it is shown that the area of a triangle with vertices (x_1, y_1), (x_2, y_2), and (x_3, y_3) is the absolute value of

$$\frac{1}{2} \begin{vmatrix} x_1 & y_1 & 1 \\ x_2 & y_2 & 1 \\ x_3 & y_3 & 1 \end{vmatrix}$$

Use this result to find the area of a triangle with given vertices $(-1, 4)$, $(4, 8)$, $(1, 1)$. _____

42. Find the area of a triangle with given vertices $(-1, 2)$, $(2, 5)$, and $(6, -3)$. (See Problem 41.) _____

43. Prove one case of Theorem 1 by expanding the left side of Formula (1) using the first row and cofactors to obtain the right side.

The Check Exercise for
this section is on
page 749.

44. Prove one case of Theorem 1 by expanding the left side of Formula (1) using the second column and cofactors to obtain the right side.

E Cramer's Rule

Now let us see how determinants arise rather naturally in the process of solving systems of linear equations. We will start by investigating two equations and two unknowns and then extend any results to three equations and three unknowns.

Instead of thinking of each system of linear equations in two unknowns as a different problem, let us see what happens when we attempt to solve the general system

$$a_{11}x + a_{12}y = k_1 \tag{1A}$$
$$a_{21}x + a_{22}y = k_2 \tag{1B}$$

once and for all in terms of the unspecified real constants a_{11}, a_{12}, a_{21}, a_{22}, k_1, and k_2.

We proceed by multiplying Equations (1A) and (1B) by suitable constants so that when the resulting equations are added, left side to left side and right side to right side, one of the variables drops out. Suppose we choose to eliminate y. What constants should we use to make the coefficients of y the same except for the signs? Multiply Equation (1A) by a_{22} and Equation (1B) by $-a_{12}$. Then add:

$$
\begin{array}{ll}
a_{11}a_{22}x + a_{12}a_{22}y = k_1a_{22} & a_{22}(1\text{A}) \\
\underline{-a_{21}a_{12}x - a_{12}a_{22}y = -k_2a_{12}} & -a_{12}(1\text{B}) \\
a_{11}a_{22}x - a_{21}a_{12}x + 0y = k_1a_{22} - k_2a_{12} & \\
(a_{11}a_{22} - a_{21}a_{12})x = k_1a_{22} - k_2a_{12} & \\
x = \dfrac{k_1a_{22} - k_2a_{12}}{a_{11}a_{22} - a_{21}a_{12}} & a_{11}a_{22} - a_{21}a_{12} \neq 0
\end{array}
$$

What do the numerator and denominator remind you of? From your experience with determinants in Appendix D you should recognize these expressions as

$$x = \frac{\begin{vmatrix} k_1 & a_{12} \\ k_2 & a_{22} \end{vmatrix}}{\begin{vmatrix} a_{11} & a_{12} \\ a_{21} & a_{22} \end{vmatrix}}$$

Similarly, starting with system (1) and eliminating x (this is left as an exercise), we obtain

$$y = \frac{\begin{vmatrix} a_{11} & k_1 \\ a_{21} & k_2 \end{vmatrix}}{\begin{vmatrix} a_{11} & a_{12} \\ a_{21} & a_{22} \end{vmatrix}}$$

These results are summarized in the following theorem, which is named after the Swiss mathematician G. Cramer (1704–1752):

THEOREM 1

> **Cramer's Rule for Two Equations and Two Unknowns**
>
> Given the system
>
> $$\begin{aligned} a_{11}x + a_{12}y &= k_1 \\ a_{21}x + a_{22}y &= k_2 \end{aligned} \qquad \text{with} \qquad D = \begin{vmatrix} a_{11} & a_{12} \\ a_{21} & a_{22} \end{vmatrix} \neq 0$$
>
> then
>
> $$x = \frac{\begin{vmatrix} k_1 & a_{12} \\ k_2 & a_{22} \end{vmatrix}}{D} \qquad \text{and} \qquad y = \frac{\begin{vmatrix} a_{12} & k_1 \\ a_{21} & k_2 \end{vmatrix}}{D}$$

The determinant D is called the **coefficient determinant**. If $D \neq 0$, then the system has exactly one solution, which is given by Cramer's rule. If, on the other hand, $D = 0$, then it can be shown that the system is either inconsistent or dependent; that is, the system either has no solutions or has an infinite number of solutions.

EXAMPLE 1   Solve using Cramer's rule:

$$\begin{aligned} 2x - 3y &= 7 \\ -3x + y &= -7 \end{aligned}$$

Solution

$$D = \begin{vmatrix} 2 & -3 \\ -3 & 1 \end{vmatrix} = -7$$

$$x = \frac{\begin{vmatrix} 7 & -3 \\ -7 & 1 \end{vmatrix}}{-7} = \frac{-14}{-7} = 2 \qquad y = \frac{\begin{vmatrix} 2 & 7 \\ -3 & -7 \end{vmatrix}}{-7} = \frac{7}{-7} = -1$$

PROBLEM 1

Solve using Cramer's rule: $\begin{aligned} 3x + 2y &= -3 \\ -4x + 3y &= -13 \end{aligned}$

Solution

Cramer's rule generalizes completely for any size of linear system that has the same number of unknowns as equations. We state without proof the rule for three equations and three unknowns.

THEOREM 2

Cramer's Rule for Three Equations and Three Unknowns

Given the system

$$a_{11}x + a_{12}y + a_{13}z = k_1$$
$$a_{21}x + a_{22}y + a_{23}z = k_2 \qquad \text{with} \qquad D = \begin{vmatrix} a_{11} & a_{12} & a_{13} \\ a_{21} & a_{22} & a_{23} \\ a_{31} & a_{32} & a_{33} \end{vmatrix} \neq 0$$
$$a_{31}x + a_{32}y + a_{33}z = k_3$$

then

$$x = \frac{\begin{vmatrix} k_1 & a_{12} & a_{13} \\ k_2 & a_{22} & a_{23} \\ k_3 & a_{32} & a_{33} \end{vmatrix}}{D} \qquad y = \frac{\begin{vmatrix} a_{11} & k_1 & a_{13} \\ a_{21} & k_2 & a_{23} \\ a_{31} & k_3 & a_{33} \end{vmatrix}}{D} \qquad z = \frac{\begin{vmatrix} a_{11} & a_{12} & k_1 \\ a_{21} & a_{22} & k_2 \\ a_{31} & a_{32} & k_3 \end{vmatrix}}{D}$$

It is easy to remember these determinant formulas for x, y, and z if you observe the following:

1. Determinant D is formed from the coefficients of x, y, and z, keeping the same relative position in the determinant as found in the system.
2. Determinant D appears in the denominators for x, y, and z.
3. The numerator for x can be obtained from D by replacing the coefficients of x—that is, a_{11}, a_{21}, and a_{31}—with the constants k_1, k_2, and k_3, respectively. Similar statements can be made for the numerators for y and z.

EXAMPLE 2   Use Cramer's rule to solve:

$$x + y = 1$$
$$3y - z = -4$$
$$x + z = 3$$

Solution

$$D = \begin{vmatrix} 1 & 1 & 0 \\ 0 & 3 & -1 \\ 1 & 0 & 1 \end{vmatrix} = 2 \quad \text{Missing variables have 0 coefficients.}$$

$$x = \frac{\begin{vmatrix} 1 & 1 & 0 \\ -4 & 3 & -1 \\ 3 & 0 & 1 \end{vmatrix}}{2} = \frac{4}{2} = 2 \qquad y = \frac{\begin{vmatrix} 1 & 1 & 0 \\ 0 & -4 & -1 \\ 1 & 3 & 1 \end{vmatrix}}{2} = \frac{-2}{2} = -1$$

$$z = \frac{\begin{vmatrix} 1 & 1 & 1 \\ 0 & 3 & -4 \\ 1 & 0 & 3 \end{vmatrix}}{2} = \frac{2}{2} = 1$$

PROBLEM 2   Use Cramer's rule to solve: $3x - z = 5$
$x - y + z = 0$
$x + y = 0$

Solution

In practice, Cramer's rule is rarely used to solve systems of order higher than 2 or 3; more efficient methods are available. Cramer's rule is, however, a valuable tool in theoretical mathematics.

We now summarize the results of the various methods of solving two equations and two unknowns in Table 1.

TABLE 1
SOLVING SECOND-ORDER SYSTEMS OF LINEAR EQUATIONS

Solutions	Graphing	Elimination or Substitution	Cramer's Rule
Exactly one	Lines intersect in exactly one point	One unique pair of numbers	$D \neq 0$
None	Lines are parallel	Contradiction occurs—such as $0 = 5$	$D = 0$
Infinite	Lines coincide	$0 = 0$	$D = 0$

Solutions to Matched Problems

1. $3x + 2y = -3$
$-4x + 3y = -13$

$$D = \begin{vmatrix} 3 & 2 \\ -4 & 3 \end{vmatrix} = 17 \qquad x = \frac{\begin{vmatrix} -3 & 2 \\ -13 & 3 \end{vmatrix}}{17} = \frac{17}{17} = 1$$

$$y = \frac{\begin{vmatrix} 3 & -3 \\ -4 & -13 \end{vmatrix}}{17} = \frac{-51}{17} = -3$$

2. $3x \qquad - z = 5$
$\quad x - y + z = 0 \qquad D = \begin{vmatrix} 3 & 0 & -1 \\ 1 & -1 & 1 \\ 1 & 1 & 0 \end{vmatrix} = -5$
$\quad x + y \qquad = 0$

$$x = \frac{\begin{vmatrix} 5 & 0 & -1 \\ 0 & -1 & 1 \\ 0 & 1 & 0 \end{vmatrix}}{-5} = \frac{-5}{-5} = 1 \qquad y = \frac{\begin{vmatrix} 3 & 5 & -1 \\ 1 & 0 & 1 \\ 1 & 0 & 0 \end{vmatrix}}{-5} = \frac{5}{-5} = -1$$

$$z = \frac{\begin{vmatrix} 3 & 0 & 5 \\ 1 & -1 & 0 \\ 1 & 1 & 0 \end{vmatrix}}{-5} = \frac{10}{-5} = -2$$

Practice Exercise Appendix E ■

Work odd-numbered problems first, check answers, then work even-numbered problems in areas of weakness. Answers to all problems are in the back of the book. Make every effort to work a problem yourself before you look at an answer.

A Solve using Cramer's rule.

1. $x + 2y = 1$
$\quad x + 3y = -1$ _____

2. $x + 2y = 3$
$\quad x + 3y = 5$ _____

3. $2x + y = 1$
$\quad 5x + 3y = 2$ _____

4. $x + 3y = 1$
$\quad 2x + 8y = 0$ _____

5. $2x - y = -3$
$\quad -x + 3y = 4$ _____

6. $2x + y = 1$
$\quad 5x + 3y = 2$ _____

B **7.** $x + y = 0$
$\quad 2y + z = -5$
$\quad -x + z = -3$ _____

8. $x + y = -4$
$\quad 2y + z = 0$
$\quad -x + z = 5$ _____

9. $x + y = 1$
$\quad 2y + z = 0$
$\quad -x + z = 0$ _____

10. $x + y = -4$
$\quad 2y + z = 3$
$\quad -x + z = 7$ _____

11. $y + z = -4$
$\quad x + 2z = 0$
$\quad x - y = 5$ _____

12. $x - z = 2$
$\quad 2x - y = 8$
$\quad x + y + z = 2$ _____

13. $2y - z = -4$
$\quad x - y - z = 0$
$\quad x - y + 2z = 6$ _____

14. $2x + y = 2$
$\quad x - y + z = -1$
$\quad x + y + z = -1$ _____

15. $2a + 4b + 3c = 6$ **16.** $3u - 2v + 3w = 11$
          $a - 3b + 2c = -7$ $2u + 3v - 2w = -5$
         $-a + 2b - c = 5$ _____ $u + 4v - w = -5$ _____

C   *It is clear that* x = 0, y = 0, z = 0 *is a solution to each of the following systems.*
       Use Cramer's rule to determine if this solution is unique. [Hint: If D ≠ 0, what
       can you conclude? If D = 0, what can you conclude?]

**The Check Exercise for
this section is on
page 751.** **17.** $x - 4y + 9z = 0$ **18.** $3x - y + 3z = 0$
                $4x - y + 6z = 0$ $5x + 5y - 9z = 0$
                $x - y + 3z = 0$ _____ $-2x + y - 3z = 0$ _____

F   Matrices I

- Dimension of a Matrix
- Matrix Addition
- Multiplication of a Matrix by a Number
- Application

In Section 10-3 we introduced the important idea of matrix. In this appendix and the next, we will develop this concept further.

■ Dimension of a Matrix

A **matrix** is any rectangular array of numbers enclosed within brackets. The **size** or **dimension of a matrix** is important relative to operations on matrices. We define *an* ***m* × *n* matrix** (read "m by n matrix") to be one with m rows and n columns. It is important to note that the number of rows is always given first. If a matrix has the same number of rows and columns, it is called a **square matrix**. A matrix with only one column is called a **column matrix**, and one with only one row is called a **row matrix**. These definitions are illustrated by the following:

$$
\begin{array}{cccc}
3 \times 2 & 3 \times 3 & 4 \times 1 & 1 \times 4 \\
\begin{bmatrix} -2 & 5 \\ 0 & -2 \\ 3 & 6 \end{bmatrix} &
\begin{bmatrix} 0.5 & 0.2 & 1.0 \\ 0.0 & 0.3 & 0.5 \\ 0.7 & 0.0 & 0.2 \end{bmatrix} &
\begin{bmatrix} 3 \\ -2 \\ 1 \\ 0 \end{bmatrix} &
[2 \quad \tfrac{1}{2} \quad 0 \quad -\tfrac{2}{3}]
\end{array}
$$

Square matrix   **Row matrix**

Column matrix

Two matrices are **equal** if they have the same dimension and their corresponding elements are equal. For example:

$$
\begin{array}{cc}
2 \times 3 & 2 \times 3 \\
\begin{bmatrix} a & b & c \\ d & e & f \end{bmatrix} = \begin{bmatrix} u & v & w \\ x & y & z \end{bmatrix}
\end{array}
\quad \text{if and only if} \quad
\begin{array}{ccc}
a = u & b = v & c = w \\
d = x & e = y & f = z
\end{array}
$$

■ Matrix Addition

The **sum of two matrices** of the same dimension is a matrix with elements that are the sums of the corresponding elements of the two given matrices. Addition is not defined for matrices with different dimensions.

EXAMPLE 1   **(A)** $\begin{bmatrix} a & b \\ c & d \end{bmatrix} + \begin{bmatrix} w & x \\ y & z \end{bmatrix} = \begin{bmatrix} (a+w) & (b+x) \\ (c+y) & (d+z) \end{bmatrix}$

(B) $\begin{bmatrix} 2 & -3 & 0 \\ 1 & 2 & -5 \end{bmatrix} + \begin{bmatrix} 3 & 1 & 2 \\ -3 & 2 & 5 \end{bmatrix} = \begin{bmatrix} 5 & -2 & 2 \\ -2 & 4 & 0 \end{bmatrix}$

PROBLEM 1   Add: $\begin{bmatrix} 3 & 2 \\ -1 & -1 \\ 0 & 3 \end{bmatrix} + \begin{bmatrix} -2 & 3 \\ 1 & -1 \\ 2 & -2 \end{bmatrix}$

Solution

Because we add two matrices by adding their corresponding elements, it follows from the properties of real numbers that matrices of the same dimension are commutative and associative relative to addition. That is, if A, B, and C are matrices of the same dimension, then

$$A + B = B + A \qquad \text{Commutative}$$
$$(A + B) + C = A + (B + C) \qquad \text{Associative}$$

A matrix with elements that are all 0's is called a **zero matrix**. For example,

$$[0 \ \ 0 \ \ 0] \qquad \begin{bmatrix} 0 & 0 \\ 0 & 0 \end{bmatrix} \qquad \begin{bmatrix} 0 \\ 0 \\ 0 \\ 0 \end{bmatrix} \qquad \begin{bmatrix} 0 & 0 & 0 & 0 \\ 0 & 0 & 0 & 0 \\ 0 & 0 & 0 & 0 \end{bmatrix}$$

are zero matrices of different dimensions. [*Note:* "0" may be used to denote the zero matrix of any dimension.] The **negative of a matrix M**, denoted by $-M$, is a matrix with elements that are the negative of the elements in M. Thus, if

$$M = \begin{bmatrix} a & b \\ c & d \end{bmatrix} \qquad \text{then} \qquad -M = \begin{bmatrix} -a & -b \\ -c & -d \end{bmatrix}$$

Note that $M + (-M) = 0$ (a zero matrix).

If A and B are matrices of the same dimension, then we define **subtraction** as follows:

$$A - B = A + (-B)$$

Thus, to subtract matrix B from matrix A, we simply subtract corresponding elements.

EXAMPLE 2

$$\begin{bmatrix} 3 & -2 \\ 5 & 0 \end{bmatrix} - \begin{bmatrix} -2 & 2 \\ 3 & 4 \end{bmatrix} = \begin{bmatrix} 3 & -2 \\ 5 & 0 \end{bmatrix} + \begin{bmatrix} 2 & -2 \\ -3 & -4 \end{bmatrix} = \begin{bmatrix} 5 & -4 \\ 2 & -4 \end{bmatrix}$$

PROBLEM 2   Subtract: $[2 \quad -3 \quad 5] - [3 \quad -2 \quad 1]$

Solution

■ Multiplication of a Matrix by a Number

Finally, the **product of a number *k* and a matrix *M***, denoted by kM, is a matrix formed by multiplying each element of M by k. This definition is partly motivated by the fact that if M is a matrix, then we would like $M + M$ to equal $2M$.

EXAMPLE 3

$$-2 \begin{bmatrix} 3 & -1 & 0 \\ -2 & 1 & 3 \\ 0 & -1 & -2 \end{bmatrix} = \begin{bmatrix} -6 & 2 & 0 \\ 4 & -2 & -6 \\ 0 & 2 & 4 \end{bmatrix}$$

PROBLEM 3   Find: $10 \begin{bmatrix} 1.3 \\ 0.2 \\ 3.5 \end{bmatrix}$

Solution

We now consider an application that uses various operations.

■ Application

EXAMPLE 4   Ms. Smith and Mr. Jones are salespeople in a new-car agency that sells only two models. August was the last month for this year's models, and next year's models were introduced in September. Gross dollar sales for each month are given in the following matrices:

| | AUGUST SALES | | | SEPTEMBER SALES | |
	Compact	Luxury		Compact	Luxury
Ms. Smith	$18,000	$36,000		$72,000	$144,000
Mr. Jones	$36,000	0		$90,000	$108,000

$$\begin{bmatrix} \$18{,}000 & \$36{,}000 \\ \$36{,}000 & 0 \end{bmatrix} = A \qquad \begin{bmatrix} \$72{,}000 & \$144{,}000 \\ \$90{,}000 & \$108{,}000 \end{bmatrix} = B$$

(For example, Ms. Smith had $18,000 in compact sales in August, and Mr. Jones had $108,000 in luxury car sales in September.)

(A) What was the combined dollar sales in August and September for each person and each model?
(B) What was the increase in dollar sales from August to September?
(C) If both salespeople receive 5% commissions on gross dollar sales, compute the commission for each person for each model sold in September.

Solution **(A)** $A + B = \begin{bmatrix} \$\ 90{,}000 & \$180{,}000 \\ \$126{,}000 & \$108{,}000 \end{bmatrix}$ Ms. Smith Mr. Jones

(columns: Compact, Luxury)

(B) $B - A = \begin{bmatrix} \$54{,}000 & \$108{,}000 \\ \$54{,}000 & \$108{,}000 \end{bmatrix}$ Ms. Smith Mr. Jones

(columns: Compact, Luxury)

(C) $0.05B = \begin{bmatrix} (0.05)(\$72{,}000) & (0.05)(\$144{,}000) \\ (0.05)(\$90{,}000) & (0.05)(\$108{,}000) \end{bmatrix}$

$= \begin{bmatrix} \$3{,}600 & \$7{,}200 \\ \$4{,}500 & \$5{,}400 \end{bmatrix}$ Ms. Smith Mr. Jones

(columns: Compact, Luxury)

In Example 4 we chose a relatively simple example involving an agency with only two salespeople and two models. Consider the more realistic problem of an agency with nine models and perhaps seven salespeople—then you can begin to see the value of matrix methods.

PROBLEM 4   Repeat Example 4 with

$$A = \begin{bmatrix} \$36{,}000 & \$36{,}000 \\ \$18{,}000 & \$36{,}000 \end{bmatrix} \quad \text{and} \quad B = \begin{bmatrix} \$90{,}000 & \$108{,}000 \\ \$72{,}000 & \$108{,}000 \end{bmatrix}$$

Solution   **(A)**

(B)

(C)

Solutions to Matched Problems

1. $\begin{bmatrix} 3 & 2 \\ -1 & -1 \\ 0 & 3 \end{bmatrix} + \begin{bmatrix} -2 & 3 \\ 1 & -1 \\ 2 & -2 \end{bmatrix} = \begin{bmatrix} 1 & 5 \\ 0 & -2 \\ 2 & 1 \end{bmatrix}$

2. $\begin{bmatrix} 2 & -3 & 5 \end{bmatrix} - \begin{bmatrix} 3 & -2 & 1 \end{bmatrix} = \begin{bmatrix} 2 & -3 & 5 \end{bmatrix} + \begin{bmatrix} -3 & 2 & -1 \end{bmatrix}$
$$= \begin{bmatrix} -1 & -1 & 4 \end{bmatrix}$$

3. $10 \begin{bmatrix} 1.3 \\ 0.2 \\ 3.5 \end{bmatrix} = \begin{bmatrix} 13 \\ 2 \\ 35 \end{bmatrix}$

4. (A) $A + B = \begin{bmatrix} \$126{,}000 & \$144{,}000 \\ \$\ 90{,}000 & \$144{,}000 \end{bmatrix}$

Compact   Luxury — Ms. Smith / Mr. Jones

(B) $B - A = \begin{bmatrix} \$54{,}000 & \$72{,}000 \\ \$54{,}000 & \$72{,}000 \end{bmatrix}$

Compact   Luxury — Ms. Smith / Mr. Jones

(C) $0.05B = \begin{bmatrix} \$4{,}500 & \$5{,}400 \\ \$3{,}600 & \$5{,}400 \end{bmatrix}$

Compact   Luxury — Ms. Smith / Mr. Jones

Practice Exercise Appendix F ∎

Work odd-numbered problems first, check answers, and then work even-numbered problems in areas of weakness. Answers to all problems are in the back of the book. Make every effort to work a problem yourself before you look at an answer.

A Problems 1–18 refer to the following matrices:

$$A = \begin{bmatrix} 2 & -1 \\ 3 & 0 \end{bmatrix} \qquad B = \begin{bmatrix} -3 & 1 \\ 2 & -3 \end{bmatrix} \qquad C = \begin{bmatrix} 2 \\ -3 \\ 0 \end{bmatrix}$$

$$D = \begin{bmatrix} 1 \\ 3 \\ 5 \end{bmatrix} \qquad E = \begin{bmatrix} -4 & 1 & 0 & -2 \end{bmatrix} \qquad F = \begin{bmatrix} 2 & -3 \\ -2 & 0 \\ 1 & 2 \\ 3 & 5 \end{bmatrix}$$

1. What are the dimensions of *B*? Of *E*? _____

2. What are the dimensions of *F*? Of *D*? _____

3. What element is in the third row and second column of matrix *F*?

_____

4. What element is in the second row and first column of matrix F? _____

5. Write a zero matrix of the same dimension as B. _____

6. Write a zero matrix of the same dimension as E. _____

7. Identify all column matrices. _____

8. Identify all row matrices. _____

9. Identify all square matrices. _____

10. How many additional columns would F have to have to be a square matrix?

_____

11. Find $A + B$. _____   **12.** Find $C + D$. _____

13. Write the negative of matrix C. _____

14. Write the negative of matrix B. _____

15. Find $D - C$. _____   **16.** Find $A - A$. _____

17. Find $5B$. _____   **18.** Find $-2E$. _____

B In Problems 19–24 perform the indicated operations.

19. $[-2 \quad 3 \quad 0] + 2[1 \quad -1 \quad 2]$ _____

20. $\begin{bmatrix} 230 \\ 120 \end{bmatrix} + 3\begin{bmatrix} 20 \\ 60 \end{bmatrix}$ _____

21. $1{,}000\begin{bmatrix} 0.25 & 0.36 \\ 0.04 & 0.35 \end{bmatrix}$ _____

22. $100\begin{bmatrix} 0.32 & 0.05 & 0.17 \\ 0.22 & 0.03 & 0.21 \end{bmatrix}$ _____

23. $2\begin{bmatrix} 1 & 2 \\ -1 & 3 \\ 0 & -2 \end{bmatrix} - \begin{bmatrix} 3 & 4 \\ -2 & 0 \\ 1 & -3 \end{bmatrix}$ _____

24. $-2\begin{bmatrix} 1 & 3 & 0 \\ -2 & -1 & 1 \end{bmatrix} - \begin{bmatrix} -3 & -1 & 1 \\ 0 & 2 & -1 \end{bmatrix}$ _____

C **25.** Find a, b, c, and d so that

$$\begin{bmatrix} a & b \\ c & d \end{bmatrix} + \begin{bmatrix} 2 & -3 \\ 0 & 1 \end{bmatrix} = \begin{bmatrix} 1 & -2 \\ 3 & -4 \end{bmatrix}$$ _____

26.   Find w, x, y, and z so that

$$\begin{bmatrix} 4 & -2 \\ -3 & 0 \end{bmatrix} + \begin{bmatrix} w & x \\ y & z \end{bmatrix} = \begin{bmatrix} 2 & -3 \\ 0 & 5 \end{bmatrix}$$

APPLICATIONS   **27.**   *Cost analysis:*   A company with two different plants manufactures guitars and banjos. Its production costs for each instrument are given in the following matrices:

	PLANT X			PLANT Y	
	Guitar	Banjo		Guitar	Banjo
Materials	$30	$25	$= A$	$36	$27
Labor	$60	$80		$54	$74

with $\begin{bmatrix} \$30 & \$25 \\ \$60 & \$80 \end{bmatrix} = A$ and $\begin{bmatrix} \$36 & \$27 \\ \$54 & \$74 \end{bmatrix} = B$

Find $\frac{1}{2}(A + B)$, the average cost of production for the two plants.

28.   *Heredity:*   Gregor Mendel (1822–1884), a Bavarian monk and botanist, made discoveries that revolutionized the science of heredity. In one experiment he crossed dihybrid yellow round peas (yellow and round are dominant characteristics; the peas also contained green and wrinkled as recessive genes) and obtained 560 peas of the types indicated in the matrix:

	Round	Wrinkled	
Yellow	319	101	$= M$
Green	108	32	

Suppose he carried out a second experiment of the same type and obtained 640 peas of the types indicated in this matrix:

	Round	Wrinkled	
Yellow	370	124	$= N$
Green	110	36	

The Check Exercise for this section is on page 753.

If the results of the two experiments are combined, write the resulting matrix $M + N$. Compute the decimal fraction of the total number of peas (1,200) in each category of the combined results. [*Hint:* Compute $(1/1,200)(M + N)$.]

G   Matrices II

■ Dot Product
■ Matrix Product
■ Multiplication Properties
■ Application

In this Appendix we are going to introduce two types of matrix multiplication that will at first seem rather strange. In spite of this apparent strangeness, these operations are well founded in the general theory of matrices and, as we will see, are extremely useful in practical problems.

■ Dot Product

We start by defining the dot product of two special matrices.

Dot Product

The **dot product** of a $1 \times n$ row matrix and an $n \times 1$ column matrix is a real number given by:

$$[a_1 \quad a_2 \quad \cdots \quad a_n] \cdot \begin{bmatrix} b_1 \\ b_2 \\ \vdots \\ b_n \end{bmatrix} = a_1 b_1 + a_2 b_2 + \cdots + a_n b_n$$

$1 \times n$ (label over row matrix), $n \times 1$ (label over column matrix), A real number

---

The dot between the two matrices is important. If the dot is omitted, the multiplication is of another type, which we will consider later.

EXAMPLE 1

$$[2 \quad -3 \quad 0] \cdot \begin{bmatrix} -5 \\ 2 \\ -2 \end{bmatrix} = (2)(-5) + (-3)(2) + (0)(-2)$$

$$= -10 - 6 + 0 = -16$$

PROBLEM 1

$$[-1 \quad 0 \quad 3 \quad 2] \cdot \begin{bmatrix} 2 \\ 3 \\ 4 \\ -1 \end{bmatrix} = ?$$

Solution

EXAMPLE 2 A factory produces a slalom water ski that requires 4 workhours in the fabricating department and 1 workhour in the finishing department. Fabricating personnel receive \$8 per hour and finishing personnel receive \$6 per hour. Total labor cost per ski is given by the dot product:

$$[4 \quad 1] \cdot \begin{bmatrix} 8 \\ 6 \end{bmatrix} = (4)(8) + (1)(6) = 32 + 6 = \$38 \text{ per ski}$$

PROBLEM 2   If the factory in Example 2 also produces a trick water ski that requires 6 workhours in the fabricating department and 1.5 workhours in the finishing department, write a dot product between appropriate row and column matrices that will give the total labor cost for this ski. Compute the cost.

Solution

■ Matrix Product

It is important to remember that the dot product of a row matrix and a column matrix is a real number and not a matrix. We now define a matrix product for certain matrices.

Matrix Product

The **product of two matrices** A and B is defined only on the assumption that the number of columns in A is equal to the number of rows in B. If A is an $m \times p$ matrix and B is a $p \times n$ matrix, then the matrix product of A and B, denoted by AB, is an $m \times n$ matrix whose element in the ith row and jth column is the dot product of the ith row matrix of A and the jth column matrix of B.

It is important to check dimensions before starting the multiplication process. If matrix A has dimension $a \times b$ and matrix B has dimension $c \times d$, then if $b = c$, the product AB will exist and will have dimension $a \times d$. This is shown schematically in Figure 1.

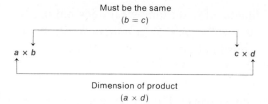

Must be the same
$(b = c)$

$a \times b$ $c \times d$

Dimension of product
$(a \times d)$

FIGURE 1

The definition is not as complicated as it might first seem. An example should help to clarify the process. For

$$A = \begin{bmatrix} 2 & 3 & -1 \\ -2 & 1 & 2 \end{bmatrix} \quad \text{and} \quad B = \begin{bmatrix} 1 & 3 \\ 2 & 0 \\ -1 & 2 \end{bmatrix}$$

A is 2×3, B is 3×2, and AB will be 2×2. The four dot products used to produce the four elements in AB (usually calculated mentally for small whole numbers) are shown in the following large matrix:

$$
\underbrace{\begin{bmatrix} 2 & 3 & -1 \\ -2 & 1 & 2 \end{bmatrix}}_{2 \times 3}
\underbrace{\begin{bmatrix} 1 & 3 \\ 2 & 0 \\ -1 & 2 \end{bmatrix}}_{3 \times 2}
=
\begin{bmatrix}
[2 \quad 3 \quad -1] \cdot \begin{bmatrix} 1 \\ 2 \\ -1 \end{bmatrix} & [2 \quad 3 \quad -1] \cdot \begin{bmatrix} 3 \\ 0 \\ 2 \end{bmatrix} \\
[-2 \quad 1 \quad 2] \cdot \begin{bmatrix} 1 \\ 2 \\ -1 \end{bmatrix} & [-2 \quad 1 \quad 2] \cdot \begin{bmatrix} 3 \\ 0 \\ 2 \end{bmatrix}
\end{bmatrix}
=
\underbrace{\begin{bmatrix} 9 & 4 \\ -2 & -2 \end{bmatrix}}_{2 \times 2}
$$

EXAMPLE 3

$$
\underbrace{\begin{bmatrix} 2 & 1 \\ 1 & 0 \\ -1 & 2 \end{bmatrix}}_{3 \times 2}
\underbrace{\begin{bmatrix} 1 & -1 & 0 & 1 \\ 2 & 1 & 2 & 0 \end{bmatrix}}_{2 \times 4}
=
\underbrace{\begin{bmatrix} 4 & -1 & 2 & 2 \\ 1 & -1 & 0 & 1 \\ 3 & 3 & 4 & -1 \end{bmatrix}}_{3 \times 4}
$$

PROBLEM 3   Find the product:

$$
\begin{bmatrix} 1 & -1 & 0 \\ 2 & 1 & 2 \end{bmatrix}
\begin{bmatrix} 2 & -1 \\ 1 & 0 \\ -1 & 2 \end{bmatrix}
$$

Solution

■ Multiplication Properties

In the arithmetic of real numbers it does not matter in which order we multiply; for example, $5 \times 7 = 7 \times 5$. In matrix multiplication it does make a difference; that is, MN does not always equal NM, even if both multiplications are defined (see Problems 11, 12, 25, 27, and 28 in Exercise G). Also, MN may be 0 with neither M nor N equal to 0.

Matrices do, however, have other general properties. We state three important properties without proof. Assuming that all products and sums are defined for the indicated matrices A, B, and C, then for k a real number:

1. $(AB)C = A(BC)$ Associative property
2. $A(B + C) = AB + AC$ Left-hand distributive property
3. $(B + C)A = BA + CA$ Right-hand distributive property
4. $k(AB) = (kA)B = A(kB)$

Since matrix multiplication is not commutative, Properties 2 and 3 must be listed as distinct properties (see Problems 27 and 28 in Exercise G).

■ Application

The next example illustrates the use of the dot and matrix product in a business application.

EXAMPLE 4   *Production scheduling:*   Let us combine the time requirements discussed in Example 2 and Problem 2 into one matrix:

$$
\begin{array}{cc}
& \text{Fabricating} \quad \text{Finishing} \\
& \text{department} \quad \text{department}
\end{array}
$$

$$
\begin{array}{c}
\text{Trick ski} \\
\text{Slalom ski}
\end{array}
\begin{bmatrix}
6 \text{ hours} & 1.5 \text{ hours} \\
4 \text{ hours} & 1 \text{ hour}
\end{bmatrix} = A
$$

Now suppose the company has two manufacturing plants X and Y in different parts of the country and that their hourly rates for each department are given in the following matrix:

$$
\begin{array}{cc}
& \text{Plant } X \quad \text{Plant } Y
\end{array}
$$

$$
\begin{array}{c}
\text{Fabricating department} \\
\text{Finishing department}
\end{array}
\begin{bmatrix}
\$8 & \$7 \\
\$6 & \$4
\end{bmatrix} = B
$$

To find the total labor costs for each ski at each factory, we multiply A and B:

$$
AB = \begin{bmatrix} 6 & 1.5 \\ 4 & 1 \end{bmatrix} \begin{bmatrix} 8 & 7 \\ 6 & 4 \end{bmatrix} = \begin{bmatrix} \$57 & \$48 \\ \$38 & \$32 \end{bmatrix}
\begin{array}{l}
\text{Trick ski} \\
\text{Slalom ski}
\end{array}
$$

Notice that the dot product of the first row matrix of A and the first column matrix of B gives us the labor costs, $57, for a trick ski manufactured at plant X; the dot product of the second row matrix of A and the second column matrix of B gives us the labor costs, $32, for manufacturing a slalom ski at plant Y; and so on.

Example 4 is, of course, oversimplified. Companies that manufacture many different items in many different plants deal with matrices that have very large numbers of rows and columns.

PROBLEM 4   Repeat Example 4 with

$$
A = \begin{bmatrix} 7 \text{ hours} & 2 \text{ hours} \\ 5 \text{ hours} & 1.5 \text{ hours} \end{bmatrix} \quad \text{and} \quad B = \begin{bmatrix} \$10 & \$8 \\ \$6 & \$4 \end{bmatrix}
$$

Solution

1. $[-1 \quad 0 \quad 3 \quad 2] \cdot \begin{bmatrix} 2 \\ 3 \\ 4 \\ -1 \end{bmatrix} = 8$ **2.** $[6 \quad 1.5] \cdot \begin{bmatrix} 8 \\ 6 \end{bmatrix} = \$57$

3. $\begin{bmatrix} 1 & -1 & 0 \\ 2 & 1 & 2 \end{bmatrix} \begin{bmatrix} 2 & -1 \\ 1 & 0 \\ -1 & 2 \end{bmatrix} = \begin{bmatrix} 1 & -1 \\ 3 & 2 \end{bmatrix}$

4. $AB = \begin{bmatrix} 7 & 2 \\ 5 & 1.5 \end{bmatrix} \overset{X}{\begin{bmatrix} 10 & 8 \\ 6 & 4 \end{bmatrix}} = \begin{bmatrix} \$82 & \$64 \\ \$59 & \$46 \end{bmatrix} \begin{matrix} \text{Trick} \\ \text{Slalom} \end{matrix}$

Practice Exercise Appendix G ■

Work odd-numbered problems first, check answers, and then work even-numbered problems in areas of weakness. Answers to all problems are in the back of the book. Make every effort to work a problem before you look at an answer.

A Find the dot products.

1. $[2 \quad 4] \cdot \begin{bmatrix} 3 \\ 1 \end{bmatrix}$ _____   **2.** $[3 \quad 1] \cdot \begin{bmatrix} 2 \\ 4 \end{bmatrix}$ _____

3. $[-3 \quad 2] \cdot \begin{bmatrix} -1 \\ -2 \end{bmatrix}$ _____   **4.** $[3 \quad -2] \cdot \begin{bmatrix} -4 \\ -1 \end{bmatrix}$ _____

Find the matrix products.

5. $[2 \quad 5] \begin{bmatrix} 1 & -1 \\ 2 & 3 \end{bmatrix}$ _____   **6.** $[1 \quad 3] \begin{bmatrix} 2 & 3 \\ 1 & -4 \end{bmatrix}$ _____

7. $\begin{bmatrix} 3 & 4 \\ -1 & -2 \end{bmatrix} \begin{bmatrix} -1 \\ 2 \end{bmatrix}$ _____   **8.** $\begin{bmatrix} -1 & 1 \\ 2 & -3 \end{bmatrix} \begin{bmatrix} 4 \\ -2 \end{bmatrix}$ _____

9. $\begin{bmatrix} 2 & -3 \\ 1 & 2 \end{bmatrix} \begin{bmatrix} 1 & -1 \\ 0 & -2 \end{bmatrix}$ _____   **10.** $\begin{bmatrix} -3 & 2 \\ 4 & -1 \end{bmatrix} \begin{bmatrix} -2 & 5 \\ -1 & 3 \end{bmatrix}$ _____

11. $\begin{bmatrix} -5 & -2 \\ 1 & -3 \end{bmatrix} \begin{bmatrix} -2 & 1 \\ 0 & -3 \end{bmatrix}$ _____

12. $\begin{bmatrix} -2 & 1 \\ 0 & -3 \end{bmatrix} \begin{bmatrix} -5 & -2 \\ 1 & -3 \end{bmatrix}$ _____

B   *Find the dot products.*

13. $[-1 \quad -2 \quad 2] \cdot \begin{bmatrix} 2 \\ -1 \\ 3 \end{bmatrix}$ _____

14. $[-2 \quad 4 \quad 0] \cdot \begin{bmatrix} -1 \\ -3 \\ 2 \end{bmatrix}$ _____

15. $[-1 \quad -3 \quad 0 \quad 5] \cdot \begin{bmatrix} 4 \\ -3 \\ -1 \\ 2 \end{bmatrix}$ _____

16. $[-1 \quad 2 \quad 3 \quad -2] \cdot \begin{bmatrix} 3 \\ -2 \\ 0 \\ 4 \end{bmatrix}$ _____

Find the matrix products.

17. $\begin{bmatrix} 2 & -1 & 1 \\ 1 & 3 & -2 \end{bmatrix} \begin{bmatrix} 1 & 3 \\ 0 & -1 \\ -2 & 2 \end{bmatrix}$ _____

18. $\begin{bmatrix} -1 & -4 & 3 \\ 2 & 0 & 1 \end{bmatrix} \begin{bmatrix} 2 & -3 \\ 1 & 2 \\ 0 & -1 \end{bmatrix}$ _____

19. $\begin{bmatrix} 1 & 3 \\ 0 & -1 \\ -2 & 2 \end{bmatrix} \begin{bmatrix} 2 & -1 & 1 \\ 1 & 3 & -2 \end{bmatrix}$ _____

20. $\begin{bmatrix} 2 & -3 \\ 1 & 2 \\ 0 & -1 \end{bmatrix} \begin{bmatrix} -1 & -4 & 3 \\ 2 & 0 & 1 \end{bmatrix}$ _____

21. $[3 \quad -2 \quad -4] \begin{bmatrix} 1 \\ 2 \\ -3 \end{bmatrix}$ _____

22. $[1 \quad -2 \quad 2] \begin{bmatrix} 2 \\ -1 \\ 1 \end{bmatrix}$ _____

23. $\begin{bmatrix} 1 \\ 2 \\ -3 \end{bmatrix} [3 \quad -2 \quad -4]$ _____

24. $\begin{bmatrix} 2 \\ -1 \\ 1 \end{bmatrix} [1 \quad -2 \quad 2]$ _____

C   *In Problems 25–28 verify each statement using the following matrices:*

$$A = \begin{bmatrix} 1 & 2 \\ 0 & 1 \end{bmatrix} \qquad B = \begin{bmatrix} 1 & 1 \\ 2 & 3 \end{bmatrix} \qquad C = \begin{bmatrix} -3 & 1 \\ -1 & 2 \end{bmatrix}$$

25. $AB \neq BA$ _____

26. $(AB)C = A(BC)$ _____

27. $A(B + C) = AB + AC$ _____   **28.** $(B + C)A = BA + CA$ _____

APPLICATIONS **29.** *Labor costs:* A company with manufacturing plants located in different parts of the country has workhour and wage requirements for the manufacturing of three types of inflatable boats as given in the following two matrices:

WORKHOURS PER BOAT

	Cutting department	Assembly department	Packaging department	
	0.6 hour	0.6 hour	0.2 hour	One-person boat
$M =$	1.0 hour	0.9 hour	0.3 hour	Two-person boat
	1.5 hours	1.2 hours	0.4 hour	Four-person boat

HOURLY WAGES

	Plant I	Plant II	
	$6	$7	Cutting department
$N =$	$8	$10	Assembly department
	$3	$4	Packaging department

(A) Find the labor costs for a one-person boat manufactured at plant I; that is, find the dot product

$$[0.6 \quad 0.6 \quad 0.2] \cdot \begin{bmatrix} 6 \\ 8 \\ 3 \end{bmatrix}$$ _____

(B) Find the labor costs for a four-person boat manufactured at plant II. Set up a dot product as in part (A) and multiply. _____

(C) What is the dimension of MN? _____

(D) Find MN and interpret. _____

30. *Nutrition:* A nutritionist for a cereal company blends two cereals in different mixes. The amounts of protein, carbohydrate, and fat (in grams per ounce) in each cereal are given by matrix M. The amounts of each cereal used in the three mixes is given by matrix N.

	Cereal A	Cereal B	
	4 grams	2 grams	Protein
$M =$	20 grams	16 grams	Carbohydrate
	3 grams	1 gram	Fat

	Mix X	Mix Y	Mix Z	
$N =$	15 ounces	10 ounces	5 ounces	Cereal A
	5 ounces	10 ounces	15 ounces	Cereal B

(A) Find the amount of protein in mix X by computing the dot product

$$[4 \quad 2] \cdot \begin{bmatrix} 15 \\ 5 \end{bmatrix}$$ _____

(B)   Find the amount of fat in mix Z. Set up a dot product as in part (A) and multiply. _____

(C)   What is the dimension of *MN*? _____

The Check Exercise for
this section is on
page 755.

(D)   Find *MN* and interpret. _____

(E)   Find $\frac{1}{20}(MN)$ and interpret. _____

Name _____ Class _____ Score _____

Check Exercise B ■

Work the following problems without looking at any text examples. Show your work in the space provided. Write your answer in the answer column.

1. _____

2. _____

3. _____

4. _____

5. _____

1. Use the portion of Table II below to find log 0.0237.

2. Use the portion of Table II below to approximate log 20,420.

3. Solve log x = 2.3456 for x using the nearest values in the portion of Table II below.

TABLE II   COMMON LOGARITHMS

x	0	1	2	3	4	5	6	7	8	9
2.0	0.3010	0.3032	0.3054	0.3075	0.3096	0.3118	0.3139	0.3160	0.3181	0.3201
2.1	0.3222	0.3243	0.3263	0.3284	0.3304	0.3324	0.3345	0.3365	0.3385	0.3404
2.2	0.3424	0.3444	0.3464	0.3483	0.3502	0.3522	0.3541	0.3560	0.3579	0.3598
2.3	0.3617	0.3636	0.3655	0.3674	0.3692	0.3711	0.3729	0.3747	0.3766	0.3784
2.4	0.3802	0.3820	0.3838	0.3856	0.3874	0.3892	0.3909	0.3927	0.3945	0.3962

4. Use the portion of Table III below to find ln 146.

5. Use the portion of Table III below to find ln 0.0135.

TABLE III   NATURAL LOGARITHMS (ln x = log$_e$ x)

ln 10 = 2.3026	5 ln 10 = 11.5130	9 ln 10 = 20.7233
2 ln 10 = 4.6052	6 ln 10 = 13.8155	10 ln 10 = 23.0259
3 ln 10 = 6.9078	7 ln 10 = 16.1181	
4 ln 10 = 9.2103	8 ln 10 = 18.4207	

x	.00	.01	.02	.03	.04	.05	.06	.07	.08	.09
1.0	0.0000	0.0100	0.0198	0.0296	0.0392	0.0488	0.0583	0.0677	0.0770	0.0862
1.1	0.0953	0.1044	0.1133	0.1222	0.1310	0.1398	0.1484	0.1570	0.1655	0.1740
1.2	0.1823	0.1906	0.1989	0.2070	0.2151	0.2231	0.2311	0.2390	0.2469	0.2546
1.3	0.2624	0.2700	0.2776	0.2852	0.2927	0.3001	0.3075	0.3148	0.3221	0.3293
1.4	0.3365	0.3436	0.3507	0.3577	0.3646	0.3716	0.3784	0.3853	0.3920	0.3988

Check Exercise C ■

Work the following problems without looking at any text examples. Show your work in the space provided. Write your answer in the answer column.

1. _____

Use linear interpolation and the portion of Table II below to find each logarithm.

2. _____

1. log 0.02468

3. _____

4. _____

5. _____

2. log 235,700

TABLE II COMMON LOGARITHMS

x	0	1	2	3	4	5	6	7	8	9
2.0	0.3010	0.3032	0.3054	0.3075	0.3096	0.3118	0.3139	0.3160	0.3181	0.3201
2.1	0.3222	0.3243	0.3263	0.3284	0.3304	0.3324	0.3345	0.3365	0.3385	0.3404
2.2	0.3424	0.3444	0.3464	0.3483	0.3502	0.3522	0.3541	0.3560	0.3579	0.3598
2.3	0.3617	0.3636	0.3655	0.3674	0.3692	0.3711	0.3729	0.3747	0.3766	0.3784
2.4	0.3802	0.3820	0.3838	0.3856	0.3874	0.3892	0.3909	0.3927	0.3945	0.3962

Use linear interpolation and the portion of Table II below to find x.

3. log x = 2.702

4. log x = −3.731

5. log x = 6.0708

TABLE II COMMON LOGARITHMS

x	0	1	2	3	4	5	6	7	8	9
5.0	0.6990	0.6998	0.7007	0.7016	0.7024	0.7033	0.7042	0.7050	0.7059	0.7067
5.1	0.7076	0.7084	0.7093	0.7101	0.7110	0.7118	0.7126	0.7135	0.7143	0.7152
5.2	0.7160	0.7168	0.7177	0.7185	0.7193	0.7202	0.7210	0.7218	0.7226	0.7235
5.3	0.7243	0.7251	0.7259	0.7267	0.7275	0.7284	0.7292	0.7300	0.7308	0.7316
5.4	0.7324	0.7332	0.7340	0.7348	0.7356	0.7364	0.7372	0.7380	0.7388	0.7396

Check Exercise D ■

Work the following problems without looking at any text examples. Show your work in the space provided. Write your answer in the answer column.

1. _____

2. _____

Evaluate each determinant.

3. _____

1. $\begin{vmatrix} 3 & 2 \\ -1 & -3 \end{vmatrix}$

4. _____

5. _____

2. $\begin{vmatrix} 2.1 & 3.2 \\ 2 & 4 \end{vmatrix}$

3. $\begin{vmatrix} 1 & 2 & 3 \\ 0 & 1 & 0 \\ -3 & 2 & -1 \end{vmatrix}$

4.
$$\begin{vmatrix} -3 & 3 & 5 \\ 2 & 0 & 1 \\ -7 & 3 & 3 \end{vmatrix}$$

5.
$$\begin{vmatrix} 0 & 1 & 0 \\ a & b & c \\ d & e & f \end{vmatrix}$$

Check Exercise E ■

ANSWER COLUMN

1. _____

2. _____

3. _____

4. _____

5. _____

Work the following problems without looking at any text examples. Show your work in the space provided. Write your answer in the answer column.

Solve using Cramer's rule.

1. $2x + 3y = -1$
$\quad\;\; x - 2y = 3$

2. $\quad x - 4y = 0$
$\quad 3x + \;\; y = 1$

3. $\quad x + 2y = 3$
$\qquad\; y + 2z = 4$
$\quad 2x + \quad z = 5$

4. $x + y + z = 3$
$x - y - z = 2$
$-x + y - z = 1$

5. $2x - 3y = 0$
$4x + 5y = 0$

Check Exercise F ■

Work the following problems without looking at any text examples. Show your work in the space provided. Write your answer in the answer column.

1. _____

In Problems 1–3, refer to the matrices

2. _____

$$A = \begin{bmatrix} 3 & 0 \\ 1 & 4 \end{bmatrix} \qquad B = \begin{bmatrix} -2 & 5 \\ 3 & -4 \end{bmatrix}$$

3. _____

1. Find $A + B$.

4. _____

5. _____

2. Find $-2B$.

3. Find $B - A$.

4. Find

$$5 \begin{bmatrix} 0 \\ 2 \\ 1 \end{bmatrix} - 3 \begin{bmatrix} 1 \\ 2 \\ 0 \end{bmatrix}$$

5. Find a matrix D so that

$$\begin{bmatrix} 2 & 3 & 5 \\ 1 & 0 & -1 \end{bmatrix} + D = \begin{bmatrix} 1 & 4 & 2 \\ -3 & 1 & 0 \end{bmatrix}$$

Check Exercise G ■

Work the following problems without looking at any text examples. Show your work in the space provided. Write your answer in the answer column.

1. _____

2. _____

In Problems 1 and 2 find the dot product.

1. $[3 \quad -1] \cdot \begin{bmatrix} 2 \\ 4 \end{bmatrix}$

3. _____

4. _____

5. _____

2. $[-1 \quad 0 \quad 3] \cdot \begin{bmatrix} 8 \\ 6 \\ 4 \end{bmatrix}$

In Problems 3–5 find the matrix product.

3. $\begin{bmatrix} 1 & 2 \\ 3 & -4 \end{bmatrix} \begin{bmatrix} 2 & 0 \\ 1 & -1 \end{bmatrix}$

4. $\begin{bmatrix} 1 & 1 & -1 \\ -1 & 0 & 1 \end{bmatrix} \begin{bmatrix} 1 & -3 \\ 2 & -2 \\ 3 & -1 \end{bmatrix}$

5. $\begin{bmatrix} 2 & 1 \\ 1 & 2 \\ 0 & 3 \end{bmatrix} \begin{bmatrix} 0 & 1 \\ 1 & 2 \end{bmatrix}$

■ Tables

TABLE I VALUES OF e^x AND e^{-x} (0.00 TO 3.00)

x	e^x	e^{-x}	x	e^x	e^{-x}	x	e^x	e^{-x}
0.00	1.0000	1.000 00	0.50	1.6487	0.606 53	1.00	2.7183	0.367 88
0.01	1.0101	0.990 05	0.51	1.6653	0.600 50	1.01	2.7456	0.364 22
0.02	1.0202	0.980 20	0.52	1.6820	0.594 52	1.02	2.7732	0.360 59
0.03	1.0305	0.970 45	0.53	1.6989	0.588 60	1.03	2.8011	0.357 01
0.04	1.0408	0.960 79	0.54	1.7160	0.582 75	1.04	2.8292	0.353 45
0.05	1.0513	0.951 23	0.55	1.7333	0.576 95	1.05	2.8577	0.349 94
0.06	1.0618	0.941 76	0.56	1.7507	0.571 21	1.06	2.8864	0.346 46
0.07	1.0725	0.932 39	0.57	1.7683	0.565 53	1.07	2.9154	0.343 01
0.08	1.0833	0.923 12	0.58	1.7860	0.559 90	1.08	2.9447	0.339 60
0.09	1.0942	0.913 93	0.59	1.8040	0.554 33	1.09	2.9743	0.336 22
0.10	1.1052	0.904 84	0.60	1.8221	0.548 81	1.10	3.0042	0.332 87
0.11	1.1163	0.895 83	0.61	1.8404	0.543 35	1.11	3.0344	0.329 56
0.12	1.1275	0.886 92	0.62	1.8589	0.537 94	1.12	3.0649	0.326 28
0.13	1.1388	0.878 10	0.63	1.8776	0.532 59	1.13	3.0957	0.323 03
0.14	1.1503	0.869 36	0.64	1.8965	0.527 29	1.14	3.1268	0.319 82
0.15	1.1618	0.860 71	0.65	1.9155	0.522 05	1.15	3.1582	0.316 64
0.16	1.1735	0.852 14	0.66	1.9348	0.516 85	1.16	3.1899	0.313 49
0.17	1.1853	0.843 66	0.67	1.9542	0.511 71	1.17	3.2220	0.310 37
0.18	1.1972	0.835 27	0.68	1.9739	0.506 62	1.18	3.2544	0.307 28
0.19	1.2092	0.826 96	0.69	1.9937	0.501 58	1.19	3.2871	0.304 22
0.20	1.2214	0.818 73	0.70	2.0138	0.496 59	1.20	3.3201	0.301 19
0.21	1.2337	0.810 58	0.71	2.0340	0.491 64	1.21	3.3535	0.298 20
0.22	1.2461	0.802 52	0.72	2.0544	0.486 75	1.22	3.3872	0.295 23
0.23	1.2586	0.794 53	0.73	2.0751	0.481 91	1.23	3.4212	0.292 29
0.24	1.2712	0.786 63	0.74	2.0959	0.477 11	1.24	3.4556	0.289 38
0.25	1.2840	0.778 80	0.75	2.1170	0.472 37	1.25	3.4903	0.286 50
0.26	1.2969	0.771 05	0.76	2.1383	0.467 67	1.26	3.5254	0.283 65
0.27	1.3100	0.763 38	0.77	2.1598	0.463 01	1.27	3.5609	0.280 83
0.28	1.3231	0.755 78	0.78	2.1815	0.458 41	1.28	3.5966	0.278 04
0.29	1.3364	0.748 26	0.79	2.2034	0.453 84	1.29	3.6328	0.275 27
0.30	1.3499	0.740 82	0.80	2.2255	0.449 33	1.30	3.6693	0.272 53
0.31	1.3634	0.733 45	0.81	2.2479	0.444 86	1.31	3.7062	0.269 82
0.32	1.3771	0.726 15	0.82	2.2705	0.440 43	1.32	3.7434	0.267 14
0.33	1.3910	0.718 92	0.83	2.2933	0.436 05	1.33	3.7810	0.264 48
0.34	1.4049	0.711 77	0.84	2.3164	0.431 71	1.34	3.8190	0.261 85
0.35	1.4191	0.704 69	0.85	2.3396	0.427 41	1.35	3.8574	0.259 24
0.36	1.4333	0.697 68	0.86	2.3632	0.423 16	1.36	3.8962	0.256 66
0.37	1.4477	0.690 73	0.87	2.3869	0.418 95	1.37	3.9354	0.254 11
0.38	1.4623	0.683 86	0.88	2.4109	0.414 78	1.38	3.9749	0.251 58
0.39	1.4770	0.677 06	0.89	2.4351	0.410 66	1.39	4.0149	0.249 08
0.40	1.4918	0.670 32	0.90	2.4596	0.406 57	1.40	4.0552	0.246 60
0.41	1.5068	0.663 65	0.91	2.4843	0.402 52	1.41	4.0960	0.244 14
0.42	1.5220	0.657 05	0.92	2.5093	0.398 52	1.42	4.1371	0.241 71
0.43	1.5373	0.650 51	0.93	2.5345	0.394 55	1.43	4.1787	0.239 31
0.44	1.5527	0.644 04	0.94	2.5600	0.390 63	1.44	4.2207	0.236 93
0.45	1.5683	0.637 63	0.95	2.5857	0.386 74	1.45	4.2631	0.234 57
0.46	1.5841	0.631 28	0.96	2.6117	0.382 89	1.46	4.3060	0.232 24
0.47	1.6000	0.625 00	0.97	2.6379	0.379 08	1.47	4.3492	0.229 93
0.48	1.6161	0.618 78	0.98	2.6645	0.375 31	1.48	4.3939	0.227 64
0.49	1.6323	0.612 63	0.99	2.6912	0.371 58	1.49	4.4371	0.225 37
0.50	1.6487	0.606 53	1.00	2.7183	0.367 88	1.50	4.4817	0.223 13

x	e^x	e^{-x}	x	e^x	e^{-x}	x	e^x	e^{-x}
1.50	4.4817	0.223 13	2.00	7.3891	0.135 34	2.50	12.182	0.082 085
1.51	4.5267	0.220 91	2.01	7.4633	0.133 99	2.51	12.305	0.081 268
1.52	4.5722	0.218 71	2.02	7.5383	0.132 66	2.52	12.429	0.080 460
1.53	4.6182	0.216 54	2.03	7.6141	0.131 34	2.53	12.554	0.079 659
1.54	4.6646	0.214 38	2.04	7.6906	0.130 03	2.54	12.680	0.078 866
1.55	4.7115	0.212 25	2.05	7.7679	0.128 73	2.55	12.807	0.078 082
1.56	4.7588	0.210 14	2.06	7.8460	0.127 45	2.56	12.936	0.077 305
1.57	4.8066	0.208 05	2.07	7.9248	0.126 19	2.57	13.066	0.076 536
1.58	4.8550	0.205 98	2.08	8.0045	0.124 93	2.58	13.197	0.075 774
1.59	4.9037	0.203 93	2.09	8.0849	0.123 69	2.59	13.330	0.075 020
1.60	4.9530	0.201 90	2.10	8.1662	0.122 46	2.60	13.464	0.074 274
1.61	5.0028	0.199 89	2.11	8.2482	0.121 24	2.61	13.599	0.073 535
1.62	5.0531	0.197 90	2.12	8.3311	0.120 03	2.62	13.736	0.072 803
1.63	5.1039	0.195 93	2.13	8.4149	0.118 84	2.63	13.874	0.072 078
1.64	5.1552	0.193 98	2.14	8.4994	0.117 65	2.64	14.013	0.071 361
1.65	5.2070	0.192 05	2.15	8.5849	0.116 48	2.65	14.154	0.070 651
1.66	5.2593	0.190 14	2.16	8.6711	0.115 33	2.66	14.296	0.069 948
1.67	5.3122	0.188 25	2.17	8.7583	0.114 18	2.67	14.440	0.069 252
1.68	5.3656	0.186 37	2.18	8.8463	0.113 04	2.68	14.585	0.068 563
1.69	5.4195	0.184 52	2.19	8.9352	0.111 92	2.69	14.732	0.067 881
1.70	5.4739	0.182 68	2.20	9.0250	0.110 80	2.70	14.880	0.067 206
1.71	5.5290	0.180 87	2.21	9.1157	0.109 70	2.71	15.029	0.066 537
1.72	5.5845	0.179 07	2.22	9.2073	0.108 61	2.72	15.180	0.065 875
1.73	5.6407	0.177 28	2.23	9.2999	0.107 53	2.73	15.333	0.065 219
1.74	5.6973	0.175 52	2.24	9.3933	0.106 46	2.74	15.487	0.064 570
1.75	5.7546	0.173 77	2.25	9.4877	0.105 40	2.75	15.643	0.063 928
1.76	5.8124	0.172 04	2.26	9.5831	0.104 35	2.76	15.800	0.063 292
1.77	5.8709	0.170 33	2.27	9.6794	0.103 31	2.77	15.959	0.062 662
1.78	5.9299	0.168 64	2.28	9.7767	0.102 28	2.78	16.119	0.062 039
1.79	5.9895	0.166 96	2.29	9.8749	0.101 27	2.79	16.281	0.061 421
1.80	6.0496	0.165 30	2.30	9.9742	0.100 26	2.80	16.445	0.060 810
1.81	6.1104	0.163 65	2.31	10.074	0.099 261	2.81	16.610	0.060 205
1.82	6.1719	0.162 03	2.32	10.176	0.098 274	2.82	16.777	0.059 606
1.83	6.2339	0.160 41	2.33	10.278	0.097 296	2.83	16.945	0.059 013
1.84	6.2965	0.158 82	2.34	10.381	0.096 328	2.84	17.116	0.058 426
1.85	6.3598	0.157 24	2.35	10.486	0.095 369	2.85	17.288	0.057 844
1.86	6.4237	0.155 67	2.36	10.591	0.094 420	2.86	17.462	0.057 269
1.87	6.4883	0.154 12	2.37	10.697	0.093 481	2.87	17.637	0.056 699
1.88	6.5535	0.152 59	2.38	10.805	0.092 551	2.88	17.814	0.056 135
1.89	6.6194	0.151 07	2.39	10.913	0.091 630	2.89	17.993	0.055 576
1.90	6.6859	0.149 57	2.40	11.023	0.090 718	2.90	18.174	0.055 023
1.91	6.7531	0.148 08	2.41	11.134	0.089 815	2.91	18.357	0.054 476
1.92	6.8210	0.146 61	2.42	11.246	0.088 922	2.92	18.541	0.053 934
1.93	6.8895	0.145 15	2.43	11.359	0.088 037	2.93	18.728	0.053 397
1.94	6.9588	0.143 70	2.44	11.473	0.087 161	2.94	18.916	0.052 866
1.95	7.0287	0.142 27	2.45	11.588	0.086 294	2.95	19.106	0.052 340
1.96	7.0993	0.140 86	2.46	11.705	0.085 435	2.96	19.298	0.051 819
1.97	7.1707	0.139 46	2.47	11.822	0.084 585	2.97	19.492	0.051 303
1.98	7.2427	0.138 07	2.48	11.941	0.083 743	2.98	19.688	0.050 793
1.99	7.3155	0.136 70	2.49	12.061	0.082 910	2.99	19.886	0.050 287
2.00	7.3891	0.135 34	2.50	12.182	0.082 085	3.00	20.086	0.049 787

TABLE II COMMON LOGARITHMS

x	0	1	2	3	4	5	6	7	8	9
1.0	0.0000	0.004321	0.008600	0.01284	0.01703	0.02119	0.02531	0.02938	0.03342	0.03743
1.1	0.04139	0.04532	0.04922	0.05308	0.05690	0.06070	0.06446	0.06819	0.07188	0.07555
1.2	0.07918	0.08279	0.08636	0.08991	0.09342	0.09691	0.1004	0.1038	0.1072	0.1106
1.3	0.1139	0.1173	0.1206	0.1239	0.1271	0.1303	0.1335	0.1367	0.1399	0.1430
1.4	0.1461	0.1492	0.1523	0.1553	0.1584	0.1614	0.1644	0.1673	0.1703	0.1732
1.5	0.1761	0.1790	0.1818	0.1847	0.1875	0.1903	0.1931	0.1959	0.1987	0.2014
1.6	0.2041	0.2068	0.2095	0.2122	0.2148	0.2175	0.2201	0.2227	0.2253	0.2279
1.7	0.2304	0.2330	0.2355	0.2380	0.2405	0.2430	0.2455	0.2480	0.2504	0.2529
1.8	0.2553	0.2577	0.2601	0.2625	0.2648	0.2673	0.2695	0.2718	0.2742	0.2765
1.9	0.2788	0.2810	0.2833	0.2856	0.2878	0.2900	0.2923	0.2945	0.2967	0.2989
2.0	0.3010	0.3032	0.3054	0.3075	0.3096	0.3118	0.3139	0.3160	0.3181	0.3201
2.1	0.3222	0.3243	0.3263	0.3284	0.3304	0.3324	0.3345	0.3365	0.3385	0.3404
2.2	0.3424	0.3444	0.3464	0.3483	0.3502	0.3522	0.3541	0.3560	0.3579	0.3598
2.3	0.3617	0.3636	0.3655	0.3674	0.3692	0.3711	0.3729	0.3747	0.3766	0.3784
2.4	0.3802	0.3820	0.3838	0.3856	0.3874	0.3892	0.3909	0.3927	0.3945	0.3962
2.5	0.3979	0.3997	0.4014	0.4031	0.4048	0.4065	0.4082	0.4099	0.4116	0.4133
2.6	0.4150	0.4166	0.4183	0.4200	0.4216	0.4232	0.4249	0.4265	0.4281	0.4298
2.7	0.4314	0.4330	0.4346	0.4362	0.4378	0.4393	0.4409	0.4425	0.4440	0.4456
2.8	0.4472	0.4487	0.4502	0.4518	0.4533	0.4548	0.4564	0.4579	0.4594	0.4609
2.9	0.4624	0.4639	0.4654	0.4669	0.4683	0.4698	0.4713	0.4728	0.4742	0.4757
3.0	0.4771	0.4786	0.4800	0.4814	0.4829	0.4843	0.4857	0.4871	0.4886	0.4900
3.1	0.4914	0.4928	0.4942	0.4955	0.4969	0.4983	0.4997	0.5011	0.5024	0.5038
3.2	0.5051	0.5065	0.5079	0.5092	0.5105	0.5119	0.5132	0.5145	0.5159	0.5172
3.3	0.5185	0.5198	0.5211	0.5224	0.5237	0.5250	0.5263	0.5276	0.5289	0.5302
3.4	0.5315	0.5328	0.5340	0.5353	0.5366	0.5378	0.5391	0.5403	0.5416	0.5428
3.5	0.5441	0.5453	0.5465	0.5478	0.5490	0.5502	0.5514	0.5527	0.5539	0.5551
3.6	0.5563	0.5575	0.5587	0.5599	0.5611	0.5623	0.5635	0.5647	0.5658	0.5670
3.7	0.5682	0.5694	0.5705	0.5717	0.5729	0.5740	0.5752	0.5763	0.5775	0.5786
3.8	0.5798	0.5809	0.5821	0.5832	0.5843	0.5855	0.5866	0.5877	0.5888	0.5899
3.9	0.5911	0.5922	0.5933	0.5944	0.5955	0.5966	0.5977	0.5988	0.5999	0.6010
4.0	0.6021	0.6031	0.6042	0.6053	0.6064	0.6075	0.6085	0.6096	0.6107	0.6117
4.1	0.6128	0.6138	0.6149	0.6160	0.6170	0.6180	0.6191	0.6201	0.6212	0.6222
4.2	0.6232	0.6243	0.6253	0.6263	0.6274	0.6284	0.6294	0.6304	0.6314	0.6325
4.3	0.6335	0.6345	0.6355	0.6365	0.6375	0.6385	0.6395	0.6405	0.6415	0.6425
4.4	0.6435	0.6444	0.6454	0.6464	0.6474	0.6484	0.6493	0.6503	0.6513	0.6522
4.5	0.6532	0.6542	0.6551	0.6561	0.6571	0.6580	0.6590	0.6599	0.6609	0.6618
4.6	0.6628	0.6637	0.6646	0.6656	0.6665	0.6675	0.6684	0.6693	0.6702	0.6712
4.7	0.6721	0.6730	0.6739	0.6749	0.6758	0.6767	0.6776	0.6785	0.6794	0.6803
4.8	0.6812	0.6821	0.6830	0.6839	0.6848	0.6857	0.6866	0.6875	0.6884	0.6893
4.9	0.6902	0.6911	0.6920	0.6928	0.6937	0.6946	0.6955	0.6964	0.6972	0.6981
5.0	0.6990	0.6998	0.7007	0.7016	0.7024	0.7033	0.7042	0.7050	0.7059	0.7067
5.1	0.7076	0.7084	0.7093	0.7101	0.7110	0.7118	0.7126	0.7135	0.7143	0.7152
5.2	0.7160	0.7168	0.7177	0.7185	0.7193	0.7202	0.7210	0.7218	0.7226	0.7235
5.3	0.7243	0.7251	0.7259	0.7267	0.7275	0.7284	0.7292	0.7300	0.7308	0.7316
5.4	0.7324	0.7332	0.7340	0.7348	0.7356	0.7364	0.7372	0.7380	0.7388	0.7396

x	0	1	2	3	4	5	6	7	8	9
5.5	0.7404	0.7412	0.7419	0.7427	0.7435	0.7443	0.7451	0.7459	0.7466	0.7474
5.6	0.7482	0.7490	0.7497	0.7505	0.7513	0.7520	0.7528	0.7536	0.7543	0.7551
5.7	0.7559	0.7566	0.7574	0.7582	0.7589	0.7597	0.7604	0.7612	0.7619	0.7627
5.8	0.7634	0.7642	0.7649	0.7657	0.7664	0.7672	0.7679	0.7686	0.7694	0.7701
5.9	0.7709	0.7716	0.7723	0.7731	0.7738	0.7745	0.7752	0.7760	0.7767	0.7774
6.0	0.7782	0.7789	0.7796	0.7803	0.7810	0.7818	0.7825	0.7832	0.7839	0.7846
6.1	0.7853	0.7860	0.7868	0.7875	0.7882	0.7889	0.7896	0.7903	0.7910	0.7917
6.2	0.7924	0.7931	0.7938	0.7945	0.7952	0.7959	0.7966	0.7973	0.7980	0.7987
6.3	0.7993	0.8000	0.8007	0.8014	0.8021	0.8028	0.8035	0.8041	0.8048	0.8055
6.4	0.8062	0.8069	0.8075	0.8082	0.8089	0.8096	0.8102	0.8109	0.8116	0.8122
6.5	0.8129	0.8136	0.8142	0.8149	0.8156	0.8162	0.8169	0.8176	0.8182	0.8189
6.6	0.8195	0.8202	0.8209	0.8215	0.8222	0.8228	0.8235	0.8241	0.8248	0.8254
6.7	0.8261	0.8267	0.8274	0.8280	0.8287	0.8293	0.8299	0.8306	0.8312	0.8319
6.8	0.8325	0.8331	0.8338	0.8344	0.8351	0.8357	0.8363	0.8370	0.8376	0.8382
6.9	0.8388	0.8395	0.8401	0.8407	0.8414	0.8420	0.8426	0.8432	0.8439	0.8445
7.0	0.8451	0.8457	0.8463	0.8470	0.8476	0.8482	0.8488	0.8494	0.8500	0.8506
7.1	0.8513	0.8519	0.8525	0.8531	0.8537	0.8543	0.8549	0.8555	0.8561	0.8567
7.2	0.8573	0.8579	0.8585	0.8591	0.8597	0.8603	0.8609	0.8615	0.8621	0.8627
7.3	0.8633	0.8639	0.8645	0.8651	0.8657	0.8663	0.8669	0.8675	0.8681	0.8686
7.4	0.8692	0.8698	0.8704	0.8710	0.8716	0.8722	0.8727	0.8733	0.8739	0.8745
7.5	0.8751	0.8756	0.8762	0.8768	0.8774	0.8779	0.8785	0.8791	0.8797	0.8802
7.6	0.8808	0.8814	0.8820	0.8825	0.8831	0.8837	0.8842	0.8848	0.8854	0.8859
7.7	0.8865	0.8871	0.8876	0.8882	0.8887	0.8893	0.8899	0.8904	0.8910	0.8915
7.8	0.8921	0.8927	0.8932	0.8938	0.8943	0.8949	0.8954	0.8960	0.8965	0.8971
7.9	0.8976	0.8982	0.8987	0.8993	0.8998	0.9004	0.9009	0.9015	0.9020	0.9025
8.0	0.9031	0.9036	0.9042	0.9047	0.9053	0.9058	0.9063	0.9069	0.9074	0.9079
8.1	0.9085	0.9090	0.9096	0.9101	0.9106	0.9112	0.9117	0.9122	0.9128	0.9133
8.2	0.9138	0.9143	0.9149	0.9154	0.9159	0.9165	0.9170	0.9175	0.9180	0.9186
8.3	0.9191	0.9196	0.9201	0.9206	0.9212	0.9217	0.9222	0.9227	0.9232	0.9238
8.4	0.9243	0.9248	0.9253	0.9258	0.9263	0.9269	0.9274	0.9279	0.9284	0.9289
8.5	0.9294	0.9299	0.9304	0.9309	0.9315	0.9320	0.9325	0.9330	0.9335	0.9340
8.6	0.9345	0.9350	0.9355	0.9360	0.9365	0.9370	0.9375	0.9380	0.9385	0.9390
8.7	0.9395	0.9400	0.9405	0.9410	0.9415	0.9420	0.9425	0.9430	0.9435	0.9440
8.8	0.9445	0.9450	0.9455	0.9460	0.9465	0.9469	0.9474	0.9479	0.9484	0.9489
8.9	0.9494	0.9499	0.9504	0.9509	0.9513	0.9518	0.9523	0.9528	0.9533	0.9538
9.0	0.9542	0.9547	0.9552	0.9557	0.9562	0.9566	0.9571	0.9576	0.9581	0.9586
9.1	0.9590	0.9595	0.9600	0.9605	0.9609	0.9614	0.9619	0.9624	0.9628	0.9633
9.2	0.9638	0.9643	0.9647	0.9652	0.9657	0.9661	0.9666	0.9671	0.9675	0.9680
9.3	0.9685	0.9689	0.9694	0.9699	0.9703	0.9708	0.9713	0.9717	0.9722	0.9727
9.4	0.9731	0.9736	0.9741	0.9745	0.9750	0.9754	0.9759	0.9763	0.9768	0.9773
9.5	0.9777	0.9782	0.9786	0.9791	0.9795	0.9800	0.9805	0.9809	0.9814	0.9818
9.6	0.9823	0.9827	0.9832	0.9836	0.9841	0.9045	0.9850	0.9854	0.9859	0.9863
9.7	0.9868	0.9872	0.9877	0.9881	0.9886	0.9890	0.9894	0.9899	0.9903	0.9908
9.8	0.9912	0.9917	0.9921	0.9926	0.9930	0.9934	0.9939	0.9943	0.9948	0.9952
9.9	0.9956	0.9961	0.9965	0.9969	0.9974	0.9978	0.9983	0.9987	0.9991	0.9996

TABLE III   NATURAL LOGARITHMS (ln $x = \log_e x$)

ln 10 = 2.3026	5 ln 10 = 11.5130	9 ln 10 = 20.7233
2 ln 10 = 4.6052	6 ln 10 = 13.8155	10 ln 10 = 23.0259
3 ln 10 = 6.9078	7 ln 10 = 16.1181	
4 ln 10 = 9.2103	8 ln 10 = 18.4207	

x	.00	.01	.02	.03	.04	.05	.06	.07	.08	.09
1.0	0.0000	0.0100	0.0198	0.0296	0.0392	0.0488	0.0583	0.0677	0.0770	0.0862
1.1	0.0953	0.1044	0.1133	0.1222	0.1310	0.1398	0.1484	0.1570	0.1655	0.1740
1.2	0.1823	0.1906	0.1989	0.2070	0.2151	0.2231	0.2311	0.2390	0.2469	0.2546
1.3	0.2624	0.2700	0.2776	0.2852	0.2927	0.3001	0.3075	0.3148	0.3221	0.3293
1.4	0.3365	0.3436	0.3507	0.3577	0.3646	0.3716	0.3784	0.3853	0.3920	0.3988
1.5	0.4055	0.4121	0.4187	0.4253	0.4318	0.4383	0.4447	0.4511	0.4574	0.4637
1.6	0.4700	0.4762	0.4824	0.4886	0.4947	0.5008	0.5068	0.5128	0.5188	0.5247
1.7	0.5306	0.5365	0.5423	0.5481	0.5539	0.5596	0.5653	0.5710	0.5766	0.5822
1.8	0.5878	0.5933	0.5988	0.6043	0.6098	0.6152	0.6206	0.6259	0.6313	0.6366
1.9	0.6419	0.6471	0.6523	0.6575	0.6627	0.6678	0.6729	0.6780	0.6831	0.6881
2.0	0.6931	0.6981	0.7031	0.7080	0.7129	0.7178	0.7227	0.7275	0.7324	0.7372
2.1	0.7419	0.7467	0.7514	0.7561	0.7608	0.7655	0.7701	0.7747	0.7793	0.7839
2.2	0.7885	0.7930	0.7975	0.8020	0.8065	0.8109	0.8154	0.8198	0.8242	0.8286
2.3	0.8329	0.8372	0.8416	0.8459	0.8502	0.8544	0.8587	0.8629	0.8671	0.8713
2.4	0.8755	0.8796	0.8838	0.8879	0.8920	0.8961	0.9002	0.9042	0.9083	0.9123
2.5	0.9163	0.9203	0.9243	0.9282	0.9322	0.9361	0.9400	0.9439	0.9478	0.9517
2.6	0.9555	0.9594	0.9632	0.9670	0.9708	0.9746	0.9783	0.9821	0.9858	0.9895
2.7	0.9933	0.9969	1.0006	1.0043	1.0080	1.0116	1.0152	1.0188	1.0225	1.0260
2.8	1.0296	1.0332	1.0367	1.0403	1.0438	1.0473	1.0508	1.0543	1.0578	1.0613
2.9	1.0647	1.0682	1.0716	1.0750	1.0784	1.0818	1.0852	1.0886	1.0919	1.0953
3.0	1.0986	1.1019	1.1053	1.1086	1.1119	1.1151	1.1184	1.1217	1.1249	1.1282
3.1	1.1314	1.1346	1.1378	1.1410	1.1442	1.1474	1.1506	1.1537	1.1569	1.1600
3.2	1.1632	1.1663	1.1694	1.1725	1.1756	1.1787	1.1817	1.1848	1.1878	1.1909
3.3	1.1939	1.1969	1.2000	1.2030	1.2060	1.2090	1.2119	1.2149	1.2179	1.2208
3.4	1.2238	1.2267	1.2296	1.2326	1.2355	1.2384	1.2413	1.2442	1.2470	1.2499
3.5	1.2528	1.2556	1.2585	1.2613	1.2641	1.2669	1.2698	1.2726	1.2754	1.2782
3.6	1.2809	1.2837	1.2865	1.2892	1.2920	1.2947	1.2975	1.3002	1.3029	1.3056
3.7	1.3083	1.3110	1.3137	1.3164	1.3191	1.3218	1.3244	1.3271	1.3297	1.3324
3.8	1.3350	1.3376	1.3403	1.3429	1.3455	1.3481	1.3507	1.3533	1.3558	1.3584
3.9	1.3610	1.3635	1.3661	1.3686	1.3712	1.3737	1.3762	1.3788	1.3813	1.3838
4.0	1.3863	1.3888	1.3913	1.3938	1.3962	1.3987	1.4012	1.4036	1.4061	1.4085
4.1	1.4110	1.4134	1.4159	1.4183	1.4207	1.4231	1.4255	1.4279	1.4303	1.4327
4.2	1.4351	1.4375	1.4398	1.4422	1.4446	1.4469	1.4493	1.4516	1.4540	1.4563
4.3	1.4586	1.4609	1.4633	1.4656	1.4679	1.4702	1.4725	1.4748	1.4770	1.4793
4.4	1.4816	1.4839	1.4861	1.4884	1.4907	1.4929	1.4951	1.4974	1.4996	1.5019
4.5	1.5041	1.5063	1.5085	1.5107	1.5129	1.5151	1.5173	1.5195	1.5217	1.5239
4.6	1.5261	1.5282	1.5304	1.5326	1.5347	1.5369	1.5390	1.5412	1.5433	1.5454
4.7	1.5476	1.5497	1.5518	1.5539	1.5560	1.5581	1.5602	1.5623	1.5644	1.5665
4.8	1.5686	1.5707	1.5728	1.5748	1.5769	1.5790	1.5810	1.5831	1.5851	1.5872
4.9	1.5892	1.5913	1.5933	1.5953	1.5974	1.5994	1.6014	1.6034	1.6054	1.6074
5.0	1.6094	1.6114	1.6134	1.6154	1.6174	1.6194	1.6214	1.6233	1.6253	1.6273
5.1	1.6292	1.6312	1.6332	1.6351	1.6371	1.6390	1.6409	1.6429	1.6448	1.6467
5.2	1.6487	1.6506	1.6525	1.6544	1.6563	1.6582	1.6601	1.6620	1.6639	1.6658
5.3	1.6677	1.6696	1.6715	1.6734	1.6752	1.6771	1.6790	1.6808	1.6827	1.6845
5.4	1.6864	1.6882	1.6901	1.6919	1.6938	1.6956	1.6974	1.6993	1.7011	1.7029

Note: ln 35,200 = ln(3.52 × 10⁴) = ln 3.52 + 4 ln 10

ln 0.008 64 = ln(8.64 × 10⁻³) = ln 8.64 − 3 ln 10

x	.00	.01	.02	.03	.04	.05	.06	.07	.08	.09
5.5	1.7047	1.7066	1.7084	1.7102	1.7120	1.7138	1.7156	1.7174	1.7192	1.7210
5.6	1.7228	1.7246	1.7263	1.7281	1.7299	1.7317	1.7334	1.7352	1.7370	1.7387
5.7	1.7405	1.7422	1.7440	1.7457	1.7475	1.7492	1.7509	1.7527	1.7544	1.7561
5.8	1.7579	1.7596	1.7613	1.7630	1.7647	1.7664	1.7681	1.7699	1.7716	1.7733
5.9	1.7750	1.7766	1.7783	1.7800	1.7817	1.7834	1.7851	1.7867	1.7884	1.7901
6.0	1.7918	1.7934	1.7951	1.7967	1.7984	1.8001	1.8017	1.8034	1.8050	1.8066
6.1	1.8083	1.8099	1.8116	1.8132	1.8148	1.8165	1.8181	1.8197	1.8213	1.8229
6.2	1.8245	1.8262	1.8278	1.8294	1.8310	1.8326	1.8342	1.8358	1.8374	1.8390
6.3	1.8405	1.8421	1.8437	1.8453	1.8469	1.8485	1.8500	1.8516	1.8532	1.8547
6.4	1.8563	1.8579	1.8594	1.8610	1.8625	1.8641	1.8656	1.8672	1.8687	1.8703
6.5	1.8718	1.8733	1.8749	1.8764	1.8779	1.8795	1.8810	1.8825	1.8840	1.8856
6.6	1.8871	1.8886	1.8901	1.8916	1.8931	1.8946	1.8961	1.8976	1.8991	1.9006
6.7	1.9021	1.9036	1.9051	1.9066	1.9081	1.9095	1.9110	1.9125	1.9140	1.9155
6.8	1.9169	1.9184	1.9199	1.9213	1.9228	1.9242	1.9257	1.9272	1.9286	1.9301
6.9	1.9315	1.9330	1.9344	1.9359	1.9373	1.9387	1.9402	1.9416	1.9430	1.9445
7.0	1.9459	1.9473	1.9488	1.9502	1.9516	1.9530	1.9544	1.9559	1.9573	1.9587
7.1	1.9601	1.9615	1.9629	1.9643	1.9657	1.9671	1.9685	1.9699	1.9713	1.9727
7.2	1.9741	1.9755	1.9769	1.9782	1.9796	1.9810	1.9824	1.9838	1.9851	1.9865
7.3	1.9879	1.9892	1.9906	1.9920	1.9933	1.9947	1.9961	1.9974	1.9988	2.0001
7.4	2.0015	2.0028	2.0042	2.0055	2.0069	2.0082	2.0096	2.0109	2.0122	2.0136
7.5	2.0149	2.0162	2.0176	2.0189	2.0202	2.0215	2.0229	2.0242	2.0255	2.0268
7.6	2.0281	2.0295	2.0308	2.0321	2.0334	2.0347	2.0360	2.0373	2.0386	2.0399
7.7	2.0412	2.0425	2.0438	2.0451	2.0464	2.0477	2.0490	2.0503	2.0516	2.0528
7.8	2.0541	2.0554	2.0567	2.0580	2.0592	2.0605	2.0618	2.0631	2.0643	2.0656
7.9	2.0669	2.0681	2.0694	2.0707	2.0719	2.0732	2.0744	2.0757	2.0769	2.0782
8.0	2.0794	2.0807	2.0819	2.0832	2.0844	2.0857	2.0869	2.0882	2.0894	2.0906
8.1	2.0919	2.0931	2.0943	2.0956	2.0968	2.0980	2.0992	2.1005	2.1017	2.1029
8.2	2.1041	2.1054	2.1066	2.1078	2.1090	2.1102	2.1114	2.1126	2.1138	2.1150
8.3	2.1163	2.1175	2.1187	2.1199	2.1211	2.1223	2.1235	2.1247	2.1258	2.1270
8.4	2.1282	2.1294	2.1306	2.1318	2.1330	2.1342	2.1353	2.1365	2.1377	2.1389
8.5	2.1401	2.1412	2.1424	2.1436	2.1448	2.1459	2.1471	2.1483	2.1494	2.1506
8.6	2.1518	2.1529	2.1541	2.1552	2.1564	2.1576	2.1587	2.1599	2.1610	2.1622
8.7	2.1633	2.1645	2.1656	2.1668	2.1679	2.1691	2.1702	2.1713	2.1725	2.1736
8.8	2.1748	2.1759	2.1770	2.1782	2.1793	2.1804	2.1815	2.1827	2.1838	2.1849
8.9	2.1861	2.1872	2.1883	2.1894	2.1905	2.1917	2.1928	2.1939	2.1950	2.1961
9.0	2.1972	2.1983	2.1994	2.2006	2.2017	2.2028	2.2039	2.2050	2.2061	2.2072
9.1	2.2083	2.2094	2.2105	2.2116	2.2127	2.2138	2.2148	2.2159	2.2170	2.2181
9.2	2.2192	2.2203	2.2214	2.2225	2.2235	2.2246	2.2257	2.2268	2.2279	2.2289
9.3	2.2300	2.2311	2.2322	2.2332	2.2343	2.2354	2.2364	2.2375	2.2386	2.2396
9.4	2.2407	2.2418	2.2428	2.2439	2.2450	2.2460	2.2471	2.2481	2.2492	2.2502
9.5	2.2513	2.2523	2.2534	2.2544	2.2555	2.2565	2.2576	2.2586	2.2597	2.2607
9.6	2.2618	2.2628	2.2638	2.2649	2.2659	2.2670	2.2680	2.2690	2.2701	2.2711
9.7	2.2721	2.2732	2.2742	2.2752	2.2762	2.2773	2.2783	2.2793	2.2803	2.2814
9.8	2.2824	2.2834	2.2844	2.2854	2.2865	2.2875	2.2885	2.2895	2.2905	2.2915
9.9	2.2925	2.2935	2.2946	2.2956	2.2966	2.2976	2.2986	2.2996	2.3006	2.3016

TABLE IV   LOGARITHMS OF FACTORIAL *n*

n	log n!	n	log n!	n	log n!	n	log n!
		50	64.483 07	100	157.970 00	150	262.756 89
1	0.000 00	51	66.190 64	101	159.974 32	151	264.935 87
2	0.301 03	52	67.906 65	102	161.982 93	152	267.117 71
3	0.778 15	53	69.630 92	103	163.995 76	153	269.302 41
4	1.380 21	54	71.363 32	104	166.012 80	154	271.489 93
5	2.079 18	55	73.103 68	105	168.033 99	155	273.680 26
6	2.857 33	56	74.851 87	106	170.059 29	156	275.873 38
7	3.702 43	57	76.607 74	107	172.086 67	157	278.069 28
8	4.605 52	58	78.371 17	108	174.122 10	158	280.267 94
9	5.559 76	59	80.142 02	109	176.159 52	159	282.469 34
10	6.559 76	60	81.920 17	110	178.200 92	160	284.673 46
11	7.601 16	61	83.705 50	111	180.246 24	161	286.880 28
12	8.680 34	62	85.497 90	112	182.295 46	162	289.089 80
13	9.794 28	63	87.297 24	113	184.348 54	163	291.301 98
14	10.940 41	64	89.103 42	114	186.405 44	164	293.516 83
15	12.116 50	65	90.916 33	115	188.466 14	165	295.734 31
16	13.320 62	66	92.735 87	116	190.530 60	166	297.954 42
17	14.551 07	67	94.561 95	117	192.598 78	167	300.177 14
18	15.806 34	68	96.394 46	118	194.670 67	168	302.402 45
19	17.085 09	69	98.233 31	119	196.746 21	169	304.630 33
20	18.386 12	70	100.078 40	120	198.825 39	170	306.860 78
21	19.708 34	71	101.929 66	121	200.908 18	171	309.093 78
22	21.050 77	72	103.787 00	122	202.994 54	172	311.329 31
23	22.412 49	73	105.650 32	123	205.084 44	173	313.567 35
24	23.792 71	74	107.519 55	124	207.177 87	174	315.807 90
25	25.190 65	75	109.394 61	125	209.274 78	175	318.050 94
26	26.605 62	76	111.275 43	126	211.375 15	176	320.296 45
27	28.036 98	77	113.161 92	127	213.478 95	177	322.544 43
28	29.484 14	78	115.054 01	128	215.586 16	178	324.794 85
29	30.946 54	79	116.951 64	129	217.696 75	179	327.047 70
30	32.423 66	80	118.854 73	130	219.810 69	180	329.302 97
31	33.915 02	81	120.763 21	131	221.927 96	181	331.560 65
32	35.420 17	82	122.677 03	132	224.048 54	182	333.820 72
33	36.938 69	83	124.596 10	133	226.172 39	183	336.083 17
34	38.470 16	84	126.520 38	134	228.299 49	184	338.347 99
35	40.014 23	85	128.449 80	135	230.429 83	185	340.615 16
36	41.570 54	86	130.384 30	136	232.563 37	186	342.884 68
37	43.138 74	87	132.323 82	137	234.700 09	187	345.156 52
38	44.718 52	88	134.268 30	138	236.839 97	188	347.430 67
39	46.309 59	89	136.217 69	139	238.982 98	189	349.707 14
40	47.911 65	90	138.171 94	140	241.129 11	190	351.985 89
41	49.524 43	91	140.130 98	141	242.278 33	191	354.266 92
42	51.147 68	92	142.094 76	142	245.430 62	192	356.550 22
43	52.781 15	93	144.063 25	143	247.585 95	193	358.835 78
44	54.424 60	94	146.036 38	144	249.744 32	194	361.123 58
45	56.077 81	95	148.014 10	145	251.905 68	195	363.413 62
46	57.740 57	96	149.996 37	146	254.070 04	196	365.705 87
47	59.412 67	97	151.983 14	147	256.237 35	197	368.000 34
48	61.093 91	98	153.974 37	148	258.407 62	198	370.297 01
49	62.784 10	99	155.970 00	149	260.580 80	199	372.595 86

■ Answers to Problems

Chapter 1   Exercise 1-1

Odd

1. T   **3.** T   **5.** T   **7.** T   **9.** T   **11.** F
13. $-3, 0, 5$ (Infinitely many more answers are possible, except for 0.)   **15.** $\frac{2}{3}, -\frac{7}{8}, 2.65$ are three of infinitely many
17. $\{6, 7, 8, 9\}$ **19.** $\{a, s, t, u\}$ **21.** $\varnothing$ **23.** $\{5\}$ **25.** $\varnothing$ **27.** $\{-2, 2\}$ **29.** **(A)** T **(B)** F **(C)** T
31. **(A)** 3 and 4 **(B)** -2 and -1 **(C)** -3 and -2 **33.** **(A)** $\{1, 2, 3, 4, 6\}$ **(B)** $\{2, 4\}$
35. **(A)** J, Q, R **(B)** Q, R **(C)** R **(D)** Q, R **37.** $\frac{1}{11}$
39. **(A)** $0.88888888\ldots$ **(B)** $0.27272727\ldots$ **(C)** $2.23606797\ldots$ **(D)** $1.37500000\ldots$ **41.** 6

Even

2. F   **4.** F   **6.** F   **8.** T   **10.** T   **12.** T
14. $-\frac{3}{4}, 0, \frac{2}{3}$ (Infinitely many more answers are possible, except for 0.)   **16.** -3 and 0 are two of infinitely many
18. $\{5, 6, 7\}$ **20.** $\{i, l, n, o, s\}$ **22.** $\varnothing$ **24.** $\{-3\}$ **26.** $\varnothing$ **28.** $\{-3, 3\}$ **30.** **(A)** F **(B)** T **(C)** T
32. **(A)** 3 and 4 **(B)** -3 and -2 **(C)** -5 and -4 **34.** **(A)** $\{-2, -1, 0, 1, 2\}$ **(B)** $\{0, 2\}$
36. **(A)** N, J, Q, R **(B)** R **(C)** Q, R **(D)** Q, R
38. $\frac{2}{11}$ **40.** **(A)** $2.16666666\ldots$ **(B)** $4.58257569\ldots$ **(C)** $0.43750000\ldots$ **(D)** $0.26126126\ldots$ **42.** 4

Exercise 1-2

Odd

1. $11 - 5 = \frac{12}{2}$ **3.** $4 > -18$ **5.** $-12 < -3$ **7.** $x \geq -8$ **9.** $-2 < x < 2$ **11.** Is equal to
13. Is greater than   **15.** Is less than   **17.** Is less than or equal to   **19.** Is not equal to
21. Is greater than or equal to   **23.** $<$ **25.** $>$ **27.** $<$ **29.** $=$ **31.** $<$ **33.** $>$ **35.** $<, <$
37. $>$ **39.** $>$ **41.** $<$ **43.** Greater than   **45.**   **47.**

49.   **51.**   **53.** $x - 8 > 0$ **55.** $x + 4 \geq 0$ **57.** $80 = 3 + 2x$

59. $-3 \leq x < 4$ **61.** $26 = x - 12$ **63.** $x < 2x - 6$ **65.** $6x = 4 + 3x$ **67.** $x - 6 = 5(7 + x)$
69. $63 \leq \frac{9}{5}C + 32 \leq 72$ **71.** $x + (x + 1) + (x + 2) = 186$ **73.** $t = -5$ **75.** $5x + 7x = 12x$ **77.** $3 - x$
79. "Is" does not translate into "equal" in this case. (The number 8 is actually an element in the set of even numbers.) The properties of equality do not apply.
81. $90 = x(2x - 3)$ **83.** $2x + 2(3x - 10) = 210$ **85.** **(A)** $A = x(200 - x)$ **(B)** $0 \leq x \leq 200$

Even

2. $\frac{36}{2} = 3 \cdot 6$ **4.** $3 > -7$ **6.** $-20 < 1$ **8.** $x \leq 12$ **10.** $0 < x < 10$ **12.** Is equal to   **14.** Is greater than
16. Is less than   **18.** Is less than or equal to   **20.** Is not equal to   **22.** Is greater than or equal to
24. $>$ **26.** $>$ **28.** $<$ **30.** $=$ **32.** $>$ **34.** $<$ **36.** $<, <$ **38.** $<$ **40.** $<$ **42.** $>$
44. Right   **46.**   **48.**   **50.**

52.   **54.** $2x - 8 < 0$ **56.** $4x + 1 \leq 0$ **58.** $18 = 3x$ **60.** $0 < x < 10$

62. $32 = x - 5$ **64.** $x > 9 + 4x$ **66.** $7x = 4x - 12$ **68.** $5 + x = 3(x - 4)$ **70.** $20 \leq \frac{5}{9}(F - 32) \leq 25$

A1

72. $x + (x + 1) + (x + 2) = 372$ **74.** $x < y$ **76.** $x > 3$ **78.** $x - 3$
80. "Is" does not translate to "equal" in this case. The properties of equality do not apply.
82. $500 = x(x + 5)$ **84.** $2x + 2(x - 7) = 186$ **86.** **(A)** $V = x(12 - 2x)(18 - 2x)$ **(B)** $0 \le x \le 6$

Exercise 1-3

Odd

1. $3 + x$ **3.** $(5 \cdot 7)z$ **5.** mn **7.** $(9 + 11) + M$ **9.** $7x$ **11.** $x + y$ **13.** Commutative axiom for addition
15. Associative axiom for addition   **17.** Commutative axiom for multiplication   **19.** Associative axiom for multiplication
21. Identity axiom for addition   **23.** Identity axiom for multiplication   **25.** $x + 9$ **27.** $20y$ **29.** $u + 15$
31. $21x$ **33.** x **35.** Commutative axiom for addition   **37.** Commutative axiom for multiplication
39. Commutative axiom for addition   **41.** Associative axiom for multiplication   **43.** $x + y + z + 12$
45. $3x + 4y + 11$ **47.** $x + y + 5$ **49.** $36mnp$ **51.** (B) and (D) are false, since $12 - 4 \ne 4 - 12$ and $12 \div 4 \ne 4 \div 12$.
53. *1.* Commutative axiom for addition *2.* Associative axiom for addition *3.* Associative axiom for addition
   4. Substitution principle for equality *5.* Commutative axiom for addition *6.* Associative axiom for addition

Even

2. $n + m$ **4.** $u(vw)$ **6.** cd **8.** $x + (7 + 5)$ **10.** $x + z$ **12.** uv **14.** Commutative axiom for addition
16. Associative axiom for addition   **18.** Commutative axiom for multiplication   **20.** Associative axiom for multiplication
22. Identity axiom for addition   **24.** Identity axiom for multiplication   **26.** $8 + m$ **28.** $48n$ **30.** $x + 17$
32. $12y$ **34.** $y + 3$ **36.** Commutative axiom for multiplication   **38.** Commutative axiom for addition
40. Associative axiom for addition   **42.** Associative axiom for addition   **44.** $18 + m + n + p$
46. $3a + 5b + 9$ **48.** $x + y + 5$ **50.** $64xyz$
52. **(A)** True **(B)** False; $(8 - 4) - 2 \ne 8 - (4 - 2)$ **(C)** True **(D)** False; $(8 \div 4) \div 2 \ne 8 \div (4 \div 2)$
54. *1.* Commutative axiom for multiplication   *2.* Associative axiom for multiplication
   3. Associative axiom for multiplication   *4.* Substitution principle for equality
   5. Associative axiom for multiplication   *6.* Commutative axiom for multiplication
   7. Associative axiom for multiplication

Exercise 1-4

Odd

1. -7 **3.** 6   **5.** 2   **7.** 27   **9.** 0   **11.** -10 **13.** 4   **15.** -6 **17.** 12   **19.** Sometimes
21. -3 **23.** -2 **25.** -6 **27.** -5 **29.** -5 **31.** -5 **33.** -1 **35.** -6 **37.** 28   **39.** -5
41. 7 or -7 **43.** -5 **45.** 5   **47.** -5 **49.** -11 **51.** -3 **53.** 6   **55.** -5 **57.** True
59. False; $(+7) - (-3) = +10, (-3) - (+7) = -10$ **61.** True   **63.** False; $|(+9) + (-3)| = +6, |+9| + |-3| = +12$
65. Commutative axiom for addition, associative axiom for addition, additive inverse axiom, identity axiom for addition
67. -29.191 **69.** 76.025   **71.** -16.179 **73.** \$23.75

Even

2. -12 **4.** 8   **6.** 9   **8.** 32   **10.** 0   **12.** -4 **14.** -4 **16.** 12   **18.** -6
20. Sometimes, since $|0| = 0$ **22.** 6   **24.** -3 **26.** -6 **28.** -5 **30.** 5   **32.** -3 **34.** -2
36. -7 **38.** -5 **40.** 8   **42.** No real number   **44.** -11 **46.** -6 **48.** -10 **50.** -6 **52.** -3
54. 4   **56.** -7 **58.** True   **60.** True   **62.** False; $[(+9) - (+6)] - (+3) = 0, (+9) - [(+6) - (+3)] = +6$
64. False; $|(+9) - (-6)| = +15, |+9| - |-6| = +3$
66. Commutative axiom for addition, associative axiom for addition, associative axiom for addition, inverse axiom for addition, identity axiom for addition, inverse axiom for addition, inverse axiom for addition
68. 10.405   **70.** 39.596   **72.** -89.037 **74.** $(20,270) - (-280) = 20,550$ feet

Exercise 1-5

Odd

1. 15 **3.** 3 **5.** −18 **7.** −3 **9.** 0 **11.** 0 **13.** Not defined **15.** Not defined **17.** −7
19. −1 **21.** 11 **23.** 9 **25.** Both are 8 **27.** −10 **29.** 3 **31.** −14 **33.** −5 **35.** −70
37. 0 **39.** 10 **41.** 40 **43.** 12 **45.** −18 **47.** −12 **49.** Not defined **51.** 56 **53.** 0
55. −6 **57.** 8 **59.** 0 **61.** −50 **63.** 0 **65.** Not defined (cannot divide by 0) **67.** 100 **69.** Never
71. Always **73.** When x and y are of opposite signs
75. *1.* Identity axiom for addition *2.* Distributive axiom *3.* Addition property for equality
    4. Inverse and associative axioms for addition *5.* Inverse axiom for addition *6.* Identity axiom for addition
    7. Symmetric property for equality

Even

2. 28 **4.** 5 **6.** −18 **8.** −3 **10.** 0 **12.** 0 **14.** Undefined **16.** Undefined **18.** 9
20. −13 **22.** −19 **24.** −1 **26.** −3 and −3 **28.** −9 **30.** −5 **32.** −8 **34.** −6 **36.** 72
38. 0 **40.** −22 **42.** 34 **44.** −2 **46.** 13 **48.** 8 **50.** 0 **52.** 6 **54.** 0 **56.** 72 **58.** −12
60. −108 **62.** −3 **64.** −9 **66.** Undefined **68.** −23 **70.** Sometimes, $\frac{0}{0}$ is undefined **72.** Never
74. When x and y have the same time
76. *1.* Identity axiom for multiplication *2.* Distributive axiom *3.* Inverse axiom for addition
    4. Theorem 4: multiplication involving 0 *5.* Inverse axiom for addition

Diagnostic (Review) Exercise 1-6

1. **(A)** T **(B)** T **(C)** F **(D)** F **(E)** T **(F)** T **(G)** T **(H)** F *(1-1)* **2.** Rational *(1-1)* **3.** 17 *(1-5)*
4. 13 *(1-5)* **5.** −5 *(1-4)* **6.** −13 *(1-4)* **7.** 6 *(1-4)* **8.** −3 *(1-4)* **9.** 3 *(1-4)* **10.** −12 *(1-4)*
11. 28 *(1-5)* **12.** −18 *(1-5)* **13.** −4 *(1-5)* **14.** 6 *(1-5)* **15.** Not defined *(1-5)* **16.** 0 *(1-5)*
17. 4 *(1-5)* **18.** −14 *(1-5)* **19.** −5 *(1-5)* **20.** −12 *(1-5)* **21.** 8 *(1-4)* **22.** 5 *(1-4)*
23. −3 *(1-4)* **24.** −2 *(1-4)* **25.** −5 *(1-4)* **26.** −5 *(1-4)* **27.** x + 10 *(1-3)* **28.** 15x *(1-3)*
29. 8xy *(1-3)* **30.** x + y + z + 12 *(1-3)* **31.** x + 2 *(1-3)* **32.** x *(1-3)* **33.** > *(1-2)* **34.** < *(1-2)*
35. < *(1-2)* **36.** < *(1-3)* **37.** > *(1-3)* **38.** < *(1-3)*
39. **(A)** F **(B)** T **(C)** T **(D)** F **(E)** T **(F)** T **(G)** F **(H)** T *(1-1)*
40. **(A)** **(B)** *(1-2)* **41.** 6 *(1-5)* **42.** 2 *(1-4)* **43.** 4 *(1-4)*

44. 1 *(1-4)* **45.** 4 *(1-4)* **46.** −26 *(1-5)* **47.** 10 *(1-5)* **48.** 35 *(1-5)* **49.** −2 *(1-5)*
50. Not defined *(1-5)* **51.** 6 *(1-4)* **52.** −2 *(1-4)* **53.** 8 *(1-4)* **54.** −6 *(1-4)* **55.** −9 *(1-5)*
56. −7 *(1-5)* **57.** −1 *(1-5)* **58.** x − 1 > 0 *(1-2)* **59.** 2x + 3 ≥ 0 *(1-2)* **60.** 50 = 2x − 10 *(1-2)*
61. x < 2x − 12 *(1-2)* **62.** −5 ≤ x < 5 *(1-2)* **63.** 8 + x = 5(x − 6) *(1-2)* **64.** x(x − 10) = 1,200 *(1-2)*
65. 2x + 2(x + 5) = 43 *(1-2)* **66.** *a* *(1-3)* **67.** 0 *(1-3)* **68.** 0 *(1-3)* **69.** *a* *(1-3)* **70.** 1 *(1-3)*
71. 1/a *(1-3)* **72.** Less than *(1-2)* **73.** Greater than *(1-2)* **74.** A = P + Prt *(1-2)* **75.** x + y < z *(1-2)*
76. Q + P *(1-3)* **77.** −x *(1-3)* **78.** u + 4 *(1-2)* **79.** 1 *(1-3)* **80.** x + (3 + 5) *(1-3)* **81.** y < 5 *(1-2)*
82. Commutative axiom for addition *(1-3)* **83.** Associative axiom for addition *(1-3)*
84. Commutative axiom for multiplication *(1-3)* **85.** Associative axiom for multiplication *(1-3)*
86. Additive identity *(1-3)* **87.** Additive inverse *(1-3)* **88.** −42 *(1-5)* **89.** 6 *(1-5)* **90.** 3, < *(1-1, 1-2)*
91. **(A)** {−1, 0, 1, 2} **(B)** ∅ *(1-1, 1-2)* **92.** **(A)** {3, 4, 5, 6, 7} **(B)** {4, 5} *(1-1)*
93. **(A)** All nonnegative real numbers **(B)** All nonpositive real numbers *(1-4)*

Practice Test Chapter 1

1. 24 *(1-5)* **2.** 38 *(1-5)* **3.** 5 *(1-4)* **4.** −2 *(1-5)* **5.** 1 *(1-5)* **6.** Not defined *(1-5)*
7. 3x > 4x − 6 *(1-2)* **8.** 2x + 4 = 3x − 5 *(1-2)* **9.** *(1-2)*
10. 54 = 2x + 2(2x + 3) *(1-2)* **11.** (C), (F) *(1-1)* **12.** (A), (B), (D) *(1-3)*
13. (A), (C) *(1-3)* **14.** {−1, 0, 1, 2, 3} *(1-1)*

Chapter 2   Exercise 2-1

Odd

1. 9   **3.** 5   **5.** 12   **7.** 2   **9.** $u^7 v^7$ **11.** 4   **13.** $\dfrac{a^8}{b^8}$ **15.** 3   **17.** 6   **19.** 2   **21.** 7   **23.** 12

25. $10x^{11}$ **27.** $3x^2$ **29.** $\dfrac{3}{4m^2}$ **31.** $x^{10}y^{10}$ **33.** $\dfrac{m^5}{n^5}$ **35.** $12y^{10}$ **37.** 35×10^{17} **39.** 10^{14} **41.** x^6

43. $m^6 n^{15}$ **45.** $\dfrac{c^6}{d^{15}}$ **47.** $\dfrac{3u^4}{v^2}$ **49.** $2^4 s^8 t^{16}$ or $16 s^8 t^{16}$ **51.** $6x^5 y^{15}$ **53.** $\dfrac{m^4 n^{12}}{p^8 q^4}$ **55.** $\dfrac{u^3}{v^9}$ **57.** $9x^4$

59. -1 **61.** $\dfrac{-1}{x^5}$ **63.** $-\dfrac{wy}{x^3}$ **65.** $\dfrac{(x-y)^2}{2(x+y)^2}$ **67.** x^7 **69.** x^n **71.** x^{2n+2} **73.** $\dfrac{u^2}{v^4}$

Even

2. 12   **4.** 4   **6.** v^6 **8.** 2   **10.** 5   **12.** 3   **14.** 4   **16.** 7   **18.** 4   **20.** 7   **22.** 4   **24.** 7

26. $6x^{10}$ **28.** $2x^2$ **30.** $\dfrac{2}{u^4}$ **32.** $c^{12}d^{12}$ **34.** $\dfrac{x^6}{y^6}$ **36.** $6x^9$ **38.** 6×10^{15} **40.** 10^{20} **42.** y^{20}

44. $x^8 y^{12}$ **46.** $\dfrac{a^{12}}{b^8}$ **48.** $\dfrac{y^6}{3x^4}$ **50.** $3^3 a^9 b^6$ **52.** $2x^8 y^4$ **54.** $\dfrac{x^6 y^3}{8w^6}$ **56.** $\dfrac{y^3}{16x^4}$ **58.** $\dfrac{-x^2}{32}$ **60.** -1

62. $\dfrac{-1}{a^8}$ **64.** $\dfrac{c}{a^2 b^4}$ **66.** $\dfrac{2}{(u-v+w)^3}$ **68.** y^{3n} **70.** x^2 **72.** x^{n^2+n} **74.** $\dfrac{y^2}{x}$

Exercise 2-2

Odd

1. 5   **3.** 4   **5.** 6   **7.** Binomial, 2   **9.** Trinomial, 6   **11.** Binomial, 3   **13.** Monomial, 8   **15.** -3
17. 3   **19.** 1   **21.** $17x$ **23.** x **25.** $8x$ **27.** $-13t$ **29.** $3x+5y$ **31.** $4+4x$ **33.** $4m-6n$
35. $9u-4v$ **37.** $-2m-24n$ **39.** $5u-6v$ **41.** $4x-3$ **43.** $x+11$ **45.** $9x-3$ **47.** $-2x-4$
49. $7x^2-x-12$ **51.** $-x+1$ **53.** $-y^2-2$ **55.** $-3x^2 y$ **57.** $3y^3+4y^2-y-3$ **59.** $3a^2-b^2$
61. $-7x+9y$ **63.** $-5x+3y$ **65.** $4x-6$ **67.** $-8x+12$ **69.** $10t-18$ **71.** $x-14$ **73.** $-m+2n$
75. $y+z$ **77.** $2x^4+3x^3+7x^2-x-8$ **79.** $-3x^3+x^2+3x-2$ **81.** $-2m^3-5$ **83.** $-t+27$
85. $2x-w$ **87.** -3 **89.** $3x-2$ **91.** $P = 2x + 2(x-5) = 4x-10$
93. Value in cents $= 5x + 10(x-5) + 25[(x-5)+2] = 40x - 125$ **95.** $x+2x+2x+3x = 8x$ **97.** $8t + 12 \cdot 2t = 32t$

Even

2. 3   **4.** 3   **6.** 8   **8.** Trinomial, 2   **10.** Binomial, 5   **12.** Monomial, 6   **14.** Trinomial, 4   **16.** -1
18. 1   **20.** 7   **22.** $10x$ **24.** $4x$ **26.** $8x$ **28.** $-2x$ **30.** $7x+3y$ **32.** $2-x$ **34.** $-5x+3y$
36. $6m-2n$ **38.** $-7x+4y$ **40.** $3x-2y$ **42.** $3y+3$ **44.** z **46.** $5x-2$ **48.** $5x-3$
50. $6x^2-x+3$ **52.** $-2x+12$ **54.** x^2-3x **56.** $-9r^3 t^3$ **58.** $2x^2+2x-3$ **60.** $-2x^2 y - 6xy + 4xy^2$
62. $-a+3b$ **64.** $-2x-y$ **66.** $2t-20$ **68.** $-3y+4$ **70.** $x-14$ **72.** $12x+4$ **74.** $-3x+y$
76. $y+z-5$ **78.** $3x^3+x^2+x+6$ **80.** $-x^3+2x^2-3x+7$ **82.** $2x^2-2xy+3y^2$ **84.** 1
86. $13x^2-26x+10$ **88.** x **90.** $3x-2$ **92.** $P = 2x + 2(2x+3) = 6x+6$
94. Value in cents $= 10x + 25(x-4) = 35x - 100$ **96.** $x + (x+3) + \{[x+(x+3)]+8\} = 4x+14$
98. $2(s) + (s-12)(1.5) = 3.5s - 18$

Exercise 2-3

Odd

1. y^5 **3.** $10y^5$ **5.** $-24x^{20}$ **7.** $6u^{16}$ **9.** $c^3 d^4$ **11.** $15x^2 y^3 z^5$ **13.** y^2+7y **15.** $10y^2-35y$
17. $3a^5+6a^4$ **19.** $2y^3+4y^2-6y$ **21.** $7m^6-14m^5-7m^4+28m^3$ **23.** $10u^4 v^3 - 15u^2 v^4$

25. $2c^3d^4 - 4c^2d^4 + 8c^4d^5$ **27.** $6y^3 + 19y^2 + y - 6$ **29.** $m^3 - 2m^2n - 9mn^2 - 2n^3$
31. $6m^4 + 2m^3 - 5m^2 + 4m - 1$ **33.** $a^3 + b^3$ **35.** $2x^4 + x^3y - 7x^2y^2 + 5xy^3 - y^4$ **37.** $x^2 + 5x + 6$
39. $a^2 + 4a - 32$ **41.** $t^2 - 16$ **43.** $m^2 - n^2$ **45.** $4t^2 - 11t + 6$ **47.** $3x^2 - 7xy - 6y^2$ **49.** $4m^2 - 49$
51. $30x^2 - 2xy - 12y^2$ **53.** $6s^2 - 11st + 3t^2$ **55.** $x^4 + x^3y + xy^3 + y^4$ **57.** $2x^3 + 6xy - x^2y^2 - 3y^3$
59. $9x^2 + 12x + 4$ **61.** $4x^2 - 20xy + 25y^2$ **63.** $36u^2 + 60uv + 25v^2$ **65.** $4m^2 - 20mn + 25n^2$
67. $x^3 + 6x^2y + 12xy^2 + 8y^3$ **69.** $-x^2 + 17x - 11$ **71.** $2x^3 - 13x^2 + 25x - 18$ **73.** $9x^3 - 9x^2 - 18x$
75. $12x^5 - 19x^3 + 12x^2 + 4x - 3$ **77.** $m + n$ **79.** Area $= y(y - 8) = y^2 - 8y$

Even

2. x^5 **4.** $6x^5$ **6.** $-35u^{16}$ **8.** $24x^9$ **10.** a^3b^3 **12.** $-6x^4y^4z^2$ **14.** $x + x^2$ **16.** $6x^2 - 15x$
18. $2m^4 + 6m^3$ **20.** $4x^3 - 6x^2 + 2x$ **22.** $6x^5 + 9x^4 - 3x^3 - 6x^2$ **24.** $8m^5n^4 - 4m^3n^5$
26. $6x^3y^4 + 12x^3y - 3x^2y^3$ **28.** $2x^3 - 7x^2 + 13x - 5$ **30.** $x^3 - 6x^2y + 10xy^2 - 3y^3$
32. $2x^4 - 5x^3 + 5x^2 + 11x - 10$ **34.** $a^3 - b^3$ **36.** $a^4 - 2a^2b^2 + b^4$ **38.** $m^2 - 5m + 6$ **40.** $m^2 - 7m - 60$
42. $u^2 - 9$ **44.** $a^2 - b^2$ **46.** $6x^2 - 7x - 5$ **48.** $2x^2 + xy - 6y^2$ **50.** $9y^2 - 4$ **52.** $6m^2 - mn - 35n^2$
54. $6x^2 - 13xy + 6y^2$ **56.** $x^5 + x^2y + x^3y^3 + y^4$ **58.** $2x^4 + x^3y^2 - 2xy^2 - y^4$ **60.** $16x^2 + 24xy + 9y^2$
62. $4x^2 - 28x + 49$ **64.** $49p^2 + 28pq + 4q^2$ **66.** $16x^2 - 8x + 1$ **68.** $8m^3 - 12m^2n + 6mn^2 - n^3$
70. $-7x^2 - x - 16$ **72.** $8x^3 - 20x^2 + 20x + 1$ **74.** $4x^3 - 14x^2 + 8x - 6$ **76.** $2x^5 - 9x^4 + 11x^3 - x^2 - x - 2$
78. m **80.** Area $= y(2y - 3) = 2y^2 - 3y$

Exercise 2-4

Odd

1. $A(2x + 3)$ **3.** $5x(2x + 3)$ **5.** $2u(7u - 3)$ **7.** $2u(3u - 5v)$ **9.** $5mn(2m - 3n)$ **11.** $2x^2y(x - 3y)$
13. $(x + 2)(3x + 5)$ **15.** $(m - 4)(3m - 2)$ **17.** $(x + y)(x - y)$ **19.** $3x^2(2x^2 - 3x + 1)$ **21.** $2xy(4x^2 - 3xy + 2y^2)$
23. $4x^2(2x^2 - 3xy + y^2)$ **25.** $(2x + 3)(3x - 5)$ **27.** $(x + 1)(x - 1)$ **29.** $(2x - 3)(4x - 1)$ **31.** $2x - 2$
33. $2x - 8$ **35.** $2u + 1$ **37.** $3x(x - 1) + 2(x - 1) = (x - 1)(3x + 2)$ **39.** $3x(x - 4) - 2(x - 4) = (x - 4)(3x - 2)$
41. $4u(2u + 1) - (2u + 1) = (2u + 1)(4u - 1)$ **43.** $(x - 1)(3x + 2)$ **45.** $(x - 4)(3x - 2)$ **47.** $(2u + 1)(4u - 1)$
49. $2m(m - 4) + 5(m - 4) = (m - 4)(2m + 5)$ **51.** $3x(2x - 3) - 2(2x - 3) = (2x - 3)(3x - 2)$
53. $3u(u - 4) - (u - 4) = (u - 4)(3u - 1)$ **55.** $3u(2u + v) - 2v(2u + v) = (2u + v)(3u - 2v)$
57. $3x(2x + y) - 5y(2x + y) = (2x + y)(3x - 5y)$ **59.** $(3a + b)(a + 3b)$ **61.** $(u - v)(w - x)$

Even

2. $M(x - 4)$ **4.** $3y(3y - 2)$ **6.** $4m(5m + 3)$ **8.** $7x(2x - 3y)$ **10.** $3uv(3u + 2v)$ **12.** $6xy^2(x - y)$
14. $(y + 3)(4y + 7)$ **16.** $(x - 1)(x - 4)$ **18.** $(m - n)(m + n)$ **20.** $2m^2(3m^2 - 4m - 1)$ **22.** $5uv(2u^2 + 4uv - 3v^2)$
24. $3m^2(3m^2 - 2mn - 2n^2)$ **26.** $(3u - 8)(2u - 3)$ **28.** $(u - 1)(3u - 1)$ **30.** $(4y - 5)(3y - 1)$ **32.** $3x + 6$
34. $3y - 15$ **36.** $3x + 5$ **38.** $2x(x + 2) + 3(x + 2) = (x + 2)(2x + 3)$ **40.** $2y(y - 5) - 3(y - 5) = (y - 5)(2y - 3)$
42. $2x(3x + 5) - (3x + 5) = (3x + 5)(2x - 1)$ **44.** $(x + 2)(2x + 3)$ **46.** $(y - 5)(2y - 3)$ **48.** $(3x + 5)(2x - 1)$
50. $5x(x - 2) + 2(x - 2) = (x - 2)(5x + 2)$ **52.** $4x(3x + 2) - 3(3x + 2) = (3x + 2)(4x - 3)$
54. $2m(3m + 2) - (3m + 2) = (3m + 2)(2m - 1)$ **56.** $2x(x - 2y) - y(x - 2y) = (x - 2y)(2x - y)$
58. $4u(u - 4v) - 3v(u - 4v) = (u - 4v)(4u - 3v)$ **60.** $(a + b)(a + b)$ or $(a + b)^2$ **62.** $(a + 3)(2b + 4)$

Exercise 2-5

Odd

1. $(x + 1)(x + 4)$ **3.** $(x + 2)(x + 3)$ **5.** $(x - 1)(x - 3)$ **7.** $(x - 2)(x - 5)$ **9.** Not factorable
11. Not factorable   **13.** $(x + 3y)(x + 5y)$ **15.** $(x - 3y)(x - 7y)$ **17.** Not factorable   **19.** $(3x + 1)(x + 2)$
21. $(3x - 4)(x - 1)$ **23.** $(x - 4)(3x - 2)$ **25.** $(3x - 2y)(x - 3y)$ **27.** $(n - 4)(n + 2)$ **29.** Not factorable
31. $(x - 1)(3x + 2)$ **33.** $(x + 6y)(x - 2y)$ **35.** $(u - 4)(3u + 1)$ **37.** $(3x + 5)(2x - 1)$ **39.** $(3s + 1)(s - 2)$
41. Not factorable   **43.** $(x - 2)(5x + 2)$ **45.** $(2u + v)(3u - 2v)$ **47.** $(4x - 3)(2x + 3)$ **49.** $(3u - 2v)(u + 3v)$
51. $(u - 4v)(4u - 3v)$ **53.** $(6x + y)(2x - 7y)$ **55.** $(12x - 5y)(x + 2y)$ **57.** $(x^2 + 1)(x^2 + 3)$ **59.** $(2x^2 - 3)(x^2 + 1)$
61. $(x^3 - 2)(x^3 + 3)$

Even

2. $(x + 1)(x + 3)$ **4.** $(x + 2)(x + 5)$ **6.** $(x - 1)(x - 4)$ **8.** $(x - 2)(x - 3)$ **10.** Not factorable
12. Not factorable   **14.** $(x + 4y)(x + 5y)$ **16.** $(x - 2y)(x - 8y)$ **18.** Not factorable   **20.** $(2x + 1)(x + 3)$
22. $(2x - 3)(x - 2)$ **24.** $(y - 5)(2y - 3)$ **26.** $(2x - 3y)(x - 2y)$ **28.** $(n + 4)(n - 2)$ **30.** Not factorable
32. $(3m + 2)(2m - 1)$ **34.** $(x - 2y)(2x + y)$ **36.** $(2u + 1)(4u - 1)$ **38.** $(m - 4)(2m + 5)$ **40.** $(2s - 1)(s + 3)$
42. Not factorable   **44.** $(6x - 1)(2x + 3)$ **46.** $(2x + y)(3x - 5y)$ **48.** $(2x - 3)(3x - 2)$ **50.** $(4m - 2n)(m + 3n)$
52. $(3x + 2y)(4x - 3y)$ **54.** $(5x - y)(3x + 4y)$ **56.** $(8x + 3y)(3x - 5y)$ **58.** $(x^2 - 2)(x^2 + 1)$
60. $(3x^2 + 1)(2x^2 - 3)$ **62.** $(2x^3 + 1)(x^3 - 4)$

Exercise 2-6

Odd

1. $(3x - 4)(x - 1)$ **3.** Not factorable   **5.** $(2x - 1)(x + 3)$ **7.** Not factorable   **9.** $(x - 4)(3x - 2)$
11. $(3x + 5)(2x - 1)$ **13.** Not factorable   **15.** $(m - 4)(2m + 5)$ **17.** $(u - 4)(3u + 1)$ **19.** $(2u + v)(3u - 2v)$
21. Not factorable   **23.** $(4x - 3)(2x + 3)$ **25.** $(4m - 2n)(m + 3n)$ **27.** Not factorable   **29.** $(u - 4v)(4u - 3v)$
31. $(6x + y)(2x - 7y)$ **33.** $(6x + 5y)(3x - 4y)$ **35.** $-13, 13, -8, 8, -7, 7$ **37.** $(2x^2 + 1)(3x^2 + 2)$
39. $(x^2 + y^2)(2x^2 + 3y^2)$ **41.** $(2x^3 - 3)(x^3 + 1)$

Even

2. $(2x - 3)(x - 2)$ **4.** Not factorable   **6.** $(3x + 1)(x - 2)$ **8.** Not factorable   **10.** $(y - 5)(2y - 3)$
12. $(x - 2)(5x + 2)$ **14.** Not factorable   **16.** $(6x - 1)(2x + 3)$ **18.** $(2u + 1)(4u - 1)$ **20.** $(2x + y)(3x - 5y)$
22. Not factorable   **24.** $(2x - 3)(3x + 2)$ **26.** $(3u - 2v)(u + 3v)$ **28.** Not factorable   **30.** $(3x + 2y)(4x - 3y)$
32. $(5x - y)(3x + 4y)$ **34.** $(5m - 6n)(3m + 4n)$ **36.** $6, 10, 12$ **38.** $(5x^2 - 1)(x^2 - 2)$ **40.** $(3x^2 - y^2)(x^2 + 2y^2)$
42. $(3x^3 + 2)(x^3 - 4)$

Exercise 2-7

Odd

1. $(v - 5)(v + 5)$ **3.** $(3x - 2)(3x + 2)$ **5.** Not factorable   **7.** $(3x - 4y)(3x + 4y)$ **9.** $(x + 1)(x^2 - x + 1)$
11. $(m - n)(m^2 + mn + n^2)$ **13.** $(2x + 3)(4x^2 - 6x + 9)$ **15.** $3uv^2(2u - v)$ **17.** $2(x - 2)(x + 2)$ **19.** $2x(x^2 + 4)$
21. $3x(2x - y)(2x + y)$ **23.** $2x(x + 1)(x^2 - x + 1)$ **25.** $6(x + 2)(x + 4)$ **27.** $3x(x^2 - 2x + 5)$
29. $(xy - 4)(xy + 4)$ **31.** $(ab + 2)(a^2b^2 - 2ab + 4)$ **33.** $2xy(2x + y)(x + 3y)$ **35.** $4(u + 2v)(u^2 - 2uv + 4v^2)$
37. $5y^2(6x + y)(2x - 7y)$ **39.** $(y + 2)(x + y)$ **41.** $(x - 5)(x + y)$ **43.** $(a - 2b)(x - y)$ **45.** $(3c - 4d)(5a + b)$
47. $(x - 2)(x + 1)(x - 1)$ **49.** $(y - x)[(y - x) - 1] = (y - x)(y - x - 1)$ **51.** $(xy + 2)(xy - 3)$ **53.** $(z^2 - 3)(z^2 + 2)$
55. $(x^4 + 2)(x^4 - 2)$ **57.** $(r^2 + s^2)(r - s)(r + s)$ **59.** $(x^2 - 4)(x^2 + 1) = (x - 2)(x + 2)(x^2 + 1)$
61. $[(x - 3) - 4y][(x - 3) + 4y] = (x - 3 - 4y)(x - 3 + 4y)$ **63.** $[(a - b) - 2(c - d)][(a - b) + 2(c - d)]$
65. $[5(2x - 3y) - 3ab][5(2x - 3y) + 3ab]$ **67.** $(x - 1)(x^2 + x + 1)(x + 1)(x^2 - x + 1)$ **69.** $(2x - 1)(x - 2)(x + 2)$
71. $[5 - (a + b)][5 + (a + b)]$ **73.** $[4x^2 - (x - 3y)][4x^2 + (x - 3y)]$ **75.** $(x - 2)(x^2 + 3)$ **77.** $(x^4 + 1)(x - 1)$
79. $(3x - 1)(x^2 + 4)$ **81.** $(x + 2 + y)(x + 2 - y)$

Even

2. $(x - 9)(x + 9)$ **4.** $(2m - 1)(2m + 1)$ **6.** Not factorable   **8.** $(5u - 2v)(5u + 2v)$ **10.** $(y - 1)(y^2 + y + 1)$
12. $(p + q)(p^2 - pq + q^2)$ **14.** $(u - 2v)(u^2 + 2uv + 4v^2)$ **16.** $2x^2y(x - 3y^2)$ **18.** $3(y - 3)(y + 3)$
20. $3x^2(x^2 + 9)$ **22.** $2uv(u - v)(u + v)$ **24.** $x(y + x)(y^2 - xy + x^2)$ **26.** $4(x - 3)(x + 2)$ **28.** $2x(x^2 - x + 4)$
30. $(mn - 6)(mn + 6)$ **32.** $(3 - xy)(9 + 3xy + x^2y^2)$ **34.** $3xy(x - 2y)(x - 3y)$ **36.** $2(3x - y)(9x^2 + 3xy + y^2)$
38. $4x^2(5x - y)(3x + 4y)$ **40.** $(x + 3)(x + y)$ **42.** $(x - 3)(x - y)$ **44.** $(x + y)(m - 2n)$ **46.** $(2m - 3n)(a + b)$
48. $(x - 2)(x^2 + 1)$ **50.** $(x - 1)^2(x + 1)$ **52.** $(ab - 3)(ab - 4)$ **54.** $(x^2 + 2)(x^2 + 2)$ or $(x^2 + 2)^2$
56. $(a^3 + 10)(a^3 - 2)$ **58.** $(4a^2 + b^2)(2a - b)(2a + b)$ **60.** $(x^2 + 2)(x - 3)(x + 3)$
62. $[(x + 2) - 3y][(x + 2) + 3y] = (x + 2 - 3y)(x + 2 + 3y)$ **64.** $[(x^2 - x) - 3(y^2 - y)][(x^2 - x) + 3(y^2 - y)]$
66. $2a[3a - 2(x + 4)][3a + 2(x + 4)]$ **68.** $(a + 2b)(a^2 - 2ab + 4b^2)(a - 2b)(a^2 + 2ab + 4b^2)$ **70.** $(y - 3)(2y - 3)(2y + 3)$

72. $(x - y + 3)(x - y - 3)$ **74.** $(x^2 - x + 2)(x + 2)(x - 1)$ **76.** $(x^2 - 5)(x + 2)$ **78.** $(x^2 + 1)(x^3 + 2)$
80. $(2x + 1)(x^2 + 2)$ **82.** $(x + y + z)(x + y - z)$

Diagnostic (Review) Exercise 2-8

1. x^5 (2-1)   **2.** $x^3 y^3$ (2-1)   **3.** $\dfrac{x^3}{y^3}$ (2-1)   **4.** $\dfrac{1}{x^5}$ (2-1)   **5.** x^{24} (2-1)   **6.** 1 (2-1)   **7.** x^{11} (2-1)

8. $-8x^3$ (2-1)   **9.** $-6x^{11}$ (2-3)   **10.** $\dfrac{3x^2}{4y^2}$ (2-1)   **11.** **(A)** 5 **(B)** 3 (2-2)   **12.** **(A)** 5 **(B)** 11 (2-3)

13. $x^2 + 2x + 1$ (2-2)   **14.** $-x^2 + 2x + 9$ (2-2)   **15.** $2x^3 + 5x^2 - 8x - 20$ (2-3)   **16.** $x^2 + 5x - 3$ (2-2)
17. $5x^2 - x + 5$ (2-2)   **18.** $6x^4 - 5x^3 + 8x^2 + 5x + 4$ (2-3)   **19.** $2x^2 - 2$ (2-2)   **20.** $2x^3 - 4x^2 + 12x$ (2-2)
21. $3x^3(x + 3)$ (2-4)   **22.** $x^2 y(2y^2 + x)$ (2-4)   **23.** $(x - 5)(2x + 1)$ (2-4)   **24.** $(x + 3)(4x - 1)$ (2-5)
25. $(2x + 5)(2x - 5)$ (2-7)   **26.** $(x + y)(x + 2y)$ (2-5)   **27.** Not factorable (2-7)
28. $(x + 2)(x + 2)$ or $(x + 2)^2$ (2-4)   **29.** $(2x + 1)(x - 3)$ (2-5)   **30.** $(x - 3)(x - 3)$ or $(x - 3)^2$ (2-4)
31. $(3x + 2)(x - 1)$ (2-4)   **32.** $(a - 2)(a^2 + 2a + 4)$ (2-7)   **33.** $(x + 2)(x + 4)$ (2-5)   **34.** $(x - 2y)(2x - y)$ (2-5)

35. $(x + 4)(x^2 - 4x + 16)$ (2-7)   **36.** $(a - 3b)(a + 3b)$ (2-7)   **37.** $\dfrac{4x^6}{y^{16}}$ (2-1)   **38.** $-x^7 y^8$ (2-3)

39. $-x^4 y^3$ (2-1)   **40.** $9x^8 y^9$ (2-3)   **41.** $\dfrac{-8x^3}{y^6}$ (2-1)   **42.** $\dfrac{9x^2}{4y^2}$ (2-1)   **43.** $(x^2 + 1)(x^4 - x^2 + 1)$ (2-7)

44. $x^2(2x + 1)(x + 3)$ (2-7)   **45.** $a^3(x - 2)^2$ (2-7)   **46.** $3xy(x + y)(x - y)$ (2-7)   **47.** $(x^2 + 1)(x - 3)$ (2-7)
48. $x(x^4 + 1)(x + 1)$ (2-7)   **49.** $2x(x + 1)(x - 2)$ (2-7)   **50.** $(x + y + 1)(x + y - 1)$ (2-7)
51. $(x^2 + 1)(x - 1)(x + 1)$ (2-7)   **52.** $(2x - 5)(4x^2 + 10x + 25)$ (2-7)   **53.** $(3x + 2)(x^2 - 5)$ (2-7)
54. $x^2(x + 3)(x - 1)$ (2-7)   **55.** $-x^3 y(xy + 1)(xy + 1)$ (2-7)   **56.** $2a(a + 1)(a^2 - a + 1)$ (2-7)

Practice Test Chapter 2

1. (A), (D) (2-2)   **2.** $\dfrac{3x^3}{2y^2}$ (2-1)   **3.** $-4x^7 y^9$ (2-1)   **4.** $6x^2 + x + 1$ (2-2)   **5.** $x^2 + 2x + 2$ (2-2)

6. $12x^3 - 7x^2 - 14x + 5$ (2-3)   **7.** $5x^2 + 15x - 10$ (2-3)   **8.** $3(x + 2)(x + 4)$ (2-5)   **9.** $3x^2 y^2(4x - y)$ (2-4)
10. $(x + y)(3x - y)$ (2-4)   **11.** $2xy^2(x - 3)(x + 1)$ (2-5)   **12.** $(3x - 2)(9x^2 + 6x + 4)$ (2-7)
13. $(2x - 3)(3x + 5)$ (2-5, 2-6)   **14.** $(x^2 + 1)(x + y)(x - y)$ (2-7)

Chapter 3   Exercise 3-1

Odd

1. $\dfrac{1}{2x^2}$ **3.** $\dfrac{2x^2}{3y}$ **5.** $\dfrac{3(x - 9)}{y}$ **7.** $\dfrac{2x - 1}{3x}$ **9.** $\dfrac{x}{2}$ **11.** $\dfrac{1}{n}$ **13.** $12xy$ **15.** $14x^3 y$ **17.** $\dfrac{x + 2}{3x}$

19. $\dfrac{x - 3}{x + 3}$ **21.** $\dfrac{2x - 3y}{2xy}$ **23.** $\dfrac{x + 2}{x + y}$ **25.** $\dfrac{x + 5}{2x}$ **27.** $\dfrac{x^2 + 2x + 4}{x + 2}$ **29.** $\dfrac{x + 3}{x - 2}$ **31.** $\dfrac{2x}{y}$ **33.** $\dfrac{x^2 + 2}{x - 2}$

35. $3x^2 + 3xy$ **37.** $x^2 - y^2$ **39.** $\dfrac{x - y}{3x}$ **41.** $\dfrac{x - y}{2x + y}$ **43.** $\dfrac{x^2 + y^2}{(x + y)^2}$ **45.** $x + 1$ **47.** $\dfrac{x^2 + y^2}{x^3 + y^3}$

49. $\dfrac{x + y}{u + v}$

Even

2. $\dfrac{3}{u}$ **4.** $\dfrac{4n^4}{3m}$ **6.** $\dfrac{x}{3}$ **8.** $\dfrac{x + 3}{2x^2}$ **10.** $\dfrac{x}{2}$ **12.** a **14.** $6x^2 y^2$ **16.** $16u^2 v^3$ **18.** $\dfrac{x + 2}{2x}$ **20.** $\dfrac{x - 2}{x + 2}$

22. $\dfrac{a + 4b}{4b}$ **24.** $\dfrac{u - 2}{u + v}$ **26.** $\dfrac{3(x - 7)}{4x^2}$ **28.** $\dfrac{y + 3}{2y}$ **30.** $\dfrac{x - 1}{x + 5}$ **32.** $\dfrac{2x + 3}{4x + 5}$ **34.** $\dfrac{x^2 + 1}{x + 1}$ **36.** $5mn - 5n^2$

38. $6x^2 + 17x + 5$ **40.** $\dfrac{2uv}{u + v}$ **42.** $\dfrac{m + n}{m - n}$ **44.** $\dfrac{(x - y)(x + y)}{x^2 + y^2}$ **46.** x **48.** $\dfrac{z}{z - 1}$ **50.** $\dfrac{xy - z}{x - yz}$

Exercise 3-2

Odd

1. $3x + 1$ **3.** $2y^2 + y - 3$ **5.** $3x + 1, R = 3$ **7.** $4x - 1$ **9.** $3x - 4, R = -1$ **11.** $x + 2$
13. $4x + 1, R = -4$ **15.** $4x + 6, R = 25$ **17.** $x - 4, R = 3$ **19.** $x^2 + x + 1$ **21.** $x^3 + 3x^2 + 9x + 27$
23. $4a + 5, R = -7$ **25.** $x^2 + 3x - 5$ **27.** $x^2 + 3x + 8, R = 27$ **29.** $3x^3 + x^2 - 2, R = -4$
31. $4x^2 - 2x - 1, R = -2$ **33.** $2x^3 + 6x^2 + 32x + 84, R = 186x - 170$ **35.** $Q = x - 3, R = x + 2$

Even

2. $2x - 3$ **4.** $x^2 - 3x - 5$ **6.** $2x + 3, R = 5$ **8.** $2x + 3$ **10.** $2x + 5, R = -2$ **12.** $y - 3$
14. $4x - 1, R = 5$ **16.** $3x + 2, R = -4$ **18.** $x + 5, R = -2$ **20.** $a^2 - 3a + 9$ **22.** $x^3 - 2x^2 + 4x - 8$
24. $5c - 2, R = 8$ **26.** $2y^2 + y - 3$ **28.** $2y^2 - 5y + 13, R = -27$ **30.** $2x^3 - 3x^2 - 5, R = 5$
32. $2x^2 - 3x + 2, R = 0$ **34.** $3x^3 - 4x + 3, R = -8x$ **36.** $Q(x) = x^2 + 2x - 3, R(x) = 2x + 5$

Exercise 3-3

Odd

1. $x^2 + 4x + 11, R = 26$ **3.** $x^2 + x + 2, R = 2$ **5.** $2x^2 + 5x + 16, R = 46$ **7.** $2x^2 - 3x + 4, R = -6$
9. $x^3 + 5x^2 + 23x + 95, R = 385$ **11.** $x^3 - x^2 + 5x - 7, R = 19$ **13.** $x^2 + 2x + 2, R = 8$ **15.** $x^2 - x - 1, R = 5$
17. $x^3 + 4x^2 + 16x + 61, R = 249$ **19.** $x^3 - 2x^2 + 4x - 11, R = 27$ **21.** 170 **23.** -2 **25.** -5
27. $-0.389\,000$ **29.** $-1.234\,625$ **31.** $-43.817\,000$ **33.** $5.297\,889$

Even

2. $x^2 + 3x + 6, R = 10$ **4.** $x^2 + 3, R = -2$ **6.** $2x^2 + x + 2, R = 0$ **8.** $2x^2 - 7x + 22, R = -68$
10. $x^3 + 3x^2 + 9x + 21, R = 47$ **12.** $x^3 - 3x^2 + 15x - 57, R = 233$ **14.** $x^2 + x - 1, R = 3$ **16.** $x^2 - 2x + 2, R = 0$
18. $x^3 + 2x^2 + 4x + 5, R = 15$ **20.** $x^3 - 4x^2 + 16x - 67, R = 273$ **22.** 11 **24.** 2 **26.** -154
28. -0.039899 **30.** -1.840269 **32.** -0.003999 **34.** -37.297494

Exercise 3-4

Odd

1. $\dfrac{8}{9}$ **3.** $\dfrac{6}{b}$ **5.** $\dfrac{y}{x}$ **7.** $\dfrac{3}{2}$ **9.** $\dfrac{3c}{a}$ **11.** $\dfrac{x}{9y^2}$ **13.** $\dfrac{16xy}{3}$ **15.** $\dfrac{9xy}{8c}$ **17.** $\dfrac{-45u^2}{16v^2}$ **19.** $\dfrac{c^3 d^2}{a^6 b^6}$

21. $\dfrac{x}{2}$ **23.** $\dfrac{x}{x - 3}$ **25.** $\dfrac{3y}{x + 3}$ **27.** $\dfrac{1}{2y}$ **29.** $t(t - 4)$ **31.** $\dfrac{1}{m}$ **33.** $-x(x - 2)$ or $2x - x^2$ **35.** $\dfrac{a^2}{2}$

37. 2 **39.** -1 **41.** $\dfrac{(x - y)^2}{y^2(x + y)}$ **43.** $x = 3$ **45.** $\dfrac{R}{S} \cdot \left(\dfrac{P}{Q} \cdot \dfrac{S}{R} \right) = \dfrac{RPS}{SQR} = \dfrac{P}{Q}$

Even

2. $\frac{2}{3}$ **4.** $\dfrac{1}{z}$ **6.** $\dfrac{3x}{2y}$ **8.** 4 **10.** y **12.** $2y^2$ **14.** $\dfrac{3ad}{2c}$ **16.** $\dfrac{3v}{2u}$ **18.** $-\dfrac{2x^2}{3y}$ **20.** $\dfrac{u^2 w^2}{25y^2}$ **22.** $\dfrac{2}{x}$

24. $a + 1$ **26.** $\dfrac{x - 2}{2x}$ **28.** $\dfrac{1}{y(x + 4)}$ **30.** $\dfrac{1}{2y - 1}$ **32.** $\dfrac{x}{x + 5}$ **34.** $-3(x - 2)$ or $6 - 3x$ **36.** $8d^6$

38. -2 **40.** $-\dfrac{1}{m}$ **42.** $\dfrac{x^2(x + y)}{(x - y)^2}$ **44.** All but one (namely, $x = 1$)

Exercise 3-5

Odd

1. $3x$ **3.** x **5.** v^3 **7.** $12x^2$ **9.** $(x+1)(x-2)$ **11.** $3y(y+3)$ **13.** $\dfrac{7x+2}{5x^2}$ **15.** 2 **17.** $\dfrac{1}{y+3}$

19. $\dfrac{3-2x}{k}$ **21.** $\dfrac{12x+y}{4y}$ **23.** $\dfrac{2+y}{y}$ **25.** $\dfrac{u^3+uv-v^2}{v^3}$ **27.** $\dfrac{9x^2+8x-2}{12x^2}$ **29.** $\dfrac{5x-1}{(x+1)(x-2)}$

31. $\dfrac{7y-6}{3y(y+3)}$ **33.** $24x^3y^2$ **35.** $75x^2y^2$ **37.** $18(x-1)^2$ **39.** $24(x-7)(x+7)^2$ **41.** $(x-2)(x+2)^2$

43. $12x^2(x+1)^2$ **45.** $\dfrac{8v-6u^2v^2+3u^3}{36u^3v^3}$ **47.** $\dfrac{15t^2+14t-6}{36t^3}$ **49.** $\dfrac{2}{t-1}$ **51.** $\dfrac{5a^2-2a-5}{(a+1)(a-1)}$

53. $\dfrac{5x+55}{12(x-5)^2(x+5)}$ **55.** $\dfrac{15x-11}{18(x-1)^2}$ **57.** $\dfrac{-4}{(x-1)(x+3)}$ **59.** $\dfrac{2s^2+s-2}{2s(s-2)(s+2)}$ **61.** $\dfrac{2(x+4)}{(x-2)(x+2)^2}$

63. $\dfrac{3}{x+3}$ **65.** $\dfrac{x+3}{(x-2)(x+7)}$ **67.** $\dfrac{(3x+1)(x+3)}{12x^2(x+1)^2}$ **69.** $\dfrac{xy^2-xy+y^2}{x^3-y^3}$ **71.** $\dfrac{7}{y-3}$ **73.** -1

75. $\dfrac{-17}{15(x-1)}$

Even

2. $4y$ **4.** y **6.** x^2 **8.** $24u^3$ **10.** $(x-2)(x+3)$ **12.** $2x(x-2)$ **14.** $\dfrac{3m+1}{2m^2}$ **16.** 5

18. $\dfrac{1}{2x-3}$ **20.** $\dfrac{1-b}{a^2}$ **22.** $\dfrac{6-x}{3x}$ **24.** $\dfrac{x^2+1}{x}$ **26.** $\dfrac{x^2-xy+y^2}{x^3}$ **28.** $\dfrac{20u^2-18u+3}{24u^3}$

30. $\dfrac{2x+1}{(x-2)(x+3)}$ **32.** $\dfrac{x+6}{2x(x-2)}$ **34.** $36u^3v^3$ **36.** $36m^4n^4$ **38.** $24(y-3)^2$ **40.** $12(x-5)^2(x+5)$

42. $(x-3)^2(x+3)$ **44.** $15m^2(m-1)^2$ **46.** $\dfrac{2y^2+9x-16x^2}{24x^3y^2}$ **48.** $\dfrac{y^2+8}{8y^3}$ **50.** $\dfrac{3x-5}{x-3}$ **52.** $\dfrac{3y^2-y-18}{(y+2)(y-2)}$

54. $\dfrac{13x-35}{24(x-7)(x+7)^2}$ **56.** $\dfrac{21-4y}{24(y-3)^2}$ **58.** $\dfrac{2x+7}{(2x-3)(x+2)}$ **60.** $\dfrac{5t-12}{3(t-4)(t+4)}$ **62.** $\dfrac{x+9}{(x-3)^2(x+3)}$

64. $\dfrac{2}{x+y}$ **66.** $\dfrac{5m^2+1}{6(m+1)^2}$ **68.** $\dfrac{17m^2+m-3}{15m^2(m-1)^2}$ **70.** $\dfrac{x^2}{x^3+y^3}$ **72.** $\dfrac{1}{x-1}$ **74.** -1 **76.** $\dfrac{7y-9x}{xy(a-b)}$

Exercise 3-6

Odd

1. $\dfrac{3}{4}$ **3.** $\dfrac{9}{10}$ **5.** $\dfrac{8}{13}$ **7.** $\dfrac{22}{51}$ **9.** xy **11.** $\dfrac{3xy}{2}$ **13.** $\dfrac{1}{x-3}$ **15.** $\dfrac{x+y}{x}$ **17.** $\dfrac{1}{y-x}$ **19.** $\dfrac{x-y}{x+y}$

21. 1 **23.** $-\dfrac{1}{2}$ **25.** $\dfrac{1}{1-x}$ **27.** $-x$ **29.** $\dfrac{3x+5}{2x+3}$ **31.** $r=\dfrac{2r_Rr_G}{r_R+r_G}$

Even

2. $\dfrac{3}{8}$ **4.** $\dfrac{8}{25}$ **6.** $\dfrac{31}{22}$ **8.** $\dfrac{4}{3}$ **10.** $\dfrac{1}{ab}$ **12.** $\dfrac{6x^2}{5y}$ **14.** $\dfrac{1}{(x+2)}$ **16.** $\dfrac{(a+b)}{b}$ **18.** $a(a+b)$

20. $\dfrac{(x-3)}{(x-1)}$ **22.** $\dfrac{-m(m+n)}{n}$ **24.** -1 **26.** $\dfrac{(x-2)}{x}$ **28.** $\dfrac{(t-1)}{t}$ **30.** $1-x$ **32.** $v=\dfrac{c^2(v_1+v_2)}{c^2+v_1v_2}$

Diagnostic (Review) Exercise 3-7

1. $\dfrac{3x^2}{2(z+3)^2}$ (3-1)   **2.** $\dfrac{x+1}{x-1}$ (3-1)   **3.** $\dfrac{3x+2}{3x}$ (3-5)   **4.** $\dfrac{2x+11}{6x}$ (3-5)   **5.** $\dfrac{2-9x^2-8x^3}{12x^3}$ (3-5)

6. $\dfrac{2xy}{ab}$ (3-4)   **7.** $2(x+1)$ (3-5)   **8.** $\dfrac{2}{m+1}$ (3-5)   **9.** $\dfrac{x+7}{(x-2)(x+1)}$ (3-5)   **10.** $\dfrac{(d-2)^2}{d+2}$ (3-4)

11. $\dfrac{-1}{(x+2)(x+3)}$ (3-5)   **12.** $\dfrac{3}{8}$ (3-6)   **13.** $\dfrac{11}{6}$ (3-6)   **14.** $\dfrac{y-2}{y+1}$ (3-6)   **15.** $x^2-2x-1,\ R=-4$ (3-2 or 3-3)

16. $x,\ R=0$ (3-2)   **17.** $x^2+2x+1,\ R=3$ (3-2)   **18.** $x^3-2x^2+5x-10,\ R=19$ (3-2 or 3-3)

19. $x^3+x^2+x+1,\ R=0$ (3-2 or 3-3)   **20.** $x^2-1,\ R=2x+1$ (3-2)   **21.** $\dfrac{12a^2b^2-40a^2-5b}{30a^2b^3}$ (3-5)

22. $\dfrac{5-2x}{2x-3}$ (3-5)   **23.** $\dfrac{2y^4}{9a^4}$ (3-4)   **24.** $\dfrac{5x-12}{3(x-4)(x+4)}$ (3-4)   **25.** $\dfrac{x}{x+1}$ (3-4)   **26.** $\dfrac{x-y}{x}$ (3-6)

27. $\dfrac{y}{x^3-y^3}$ (3-5)   **28.** $\dfrac{x}{y(x+y)}$ (3-6)   **29.** $x+1$ (3-4)   **30.** $\dfrac{x^2+24x-9}{12x(x-3)(x+3)^2}$ (3-5)   **31.** $\dfrac{-1}{s+2}$ (3-5)

32. -1 (3-4)   **33.** $\dfrac{y^2}{x}$ (3-4)   **34.** $\dfrac{x-y}{x+y}$ (3-6)   **35.** $\dfrac{(x+1)(x-2)}{2x}$ (3-6)

Practice Test Chapter 3

1. $x^2-3x,\ R=x+1$ (3-2)   **2.** $2x^2+8x+29,\ R=121$ (3-2)   **3.** $\dfrac{5x-5}{x^2+1}$ (3-5)   **4.** $\dfrac{2x-8}{x-2}$ (3-5)

5. $\dfrac{2x}{x^2+5x+6}$ (3-5)   **6.** $\dfrac{2x-2}{x-4}$ (3-4)   **7.** $-\dfrac{2a^2}{9}$ (3-4)   **8.** $\dfrac{2x+3y+z}{xyz}$ (3-5)   **9.** $\dfrac{3}{x+1}$ (3-5)

10. $\dfrac{a-2b}{a}$ (3-4)   **11.** $\dfrac{a^2}{b^2}$ (3-6)   **12.** x (3-6)   **13.** $\dfrac{x+1}{x}$ (3-6)   **14.** $\dfrac{a^2+2ab}{ab-b^2}$ (3-4)

Chapter 4   Exercise 4-1

Odd

1. 18   **3.** All real numbers   **5.** 9   **7.** No solution   **9.** 10   **11.** 1   **13.** Solution set $= R$ **15.** 4
17. No solution   **19.** 13   **21.** 8   **23.** -6 **25.** $\frac{-1}{12}$ **27.** 30   **29.** 20   **31.** 10   **33.** 3   **35.** $\frac{-7}{4}$
37. 3   **39.** 10   **41.** -9 **43.** 4   **45.** No solution   **47.** 5   **49.** $\frac{53}{11}$ **51.** No solution   **53.** 1
55. $\frac{31}{24}$ **57.** -4 **59.** No solution   **61.** $\frac{2}{3}$

Even

2. 4   **4.** All real numbers   **6.** $\frac{1}{3}$ **8.** No solution   **10.** 16   **12.** 4   **14.** Solution set $= R$ **16.** 6
18. No solution   **20.** 8   **22.** 12   **24.** 36   **26.** $\frac{-4}{3}$ **28.** 600   **30.** 30   **32.** -9 **34.** 4   **36.** 15
38. -3 **40.** 150   **42.** 8   **44.** -4 **46.** No solution   **48.** $\frac{-6}{5}$ **50.** 2   **52.** No solution   **54.** 8
56. $\frac{8}{5}$ **58.** 8   **60.** No solution   **62.** -4

Exercise 4-2

Odd

1. $2x$ **3.** $x-3$ **5.** $\dfrac{x}{3}$ **7.** $\dfrac{3x}{4}$ **9.** $\dfrac{2x}{3}-5$ **11.** $\dfrac{3}{2}$ **13.** $\dfrac{5}{2}$ **15.** $\dfrac{25}{2}$ **17.** 20   **19.** 7

21. $7x=4x-12,\ x=-4$ **23.** $3+\dfrac{x}{6}=\dfrac{2}{3};\ -14$ **25.** $x+(x+1)+(x+2)=96;\ 31,\ 32,\ 33$

27. $x+(x+2)+(x+4)=42;\ 12,\ 14,\ 16$ **29.** $(2x-3)+x=12;\ 5\ \text{ft and }7\ \text{ft}$ **31.** $2x+2(x-6)=36;\ 12\times6\ \text{ft}$

33. 60 quarters   **35.** 40 m   **37.** 162 km   **39.** $2(2x + 3) + 2x = 66$; 23×10 cm   **41.** $\dfrac{x}{6} - 2 = \dfrac{x}{4} + 1$; -36

43. $\dfrac{x}{52} = \dfrac{9}{46}$; 10.17 ml   **45.** $x + (x + 2) = (x + 4) + 5$; 7, 9, 11   **47.** $\dfrac{x}{4} = \dfrac{1}{1.06}$; 3.77 liters   **49.** $\dfrac{x}{1} = \dfrac{1}{0.62}$; 1.61 km

51. $2x + 2 \cdot \dfrac{x}{6} = 84$; 36×6 m   **53.** $\dfrac{x}{500} = \dfrac{240}{200}$; \$600   **55.** $\dfrac{x}{23} = \dfrac{10}{35}$; 6.57 in.   **57.** $\dfrac{3x}{5} - 4 = \dfrac{x}{3} + 8$; 45

59. $\dfrac{x}{5} = \dfrac{1}{0.26}$; 19.23 liters   **61.** $\dfrac{2P}{5} + 70 + \dfrac{P}{4} = P$; 200 cm   **63.** $\dfrac{x}{300} = \dfrac{250}{25}$; 3,000 trout   **65.** $\dfrac{D}{3} + 6 + \dfrac{D}{2} = D$; 36 km

67. 3 kg

Even

2. $3x$ **4.** $2 + x$ **6.** $\dfrac{x}{5}$ **8.** $\dfrac{2x}{5}$ **10.** $\dfrac{3x}{4} - 11$ **12.** $\frac{1}{3}$ **14.** $\frac{5}{3}$ **16.** $\frac{80}{3}$ **18.** 8   **20.** 18

22. $6x = 24 + 3x$; $x = 8$ **24.** $2 + \dfrac{x}{4} = \dfrac{1}{2}$; -6 **26.** $x + (x + 1) + (x + 2) = 78$; 25, 26, 27

28. $x + (x + 2) + (x + 4) = 54$; 16, 18, 20   **30.** $(3x + 4) + x = 32$; 7 cm and 25 cm   **32.** $2x + 2(x + 7) = 54$; 17×10 m

34. 169 pennies   **36.** 32 cm   **38.** 60 mi   **40.** $2(4x - 6) + 2x = 128$; 50×14 m   **42.** $\dfrac{x}{2} - 5 = \dfrac{x}{3} + 3$; 48

44. $\dfrac{x}{9} = \dfrac{0.75}{6}$; 1.13 cups   **46.** $(x + 2) + (x + 4) = 3x + 1$; 5, 7, 9   **48.** $\dfrac{x}{10} = \dfrac{1}{2.2}$; 4.55 kg   **50.** $\dfrac{x}{1} = \dfrac{1}{0.91}$; 1.10 yd

52. $2x + 2 \cdot \dfrac{x}{3} = 72$; 27×9 cm   **54.** \$197.40   **56.** $\dfrac{x}{4.25} = \dfrac{12}{3.25}$; 15.69 in.   **58.** $\dfrac{2x}{3} + 5 = \dfrac{x}{4} - 10$; -36

60. $\dfrac{z}{10} = \dfrac{1}{0.94}$; 10.64 qt   **62.** $\dfrac{P}{4} + 3 + \dfrac{P}{3} = P$; 7.20 m   **64.** $\dfrac{x}{400} = \dfrac{264}{24}$; 4,400 trout   **66.** $\dfrac{H}{5} + 6 + \dfrac{H}{2} = H$; 20 m

68. $\dfrac{C}{500} = \dfrac{360}{7.5}$; 24,000 mi

Exercise 4-3

Odd

1. 5 hr   **3.** 6 hr   **5.** 65 min   **7.** 30 quarters and 70 dimes   **9.** 700 \$2 tickets; 2,800 \$4 tickets   **11.** 20 dl
13. 5 hr   **15.** 20 liters   **17.** 71.25 min   **19.** 60 dl of 20% solution, 30 dl of 50% solution   **21.** 3.43 hr
23. 25 kg of \$5/kg tea and 50 kg of \$6.50/kg tea   **25.** 7.5 days   **27.** \$8,000 at 10% and \$12,000 at 15%   **29.** 3.6 liters
31. 670 m

Even

2. 2.6 hr   **4.** 4.5 hr   **6.** 32 min   **8.** 30 nickels and 20 dimes   **10.** 3,000 \$10 tickets; 5,000 \$6 tickets   **12.** 2 ml
14. 3 hr   **16.** 100 cl   **18.** 40 min   **20.** 75 liters of 30% solution; 25 liters of 70% solution   **22.** 2.92 hr
24. 60 kg of \$7.00/kg coffee; 40 kg of \$9.50/kg coffee   **26.** 10 A.M.; 24 km   **28.** \$8,000 at 11% and \$2,000 at 16%
30. 6 liters   **32.** Air: 1,101 ft/sec; water: 4,954 ft/sec

Exercise 4-4

Odd

1. 180 cm   **3.** 85.8 ft   **5.** Solve $\dfrac{N}{600} = \dfrac{500}{60}$; $N = 5,000$ chipmunks   **7.** 205 points   **9.** \$6,400

11. **(A)** 216 mi   **(B)** 225 mi

13. **(A)** 15 in. **(B)** 20 in. **(C)** 22.5 in. **(D)** 24 in. **(E)** 25 in. **(F)** 18 in. **(G)** 18.75 in.

15. Solve $x - 0.2x = 160$; $x = \$200$ **17.** 5,300 copies **19.** Solve $(x + 0.1x) - 0.1(x + 0.1x) = 99$; $x = \$100$

21. 8 sec **23.** 7 cm^2 **25.** Solve $20x = 50(180)$; $x = 450$ kg **27.** 200,000 mi/sec

29. Solve $\dfrac{x}{5} + 4 + \dfrac{2x}{3} = x$; $x = 30$ m **31.** 15 min or $\frac{1}{4}$ hr

---

Even

2. 158 cm **4.** 66 ft **6.** Solve $\dfrac{N}{200} = \dfrac{200}{8}$; $N = 5,000$ trout **8.** 192 points **10.** Solve $\dfrac{C}{3} - \dfrac{C}{4} = 400$; $C = \$4,800$

12. **(A)** 504 km **(B)** 525 km **14.** 330 hertz, 396 hertz **16.** 112 mi **18.** \$1,200 **20.** \$10,000,000

22. 10.8 sec **24.** 30 kg **26.** Solve $8(150) + 6x = 1,920(1)$; $x = 120$ lb **28.** 90 mi **30.** 84 yr

32. 5 and $\frac{5}{11}$ min after 1 P.M.

Exercise 4-5

Odd

1. $r = d/t$ **3.** $r = C/2\pi$ **5.** $\pi = C/D$ **7.** $x = -b/a$, $a \neq 0$ **9.** $x = \dfrac{y + 5}{2}$ **11.** $y = \frac{3}{4}x - 3$

13. $R = E/I$ **15.** $B = \dfrac{CL}{100}$ **17.** $G = \dfrac{Fd^2}{m_1 m_2}$ **19.** $C = \frac{5}{9}(F - 32)$ **21.** $f = \dfrac{ab}{a + b}$ **23.** $n = \dfrac{a_n - a_1 + d}{d}$

25. $T_2 = \dfrac{T_1 P_2 V_2}{P_1 V_1}$ **27.** $x = \dfrac{5y - 3}{3y - 2}$

Even

2. $t = \dfrac{d}{1,100}$ **4.** $t = \dfrac{I}{Pr}$ **6.** $m = \dfrac{e}{c^2}$ **8.** $a = \dfrac{p - 2b}{2}$ **10.** $m = \dfrac{y - b}{x}$, $x \neq 0$ **12.** $y = -\dfrac{A}{B}x - \dfrac{C}{B}$, $B \neq 0$

14. $a = \dfrac{b}{m}$ **16.** $(CA) = \dfrac{100(MA)}{(IQ)}$ **18.** $m_1 = \dfrac{Fd^2}{Gm_2}$ **20.** $F = \frac{9}{5}C + 32$ **22.** $R = \dfrac{R_1 R_2}{R_1 + R_2}$ **24.** $d = \dfrac{a_n - a_1}{n - 1}$

26. $V_1 = \dfrac{T_1 P_2 V_2}{P_1 T_2}$ **28.** $x = \dfrac{4y + 2}{2y - 3}$

Exercise 4-6

Odd

1. $-8 \leq x \leq 7$ **3.** $-6 \leq x < 6$

5. $x \geq -6$ **7.** $(-2, 6]$ **9.** $(-7, 8)$

11. $(-\infty, -2]$ **13.** $[-7, 2)$; $-7 \leq x < 2$ **15.** $(-\infty, 0]$; $x \leq 0$

17. $x < 5$ or $(-\infty, 5)$ **19.** $x \geq 3$ or $[3, \infty)$ **21.** $N < -8$ or $(-\infty, -8)$

23. $t > 2$ or $(2, \infty)$ **25.** $m > 3$ or $(3, \infty)$ **27.** $B \geq -4$ or $[-4, \infty)$

29. $-2 < t \leq 3$ or $(-2, 3]$ **31.** $2x - 3 \geq -6$; $x \geq \frac{-3}{2}$ **33.** $15 - 3x < 6$; $x > 3$

35. $q < -14$ or $(-\infty, -14)$ **37.** $x \geq 4.5$ or $[4.5, \infty)$

39. $-20 \leq x \leq 20$ or $[-20, 20]$ **41.** $-30 \leq x < 18$ or $[-30, 18)$

43. $-8 \le x < -3$ or $[-8, -3)$ **45.** $-14 < x \le 11$ or $(-14, 11]$ **47.** Positive

49. **(A)** F   **(B)** T   **(C)** T   **51.** $8{,}000 \le h \le 20{,}000$ or $[8{,}000, 20{,}000]$ **53.** $x > 600$

55. If r is the worker's maximum running rate and R is the train's rate, then he will escape running toward the train if $r > R/3 = 7R/21$, and he will escape running away from the train if $r > 3R/7 = 9R/21$. Thus, his chances are better if he runs toward the train!

Even

2. $-4 < x < 8$ **4.** $-3 < x \le 3$

6. $x < 7$ **8.** $[-5, 5]$ **10.** $[-4, 5)$

12. $(3, \infty)$ **14.** $[-5, 6]$; $-5 \le x \le 6$ **16.** $(1, \infty)$; $x > 1$

18. $x \ge -3$ or $[-3, \infty)$ **20.** $x < 2$ or $(-\infty, 2)$ **22.** $M \ge 6$ or $[6, \infty)$

24. $n \le -3$ or $(-\infty, -3]$ **26.** $u \le \frac{2}{7}$ or $(-\infty, \frac{2}{7}]$

28. $y < -7$ or $(-\infty, -7)$ **30.** $3 \le m < 7$ or $[3, 7)$ **32.** $2x + 5 \le 7$; $x \le 1$

34. $3x - 5 \le 4x$; $x \ge -5$ **36.** $p \ge 12$ or $[12, \infty)$ **38.** $x > -4\frac{2}{9}$ or $(-4\frac{2}{9}, \infty)$

40. $-9 \le A \le 9$ or $[-9, 9]$ **42.** $41 \le x < 59$ or $[41, 59)$

44. $5 \ge x > 2$ or $(2, 5]$ **46.** $-20 \ge x \ge -35$ or $[-35, -20]$ **48.** Negative

50. When both sides are divided by $n - m$, the sense of the inequality should be changed, because $n - m$ is negative.

52. $2 \le I \le 25$ or $[2, 25]$ **54.** $9.6 \le MA \le 16.8$ or $[9.6, 16.8]$

Exercise 4-7

Odd

1. $\sqrt{5}$ **3.** 4   **5.** $5 - \sqrt{5}$ **7.** $5 - \sqrt{5}$ **9.** 12   **11.** 12   **13.** 9   **15.** 4   **17.** 4   **19.** 9

21. $x = \pm 7$ **23.** $-7 \le x \le 7$ **25.** $x \le -7$ or $x \ge 7$

27. $y = 2$ or 8 **29.** $2 < y < 8$ **31.** $y < 2$ or $y > 8$

33. $u = -11$ or -5 **35.** $-11 \le u \le -5$ **37.** $u \le -11$ or $u \ge -5$

39. $x = -4, \frac{4}{3}$ **41.** $-\frac{9}{5} \le x \le 3$ **43.** $y < 3$ or $y > 5$ **45.** $t = -\frac{4}{5}, \frac{18}{5}$ **47.** $-\frac{5}{7} < u < \frac{23}{7}$

49. $x \le -6$ or $x \ge 9$ **51.** $-35 < C < -\frac{5}{9}$ **53.** $x \ge 5$ **55.** $x \le -8$ **57.** $x \ge -\frac{3}{4}$ **59.** $x \le \frac{2}{5}$

61. Case 1: $a = b$; $|b - a| = |0| = 0$; $|a - b| = |0| = 0$

     Case 2: $a > b$; $|b - a| = -(b - a) = a - b$

                  $|a - b| = a - b$

     Case 3: $b > a$; $|b - a| = b - a$

                  $|a - b| = -(a - b) = b - a$

Even

2. $\frac{3}{4}$ **4.** 4   **6.** $\sqrt{7} - 2$ **8.** $\sqrt{7} - 2$ **10.** 9   **12.** 9   **14.** 8   **16.** 5   **18.** 5   **20.** 3

22. $x = \pm 5$ **24.** $-5 \le t \le 5$ **26.** $x \le -5$ or $x \ge 5$

28. $t = -1$ or 7 **30.** $-1 < t < 7$ **32.** $t < -1$ or $t > 7$

34. $x = -6$ or 4 **36.** $-6 \le x \le 4$ **38.** $x \le -6$ or $x \ge 4$

40. $x = -1, 4$ **42.** $-1 \le x \le 4$ **44.** $u < -\frac{7}{3}$ or $u > -\frac{1}{3}$ **46.** $m = -\frac{11}{3}, \frac{2}{3}$ **48.** $-\frac{8}{9} < M < \frac{22}{9}$

50. $x \le -16$ or $x \ge 8$ **52.** $-40 < F < 104$ **54.** $x \ge -7$ **56.** $x \le 11$ **58.** $x \ge \frac{9}{5}$ **60.** $x \le -\frac{7}{3}$

Diagnostic (Review) Exercise 4-8

1. $x = -2$ *(4-1)* **2.** $x = 2$ *(4-1)* **3.** $x < -2$ *(4-6)* **4.** $1 < x < 6$ *(4-6)* **5.** $x = \pm 6$ *(4-7)*

6. $-6 < x < 6$ *(4-7)* **7.** $x < -6$ or $x > 6$ *(4-7)* **8.** $-14, -4$ *(4-7)* **9.** $-14 < y < -4$ *(4-7)*

10. $y < -14$ or $y > -4$ *(4-7)* **11.** 9 *(4-1)* **12.** $\frac{-10}{9}$ *(4-1)* **13.** 60 *(4-1)* **14.** $x \le -12$ *(4-6)*

15. $b = \dfrac{2A}{h}$ *(4-5)* **16.** **(A)** 6 **(B)** 6 *(4-7)* **17.** $-4 \le x < 3$ *(4-6)*

18. $1 < x \le 4$ *(4-6)* **19.** $x \ge 1$ *(4-6)* **20.** 41 *(4-1)* **21.** -12 *(4-1)*

22. 5 *(4-1)* **23.** No solution *(4-1)* **24.** $\frac{1}{2}, 3$ *(4-7)* **25.** $\frac{1}{2} \le x \le 3$ *(4-7)*

26. $x < \frac{1}{2}$ or $x > 3$ *(4-7)* **27.** 11 *(4-1)* **28.** $x \ge -19$ *(4-6)*

29. $-6 < x \le -1$ *(4-6)* **30.** -2 *(4-1)* **31.** -5 *(4-1)* **32.** $\frac{-3}{5}$ *(4-1)* **33.** No solution *(4-1)*

34. $\frac{3}{4}$ *(4-1)* **35.** $L = \dfrac{2S}{n} - a$ or $L = \dfrac{2S - an}{n}$ *(4-5)* **36.** $M = \dfrac{P}{1 - dt}$ *(4-5)* **37.** **(A)** T **(B)** T *(4-6)*

38. $-1 \le x \le 4$ or $[-1, 4]$ *(4-7)* **39.** $\frac{-13}{5}$ *(4-1)* **40.** $x \ge \frac{25}{7}$ *(4-6)*

41. $-3 \le x \le 6$ *(4-6)* **42.** -15 *(4-1)* **43.** No solution *(4-7)* **44.** $x = \dfrac{5y + 3}{2y - 4}$ *(4-5)*

45. $f_1 = \dfrac{ff_2}{f_2 - f}$ *(4-5)* **46.** $x \ge \frac{3}{2}$ *(4-5)* **47.** $x \le \frac{3}{2}$ *(4-5)* **48.** Solve $2x + 2\left(\dfrac{3x}{5} - 2\right) = 76$; 25 by 13 cm *(4-2)*

49. Solve $56(x + 1.5) = 76x$; $x = 4.2$ hr *(4-3)* **50.** Solve $x - 0.3x = 210$; $x = \$300$ *(4-4)*

51. Solve $45x + 55(x - 10) = 3{,}000$; $x = 35.5$ min *(4-3)* **52.** Solve $\dfrac{x}{127} = \dfrac{1}{2.54}$; $x = 50$ in. *(4-2)*

53. Solve $\dfrac{x}{70} = \dfrac{18}{50}$; $x = 25.2$ ml of alcohol *(4-2)* **54.** Solve $\dfrac{65 + 80 + x}{3} \ge 75$; $x \ge 80$ *(4-4)*

55. Solve $0.3(60) = 0.25(x + 60)$; $x = 12$ ml *(4-3)*

56. Solve $0.5x + 0.8(36 - x) = 0.6(36)$; 24 dl of 50% solution, 12 dl of 80% solution *(4-3)*

57. Solve $\dfrac{x}{360} = \dfrac{55}{6}$; $x = 3{,}300$ squirrels *(4-2)* **58.** Solve $10 \le \frac{5}{9}(F - 32) \le 15$; $50° \le F \le 59°$ *(4-6)*

59. Solve $0.4(24 - x) + x = 0.5(24)$; $x = 4$ dl *(4-3)*

Practice Test Chapter 4

1. 0 *(4-1)* **2.** $\frac{1}{4}$ *(4-1)* **3.** $\dfrac{y + 3}{7y - 2}$ *(4-5)* **4.** $2, -\frac{10}{3}$ *(4-7)* **5.** $x \le 5$ or $(-\infty, 5]$ *(4-6)*

6. $-1 < x < 4$ or $(-1, 4)$ *(4-7)* **7.** $-2 \le x < 4$ or $[-2, 4)$ *(4-6)* **8.** 18 *(4-2)*

9. 160 km *(4-2)* **10.** 30 hr *(4-3)* **11.** 60 m *(4-2)* **12.** 32 lb of 30%; 48 lb of 10% *(4-3)* **13.** 1 gal *(4-3)*

14. 12 hr *(4-3)*

Chapter 5 Exercise 5-1

Odd

1. 1 **3.** 1 **5.** $\dfrac{1}{3^3}$ **7.** $\dfrac{1}{m^7}$ **9.** 4^3 **11.** y^5 **13.** 10^2 **15.** y **17.** 1 **19.** 10^{10} **21.** x^{11}

23. $\dfrac{1}{z^5}$ **25.** $\dfrac{1}{10^7}$ **27.** 10^{12} **29.** y^8 **31.** $u^{10}v^6$ **33.** $\dfrac{x^4}{y^6}$ **35.** $\dfrac{x^2}{y^3}$ **37.** 1 **39.** 10^2 **41.** y

43. 10 **45.** 3×10^{16} **47.** y^9 **49.** $3^2m^2n^2$ **51.** $\dfrac{2^3m^3}{n^9}$ **53.** $\dfrac{n^{15}}{m^{12}}$ **55.** $\dfrac{3^3}{2^2}$ **57.** 1 **59.** $\dfrac{4y^3}{3x^5}$

61. $\dfrac{a^9}{8b^4}$ **63.** $\dfrac{1}{x^7}$ **65.** $\dfrac{n^8}{m^{12}}$ **67.** $\dfrac{m^3n^3}{8}$ **69.** $\dfrac{t^2}{x^2y^{10}}$ **71.** 4 **73.** $\dfrac{1}{a^2 - b^2}$ **75.** $\dfrac{1}{xy}$ **77.** $-cd$

79. $\dfrac{xy}{x + y}$ **81.** $\dfrac{(y - x)^2}{x^2y^2}$ **83.** $\dfrac{y - x}{y}$

Even

2. 1 **4.** 1 **6.** $\dfrac{1}{2^2}$ **8.** $\dfrac{1}{x^4}$ **10.** 3^2 **12.** x^3 **14.** 10^2 **16.** x^4 **18.** 1 **20.** 10^{11} **22.** a^{12}

24. $\dfrac{1}{b^8}$ **26.** $\dfrac{1}{10^6}$ **28.** 2^6 **30.** x^{10} **32.** x^3y^2 **34.** $\dfrac{y^6}{x^4}$ **36.** $\dfrac{y^3}{x^2}$ **38.** 1 **40.** 10^2 **42.** $\dfrac{1}{x}$

44. 10^{17} **46.** 4×10^2 **48.** x^6 **50.** $\dfrac{1}{2^3c^3d^6}$ **52.** $\dfrac{3^2x^6}{y^4}$ **54.** $\dfrac{x^6}{y^4}$ **56.** $\dfrac{2^6}{3^4}$ **58.** $\dfrac{1}{10^4}$

60. $\dfrac{3n^4}{4m^3}$ **62.** $\dfrac{2x}{y^2}$ **64.** n^2 **66.** $\dfrac{x^{12}}{y^8}$ **68.** $\dfrac{4x^8}{y^6}$ **70.** $\dfrac{w^{12}}{u^{20}v^4}$ **72.** $\dfrac{27y^3}{2x^3}$ **74.** $\dfrac{1}{(x + 2)^2}$ **76.** $\dfrac{1}{30}$

78. $\dfrac{144}{7}$ **80.** $\dfrac{36}{13}$ **82.** $\dfrac{1,000}{11}$ **84.** $\dfrac{v - u}{v + u}$

Exercise 5-2

Odd

1. 7×10 **3.** 8×10^2 **5.** 8×10^4 **7.** 8×10^{-3} **9.** 8×10^{-8} **11.** 5.2×10 **13.** 6.3×10^{-1}
15. 3.4×10^2 **17.** 8.5×10^{-2} **19.** 6.3×10^3 **21.** 6.8×10^{-6} **23.** 800 **25.** 0.04 **27.** 300,000
29. 0.0009 **31.** 56,000 **33.** 0.0097 **35.** 430,000 **37.** 0.000 000 38 **39.** 5.46×10^9 **41.** 7.29×10^{-8}
43. 10^{13} **45.** 10^{-5} **47.** 83,500,000,000 **49.** 0.000 000 000 006 14 **51.** 865,000
53. 0.000 000 000 000 000 000 000 001 7 **55.** 9×10^4 **57.** 6×10^{-4} **59.** 3×10^5 **61.** 5×10^4
63. 3×10 or 30 **65.** 3×10^{-4} or 0.0003 **67.** 6.6×10^{21} tons **69.** 10^7; 6×10^8 **71.** 562

Even

2. 5×10 **4.** 6×10^2 **6.** 6×10^5 **8.** 6×10^{-2} **10.** 6×10^{-5} **12.** 3.5×10 **14.** 7.2×10^{-1}
16. 2.7×10^2 **18.** 3.2×10^{-2} **20.** 5.2×10^3 **22.** 7.2×10^{-4} **24.** 500 **26.** 0.08 **28.** 6,000,000
30. 0.000 02 **32.** 7,100 **34.** 0.000 86 **36.** 8,800,000 **38.** 0.000 006 1 **40.** 4.27×10^7 **42.** 7.23×10^{-5}
44. 5.87×10^{12} **46.** 3×10^{-23} **48.** 3,460,000,000 **50.** 0.000 000 623 **52.** 93,000,000 **54.** 0.000 075
56. 8×10^2 **58.** 8×10^{-3} **60.** 3×10^3 **62.** 3×10^7 **64.** 2×10^4 or 20,000 **66.** 2×10^{-4} or 0.0002
68. 3.3×10^{18} lb **70.** 0.0186 mi or 98.2 ft **72.** 64

Exercise 5-3

Odd

1. 5   **3.** Not a real number   **5.** 2   **7.** -2 **9.** -2 **11.** 64   **13.** 4   **15.** x **17.** $\dfrac{1}{x^{1/5}}$

19. x^2 **21.** ab^3 **23.** $\dfrac{x^3}{y^4}$ **25.** x^2y^3 **27.** $\frac{2}{5}$ **29.** $\frac{8}{125}$ **31.** $\frac{1}{4}$ **33.** $\frac{1}{6}$ **35.** $\frac{1}{125}$ **37.** 25   **39.** $\frac{1}{9}$

41. $\dfrac{1}{x^{1/2}}$ **43.** $n^{1/12}$ **45.** x^4 **47.** $\dfrac{2v^2}{u}$ **49.** $\dfrac{1}{x^2y^3}$ **51.** $\dfrac{x^4}{y^3}$ **53.** $\frac{5}{4}x^4y^2$ **55.** $64y^{1/3}$

57. $12m - 6m^{35/4}$ **59.** $2x + 3x^{1/2}y^{1/2} + y$ **61.** $x + 2x^{1/2}y^{1/2} + y$ **63.** Not defined   **65.** $\dfrac{2}{a} + \dfrac{5}{a^{1/2}b^{1/2}} - \dfrac{3}{b}$

67. $a^{1/2}b^{1/3}$ **69.** x **71.** $\dfrac{1}{x^m}$ **73.** **(A)** Any negative number   **(B)** n even and x any negative number

Even

2. 6   **4.** Not a real number   **6.** 3   **8.** -3 **10.** -3 **12.** 125   **14.** 9   **16.** $y^{3/5}$ **18.** $a^{1/3}$

20. y^2 **22.** x^2y **24.** $\dfrac{m^3}{n^4}$ **26.** $\dfrac{u^6}{v^4}$ **28.** $\frac{3}{2}$ **30.** $\frac{27}{8}$ **32.** $\frac{1}{9}$ **34.** $\frac{1}{5}$ **36.** $\frac{1}{64}$ **38.** 49

40. $\frac{1}{8}$ **42.** d **44.** $m^{1/6}$ **46.** $\dfrac{1}{y^{1/2}}$ **48.** $\dfrac{2x}{y^2}$ **50.** $16xy^3$ **52.** $m^{1/2}n^{1/3}$ **54.** $\dfrac{2b^2}{3a^2}$ **56.** $3x^{1/2}$

58. $6x - 2x^{19/3}$ **60.** $x - y$ **62.** $x - 2x^{1/2}y^{1/2} + y$ **64.** $-\frac{1}{64}$ **66.** $\dfrac{1}{x} - \dfrac{2}{x^{1/2}y^{1/2}} + \dfrac{1}{y}$ **68.** $a^{1/n}b^{1/m}$

70. a **72.** y^{m+1} **74.** All real numbers

Exercise 5-4

Odd

1. $\sqrt{11}$ **3.** $\sqrt[3]{5}$ **5.** $\sqrt[5]{u^3}$ **7.** $4\sqrt[7]{y^3}$ **9.** $\sqrt[7]{(4y)^3}$ **11.** $\sqrt[5]{(4ab^3)^2}$ **13.** $\sqrt{a+b}$ **15.** $6^{1/2}$ **17.** $m^{1/4}$

19. $y^{3/5}$ **21.** $(xy)^{3/4}$ **23.** $(x^2 - y^2)^{1/2}$ **25.** $-5\sqrt[5]{y^2}$ **27.** $\sqrt[7]{(1 + m^2n^2)^3}$ **29.** $\dfrac{1}{\sqrt[3]{w^2}}$ **31.** $\dfrac{1}{\sqrt[5]{(3m^2n^3)^3}}$

33. $\sqrt{a} + \sqrt{b}$ **35.** $\sqrt[3]{(a^3 + b^3)^2}$ **37.** $(a + b)^{2/3}$ **39.** $-3x(a^3b)^{1/4}$ **41.** $(-2x^3y^7)^{1/9}$ **43.** $\dfrac{3}{y^{1/3}}$ or $3y^{-1/3}$

45. $\dfrac{-2x}{(x^2 + y^2)^{1/2}}$ or $-2x(x^2 + y^2)^{-1/2}$ **47.** $m^{2/3} - n^{1/2}$ **49.** $\sqrt{2^2 + 3^2} = \sqrt{13} \neq 2 + 3 = 5$

51. **(A)** Any negative number   **(B)** n even and x any negative number

Even

2. $\sqrt{7}$ **4.** $\sqrt[6]{6}$ **6.** $\sqrt[4]{x^3}$ **8.** $5\sqrt[3]{m^2}$ **10.** $\sqrt[3]{(5m)^2}$ **12.** $\sqrt[3]{(7x^2y)^2}$ **14.** $\sqrt{a^2 + b^2}$ **16.** $3^{1/2}$

18. $m^{1/7}$ **20.** $a^{2/3}$ **22.** $(7m^3n^3)^{4/5}$ **24.** $(1 + y^2)^{1/2}$ **26.** $-3\sqrt{x}$ **28.** $\sqrt[5]{(x^2y^2 - w^3)^4}$ **30.** $\dfrac{1}{\sqrt[3]{y^3}}$

32. $\dfrac{1}{\sqrt[3]{(2xy)^2}}$ **34.** $\dfrac{1}{\sqrt{x}} + \dfrac{1}{\sqrt{y}}$ **36.** $\sqrt[3]{\sqrt{x} + \dfrac{1}{\sqrt{y}}}$ **38.** $(x - y)^{2/5}$ **40.** $-5(2x^2y^2)^{1/3}$ **42.** $(-4m^2n^3)^{1/5}$

44. $\dfrac{2x}{y^{1/2}}$ or $2xy^{-1/2}$ **46.** $\dfrac{2}{x^{1/2}} + \dfrac{3}{y^{1/2}}$ or $2x^{-1/2} + 3y^{-1/2}$ **48.** $\dfrac{-5u^2}{u^{1/2} + v^{3/5}}$ **50.** $\sqrt{3^2 + 1^2} = \sqrt{10} \neq 3 + 1 = 4$

52. Both are true

Exercise 5-5

Odd

1. y **3.** $2u$ **5.** $7x^2y$ **7.** $3\sqrt{2}$ **9.** $m\sqrt{m}$ **11.** $2x\sqrt{2x}$ **13.** $\frac{1}{3}$ **15.** $\frac{1}{y}$ **17.** $\frac{\sqrt{5}}{5}$ **19.** $\frac{\sqrt{5}}{5}$

21. $\frac{\sqrt{y}}{y}$ **23.** $\frac{\sqrt{y}}{y}$ **25.** $3xy^2\sqrt{xy}$ **27.** $3x^4y^2\sqrt{2y}$ **29.** $\frac{\sqrt{2x}}{2x}$ **31.** $2x\sqrt{3x}$ **33.** $\frac{3\sqrt{2ab}}{2b}$ **35.** $\frac{\sqrt{42xy}}{7y}$

37. $\frac{3m^2\sqrt{2mn}}{2n}$ **39.** $2x^2y$ **41.** $2xy^2\sqrt[3]{2xy}$ **43.** $\sqrt{x}$ **45.** 4 **47.** $6m^3n^3$ **49.** $2\sqrt[3]{9}$ **51.** $\frac{2a\sqrt{3ab}}{3b}$

53. Is in the simplest radical form **55.** $\frac{2x}{3y^2}$ **57.** $-3m^2n^2\sqrt[5]{3m^2n}$ **59.** $\sqrt[3]{x^2(x-y)}$ **61.** $x^2y\sqrt[3]{6xy}$

63. $2x^2y\sqrt[3]{4x^2y}$ **65.** $-\sqrt[3]{6x^2y^2}$ **67.** $\sqrt[3]{(x-y)^2}$ **69.** $\frac{\sqrt[4]{12x^3y^3}}{2x}$ **71.** $-x\sqrt{x^2+2}$ **73.** $4x^9y\sqrt[3]{2y}$

75. $mn\sqrt[12]{3^7m^5n^2}$ **77.** $x^n(x+y)^{n+2}$ **79.** **(A)** $6x$ **(B)** $-2x$ **81.** **(A)** $2x$ **(B)** 0 **83.** **(A)** x **(B)** $-5x$

Even

2. x **4.** $3m$ **6.** $5xy^2$ **8.** $2\sqrt{2}$ **10.** $x\sqrt{x}$ **12.** $3y\sqrt{2y}$ **14.** $\frac{1}{2}$ **16.** $\frac{1}{x}$ **18.** $\frac{\sqrt{3}}{3}$ **20.** $\frac{\sqrt{3}}{3}$

22. $\frac{\sqrt{x}}{x}$ **24.** $\frac{\sqrt{x}}{x}$ **26.** $2x^2y\sqrt{xy}$ **28.** $2x^3y^3\sqrt{2x}$ **30.** $\frac{\sqrt{3y}}{3y}$ **32.** $2x\sqrt{2y}$ **34.** $\frac{2}{3}x\sqrt{3xy}$ **36.** $\frac{\sqrt{6mn}}{2n}$

38. $\frac{2a\sqrt{3ab}}{3b}$ **40.** $2mn^3$ **42.** $2ab^2\sqrt[4]{a}$ **44.** $\sqrt[5]{x^3}$ **46.** 3 **48.** $3xy$ **50.** $\sqrt[3]{4}$ **52.** $\frac{3m^2\sqrt{2mn}}{2n}$

54. Is in the simplest radical form **56.** $\frac{a^2b}{2c^3}$ **58.** $-4x^3y^4\sqrt[3]{x^2y}$ **60.** $\sqrt[4]{2^3(x+y)^3}$ **62.** $u^2v\sqrt[4]{24v}$

64. $4u^2v^4\sqrt[3]{2uv}$ **66.** $2\sqrt[3]{6abc}$ **68.** $\frac{\sqrt[3]{x-y}}{x-y}$ **70.** $\frac{\sqrt[5]{8m^2n^2}}{2m}$ **72.** $m\sqrt[4]{1+4m^2}$ **74.** $2x\sqrt[3]{4y^2}$

76. $2x^4y^3\sqrt[6]{2^5x^5y^5}$ **78.** x^2y^{n+1} **80.** **(A)** $-2x$ **(B)** $6x$ **82.** **(A)** $2x$ **(B)** 0 **84.** **(A)** $2x$ **(B)** $8x$

Exercise 5-6

Odd

1. $9\sqrt{3}$ **3.** $-5\sqrt{a}$ **5.** $-5\sqrt{n}$ **7.** $4\sqrt{5}-2\sqrt{3}$ **9.** $\sqrt{m}-3\sqrt{n}$ **11.** $4\sqrt{2}$ **13.** $-6\sqrt{2}$
15. $7-2\sqrt{7}$ **17.** $3\sqrt{2}-2$ **19.** $y-8\sqrt{y}$ **21.** $4\sqrt{n}-n$ **23.** $3+3\sqrt{2}$ **25.** $3-\sqrt{3}$, **27.** $9+4\sqrt{5}$
29. $m-7\sqrt{m}+12$ **31.** $\sqrt{5}-2$ **33.** $\frac{\sqrt{5}-1}{2}$ **35.** $\frac{\sqrt{5}+\sqrt{2}}{3}$ **37.** $\frac{y-3\sqrt{y}}{y-9}$ **39.** $8\sqrt{2mn}$

41. $6\sqrt{2}-2\sqrt{5}$ **43.** $-\sqrt[5]{a}$ **45.** $5\sqrt[3]{x}-\sqrt{x}$ **47.** $\frac{9\sqrt{2}}{4}$ **49.** $\frac{-3\sqrt{6uv}}{2}$ **51.** $38-11\sqrt{3}$ **53.** $x-y$

55. $10m-11\sqrt{m}-6$ **57.** $5+\sqrt[3]{18}+\sqrt[3]{12}$ **59.** $(3-\sqrt{2})^2-6(3-\sqrt{2})+7=9-6\sqrt{2}+2-18+6\sqrt{2}+7=0$

61. $-7-4\sqrt{3}$ **63.** $5+2\sqrt{6}$ **65.** $\frac{x+5\sqrt{x}+6}{x-9}$ **67.** $\frac{6x+9\sqrt{x}}{4x-9}$ **69.** $3\sqrt{3}$ **71.** $\frac{10}{3}\sqrt[3]{9}$

73. $x+2\sqrt[3]{xy}-\sqrt[3]{x^2y^2}-2y$ **75.** $x+y$ **77.** $\frac{8x-22\sqrt{xy}+15y}{16x-25y}$ **79.** $\frac{\sqrt[3]{x^2}-\sqrt[3]{x}\sqrt[3]{y}+\sqrt[3]{y^2}}{x+y}$

81. $\frac{(\sqrt{x}+\sqrt{y}+\sqrt{z})[(x+y-z)-2\sqrt{xy}]}{(x+y-z)^2-4xy}$

Even

2. $8\sqrt{2}$ **4.** $-3\sqrt{y}$ **6.** $4\sqrt{x}$ **8.** $2\sqrt{2} - 2\sqrt{3}$ **10.** $2\sqrt{x} + 2\sqrt{y}$ **12.** $\sqrt{2}$ **14.** $-3\sqrt{3}$ **16.** $5 - 2\sqrt{5}$
18. $2\sqrt{3} - 3$ **20.** $x - 3\sqrt{x}$ **22.** $3\sqrt{m} - m$ **24.** $5\sqrt{2} + 5$ **26.** $2\sqrt{2} - 1$ **28.** $12 - 6\sqrt{3}$
30. $x - \sqrt{x} - 6$ **32.** $\dfrac{\sqrt{11} + 3}{2}$ **34.** $2\sqrt{6} + 4$ **36.** $\sqrt{3} - \sqrt{2}$ **38.** $\dfrac{x + 2\sqrt{x}}{x - 4}$ **40.** $-\sqrt{x}$ **42.** $2\sqrt{6} + \sqrt{3}$
44. $-\sqrt[3]{u}$ **46.** $3\sqrt[5]{y} + 3\sqrt[4]{y}$ **48.** $\dfrac{-\sqrt{6}}{6}$ **50.** $\dfrac{5\sqrt{2xy}}{2}$ **52.** 25 **54.** $4x - 9$ **56.** $6u + 8\sqrt{u} - 8$
58. -2 **60.** $(2 + \sqrt{3})^2 - 4(2 + \sqrt{3}) + 1 = 4 + 4\sqrt{3} + 3 - 8 - 4\sqrt{3} + 1 = 0$ **62.** $3 - 2\sqrt{2}$ **64.** $\dfrac{6 + \sqrt{a} - a}{a - 4}$
66. $\dfrac{7 - 2\sqrt{10}}{3}$ **68.** $\dfrac{15\sqrt{a} + 10a}{9 - 4a}$ **70.** $3\sqrt{2}$ **72.** $\frac{3}{2}\sqrt[4]{2}$ **74.** $u + \sqrt[5]{u^2 v^2} - \sqrt[5]{u^3 v^3} - v$ **76.** $x - y$
78. $\dfrac{6x + 19\sqrt{xy} + 10y}{4x - 25y}$ **80.** $\dfrac{\sqrt[3]{x^2} + \sqrt[3]{x}\sqrt[3]{y} + \sqrt[3]{y^2}}{x - y}$ **82.** $\dfrac{(\sqrt{x} - \sqrt{y} - \sqrt{z})[(x + y - z) + 2\sqrt{xy}]}{(x + y - z)^2 - 4xy}$

Exercise 5-7

Odd

1. $8 + 3i$ **3.** $-5 + 3i$ **5.** $5 + 3i$ **7.** $6 + 13i$ **9.** $3 - 2i$ **11.** -15 or $-15 + 0i$ **13.** $-6 - 10i$
15. $15 - 3i$ **17.** $-4 - 33i$ **19.** 65 or $65 + 0i$ **21.** $\frac{2}{5} - \frac{1}{5}i$ **23.** $\frac{3}{13} + \frac{11}{13}i$ **25.** $5 + 3i$ **27.** $7 - 5i$
29. $-3 + 2i$ **31.** $8 + 25i$ **33.** $\frac{5}{3} - \frac{2}{3}i$ **35.** $\frac{2}{13} + \frac{3}{13}i$ **37.** $-\frac{2}{5}i$ or $0 - \frac{2}{5}i$ **39.** $\frac{3}{2} - \frac{1}{2}i$ **41.** $4 - 7i$
43. 0 or $0 + 0i$ **45.** 0 **47.** $-1, -i, 1, i, -1, -i, 1$ **49.** $(a + c) + (b + d)i$ **51.** $a^2 + b^2$ or $(a^2 + b^2) + 0i$
53. $(ac - bd) + (ad + bc)i$ **55.** 1 **57.** $\pm 6i$ **59.** $9 \pm 3i$ **61.** For $x \geq 10$ **63.** 1

Even

2. $8 + 4i$ **4.** $7 - 5i$ **6.** $7 + 2i$ **8.** $-3 + 2i$ **10.** $17 - 2i$ **12.** -8 or $-8 + 0i$ **14.** $-12 - 6i$
16. $-21 + i$ **18.** $8 + i$ **20.** 34 or $34 + 0i$ **22.** $\frac{3}{10} + \frac{1}{10}i$ **24.** $\frac{4}{13} - \frac{7}{13}i$ **26.** $3 + 3i$ **28.** $-5 + 3i$
30. $6 + 13i$ **32.** $13 + i$ **34.** $3 - 4i$ **36.** $\frac{3}{25} + \frac{4}{25}i$ **38.** $-\frac{1}{3}i$ or $0 - \frac{1}{3}i$ **40.** $-\frac{1}{3} - \frac{2}{3}i$ **42.** $-6i$ or $0 - 6i$
44. 0 **46.** 0 **48.** $1, i, -1, -i, 1$ **50.** $(a - c) + (b - d)i$ **52.** $u^2 + v^2$ or $(u^2 + v^2) + 0i$
54. $\dfrac{ac + bd}{c^2 + d^2} + \dfrac{bc - ad}{c^2 + d^2}i$ **56.** 1 **58.** $\pm 5i$ **60.** $3 \pm 2i$
62. Real if $b^2 - 4ac \geq 0$; nonreal complex if $b^2 - 4ac < 0$.

Diagnostic (Review) Exercise 5-8

1. 1 *(5-1)*   **2.** $\frac{1}{9}$ *(5-1)*   **3.** 8 *(5-1)*   **4.** $\frac{1}{2}$ *(5-3)*   **5.** Not a real number *(5-3)*   **6.** 4 *(5-3)*
7. **(A)** 4.28×10^9 **(B)** 3.18×10^{-5} *(5-2)*   **8.** **(A)** $729{,}000$ **(B)** $0.000\,603$ *(5-2)*   **9.** $6x^4 y^7$ *(5-1)*
10. $\dfrac{3u^4}{v^2}$ *(5-1)*   **11.** $6x^5 y^{15}$ *(5-1)*   **12.** $\dfrac{c^6}{d^{15}}$ *(5-1)*   **13.** $\dfrac{4x^4}{9y^6}$ *(5-1)*   **14.** x^{12} *(5-1)*   **15.** y^2 *(5-1)*
16. $\dfrac{y^3}{x^2}$ *(5-1)*   **17.** x^3 *(5-3)*   **18.** $\dfrac{1}{x^2}$ *(5-3)*   **19.** $\dfrac{1}{x^{1/3}}$ *(5-3)*   **20.** u *(5-3)*
21. **(A)** $\sqrt{3m}$ **(B)** $3\sqrt{m}$ *(5-4)*   **22.** **(A)** $(2x)^{1/2}$ **(B)** $(a + b)^{1/2}$ *(5-4)*   **23.** $2xy^2$ *(5-5)*   **24.** $\dfrac{5}{y}$ *(5-5)*
25. $6x^2 y^3 \sqrt{y}$ *(5-5)*   **26.** $\dfrac{\sqrt{2y}}{2y}$ *(5-5)*   **27.** $2b\sqrt{3a}$ *(5-5)*   **28.** $6x^2 y^3 \sqrt{xy}$ *(5-5)*   **29.** $\dfrac{\sqrt{2xy}}{2x}$ *(5-5)*
30. $-3\sqrt{x}$ *(5-6)*   **31.** $\sqrt{7} - 2\sqrt{3}$ *(5-6)*   **32.** $5 + 2\sqrt{5}$ *(5-6)*   **33.** $1 + \sqrt{3}$ *(5-6)*   **34.** $\dfrac{5 + 3\sqrt{5}}{4}$ *(5-6)*
35. $3 - 6i$ *(5-7)*   **36.** $15 + 3i$ *(5-7)*   **37.** $2 + i$ *(5-7)*   **38.** $-\frac{1}{2} - i$ *(5-7)*   **39.** 2×10^{-3} or 0.002 *(5-2)*
40. $\dfrac{m^2}{2n^5}$ *(5-1)*   **41.** $\dfrac{x^6}{y^4}$ *(5-1)*   **42.** $\dfrac{4x^4}{y^6}$ *(5-1)*   **43.** $\dfrac{c}{a^2 b^4}$ *(5-1)*   **44.** $\frac{1}{4}$ *(5-1)*   **45.** $\dfrac{n^{10}}{9m^{10}}$ *(5-1)*

46. $\dfrac{1}{(x-y)^2}$ *(5-1)*   **47.** $\dfrac{3a^2}{b}$ *(5-3)*   **48.** $\dfrac{3x^2}{2y^2}$ *(5-3)*   **49.** $\dfrac{1}{m}$ *(5-3)*   **50.** $6x^{1/6}$ *(5-3)*   **51.** $\dfrac{x^{1/12}}{2}$ *(5-3)*

52. $\frac{5}{9}$ *(5-1)*   **53.** $x + 2x^{1/2}y^{1/2} + y$ *(5-3)*   **54.** $a^2 = b$ *(5-3)*   **55.** **(A)** $\sqrt[3]{4m^2n^2}$ **(B)** $3\sqrt[5]{x^2}$ *(5-4)*

56. **(A)** $x^{5/7}$ **(B)** $-3(xy)^{2/3}$ *(5-4)*   **57.** $2x^2y$ *(5-5)*   **58.** $3x^2y\sqrt[3]{x^2y}$ *(5-5)*   **59.** $\dfrac{n^2\sqrt{6m}}{3}$ *(5-5)*

60. $\sqrt[4]{y^3}$ *(5-5)*   **61.** $-6x^2y^2\sqrt[5]{3x^2y}$ *(5-5)*   **62.** $x\sqrt[3]{2x^2}$ *(5-5)*   **63.** $\dfrac{\sqrt[5]{12x^3y^2}}{2x}$ *(5-6)*

64. $2x - 3\sqrt{xy} - 5y$ *(5-6)*   **65.** $\dfrac{x - 4\sqrt{x} + 4}{x - 4}$ *(5-6)*   **66.** $\dfrac{6x + 3\sqrt{xy}}{4x - y}$ *(5-6)*   **67.** $\dfrac{5\sqrt{6}}{6}$ *(5-6)*

68. $-1 - i$ *(5-7)*   **69.** $\frac{4}{13} - \frac{7}{13}i$ *(5-7)*   **70.** $5 + 4i$ *(5-7)*   **71.** $\dfrac{xy}{x+y}$ *(5-1)*   **72.** $\dfrac{a^2b^2}{a^3+b^3}$ *(5-1)*

73. $y\sqrt[3]{2x^2y}$ *(5-5)*   **74.** 0 *(5-6)*   **75.** **(A)** x **(B)** $5x$ *(5-5)*

Practice Test Chapter 5

1. $\dfrac{9u^2}{v^6}$ *(5-1)*   **2.** $\dfrac{x^2}{2y^3}$ *(5-3)*   **3.** xy *(5-3)*   **4.** $\dfrac{x + 2x^{1/2}y^{1/2} + y}{xy}$ *(5-3)*   **5.** 3×10^{-6} *(5-2)*   **6.** $-x$ *(5-5)*

7. $-2\sqrt[3]{2}$ *(5-5)*   **8.** $\sqrt[3]{4x^2y^2}$ *(5-5)*   **9.** 0 *(5-6)*   **10.** $\dfrac{x + 2\sqrt{xy} + y}{x - y}$ *(5-6)*   **11.** $13 + 3i$ *(5-7)*

12. $37i$ *(5-7)*   **13.** $-i$ *(5-7)*   **14.** $\sqrt[6]{yz^2}$ *(5-5)*

Chapter 6   Exercise 6-1

Odd

1. ± 4 **3.** $\pm 4i$ **5.** $\pm 3\sqrt{5}$ **7.** $\pm \frac{3}{2}$ **9.** $\pm \frac{3}{4}$ **11.** $0, -5$ **13.** $0, -4$ **15.** $12, -1$ **17.** $1, -5$

19. $-\frac{2}{3}, 4$ **21.** $\pm\sqrt{2}$ **23.** $\pm\frac{3}{4}i$ **25.** $\pm\sqrt{\frac{7}{9}}$ or $\pm\dfrac{\sqrt{7}}{3}$ **27.** $8, -2$ **29.** $-1 \pm 3i$ **31.** $-\frac{1}{3}, 1$

33. $-1, 3$ **35.** $-\frac{2}{3}, 1$ **37.** Not factorable in the integers   **39.** $-\frac{1}{2}, 3$ **41.** $-2, 2$ **43.** $3, -4$

45. $-\frac{1}{2}, 2$ **47.** $\frac{1}{2}, 2$ **49.** $2, -3$ **51.** 11×3 in.   **53.** $\dfrac{-5 \pm \sqrt{10}}{2}$ **55.** $2 \pm i$ **57.** $-1, 1$

59. No solution   **61.** $a = \sqrt{c^2 - b^2}$ **63.** 90¢/gal

Even

2. ± 5 **4.** $\pm 5i$ **6.** $\pm 2\sqrt{3}$ **8.** $\pm \frac{4}{3}$ **10.** $\pm \frac{2}{3}$ **12.** $0, 3$ **14.** $0, 2$ **16.** $1, 5$ **18.** $-2, 6$

20. $\frac{1}{2}, -8$ **22.** $\pm\sqrt{3}$ **24.** $\pm\frac{5}{2}i$ **26.** $\pm\sqrt{\frac{3}{4}}$ or $\dfrac{\pm\sqrt{3}}{2}$ **28.** $-2, -8$ **30.** $3 \pm 2i$ **32.** $-1, 2$

34. $3, -5$ **36.** $\frac{1}{2}, -3$ **38.** Not factorable in the integers   **40.** $\frac{2}{3}, 2$ **42.** $-3, 3$ **44.** $-2, 4$

46. $5, -3$ **48.** $-2, 5$ **50.** $1, -2$ **52.** $h = 1$ ft, $b = 4$ ft   **54.** $\dfrac{3 \pm \sqrt{6}}{2}$ **56.** $-\dfrac{1}{2} \pm \dfrac{\sqrt{3}}{2}i$

58. No solution   **60.** 1 **62.** $t = \sqrt{\dfrac{2s}{g}}$ **64.** 2.83 ft/sec

Exercise 6-2

Odd

1. $x^2 + 4x + 4 = (x + 2)^2$ **3.** $x^2 - 6x + 9 = (x - 3)^2$ **5.** $x^2 + 12x + 36 = (x + 6)^2$ **7.** $-2 \pm \sqrt{2}$

9. $3 \pm 2\sqrt{3}$ **11.** $x^2 + 3x + \frac{9}{4} = (x + \frac{3}{2})^2$ **13.** $u^2 - 5u + \frac{25}{4} = (u - \frac{5}{2})^2$ **15.** $\dfrac{-1 \pm \sqrt{5}}{2}$ **17.** $\dfrac{5 \pm \sqrt{17}}{2}$

19. $2 \pm 2i$ **21.** $\dfrac{2 \pm \sqrt{2}}{2}$ **23.** $\dfrac{-3 \pm \sqrt{17}}{4}$ **25.** $\dfrac{3 \pm i\sqrt{7}}{4}$ **27.** $\dfrac{-1 \pm i\sqrt{3}}{2}$ **29.** $-\sqrt{2} \pm 2$ **31.** $2\sqrt{3} \pm i$

33. $i \pm \sqrt{3}$ **35.** $x = \dfrac{-m \pm \sqrt{m^2 - 4n}}{2}$

Even

2. $x^2 + 8x + 16 = (x + 4)^2$ **4.** $x^2 - 10x + 25 = (x - 5)^2$ **6.** $x^2 + 2x + 1 = (x + 1)^2$ **8.** $-4 \pm \sqrt{13}$

10. $5 \pm 2\sqrt{7}$ **12.** $x^2 + x + \frac{1}{4} = (x + \frac{1}{2})^2$ **14.** $m^2 - 7m + \frac{49}{4} = (m - \frac{7}{2})^2$ **16.** $\dfrac{-3 \pm \sqrt{13}}{2}$ **18.** $\dfrac{3 \pm \sqrt{13}}{2}$

20. $1 \pm i\sqrt{2}$ **22.** $\dfrac{3 \pm \sqrt{3}}{2}$ **24.** $\dfrac{-1 \pm \sqrt{13}}{6}$ **26.** $\dfrac{5 \pm i\sqrt{11}}{6}$ **28.** $\dfrac{3 \pm i\sqrt{23}}{4}$ **30.** $\sqrt{5}$ **32.** $-\sqrt{2} \pm i$

34. $-i \pm i\sqrt{3}$ **36.** $x = \dfrac{-b \pm \sqrt{b^2 - 4ac}}{2a}$

Exercise 6-3

Odd

1. $a = 2, b = -5, c = 3$ **3.** $a = 3, b = 1, c = -1$ **5.** $a = 3, b = 0, c = -5$ **7.** $-4 \pm \sqrt{13}$ **9.** $5 \pm 2\sqrt{7}$

11. $\dfrac{-3 \pm \sqrt{13}}{2}$ **13.** $1 \pm i\sqrt{2}$ **15.** $\dfrac{3 \pm \sqrt{3}}{2}$ **17.** $\dfrac{-1 \pm \sqrt{13}}{6}$ **19.** Two real roots   **21.** One real root

23. Two nonreal complex roots   **25.** $5 \pm \sqrt{7}$ **27.** $-1 \pm \sqrt{3}$ **29.** $0, -\frac{3}{2}$ **31.** $2 \pm 3i$ **33.** $5 \pm 2\sqrt{7}$

35. $\dfrac{2 \pm \sqrt{2}}{2}$ **37.** $1 \pm i\sqrt{2}$ **39.** $\frac{2}{5}, 3$ **41.** $t = \sqrt{2d/g}$ **43.** $r = -1 + \sqrt{A/P}$ **45.** $\dfrac{\sqrt{7} \pm i}{2}$

47. $-\sqrt{3}, \dfrac{-\sqrt{3}}{3}$ **49.** $\frac{1}{2}i, -2i$ **51.** $\dfrac{\pm\sqrt{3} - i}{2}$ **53.** $\dfrac{y \pm \sqrt{5y^2}}{2}$ **55.** $y \pm \sqrt{2y^2}$ **57.** $\frac{9}{8}$

59. $\left(\dfrac{-b + \sqrt{b^2 - 4ac}}{2a}\right)\left(\dfrac{-b - \sqrt{b^2 - 4ac}}{2a}\right) = \dfrac{b^2 - (b^2 - 4ac)}{4a^2} = \dfrac{c}{a}$ **61.** $1.35, 0.48$ **63.** $-1.05, 0.63$

65. Has real solutions, since discriminant is positive   **67.** Has no real solutions, since discriminant is negative

---

Even

2. $a = 3, b = -2, c = 1$ **4.** $a = 2, b = 3, c = -1$ **6.** $a = 2, b = -5, c = 0$ **8.** $-2 \pm \sqrt{2}$ **10.** $3 \pm 2\sqrt{3}$

12. $\dfrac{-1 \pm \sqrt{5}}{2}$ **14.** $2 \pm 2i$ **16.** $\dfrac{2 \pm \sqrt{2}}{2}$ **18.** $\dfrac{-3 \pm \sqrt{17}}{4}$ **20.** Two nonreal complex roots

22. One real root   **24.** Two real roots   **26.** $-4 \pm \sqrt{11}$ **28.** $\dfrac{3 \pm \sqrt{13}}{2}$ **30.** $0, 2$ **32.** $-3 \pm 2i$

34. $3 \pm 2\sqrt{3}$ **36.** $\dfrac{3 \pm \sqrt{3}}{2}$ **38.** $2 \pm 2i$ **40.** $-50, 2$ **42.** $a = \sqrt{c^2 - b^2}$ **44.** $I = \dfrac{E \pm \sqrt{E^2 - 4RP}}{2R}$

46. $\dfrac{\sqrt{15}}{3}$ **48.** $\dfrac{-\sqrt{6} \pm \sqrt{2}}{2}$ **50.** $-2i, 3i$ **52.** $\frac{-1}{3}i, 2i$ **54.** $\dfrac{1 \pm \sqrt{1 + 4y - 4y^2}}{2}$ **56.** $\dfrac{2 - 3y \pm \sqrt{5y^2 - 16y}}{2}$

58. $a < \frac{9}{8}$ **60.** $\dfrac{-b + \sqrt{b^2 - 4ac}}{2a} + \dfrac{-b - \sqrt{b^2 - 4ac}}{2a} = \dfrac{-2b}{2a} = -\dfrac{b}{a}$ **62.** $6.25, 0.77$ **64.** $-1.87, 0.45$

66. Has real solutions, since discriminant is positive   **68.** Has no real solutions, since discriminant is negative

Exercise 6-4

Odd

1. $12, 14$ **3.** $0, 2$ **5.** 127 mi   **7.** 5.12 by 3.12 cm   **9.** 1 ft   **11.** 5.66 ft/sec   **13.** 50 mi/hr

15. 1.41 min   **17.** 2 hr; 3 hr   **19.** 2 km/hr   **21.** $\$6$

Even

2. 8, 13   **4.** 3 or $\frac{1}{3}$ **6.** $h = 1$ m, $b = 4$ m [*Note:* $h = -4$ must be discarded.]   **8.** 4.61 by 2.61   **10.** 70 mph
12. 60 mph   **14.** **(A)** $t = 0, 11$; y is zero at the beginning and end of the flight   **(B)** 10.91 sec, 0.09 sec
16. 13.09 hr; 8.09 hr   **18.** 5 km/hr; 12 km/hr   **20.** 20%

Exercise 6-5

Odd

1. 4   **3.** 18   **5.** $\pm 1, \pm 3$ **7.** $\pm 3, \pm i\sqrt{2}$ **9.** $-1, 4$ **11.** 9, 16   **13.** 4   **15.** No solution
17. No solution   **19.** No solution   **21.** 5, 13   **23.** $-1, 2$ **25.** $\pm 1, \pm 2, \pm 2i, \pm i$ **27.** $-8, 125$ **29.** 1, 16
31. $\frac{2}{3}, -\frac{3}{2}$ **33.** $\pm 2, \pm \frac{1}{2}$ **35.** $-2, 3, \frac{1}{2} \pm \dfrac{\sqrt{7}}{2}i$ **37.** $3 \pm 2i, 4, 2$ **39.** 4   **41.** 11   **43.** 1, 5   **45.** 4, 1.4

Even

2. 9   **4.** 8   **6.** $\pm 2, \pm 3$ **8.** $\pm 2, \pm i\sqrt{2}$ **10.** $-9, 1$ **12.** 4, 81   **14.** 6   **16.** 0, 4   **18.** 9   **20.** 1
22. $\frac{9}{4}, 3$ **24.** $-\sqrt[3]{5}, \sqrt[3]{2}$ **26.** $\pm\sqrt{3}$ or $\pm\dfrac{\sqrt{21}}{3}, \pm i$ **28.** $\frac{1}{8}, -8$ **30.** 16, 81   **32.** $-\frac{3}{4}, \frac{1}{5}$ **34.** $\pm 1, \pm 3$
36. $1, -3, -1 \pm i$ **38.** 2, 3, 7, 8   **40.** 2   **42.** 4   **44.** 1, 2   **46.** 1, 4

Exercise 6-6

Odd

1. $(-4, 3)$ **3.** $(-\infty, -4] \cup [3, \infty)$ **5.** $(-4, 3)$

7. $(-\infty, 3) \cup (7, \infty)$ **9.** $\{1, 2, 3, 4, 5\}$ **11.** $\{3\}$ **13.** $\varnothing$ **15.** $[-2, 0]$ **17.** $(-1, \infty)$

19. $(-\infty, 5]$ **21.** $(-\infty, -6] \cup [0, \infty)$ **23.** $(-\infty, -3] \cup [3, \infty)$

25. $(-2, 5]$ **27.** $(-\infty, -2) \cup (5, \infty)$ **29.** $(-\infty, -2) \cup (0, 4]$

31. $(-\infty, 0) \cup (\frac{1}{4}, \infty)$ **33.** All real numbers; graph is whole real line   **35.** No solution

37. $(-\infty, -\sqrt{3}] \cup [\sqrt{3}, \infty)$ **39.** $[2, 3)$ **41.** $(-2, 3)$

43. $(-3, 0) \cup (1, \infty)$ **45.** $(-1, \frac{1}{2}] \cup (2, \infty)$ **47.** $(-3, 1)$

49. $x \geq 2$ or $x \leq 1$ **51.** $2 \leq t \leq 8$ or $[2, 8]$ **53.**

$ax + b = 0$	$x < -b/a$	$x > -b/a$
$ax = -b$	$ax < -b$	$ax > -b$
$x = -b/a$	$ax + b < 0$	$ax + b > 0$
(critical point)		

Even

2. $(-2, 4)$ **4.** $(-\infty, -2) \cup (4, \infty)$ **6.** $(-5, 2)$

8. $(-\infty, -5) \cup (-2, \infty)$ **10.** $\{3, 4, 5, 6, 7\}$ **12.** $\{3, 4\}$ **14.** $\varnothing$ **16.** $(-3, 8]$

18. $[-3, -1)$ **20.** $[0, 5)$ **22.** $[0, 8]$ **24.** $(-\infty, -2) \cup (2, \infty)$

26. $(-2, 3)$ **28.** $(-\infty, -2] \cup (3, \infty)$ **30.** $[-5, 0] \cup (3, \infty)$

32. $(0, \frac{5}{3})$ ⟶ x   **34.** All real numbers; graph is whole real line   **36.** No solution

38. $(-\sqrt{2}, \sqrt{2})$ ⟶ x   **40.** $(-\infty, -3) \cup [3, \infty)$ ⟶ x   **42.** $(-1, 2) \cup [5, \infty)$ ⟶ x

44. $(-\infty, -2) \cup (0, 3)$ ⟶ x   **46.** $(-2, 2) \cup (4, \infty)$ ⟶ x

48. $(-\infty, -3) \cup (-2, 0) \cup (1, \infty)$ ⟶ x   **50.** $x < -5$ or $x \geq 3$ **52.** $0 \leq t \leq 10$

54. $ax + b = 0$ $x < -b/a$ $x > -b/a$
      $ax = -b$ $ax > -b$ $ax < -b$
      $x = -b/a$ $ax + b > 0$ $ax + b < 0$
          (critical point)

Diagnostic (Review) Exercise 6-7

1. 0, 3 *(6-1)*   **2.** ± 5 *(6-1)*   **3.** 2, 3 *(6-1)*   **4.** $-3, 5$ *(6-1)*   **5.** $\pm\sqrt{7}$ *(6-1)*

6. $a = 3, b = 4, c = -2$ *(6-3)*   **7.** $x = \dfrac{-b \pm \sqrt{b^2 - 4ac}}{2a}$ *(6-3)*   **8.** $\dfrac{-3 \pm \sqrt{5}}{2}$ *(6-3)*   **9.** 3, 9 *(6-4)*

10. -19, two nonreal complex roots *(6-3)*   **11.** $\{1, 2, 3, 4, 5, 7, 9\}$ *(6-7)*   **12.** $\{1, 3, 5\}$ *(6-5)*   **13.** $(-1, 5)$ *(6-7)*
14. $[0, 3]$ *(6-7)*   **15.** $(-5, 4)$ ⟶ x *(6-6)*   **16.** $(-\infty, -5] \cup [4, \infty)$ ⟶ x *(6-6)*   **17.** 0, 2 *(6-1)*

18. $\pm 2\sqrt{3}$ *(6-1)*   **19.** $\pm 3i$ *(6-1)*   **20.** $-2, 6$ *(6-1)*   **21.** $\dfrac{-1}{3}, 3$ *(6-1)*   **22.** $\frac{1}{2}, -3$ *(6-1)*

23. $\dfrac{1 \pm \sqrt{7}}{3}$ *(6-3)*   **24.** $\dfrac{1 \pm \sqrt{7}}{2}$ *(6-3)*   **25.** $-4, 5$ *(6-1)*   **26.** $\frac{3}{4}, \frac{5}{2}$ *(6-1)*   **27.** $\dfrac{-1 \pm \sqrt{5}}{2}$ *(6-3)*

28. $1 \pm i\sqrt{2}$ *(6-3)*   **29.** 2, 3 *(6-5)*   **30.** 9, 25 *(6-5)*   **31.** $\pm 2, \pm 3i$ *(6-5)*   **32.** $64, \dfrac{-27}{8}$ *(6-5)*

33. $(-\infty, -3] \cup [7, \infty)$ ⟶ x *(6-6)*   **34.** $(-\infty, 0) \cup (\frac{1}{2}, \infty)$ ⟶ x *(6-6)*

35. No solution *(6-6)*   **36.** All real numbers; graph is the real line *(6-6)*   **37.** 6×5 in. *(6-4)*   **38.** $3 \pm 2\sqrt{3}$ *(6-2)*

39. $\dfrac{3 \pm i\sqrt{6}}{2}$ or $\dfrac{3}{2} \pm \dfrac{\sqrt{6}}{2}i$ *(6-5)*   **40.** $\dfrac{-3 \pm \sqrt{57}}{6}$ *(6-3)*   **41.** $\pm 1, \pm 2, \pm 2i, \pm i$ *(6-5)*   **42.** $3, \frac{9}{4}$ *(6-5)*

43. $(-\infty, 1] \cup (3, 4)$ ⟶ x *(6-6)*   **44.** 9 cm and 12 cm *(6-4)*   **45. (A)** 2,000 and 8,000   **(B)** 5,000 *(6-4)*

Practice Test Chapter 6

1. 0, 5 *(6-1)*   **2.** $\pm 3i$ *(6-1)*   **3.** $\dfrac{-1 \pm \sqrt{5}}{3}$ *(6-1)*   **4.** 2, 3 *(6-1)*   **5.** $\dfrac{-5 \pm \sqrt{13}}{2}$ *(6-3)*   **6.** 3, 7 *(6-5)*

7. $\pm i\sqrt{3}, \pm\sqrt{2}$ *(6-5)*   **8.** 1, 16 *(6-5)*   **9.** 10 *(6-5)*   **10.** $5 \leq x \leq 6$ or $[5, 6]$ ⟶ x *(6-6)*

11. $x < -5$ or $-3 < x < -1$ or $(-\infty, -5) \cup (-3, -1)$ ⟶ x *(6-6)*   **12.** 4 sec *(6-4)*

13. $k > 3$ *(6-3)*   **14.** $\dfrac{-Q \pm \sqrt{Q^2 - 4PR}}{2P}$ *(6-5)*

Chapter 7   Exercise 7-1

Odd

1. $A(-10, 10)$, $B(10, -10)$, $C(16, 14)$, $D(16, 0)$, $E(-14, -16)$, $F(0, 4)$, $G(6, -16)$, $H(-14, 0)$

3.
5.

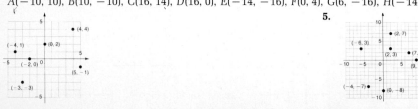

7. $A(-3\frac{1}{2}, 2)$, $B(-2, -4\frac{1}{2})$, $C(0, -2\frac{1}{2})$, $D(2\frac{3}{4}, 0)$, $E(3\frac{3}{4}, 2\frac{3}{4})$, $F(4\frac{1}{4}, -3\frac{1}{2})$

9.

11.

13.

15.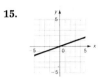

17.

19.

21.

23.

25.

27.

29.

31.

33.

35.

37.

39.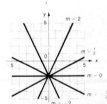

41.

43.

45. $y = 2x - 4$

47. $x + 2y = -6$

49.

51.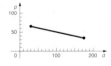

53.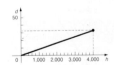

55.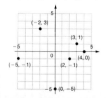

57.

59.

61.

Even

2. $A(5, 5)$, $B(8, 2)$, $C(-5, 5)$, $D(-3, 8)$, $E(-5, -6)$, $F(-7, -7)$, $G(5, -5)$, $H(2, -2)$, $I(7, 0)$, $J(-2, 0)$, $K(0, -9)$, $L(0, 4)$

4.

6.

8. $A(2\frac{1}{2}, 1)$, $B(-2\frac{1}{2}, 3\frac{1}{2})$, $C(-2, -4\frac{1}{2})$, $D(3\frac{1}{4}, -3)$, $E(1\frac{1}{4}, 2\frac{1}{4})$, $F(-3\frac{1}{4}, 0)$, $G(1\frac{1}{2}, -4\frac{1}{2})$

10.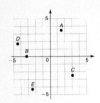

12.

14.

16.

18.

20.

22.

24.

26.

28.

30.

32.

34.

36.

38.

40.

42.

44.

46. $y = 2x + 3$

48. $2x - y = 8$

50.

52.

54.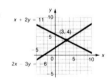

56.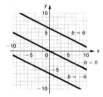

58.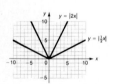

60.

Exercise 7-2

Odd

1. Slope: 2
y intercept: -3

3. Slope: -1
y intercept: 2

5. $y = 5x - 2$

7. $y = -2x + 4$

9. $y - 4 = 2(x - 5)$

11. $y - 1 = -2(x - 2)$

13. 2

15. $\frac{1}{2}$

17. $y - 6 = 2(x - 5)$ or $y - 2 = 2(x - 3)$

19. $y - 5 = \frac{1}{2}(x - 10)$ or $y - 1 = \frac{1}{2}(x - 2)$ **21.** Slope: $-\frac{1}{3}$
y intercept: 2

23. Slope: $-\frac{1}{2}$
y intercept: 2

25. Slope: $-\frac{2}{3}$
y intercept: 2

27. $y = -\dfrac{x}{2} - 2$ **29.** $y = \frac{2}{3}x + \frac{3}{2}$ **31.** $y - 2 = -2(x + 3)$, $y = -2x - 4$ **33.** $y - 3 = \frac{1}{2}(x + 4)$, $y = \dfrac{x}{2} + 5$

35. $\frac{1}{3}$ **37.** $-\frac{1}{4}$ **39.** $y - 4 = \frac{1}{3}(x + 6)$ or $y - 7 = \frac{1}{3}(x - 3)$, $y = \dfrac{x}{3} + 6$

41. $y = -\frac{1}{4}(x + 4)$ or $y + 2 = -\frac{1}{4}(x - 4)$, $y = -\dfrac{x}{4} - 1$ **43.** $x = -3$, $y = 5$ **45.** $x = -1$, $y = 22$

47. **(A)** $y = \frac{3}{5}x + \frac{17}{5}$ **(B)** $y = -\frac{5}{3}x + \frac{17}{3}$ **49.** **(A)** $y = -3x + 2$ **(B)** $y = \frac{1}{3}x + 2$

51. **(A)** $y = -x + 2$ **(B)** $y = x$ **53.** **(A)** $y = 5$ **(B)** $x = -2$

55. **(A)** $R = \frac{3}{2}C + 3$ **(B)**   **(C)** \$158   **57.** $F = \frac{9}{5}C + 32$

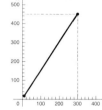

59. **(A)** $V = -1{,}800t + 20{,}000$ **(B)** \$12,800, \$5,600 **(C)** $-1{,}800$ **(D)**

Even

2. Slope: 1
y intercept: 1

4. Slope: -2
y intercept: 1

6. $y = 3x - 5$ **8.** $y = -x + 2$ **10.** $y - 5 = 3(x - 2)$ **12.** $y - 3 = -3(x - 1)$ **14.** 1   **16.** $\frac{1}{3}$

18. $y - 4 = x - 2$ or $y - 3 = x - 1$ **20.** $y - 5 = \frac{1}{3}(x - 7)$ or $y - 3 = \frac{1}{3}(x - 1)$

22. Slope: $-\frac{1}{4}$
y intercept: -1

24. Slope: $\frac{1}{3}$
y intercept: 2

26. Slope: $-\frac{3}{4}$
y intercept: 3

28. $y = -\dfrac{x}{3} - 5$ **30.** $y = -\frac{3}{2}x + \frac{5}{2}$ **32.** $y + 1 = -3(x - 4)$, or $y = -3x + 11$ **34.** $y + 5 = \frac{2}{3}(x + 6)$, or $y = \frac{2}{3}x - 1$

36. Undefined   **38.** $-\frac{1}{3}$ **40.** $y + 4 = -\frac{1}{5}(x - 5)$, or $y + 2 = -\frac{1}{5}(x + 5)$; $y = -\dfrac{x}{5} - 3$

42. $y + 4 = -\frac{2}{3}(x - 3)$, or $y = -\frac{2}{3}(x + 3)$; $y = -\frac{2}{3}x - 2$ **44.** $x = 6$, $y = -2$ **46.** $x = 5$, $y = 0$

48. **(A)** $y = -\frac{2}{3}x - \frac{8}{3}$ **(B)** $y = \frac{3}{2}x - \frac{1}{2}$ **50.** **(A)** $y = 5x + 10$ **(B)** $y = -\frac{1}{5}x - \frac{2}{5}$

52. **(A)** $y = x + 1$ **(B)** $y = -x + 5$ **54.** **(A)** $x = 3$ **(B)** $y = -2$ **56.** **(A)** $C = \dfrac{x}{2} + 200$ **(B)**

58. **(A)** $s = \dfrac{1}{10}w$ **60.** $0.2x + 0.1y = 20$, or $2x + y = 200$

     (B) 1.5 in., 3 in.

     (C) Slope = $\frac{1}{10}$

     (D)

Exercise 7-3

Odd

1.

3.

5.

7.

9.

11.

13.

15.

17.

19.

21.

23.

25.

27. $6x + 8y \le 120$
       $x + 3y \le 30$
           $x \ge 0$
           $y \ge 0$

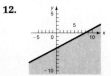

Even

2.

4.

6.

8.

10.

12.

14.

16.

18.

20.

22.

24.

26.

Exercise 7-4

Odd

1. $\sqrt{5}$ **3.** $\sqrt{89}$ **5.** $\sqrt{18}$ or $3\sqrt{2}$ **7.**   **9.**   **11.** $x^2 + y^2 = 49$

13. $x^2 + y^2 = 5$ **15.** Yes, since two sides have length $\sqrt{17}$.   **17.**   **19.**

21.   **23.** $(x - 3)^2 + (y - 5)^2 = 49$ **25.** $(x + 3)^2 + (y - 3)^2 = 64$ **27.** $(x + 4)^2 + (y + 1)^2 = 3$

29. $(x - 2)^2 + (y - 3)^2 = 9$ **31.** $(x - 3)^2 + (y + 3)^2 = 16$ **33.** $(x + 3)^2 + (y + 2)^2 = 9$ **35.** $x = -3, 7$

37. $x^2 + y^2 = 25$ **39.** $x^2 + y^2 = 50^2$

Even

2. $\sqrt{10}$ **4.** $\sqrt{41}$ **6.** $\sqrt{52}$ or $2\sqrt{13}$ **8.**   **10.**   **12.** $x^2 + y^2 = 64$

14. $x^2 + y^2 = 10$ **16.** No, the length of the third side is $\sqrt{18}$, while the other two sides have length $\sqrt{17}$.

18.   **20.**   **22.**

24. $(x - 4)^2 + (y - 1)^2 = 4$ **26.** $(x - 5)^2 + (y + 2)^2 = 36$ **28.** $(x + 7)^2 + (y + 5)^2 = 14$

30. $(x - 3)^2 + (y - 2)^2 = 9$ **32.** $(x + 3)^2 + (y - 2)^2 = 16$ **34.** $(x + 2)^2 + (y + 2)^2 = 16$

36. $x - y = 3$ **38.** $(x - 4)^2 + (y + 3)^2 = 25$ **40.** 50 cm

Exercise 7-5

Odd

1.   **3.**   **5.** 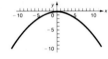   **7.**

9.   **11.**   **13.** 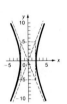   **15.**

17. $y^2 = 2x$ **19.** $\dfrac{x^2}{9} + \dfrac{y^2}{4} = 1$ **21.** $\dfrac{x^2}{9} - \dfrac{y^2}{4} = 1$ **23.** $(1, 0),\ x = -1$

25. $\left(0, -\sqrt{21}\right),\ \left(0, \sqrt{21}\right)$ **27.** $(5, 0),\ (-5, 0)$ **29.** $(x - 2)^2 = -4(y - 3)$ or $x^2 - 4x + 4y - 8 = 0$ **31.** $x^2 = -100y$

Even

2.   **4.**   **6.**   **8.**

10.   **12.**   **14.**   **16.**

18. $x^2 = -3y$

20. $\dfrac{x^2}{25} + \dfrac{y^2}{4} = 1$

22. $\dfrac{y^2}{4} - \dfrac{x^2}{25} = 1$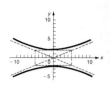

24. $(-3, 0)$, $x = 3$ **26.** $\left(-\sqrt{5}, 0\right)$, $\left(\sqrt{5}, 0\right)$ **28.** $\left(0, -\sqrt{29}\right)$, $\left(0, \sqrt{29}\right)$

30. $(y - 4)^2 = 8(x - 4)$ or $y^2 - 8x - 8y + 48 = 0$ **32.** $x^2 = 20y$

Diagnostic (Review) Exercise 7-6

1. *(7-1)*

2. *(7-1)*

3. *(7-3)*

4. *(7-3)*

5. *(7-4)*

6. Slope $= -2$, y intercept $= -3$ *(7-2)*   **7.** $2x + y = 8$ *(7-2)*   **8.** 2 *(7-2)*   **9.** $y = 2x + 1$ *(7-2)*

10. $\sqrt{20}$ or $2\sqrt{5}$ *(7-4)*   **11.** $x^2 + y^2 = 25$ *(7-4)*

12. *(7-1)*

13. *(7-1)*

14. *(7-1)*

15. *(7-3)*

16. *(7-3)*

17. *(7-3)*

18. *(7-3)*

19. *(7-3)*

20. *(7-4)*

21. *(7-5)*

22. *(7-5)*

23. *(7-5)*

24. Slope $= -\frac{1}{2}$, y intercept $= -3$ *(7-2)*   **25.** $y = -\frac{1}{3}x + 1$ *(7-2)*   **26.** $2x + 3y = 0$ *(7-2)*   **27.** $y = 2x - 10$ *(7-2)*

28. Vertical: $x = 5$; horizontal: $y = -2$ *(7-2)*   **29.** $(x + 3)^2 + (y - 4)^2 = 49$ *(7-4)*   **30.** $x^2 + y^2 = 169$ *(7-4)*

31. $y = \frac{3}{2}x + 11$ *(7-2)*   **32.** $(x + 3)^2 + (y - 4)^2 = 25$; radius $= 5$, center $= (-3, 4)$ *(7-4)*   **33.** *(7-3)*

Practice Test Chapter 7

1. *(7-1)*

2. *(7-3)*

3. *(7-4)*

4. *(7-5)*

5. *(7-5)*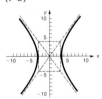

6. *(7-1)*

7. $y = -3x + 5$ *(7-1)*   **8.** $y = \frac{1}{2}x - 2$ *(7-1)*   **9.** $y = -2$ *(7-1)*   **10.** $y = -\frac{2}{3}x + 2$ *(7-1)*

11. $(x - 4)^2 + (y + 3)^2 = 4^2$, center $(4, -3)$, radius 4 *(7-5)*   **12.** $x^2 + y^2 = 68$ *(7-5)*

13. *(7-5)*

14. *(7-3)*

Chapter 8   Exercise 8-1

Odd

1. Function   **3.** Not a function   **5.** Function   **7.** Function   **9.** Not a function   **11.** Function

13. Function   **15.** Function   **17.** Not a function   **19.** Not a function   **21.** Function   **23.** Function

25. Domain = $\{1, 2, 3\}$
Range = $\{1, 2, 3\}$
Not a function

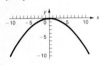

27. Domain = $\{-1, 0, 1, 2, 3, 4\}$
Range = $\{-2, -1, 0, 1, 2\}$
A function

29. Domain = $\{0, 1, 2, 3, 4\}$
Range = $\{-2, 0, 2, 4, 6\}$
A function

31. Domain = $\{0, 1, 4\}$
Range = $\{-2, -1, 0, 1, 2\}$
Not a function

33. Domain = $\{-2, 0, 2\}$
Range = $\{-2, 0, 2\}$
Not a function

35. Domain = $\{-2, 0, 2\}$
Range = $\{0, 2\}$
A function

37. R   **39.** R   **41.** $x \geq -3$ or $[-3, \infty)$ **43.** $-8 \leq x \leq 8$ or $[-8, 8]$

45. All real numbers except $x = -3$ and $x = \frac{1}{2}$ **47.** $x \leq 4$ or $(-\infty, 4]$ **49.** $x < -3$ or $x \geq 1$ or $(-\infty, -3) \cup [1, \infty)$

51. Domain = $\{-4, -2, 0, 2, 4\}$
Range = $\{-2, -1, 0, 1, 2\}$
A function

53. Domain = $\{0, 1, 2, 3\}$
Range = $\{0, 1, 2, 3\}$
Not a function

55. (A)

PARTIAL TABLE	
t	d
0	0
1	144
2	256
3	336
4	384
5	400
6	384
7	336
8	256
9	144
10	0

(B)

DOMAIN	RANGE
Set of all real t	Set of all real d
$0 \le t \le 10$	$0 \le d \le 400$

(C) The relation is a function.

Even

2. Function   **4.** Not a function   **6.** Function   **8.** Function   **10.** Not a function   **12.** Not a function
14. Function   **16.** Function   **18.** Not a function   **20.** Not a function   **22.** Function   **24.** Function
26. Domain = $\{2, 4\}$
Range = $\{-2, 0, 2, 4\}$
Not a function

28. Domain = $\{-2, 0, 2\}$
Range = $\{0, 2\}$
A function

30. Domain = $\{0, 1, 2, 3, 4\}$
Range = $\{-4, -3\frac{1}{2}, -3, -2\frac{1}{2}, -2\}$
A function

32. Domain = $\{-2, 0, 2\}$
Range = $\{0, 4\}$
A function

34. Domain = $\{-3, 0, 3\}$
Range = $\{-3, 0, 3\}$
Not a function

36. Domain = $\{0, 1, 4\}$
Range = $\{-4, -1, 0, 1, 4\}$
Not a function

38. R **40.** R **42.** R **44.** $-1 \le x \le 1$ or $[-1, 1]$ **46.** All real numbers, except $x = -4, 3$
48. $x \ge 5$ or $[5, \infty)$ **50.** $x \le -5$ or $x \ge 2$, or $(-\infty, -5] \cup [2, \infty)$
52. Domain = $\{-3, -1, 0, 2\}$
Range = $\{0, 2, 3, 5\}$
A function

54. Domain = $\{-2, -1, 1, 2\}$
Range = $\{0, 1\}$
Not a function

56. (A)

t	s
0	0
1	16
2	64
3	144
4	256
5	400

(B) Both are the set of all nonnegative real numbers.
(C) The relation is a function.

Exercise 8-2

Odd

1. 4   **3.** -8 **5.** -2 **7.** -2 **9.** -12 **11.** -6 **13.** -27 **15.** 2   **17.** 6   **19.** 25
21. 22   **23.** -91 **25.** $10 - 2u$ **27.** $3a - 9a^2$ **29.** $2 - 2h$ **31.** -2 **33.** 10   **35.** 48   **37.** 0
39. -7 **41.** $\frac{1}{5}$, $-\frac{3}{5}$, not defined   **43.** 10   **45.** **(A)** Yes **(B)** Yes **(C)** No   **47.** $C(x) = 5x$
49. $C(F) = \frac{5}{9}(F - 32)$ **51.** **(A)** 30 mi, 300 mi **(B)** 30

Even

2. 1   **4.** -5 **6.** 10   **8.** 0   **10.** -20 **12.** -2 **14.** 3   **16.** -12 **18.** 0   **20.** 6
22. -36 **24.** $-\frac{3}{4}$ **26.** $10v + 3$ **28.** $12c^2$ **30.** $3(2 + h)^2$ **32.** $12 + 3h$ **34.** $-3 - h$ **36.** -6
38. 6   **40.** -2 **42.** 3, 0, not defined   **44.** -2 **46.** **(A)** No **(B)** No **(C)** Yes

48. $C(x) = 800 + 60x$ **50.** $P(d) = 15\left(1 + \dfrac{d}{33}\right)$

52. **(A)** 0, 16, 64, 144 **(B)** $64 + 16h$; tends to 64 as h tends to 0. The average speed from $t = 2$ to $t = 2 + h$ tends to 64 as h tends to 0.

Exercise 8-3

Odd

1. Slope: 2
y intercept: -4

3. Slope: -2
y intercept: 4

5. Slope: $-\frac{2}{3}$
y intercept: 4

7. Vertex: $(-4, 0)$
Min: $f(-4) = 0$
Axis: $x = -4$

9. Vertex: $(1, 3)$
Min: $f(1) = 3$
Axis: $u = 1$

11. Vertex: $(2, 6)$
Max: $h(2) = 6$
Axis: $x = 2$

13. Vertex: $(3, 9)$
Max: $f(3) = 9$
Axis: $x = 3$

15. Vertex: $(0, -4)$
Min: $F(0) = -4$
Axis: $s = 0$

17. Vertex: $(0, 4)$
Max: $F(0) = 4$
Axis: $x = 0$

19. Vertex: $(3.5, -2.25)$
Min: $f(3.5) = -2.25$
Axis: $x = 3.5$

21. Vertex: $(1.5, 6.25)$
Max: $g(1.5) = 6.25$
Axis: $t = 1.5$

23. Vertex: $(-2, -2)$
Min: $f(x) = f(-2) = -2$
Axis: $x = -2$

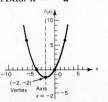

25. Vertex: $(-2, 6)$
Max: $f(x) = f(-2) = 6$
Axis: $x = -2$

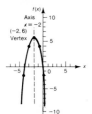

27.

29.

31.

33.

35.

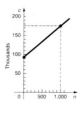

37. **(A)** $A(x) = x(50 - x)$
 (B) Domain: $0 \le x \le 50$
 (C)

 (D) $A(25) = 625 \text{ m}^2$ is maximum for a 25 by 25 m rectangle (a square)

39. **(A)** $V(x) = (12 - 2x)(8 - 2x)x$
$= 4x^3 - 40x^2 + 96x$
 (B) Domain: $0 < x < 4$. [*Note:* At $x = 0$ and $x = 4$, we have zero volume.]
 (C)

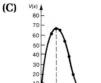

 (D) Max: $V(x) \approx V(1.5) \approx 67.5 \text{ in.}^3$
A 1.5-in. square should be cut from each corner. Dimensions of box are $5 \times 9 \times 1.5$ inches.

Even

2. Slope: $\frac{1}{2}$
y intercept: 0

4. Slope: $-\frac{1}{2}$
y intercept: 3

6. Slope: 0
y intercept: 3

8. Vertex: $(1, -4)$
Min: $h(1) = -4$
Axis: $x = 1$

10. Vertex: $(5, 0)$
Min: $f(5) = 0$
Axis: $x = 5$

12. Vertex: $(-3, 5)$
Max: $g(-3) = 5$
Axis: $x = -3$

14. Vertex: (4, 32)
Max: G(4) = 32
Axis: x = 4

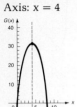

16. Vertex: (0, 4)
Min: g(0) = 4
Axis: t = 0

18. Vertex: (0, 9)
Max: G(0) = 9
Axis: x = 0

20. Vertex: (2.5, −4.25)
Min: g(2.5) = −4.25
Axis: t = 2.5

22. Vertex: (−2.5, 8.25)
Max: h(−2.5) = 8.25
Axis: x = −2.5

24. Vertex: (3, −4)
Min: f(x) = f(3) = −4
Axis: x = 3

26. Vertex: (4, 4)
Max: f(x) = f(4) = 4
Axis: x = 4

28.

30.

32.

34.

36.

38. **(A)** $A(x) = x(100 - 2x)$
(B) Domain: $0 \le x \le 50$
(C)

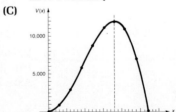

(D) $A(25) = 1{,}250$ m²; maximum for a 25 m by 50 m rectangle

40. **(A)** $V(x) = x^2(108 - 4x)$
(B) Domain: $0 \le x \le 27$ [*Note*: At $x = 0$ and $x = 27$, we have 0 volume.]
(C)

(D) Max: $V(x) \approx V(18) \approx 11{,}664$ in.³. Dimensions of box are 18 × 18 × 36 inches.

Exercise 8-4

Odd

1. $R^{-1} = \{(1, -2), (3, 0), (2, 2)\}$ **3.** $G^{-1} = \{(4, -2), (1, -1), (0, 0), (1, 1), (4, 2)\}$ **5.**

7.   **9.** Both are functions.   **11.** G is a function; G^{-1} is not.

13. f^{-1}: $x = 3y - 2$ or $y = (x + 2)/3$ **15.** F^{-1}: $x = y/3 - 2$ or $y = 3(x + 2)$ **17.** h^{-1}: $x = y^2/2$ or $y = \pm\sqrt{2x}$

19. Both are functions; both are one-to-one.   **21.** h is a function; h^{-1} is not a function; neither is one-to-one.

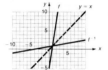

23. **(A)** $f^{-1}(x) = \dfrac{x + 2}{3}$ **(B)** $\frac{4}{3}$ **(C)** 3 **25.** **(A)** $F^{-1}(x) = 3(x + 2)$ **(B)** 3 **(C)** 4

27. **(A)** $f^{-1}(x) = 3(x - 2)$ **(B)** a **29.** x **31.** **(A)**   **(B)** Both G and G^{-1} are functions.

Even

2. $F^{-1} = \{(-1, -3), (1, 0), (2, 3)\}$ **4.** $H^{-1} = \{(0, -5), (1, -2), (0, 0), (1, 2), (0, 5)\}$

6.   **8.**

10. Both are functions.   **12.** H is a function; H^{-1} is not.   **14.** g^{-1}: $x = 2y + 3$ or $y = \dfrac{x - 3}{2}$

16. G^{-1}: $x = \dfrac{y}{2} + 5$ or $y = 2(x - 5)$ **18.** H^{-1}: $x = |2y|$

20. Both are functions; both are one-to-one.   **22.** H is a function; H^{-1} is not a function; neither is one-to-one.

24. **(A)** $g^{-1}(x) = \dfrac{x-3}{2}$ **(B)** 1 **(C)** 4 **26.** **(A)** $G^{-1}(x) = 2(x-5)$ **(B)** 6 **(C)** -4

28. **(A)** $g^{-1}(x) = \dfrac{x-2}{4}$ **(B)** a **30.** x **32.** **(A)** $H^{-1}(x) = x^2,\ x \geq 0$ **(B)** **(C)** 9; x

Exercise 8-5

Odd

1. $F = kv^2$ **3.** $f = k\sqrt{T}$ **5.** $y = k/\sqrt{x}$ **7.** $t = k/T$ **9.** $R = kSTV$ **11.** $V = khr^2$ **13.** 4 **15.** $9\sqrt{3}$
17. $U = k(ab/c^3)$ **19.** $L = k(wh^2/l)$ **21.** -12 **23.** 83 lb **25.** 20 amp
27. The new horsepower must be eight times the old. **29.** No effect **31.** $t^2 = kd^3$ **33.** 1.47 hr (approx.)
35. 20 days **37.** Quadrupled **39.** 540 lb **41.** **(A)** $\Delta S = kS$ **(B)** 10 oz **(C)** 8 candlepower
43. 32 times/sec **45.** $N = k(F/d)$ **47.** 1.2 mi/sec **49.** 20 days **51.** The volume is increased by a factor of 8.

Even

2. $u = kv$ **4.** $P = kv^6$ **6.** $I = \dfrac{k}{t}$ **8.** $P = \dfrac{k}{n}$ **10.** $g = kxy^2$ **12.** $Q = ktRI^2$ **14.** 125 **16.** $\frac{1}{2}$

18. $w = k\dfrac{x^2}{\sqrt{y}}$ **20.** $n = k\dfrac{P_1 P_2}{d}$ **22.** 3 **24.** 60 mph or more **26.** 90 **28.** $\frac{1}{4}$ of the original

30. Under the new conditions, it would have 8 times the destructive force of the old.

32. F is cut in half. **34.** N is quadrupled. **36.** 45 mph **38.** $t = k\dfrac{wd}{P}$ **40.** Quadrupled **42.** 140

44. $f = k\dfrac{\sqrt{T}}{ld}$ **46.** The f-stop number should be cut in half. **48.** $v = k\sqrt{\dfrac{T}{w}}$ **50.** \$9

52. The area is reduced by a factor of $\frac{1}{4}$.

Diagnostic (Review) Exercise 8-6

1. Not a function *(8-1)* **2.** A function *(8-1)* **3.** A function *(8-1)* **4.** A function *(8-1)*
5. Not a function *(8-1)* **6.** A function *(8-1)* **7.** A function *(8-1)* **8.** Not a function *(8-1)*
9. Not a function *(8-1)* **10.** Not a function *(8-1)* **11.** A function *(8-1)* **12.** A function *(8-1)*
13. Domain $= \{0, 1\}$, Range $= \{-2, 2, 3\}$ *(8-1)* **14.** Domain $= \{-1, 1, 2\}$, Range $= \{2, 3, 4\}$ *(8-1)*
15. Domain $= \{-2, 0, 2\}$, Range $= \{3\}$ *(8-1)* **16.** **(A)** 0 **(B)** 6 **(C)** 9 **(D)** $6 - m$ *(8-2)*
17. **(A)** -6 **(B)** 0 **(C)** -3 **(D)** $c - 2c^2$ *(8-2)*
18. Slope $= 2$ **19.** *(8-3)* **20.** M is not a function;
y intercept: -4 *(8-3)* M^{-1} is a function *(8-4)*

  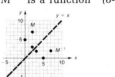

21. Domain $= \{3, 5, 7\}$, Range $= \{0, 2\}$ *(8-4)* **22.** $m = kn^2$ *(8-5)* **23.** $P = \dfrac{k}{Q^3}$ *(8-5)* **24.** $A = kab$ *(8-5)*

25. $y = k\dfrac{x^3}{\sqrt{z}}$ *(8-5)*

26. No. A function is a relation with the added restriction that each domain element corresponds to exactly one range element. *(8-1)*
27. **(A)** 7 **(B)** 1 **(C)** $x^2 + x - 7$ **(D)** 3 *(8-2)* **28.** **(A)** $3 + 2h$ **(B)** 2 *(8-2)*

29. Slope: $-\frac{3}{2}$
    y intercept: 6   *(8-3)*

30. Vertex: (2, 1)
    Min: $f(2) = 1$
    Axis: $x = 2$ *(8-3)*

31. $P(x) = [(x - 2)x - 5]x + 6$ *(8-3)*

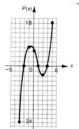

32. Problem 3   *(8-4)*   **33.** Problem 6   *(8-4)*   **34.** None   *(8-4)*   **35.** Problem 11   *(8-4)*   **36.** Problem 6   *(8-4)*

37. Problem 3   *(8-4)*   **38.** $(-\infty, -2] \cup (5, \infty)$ *(8-1)*   **39.** **(A)** $M^{-1}(x) = 2x - 3$ **(B)** Yes   **(C)** 1   **(D)** 3   *(8-4)*

40. **(A)** $y = k\dfrac{x}{z}$ **(B)** $y = \frac{4}{3}$ *(8-5)*   **41.** **(A)** $-3 - 4h - h^2$ **(B)** $-4 - h$ *(8-2)*

42. Vertex: (3, 144)
    Max: $g(3) = 144$
    Axis: $t = 3$ *(8-3)*

43. **(A)** $f^{-1}(x) = \sqrt{x}, x \geq 0$ *(8-4)*
    (B)

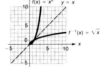

(C) $f^{-1}(9) = 3, f^{-1}[f(x)] = x, x \geq 0$

44. **(A)**

(B) Both are functions.   *(8-4)*

45. *(8-3)*

46. **(A)**

(B) $p = \$100$, maximum $R = f(100) = \$300,000$ *(8-3)*

47. $t = k\dfrac{wd}{p}, t = 24$ sec   *(8-5)*   **48.** The total force is doubled.   *(8-5)*

Practice Test Chapter 8

1. (A), (C)   *(8-1)*   **2.** $\{-1, 1, 5\}$ *(8-1)*   **3.** -14 *(8-2)*   **4.** 2   *(8-2)*   **5.** $4 - m^2$ *(8-2)*

6. $(-\infty, -2) \cup (1, \infty)$ *(8-2)*   **7.** $f^{-1}(x) = 4x + 2, f^{-1}(1) = 6$ *(8-4)*   **8.** $m = \dfrac{kn^{2/3}}{p^4}$ *(8-5)*   **9.** 54   *(8-5)*

10. $-\frac{1}{2}, 4$ *(8-3)*   **11.** Minimum is -1.   *(8-3)*   **12.** f is a function, f^{-1} is not.   *(8-4)*

13. *(8-3)*

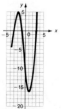

14. $(-\infty, -1) \cup [1, \infty)$ *(8-2)*

Chapter 9   Exercise 9-1

Odd

1. $y = 3^x$

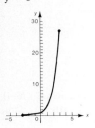

3. $y = 3^{-x}$

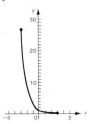

5. $y = 4 \cdot 3^x$

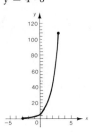

7. $y = 2^{x+3}$

9. $y = 7 \cdot 2^{-2x}$

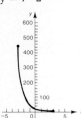

11. $y = e^x$

13. $y = 10e^{-0.12x}$

15. $y = 10e^{-x^2}$

17. $y = y_0 2^x$

19.

21. $f(n) = 2^n$

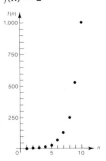

23. $P = 14.7e^{-0.21x}$

25. $A = 100(\frac{1}{2})^{t/28}$

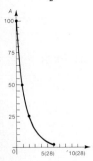

27. $N(i) = 100e^{-0.11(i-1)}$

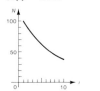

Even

2. $y = 2^x$

4. $y = 2^{-x}$

6. $y = 5 \cdot 2^x$

8. $y = 3^{x+1}$

10. $y = 11 \cdot 2^{-2x}$

12. $y = e^{-x}$
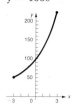

14. $y = 100e^{0.25x}$

16. $y = e^{-x^2}$
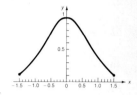

18. $y = y_0 e^{-0.22x}$
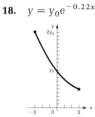

20.

22. $A = 10(1.1)^t$

24. $N = N_0 2^n$

26. $A = 100(\frac{1}{2})^{t/4}$

Exercise 9-2

Odd

1. $9 = 3^2$ **3.** $81 = 3^4$ **5.** $1{,}000 = 10^3$ **7.** $1 = e^0$ **9.** $\log_8 64 = 2$ **11.** $\log_{10} 10{,}000 = 4$
13. $\log_v u = x$ **15.** $\log_{27} 9 = \frac{2}{3}$ **17.** 5 **19.** -4 **21.** 2 **23.** 3 **25.** $x = 4$ **27.** $y = 2$
29. $b = 4$ **31.** $0.001 = 10^{-3}$ **33.** $3 = 81^{1/4}$ **35.** $16 = (\frac{1}{2})^{-4}$ **37.** $N = a^e$ **39.** $\log_{10} 0.01 = -2$
41. $\log_e 1 = 0$ **43.** $\log_2(\frac{1}{8}) = -3$ **45.** $\log_{81}(\frac{1}{3}) = -\frac{1}{4}$ **47.** $\log_{49} 7 = \frac{1}{2}$ **49.** u **51.** $\frac{1}{2}$ **53.** $\frac{3}{2}$ **55.** 0
57. $\frac{3}{2}$ **59.** $x = 2$ **61.** $y = -2$ **63.** $b = 100$ **65.** Any positive real number except 1
67. Domain of f in the set of all real numbers; range of f is 1. Domain of f^{-1} is 1; range of f^{-1} is the set of all real numbers. No, f^{-1} is not.
69. **(A)**

(B) Domain of f is the set of real numbers; range of f is the set of positive real numbers. Domain of f is the range of f^{-1}; range of f is the domain of f^{-1}.   **71.** -3
(C) f^{-1} is called the logarithmic function with base 10.

Even

2. $4 = 2^2$ **4.** $125 = 5^3$ **6.** $100 = 10^2$ **8.** $1 = 8^0$ **10.** $\log_5 25 = 2$ **12.** $\log_{10} 1{,}000 = 3$ **14.** $\log_b a = c$
16. $\log_4 8 = \frac{3}{2}$ **18.** 3 **20.** -7 **22.** 2 **24.** -3 **26.** $x = 9$ **28.** $y = 2$ **30.** $b = 10$
32. $0.01 = 10^{-2}$ **34.** $2 = 4^{1/2}$ **36.** $27 = (\frac{1}{3})^{-3}$ **38.** $u = k^v$ **40.** $\log_{10} 0.001 = -3$ **42.** $\log_{1/2} 1 = 0$
44. $\log_{1/2}(\frac{1}{8}) = 3$ **46.** $\log_{32}(\frac{1}{2}) = -\frac{1}{5}$ **48.** $\log_{121} 11 = \frac{1}{2}$ **50.** uv **52.** -3 **54.** $\frac{1}{3}$ **56.** 0 **58.** -1
60. $x = 5$ **62.** $y = -\frac{1}{2}$ **64.** $b = 8$ **66.** Any positive real number except 1
68. If $\log_1 x = y$, then $1^y = x$; that is, $1 = x$ for arbitrary positive x, which is impossible.

Exercise 9-3

Odd

1. $\log_b u + \log_b v$ **3.** $\log_b A - \log_b B$ **5.** $5 \log_b u$ **7.** $\frac{3}{5} \log_b N$ **9.** $\frac{1}{2} \log_b Q$ **11.** $\log_b u + \log_b v + \log_b w$
13. $\log_b AB$ **15.** $\log_b \dfrac{X}{Y}$ **17.** $\log_b \dfrac{wx}{y}$ **19.** 3.40 **21.** -0.92 **23.** 3.30 **25.** $2 \log_b u + 7 \log_b v$

27. $-\log_b a$ **29.** $\frac{1}{3}\log_b N - 2\log_b p - 3\log_b q$ **31.** $\frac{1}{4}(2\log_b x + 3\log_b y - \frac{1}{2}\log_b z)$ **33.** $\log_b \dfrac{x^2}{y}$

35. $\log_b \dfrac{x^3 y^2}{z^4}$ **37.** $\log_b \sqrt[5]{x^2 y^3}$ **39.** 2.02   **41.** 0.23   **43.** -0.05 **45.** 8   **47.** $y = cb^{-kt}$

49. Let $u = \log_b M$ and $v = \log_b N$; then $M = b^u$ and $N = b^v$. Thus, $\log_b \dfrac{M}{N} = \log_b \dfrac{b^u}{b^v} = \log_b b^{u-v} = u - v = \log_b M - \log_b N$.

51. $MN = b^{\log_b M} b^{\log_b N} = b^{\log_b M + \log_b N}$; hence, by definition of logarithm, $\log_b MN = \log_b M + \log_b N$.

Even

2. $\log_b r + \log_b t$ **4.** $\log_b p - \log_b q$ **6.** $25\log_b w$ **8.** $-\frac{2}{3}\log_b u$ **10.** $\frac{1}{5}\log_b M$ **12.** $\log_b u - \log_b v - \log_b w$

14. $\log_b PQR$ **16.** $\log_b \dfrac{x^2}{y^3}$ **18.** $\log_b \dfrac{w}{xy}$ **20.** 1.79   **22.** 0.51   **24.** 2.76   **26.** $\frac{1}{2}\log_b u + \frac{1}{3}\log_b v$

28. $-3\log_b M$ **30.** $5\log_b m + 3\log_b n - \frac{1}{2}\log_b p$ **32.** $\frac{3}{5}(\log_b x - 4\log_b y - 9\log_b z)$ **34.** $\log_b \dfrac{m}{\sqrt{n}}$

36. $\log_b \dfrac{\sqrt[3]{w}}{x^3 y^5}$ **38.** $\log_b \sqrt[3]{\dfrac{x}{y}}$ **40.** 0.41   **42.** 0.55   **44.** 0.137   **46.** 2   **48.** $x = 100e^{-0.08t}$

50. Let $u = \log_b M$; then $M = b^u$. Thus, $\log_b M^p = \log_b(b^u)^p = \log_b b^{pu} = pu = p\log_b M$.

52. $\dfrac{M}{N} = \dfrac{b^{\log_b M}}{b^{\log_b N}} = b^{\log_b M - \log_b N}$; hence, by definition of logarithm, $\log_b \dfrac{M}{N} = \log_b M - \log_b N$.

Exercise 9-4

Odd

1. 4.9177   **3.** -2.8419 **5.** 3.7623   **7.** -2.5128 **9.** 200,800   **11.** 0.000 664 8   **13.** 47.73   **15.** 0.6760

17. 4.959   **19.** 7.861   **21.** 3.301   **23.** 3.6776   **25.** -1.6094 **27.** -1.7372

29.   **31.**   **33.**

35. 12.725   **37.** -25.715 **39.** 1.1709×10^{32} **41.** 4.2672×10^{-7}

Even

2. 5.9260   **4.** -1.4485 **6.** 14.8604   **8.** -10.3374 **10.** 82.57   **12.** 0.009 097   **14.** 162.1

16. 0.016 44   **18.** 6.116   **20.** 7.604   **22.** 3.822   **24.** 2.2618   **26.** -7.5224 **28.** 2.4455

30.   **32.**   **34.**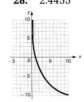

36. 17.322   **38.** -17.027 **40.** 1.8551×10^{-13} **42.** 1.6007×10^8

Exercise 9-5

Odd

1. 1.46   **3.** 0.321   **5.** 1.29   **7.** 3.50   **9.** 1.80   **11.** 2.07   **13.** 20   **15.** 5   **17.** 14.2

19. -1.83 **21.** 11.7   **23.** 5   **25.** $1, e^2, e^{-2}$ **27.** $x = e^e$ **29.** 100, 0.1   **31.** $x = -(1/k)\ln(I/I_0)$

33. $I = I_0 10^{N/10}$ **35.** $t = (-L/R)\ln[1 - (RI/E)]$
37. Inequality sign should have been reversed when both sides were multiplied by $\log \frac{1}{2}$, a negative quantity.
39. 5 years to the nearest year   **41.** Approx. 3.8 hr   **43.** Approx. 35 years   **45.** Approx. 28 years
47. Divide both sides by I_0, take logs of both sides, and then multiply both sides by 10.   **49.** 95 ft, 489 ft

Even

2. 1.16   **4.** 0.908   **6.** 4.25   **8.** -0.610 **10.** 1.26   **12.** 2.37   **14.** 80   **16.** 10   **18.** 18.9
20. 20.2   **22.** 3.14   **24.** $\frac{2}{3}$ **26.** $1, 10^2, 10^{-2}$ **28.** $x = 10^{10}$ **30.** $x = 10^{(\log 3)/(\log 3 - 1)}$

32. $n = \dfrac{\log A - \log P}{\log(1 + i)}$ **34.** $A = A_0 e^{-kt}$ **36.** $n = \dfrac{\log\left(\dfrac{iS}{R} + 1\right)}{\log(1 + i)}$

38. The inequality sign should have been reversed when both sides were divided by $\ln e^{-1}$, a negative number.
40. 8 years to the nearest year   **42.** (A) Approx. 12.8 (B) Approx. 206 in.   **44.** Approx. 533 years
46. Approx. 18,600 years old   **48.** (A) 30 decibels (B) 65 decibels (C) 80 decibels (D) 150 decibels   **50.** 17 wk

Diagnostic (Review) Exercise 9-6

1. $n = \log_{10} m$ (9-2)   **2.** $x = 10^y$ (9-2)   **3.** 8 (9-2)   **4.** 5 (9-2)   **5.** 3 (9-2)   **6.** 1.24 (9-5)
7. 11.9 (9-5)   **8.** 900 (9-3, 9-5)   **9.** 5 (9-3, 9-5)   **10.** $y = e^x$ (9-2)   **11.** $y = \ln x$ (9-2)   **12.** -2 (9-2)
13. $\frac{1}{3}$ (9-2)   **14.** 64 (9-2)   **15.** e (9-2)   **16.** 33 (9-2)   **17.** 1 (9-2)   **18.** 2.32 (9-5)   **19.** 3.92 (9-5)
20. 92.1 (9-5)   **21.** 300 (9-3, 9-5)   **22.** 2 (9-3, 9-5)   **23.** $1, 10^3, 10^{-3}$ (9-3, 9-5)   **24.** 10^e (9-3, 9-5)
25. 1.95 (9-4)
26. (9-1)   **27.** (9-1)   **28.** (9-4)   **29.** (9-4)

30. $y = ce^{-5t}$ (9-3, 9-5)   **31.** Domain $f = (0, \infty) = $ Range f^{-1}; Range $f = R = $ Domain f^{-1} (9-2)

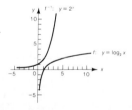

32. If $\log_1 x = y$, then we would have to have $1^y = x$; that is, $1 = x$ for arbitrary positive x, which is impossible. (9-2)
33. Let $u = \log_b M$ and $v = \log_b N$; then $M = b^u$ and $N = b^v$. Thus,
$\log_b(M/N) = \log_b(b^u/b^v) = \log_b b^{u-v} = u - v = \log_b M - \log_b N$. (9-3)
34. 23.4 years (9-5)   **35.** 23.1 years (9-5)   **36.** 37,100 years (9-5)   **37.** $I = I_0 e^{-kx}$ (9-3, 9-5)
38. $n = -\log[1 - (Pi/R)]/\log(1 + i)$ (9-3, 9-5)

Practice Test Chapter 9

1. $8 = 16^{3/4}$ (9-2)   **2.** $y = 10^{0.2x}$ (9-2)   **3.** $\ln A = 0.08t$ (9-2)   **4.** 16 (9-2)   **5.** $\frac{1}{2}$ (9-2)   **6.** $\frac{1}{4}$ (9-2)
7. 10 (9-5)   **8.** 0.701 (9-5)   **9.** 1.63 (9-5)   **10.** 5.76 (9-5)   **11.** 18 years (9-5)   **12.** 11,200 years (9-5)
13. (9-4)   **14.** (9-1)

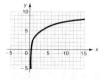   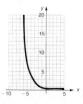

Chapter 10   Exercise 10-1

Odd

1. $(2, -3)$ **3.** No solution   **5.** $(1, 4)$ **7.** $(-2, -3)$ **9.** $(-\frac{4}{3}, 1)$
11. Infinite number of solutions, $(x, 3x - 9)$ for any real number x **13.** $(-2, 3)$ **15.** $(1, -5)$ **17.** $(\frac{1}{3}, -2)$
19. Infinite number of solutions, $(x, \frac{1}{2}x + 3)$ for any real number x **21.** $(-2, 2)$ **23.** $(1.1, 0.3)$ **25.** $(8, 6)$
27. 14 nickels and 8 dimes   **29.** $52°; 38°$ **31.** $84\frac{1}{4}$-lb packages; $60\frac{1}{2}$-lb packages
33. 60 ml of 80% solution and 40 ml of 50% solution   **35.** **(A)** $1\frac{9}{13}$ sec, $7\frac{9}{13}$ sec   **(B)** Approximately 8,462 ft
37. Both companies pay $136 on sales of $1,700. The straight-commission company pays better to the right of this point; the other company pays better to the left.
39. At $2 per record, $d = s = 3,000$.

Even

2. $(-5, -1)$ **4.** Infinite number of solutions (lines coincide)   **6.** No solution   **8.** $(-3, -5)$ **10.** $(-2, -3)$
12. $\frac{3}{2}, -\frac{2}{3}$ **14.** $(1, -1)$ **16.** $(2, -4)$ **18.** $(-2, -3)$ **20.** No solution   **22.** $(2, 3)$ **24.** $(1, 0.2)$
26. $(-6, 12)$ **28.** 18 4¢ stamps; 12 5¢ stamps   **30.** 16×20 in.   **32.** 60 oz of mix A and 80 oz of mix B
34. Total time: 2 hr; faster machine: 2 hr, slower machine: 1.5 hr   **36.** 40 sec; 24 sec; 120 mi
38. Both companies pay $175 on sales of $2,500. The straight-commission company pays better to the right of this point; the other company pays better to the left.

Exercise 10-2

Odd

1. $x = -1, y = -1, z = -2$ **3.** $x = 0, y = -2, z = 5$ **5.** $x = 2, y = 0, z = -1$ **7.** $a = -1, b = 2, c = 0$
9. $x = 0, y = 2, z = -3$ **11.** $x = 1, y = -2, z = 1$ **13.** No solution   **15.** $w = 1, x = -1, y = 0, z = 2$
17. $D = -4, E = -4, F = -17$ **19.** 1,200 style A; 800 style B; 2,000 style C
21. 60 grams of mix A, 50 grams of mix B, 40 grams of mix C
23. Oldest press: 12 hr; middle press: 6 hr; newest press: 4 hr

Even

2. $x = -4, y = -3, z = 2$ **4.** $x = -3, y = 2, z = 1$ **6.** $x = 3, y = -1, z = -2$ **8.** $u = 1, v = -1, w = 2$
10. $x = -2, y = 1, z = 0$ **12.** $x = -1, y = 1, z = 2$ **14.** Infinitely many solutions
16. $r = 2, s = 0, t = 0, u = -1$ **18.** $D = -6, E = 8, F = 0$ **20.** 900 style A; 1,300 style B; 1,600 style C
22. 50 g mix A, 40 g mix B, 30 g mix C

Exercise 10-3

Odd

1. $\begin{bmatrix} 4 & -6 & \vert & -8 \\ 1 & -3 & \vert & 2 \end{bmatrix}$ **3.** $\begin{bmatrix} -4 & 12 & \vert & -8 \\ 4 & -6 & \vert & -8 \end{bmatrix}$ **5.** $\begin{bmatrix} 1 & -3 & \vert & 2 \\ 8 & -12 & \vert & -16 \end{bmatrix}$ **7.** $\begin{bmatrix} 1 & -3 & \vert & 2 \\ 0 & 6 & \vert & -16 \end{bmatrix}$

9. $\begin{bmatrix} 1 & -3 & \vert & 2 \\ 2 & 0 & \vert & -12 \end{bmatrix}$ **11.** $\begin{bmatrix} 1 & -3 & \vert & 2 \\ 3 & -3 & \vert & -10 \end{bmatrix}$ **13.** $x = 3, y = 2$ **15.** $x = 3, y = 1$ **17.** $x = 2, y = 1$

19. $x = 2, y = 4$ **21.** No solution   **23.** $x = 1, y = 4$
25. Infinitely many solutions: $y = s, x = 2s - 3$ for any real number s
27. Infinitely many solutions: $y = s, x = \frac{1}{2}s + \frac{1}{2}$ for any real number s **29.** $x = 2, y = -1$ **31.** $x = 2, y = -1$
33. $x = 1.1, y = 0.3$ **35.** $x = -2, y = 3, z = 1$ **37.** $x = 0, y = -2, z = 2$

Even

2. $\begin{bmatrix} 1 & -3 & | & 2 \\ 2 & -3 & | & -4 \end{bmatrix}$ **4.** $\begin{bmatrix} -2 & 6 & | & -4 \\ 4 & -6 & | & -8 \end{bmatrix}$ **6.** $\begin{bmatrix} 1 & -3 & | & 2 \\ -4 & 6 & | & 8 \end{bmatrix}$ **8.** $\begin{bmatrix} -1 & 0 & | & 6 \\ 4 & -6 & | & -8 \end{bmatrix}$

10. $\begin{bmatrix} 1 & -3 & | & 2 \\ 1 & 3 & | & -14 \end{bmatrix}$ **12.** $\begin{bmatrix} 1 & -3 & | & 2 \\ 5 & -9 & | & -6 \end{bmatrix}$ **14.** $x = 4, y = 2$ **16.** $x = 4, y = -2$

18. $x = -2, y = 1$ **20.** $x = -1, y = 2$ **22.** No solution **24.** $x = -1, y = 2$

26. Infinitely many solutions: $y = s, x = 2s - 1$ for s any real number

28. Infinitely many solutions: $y = s, x = \frac{1}{3}s - \frac{2}{3}$ for s any real number **30.** $x = -1, y = 2$ **32.** $x = 5, y = 2$

34. $x = 1.2, y = 0.3$ **36.** $x = -2, y = 0, z = 1$ **38.** $x = 1, y = -2, z = 0$

Exercise 10-4

Odd

1.

3.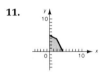

5.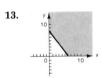

7.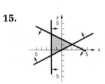

9.

11.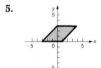

13.

15.

17.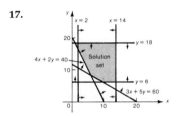

19. $6x + 4y \leq 108$
$x + y \leq 24$
$x \geq 0$
$y \geq 0$

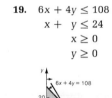

Even

2.

4.

6.

8.

10.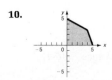

12.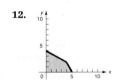

14.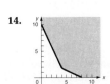

16.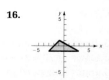

18.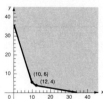

20. $30x + 10y \geq 360$
$10x + 10y \geq 160$
$10x + 30y \geq 240$
$x \geq 0$
$y \geq 0$

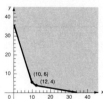

Exercise 10-5

Odd

1. $(-3, -4), (3, -4)$ **3.** $(\frac{1}{2}, 1)$ **5.** $(4, -2), (-4, 2)$ **7.** $(3 - i, 4 - 3i), (3 + i, 4 + 3i)$
9. $(2, 1), (2, -1), (-2, 1), (-2, -1)$ **11.** $(-3, -2), (-3, 2), (3, -2), (3, 2)$
13. $(2 + \sqrt{10}, -2 + \sqrt{10}), (2 - \sqrt{10}, -2 - \sqrt{10})$ **15.** $(-1, 2), (4, 7)$ **17.** $(i, 2i), (i, -2i), (-i, 2i), (-i, -2i)$
19. $(2, 4), (-2, 4), (i\sqrt{5}, -5), (-i\sqrt{5}, -5)$ **21.** $(4, 0), (-3, \sqrt{7}), (-3, -\sqrt{7})$ **23.** 2 by 16 ft
25. $(2\sqrt{2}, \sqrt{2}), (-2\sqrt{2}, -\sqrt{2}), (1, 4), (-1, -4)$ **27.** $(3, 3), (-3, -3), (0, 6), (0, -6)$
29. $(4, 4), (-4, -4), (\frac{4}{5}\sqrt{5}, -\frac{4}{5}\sqrt{5}), (-\frac{4}{5}\sqrt{5}, \frac{4}{5}\sqrt{5})$

Even

2. $(-12, 5), (-12, -5)$ **4.** $(2, 4), (-2, -4)$ **6.** $(5i, -5i), (-5i, 5i)$ **8.** $(3 + 4i, -1 + 2i), (3 - 4i, -1 - 2i)$
10. $(2, 4), (2, -4), (-2, 4), (-2, -4)$ **12.** $(1, 3), (1, -3), (-1, 3), (-1, -3)$
14. $(-1 + \sqrt{3}i, 1 + \sqrt{3}i), (-1 - \sqrt{3}i, 1 - \sqrt{3}i)$ **16.** $(0, -1), (-4, -3)$ **18.** $(2, 2i), (2, -2i), (-2, 2i), (-2, -2i)$
20. $(2, \sqrt{2}), (2, -\sqrt{2}), (-1, i), (-1, -i)$ **22.** $(2, 1), (-2, 1), (2i, 3), (-2i, 3)$ **24.** $\frac{1}{2} + \frac{\sqrt{3}}{2}i$ and $\frac{1}{2} - \frac{\sqrt{3}}{2}i$
26. $(2, 1), (-2, -1), (i, -2i), (-i, 2i)$ **28.** $(2, 2), (-2, -2), (\sqrt{2}, -\sqrt{2}), (-\sqrt{2}, \sqrt{2})$
30. $(-3, 1), (3, -1), (-i, i), (i, -i)$

Diagnostic (Review) Exercise 10-6

1. $x = 6, y = 1$ *(10-1)*   **2.** *(10-4)*    **3.** $x = 2, y = 1$ *(10-1)*   **4.** $x = 2, y = 1$ *(10-1)*

5. $x = -1, y = 2, z = 1$ *(10-2)*   **6.** $(1, -1), (\frac{7}{5}, -\frac{1}{5})$ *(10-5)*   **7.** $(4, 3), (-4, 3), (4, -3), (-4, -3)$ *(10-5)*
8. $(2, 1, -1)$ *(10-2)*   **9.** *(10-4)*   **10.** No solution *(10-1)*   **11.** $(2, -1, 2)$ *(10-2)*

12. $(1, 3), (1, -3), (-1, 3), (-1, -3)$ *(10-5)*   **13.** $(1 + i, 2i), (1 - i, -2i)$ *(10-5)*
14. Lines coincide—infinitely many solutions; $(x, \frac{1}{3}x + \frac{1}{2})$ for all real x *(10-1)*
15. $(0, 2), (0, -2), (i\sqrt{2}, i\sqrt{2}), (-i\sqrt{2}, -i\sqrt{2})$ *(10-5)*   **16.** $x = -1, y = 3$ *(10-3)*
17. $x = -1, y = 2, z = 1$ *(10-3)*   **18.** $x = 2, y = 1, z = -1$ *(10-3)*   **19.** $x = -12, y = 7, z = 0$ *(10-3)*
20. 16 dimes and 14 nickels   *(10-1)*   **21.** $4,000 at 6% and $2,000 at 10% *(10-1)*   **22.** 6 by 5 cm *(10-1)*
23. 30 grams of 70% solution and 70 grams of 40% solution *(10-1)*   **24.** 48 $\frac{1}{2}$-lb packages, 72 $\frac{1}{3}$-lb packages *(10-1)*
25. 40 grams mix A, 60 grams mix B, 30 grams mix C *(10-2, 10-3)*

Practice Test Chapter 10

1. $x = 4, y = 3$ *(10-1)*

2. *(10-4)*

3. *(10-4)*

4. $x = -2, y = 3$ *(10-1)* **5.** $(x, 3x + 2)$ for any x *(10-1)* **6.** No solution *(10-1)*

7. $x = -1, y = 2, z = -3$ *(10-2)* **8.** $(0, -1)$ and $(\frac{8}{5}, \frac{3}{5})$ *(10-5)*

9. $(\sqrt{3}, \sqrt{2}), (-\sqrt{3}, \sqrt{2}), (-\sqrt{3}, -\sqrt{2}), (\sqrt{3}, -\sqrt{2})$ *(10-5)* **10.** -4 *(10-3)*

11. 31 nickels and 25 dimes *(10-1)* **12.** 300 lb mix A, 1,000 lb mix B *(10-1)*

13. $\left(\dfrac{4\sqrt{17}}{17}, \dfrac{2\sqrt{17}}{17}\right), \left(\dfrac{4\sqrt{17}}{17}, \dfrac{-2\sqrt{17}}{17}\right), \left(\dfrac{-4\sqrt{17}}{17}, \dfrac{2\sqrt{17}}{17}\right), \left(\dfrac{-4\sqrt{17}}{17}, \dfrac{-2\sqrt{17}}{17}\right)$ *(10-5)*

14. $0.2x + 0.4y + 0.3z = 1,160$
$0.3x + 0.5y + 0.4z = 1,560$
$0.1x + 0.2y + 0.1z = 480$ *(10-2)*

Chapter 11   Exercise 11-1

Odd

1. $-1, 0, 1, 2$ **3.** $0, \frac{1}{3}, \frac{1}{2}, \frac{3}{5}$ **5.** $4, -8, 16, -32$ **7.** 6 **9.** $\frac{99}{101}$ **11.** $S_5 = 1 + 2 + 3 + 4 + 5$

13. $S_3 = \frac{1}{10} + \frac{1}{100} + \frac{1}{1000}$ **15.** $S_4 = -1 + 1 - 1 + 1$ **17.** $1, -4, 9, -16, 25$ **19.** $0.3, 0.33, 0.333, 0.3333, 0.33333$

21. $7, 3, -1, -5, -9$ **23.** $4, 1, \frac{1}{4}, \frac{1}{16}, \frac{1}{64}$ **25.** $a_n = n + 3$ **27.** $a_n = 3n$ **29.** $a_n = \dfrac{n}{n+1}$ **31.** $a_n = (-1)^{n+1}$

33. $a_n = (-2)^n$ **35.** $a_n = \dfrac{x^n}{n}$ **37.** $\frac{4}{1} - \frac{8}{2} + \frac{16}{3} - \frac{32}{4}$ **39.** $S_3 = x^2 + \dfrac{x^3}{2} + \dfrac{x^4}{3}$

41. $x - \dfrac{x^2}{2} + \dfrac{x^3}{3} - \dfrac{x^4}{4} + \dfrac{x^5}{5}$ **43.** $S_4 = \displaystyle\sum_{k=1}^{4} k^2$ **45.** $S_5 = \displaystyle\sum_{k=1}^{5} 1/2^k$ **47.** $S_n = \displaystyle\sum_{k=1}^{n} 1/k^2$

49. $S_n = \displaystyle\sum_{k=1}^{n} (-1)^{k+1} k^2$ **53. (A)** $3, 1.833, 1.462, 1.415$ **(B)** Table, $\sqrt{2} = 1.414$ **(C)** $1, 1.5, 1.417, 1.414$

Even

2. $4, 5, 6, 7$ **4.** $2, (\frac{3}{2})^2, (\frac{4}{3})^3, (\frac{5}{4})^4$ **6.** $1, -\frac{1}{4}, \frac{1}{9}, -\frac{1}{16}$ **8.** 13 **10.** $(\frac{201}{200})^{200}$ **12.** $S_4 = 1 + 4 + 9 + 16$

14. $S_5 = \frac{1}{3} + \frac{1}{9} + \frac{1}{27} + \frac{1}{81} + \frac{1}{243}$ **16.** $S_6 = 1 - 2 + 3 - 4 + 5 - 6$ **18.** $\frac{1}{2}, -\frac{1}{4}, \frac{1}{8}, -\frac{1}{16}, \frac{1}{32}$ **20.** $2, 0, 6, 0, 10$

22. $1, 1, 2, 3, 5$ **24.** $2, 4, 8, 16, 32$ **26.** $a_n = n - 3$ **28.** $a_n = -2n$ **30.** $a_n = \dfrac{2n - 1}{2n}$ **32.** $a_n = (-1)^{n+1} n$

34. $a_n = (-1)^{n+1}(2n - 1)$ **36.** $a_n = (-1)^{n+1} x^{2n-1}$ **38.** $1^2 - 3^2 + 5^2 - 7^2 + 9^2$ **40.** $S_5 = 1 + x + x^2 + x^3 + x^4$

42. $x - \dfrac{x^3}{3} + \dfrac{x^5}{5} - \dfrac{x^7}{7} + \dfrac{x^9}{9}$ **44.** $S_5 = \displaystyle\sum_{k=1}^{5} (k + 1)$ **46.** $S_4 = \displaystyle\sum_{k=1}^{4} \dfrac{(-1)^{k+1}}{k}$ **48.** $S_n = \displaystyle\sum_{k=1}^{n} \dfrac{k+1}{k}$

50. $S_n = \displaystyle\sum_{k=1}^{n} \dfrac{(-1)^{k+1}}{2^k}$ **54. (A)** $2, 2.25, 2.236, 2.236$ **(B)** Table, $\sqrt{5} = 2.236$ **(C)** $3, 2.333, 2.238, 2.236$

Exercise 11-2

Odd

1. (B) $d = -0.5; 5.5, 5$ **(C)** $d = -5; -26, -31$ **3.** $a_2 = -1, a_3 = 3, a_4 = 7$ **5.** $a_{15} = 67; S_{11} = 242$

7. $S_{21} = 861$ **9.** $a_{15} = -21$ **11.** $d = 6; a_{101} = 603$ **13.** $S_{40} = 200$ **15.** $a_{11} = 2, S_{11} = \frac{77}{6}$ **17.** $a_1 = 1$

19. $S_{51} = 4,131$ **21.** $-1,071$ **23.** $4,446$ **27.** $a_n = -3 + (n - 1)3$ **29. (A)** 336 ft **(B)** 1,936 ft **(C)** $16t^2$

Even

2. (A) $d = -6$; -13, -19 **(D)** $d = 32$; 112, 144 **4.** $a_2 = -15$, $a_3 = -12$, $a_4 = -9$ **6.** $a_{22} = 87$; $S_{21} = 903$
8. $S_{11} = 385$ **10.** $a_{10} = -39$ **12.** $d = 3$; $a_{25} = 79$ **14.** $S_{24} = -48$ **16.** $a_{19} = \frac{5}{3}$; $S_{19} = \frac{209}{12}$

18. $a_1 = -42$ **20.** $S_{40} = 1{,}520$ **22.** 320 **24.** $60{,}000$ **28.** $S_n = \dfrac{n}{2}[-6 + (n-1)3]$ **30.** Firm B; $216{,}000

Exercise 11-3

Odd

1. (A) $r = -2$; -16, 32 **(D)** $r = \frac{1}{3}$; $\frac{1}{54}$, $\frac{1}{162}$ **3.** $a_2 = 3$, $a_3 = -\frac{3}{2}$, $a_4 = \frac{3}{4}$ **5.** $a_{10} = \frac{1}{243}$ **7.** $S_7 = 3{,}279$
9. $r = 0.398$ **11.** $S_{10} = -1{,}705$ **13.** $a_2 = 6$, $a_3 = 4$ **15.** $S_7 = 547$ **17.** $\frac{1023}{1024}$ **19.** $x = 2\sqrt{3}$ **21.** $S_\infty = \frac{9}{2}$
23. No sum   **25.** $S_\infty = \frac{8}{5}$ **27.** $\frac{7}{9}$ **29.** $\frac{6}{11}$ **31.** $3\frac{8}{37}$ or $\frac{119}{37}$ **33.** $A = P(1 + r)^n$; approx. 12 years
35. 900 **37.** $4{,}000{,}000

Even

2. (B) $r = \frac{1}{3}$; $\frac{4}{9}$, $\frac{4}{27}$ **4.** $a_2 = 8$, $a_3 = \frac{16}{3}$, $a_4 = \frac{32}{9}$ **6.** $a_{13} = \frac{1}{64}$ **8.** $S_7 = 547$ **10.** $r = \sqrt[9]{3} \approx 1.13$
12. $S_{10} = 3{,}069$ **14.** $a = -4$, $a_3 = \frac{4}{3}$ **16.** $S_7 = 3{,}279$ **18.** $2{,}046$ **20.** $x = 4\sqrt{3}$ **22.** $S_\infty = \frac{64}{3}$
24. No sum   **26.** $S_\infty = \frac{147}{8}$ **28.** $\frac{5}{9}$ **30.** $\frac{3}{11}$ **32.** $5\frac{7}{11}$ or $\frac{62}{11}$ **34.** $A = A_0(1 + r)^t$; approx. 35 years
36. 100 in.   **38.** 1 min

Exercise 11-4

Odd

1. 720 **3.** 20 **5.** 720 **7.** 15 **9.** 1 **11.** 28 **13.** $9!/8!$ **15.** $8!/5!$ **17.** 126 **19.** 6
21. 1 **23.** $2{,}380$ **25.** $u^5 + 5u^4v + 10u^3v^2 + 10u^2v^3 + 5uv^4 + v^5$ **27.** $y^4 - 4y^3 + 6y^2 - 4y + 1$
29. $32x^5 - 80x^4y + 80x^3y^2 - 40x^2y^3 + 10xy^4 - y^5$ **31.** $5{,}005u^9v^6$ **33.** $264m^2n^{10}$ **35.** $924w^6$ **37.** 1.1046
39. $\dbinom{n}{r} = \dfrac{n!}{r!(n-r)!} = \dfrac{n!}{(n-r)![n-(n-r)]!} = \dbinom{n}{n-r}$

Even

2. 24 **4.** 5 **6.** 504 **8.** 10 **10.** 1 **12.** 35 **14.** $\dfrac{12!}{11!}$ **16.** $\dfrac{12!}{8!}$ **18.** 10 **20.** 7 **22.** 1

24. $4{,}845$ **26.** $x^4 + 4x^3y + 6x^2y^2 + 4xy^3 + y^4$ **28.** $x^5 - 10x^4 + 40x^3 - 80x^2 + 80x - 32$
30. $m^6 + 12m^5n + 60m^4n^2 + 160m^3n^3 + 240m^2n^4 + 192mn^5 + 64n^6$ **32.** $495a^8b^4$ **34.** $760x^{18}y^2$ **36.** $-3{,}240x^7$
38. 0.9415 **40.** 1   5   10   10   5   1   and   1   6   15   20   15   6   1

Diagnostic (Review) Exercise 11-5

1. Arithmetic: (B) and (C); Geometric: (A) and (E) *(11-2, 11-3)*
2. (A) 5, 7, 9, 11 **(B)** $a_{10} = 23$ **(C)** $S_{10} = 140$ *(11-2)*
3. (A) 16, 8, 4, 2 **(B)** $a_{10} = \frac{1}{32}$ **(C)** $S_{10} = 31\frac{31}{32}$ *(11-3)*
4. (A) -8, -5, -2, 1 **(B)** $a_{10} = 19$ **(C)** $S_{10} = 55$ *(11-2)*
5. (A) -1, 2, -4, 8 **(B)** $a_{10} = 512$ **(C)** $S_{10} = 341$ *(11-3)*   **6.** $S_\infty = 32$ *(11-3)*   **7.** 720 *(11-4)*
8. $20 \cdot 21 \cdot 22 = 9{,}240$ *(11-4)*   **9.** 21 *(11-4)*   **10.** $S_{10} = -6 - 4 - 2 + 0 + 2 + 4 + 6 + 8 + 10 + 12 = 30$ *(11-2)*
11. $S_7 = 8 + 4 + 2 + 1 + \frac{1}{2} + \frac{1}{4} + \frac{1}{8} = 15\frac{7}{8}$ *(11-3)*   **12.** $S_\infty = \frac{81}{5}$ *(11-3)*   **13.** $S_n = \displaystyle\sum_{k=1}^{n} \dfrac{(-1)^{k+1}}{3^k}$, $S_\infty = \frac{1}{4}$ *(11-3)*

14. $d = 3, a_5 = 25$ *(11-2)*   **15.** 190 *(11-4)*   **16.** 1,820 *(11-4)*   **17.** 1 *(11-4)*
18. $x^5 - 5x^4y + 10x^3y^2 - 10x^2y^3 + 5xy^4 - y^5$ *(11-4)*   **19.** $-1,760x^3y^9$ *(11-4)*   **20.** $\frac{8}{11}$ *(11-3)*
21. $49g/2$ ft; $625g/2$ ft *(11-2)*   **22.** $x^6 + 6ix^5 - 15x^4 - 20ix^3 + 15x^2 + 6ix - 1$ *(11-4)*

Practice Test Chapter 11

1. $2, -1, -4, -7$ *(11-2)*   **2.** -130 *(11-2)*   **3.** $-2,880$ *(11-2)*   **4.** 0.0002 *(11-3)*   **5.** 0.222 22 *(11-3)*

6. $\frac{2}{9}$ *(11-3)*   **7.** $0 - 3 + 8 - 15 + 24$ *(11-1)*   **8.** $\sum_{k=1}^{5} \frac{2k}{2k+1}$ *(11-1)*   **9.** $6\sqrt{2}$ *(11-3)*   **10.** 9 *(11-3)*

11. 165 *(11-4)*   **12.** $165x^3$ *(11-4)*   **13.** -330 *(11-1, 11-2, 11-3)*   **14.** $\dfrac{\log 3}{\log 1.06}$ *(11-3)*

Appendix   Exercise B

Odd

1. 0.8627   **3.** 3.3096   **5.** -1.3840 **7.** 5.4843   **9.** -2.3215 **11.** 2.74×10^7 **13.** 1.58×10^{-5}
15. 0.8544   **17.** 13.3114   **19.** -5.7541

Even

2. 0.8041   **4.** 2.5145   **6.** -3.0329 **8.** 4.9175   **10.** -1.2083 **12.** 1.30×10^9 **14.** 3.87×10^{-4}
16. 1.9488   **18.** 8.5622   **20.** -2.6409

Exercise C

Odd

1. 0.3649   **3.** 5.8472   **5.** 1.8131   **7.** $0.6027 - 3$ **9.** $0.9546 - 1$ **11.** 5.8403   **13.** 5.204
15. $3.514 \times 10^5 = 351,400$ **17.** $2.693 \times 10^3 = 2,693$ **19.** $4.016 \times 10^{-3} = 0.004\ 016$ **21.** $6.906 \times 10^{-1} = 0.6906$
23. $1.474 \times 10^{-1} = 0.1474$

Even

2. 0.7112   **4.** 4.4538   **6.** 1.3086   **8.** $0.5697 - 2$ **10.** $0.8071 - 1$ **12.** 4.9277   **14.** 2.562
16. $2.975 \times 10^2 = 297.5$ **18.** $8.206 \times 10 = 82.06$ **20.** $1.651 \times 10^{-1} = 0.1651$ **22.** $7.094 \times 10^{-4} = 0.000\ 709\ 4$
24. $2.162 \times 10^{-3} = 0.002\ 162$

Exercise D

Odd

1. -14 **3.** 2   **5.** -2.6 **7.** $\begin{vmatrix} a_{22} & a_{23} \\ a_{32} & a_{33} \end{vmatrix}$ **9.** $\begin{vmatrix} a_{11} & a_{12} \\ a_{31} & a_{32} \end{vmatrix}$ **11.** $(-1)^{1+1}\begin{vmatrix} a_{22} & a_{23} \\ a_{32} & a_{33} \end{vmatrix}$ **13.** $(-1)^{2+3}\begin{vmatrix} a_{11} & a_{12} \\ a_{31} & a_{32} \end{vmatrix}$

15. $\begin{vmatrix} 1 & -2 \\ -4 & 8 \end{vmatrix}$ **17.** $\begin{vmatrix} -2 & 0 \\ 5 & -2 \end{vmatrix}$ **19.** $(-1)^{1+1}\begin{vmatrix} 1 & -2 \\ -4 & 8 \end{vmatrix} = 0$ **21.** $(-1)^{3+2}\begin{vmatrix} -2 & 0 \\ 5 & -2 \end{vmatrix} = -4$ **23.** 10

25. -21 **27.** -120 **29.** -40 **31.** -12 **33.** 0   **35.** -8 **37.** 48
39. Expand the determinant about the first row to obtain $x - 3y + 7 = 0$; then show that the two points satisfy this linear equation.
41. $\frac{23}{2}$

Even

2. 8   **4.** -20 **6.** 4.3   **8.** $\begin{vmatrix} a_{11} & a_{12} \\ a_{21} & a_{22} \end{vmatrix}$ **10.** $\begin{vmatrix} a_{11} & a_{13} \\ a_{31} & a_{33} \end{vmatrix}$ **12.** $(-1)^{3+3}\begin{vmatrix} a_{11} & a_{12} \\ a_{21} & a_{22} \end{vmatrix}$ **14.** $(-1)^{2+2}\begin{vmatrix} a_{11} & a_{13} \\ a_{31} & a_{33} \end{vmatrix}$

16. $\begin{vmatrix} -2 & 0 \\ 7 & 8 \end{vmatrix}$ **18.** $\begin{vmatrix} 3 & 0 \\ -4 & 8 \end{vmatrix}$ **20.** $(-1)^{2+2}\begin{vmatrix} -2 & 0 \\ 7 & 8 \end{vmatrix} = -16$ **22.** $(-1)^{2+1}\begin{vmatrix} 3 & 0 \\ -4 & 8 \end{vmatrix} = -24$ **24.** -24 **26.** -40

28. 22   **30.** -43 **32.** -1 **34.** -18 **36.** 6   **38.** 60

40. Expand the determinant about the first row to obtain $(y_1 - y_2)x - (x_1 - x_2)y + (x_1y_2 - x_2y_1) = 0$; then show that the two points satisfy this linear equation.

42. 18

Exercise E

Odd

1. $x = 5, y = -2$ **3.** $x = 1, y = -1$ **5.** $x = -1, y = 1$ **7.** $x = 2, y = -2, z = -1$ **9.** $x = 2, y = -1, z = 2$

11. $x = 2, y = -3, z = -1$ **13.** $x = 1, y = -1, z = 2$ **15.** $a = -1, b = 2, c = 0$

17. $D = 0$; infinitely many solutions

Even

2. $x = -1, y = 2$ **4.** $x = 4, y = -1$ **6.** $x = 1, y = -1$ **8.** $x = -3, y = -1, z = 2$ **10.** $x = -4, y = 0, z = 3$

12. $x = 3, y = -2, z = 1$ **14.** $x = 1, y = 0, z = -2$ **16.** $u = 1, v = -1, w = 2$

18. Since $D \neq 0$, $(0, 0, 0)$ is the only solution.

Exercise F

Odd

1. $2 \times 2, 1 \times 4$ **3.** 2   **5.** $\begin{bmatrix} 0 & 0 \\ 0 & 0 \end{bmatrix}$ **7.** C, D **9.** A, B **11.** $\begin{bmatrix} -1 & 0 \\ 5 & -3 \end{bmatrix}$ **13.** $\begin{bmatrix} -2 \\ 3 \\ 0 \end{bmatrix}$ **15.** $\begin{bmatrix} -1 \\ 6 \\ 5 \end{bmatrix}$

17. $\begin{bmatrix} -15 & 5 \\ 10 & -15 \end{bmatrix}$ **19.** $[0 \quad 1 \quad 4]$ **21.** $\begin{bmatrix} 250 & 360 \\ 40 & 350 \end{bmatrix}$ **23.** $\begin{bmatrix} -1 & 0 \\ 0 & 6 \\ -1 & -1 \end{bmatrix}$ **25.** $a = -1, b = 1, c = 3, d = -5$

27.
$$\begin{array}{cc} \text{Guitar} & \text{Banjo} \\ \begin{bmatrix} \$33 & \$26 \\ \$57 & \$77 \end{bmatrix} & \begin{array}{l} \text{Materials} \\ \text{Labor} \end{array} \end{array}$$

Even

2. $4 \times 2, 3 \times 1$ **4.** -2 **6.** $[0 \quad 0 \quad 0 \quad 0]$ **8.** E **10.** 2   **12.** $\begin{bmatrix} 3 \\ 0 \\ 5 \end{bmatrix}$ **14.** $\begin{bmatrix} 3 & -1 \\ -2 & 3 \end{bmatrix}$ **16.** $\begin{bmatrix} 0 & 0 \\ 0 & 0 \end{bmatrix}$

18. $[8 \quad -2 \quad 0 \quad 4]$ **20.** $\begin{bmatrix} 290 \\ 300 \end{bmatrix}$ **22.** $\begin{bmatrix} 32 & 5 & 17 \\ 22 & 3 & 21 \end{bmatrix}$ **24.** $\begin{bmatrix} 1 & -5 & -1 \\ 4 & 0 & -1 \end{bmatrix}$ **26.** $w = -2, x = -1, y = 3, z = 5$

28. $M + N = \begin{bmatrix} 689 & 225 \\ 218 & 68 \end{bmatrix}$; $\dfrac{1}{1,200}(M + N) = \begin{bmatrix} 0.57 & 0.19 \\ 0.18 & 0.06 \end{bmatrix}$

Exercise G

Odd

1. 10 **3.** -1 **5.** [12 13] **7.** $\begin{bmatrix} 5 \\ -3 \end{bmatrix}$ **9.** $\begin{bmatrix} 2 & 4 \\ 1 & -5 \end{bmatrix}$ **11.** $\begin{bmatrix} 10 & 1 \\ -2 & 10 \end{bmatrix}$ **13.** 6 **15.** 15

17. $\begin{bmatrix} 0 & 9 \\ 5 & -4 \end{bmatrix}$ **19.** $\begin{bmatrix} 5 & 8 & -5 \\ -1 & -3 & 2 \\ -2 & 8 & -6 \end{bmatrix}$ **21.** [11] **23.** $\begin{bmatrix} 3 & -2 & -4 \\ 6 & -4 & -8 \\ -9 & 6 & 12 \end{bmatrix}$ **25.** $AB = \begin{bmatrix} 5 & 7 \\ 2 & 3 \end{bmatrix}, BA = \begin{bmatrix} 1 & 3 \\ 2 & 7 \end{bmatrix}$

27. Both sides equal $\begin{bmatrix} 0 & 12 \\ 1 & 5 \end{bmatrix}$ **29.** **(A)** \$9 per boat **(B)** $[1.5 \quad 1.2 \quad 0.4] \cdot \begin{bmatrix} 7 \\ 10 \\ 4 \end{bmatrix} = \$24.10$ **(C)** 3×2

(D)
	I	II	
	\$9.00	\$11.00	One-person
	\$14.10	\$17.20	Two-person
	\$19.80	\$24.10	Four-person

Labor costs per boat
at each plant

Even

2. 10 **4.** -10 **6.** [5 -9] **8.** $\begin{bmatrix} -6 \\ 14 \end{bmatrix}$ **10.** $\begin{bmatrix} 4 & -9 \\ -7 & 17 \end{bmatrix}$ **12.** $\begin{bmatrix} 11 & 1 \\ -3 & 9 \end{bmatrix}$ **14.** -10

16. -15 **18.** $\begin{bmatrix} -6 & -8 \\ 4 & -7 \end{bmatrix}$ **20.** $\begin{bmatrix} -8 & -8 & 3 \\ 3 & -4 & 5 \\ -2 & 0 & -1 \end{bmatrix}$ **22.** [6] **24.** $\begin{bmatrix} 2 & -4 & 4 \\ -1 & 2 & -2 \\ 1 & -2 & 2 \end{bmatrix}$

26. Both sides equal $\begin{bmatrix} -22 & 19 \\ -9 & 8 \end{bmatrix}$ **28.** Both sides equal $\begin{bmatrix} -2 & -2 \\ 1 & 7 \end{bmatrix}$

30. **(A)** 70 g **(B)** $[3 \quad 1] \cdot \begin{bmatrix} 5 \\ 15 \end{bmatrix} = 30$ g **(C)** 3×3

(D)
Mix X	Mix Y	Mix Z	
70 g	60 g	50 g	Protein
380 g	360 g	340 g	Carbohydrate
50 g	40 g	30 g	Fat

Total amounts in grams of
protein, carbohydrate, and
fat in 20 oz of each mix

(E)
Mix X	Mix Y	Mix Z	
3.5 g	3.0 g	2.5 g	Protein
19.0 g	18.0 g	17.0 g	Carbohydrate
2.5 g	2.0 g	1.5 g	Fat

Total amounts of protein,
carbohydrate, and fat in grams
per ounce of each mix

■ Index